CZECHOSLOVAK ACADEMY OF SCIENCES

TRANSACTIONS
of the
SEVENTH PRAGUE CONFERENCE
on
INFORMATION THEORY, STATISTICAL DECISION FUNCTIONS, RANDOM PROCESSES
and of the
1974 EUROPEAN MEETING OF STATISTICIANS

held at

Prague, from August 18 to 23, 1974

VOLUME A

1977

D. REIDEL PUBLISHING COMPANY

DORDRECHT – HOLLAND / BOSTON – U.S.A.

Sold and distributed in the U.S.A., Canada and Mexico by D. Reidel Publishing Company, Inc.
Lincoln Building, 160 Old Derby Street, Hingham, Mass. 02043, U.S.A.

Published by D. Reidel Publishing Company, P.O. Box 17, Dordrecht, Holland, in co-edition with ACADEMIA, Publishing House of the Czechoslovak Academy of Sciences, Prague

ISBN 978-94-010-9912-7 ISBN 978-94-010-9910-3 (eBook)
DOI 10.1007/978-94-010-9910-3

CONTENTS

Preface 11

Invited Papers

Csiszár I.: Information Measures: a Critical Survey . . . . vol. B
Durbin J.: Goodness-of-Fit Tests Based on the Order Statistics . . . . . . vol. B
Gani J.: A Markov Chain for Type Counts in Vocabulary Subgroups . . . 155
Jacobs K.: Recent Developments in Ergodic Theory 283
Lukacs E.: A Stability Theorem for a Characterization of the Wiener Process 375
Mandl P.: Some Results in the Adaptive Control of Linear Systems 399
Métivier M.: Advances in the Theory of Stochastic Integration and Applications 421
Rajbman N. S.: Dispersion Functions of Random Functions and Some their Applications . . . . vol. B
Schmetterer L.: Einige Resultate aus der Theorie der erwartungstreuen Schätzungen . . . . vol. B
Šidák Z.: Miscellaneous Topics in Markov Chains with a General State Space 531
Ullrich M.: The Use of Boolean Functions for Generation of Stochastic Processes . . . . vol. B

Communications

Ahmad R.: An Algebraic Treatment of Markov Processes. 13
Aksland M.: On Interconnected Birth and Death Processes with Immigration 23
Anděl J.: On the Backward Extrapolation of Non-Stationary Autoregressive Series 29
Anderson T. W.: On Maximum Likelihood Estimation of Parameters of Autoregressive Moving Average Processes 37
Bandemer H.: Experimental Design in Decision Models . . . . vol. B
Bernardo J. M.: Information Theory and Bayesian Design of Experiments . . vol. B
Besag J.: On Spatial-Temporal Models and Markov Fields 47
Bouchon B.: Quelques résultats de la théorie des questionnaires réalisables. . vol. B
Bräuer D.: Zur Auswertung des Elektromyogramms mit spektralanalytischen Methoden 57
Broussolle C.: L'utilisation de la théorie des phénomènes d'attente dans l'analyse des systèmes de production agricole 63

Butkiewicz J., Hys E.: On a Class of Bi- and Multivariate Distributions Generated by Marginal Weibull Distributions 73
Çapar U.: Random Fixed Point Theorems and Sequential Decision Making in Abstract Spaces . . . . vol. B
Capasso V.: On the General Stochastic Epidemic with Immigration in Large Populations 83
Clifford P.: On the Age Structure of Cell-Size Dependent Branching Processes 97
Cochlar J.: Smoothing and Extrapolation of Incomplete Time Series . . . . vol. B
Conradsen K.: Some Aspects of the Analysis of Air Pollution Data . . . . . vol. B
Csibi S.: More on Computationally Constrained Learning Processes . . . . vol. B
Csörgő M., Komlós J., Major P., Révész P., Tusnády G.: On the Empirical Process when Parameters are Estimated . . . . vol. B
Drygas H.: Admissibility of Certain ANOVA-Estimators in Simple Variance-Component Models . . . . vol. B
Dziubdziela W.: On Convergence Rates in the Limit Laws for Extreme Order Statistics . . . . vol. B
Ephremides A.: Information Patterns in Stochastic Point Process Differential Games 103
Fahrmeir L., Beeck H.: Zur Simulation stetiger stochastischer Wirtschaftsmodelle 113
Fichtner K. H.: Poissonsche zufällige Punktfolgen und ortsabhängige Verdünnungen 123
Fieger W.: Die Anzahl der Kreuzungspunkte einer Funktion mit einem Gaußschen Prozeß 135
Fleischmann K.: On the Stability of Spatially Homogeneous Cluster Fields . 147
Gänssler P.: On the Realization of Stochastic Processes by Probability Distributions in Function Spaces 163
Gasser T.: Tests of Goodness-of-Fit for Stationary Processes 173
Ghosal A.: Cybernetic Modelling of Queueing & Storage Systems 181
Giguère M. A., Styan G. P. H.: Multivariate Normal Estimation with Missing Data on Several Variates . . . . vol. B
Gittins J., Nash P.: Scheduling, Queues and Dynamic Allocation Indices . . 191
Глонти О. А.: Передача гауссовских сигналов по каналу с обратной связью . . . . vol. B
Goel A. L., Joglekar A. M.: Statistical Modelling of Computer Systems: a Review . . . . vol. B
Govindarajulu Z.: Stopping Time of a c-Sample Rank-Order Sequential Probability Ratio Test . . . . vol. B
Grimmett G.: Random Graph Theorems 203
Гусак Д. В.: Распределение абсолютного максимума пуассоновских и винеровских процессов на цепи Маркова 211

Guttman I., Dutter R.: A Bayesian Approach to the Detection of Spuriousness and Estimation in the General Univariate Linear Model when Outliers are Present . . . . vol. B
Hampel F. R.: Rejection Rules and Robust Estimates of Location: an Analysis of Some Monte Carlo Results . . . . vol. B
Härtler G.: A Rough Test for Multivariate Independence . . . . vol. B
Havránek T.: Some Aspects of Automatic Systems of Statistical Inference . . 221
Hensz E., Jajte R.: Random Measures of Gleason Type 231
Hida T.: Functionals of Brownian Motion 239
Hinderer K., Hübner G.: An Improvement of J. F. Shapiro's Turnpike Theorem for the Horizon of Finite Stage Discrete Dynamic Programs . . . 245
Hübner G.: Improved Procedures for Eliminating Suboptimal Actions in Markov Programming by the Use of Contraction Properties 257
Idzik A.: On Markov Policies in Continuous Time Discounted Dynamic Programming 265
Ihara S.: Coding Theory in Gaussian Channel with Feedback . . . . vol. B
Imhof J. P.: Oscillations of Monotone Amplitude and Application 277
Irle A., Schmitz N.: Decision Theory for Continuous Observations I: Bayes Solutions . . . . vol. B
Jílek M.: A Note on the Generalized Gamma Distribution: the Statistic $U = (X - Y)/(X + Y)$ 309
Jílovec S.: Uniformly Consistent Estimates . . . . vol. B
Jurečková J.: Asymptotic Relations of Least-Squares Estimate and of Two Robust Estimates of Regression Parameter Vector . . . . vol. B
Kaňková V.: Optimum Solution of a Stochastic Optimization Problem with Unknown Parameters . . . . vol. B
Kannappan Pl.: An Axiomatic Characterization of the Measure Generalized Directed Divergence . . . . vol. B
Kannappan Pl., Rathie P. N.: On a Generalized Directed Divergence and Related Measures . . . . vol. B
Kanter M.: On the Boundedness of Stable Processes 317
Кемпе Ф., Альберендт Н.: Экстраполяция и идентификация случайных процессов на основе векторных марковских моделей . . . . vol. B
Kligienė N.: On the Estimation of the Change Point in the Autoregressive Sequence 325
König D., Jansen U.: Stochastic Processes and Properties of Invariance for Queueing Systems with Speeds and Temporary Interruptions 335
Kozek A.: A Construction of UMV Estimators and Some Problems of Partial Differential Equations . . . . vol. B
Kramosil I.: Gentzen-Like Random Axiomatic Systems 345
Krishnaiah P. R., Schuurmann F. J.: Approximations to the Traces of Complex Multivariate Beta and F Matrices . . . . vol. B

Kubát L.: On Diagonstic Methods for Multi-Level Classification of Operation of Technical Systems . . . . vol. B
Lanke J.: Asymptotic Normality in Simple Random Sampling . . . . vol. B
Le Breton A.: Parameter Estimation in a Linear Stochastic Differential Equation 353
Lee L., Thompson W. A., Jr.: Reliability of Multiple Component Systems . . vol. B
Леонов Ю. П.: Распределение вероятностей типа CR 367
Liese F.: Charakterisierung einer Klasse von Entropiemaximalfolgen . . . . vol. B
Macchi O.: The Fermion Process – a Model of Stochastic Point Process with Repulsive Points 391
Mareš M.: Competitive Equilibrium in a Market with Limited Cooperation . 411
Marosi J., Tusnády G.: On the Constructions of Confidence Intervals for the Expectation of a Poisson Distribution on the Basis of Censored Data . . . vol. B
Mathai A. M.: An Axiomatic Foundation for the Multivariate Measure of Affinity with Applications in Social Statistics . . . . vol. B
Michálek J.: Second Order Stochastic Processes and Self-Adjoint Operators . 433
Mihoc I.: On Measures of Information . . . . vol. B
Mikiewicz J.: Statistical Selection Method of the Best Objects . . . . . . vol. B
Mollison D.: The Velocity of Stochastic Processes 441
Nåsell I.: A Mathematical Model of Helminthiasis with External Infection . . 451
Nemetz T.: On a Finite Decision Model . . . . vol. B
Nollau V.: Zur operatorentheoretischen Charakterisierung symmetrisch stabiler Prozesse 461
Österreicher F.: On the Existence of Least Favourable Pairs of Distributions vol. B
Pázman A.: A Contribution to the Time Series Design Problems 467
Piccinato L.: Predictive Distributions and Noninformative Priors . . . . . vol. B
Plachky D.: A Characterization of Bounded Completeness in the Undominated Case 477
Plackett R. L.: The Information in Marginal Totals . . . . vol. B
Pop-Stojanovic Z. R.: Continuous Filtering Problem for Itô-McShane Stochastic Differential Equation 481
Pötschke D.: A Statistical Interpretation of the B-Rate of Information Theory vol. B
Radcliffe J.: Host-Vector Epidemics 489
Raudys Š.: On the Accuracy of Some Estimates of Multivariate Density Function . . . . vol. B
Richardson D.: Replication and Stochastic Processes in Space 499
Rootzén H.: Rényi-Mixing and Functional Limit Theorems for Martingales 505
Rukhin A. L.: On the Estimation of a Transformation Parameter of a Distribution Given on a Finite Group . . . . vol. B
Rüschendorf L.: On One-Sample Rank Order Statistics for Dependent Random Variables . . . . vol. B
Rusu I., Coţiu F.: Sur un problème de programmation linéaire stochastique . vol. B

Saleh A. K. Md. E.: Behrens-Fisher Problem for the Exponential Distribution in Censored and Uncensored Samples . . . . vol. B
Sayed Y., Bogr J.: Measurement of Doppler Signals with Gaussian Spectra Using Phase-Locked Loop . . . . vol. B
Schlee W.: Asymptotische Verteilung des Kolmogoroff-Mehrstichproben-Tests bei unstetiger Verteilungsfunktion . . . . vol. B
Sendler W.: On the Moments of a Certain Class of Random Permanents . . 515
Shahani A. K.: A Trichotomous Process with Applications in Quality Control vol. B
Shpilewski E.: Estimation of the Parameters and Recognition of Stochastic Processes 523
Sibson R.: Distance Distributions 527
Skala H. J.: Not Necessarily Additive Realizations of Comparative Probability Relations – a Basis of Decision Theory 545
Sladký K.: On the Optimality Conditions for Semi-Markov Decision Processes 555
Streit F.: On Generalized Inverse Matrices and Multivariate Statistical Distributions . . . . vol. B
Székely G. J.: The Brownian Taboo Bridge with an Application in the Theory of Nonparametric Statistics 563
Thyregod P.: On Bayes Decision Rules for Symmetric Finite Populations . . vol. B
Vízková Z.: A Local Limit Theorem for Dwass Statistic . . . . vol. B
Watanabe S.: MOSS – a Subspace Representation of Classes in Pattern Recognition . . . . vol. B
Weron A.: Stochastic Processes of Second Order with Values in Banach Spaces 567
Wessels J.: Stopping Times and Markov Programming 575
Winkelbauer K.: Information Channels with Memoryless Components . . vol. B
Yeo G.: Dams with Variable Release Rates 587
Zeliński R.: A Randomized Kiefer-Wolfowitz Procedure . . . . vol. B
Ziezold H.: On Expected Figures and a Strong Law of Large Numbers for Random Elements in Quasi-Metric Spaces 591

PREFACE

The Prague Conferences on Information Theory, Statistical Decision Functions, and Random Processes have been organized every three years since 1956. During the eighteen years of their existence the Prague Conferences developed from a platform for presenting results obtained by a small group of researchers into a probabilistic congress, this being documented by the increasing number of participants as well as of presented papers.

The importance of the Seventh Prague Conference has been emphasized by the fact that this Conference was held jointly with the eighth European Meeting of Statisticians. This joint meeting was held from August 18 to 23, 1974 at the Technical University of Prague. The Conference was organized by the Institute of Information Theory and Automation of the Czechoslovak Academy of Sciences and was sponsored by the Czechoslovak Academy of Sciences, by the Committee for the European Region of the Institute of Mathematical Statistics, and by the International Association for Statistics in Physical Sciences.

More than 300 specialists from 25 countries participated in the Conference. In 57 sessions 164 papers (including 17 invited papers) were read, 128 of which are published in the present two volumes of the Transactions of the Conference. Volume A includes papers related mainly to probability theory and stochastic processes, whereas the papers of Volume B concern mainly statistics and information theory.

It gives us a pleasure to express our gratitude to all who have contributed to the success of the Conference, especially to the sponsoring institutions and their representatives, and to those who read papers and delivered their manuscripts for publication in the Transactions.

Sincere thanks are due to Academician Jaroslav Kožešník, the scientific editor of these volumes, to the editorial board the members of which read all published manuscripts, and also to all employees of the Institute of Information Theory and Automation who participated in the organization of the Conference.

The Organizing Committee is very sorry that the premature death of two of its members, who participated in the preparation of this volume, prevented them from seeing the fruits of their labour. Ing. Libor Kubát died on 28th December, 1975, at the age of fourty-six and Dr. Zdeněk Koutský died on 3rd June, 1976, at the age of fifty-two. Ing. Kubát, as the executive editor, and Dr. Koutský, as one of the main organizers, contributed invaluably to the success of this Conference, as well as of all our previous conferences. The function of the executive editor of this volume was assumed by Dr. Milan Ullrich.

ORGANIZING COMMITTEE

AN ALGEBRAIC TREATMENT OF MARKOV PROCESSES

RASHID AHMAD

GLASGOW

The purpose of this paper is to introduce some algebraic structure, as in Heller [12], in the class of Markovian stochastic processes in discrete time and countable state space. By this method the algebraic structure theory of stochastic modules is developed, and consequently some results are obtained about the reduced modules through the dimension of various subspaces of these modules when viewed as vector spaces over the real line. These results turn out to be conceptually simpler and elegant in form. Finally, the stochastic module theory developed is employed in the study of the decomposition of the reduced module of a stationary Markov chain, and the relationship between probability measures and linear functionals is studied.

1. INTRODUCTION

It is well-known that a very important class of stochastic processes with a wide range of applications is the Markov processes family. Furthermore, from a mathematical viewpoint, this family is one of the most highly developed branches of generalized stochastic random variables. The applied literature in probability theory abounds with many interesting examples of Markov processes. Sometimes it happens that a process is not a Markovian process. However, there exist transformations which reduce this process to be a function of a Markov chain or process. The earliest work in this direction goes back to Blackwell and Koopman [3], Gilbert [11], Burk and Rosenblatt [5], Rosenblatt [16], Dharmadhikari [6], [7], [8] and Heller [12]. The classification and identification problems are studied by Leysieffer [14], [15], Boudreau [4] and Erickson [9] by employing essentially Dharmadhikari's approach. Other authors who worked on related but slightly different problems are listed in the references of the papers cited above.

Heller [12] completed the problem of characterizing processes which are functionals of Markov chains. In particular, Heller employed algebraic techniques, thus treating the problem in a simple and an elegant way. The algebraic approach, still unconventional perhaps in probability theory, provides better and clearer insight into the behaviour and structure of such processes. It is for this reason that we have chosen to examine this elegant and novel theory and look for generalizations and its applications to various other aspects of stochastic processes. To this end, we have relaxed the restrictions to chains with a finite state space to chains with countable state space. In the sequel a stochastic process will be regarded as a countable set of jointly distributed random variables (with a finite dimensional distribution) each of which takes values on a countable state space, $\boldsymbol{S} = \{\boldsymbol{x}_i\}$. Heller dealt with a finite state space. However, most of the theory generalizes easily to the countable case.

2. ALGEBRAIC STRUCTURE THEORY OF STOCHASTIC MODULES

In this section we shall define the process as a linear functional from an algebra generated by the state space into the reals. Consequently for convenience we redefine this as a linear functional from a module over this algebra into the reals. After some general results about these algebraic structures are established, we shall examine Markovian processes, and obtain some particular results for these processes.

Consider a stochastic process $\{\boldsymbol{X}_t, t \geqq 1\}$ where each $\boldsymbol{X}_t$ takes values on a countable state space $\boldsymbol{S}$. For some integer n, let $\{\boldsymbol{x}_{i_1}, \ldots, \boldsymbol{x}_{i_n}\}$ be a finite sequence of states for some finite set of integers $\{i_1, \ldots, i_n\}$. Let $\boldsymbol{s} = \boldsymbol{x}_{i_1} \cdot \ldots \cdot \boldsymbol{x}_{i_n}$, so that $\boldsymbol{s}$ represents a finite sequence of states. This multiplication of the $\boldsymbol{x}_j$'s is associative but it is not commutative, for rearrangements of the $\boldsymbol{x}_j$'s in a product could possibly yield a different event. If $\boldsymbol{s}_1, \ldots, \boldsymbol{s}_q$ are sequences of states involving finitely many of the $\boldsymbol{x}_j$'s we can interpret $\sum_{j=1}^{q} \boldsymbol{s}_j$ as the event corresponding to the disjoint union of the $\boldsymbol{s}_j$'s, so that probability distributes linearly over the sum. This addition is clearly associative and commutative. Notice that an infinite product or sum, would involve us with an event requiring an infinite period of time for realization. We observe at this point that the process is completely determined by the probabilities of all finite sequences of states as $\boldsymbol{s}$ above. The above discussion motivates us to introduce the idea of a *free associative polynomial algebra* (see Sikorski [17]) generated over the reals by the state space $\boldsymbol{S}$. This algebra, $\mathscr{A}_S$, is a ring of polynomials in the infinite set of noncommuting states $\{\boldsymbol{x}_i\}$. If $\boldsymbol{s}$ is a finite sequence of states we define the length of $\boldsymbol{s}$, $\lambda(\boldsymbol{s})$ to be the degree of $\boldsymbol{s}$ as a polynomial in $\mathscr{A}_S$. We interpret a sequence $\boldsymbol{s}$ in $\mathscr{A}_S$ with $\lambda(\boldsymbol{s}) = n$ as being the states occupied at times $1, \ldots, n$, that is, the values of $\boldsymbol{X}_1, \ldots, \boldsymbol{X}_n$.

Clearly $\mathscr{A}_S$ is a vector space over $\mathscr{R}$ with a basis consisting of all finite products of $\boldsymbol{x}_j$'s together with 1, the identity in $\mathscr{R}$. Note that 1 in $\mathscr{A}_S$ is interpreted as a sequence of states of length zero. We now define a generalized stochastic functional p on $\mathscr{A}_S$ by defining p on this basis and extending the definition by $\mathscr{R}$-linearity to the whole of $\mathscr{A}_S$. Thus, if $f(x_{i_1}, \ldots, x_{i_n})$ is a density with respect to the Lebesgue measure, we set $p(x_{i_1} \cdot \ldots \cdot x_{i_n}) = f(x_{i_1}, \ldots, x_{i_n})$, and $p(1) = 1$. Extending p by $\mathscr{R}$-linearity to all of $\mathscr{A}_S$, we have $p : \mathscr{A}_S \to \mathscr{R}$ is an $\mathscr{R}$-linear map, that is, $p(as + bs^*) = ap(s) +$ $+ bp(s^*)$ for any real numbers a and b. Now let $\mathscr{A}_S^+$ be the set of all polynomials in $\mathscr{A}_S$ with nonnegative coefficients. Then the definition of a stochastic process can be reformulated as follows.

Definition 2.1. A linear functional $p : \mathscr{A}_S \to \mathscr{R}$ is a stochastic process if and only if it satisfies:

(i) $$p(\mathscr{A}_S^+) \subset [0, \infty);$$

(ii) $$p(1) = 1;$$

(iii) for every ξ in $\mathscr{A}_S$,

$$\sum_{j=1}^{\infty} p(\xi x_j) = p(\xi).$$

These conditions are essentially the same as Heller's P0, P1 and P2, only P2 has to be altered for the infinite case. Heller defines the 'certain event' $\sigma = \sum_{x \in S} x$ for finite state space S. There is no analogue for σ in the infinite case since it would be a polynomial with an infinite number of terms and indeterminates and hence would not be in $\mathscr{A}_S$. The fact that the functional p is equivalent to the usual idea of a stochastic process can be seen from the following result.

Theorem 2.1. *There is* $1-1$ *correspondence between the set of linear functionals* $p : \mathscr{A}_S \to \mathscr{R}$ *satisfying* (i), (ii) *and* (iii) *and the set of probability densities* $\{f_\alpha\}_{\alpha=1}^{\infty}$ *on* S *given by* $f(x_{i_\alpha}) = \sum p(sx_{i_\alpha})$ *for every* α, i_α, *where the sum is over all* s *such that* $\lambda(s) = \alpha - 1$.

Proof. If one starts from a process and defines the extended function p, then (i) and (ii) will be satisfied. To prove (iii) it is sufficient to take the case where $\xi = s$ is a finite sequence of states, the general case is given by $\mathscr{R}$-linearity. Clearly,

$$\sum_{j=1}^{\infty} p(sx_j) = \sum_{j=1}^{\infty} p(x_j \mid s)\, p(s) = p(s) \sum_{j=1}^{\infty} p(x_j \mid s) = p(s),$$

since $\{x_j \mid s\}_{j=1}^{\infty}$, for $\lambda(sx_j) = n$, form a mutually exclusive and exhaustive set. Conversely, suppose that a functional p satisfies (i), (ii) and (iii). Then one has to show that p generates unique probability distributions for the family of random variables $\{X_j\}_{j=1}^{\infty}$. It can be shown by induction that for all positive integers n, i_n, one can define $f_n(x_{i_n}) = \sum p(sx_{i_n})$. From (i) the R.H.S. is clearly nonnegative for all n. Therefore it only remains to show that for all n, $\sum_{i_n=1}^{\infty} \mathfrak{p}(sx_{i_n}) = 1$. But this follows immediately from (i), (ii) and by induction argument. Q.E.D.

It may be remarked here that the treatment of a stochastic process given above is quite general, no assumptions are made regarding stationarity or the Markovian property. We can now reduce the algebra $\mathscr{A}_S$ by '*factoring out*', in the ring theory sense, certain polynomials having probability zero. In essence, the sequences we remove are those that can never occur in any stage in time. The structure obtained after this factoring will be called the *reduced module* associated with the process, and in the sequel will be refered as the *module*. Let ker $\mathfrak{p}$ denote the kernel of $\mathfrak{p}$, that is, the set $\{\xi \in \mathscr{A}_S \mid \mathfrak{p}(\xi) = 0\}$. Clearly, $\mathscr{N} = \{\xi \in \mathscr{A}_S \mid \mathfrak{p}(\eta\xi) = 0, \ \forall \eta \in \mathscr{A}_S\}$ is the largest left ideal which is wholly contained in ker $\mathfrak{p}$, so one can factor $\mathfrak{p}$ through $\mathscr{A}_S \mid \mathscr{N} = \mathscr{Q}$, say. Thus we have the following situation: $\mathfrak{p} : \mathscr{A}_S \to \mathscr{R}$, $\phi : \mathscr{A}_S \to \mathscr{Q}$, and $\mathfrak{q} : \mathscr{Q} \to \mathscr{R}$. That is, $\mathfrak{p} = \mathfrak{q}\phi$, where ϕ is the canonical ring homomorphism so that $\phi(\xi) = \xi + \mathscr{N}$, $\mathfrak{q}(\xi + \mathscr{N}) = \mathfrak{p}(\xi)$ for every $\xi \in \mathscr{A}_S$. The left $\mathscr{A}_S$-module structure of $\mathscr{Q}$ is given by $\xi_1(\eta_1 + \mathscr{N}) + \xi_2(\eta_2 + \mathscr{N}) = (\xi_1\eta_1 + \xi_2\eta_2) + \mathscr{N}$ for every $\xi_1, \xi_2, \eta_1, \eta_2$ in $\mathscr{A}_S$. Note that $\mathscr{Q}$ is generated over $\mathscr{A}_S$ by $\mathfrak{m} = 1 + \mathscr{N}$. Furthermore, $\mathfrak{q}$ is $\mathscr{R}$-linear since $\mathfrak{p}$ is, also ker $\mathfrak{q}$ contains no nonzero submodules, and finally since $\mathfrak{p} = \mathfrak{q}\phi$, we have $\mathfrak{p}(\xi) = \mathfrak{q}\phi(\xi) = \mathfrak{q}(\xi\mathfrak{m})$ for every $\xi \in \mathscr{A}_S$.

On the basis of above development the triple $(\mathscr{Q}, \mathfrak{q}, \mathfrak{m})$ is called the reduced $\mathscr{A}_S$-module associated with the stochastic process $\mathfrak{p}$. All our modules are left modules, so we drop the term left. As shown above, $(\mathscr{Q}, \mathfrak{q}, \mathfrak{m})$ has the following important properties: (i) $\mathfrak{q}$ is $\mathscr{R}$-linear and $\mathfrak{q}(\xi\mathfrak{m}) = \mathfrak{p}(\xi)$ for every $\xi \in \mathscr{A}_S$; (ii) $\mathscr{Q} = \mathscr{A}_S\mathfrak{m} = \{\xi\mathfrak{m} \mid \xi \in \mathscr{A}_S\}$; (iii) if $\mathscr{Q}'$ is a submodule of $\mathscr{Q}$, then $\mathscr{Q}' \subset$ ker $\mathfrak{q}$ implies $\mathscr{Q}' = 0$. The stochastic $\mathscr{A}_S$-modules $(\mathscr{Q}, \mathfrak{q}, \mathfrak{m})$ and $(\mathscr{Q}^*, \mathfrak{q}^*, \mathfrak{m}^*)$ are said to be equivalent if $\mathscr{Q}$ and $\mathscr{Q}^*$ are isomorphic and the associated stochastic processes are the same.

THEOREM 2.2. *The reduced stochastic $\mathscr{A}_S$-modules $(\mathscr{Q}, \mathfrak{q}, \mathfrak{m})$ and $(\mathscr{Q}^*, \mathfrak{q}^*, \mathfrak{m}^*)$ are equivalent if and only if there exists an $\mathscr{A}_S$-module isomorphism, $\mathbf{g} : \mathscr{Q} \to \mathscr{Q}^*$ such that*

(a) $$\mathbf{g}(\mathfrak{m}) = \mathfrak{m}^*,$$

(b) $$\mathfrak{q}^*\mathbf{g} = \mathfrak{q}.$$

Proof. Assume $(\mathscr{Q}, \mathfrak{q}, \mathfrak{m}) \equiv (\mathscr{Q}^*, \mathfrak{q}^*, \mathfrak{m}^*)$, and define $\mathbf{g} : \mathscr{Q} \to \mathscr{Q}^*$ by $\mathbf{g}(\xi\mathfrak{m}) = \xi\mathfrak{m}^*$, for every $\xi \in \mathscr{A}_S$. Clearly $\mathbf{g}$ is a surjective homomorphism, and it is also injective, for suppose $\eta\mathfrak{m}^* = 0$, then $\xi\eta\mathfrak{m}^* = 0$, for every $\xi \in \mathscr{A}_S$. Therefore $\mathfrak{q}^*(\xi\eta\mathfrak{m}^*) = \mathfrak{q}(\xi\eta\mathfrak{m}) = 0$, for every $\xi \in \mathscr{A}_S$, hence $\eta\mathfrak{m} = 0$. Now (a) is obvious, for (b) observe that for every $\xi \in \mathscr{A}_S$, we have $\mathfrak{q}(\xi\mathfrak{m}) = \mathfrak{q}^*(\xi\mathfrak{m}^*) = \mathfrak{q}^*\mathbf{g}(\xi\mathfrak{m})$, and $\mathfrak{q} = \mathfrak{q}^*\mathbf{g}$. Conversely, suppose (a) and (b) hold, then $\mathfrak{q}^*(\xi\mathfrak{m}^*) = \mathfrak{q}^*\mathbf{g}(\xi\mathfrak{m}) = \mathfrak{q}(\xi\mathfrak{m})$ for every $\xi \in \mathscr{A}_S$, and hence the associated processes are the same. Q.E.D.

If $\mathfrak{p}$ is an arbitrary $\mathscr{R}$-linear functional, $\mathfrak{p} : \mathscr{A}_S \to \mathscr{R}$ and if the triple $(\mathscr{Q}, \mathfrak{q}, \mathfrak{m})$ is such that $\mathfrak{p}(\xi) = \mathfrak{q}(\xi\mathfrak{m})$ for every $\xi \in \mathscr{A}_S$, where $\mathfrak{q}$ is $\mathscr{R}$-linear, then we can redefine a stochastic process in terms of $(\mathscr{Q}, \mathfrak{q}, \mathfrak{m})$.

DEFINITION 2.2. Let $\mathscr{Q}$ be an $\mathscr{A}_S$-module, $\mathfrak{m} \in \mathscr{Q}$ and $\mathfrak{q} : \mathscr{Q} \to \mathscr{R}$ be $\mathscr{R}$-linear. Define $\mathfrak{p} : \mathscr{A}_S \to \mathscr{R}$ by $\mathfrak{p}(\xi) = \mathfrak{q}(\xi\mathfrak{m})$ for every $\xi \in \mathscr{A}_S$. Then $\mathfrak{p}$ is a stochastic process if and only if the following conditions are satisfied:

(i) $$\mathfrak{q}(\mathscr{A}_S^+ \mathfrak{m}) \subset [0, \infty)\,;$$

(ii) $$\mathfrak{q}(\mathfrak{m}) = 1\,;$$

(iii) $$\text{for every} \quad \xi \in \mathscr{A}_S\,, \quad \sum_{j=1}^{\infty} \mathfrak{q}(\xi x_j \mathfrak{m}) = \mathfrak{q}(\xi\mathfrak{m})\,.$$

This is clearly equivalent to our previous definition (2.1) of a stochastic process. So far, our treatment of stochastic processes has been quite general. Now, we look at processes with the Markovian property and examine the reduced modules structure. We shall see that the special nature of such processes enables us to obtain some strong results about the reduced modules. These results, as shown by Heller [12], are very useful in the theory of functions of Markov chains. We shall be concerned with stationary Markov chains and omit the term stationary from now on.

To develop our theory, we look for a new definition in terms of the map $\mathfrak{p} : \mathscr{A}_S \to \mathscr{R}$. Let $\mathfrak{p}$ be a Markov chain of order r, then if s and u are any sequences of states and w is a sequence with $\lambda(w) = r$, we have $\mathfrak{p}(swu) = \mathfrak{p}(sw)\,\mathfrak{p}(swu \mid sw)$. Now since $\lambda(w) = r$, it follows that $\mathfrak{p}(swu \mid sw)$ is independent of the actual states in s and because the process is stationary it is also independent of $\lambda(s)$. Therefore we can write $\mathfrak{p}(swu \mid sw) = \tau(w, u)$, a function of w and u only. By the $\mathscr{R}$-linearity of $\mathfrak{p}$ we can generalize from a sequence s to any polynomial, $\xi \in \mathscr{A}_S$, obtaining $\mathfrak{p}(\xi wu) = \mathfrak{p}(\xi w)\,\tau(w, u)$. Let S^r be the set of all sequences of length r, and S^∞ be the set of all finite sequences of states, then we have the following definition.

DEFINITION 2.3. A stochastic process $\mathfrak{p} : \mathscr{A}_S \to \mathscr{R}$ is a Markov chain of order r if and only if there exists a map $\tau : S^r \times S^\infty \to \mathscr{R}$ such that for every $\xi \in \mathscr{A}_S$ and sequences w with $\lambda(w) = r$, and u of arbitrary length,

(*) $$\mathfrak{p}(\xi(wu - \tau(w, u)\, w)) = 0\,.$$

Our results for Markov chains concern the dimension of various subspaces of the reduced module when viewed as a vector space over $\mathscr{R}$. The result below is a generalization of Heller's proposition 3.2.

THEOREM 2.3. *Let* $(\mathcal{Q}, \mathfrak{q}, \mathfrak{m})$ *be a reduced module with associated process* $\mathfrak{p}$. *Then* $\mathfrak{p}$ *is a Markov chain of order* r *if and only if for every sequence of states* $\mathbf{w}$ *with* $\lambda(\mathbf{w}) = r$, $\mathbf{w}\mathcal{Q} = \{\mathbf{w}\xi\mathfrak{m} \mid \xi \in \mathcal{A}_S\}$ *viewed as a vector space over* $\mathcal{R}$ *has dimension* $\dim_{\mathcal{R}} \mathbf{w}\mathcal{Q} \leqq 1$.

Proof. Assume $\mathfrak{p}$ is a Markov chain of order r, then (*) holds. If τ is extended to $S^r \times \mathcal{A}_S$ by $\mathcal{R}$-linearity, equation (*) becomes

$$(**) \qquad \mathfrak{p}(\xi(\mathbf{w}\eta - \tau(\mathbf{w}, \eta)\,\mathbf{w})) = 0 \quad \text{for every} \quad \xi, \eta \in \mathcal{A}_S .$$

Since $\mathfrak{p}(\xi) = \mathfrak{q}(\xi\mathfrak{m})$, for $\xi \in \mathcal{A}_S$, it follows that for each $\eta \in \mathcal{A}_S$, $\mathfrak{q}$ vanishes on the submodules $\mathcal{A}_S(\mathbf{w}\eta - \tau(\mathbf{w}, \eta)\,\mathbf{w})$ and since the stochastic module is reduced $(\mathbf{w}\eta - \tau(\mathbf{w}, \eta)\,\mathbf{w})\,\mathfrak{m} = 0$ for each η. Therefore, $\mathbf{w}\eta\mathfrak{m} = \tau(\mathbf{w}, \eta)\,\mathbf{w}\mathfrak{m}$ for every η. But $\tau(\mathbf{w}, \eta) \in \mathcal{R}$, hence $\mathbf{w}\mathcal{Q} \subset \mathcal{R}\mathbf{w}\mathfrak{m}$ and $\dim_{\mathcal{R}} \mathbf{w}\mathcal{Q} \leqq 1$. Conversely, suppose that for every $\mathbf{w} \in S^r$, $\dim_{\mathcal{R}} \mathbf{w}\mathcal{Q} \leqq 1$. If $\dim_{\mathcal{R}} \mathbf{w}\mathcal{Q} = 0$, then $\mathbf{w}\mathfrak{m} = 0$, that is, the sequence $\mathbf{w}$ can never occur; thus (**) is trivially satisfied. Assume $\dim_{\mathcal{R}} \mathbf{w}\mathcal{Q} = 1$, so there exists η in $\mathcal{A}_S$ such that $\mathbf{w}\eta\mathfrak{m} \neq 0$, $\mathbf{w}\eta \notin \mathcal{N}$, and $\mathbf{w} \notin \mathcal{N}$. Thus $\mathbf{w}\mathfrak{m} \neq 0$ and one may take $\{\mathbf{w}\mathfrak{m}\}$ as a basis for $\mathbf{w}\mathcal{Q}$ over $\mathcal{R}$, that is, for every $\eta \in \mathcal{A}_S$ there exists $\tau(\mathbf{w}, \eta)$ in $\mathcal{A}_S$ such that $\mathbf{w}\eta\mathfrak{m} = \tau(\mathbf{w}, \eta)\,\mathbf{w}\mathfrak{m}$. Hence $\xi(\mathbf{w}\eta - \tau(\mathbf{w}, \eta)\,\mathbf{w})\,\mathfrak{m} = 0$ for every $\xi \in \mathcal{A}_S$. Therefore,

$$\mathfrak{q}(\xi(\mathbf{w}\eta - \tau(\mathbf{w}, \eta)\,\mathbf{w})\,\mathfrak{m}) = \mathfrak{p}(\xi(\mathbf{w}\eta - \tau(\mathbf{w}, \eta)\,\mathbf{w})) = 0$$

for every $\eta, \xi \in \mathcal{A}_S$, and this is (**), so $\mathfrak{p}$ is a Markov chain of order r. Q.E.D.

COROLLARY 2.1. (i) *The process* $\mathfrak{p}$ *is independent process if and only if* $\mathcal{Q} = \mathcal{R}\mathfrak{m}$. *In this case* $\dim_{\mathcal{R}} \mathcal{Q} = 1$, *for any state space.*

(ii) *The process* $\mathfrak{p}$ *is a Markov chain if and only if for every* $\mathbf{x} \in S$, $\dim_{\mathcal{R}} \mathbf{x}\mathcal{Q} \leqq 1$. *In this case, if* S *is finite* $\dim_{\mathcal{R}} \mathcal{Q} \leqq M < \infty$.

In the above discussion the strongest condition is that of independence $(r = 0)$ and this gives us the simplest reduced module as regards vector space structure over $\mathcal{R}$, that is, the space has a one element basis. Weakening the independence property by the Markov property $(r \geqq 1)$ increases the dimension of the underlying vector space. Heller has called processes with finite dimensional reduced module, *finitary*, and has shown that processes which are functions of a finite Markov chain are finitary. However, there exist finitary processes which are not induced from a Markov chain, see Heller [12].

3. SOME APPLICATIONS OF THE THEORY

In this section, we shall apply the abstract algebraic theory developed above to (i) the decomposition of the reduced module of a stationary Markov chain, and (ii) in the study of the relationship between probability measures and linear functionals.

If a process $\mathbf{p}$ has reduced module $(\mathcal{Q}, \mathfrak{q}, \mathfrak{m})$ the decomposition will consist of a direct sum decomposition of $\mathcal{Q}$ and $\mathfrak{q}$ will be expressed as a sum of $\mathcal{R}$-linear maps. We approach this problem through the theory of closed sets of states as in Feller [10]. It is well-known that the state space $\mathbf{S}$ of a Markov chain can be uniquely expressed as $\mathbf{S} = \mathbf{T} \cup (\bigcup_j \mathbf{S}_j)$ where $\mathbf{T}$ is the set of all transient states and the $\mathbf{S}_j$ are the irreducible components, that is, the minimal disjoint closed sets of recurrent (or persistent) states. If $\mathbf{S}$ is finite recurrent states always exist, but in the infinite case all states may be transient.

Let $\mathbf{p} : \mathcal{A}_\mathbf{S} \to \mathcal{R}$ be a Markov chain on the state space $\mathbf{S}$, which may be finite or countably infinite. We restrict ourselves to the case where there are only a finite number of irreducible components, that is, $\mathbf{S} = \mathbf{T} \cup (\bigcup_{j=1}^{k} \mathbf{S}_j)$ for some integer $k > 0$ (if $\mathbf{S} = \mathbf{T}$, the decomposition will be trivial). For each j, let $\mathcal{R}(\mathbf{S}_j)$ be the set of all polynomials in $\mathcal{A}_\mathbf{S}$ such that each term contains at least one state from $\mathbf{S}_j$ but no other states from any other $\mathbf{S}_i$, $j \neq i$. Thus the elements in $\mathcal{R}(\mathbf{S}_j)$ are polynomials in states from $\mathbf{T} \cup \mathbf{S}_j$, and clearly $\mathcal{R}(\mathbf{S}_j)$ is a subring of $\mathcal{A}_\mathbf{S}$ and hence a ring. For each j, let $\mathbf{p}_j : \mathcal{R}(\mathbf{S}_j) \to \mathcal{R}$ be the restriction of $\mathbf{p}$ to $\mathcal{R}(\mathbf{S}_j)$. Note that $\mathbf{p}_j$ is $\mathcal{R}$-linear since $\mathbf{p}$ is, and $\mathbf{p}_j$ satisfies our conditions (i) and (ii) for a stochastic process, but not (iii) as we may have $\sum \mathbf{p}(\xi \mathbf{x}_i) < \mathbf{p}(\xi)$, where the sum is over all $\mathbf{x}_i \in \mathbf{T} \cup \mathbf{S}_j$.

DEFINITION 3.1. An $\mathcal{R}$-linear map $\mathbf{p} : \mathcal{A} \subset \mathcal{A}_\mathbf{S} \to \mathcal{R}$ is called a *quasi-process* if conditions (i), (ii) are satisfied and $\mathcal{A}$ is a subalgebra of $\mathcal{A}_\mathbf{S}$. If $\mathbf{p}$ is a quasi-process and $\mathcal{Q}$ is a $\mathcal{A}$-module, $\mathfrak{m} \in \mathcal{Q}$ is such that $\mathcal{Q} = \mathcal{A}\mathfrak{m}$, $\mathfrak{q} : \mathcal{Q} \to \mathcal{R}$ is $\mathcal{R}$-linear and such that $\mathbf{p}(\xi) = \mathfrak{q}(\xi \mathfrak{m})$ for every ξ in $\mathcal{A}$, also $\ker \mathfrak{q}$ contains no nonzero submodules, then the triple $(\mathcal{Q}, \mathfrak{q}, \mathfrak{m})$ is a *quasi-reduced module* with associated quasi-process $\mathbf{p}$.

The structural results similar to those for stochastic processes are also valid for quasi-processes. We denote the quasi-reduced module of $\mathbf{p}_j$ by $(\mathcal{Q}_j, \mathfrak{q}_j, \mathfrak{m}_j)$ for $j = 1, 2, \ldots, k$. If $\mathbf{S}^* \subset \mathbf{S}$, then $\mathcal{A}_{\mathbf{S}^*}$ is a subring of $\mathcal{A}_\mathbf{S}$, and if $\mathbf{p} : \mathcal{A}_\mathbf{S} \to \mathcal{R}$ is a stochastic process, the restriction of $\mathbf{p}$ to $\mathcal{A}_{\mathbf{S}^*}$ is clearly a quasi-process. In particular, $(\mathcal{Q}_T, \mathfrak{q}_T, \mathfrak{m}_T)$ is the quasi-reduced module associated with a quasi-process $\mathbf{p}_T : \mathcal{A}_T \to \mathcal{R}$. If $\mathcal{Q}_1, \ldots, \mathcal{Q}_k$ are $\mathcal{A}_\mathbf{S}$-modules then $\bigoplus_{j=1}^{k} \mathcal{Q}_j$ is the set $\prod_{j=1}^{k} \mathcal{Q}_j$ with $\mathcal{A}_\mathbf{S}$-module structure given componentwise, that is, $\xi(\mathfrak{m}_1, \ldots, \mathfrak{m}_k) = (\xi\mathfrak{m}_1, \ldots, \xi\mathfrak{m}_k)$ and $(\mathfrak{m}_1, \ldots, \mathfrak{m}_k) + (\mathfrak{m}_1^*, \ldots, \mathfrak{m}_k^*) = (\mathfrak{m}_1 + \mathfrak{m}_1^*, \ldots, \mathfrak{m}_k + \mathfrak{m}_k^*)$.

THEOREM 3.1. *Let* $\mathfrak{p} : \mathscr{A}_S \to \mathscr{R}$ *be a Markov chain such that* $\mathbf{S} = \mathbf{T} \cup \left(\bigcup_{j=1}^{k} \mathbf{S}_j \right)$ *where* $\mathbf{T}$ *is the set of transient states and the* $\mathbf{S}_j$ *are irreducible components. Let* $(\mathscr{Q}_j, \mathfrak{q}_j, \mathfrak{m}_j)$ *and* $(\mathscr{Q}_T, \mathfrak{q}_T, \mathfrak{m}_T)$ *be as defined above. Then there is an* $\mathscr{A}_S$*-module structure for* $\mathscr{Q}_T$ *and each* $\mathscr{Q}_j$ *such that if* $\mathscr{Q} = \mathscr{Q}_T \oplus \left(\bigoplus_{j=1}^{k} \mathscr{Q}_j \right)$, $\mathfrak{q} : \mathscr{Q} \to \mathscr{R}$ *is given by* $\mathfrak{q} = \mathfrak{q}_T + \mathfrak{q}_1 + \ldots + \mathfrak{q}_k$ *where* $\mathfrak{q}_T$, $\mathfrak{q}_j$*'s are regarded as compositions with the projections from* $\mathscr{Q}$ *to each summond,* $\mathfrak{m} = (\mathfrak{m}_T, \mathfrak{m}_1, \ldots, \mathfrak{m}_k)$, *then* $(\mathscr{Q}^{\sim}, \mathfrak{q}, \mathfrak{m})$ *is a reduced module with associated process* $\mathfrak{p}$ *and thus* $\mathscr{Q}^{\sim} \equiv \mathscr{Q}_T \oplus \left(\bigoplus_{j=1}^{k} \mathscr{Q}_j \right)$, *where* $\mathscr{Q}$ *and* $\mathscr{Q}^{\sim}$ *are equivalent modulo* $\mathscr{N}$.

Proof. The detailed proof of the above decomposition can be found in Ahmad and Hickey [1]. For completeness we shall briefly outline the main arguments involved. To establish the result, first one constructs $\mathscr{A}_S$-module structures for $\mathscr{Q}_T$ and each $\mathscr{Q}_j$ by extending the $\mathscr{A}_{T,S_j}$-module and structures, respectively, by $\mathscr{R}$-linearity to all of $\mathscr{A}_S$. Then one shows that $\mathscr{Q}^{\sim}$ is cyclic with generator $\mathfrak{m}$, that is, $\mathscr{Q}^{\sim} = \mathscr{A}_S\mathfrak{m}$. Next, one proves that if $\mathfrak{q} = \mathfrak{q}_T + \mathfrak{q}_1 + \ldots + \mathfrak{q}_k$, then $\mathfrak{p}(\xi) = \mathfrak{q}(\xi\mathfrak{m})$ for every $\xi \in \mathscr{A}_S$. Finally, one completes the proof by showing that $\mathfrak{q}$ has no nonzero submodules. Q.E.D.

In the previous terminology if $\mathfrak{p}_T$, $\mathfrak{p}_j$, $(j = 1, 2, \ldots, k)$ are the quasiprocesses associated with $(\mathscr{Q}_T, \mathfrak{q}_T, \mathfrak{m}_T)$, $(\mathscr{Q}_j, \mathfrak{q}_j, \mathfrak{m}_j)$, respectively, then $\mathfrak{p}$ can be defined as a direct sum of these quasi-processes by $\mathfrak{p} = \mathfrak{p}_T \oplus \mathfrak{p}_1 \oplus \ldots \oplus \mathfrak{p}_k$.

To study the relationship between probability measures and linear functionals first we observe that the stochastic algebraic space $(\mathscr{A}_S, \mathfrak{p})$ generates the underlying probability space $(\mathscr{F}, \mathbf{P})$, where $\mathscr{F}$ is the smallest sigma-field. We say that $(\mathscr{A}_S, \mathfrak{p})$ is *isometric* to $(\mathscr{A}_{S^*}, \mathfrak{p}^*)$ if and only if there exists an isomorphic map $\mathbf{h}$ of $\mathscr{A}_S$ onto $\mathscr{A}_{S^*}$ such that $\mathfrak{p}(\xi) = \mathfrak{p}^*(\mathbf{h}(\xi))$ for all $\xi \in \mathscr{A}_S$. Clearly, if we have independent random variables, then $(\mathscr{A}_S, \mathfrak{p})$ is isometric to the usual product probability space. Now, with a suitably chosen metric distance $\varrho(\mathbf{x}, \mathbf{y})$ for all $(\mathbf{x}, \mathbf{y}) \in \mathscr{A}_S \times \mathscr{A}_S$, we can consider $\mathscr{A}_S$ as a *metrizable topological space.* Then the concept of metric convergence can be introduced for various limit theorems. An alternative but equivalent approach is given by Kappos [13], who has studied the so called probability algebras and stochastic spaces on the basis of lattice theory. In a similar way, it is possible to study the algebraic and topological structure in the space of all generalized random variables over $(\mathscr{A}_S, \mathfrak{p})$ with values in any abstract space.

Let $\mathscr{X}$ be a metric space and $\mathfrak{B}(\mathscr{X})$ the space of all bounded real-valued continuous functions on this space. Then the following is an easy consequence of the above development.

THEOREM 3.2. *Let* ψ *be a nonnegative linear functional on* $\mathfrak{B}(\mathscr{X})$ *such that* $\psi(1) = 1$. *Then there exists a unique finitely additive regular measure* μ *on* $\mathscr{F}(\mathscr{X})$, *the algebra generated by all the open subsets of* $\mathscr{X}$, *such that* $\psi(\mathbf{g}) = \int \mathbf{g}\, d\mu$,

$g \in \mathfrak{B}(\mathscr{X})$. Conversely, if μ is a finitely additive measure on $\mathscr{F}(\mathscr{X})$, then the map $\psi : g \to \int g \, d\mu$ is nonnegative, linear, and $\psi(1) = 1$. Furthermore, if $\mathscr{X}$ be a compact metric space and ψ be a nonnegative linear map on $\mathfrak{B}(\mathscr{X})$ with $\psi(1) = 1$, then there exists a unique measure on the Borel sigma-field $\mathscr{B}(\mathscr{X})$ such that $\psi(g) = \int g \, d\mu$, $g \in \mathfrak{B}(\mathscr{X})$.

An excellent account of interrelationships between various conditions related to stochastic independence was given by Bell [2]. In discussing the structure of stochastic independence Bell points out that any space exhibiting independence is in some way an image of a product space. We shall here mention an applicability of these ideas in the case of stochastic modules for r-dependent processes. We say $\{\mathbf{X}_j, j \geqq 1\}$ is *r-dependent process* if $\{\mathbf{X}_1, \mathbf{X}_2, \ldots, \mathbf{X}_k\}$ and $\{\mathbf{X}_n, \mathbf{X}_{n+1}, \ldots, \mathbf{X}_{n+t}\}$ are stochastically independent if $n - k > r$. Notice that a rth order Markov process is clearly distinct from a r-dependent process. By using Bell's ideas this definition of r-dependence can be extended if we replace the '*stochastic independence*' by '*σ-independence*', '*almost-σ-independence*', and '*quasi-σ-independence*', respectively. Now, for these extended r-dependent processes, either by considering the induced countable family of probability spaces or the underlying stochastic modules, one can develop some relationships as in Bell [2] between various conditions related to stochastic r-dependence.

Finally, we may remark that the above approach can also be employed in connection with controlled-Markov processes, semi-Markov processes and semilinear-Markov processes to develop various structural results. Notice that the theory of semilinear-Markov processes is, in a sense, a generalization of classical renewal theory. One interesting area of further research is to use this abstract algebraic approach to establish various local and global limit theorems for the stochastic processes mentioned above.

REFERENCES

[1] R. Ahmad and R. J. Hickey: On the theory of stochastic R[S]-modules. Manuscript, 1973.

[2] C. B. Bell: On the structure of stochastic independence. Illinois J. Math. *2* (1958), 415—424.

[3] D. Blackwell and L. Koopman: On the identifiability problem for functions of finite Markov chains. Ann. Math. Statist. *28* (1957), 1011—1015.

[4] P. E. Boudreau: Functions of finite Markov chains and exponential type process. Ph. D. Thesis, The University of Michigan 1965.

[5] C. J. Burke and M. Rosenblatt: A Markovian function of a Markov chain. Ann. Math. Statist. *29* (1958), 1111—1122.

[6] S. W. Dharmadhikari: Functions of finite Markov chains. Ann. Math. Statist. *34* (1963), 1022—1032.

[7] S. W. Dharmadhikari: Sufficient conditions for a stationary process to be a function of a finite Markov chain. Ann. Math. Statist. *34* (1963), 1033—1041.

[8] S. W. Dharmadhikari: A characterization of a class of functions of finite Markov chains. Ann. Math. Statist. *36* (1965), 524—528.

[9] R. V. Erickson: Functions of finite Markov chains. Ph. D. Thesis, The University of Michigan 1968.
[10] W. Feller: An Introduction to Probability Theory and Its Applications, 1, 2. Wiley, New York 1968, 1966.
[11] E. J. Gilbert: On the identifiability problem for functions of finite Markov chains. Ann. Math. Statist. *30* (1959), 688–697.
[12] A. Heller: On stochastic processes derived from Markov chains. Ann. Math. Statist. *36* (1965), 1286–1291.
[13] D. A. Kappos: Probability Algebras and Stochastic Spaces. Academic Press, London 1969.
[14] F. W. Leysieffer: Functions of finite Markov chains. Ph. D. Thesis, The University of Michigan, 1964.
[15] F. W. Leysieffer: Functions of finite Markov chains. Ann. Math. Statist., *38* (1967), 206 to 212.
[16] M. Rosenblatt: Functions of a Markov process that are Markovian. J. Math. Mech. *8* (1959), 585–596.
[17] R. Sikorski: Boolean Algebra. Second edition. Springer Verlag, Berlin 1964.
[18] R. Sikorski: Independent fields and cartesian products. Studia Math. *11* (1950), 171–184.
[19] R. Sikorski: Cartesian products of Boolean algebras. Fund. Math., *37* (1950), 25–54.
[20] R. E. Zink: On the structure of measure spaces. Acta Math. *107* (1962), 53–71.

University of Strathclyde

Department of Mathematics

ON INTERCONNECTED BIRTH AND DEATH PROCESSES WITH IMMIGRATION

MAGNAR AKSLAND

BERGEN

A finite number of populations each subject to a simple birth, death and immigration process is studied under the assumption of migration between the populations. Stochastic solutions for the number of individuals in each population at time t, $X_1(t), X_2(t), \ldots, X_n(t)$, are given for some special cases including the death, migration and immigration process and the case where migration occurs from one population only constituting individuals with zero birth rate.

For the general case a sequence of functions is given which converges uniformly to the p.g.f. of the joint probabilities for $X_1(t), X_2(t), \ldots, X_n(t)$.

The limiting behavior of the process without immigration is considered, and a set of algebraic equations for the extinction probabilities are given.

A sequence with recursively computable terms is given which converges to the probability of extinction in the general case, and finally a necessary and sufficient condition that the extinction probability be one is obtained.

I

Consider n colonies of particles and let $X_i(t)$ be the number of particles in colony i at time t where we have numbered the colonies from 1 to n.

The particles in colony i undergoes a linear birth, death process with birth rate λ_i and death rate μ_i. In addition they migrates to colony j with rate ν_{ij}. Further particles immigrates to colony i from outside with rate ϱ_i. The rates λ_i, μ_i, ϱ_i and ν_{ij}, $i = 1, 2, \ldots, n$, $j = 1, 2, \ldots, n$, are timedependent unless otherwise specified, and $\nu_{ii} \equiv 0$. Furthermore the necessary statements about independent behaviour of the particles to make $X(t) = (X_1(t), \ldots, X_n(t))$ an n-dimensional Markov-process are assumed.

Special cases of this model, especially the case $n = 2$, has been studied and applied in biology, for instance to describe bacterial growth and mutation. (See Armitage [1].)

The n-dimensional model is later studied by Puri [2] and Renshaw [3].

Let $G(\boldsymbol{u}; t)$ where $\boldsymbol{u} = (u_1, u_2, \ldots, u_n)$, be the probability-generating function (p.g.f.) of the joint probabilities of $(X_1(t), \ldots, X_n(t))$.

The Kolmogorov forward partial differential equation for G is obtained by the standard method and yields.

$$\frac{\partial G}{\partial t} + \sum_{i=1}^{n} \{(1 - u_i)(\lambda_i u_i - \mu_i) + \sum_{j=1}^{n} \nu_{ij}(u_i - u_j)\} \frac{\partial G}{\partial u_i} + \\ + \sum_{i=1}^{n} \varrho_i (1 - u_i) G = 0 . \tag{1}$$

This equation is intractable even if all the rates are independent of time.

However, it can be used to obtain fitst and second-order moments.

II

Let $g_i(\boldsymbol{u}; t)$ be the p.g.f. given that $X_i(0) = 1$, $X_j(0) = 0$, $j \neq i$, $i = 1, 2, \ldots, n$, and $\varrho_k = 0$, $k = 1, 2, \ldots, n$, and let $g_0(\boldsymbol{u}; t)$ be the p.g.f. given that $X_i(0) = 0$, $i = 1, 2, \ldots, n$. Assuming constant rates, a system of ordinary differential equations for $g_0, g_1, \ldots, g_n$ can be obtained.

Following a method used by Puri [2] we end up with

$$\frac{\mathrm{d}g_i}{\mathrm{d}t} = \lambda_i g_i^2 - \theta_i g_i + \sum_{j=1}^{n} \nu_{ij} g_j + \mu_i , \tag{2}$$

$$g_i(\boldsymbol{u}; 0) = u_i , \quad i = 1, 2, \ldots, n ,$$

where

$$\theta_i = \lambda_i + \mu_i + \sum_{j=1}^{n} \nu_{ij}$$

and

$$\frac{\mathrm{d}g_0}{\mathrm{d}t} = - \sum_{k=1}^{n} \varrho_k (1 - g_k) g_0 , \tag{3}$$

$$g_0(\boldsymbol{u}; 0) = 1 .$$

Because of the independent behaviour of the particles we must have:

$$G(\boldsymbol{u}; t) = g_0(\boldsymbol{u}; t) \, G(g_1(\boldsymbol{u}; t), g_2(\boldsymbol{u}; t), \ldots, g_n(\boldsymbol{u}; t); 0) .$$

Hence, there is no loss of generality in studying g_i, $i = 0, 1, 2, \ldots, n$, only.

Being a system of Riccati equations, the general form of (2) is intractable. However, putting $\lambda_i = 0$, $i = 1, 2, \ldots, n$, the solution is:

$$g_k(t) = 1 - \sum_{i=1}^{n} q_i^{(k)} \left(\sum_{j=1}^{n} r_i^{(j)} \right) e^{\omega_i t} + \sum_{j=1}^{n} \left(\sum_{i=1}^{n} r_i^{(j)} q_i^{(k)} e^{\omega_i t} \right) u_j \,. \tag{4}$$

From (3) and (4) it follows that

$$g_0(t) = b(t) \exp\left\{ \sum_{j=1}^{n} a_j(t)\, u_j \right\} = \tag{5}$$

$$= b(t) \sum_{(k_1,\ldots,k_n)} \frac{(a_1(t))^{k_1} (a_2(t))^{k_2} \ldots (a_n(t))^{k_n}}{k_1!\, k_2! \ldots k_n!} u_1^{k_1} u_2^{k_2} \ldots u_n^{k_n}$$

where the sum is over all n-tuples with k_i, $i = 1, 2, \ldots, n$, nonnegative integers,

$$a_j(t) = \sum_{i=1}^{n} \frac{r_i^{(j)}}{\omega_i} (e^{\omega_i t} - 1) \sum_{k=1}^{n} \varrho_k q_i^{(k)}$$

and

$$b(t) = \exp\left\{ - \sum_{j=1}^{n} a_j(t) \right\} .$$

In the case $\omega_i = 0$ for some i, $(e^{\omega_i t} - 1)/\omega_i$ takes the value $\lim_{\omega \to 0} [(e^{\omega t} - 1)/\omega] = t$. Here ω_i, $i = 1, 2, \ldots, n$, are the eigenvalues (assumed distinct) and $(r_i^{(1)}, \ldots, r_i^{(n)})$ and $\begin{Bmatrix} q_i^{(1)} \\ \vdots \\ q_i^{(n)} \end{Bmatrix}$ the corresponding row and column eigenvectors of the matrix

$$\begin{Bmatrix} -\mu_1 + \sum_{j=1}^{n} \nu_{1j} & \nu_{12} & \ldots & \nu_{1n} \\ \nu_{21} & -\mu_2 + \sum_{j=1}^{n} \nu_{2j} & \ldots & \nu_{2n} \\ \nu_{n1} & \nu_{n2} & \ldots & -\mu_n + \sum_{j=1}^{n} \nu_{nj} \end{Bmatrix} .$$

The following special case is a timeinhomogeneous process. Let $\lambda_1 = \nu_{kj} = 0$, $k = 2, 3, \ldots, n$; $j = 1, 2, \ldots, n$, and the remaining rates timedependent.

Equation (1) then takes the form:

$$\frac{\partial G}{\partial t} + \left\{ -\mu_1(t)(1 - u_1) + \sum_{j=1}^{n} \nu_j(t)(u_1 - u_j) \right\} \frac{\partial G}{\partial u_1} +$$

$$+ \sum_{i=2}^{n} (1 - u_i)(\lambda_i(t) u_i - \mu_i(t)) \frac{\partial G}{\partial u_i} + \sum_{i=1}^{n} \varrho_i(t)(1 - u_i)\, G = 0$$

where we write $\nu_j(t)$ instead of $\nu_{1j}(t)$.

Solving this equation subject to the initial condition $G(\boldsymbol{u}; 0) = 1$ gives:

$$G(\boldsymbol{u}; t) = \exp\{-S(\boldsymbol{u}; t)\}$$

where

(6)
$$S(\boldsymbol{u}; t) = \left[(1 - u_1)\, e^{\beta(t)} + \int_0^t \sum_{j=2}^n v_j(\tau)\, H_j(u_j, \tau, t)\, e^{\beta(\tau)}\, d\tau\right] \cdot$$

$$\cdot \int_0^t \varrho_1(\tau)\, e^{-\beta(\tau)}\, d\tau - \int_0^t \left[\varrho_1(\tau)\, e^{-\beta(\tau)} \int_0^\tau \sum_{j=2}^n v_j(s)\, H_j(u_j, s, \tau)\, e^{\beta(s)}\, ds\right] d\tau +$$

$$+ \int_0^t \sum_{j=2}^n \varrho_j(\tau)\, H_j(u_j, \tau, t)\, d\tau\,,$$

$$H_j(u, \tau, t) = \frac{e^{\alpha_j(\tau)}(1 - u)}{(1 - u)\displaystyle\int_\tau^t \lambda_j(s)\, e^{\alpha_j(s)}\, ds + e^{\alpha_j(t)}}\,,$$

$$\alpha_j(t) = \int_0^t (\mu_j(\tau) - \lambda_j(\tau))\, d\tau\,,$$

$$\beta(t) = -\int_0^t \Big(\mu_1(\tau) + \sum_{j=2}^n v_j(\tau)\Big)\, d\tau\,.$$

Now we give a method of approximation to the solution of (2) in the general case. Put

$$\boldsymbol{g}(\boldsymbol{u}; t) = (g_1(\boldsymbol{u}; t), \ldots, g_n(\boldsymbol{u}; t))$$

and define the functions

$$\boldsymbol{f}^{(0)}, \boldsymbol{f}^{(1)}, \ldots$$

by

$$\boldsymbol{f}^{(0)} = \boldsymbol{u}\,,$$

$$f_i^{(m+1)} = u_i \exp\left\{\lambda_i \int_0^t f_i^{(m)}(\boldsymbol{u}; \tau)\, d\tau - \theta_i t\right\} +$$

$$+ \int_0^t \exp\left\{\lambda_i \int_\tau^t f_i^{(m)}(\boldsymbol{u}; s)\, ds - \theta_i(t - \tau)\right\} \Big(\sum_{j=1}^n v_{ij} f_j^{(m)}(\boldsymbol{u}; \tau) + \mu_i\Big)\, d\tau\,,$$

$$i = 1, 2, \ldots, n\,, \quad m = 0, 1, 2, \ldots\,,$$

where

$$\boldsymbol{f}^{(m)} = (f_1^{(m)}, f_2^{(m)}, \ldots, f_n^{(m)})\,.$$

Let

$$[\boldsymbol{0}, \boldsymbol{1}] = \mathop{\times}_{j=1}^{n} [0, 1]\,.$$

It can be proved that the sequence $\{f^{(m)}\}$ converges to $g(u; t)$ uniformly for $u \in [0, 1]$ and $t \in [0, \infty)$. Further the functions $f_i^{(m)}(u; t)$, $i = 1, 2, \ldots, n$; $m = 0, 1, 2, \ldots$, are p.g.f's.

Actually the uniform convergence of $\{f^{(m)}\}$ to g holds for any integrable $f^{(0)}(u; t)$ satisfying $0 \leqq f_i^{(0)} \leqq 1$ and $f_i^{(0)} \not\to 1$ when $t \to \infty$, $i = 1, 2, \ldots, n$, for $u \in [0, 1)$.

III

It can be proved that $q(u) = \lim_{t \to \infty} g(u; t)$ exists. With $q = (q_1, q_2, \ldots, q_n)$ it follows from (2) that

$$\lambda_i q_i^2 - \theta_i q_i + \sum_{j=1}^{n} \nu_{ij} q_j + \mu_i = 0\,, \quad i = 1, 2, \ldots, n\,. \tag{7}$$

We will discuss equation (7) under the following assumption.

For every pair of colonies considered, the probability that a particle belonging to one of them migrates, not necessarily directly, to the other, is positive.

Let $[0, 1] = \bigtimes_{i=1}^{n} [0, 1)$. Then q is constant on $[0, 1)$ and q_i is the probability of extinction of $X(t)$ given that $X_i(0) = 1$, $X_j(0) = 0$, $j \neq i$.

From now on let q denote the constant value of $q(u)$ on $[0. 1)$. The above assumption implies that either $q = 1 = (1, 1, \ldots, 1)$ or $q \in [0, 1)$.

Equation (7) cannot be solved, but the following can be proved.

The sequence $\{x^{(m)}\}$ where $x^{(m)} = (x_1^{(m)}, x_2^{(m)}, \ldots, x_n^{(m)})$ given by

$$x_i^{(m+1)} = \frac{\mu_i + \sum_{j=1}^{n} \nu_{ij} x_j^{(m)}}{\theta_i - \lambda_i x_i^{(m)}}\,, \qquad i = 1, 2, \ldots, n\,, \quad m = 0, 1, 2, \ldots\,, \tag{8}$$

is confined in $[0, 1)$ and converges to the point q for each $x^{(0)} \in [0, 1)$.

A necessary and sufficient condition that $q = 1$ (i.e. $X(t)$ dies out with probability one) is now that $\{x^{(m)}\}$ converges to 1 for an $x^{(0)} \in [0, 1)$.

Using this, the following theorem can be proved.

THEOREM. Let $C = \{c_{ij}\}$ be the $n \times n$ matrix where

$$c_{ij} = \nu_{ij}\,, \quad i \neq j\,,$$

$$c_{ii} = \lambda_i - \mu_i - \sum_{k=1}^{n} \nu_{ik}\,.$$

It is necessary and sufficient for $\boldsymbol{q} = \boldsymbol{1}$ that there exists a vector $\boldsymbol{\varepsilon} = (\varepsilon_1, \varepsilon_2, \ldots, \varepsilon_n)$ where $\varepsilon_i > 0$, $i = 1, 2, \ldots, n$, with the property that each component of $\boldsymbol{C\varepsilon}$ is non-positive.

REFERENCES

[1] P. ARMITAGE: The Statistical Theory of Bacterial Population Subject to Mutation. J. R. Statist. Soc. B *14* (1952), 1—33.
[2] PREM. S. PURI: Interconnected Birth and Death processes. J. Appl. Prob. *5* (1968), 334—349.
[3] E. RENSHAW: Birth, death and migration processes. Biometrica *59* (1972), 1—14.

UNIVERSITY OF BERGEN

DEPARTMENT OF APPLIED MATHEMATICS

ON THE BACKWARD EXTRAPOLATION OF NON-STATIONARY AUTOREGRESSIVE SERIES

JIŘÍ ANDĚL

PRAGUE

Let $X_1, \ldots, X_N$ be a non-stationary p-dimensional autoregressive series of the order n, where $n < N$. Suppose that m and k are integers such that $1 \leqq m \leqq N - n$, $0 \leqq k \leqq m - 1$. It is proved that the best linear extrapolation of X_{m-k} based on $X_{m+1}, \ldots, X_N$ does not depend on $X_{m+n+1}, \ldots, X_N$. Some formulas for the covariance function of the non-stationary autoregressive series are given. Several special stationary cases are discussed as a consequence of the presented general theorem.

1. PRELIMINARIES

Let $X_1, \ldots, X_n$ be p-dimensional random vectors with vanishing means and finite second moments. Denote

(1) $$B = \operatorname{Var}(X_1', \ldots, X_n')'$$

the variance matrix of the random vector $(X_1', \ldots, X_n')'$. We suppose B to be regular.

Let $Y_{n+1}, \ldots, Y_N$ be p-dimensional random vectors such that

$$\mathrm{E}Y_t = 0\,, \quad \operatorname{Var} Y_t = I\,, \quad \operatorname{Cov}(Y_s, Y_t) = 0\,, \quad \operatorname{Cov}(X_k, Y_t) = 0$$

for $1 \leqq k \leqq n < s \neq t \leqq N$.

Define vectors $X_{n+1}, \ldots, X_N$ by the formula

(2) $$\sum_{j=0}^{n} A_j X_{t-j} = Y_t\,, \quad n < t \leqq N\,,$$

where $A_0, \ldots, A_n$ are given matrices of the type $p \times p$ with real elements such that $|A_0| \neq 0$, $A_n \neq 0$.

The series $X_1, \ldots, X_N$ is called general p-dimensional autoregressive series of the order n. Putting

(3) $$U_j = -A_0^{-1}A_j\,, \quad W_t = A_0^{-1}Y_t\,,$$

we obtain

(4) $$X_t = \sum_{j=1}^{n} U_j X_{t-j} + W_t\,, \quad t = n+1, \ldots, N\,.$$

Obviously,

(5) $$\mathrm{E}X_s Y_t' = 0$$

holds for any $1 \leqq s < t$ such, that $n < t \leqq N$ (not only for $s = 1, 2, \ldots, n$, which is assured by our assumptions).

Denote $R(s, t) = \mathrm{E}X_s X_t'$ $(1 \leqq s,\ t \leqq N)$ the covariance function of the series $X_1, \ldots, X_N$.

Lemma 1. *Let $\eta_1, \ldots, \eta_p$ and $\xi_1, \ldots, \xi_q$ be random variables with finite second moments. Denote $\mathfrak{M} = \mathfrak{M}(\xi_1, \ldots, \xi_q)$ the linear space based on $\xi_1, \ldots, \xi_q$. Denote $\eta = (\eta_1, \ldots, \eta_p)'$, $\xi = (\xi_1, \ldots, \xi_q)'$. Let $\hat{\eta}_1, \ldots, \hat{\eta}_p$ be the projections of $\eta_1, \ldots, \eta_p$ onto $\mathfrak{M}$ when the scalar product is defined by the covariance. If the matrix $G = \mathrm{Var}\,\xi$ is regular, then*

(6) $$\hat{\eta} = \mathrm{Cov}(\eta, \xi)\, G^{-1}\xi\,,$$

where $\hat{\eta} = (\hat{\eta}_1, \ldots, \hat{\eta}_p)'$.

Proof is clear.

2. BACKWARD EXTRAPOLATION

Let m be a natural number such that $1 \leqq m \leqq N - n$. Denote

(7) $$B^* = \mathrm{Var}\,(X_{m+1}', \ldots, X_{m+n}')'\,, \quad E^* = B^{*-1}\,.$$

Introduce following block matrices:

$$M = \begin{Vmatrix} 0 & I & 0 & \ldots & 0 \\ 0 & 0 & I & \ldots & 0 \\ \ldots & \ldots & \ldots & \ldots & \ldots \\ 0 & 0 & 0 & \ldots & I \\ U_n & U_{n-1} & U_{n-2} & \ldots & U_1 \end{Vmatrix}, \quad \Lambda = \begin{Vmatrix} 0 & 0 & \ldots & 0 \\ 0 & 0 & \ldots & 0 \\ \ldots & \ldots & \ldots & \ldots \\ 0 & 0 & \ldots & 0 \\ 0 & 0 & \ldots & (A_0'A_0)^{-1} \end{Vmatrix},$$

$$C = \left\| \begin{array}{cccc|cccccc} & E^{1/2} & & & & & 0 & & & \\ \hline A_n & A_{n-1} & \dots & A_1 & A_0 & 0 & \dots & 0 & \dots & 0 \\ 0 & A_n & \dots & A_2 & A_1 & A_0 & \dots & 0 & \dots & 0 \\ \dots & \dots & \dots & \dots & \dots & \dots & \dots & \dots & \dots & \dots \\ 0 & 0 & \dots & 0 & 0 & 0 & \dots & A_n & \dots & A_0 \end{array} \right\| .$$

The matrices M and A are of the type $np \times np$, the matrix C is of the type $(N - m)\, p \times (N - m)\, p$.

THEOREM 2. *The matrix* $G = \operatorname{Var}(X'_{m+1}, \dots, X'_N)'$ *is regular and*

$$G^{-1} = C'C \tag{8}$$

holds.

Proof is analogous to that of Theorem 6 in [1].

THEOREM 3. *Let k be an integer, $0 \leqq k \leqq m - 1$. The best linear extrapolation $\hat{X}_{m-k}$ of the random vector X_{m-k} based on $X_{m+1}, \dots, X_N$ is of the form*

$$\hat{X}_{m-k} = \sum_{j=1}^{n} V_{m,k,j} X_{m+j}\,, \tag{9}$$

where $V_{m,k,j}$ are matrices of the type $p \times p$.

Proof. According to Lemma 1 and Theorem 2 we get

$$\hat{X}_{m-k} = \operatorname{Cov}(X_{m-k}, [X'_{m+1}, \dots, X'_N]')\, G^{-1}[X'_{m+1}, \dots, X'_N]' =$$

$$= [R(m - k, m + 1), \dots, R(m - k, N)]\, C'C[X'_{m+1}, \dots, X'_N]' .$$

However, we have

$$C \left\| \begin{array}{c} R(m+1, m-k) \\ \dots\dots\dots \\ R(N, m-k) \end{array} \right\| = \left\| \begin{array}{cc} E^{*1/2} & \left\| \begin{array}{c} R(m+1, m-k) \\ \dots\dots\dots \\ R(m+n, m-k) \end{array} \right\| \\ & 0 \end{array} \right\|$$

because for $1 \leqq j \leqq N - m - n$

$$A_n R(m + j, m - k) + \dots + A_0 R(m + j + n, m - k) =$$

$$= \operatorname{Cov}\Big(\sum_{i=0}^{n} A_i X_{m+j+n-i}, X_{m-k}\Big) =$$

$$= \operatorname{Cov}(Y_{m+j+n}, X_{m-k}) = 0$$

in view of (5).

Write the matrix C in a new block form

$$C = \begin{Vmatrix} E^{*1/2} & 0 \\ C_1 & C_2 \end{Vmatrix} .$$

Then we obtain

$$\hat{X}_{m-k} = \{[R(m-k, m+1), \ldots, R(m-k, m+n)] \, E^{*1/2}, 0\} \,.$$

$$. \begin{Vmatrix} E^{*1/2} & 0 \\ C_1 & C_2 \end{Vmatrix} \begin{Vmatrix} X_{m+1} \\ \cdots\cdots \\ X_N \end{Vmatrix} =$$

$$= [R(m-k, m+1), \ldots, R(m-k, m+n)] \, E^*(X'_{m+1}, \ldots, X'_{m+n})' \,. \tag{10}$$

The proof is finished.

The meaning of formula (9) is that the backward extrapolation in general autoregressive series uses only n next terms which are known. That is quite analogous to the usual (forward) extrapolation. The result of Theorem 3 has been known in the stationary case; explicit formulas for $V_{m,1,j}$ are given in [3] (cf. Section 4 of the present paper).

3. EVALUATION OF THE COVARIANCE FUNCTION

A real use of formula (10) needs the covariance function $R(s, t)$ to be known. The following assertions may be useful for its evaluation.

THEOREM 4. *Let j, k be natural numbers such that $1 \leqq j \leqq k \leqq N + 1 - n$. Then*

$$\begin{Vmatrix} R(j,k), & \ldots, & R(j, k+n-1) \\ \cdots & \cdots & \cdots \\ R(j+n-1, k), & \ldots, & R(j+n-1, k+n-1) \end{Vmatrix} =$$

$$= \begin{Vmatrix} R(j,j), & \ldots, & R(j, j+n-1) \\ \cdots & \cdots & \cdots \\ R(j+n-1, j), & \ldots, & R(j+n-1, j+n-1) \end{Vmatrix} M'^{k-j} ,$$

$$\begin{Vmatrix} R(k,k), & \ldots, & R(k, k+n-1) \\ \cdots & \cdots & \cdots \\ R(k+n-1, k), & \ldots, & R(k+n-1, k+n-1) \end{Vmatrix} =$$

$$= M^{k-j} \begin{Vmatrix} R(j,j), & \ldots, & R(j, j+n-1) \\ \cdots & \cdots & \cdots \\ R(j+n-1, j), & \ldots, & R(j+n-1, j+n-1) \end{Vmatrix} .$$

$$. M'^{k-j} + \sum_{i=0}^{k-j-1} M^i \Lambda M'^i \,.$$

Proof. By the repeated use the formula

$$\begin{Vmatrix} X_t \\ \cdots\cdots \\ X_{t+n-2} \\ X_{t+n-1} \end{Vmatrix} = M \begin{Vmatrix} X_{t-1} \\ \cdots\cdots \\ X_{t+n-3} \\ X_{t+n-2} \end{Vmatrix} + \begin{Vmatrix} 0 \\ \cdots\cdots\cdots \\ 0 \\ A_0^{-1} Y_{t+n-1} \end{Vmatrix}$$

we obtain

$$\begin{Vmatrix} X_k \\ \cdots\cdots \\ X_{k+n-1} \end{Vmatrix} = M^{k-j} \begin{Vmatrix} X_j \\ \cdots\cdots \\ X_{j+n-1} \end{Vmatrix} + \sum_{i=0}^{k-j-1} M^i \begin{Vmatrix} 0 \\ \cdots\cdots\cdots\cdots \\ A_0^{-1} Y_{k+n-1-i} \end{Vmatrix}.$$

Now, the evaluation of covariance and variance matrices leads to formulas given in Theorem 4.

Note, that the first formula in Theorem 4 implies

$$(11) \qquad [R(m-k, m+1), \ldots, R(m-k, m+n)] = \\ = (I, 0, \ldots, 0) \operatorname{Var} (X'_{m-k}, \ldots, X'_{m-k+n-1})' M'^{k+1} .$$

4. SPECIAL CASES

In this Section we consider only one-step backward extrapolation (i.e. the case $k = 0$).

A. Multiple stationary autoregressive series

Suppose $X_1, \ldots, X_N$ to be stationary. The necessary and sufficient conditions for the stationarity are given in [1]. Denote $B^{-1} = E = \|E_{st}\|_{s,t=1}^n$, where E_{st} are $p \times p$ blocks. Obviously, in stationary case it holds $B^* = B$ and $E^* = E$.

COROLLARY 5. *If $X_1, \ldots, X_N$ is stationary, then*

$$(12) \qquad \hat{X}_m = \sum_{j=1}^{n} V_j X_{m+j} ,$$

where

$$(13) \qquad V_j = -(E_{11} + A_n' A_n)^{-1} (E_{1,j+1} + A_n' A_{n-j}) , \quad j = 1, 2, \ldots, n .$$

Proof. These formulas are proved in [3] by means of reversed autoregressive series (see [2]). Here we give another proof which is based on Theorem 3.

Formulas (10) and (11) imply

$$\hat{X}_m = (I, 0, \ldots, 0)\, BM'EX\,, \tag{14}$$

where

$$X = (X'_{m+1}, \ldots, X'_{m+n})'\,. \tag{15}$$

If $X_1, \ldots, X_N$ is stationary, then

$$B = MBM' + \Lambda \tag{16}$$

(see [1]). Suppose first, that the matrix U_n is regular. Then M is regular, too, and

$$M^{-1} = \begin{Vmatrix} A & U_n^{-1} \\ D_I & 0 \end{Vmatrix} \tag{17}$$

where

$$A = -U_n^{-1}(U_{n-1}, \ldots, U_1)\,, \quad D_I = \operatorname{diag}(I, \ldots, I)\,. \tag{18}$$

From (16) we get

$$BM'E = M^{-1} - M^{-1}\Lambda E\,. \tag{19}$$

Using (17) we see that

$$M^{-1} - M^{-1}\,\Lambda\, E = \begin{Vmatrix} W_1, & \ldots, & W_n \\ 0, & \ldots, & 0 \\ \ldots & \ldots & \ldots \\ 0, & \ldots, & 0 \end{Vmatrix}\,, \tag{20}$$

where

$$W_j = -U_n^{-1}[U_{n-j} + (A'_0A_0)^{-1}\, E_{nj}]\,, \quad j = 1, 2, \ldots, n\,. \tag{21}$$

(We put $U_0 = -I$.)

It is known that

$$A'_{n-s}A_{n-t} = E_{st} - E_{s+1,t+1} + \tag{22}$$

$$+ (E_{s+1,1} + A'_{n-s}A_n)(E_{11} + A'_nA_n)^{-1}(E_{1,t+1} + A'_nA_{n-t})$$

for $s, t = 1, 2, \ldots, n$, where $E_{st} = 0$ if $\max(s, t) > n$.

The proof of (22) is given in [2]. Putting $s = n$, $t = j$, we obtain

(23) $$A_0' A_{n-j} = E_{nj} + A_0' A_n (E_{11} + A_n' A_n)^{-1} (E_{1,j+1} + A_n' A_{n-j}) .$$

An easy modification of (23) gives $W_j = V_j$. Thus

$$(I, 0, \ldots, 0) BM'E = (I, 0, \ldots, 0)(M^{-1} - M^{-1} A E) = (V_1, \ldots, V_n) .$$

If U_n is singular, we obtain this result by a limit procedure choosing an appropriate sequence of regular matrices $U_n^{(k)}$ which converges to U_n for $k \to \infty$.

B. Symmetric stationary autoregressive series

Denote $R(t)$ the covariance function of a stationary autoregressive series. If $R(t) = R(-t)$ for any integer t, the series is called symmetric. The series $X_1, \ldots, X_N$ is symmetric if and only if

(24) $$\sum_{k=0}^{n-k} A_{h+k}' A_k = \sum_{k=0}^{n-h} A_k' A_{k+h} , \quad h = 1, 2, \ldots, n .$$

Note, that (24) holds trivially for $h = 0$, too. If $X_1, \ldots, X_N$ is symmetric, then

(25) $$E_{st} = \sum_{k=1}^{\min(s,t)} (A_{s-k}' A_{t-k} - A_{n+k-s}' A_{n+k-t})$$

for $s, t = 1, 2, \ldots, n$. A proof of (24) and (25) is given in [2].

COROLLARY 6. *If $X_1, \ldots, X_N$ is a symmetric stationary autoregressive series, then*

(26) $$\hat{X}_m = \sum_{j=1}^{n} U_j X_{m+j} ,$$

where U_j are given in (3).

Proof. According to (25) we get

(27) $$E_{1,j+1} = A_0' A_j - A_n' A_{n-j} , \quad j = 1, 2, \ldots, n-1 .$$

In view of (13)

$$V_j = -(A_0' A_0)^{-1} (A_0' A_j - A_n' A_{n-j} + A_n' A_{n-j}) = -A_0^{-1} A_j = U_j$$

for $j = 1, 2, \ldots, n-1$;

$$V_n = -(A_0' A_0)^{-1} A_n' A_0 = -A_0^{-1} A_0'^{-1} A_n' A_0 = -A_0^{-1} A_0'^{-1} A_0' A_n = U_n$$

because (24) implies $A_n' A_0 = A_0' A_n$.

C. One-dimensional stationary autoregressive series

Any one-dimensional stationary autoregressive series is symmetric. It is the well-known property and it can be easily checked by (24), because $\boldsymbol{A}_j$ are 1×1 matrices. Therefore, Corollary 6 is applicable and the one-step backward extrapolation is given by (26).

REFERENCES

[1] J. Anděl: On the multiple autoregressive series. Ann. Math. Statist. *42* (1971), 755—759.
[2] J. Anděl: Symmetric and reversed multiple stationary autoregressive series. Ann. Math. Statist. *43* (1972), 1197—1203.
[3] J. Anděl: On multiple random sequences. In: Proc. Prague Conf. on Asympt. Methods of Statist., Vol. I (1974), 15—23.

Charles University

Department of Statistics

ON MAXIMUM LIKELIHOOD ESTIMATION OF PARAMETERS OF AUTOREGRESSIVE MOVING AVERAGE PROCESSES*

T. W. ANDERSON

STANFORD

1. INTRODUCTION

A stationary stochastic process that serves as a useful model for time series analysis is the autoregressive process with moving average residuals $\{y_t\}$ which satisfies

$$\sum_{j=0}^{p} \beta_j y_{t-j} = \sum_{g=0}^{q} \alpha_g v_{t-g}, \tag{1}$$

$t = \ldots, -1, 0, 1, \ldots$, where the sequence $\{v_t\}$ consists of independently identically distributed (unobservable) random variables. To avoid indeterminancy $\beta_0 = \alpha_0 = 1$. The mean of v_t is independent of t and is taken to be 0 for convenience; the variance of v_t is σ^2. We shall assume that the v_t's are normally distributed, that is, that the process is Gaussian.

If we let the right-hand side of (1) be u_t, the covariances of the (unobservable) process $\{u_t\}$ are

$$\begin{aligned} \mathscr{E} u_t u_{t+s} = \sigma_s &= \sigma^2 \sum_{k=0}^{q-s} \alpha_k \alpha_{k+s}, \qquad s = 0, 1, \ldots, q, \\ &= 0, \qquad\qquad\qquad s > q. \end{aligned} \tag{2}$$

In this paper we consider the estimation of the parameters $\beta_1, \ldots, \beta_p, \sigma_0, \sigma_1, \ldots, \sigma_q$ on the basis of a set of $T(>p + q)$ observations.

The method of maximum likelihood can be considered, but in general an explicit solution cannot be found. The approach of this paper is to modify the model slightly so that the derivatives of the likelihood function set equal to 0 yield relatively simple

* Research supported by the U.S. Office of Naval Research under Contract Number N00014-67-A-0112-0030. The author thanks Paul Shaman for helpful discussions.

equations. Since these equations are nonlinear, an iterative procedure is proposed that yields asymptotically efficient estimates at the first step (as $T \to \infty$).

For the model (1) the estimates suggested by Durbin (1959), (1960) and Walker (1961), (1962) are not asymptotically efficient as $T \to \infty$. Box and Jenkins (1970) have suggested maximizing the likelihood function by numerical means. Clevenson (1970) and Parzen (1971) and Hannan (1969) have proposed estimation methods based on the sample spectral density (the so-called periodogram). The complete references are given by Anderson in [2] and [3]. In these latter two papers the maximum likelihood equations are obtained (in the time domain) and iterative procedures suggested (pointed out by J. N. K. Rao as the method of scoring).

2. ESTIMATES IN A MODIFIED MODEL

Let $\boldsymbol{u}$ be the (column) vector $(u_1, \ldots, u_T)'$. Its covariance matrix is

$$\mathscr{C}(\boldsymbol{u}) = \mathscr{E}\boldsymbol{u}\boldsymbol{u}' = \Sigma = \sum_{g=0}^{q} \sigma_g G_g , \tag{3}$$

where $G_0 = I$, $G_g = L^g + L'^g$, $g = 1, \ldots, q$, and L is the $T \times T$ matrix with 1's in the $(t+1, t)$ position, $t = 1, \ldots, T-1$, and 0's elsewhere. Define the random vector $\boldsymbol{y} = (y_1, \ldots, y_T)'$ by

$$\sum_{j=0}^{p} \beta_j L^j \boldsymbol{y} = \boldsymbol{u} . \tag{4}$$

The components of $\boldsymbol{y}$ are generated by (1) for $t = 1, \ldots, T$ if $y_0, y_{-1}, \ldots, y_{-p+1}$ are taken as 0. When $\boldsymbol{u}$ is normally distributed, the derivatives of the likelihood set equal to 0 lead to solving the following equations for the maximum likelihood estimates:

$$\operatorname{tr} \Big(\sum_{g=0}^{q} \hat{\sigma}_g G_g\Big)^{-1} G_f = \sum_{k,j=0}^{p} \hat{\beta}_k \hat{\beta}_j \boldsymbol{y}' L'^k \Big(\sum_{g=0}^{q} \hat{\sigma}_g G_g\Big)^{-1} G_f \Big(\sum_{g=0}^{q} \hat{\sigma}_g G_g\Big)^{-1} L^j \boldsymbol{y} , \tag{5}$$

$$f = 0, 1, \ldots, q ,$$

$$\sum_{k=1}^{p} \hat{\beta}_k \boldsymbol{y}' L'^k \Big(\sum_{g=0}^{q} \hat{\sigma}_g G_g\Big)^{-1} L^j \boldsymbol{y} = -\boldsymbol{y}' \Big(\sum_{g=0}^{q} \hat{\sigma}_g G_g\Big)^{-1} L^j \boldsymbol{y} , \quad j = 1, \ldots, p . \tag{6}$$

Except in very special cases these equations cannot be solved explicitly. Hence, we develop an iterative procedure based on the method of scoring. Let $\hat{\sigma}_0^{(i)}, \hat{\sigma}_1^{(i)}, \ldots$ $\ldots, \hat{\sigma}_q^{(i)}, \hat{\beta}_1^{(i)}, \ldots, \hat{\beta}_p^{(i)}$ be the estimated at the i-th iteration, $i = 1, 2, \ldots$. ($\hat{\beta}_0^{(i)} \equiv 1$.)

The linear equations to be solved are

$$(7)\qquad \sum_{h=0}^{q} \operatorname{tr} (\Sigma_{i-1}^{e})^{-1} G_f(\Sigma_{i-1}^{e})^{-1} G_h\hat{\sigma}_h^{(i)} - 2\sum_{k=1}^{p} \operatorname{tr} G_f(\Sigma_{i-1}^{e})^{-1} L^k(B_{i-1}^{e})^{-1} \hat{\beta}_k^{(i)} =$$

$$= -2 \operatorname{tr} G_f(\Sigma_{i-1}^{e})^{-1} + y'(B_{i-1}^{e})' (\Sigma_{i-1}^{e})^{-1} G_f(\Sigma_{i-1}^{e})^{-1} B_{i-1}^{e} y +$$

$$+ 2 \operatorname{tr} G_f(\Sigma_{i-1}^{e})^{-1} (B_{i-1}^{e})^{-1}, \quad f = 0, 1, \ldots, q,$$

$$(8)\qquad -2\sum_{h=0}^{q} \operatorname{tr} G_h(\Sigma_{i-1}^{e})^{-1} L^j(B_{i-1}^{e})^{-1} \hat{\sigma}_h^{(i)} +$$

$$+ 2\sum_{k=1}^{p} \operatorname{tr} (B_{i-1}^{e})^{-1} \Sigma_{i-1}^{e}(B_{i-1}^{e})'^{-1} L'^k(\Sigma_{i-1}^{e})^{-1} L^j\hat{\beta}_k^{(i)} =$$

$$= -2y'(B_{i-1}^{e})' (\Sigma_{i-1}^{e})^{-1} L^j y - 2 \operatorname{tr} (B_{i-1}^{e})^{-1} \Sigma_{i-1}^{e}(B_{i-1}^{e})'^{-1} (\Sigma_{i-1}^{e})^{-1} L^j,$$

$$j = 1, \ldots, p,$$

where

$$\Sigma_{i-1}^{e} = \sum_{g=0}^{q} \hat{\sigma}_g^{(i-1)} G_g$$

and

$$B_{i-1}^{e} = \sum_{k=0}^{p} \hat{\beta}_k^{(i-1)} L^k .$$

There are $p + q + 1$ linear equations in $p + q + 1$ unknowns. If the initial estimates are consistent and of order $T^{-1/2}$, then the solution for $i = 1$ is consistent, asymptotically normal, and asymptotically efficient as $T \to \infty$. The information matrix divided by T, which is the inverse of the covariance matrix of the asymptotic distribution is estimated consistently by the coefficients on the left-hand sides of (7) and (8) for $i = 1$.

3. COMPUTATION OF QUADRATIC FORMS

At each step of the iteration the observed time series y enters into an equation in terms of a quadratic form. The t-th component of the vector $L^j y$ is 0 for $t = 1, \ldots, j$ and is y_{t-j} for $t = j + 1, \ldots, T$. The other vector involved is

$$(9)\qquad \hat{u}_{i-1} = B_{i-1}^{e} y = \sum_{k=0}^{p} \hat{\beta}_k^{(i-1)} L^k y ;$$

the components of $\hat{u}_{i-1}$ are linear combinations of at most $p + 1$ successive components of y.

The rest of the computation involves $(\Sigma_{i-1}^{e})^{-1}\hat{u}_{i-1}$ or $(\Sigma_{i=1}^{e})^{-1}(L^j y)$. In either case the desired vector is the solution of an equation of the type

(10) $$\Sigma z = x\,,$$

where Σ has the form (3). Let $\varrho_h = \sigma_h/\sigma_0$, and let the $T \times T$ matrix $(1/\sigma_0)\,\Sigma$ be

(11) $$R = \sum_{g=0}^{q} \varrho_g G_g\,.$$

This matrix has 1 on the main diagonal, ϱ_1 on the diagonals directly above and below the main diagonal, ..., ϱ_q on the diagonals q above and below the main diagonal and 0's elsewhere. The forward solution of a method of pivotal condensation or successive elimination can be represented as the multiplication of R on the left by $F = F_T F_{T-1} \ldots F_2 F_1$, where $F_1 = I$. Here F_2 represents subtracting ϱ_1 times the first row of $F_1 R = R$ from the second, ϱ_2 times the first row from the third, ..., and ϱ_q times the first row from the $(q+1)$st. Then $F_2 F_1 R$ has all 0's in the first column below the first entry. Before the j-th step $F_{j-1} \ldots F_1 R$ has all 0's in the first $j-2$ columns below the main diagonal. F_j represents subtracting multiples of the $(j-1)$st row of $F_{j-1} \ldots F_1 R$ from the j-th, $(j+1)$st, ..., $(j+q-1)$st rows in order that the $(j-1)$st column have only 0's below the main diagonal. Let $\varrho_{st}^{(j-1)}$ be the s, t – th element of $R^{(j-1)} = F_{j-1} \ldots F_1 R$. Then

(12) $$F_j = \begin{pmatrix} I & 0 & 0 & 0 \\ 0 & 1 & 0 & 0 \\ 0 & f_j & I & 0 \\ 0 & 0 & 0 & I \end{pmatrix},$$

where the I's are of order $j-2$, q, and $T+1-j-q$, respectively, and

(13) $$f_j' = \left(-\frac{\varrho_1}{\varrho_{j-1,j-1}^{(j-1)}}, \ldots, -\frac{\varrho_q}{\varrho_{j-1,j-1}^{(j-1)}}\right), \quad j = 2, \ldots, T-q+1\,.$$

For $j = T-q+2, \ldots, T$, f_j consists of the first $T-j+1$ components of (13). Thus F_j consists of the identity with (at most) q nonzero elements below the main diagonal in the $(j-1)$st column.

The calculation of $w = Fx = F_T F_{T-1} \ldots F_2 F_1 x$ is done recursively. Let $w^{(j)} = F_j w^{(j-1)}$. Then $w = w^{(T)}$. In the first sequence of operations $w_t^{(j)} = w_t^{(j-1)}$ for $t = 1, \ldots, j-1$,

(14) $$w_t^{(j)} = w_t^{(j-1)} + f_{t,j-1}^{(j)} w_{j-1}^{(j-1)}\,, \quad t = j, \ldots, j+q-1\,,$$

$w_t^{(j)} = w_t^{(j-1)} = x_t$, $t = j+q, \ldots, T$. For $j = T-q+2, \ldots, T$ (14) holds for $t = j, \ldots, T$.

The matrix $\boldsymbol{FR}$ has 0's below the main diagonal; the elements above the main diagonal are the same as those of $\boldsymbol{R}$. Since $\boldsymbol{FRF}'$ is symmetric, it must be diagonal say $\boldsymbol{D}$. Hence, $\boldsymbol{R}^{-1}\boldsymbol{x} = \boldsymbol{F}'\boldsymbol{D}^{-1}\boldsymbol{F}\boldsymbol{x} = \boldsymbol{F}'\boldsymbol{D}^{-1}\boldsymbol{w}$. This vector is computed as $\boldsymbol{F}_1'\boldsymbol{F}_2' \ldots \ldots \boldsymbol{F}_{T-1}'\boldsymbol{F}_T'(\boldsymbol{D}^{-1}\boldsymbol{w})$. Then z, the solution to (10), is $1/\sigma_0$ times this vector.

4. COMPUTATION OF THE COEFFICIENTS

We shall concentrate on the computation of the coefficients of $\hat{\sigma}_0^{(i)}, \hat{\sigma}_1^{(i)}, \ldots, \hat{\sigma}_q^{(i)}$ in (7) because space in this paper is limited. These are the most difficult to compute and are needed for the pure moving average process ($p = 0$, $\boldsymbol{B}_{i-1}^{e} = \boldsymbol{I}$). The problem is to calculate $\operatorname{tr} \boldsymbol{\Sigma}_{\mathrm{MA}}^{-1}\boldsymbol{G}_j\boldsymbol{\Sigma}_{\mathrm{MA}}^{-1}\boldsymbol{G}_h$, where $\boldsymbol{\Sigma}_{\mathrm{MA}}$ [which is σ_0 times (11)] is the covariance matrix of $(u_1, \ldots, u_T)'$ from a moving average process $u_t = \sum_{j=0}^{q} \alpha_j v_{t-j}$. This covariance matrix differs only in the $q \times q$ upper left-hand corner and $q \times q$ lower right-hand corner from $\boldsymbol{\Sigma}_{\mathrm{AR}}^{-1}$, which is the inverse of the $T \times T$ covariance matrix of T successive observations from an autoregressive process with coefficients $1, \alpha_1, \ldots, \alpha_q$ and disturbance variance $1/\sigma^2$. (See Section 6.2 of [1], for example.) This suggests that $\boldsymbol{\Sigma}_{\mathrm{AR}}$ is a good approximation to $\boldsymbol{\Sigma}_{\mathrm{MA}}^{-1}$.

It has been shown by Shaman [4] that if $\boldsymbol{\Sigma}$ is the $T \times T$ covariance matrix of a stationary stochastic process with continuous positive spectral density $f(\lambda)$ and $\boldsymbol{\Sigma}^*$ is the $T \times T$ covariance matrix of a stationary stochastic process with spectral density $1/[f(\lambda)(2\pi)^2]$, then $\boldsymbol{\Sigma} - \boldsymbol{\Sigma}^{*-1}$ and $\boldsymbol{\Sigma}^* - \boldsymbol{\Sigma}^{-1}$ are positive semidefinite. Thus $\boldsymbol{\Sigma}_{\mathrm{MA}} - \boldsymbol{\Sigma}_{\mathrm{AR}}^{-1}$ and $\boldsymbol{\Sigma}_{\mathrm{AR}} - \boldsymbol{\Sigma}_{\mathrm{MA}}^{-1}$ are positive semidefinite.

THEOREM 4.1. *If $\boldsymbol{P}$ is positive definite, $\boldsymbol{Q}$ and $\boldsymbol{P} - \boldsymbol{Q}$ are positive semidefinite, and $\boldsymbol{H}$ is symmetric,* $\operatorname{tr} \boldsymbol{PHPH} \geqq \operatorname{tr} \boldsymbol{QHQH}$.

Proof. There is a nonsingular matrix $\boldsymbol{M}$ such that $\boldsymbol{P} = \boldsymbol{MM}'$ and $\boldsymbol{Q} = \boldsymbol{MDM}'$, where $\boldsymbol{D}$ is diagonal. Since $\boldsymbol{P} - \boldsymbol{Q}$ is positive semidefinite, $d_{ii} \leqq 1$. Then

$$(15) \quad \operatorname{tr} \boldsymbol{PHPH} - \operatorname{tr} \boldsymbol{QHQH} = \operatorname{tr} \boldsymbol{MM'HMM'H} - \operatorname{tr} \boldsymbol{MDM'HMDM'H} =$$

$$= \operatorname{tr} \boldsymbol{M'HMM'HM} - \operatorname{tr} \boldsymbol{DM'HMDM'HM} =$$

$$= \operatorname{tr} \boldsymbol{K}^2 - \operatorname{tr} (\boldsymbol{DK})^2 = \sum_{s,t=1}^{T} k_{st}^2 - \sum_{s,t=1}^{T} d_{ss} d_{tt} k_{st}^2 = \sum_{s,t=1}^{T} k_{st}^2 (1 - d_{ss} d_{tt}) \geqq 0 . \quad \square$$

COROLLARY 4.1. *The matrix with components* $\operatorname{tr} \boldsymbol{\Sigma}^*\boldsymbol{G}_f\boldsymbol{\Sigma}^*\boldsymbol{G}_h - \operatorname{tr} \boldsymbol{\Sigma}^{-1}\boldsymbol{G}_f\boldsymbol{\Sigma}^{-1}\boldsymbol{G}_h$ *is positive semidefinite.*

Proof. Let $\boldsymbol{P} = \boldsymbol{\Sigma}^*$, $\boldsymbol{Q} = \boldsymbol{\Sigma}^{-1}$, and $\boldsymbol{H} = \sum_{g=0}^{q} y_g \boldsymbol{G}_g$ in Theorem 4.1. $\square$

For large T the following theorem justifies the substitution and gives another approximation for the coefficients.

THEOREM 4.2. *Let Σ be the $T \times T$ covariance matrix of a stationary stochastic process with positive continuous spectral density $f(\lambda)$. Then*

$$(16)\quad \lim_{T\to\infty} \frac{1}{T} \operatorname{tr} \Sigma^{-1} G_g \Sigma^{-1} G_h = \frac{n_g n_h}{(2\pi)^3} \int_{-\pi}^{\pi} \frac{\cos \lambda g \cos \lambda h}{f^2(\lambda)} \, d\lambda\,, \quad g, h = 0, 1, \ldots, q\,,$$

where $n_0 = 1$ and $n_g = 2$, $g = 1, \ldots, q$.

Proof. From Corollary 4.1 we have

$$(17)\qquad \sum_{g,h=0}^{q} \operatorname{tr} \Sigma^{-1} G_g \Sigma^{-1} G_h y_g y_h \leqq \sum_{g,h=0}^{q} \operatorname{tr} \Sigma^* G_g \Sigma^* G_h y_g y_h =$$

$$= \sum_{g,h=0}^{q} y_g^* y_h^* \sum_{r,s,t,u=1}^{T} \sigma^*(r-s) \left(\delta_{s,t-g} + \delta_{s-g,t}\right) \sigma^*(t-u) \left(\delta_{u,r-h} + \delta_{u-h,r}\right),$$

where $y_0^* = \frac{1}{2} y_0$, $y_g^* = y_g$, $g = 1, \ldots, q$, $\delta_{s,t} = 1$, $s = t = 1, \ldots, T$, and $\delta_{s,t} = 0$ otherwise, and $\sigma^*(r-s)$ is the r, s-th element of Σ^*. The sum is

$$(18)\qquad \frac{1}{(2\pi)^4} \sum_{g,h=0}^{q} y_g^* y_h^* \int_{-\pi}^{\pi} \int_{-\pi}^{\pi} \frac{\sum_{r,s,t,u=1}^{T} e^{i\lambda(r-s) + i\mu(t-u)}}{f(\lambda) f(\mu)} \,\cdot$$

$$\cdot \left(\delta_{s,t-g} + \delta_{s-g,t}\right) \left(\delta_{u,r-h} + \delta_{u-h,r}\right) d\lambda \, d\mu =$$

$$= \frac{1}{(2\pi)^4} \sum_{g,h=0} y_g^* y_h^* \int_{-\pi}^{\pi} \int_{-\pi}^{\pi} \frac{\sum_{s=1}^{T-g} \sum_{u=1}^{T-h} e^{i(\mu-\lambda)s + i(\lambda-\mu)u}}{f(\lambda) f(\mu)} \times$$

$$\times \left[e^{i\mu g + i\lambda h} + e^{-i\mu g + i\lambda h} + e^{i\mu g - i\lambda h} + e^{-i\mu g - i\lambda h}\right] d\lambda \, d\mu =$$

$$= \frac{4}{(2\pi)^4} \sum_{g,h=0}^{q} y_g^* y_h^* \int_{-\pi}^{\pi} \int_{-\pi}^{\pi} \frac{\sin \frac{1}{2}(\lambda-\mu)(T-g)}{\sin \frac{1}{2}(\lambda-\mu)} \cdot \frac{\sin \frac{1}{2}(\lambda-\mu)(T-h)}{\sin \frac{1}{2}(\lambda-\mu)} \,\cdot$$

$$\cdot \frac{e^{i(\lambda-\mu)(g-h)/2} \cos \mu g \cos \lambda h}{f(\lambda) f(\mu)} \, d\lambda \, d\mu\,.$$

The limit of $1/T$ times the right-hand side of (18) is

$$(19)\qquad \frac{4}{(2\pi)^3} \sum_{g,h=0}^{q} y_g^* y_h^* \int_{-\pi}^{\pi} \frac{\cos \lambda g \cos \lambda h}{f^2(\lambda)} \, d\lambda$$

by the argument leading to Theorem 8.3.3 of [1].

The other part of the proof is along the lines of the proof of Theorem 10.2.7 of [1]. Since the spectral density $f(\lambda)$ is continuous and positive, for arbitrary $\varepsilon > 0$ there exists an autoregressive process with covariance matrix $\boldsymbol{\Sigma}_{\mathrm{U}}$ and (positive) spectral density $f_{\mathrm{U}}(\lambda)$ such that $f(\lambda) \leqq f_{\mathrm{U}}(\lambda)$, $-\pi \leqq \lambda \leqq \pi$, and

$$\frac{1}{f(\lambda)} - \frac{1}{f_{\mathrm{U}}(\lambda)} < \varepsilon\,, \qquad -\pi \leqq \lambda \leqq \pi\,. \tag{20}$$

Then $\boldsymbol{\Sigma}_{\mathrm{U}} - \boldsymbol{\Sigma}$ and $\boldsymbol{\Sigma}^{-1} - \boldsymbol{\Sigma}_{\mathrm{U}}^{-1}$ are positive semidefinite (Lemma 10.2.6 of [1]). For $\boldsymbol{B}$ defined by (87) of Section 10.2 of [1]

$$\begin{aligned}
&\sum_{g,h=0}^{q} \operatorname{tr} \boldsymbol{\Sigma}^{-1}\boldsymbol{G}_g\boldsymbol{\Sigma}^{-1}\boldsymbol{G}_h y_g y_h \geqq \sum_{g,h=0}^{q} \operatorname{tr} \boldsymbol{\Sigma}_{\mathrm{U}}^{-1}\boldsymbol{G}_g\boldsymbol{\Sigma}_{\mathrm{U}}^{-1}\boldsymbol{G}_h y_g y_h = \\
&= \sum_{g,h=0}^{q} \operatorname{tr} \boldsymbol{B}'\boldsymbol{B}\boldsymbol{G}_g\boldsymbol{B}'\boldsymbol{B}\boldsymbol{G}_h y_g y_h = \sum_{g,h=0}^{q} \sum_{\substack{i,j,k,l,\\ m,n=1}}^{T} b_{ji} b_{jk} g_{kl}^{(g)} b_{ml} b_{mn} g_{ni}^{(h)} y_g y_h = \\
&= \sum_{j,m=1}^{T} \Big[\sum_{g=0}^{q} \sum_{k,l=1}^{T} b_{jk} g_{kl}^{(g)} b_{ml} y_g\Big]^2 \geqq \sum_{j,m=K+1}^{T} \Big[\sum_{g=0}^{q} \sum_{k,l=1}^{T} b_{jk} g_{kl}^{(g)} b_{ml} y_g\Big]^2 = \\
&= \sum_{j,m=K+1}^{T} \Big[\sum_{g=0}^{q} \sum_{k',l'=0}^{K} b_{k'} b_{l'} g_{j-k',m-l'}^{(g)} y_g\Big]^2 ,
\end{aligned} \tag{21}$$

where K is the order of the approximating autoregressive process, $b_1, \ldots, b_K$ are the coefficients, and $g_{kl}^{(h)}$ is the k, l-th element of $\boldsymbol{G}_h$. Since

$$g_{st}^{(h)} = \frac{n_h}{4\pi} \int_{-\pi}^{\pi} \big[\mathrm{e}^{\mathrm{i}\lambda(s-t+h)} + \mathrm{e}^{\mathrm{i}\lambda(s-t-h)}\big]\, \mathrm{d}\lambda\,, \qquad h = 0, 1, \ldots, q\,, \tag{22}$$

we can write $1/T$ times the right-hand side of (21) as

$$\begin{aligned}
&\frac{1}{T} \sum_{j,m=K+1}^{T} \Bigg[\sum_{g=0}^{q} \sum_{k',l'=0}^{K} b_{k'} b_{l'} y_g^* \frac{1}{2\pi} \int_{-\pi}^{\pi} \{\mathrm{e}^{\mathrm{i}\lambda[(j-k')-(m-l')+g]} + \\
&\qquad + \mathrm{e}^{\mathrm{i}\lambda[(j-k')-(m-l')-g]}\}\, \mathrm{d}\lambda\Bigg]^2 = \\
&= \frac{1}{(2\pi)^2\, T} \sum_{j,m=K+1}^{T} \Bigg[\int_{-\pi}^{\pi} \mathrm{e}^{\mathrm{i}\lambda(j-m)} \sum_{k',l'=0}^{K} b_{k'} \mathrm{e}^{-\mathrm{i}\lambda k'} b_{l'} \mathrm{e}^{\mathrm{i}\lambda l'} \sum_{g=0}^{q} y_g^* \{\mathrm{e}^{\mathrm{i}\lambda g} + \mathrm{e}^{-\mathrm{i}\lambda g}\}\, \mathrm{d}\lambda\Bigg]^2 = \\
&= \frac{1}{(2\pi)^2\, T} \sum_{j,m=K+1}^{T} \Bigg[2 \sum_{g=0}^{q} y_g^* \int_{-\pi}^{\pi} \mathrm{e}^{\mathrm{i}\lambda(j-m)} \frac{\cos \lambda g}{2\pi f_{\mathrm{U}}(\lambda)}\, \mathrm{d}\lambda\Bigg]^2
\end{aligned} \tag{23}$$

because $f_U(\lambda) = 1/[2\pi|\sum_{k=0}^{K} b_k e^{i\lambda k}|^2]$. Then the limit of (23) is

$$(24)\quad \frac{4}{(2\pi)^4} \sum_{g,h=0}^{q} y_g^* y_h^* \lim_{T\to\infty} \int_{-\pi}^{\pi} \int_{-\pi}^{\pi} \sum_{j,m=1}^{T-K} \frac{1}{T} e^{i(\lambda+\nu)(j-m)} \frac{\cos \lambda g \cos \nu h}{f_U(\lambda) f_U(\nu)} \, d\lambda \, d\nu =$$

$$= \frac{4}{(2\pi)^4} \sum_{g,h=0}^{q} y_g^* y_h^* \lim_{T\to\infty} \int_{-\pi}^{\pi} \int_{-\pi}^{\pi} \frac{\sin^2 \frac{1}{2}(\lambda+\nu)(T-K)}{T \sin^2 \frac{1}{2}(\lambda+\nu)} \frac{\cos \lambda g \cos \nu h}{f_U(\lambda) f_U(\nu)} \, d\lambda \, d\nu .$$

The limit is (19) with $f(\lambda)$ replaced by $f_U(\lambda)$. Since ε is arbitrary,

$$(25)\quad \lim_{T\to\infty} \frac{1}{T} \sum_{g,h=0}^{q} y_g y_h \operatorname{tr} \Sigma^{-1} G_g \Sigma^{-1} G_h = \frac{1}{(2\pi)^3} \sum_{g,h=0}^{q} y_g y_h n_g n_h \int_{-\pi}^{\pi} \frac{\cos \lambda g \cos \lambda h}{f^2(\lambda)} \, d\lambda .$$

Since (25) holds for every set of $y_0, y_1, \ldots, y_q$ the theorem follows. □

It should be noted that Theorem 4.2 holds for each stationary process with continuous positive spectral density, not only for the autoregressive process with moving average residual of order q. This is a more general result than that of Whittle [5].

COROLLARY 4.2.

$$(26)\quad \lim_{T\to\infty} \frac{1}{T} \operatorname{tr} \Sigma_{MA}^{-1} G_g \Sigma_{MA}^{-1} G_h = \lim_{T\to\infty} \frac{1}{T} \operatorname{tr} \Sigma_{AR} G_g \Sigma_{AR} G_h .$$

The case of $q = 1$ can be analyzed further. The matrix G_1 can be written as $G_1 = PA_1P'$, where the orthogonal matrix P has $[2/(T+1)]^{1/2} \sin[\pi st/(T+1)]$ as its s, t – th element and the diagonal matrix Λ_1 has $2\cos[\pi t/(T+1)]$ as its t – th diagonal element. (See Section 6.5.4 of [1].) Then

$$(27)\quad \operatorname{tr} \Sigma^{-1} G_f \Sigma^{-1} G_h = \operatorname{tr} (\sigma_0 I + \sigma_1 \Lambda_1)^{-1} \Lambda_f (\sigma_0 I + \sigma_1 \Lambda_1)^{-1} \Lambda_h , \quad f, h = 0, 1 ,$$

where $\Lambda_0 = I$. The sums in (27) can be approximated by integrals with limits $\frac{1}{2}\pi/(T+1)$ and $\pi - \frac{1}{2}\pi/(T+1)$. The matrix of coefficients is approximated by

$$(28)\quad \frac{T+1}{\sigma_0^2(1-4\varrho_1^2)^{3/2}} \begin{bmatrix} 1 & -4\varrho_1 \\ -4\varrho_1 & \dfrac{(1-4\varrho_1^2)^{3/2} - 1 + 8\varrho_1^2}{\varrho_1^2} \end{bmatrix} - \frac{1}{\sigma_0^2(1-4\varrho_1^2)^2} \begin{bmatrix} 1+4\varrho_1^2 & -8\varrho_1 \\ -8\varrho_1 & 4(1+4\varrho_1^2) \end{bmatrix}$$

for $\varrho_1 \neq 0$; the lower right-hand element in the first matrix is 2 for $\varrho_1 = 0$.

In the computations $\sigma_0, \sigma_0\varrho_1 = \sigma_1, \ldots, \sigma_0\varrho_q = \sigma_q$ are replaced by the estimates $\hat\sigma_0^{(i-1)}, \hat\sigma_1^{(i-1)}, \ldots, \hat\sigma_q^{(i-1)}$. In case of the purely moving average process initial estimates are the sample covariances $c_h = y'L^h y/T$, $h = 0, 1, \ldots, q$, which are consistent estimates of order $T^{-1/2}$. The integrals (16) can be approximated by sums.

REFERENCES

[1] T. W. ANDERSON: The statistical analysis of time series. Wiley, New York 1971.

[2] T. W. ANDERSON: Asymptotically efficient estimation of covariance matrices with linear structure. Ann. Statistics *1* (1973), 135—141.

[3] T. W. ANDERSON: Maximum likelihood estimation of parameters of autoregressive process with moving average residuals and other covariance matrices with linear structure. Ann. Statistics *3* (1975), 1283—1304.

[4] P. SHAMAN: Approximations for stationary covariance matrices and their in verses with application to ARMA models. Ann. Statistics *4* (1976), 292—301.

[5] P. WHITTLE: Estimation and information in stationary time series. Ark. Mat. Fys. Ast. *2* (1953), 423—434.

STANFORD UNIVERSITY

DEPARTMENTS OF STATISTICS AND ECONOMICS

ON SPATIAL-TEMPORAL MODELS AND MARKOV FIELDS

JULIAN BESAG

LIVERPOOL

A short account of the Hammersley-Clifford theorem in the theory of Markov random fields is given. A class of time-reversible spatial-temporal processes is defined and is shown to generate the class of positive discrete Markov fields as limiting spatial distribution. Some examples are given. Spatial-temporal autoregressions for Gaussian variates are discussed and it is shown that these are often consistent with the auto-normal spatial fields introduced previously by the author.

1. INTRODUCTION

There has recently been considerable interest in the specification of spatial stochastic processes and particularly in the concept of spatial Markovity. The usual situation may be described as follows. Suppose we consider a finite system of n sites, labelled $1, \ldots, n$ which has associated with it a collection of random variables $X_1, \ldots, X_n$. We shall suppose for the moment that the variables are discrete. For each site i, we stipulate that the conditional distribution of the associated random variable X_i, given all other site values, is allowed to depend only upon the values at some specified subset $\mathcal{N}_i$ of the remaining sites. $\mathcal{N}_i$ is termed the neighbourhood of site i. A complete set of neighbourhoods $\mathcal{N}_1, \ldots, \mathcal{N}_n$ clearly generates an associated class of valid probability distributions for the random vector $X \equiv (X_1, \ldots, X_n)$. Members of this class are termed Markov (random) fields. However, the identification of the class is no obvious matter since the conditional distribution at each site is subject to awkward consistency conditions. Indeed, the problem remained largely unresolved until the advent of an unpublished paper by Hammersley and Clifford in 1971. Whilst other authors, including Brook [6], Dobrushin [7], Averintsev [1], Spitzer [12] and Besag [3], restrict their attention to the k-dimensional cubic lattice with the neighbours of any (internal) site defined as those adjacent to it, Hammersley and Clifford tackle the much wider and more interesting set-up associated with an arbitrary finite graph and described above. Each of the cited authors made the assumption of positivity which demands that if $x_1, \ldots, x_n$ are values which can individually occur at sites $1, \ldots, n$, respectively, then they can occur collectively as a single realisation of the random vector X. This condition has recently been slightly relaxed in a paper by Preston [11] and, more comprehensively, by Moussouris [10]. Nevertheless, in most practical data-analytic situations, the assumption of positivity is a reasonable one. Its validity will be assumed throughout the present paper. The original proof of the Hammersley-Clifford theorem is circuitous, involving the construction of an operational calculus (the "blackening algebra"), and short proofs have subsequently been given by Grimmett [8] and by Besag [4]. When stating the theorem in Section 2, we shall adopt the approach of [4].

In practical applications of Markov field theory, the sites usually denote points or regions of Euclidean space and the data represent a single instantaneous observation on what is, in reality, a spatial-temporal process $\{X(t)\}$, where t denotes time. For example, observations might be the yields of plots of wheat in a field at harvest or the incidence of infection in a spatially distributed population at one particular time point. In such situations, although the data is only available at one point in time, there may be objections to constructing a spatial "model" which is devoid of any temporal (and, hence, causal) considerations. An alternative approach would be to set up a model for the spatial-temporal process itself and thence derive the corresponding instantaneous spatial structure. This is, of course, more easily said than done. Whilst it may well be feasible to produce a reasonable approximation to the temporal process, it is rarely possible to obtain the associated spatial structure. Nevertheless, if an assumption of temporal stationarity is appropriate, there are some situations where progress can be made. In Section 3 of the paper, some preliminary results on time-reversible processes are presented, whilst Section 4 deals exclusively with Gaussian autoregressions and their relationship to the auto-normal schemes suggested in [4].

Statistical analysis of spatial schemes will not be considered in the present paper. Some methods are given in [4] and [5].

2. HAMMERSLEY-CLIFFORD THEOREM

For integer-valued discrete site variables $X_1, \ldots, X_n$, let

$$\Omega_i = \{x_i : \Pr(X_i = x_i) > 0\}$$

and define the Cartesian product space $\Omega = \Omega_1 \times \ldots \times \Omega_n$, with typical elements

$x \equiv (x_1, \ldots, x_n)$. Without loss of generality, we assume that the value zero can occur at any site. Then, subject to positivity, $\Pr(X = x) > 0$ for all $x \in \Omega$ and we may legitimately define the function

$$Q(x) \equiv \ln \{\Pr(X = x)/\Pr(X = 0)\}$$

such that $Q(x)$ is finite for all $x \in \Omega$. It is easily established that any such function has the representation,

$$Q(x) = \sum_{1 \leqq i \leqq n} x_i G_i(x_i) + \sum_{1 \leqq i < j \leqq n} \sum x_i x_j G_{i,j}(x_i, x_j) + \ldots \tag{1}$$

$$\ldots + x_1 x_2 \ldots x_n G_{1,\ldots,n}(x_1, \ldots, x_n),$$

where the G-functions are uniquely defined. The result of the Hammersley-Clifford theorem is to identify the constraints which are imposed upon the G-functions by the neighbourhoods $\mathcal{N}_1, \ldots, \mathcal{N}_n$. This is done in terms of the cliques which are generated by the neighbourhoods. Here, a clique is defined as any set of sites which either consists of a single site or else in which every site is a neighbour of every other site in the set. Then it is easily shown (see [4]) that the only G-functions which may be non-null in the expansion (1) are those associated with cliques. That is, for any $i < j < \ldots < s$, $G_{i,j,\ldots,s} \equiv 0$ unless the sites $i, j, \ldots, s$ are all neighbours of each other. Furthermore, if Ω is finite, this restriction is sufficient to ensure a satisfactory definition of the field; that is, the G-functions associated with cliques may be chosen arbitrarily (finite on Ω). If Ω is denumerably infinite, we additionally require the summability condition that $\Sigma Q(x)$ be finite, where Σ denotes summation over all $x \in \Omega$.

For continuous variates, the analogous result holds with the appropriate change of terminology from probabilities to probability densities and with the summability condition replaced by the corresponding one of integrability. For some examples of Markov fields on discrete or continuous state spaces, see [4].

3. MARKOV FIELDS AND TIME-REVERSIBLE MARKOV PROCESSES

In this section, we examine the relationship between the form of the intensity function and that of the limiting spatial distribution for a certain class of spatial-temporal Markov processes. A typical member of this class will be denoted by $\{X(t) : t \geqq 0\}$, where $X(t)$ is supposed to have n components $X_1(t), \ldots, X_n(t)$, these being associated with sites $1, \ldots, n$, respectively. $X_i(t)$ is assumed to take only integer values and to generate a minimal state space Ω_i. We denote by Ω the corresponding minimal state space for $X(t)$. It is assumed that the process $\{X(t)\}$ is uniquely defined in terms of the intensity function h, where h satisfies

$$\Pr\{X(t+\Delta t)=\boldsymbol{y} \mid X(t)=\boldsymbol{x}, \text{ all previous history}\} =$$

$$= \begin{cases} h(\boldsymbol{x},\boldsymbol{y})\,\Delta t + o(\Delta t) & \text{for} \quad \boldsymbol{y} \neq \boldsymbol{x},\ \boldsymbol{x},\boldsymbol{y} \in \Omega\,, \\ 1 + h(\boldsymbol{x},\boldsymbol{x})\,\Delta t + o(\Delta t) & \text{for} \quad \boldsymbol{y} = \boldsymbol{x} \in \Omega\,, \\ o(\Delta t) \quad \text{otherwise} & \end{cases}$$

where $0 \leqq h(\boldsymbol{x},\boldsymbol{y}) < \infty$ for any $\boldsymbol{y} \neq \boldsymbol{x}$ and $\Sigma h(\boldsymbol{x},\boldsymbol{y}) = 0$, the summation being over all $\boldsymbol{y}$. In general, such a process will be intractable to detailed discussion; however, we make three simplifying assumptions.

Assumption 1. $\Omega = \Omega_1 \times \ldots \times \Omega_n$ (the Cartesian product of the individual spaces Ω_i). Furthermore, all states in Ω are to be positive.

Assumption 1 ensures that there exists a unique limiting distribution $\{\pi(\boldsymbol{x}) : \boldsymbol{x} \in \Omega\}$ with all positive elements, these being determined, apart from scale factor, by the usual set of equations,

$$\sum_{\boldsymbol{x} \in \Omega} \pi(\boldsymbol{x})\, h(\boldsymbol{x},\boldsymbol{y}) = 0 \quad \text{for all} \quad \boldsymbol{y} \in \Omega\,. \tag{2}$$

More conveniently, we write equation (2) as

$$\sum_{\boldsymbol{x} \neq \boldsymbol{y}} \pi(\boldsymbol{x})\, h(\boldsymbol{x},\boldsymbol{y}) = \pi(\boldsymbol{y}) \sum_{\boldsymbol{x} \neq \boldsymbol{y}} h(\boldsymbol{y},\boldsymbol{x})\,. \tag{3}$$

Assumption 2. The process $\{X(t)\}$ is time-reversible (Kendall [9]).

Presently the important consequence of assumption 2 is that $\{\pi(\boldsymbol{x})\}$ satisfies equation (3) term by term; that is

$$\pi(\boldsymbol{x})\, h(\boldsymbol{x},\boldsymbol{y}) = \pi(\boldsymbol{y})\, h(\boldsymbol{y},\boldsymbol{x})\,, \tag{4}$$

for all $\boldsymbol{x},\boldsymbol{y} \in \Omega$. Bartlett [2], when finding a convenient intensity function for a given spatial distribution, calls the comparable result a "local" solution, as opposed to an otherwise "global" one.

Assumption 3. Adopting the notation,

$$\boldsymbol{x}_i \equiv (x_1, \ldots, x_{i-1}, x_i - 1, x_{i+1}, \ldots, x_n)\,,$$

$$\boldsymbol{x}^i \equiv (x_1, \ldots, x_{i-1}, x_i + 1, x_{i+1}, \ldots, x_n)\,,$$

we assume that

$$h(\boldsymbol{x},\boldsymbol{x}_i) > 0 \quad \text{whenever} \quad \boldsymbol{x},\boldsymbol{x}_i \in \Omega \quad \text{and}$$

$$h(\boldsymbol{x},\boldsymbol{x}^i) > 0 \quad \text{whenever} \quad \boldsymbol{x},\boldsymbol{x}^i \in \Omega\,.$$

The transitions $\boldsymbol{x}$ to $\boldsymbol{x}_i$ and $\boldsymbol{x}$ to $\boldsymbol{x}^i$ may conveniently be thought of as representing a death or birth, respectively, occurring at site i. Assumption 3 is unnecessarily severe; its relaxation will be commented upon in Section 3.5.

Making assumptions 1 to 3, it follows that

$$\pi(\boldsymbol{x})/\pi(\boldsymbol{x}_i) = h(\boldsymbol{x}_i, \boldsymbol{x})/h(\boldsymbol{x}, \boldsymbol{x}_i) \tag{5}$$

for all $\boldsymbol{x}, \boldsymbol{x}_i \in \Omega$ and hence that the neighbours of each site i are determined by the functional form of $h(\boldsymbol{x}, \boldsymbol{x}_i)$ and $h(\boldsymbol{x}, \boldsymbol{x}^i)$ alone. Specifically, in regard to the limiting spatial distribution $\{\pi(\boldsymbol{x})\}$, it is clear that site j is a neighbour of site i if and only if $h(\boldsymbol{x}_i, \boldsymbol{x})/h(\boldsymbol{x}, \boldsymbol{x}_i)$, for $\boldsymbol{x}, \boldsymbol{x}_i \in \Omega$, is functionally dependent upon x_j.

It is evident that any given (positive) Markov field on the integers may be generated in the above manner and that there will be a close relationship between the intensity function and the limiting conditional probability distribution at each site. This generalises a result for binary fields in Preston [11]. Note also that the possibility of further types of transition is not precluded by assumption 3 but that such transitions must be reversible and that the corresponding values of the intensity function must be chosen in accordance with equation (5) in order to ensure that time-reversibility is preserved.

We now consider some simple examples. It is stressed that these have been selected primarily for illustrative purposes and in accordance with the suggestions in [4]. We use $p_i(x_i; \ldots)$ to denote the conditional distribution at site i associated with the limiting spatial distribution $\{\pi(\boldsymbol{x})\}$ and conveniently define

$$\beta_i(\boldsymbol{x}) \equiv \exp\left(b_i + \sum b_{i,j} x_j\right),$$

$$\gamma_i(\boldsymbol{x}) \equiv \exp\left(c_i + \sum c_{i,j} x_j\right),$$

$$\delta_i(\boldsymbol{x}) \equiv \exp\left(d_i + \sum d_{i,j} x_j\right),$$

where $\{b_{i,j}\}$ and $\{c_{i,j} - d_{i,j}\}$ are symmetric in i and j and the summations are over all $j \neq i$.

3.1. Binary fields

For 0, 1 variables, the function $G_{i,j,\ldots,s}$ in equation (1) may be replaced by an arbitrary parameter $b_{i,j,\ldots,s}$, where $b_{i,j,\ldots,s} = G_{i,j,\ldots,s}(1, \ldots, 1)$. If, as a first step away from independence, we truncate the expansion at the bilinear term, we obtain the auto-logistic model (see [4]), for which

$$p_i(1; \ldots)/p_i(0; \ldots) = \beta_i(\boldsymbol{x}).$$

Now consider the temporal process defined on $\Omega \equiv \{0, 1\}^n$ and for which

$$h(x, x^i) = \gamma_i(x) \quad \text{for} \quad x, x^i \in \Omega ,$$
$$h(x, x_i) = \delta_i(x) \quad \text{for} \quad x, x_i \in \Omega ,$$
$$h(x, y) = 0 \quad \text{otherwise (unless } y = x) .$$

Such a model might be applied to describe the spread of recurrent infection over an array of plants, with 0 at a particular site and time denoting that the plant there is healthy and 1 denoting that it is infected (see [3]). For this model the limiting spatial distribution is clearly auto-logistic with $b_i = c_i - d_i$ and $b_{i,j} = c_{i,j} - d_{i,j}$. (We assume henceforth that these relations hold.)

3.2. Auto-binomial schemes

Suppose $\Omega_i = \{0, \ldots, m_i\}$ and consider the temporal process specified by

$$h(x, x^i) = (m_i - x_i)\,\gamma_i(x) \quad \text{for} \quad x, x^i \in \Omega ,$$
$$h(x, x_i) = x_i\,\delta_i(x) \quad \text{for} \quad x, x_i \in \Omega ,$$
$$h(x, y) = 0 \quad \text{otherwise (unless } y = x) .$$

The the limiting spatial distribution is auto-binomial (see [4]); that is, the limiting conditional distribution at site i is binomial with parameters m_i and θ_i where

$$\theta_i = \beta_i(x)/\{1 + \beta_i(x)\} .$$

3.3. Auto-Poisson schemes

Suppose $\Omega_i = \{0, 1, \ldots\}$ and consider the process specified by

$$h(x, x^i) = \gamma_i(x) \quad \text{for} \quad x \in \Omega ,$$
$$h(x, x_i) = x_i\,\delta_i(x) \quad \text{for} \quad x_i \in \Omega ,$$
$$h(x, y) = 0 \quad \text{otherwise (unless } y = x) .$$

The process is positive if and only if $b_{i,j} \leqq 0$ for each i and j and then the limiting conditional distribution at site i is Poisson with mean $\beta_i(x)$. It is shown independently in [4] that $b_{i,j} \leqq 0$ is a necessary and sufficient condition for the valid definition of an auto-Poisson distribution.

3.4. Simple immigration, death and migration

Suppose we firstly consider n independent simple death – immigration processes running in parallel; that is, for $\Omega_i = \{0, 1, \ldots\}$,

$$h(\boldsymbol{x}, \boldsymbol{x}^i) = \lambda \quad \text{for} \quad \boldsymbol{x} \in \Omega\,,$$

$$h(\boldsymbol{x}, \boldsymbol{x}_i) = \mu x_i \quad \text{for} \quad \boldsymbol{x}_i \in \Omega\,,$$

$$h(\boldsymbol{x}, \boldsymbol{y}) = 0 \quad \text{otherwise (unless } \boldsymbol{y} = \boldsymbol{x}).$$

Then, of course, the limiting distribution consists of n independent Poisson distributions each with mean λ/μ, provided $\lambda < \mu$. If we now allow *reversible* migration between some or all of the sites according to

$$h(\boldsymbol{x}, \boldsymbol{x}_i^j) = \nu x_i\,,$$

time-reversibility is maintained; that is, the limiting spatial distribution is again that of independent Poisson variates with means λ/μ.

3.5. More general results

It is possible to extend the present discussion to a wider class of time-reversible processes. In particular it may be shown that the limiting spatial structure is related in a simple manner to the form of the intensity function at least in situations where it is possible to reach any given $\boldsymbol{y} \in \Omega$ from any given $\boldsymbol{x} \in \Omega$ by means of a finite sequence of transitions, none of which involves an intermediate change of state at any site i for which $y_i = x_i$ already. This condition may be viewed as a rather strict form of irreducibility. The extended result is reasonably self-evident but nevertheless can involve irksome notation. The present narrower approach has been adopted both in the interests of clarity and to avoid proofs.

4. GAUSSIAN AUTOREGRESSIONS IN CONTINUOUS TIME

In this section, we suppose that the spatial-temporal process $\{\boldsymbol{X}(t)\}$ is a stationary Gaussian autoregression defined by the relationship

$$\mathrm{d}\boldsymbol{X}(t) + \boldsymbol{B}\boldsymbol{X}(t)\,\mathrm{d}t = \mathrm{d}\boldsymbol{Z}(t)$$

where $\boldsymbol{B}$ is a suitable $n \times n$ matrix and $\boldsymbol{Z}(t)$ represents n independent Brownian motions each with variance σ^2 per unit time. Let $\boldsymbol{\Gamma} \equiv \mathrm{E}\{\boldsymbol{X}(t)\,\boldsymbol{X}(t)^{\mathrm{T}}\}$ denote the spatial autocovariance matrix for $\{\boldsymbol{X}(t)\}$. Then $\boldsymbol{\Gamma}$ satisfies the equation,

$$\boldsymbol{B\Gamma} + \boldsymbol{\Gamma B}^{\mathrm{T}} = \sigma^2 \boldsymbol{I}$$

and, provided $\boldsymbol{B}$ and $\boldsymbol{B}^{\mathrm{T}}$ commute,

$$\boldsymbol{\Gamma} = \sigma^2(\boldsymbol{B} + \boldsymbol{B}^{\mathrm{T}})^{-1}\,.$$

Note that commutability is satisfied at least for homogeneous torus lattice systems, for the analogous infinite regular lattice systems and, of course, for $\boldsymbol{B}$ symmetric. Writing $\boldsymbol{C} = \boldsymbol{B} + \boldsymbol{B}^{\mathrm{T}}$, we now define the random process $\{\boldsymbol{U}(t)\}$ by

$$\boldsymbol{U}(t) = \boldsymbol{C}\boldsymbol{X}(t)\,.$$

Omitting the temporal parameter t when variates are considered contemporaneously, we obtain

$$\mathrm{E}(\boldsymbol{U}\boldsymbol{X}^{\mathrm{T}}) = \boldsymbol{I} \quad \text{and} \quad \mathrm{E}(\boldsymbol{U}\boldsymbol{U}^{\mathrm{T}}) = \sigma^2 \boldsymbol{C}\,.$$

Since the process is Gaussian, the component variates X_i and U_j must be independent for $j \neq i$. Furthermore, it follows that

$$\mathrm{E}(\Sigma c_{i,j} X_j \,|\, X_j,\ j \neq i) = 0$$

and that

$$\mathrm{Var}\,(X_i \,|\, X_j,\ j \neq i) = \sigma^2\,,$$

where $c_{i,j}$ denotes the (i, j) element of $\boldsymbol{C}$ and the summation is over all $j \neq i$. Thus, the spatial conditional probability structure of the process is of a particularly simple form and is consistent with the auto-normal schemes suggested in [4].

REFERENCES

[1] M. B. Averintsev: On a method of describing discrete parameter random fields. Problemy Peredachi Informatsii *6* (1970), 100—109.

[2] M. S. Bartlett: Physical nearest-neighbour models and non-linear time series. J. Appl. Prob. *8* (1971), 222—232.

[3] J. E. Besag: Nearest-neighbour systems and the auto-logistic model for binary data. J. R. Statist. Soc. B *34* (1972), 75—83.

[4] J. E. Besag: Spatial interaction and the statistical analysis of lattice systems. J. R. Statist. Soc. B *36* (1974) (with Discussion).

[5] J. E. Besag: Statistical analysis of non-lattice data. The Statisticians *24* (1975), 179—195.

[6] D. Brook: On the distinction between the conditional probability and the joint probability approaches in the specification of nearest-neighbour systems. Biometrika *51* (1964), 481 to 483.

[7] R. L. Dobrushin: Gibbsian random fields for lattice systems with pairwise interactions. Funktional'nyi Analiz *2* (1968), 292—301.

[8] G. R. Grimmett: A theorem about random fields. Bull. London Math. Soc. *5* (1973), 81—84.

[9] D. G. Kendall: Unitary dilations of one-parameter semi-groups of Markov transition operators, and the corresponding integral representations for Markov processes with a countable infinity of states. Proc. London Math. Soc. (3) *9* (1959), 417—431.

[10] J. Moussouris: Gibbs and Markov random systems with constraints. J. Statist. Phys. *10* (1976), 11—33.

[11] C. J. Preston: Generalised Gibbs states and Markov random fields. Adv. Appl. Prob. *5* (1973), 242—261.

[12] F. Spitzer: Markov random fields and Gibbs ensembles. Am. Math. Monthly *78* (1971), 142—154.

University of Liverpool

Department of Computational & Statistical Science

ZUR AUSWERTUNG DES ELEKTROMYOGRAMMS MIT SPEKTRALANALYTISCHEN METHODEN

DIETER BRÄUER

BERLIN

Die Erfassung der bioelektrischen Aktivität des tätigen Muskels erfolgt bei arbeitsphysiologischen Untersuchungen aus Gründen einer beeinflussungsarmen Messung in der Regel mittels Oberflächenelektroden. Bei weitgehender Standardisierung einer Reihe ableittechnischer Variablen wie Fläche, Abstand und Lokalisation der Elektroden können bei nicht zu geringen Belastungen gut reproduzierbare Meßergebnisse erhalten werden. Die Auswertung des Oberflächen-Elektromyogramms (Bild 1) verfolgt den Zweck, Aussagen über die Aktivität des arbeitenden Muskels und den Einfluß arbeitsbedingter Faktoren zu ermöglichen, regulative Beziehungen zwischen verschiedenen Muskelgruppen aufzuzeigen, den biologischen Aufwand für vorgegebene Arbeitsaufgaben einzuschätzen und Ermüdungszustände möglichst frühzeitig zu erkennen. Dazu müssen Rückschlüsse aus den Parametern des Oberflächen-EMG auf physiologische Vorgänge im Muskel gezogen werden, dies kann auf der Grundlage eines Modelles der Entstehung des EMG geschehen.

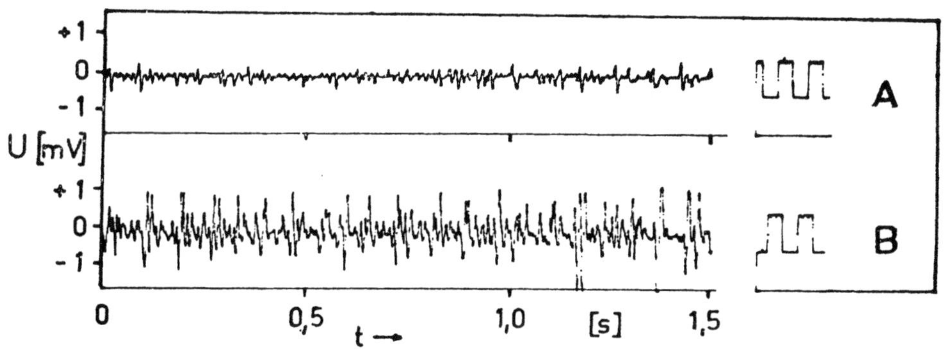

Bild 1. Oberflächen-Elektromyogramm vom M. biceps brach. eines Probanden bei ermüdender statischer Arbeit (A — Beginn der Belastung, B — Ende der Belastung).

Bei der Aufstellung eines stochastischen Modelles für das Elektromyogramm [1], [3], [4], [7], [8], [9] muß von den funktionellen Einheiten des Muskels, den sogenannten motorischen Einheiten (ME), ausgegangen werden. Sie bestehen aus jeweils einem Motoneuron und den von ihm versorgten Muskelfasern. Beim Eintreffen eines Nervenimpulses an den zu einer ME gehörenden neuromuskulären Synapsen werden Aktionspotentiale (AP) und eine mechanische Kontraktion ausgelöst. Die Summe aller AP einer ME stellt die Impulsantwort der motorischen Einheit dar. Es sei $z_k(t)$ eine Folge von Nervenimpulsen $\delta(t)$, die zu Zeitpunkten t_m, $m = 0, 1, \ldots$ an der k-ten ME eintreffen.

(1) $$z_k(t) = \sum_{m=0}^{M} \delta(t - t_m)\,, \quad t_0 = 0\,.$$

$h_k(t)$ sei die Gewichtsfunktion und $x_k(t)$ das EMG der aktivierten motorischen Einheit, dann ist

(2) $$x_k(t) = \int_0^t h_k(u)\, z_k(t-u)\, \mathrm{d}u$$

oder nach Transformation in den Frequenzbereich

(3) $$X_k(f) = H_k(f)\, Z_k(f)$$

mit X, H, Z als Fouriertransformierte von x, h, z.

(4) $$Z_k(f) = \sum_{m=0}^{M} \exp\{-\mathrm{i}2\pi f t_m\}\,.$$

Unter der Annahme (vgl. [1]), daß die t_m eine Summe von m unabhängig nach $N(\mu, \sigma^2)$ verteilten Zufallsgrößen sind, ist der Erwartungswert

(5) $$E\, Z_k(f) = \sum_{m=0}^{M} \int_{-\infty}^{\infty} \exp\{-\mathrm{i}2\pi f t_m\}\, \varphi(t_m)\, \mathrm{d}t_m = \sum_{m=0}^{M} \exp\{-2\pi f m(\pi f \sigma^2 + \mathrm{i}\mu)\}\,.$$

Die Impulsantwort der motorischen Einheit kann in erster Näherung durch eine gedämpfte Schwingung approximiert werden. Wir betrachten dann die ME als ein lineares System zweiter Ordnung, das als Ausgang einen Prozeß 2. Ordnung erzeugt.

(6) $$x_k(t) = a_1\, x_k(t-1) + a_2\, x_k(t-2) + \varepsilon(t)\,.$$

Die Frequenzantwort, die für alle ME eines Muskels als gleich angesehen wird, ist dann

(7) $$H(f) = \frac{1}{1 - a_1 \exp\{-\mathrm{i}2\pi f\} - a_2 \exp\{-4\mathrm{i}\pi f\}}\,, \quad -0{,}5 \leqq f < 0{,}5\,.$$

Und für das Spektrum gilt

(8) $$S_{x_k}(f) = |H(f)|^2 \, S_\varepsilon(f) \,.$$

Arbeiten die N aktivierten ME eines Muskels unabhängig voneinander, so gilt für den gesamten Muskel

(9) $$S_x(f) = |H(f)|^2 \, N \, S_\varepsilon(f) \,.$$

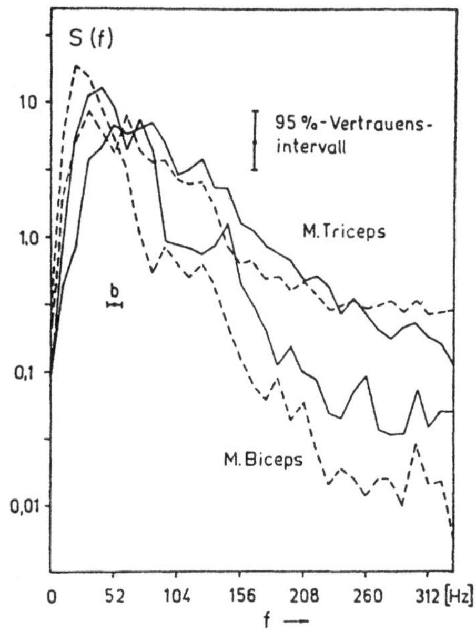

Bild 2. Leistungsspektren (Tukey-Fenster) des Elektromyogramms vom M. triceps brach. und M. biceps brach. eines Probanden bei ermüdender statischer Arbeit (A — Anfang der Belastung, B — Ende der Belastung; b — Bandbreite; Abtastintervall $\Delta t = 1{,}5$ ms, Auswertungsintervall $T = 1200$ ms.

Die Schätzung der Spektraldichte des Oberflächen-EMG erfolgte nach Digitalisierung ($\Delta t = 1{,}5$ ms) und Berechnung der Autokorrelationsfunktion $\hat{r}_x(l)$ mittels einer Gewichtsfunktion nach Tukey [5]:

(10) $$\hat{S}_x(f) = \sum_{l=-L+1}^{L-1} w(l) \, \hat{r}_x(l) \exp\{-i2\pi f l\} \,, \quad -0{,}5 \leqq f < 0{,}5 \,,$$

(11) $$w(l) = \begin{cases} 0{,}5(1 + \cos \pi l/L) & \text{für } |l| \leqq L. \\ 0 & \text{sonst.} \end{cases}$$

Ein Beispiel für die Auswertung des Elektromyogramms zweier Muskelgruppen bei einem Probanden unter ermüdender statischer Haltearbeit gibt Bild 2. Es bestehen Unterschiede zwischen den beiden Muskelgruppen (M. triceps brach., M. biceps brach.) und zwischen den Phasen des Versuches (A – Anfang der Belastung,

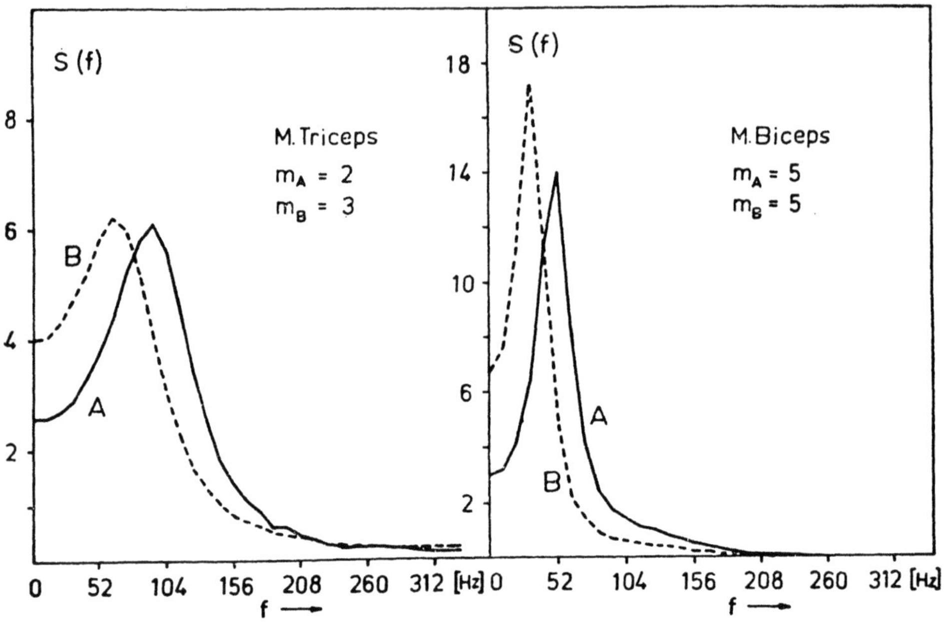

Bild 3. Leistungsspektren nach Anpassung autoregressiver Modelle an die ursprünglichen Zeitreihen des Bildes 2. m_A, m_B — Ordnung der angepaßten Modelle.

B – Ende der Belastung). Die Anpassung autoregressiver Modelle (auch höherer als zweiter Ordnung) erbrachte die in Bild 3 dargestellten Spektraldichteverteilungen.

$$S_x(f) = \frac{\hat{s}_\varepsilon^2}{\left|1 - \sum_{l=1}^{L} \hat{a}_L(l) \exp\{-i2\pi fl\}\right|^2} . \tag{12}$$

Die hier dargestellten und in weiteren Experimenten beobachteten Unterschiede bedürfen noch einer eingehenden physiologischen Deutung.

LITERATUR

[1] G. C. AGARWAL, G. L. GOTTLIEB: Fourier analysis of simulated and recorded EMG. ASCE National Transportation Engineering Meeting, Tulsa 1973. Meeting Preprint 2035.

[2] H. AKAIKE: On a semi-automatic power spectrum estimation procedure. In: Proc. of the Third Hawaii Internat. Conf. on System Sciences, Hawaii 1970, 974—977.

[3] H. P. CLAMANN: Statistical analysis of motor unit firing patterns in a human skeletal muscle. Biophysical J. *9* (1969), 1233—1251.

[4] J. C. COGGSHALL, G. A. BEKEY: A stochastic model of skeletal muscle based on motor unit properties. Math. Biosci. *7* (1970), 405—419.

[5] G. M. JENKINS, D. G. WATTS: Spectral analysis and its applications. Holden-Day, San Francisco 1968.

[6] F. H. LANGE: Signale und Systeme. Bd. 3. Verlag Technik, Berlin 1971.

[7] R. S. PERSON, M. S. LIBKIND: Modelirovanie interferentsionnoj bioelektričeskoj aktivnosti. Biofizika *12* (1967), 1, 127—134.

[8] M. S. LIBKIND: Modelirovanie interferentsionnoj bioelektričeskoj aktivnosti. Biofizika *14* (1969), 2, 369—372.

[9] R. S. PERSON, M. S. LIBKIND: Simulation of electromyograms showing interference patterns. Electroenceph. clin. Neurophysiol. *28* (1970), 625—632.

ZENTRALINSTITUT FÜR ARBEITSMEDIZIN DER DDR

L'UTILISATION DE LA THEORIE DES PHENOMENES D'ATTENTE DANS L'ANALYSE DES SYSTEMES DE PRODUCTION AGRICOLE

CLAUDE BROUSSOLLE

RENNES

L'étude de l'organisation des productions animales dans des coopératives agricoles montre que les différentes phases des processus de production, de transformation, et de vente sont soumises à d'importantes fluctuations dues aux variations de la demande qui se manifeste pour ces produits, ainsi qu'à la nature des phénomènes biologiques encore mal maitrisés sur lesquels reposent ces activités. La connaissance des distributions de probabilité qui caractérisent les principales variables permet en utilisant la théorie des phénomènes d'attente, d'étudier l'adaptation des structures de production à une demande croissante de viande bovine.

Dans la première partie de l'étude on déterminera, pour une structure de production donnée, le flux des animaux à mettre en place dans les différentes catégories d'ateliers. Dans la seconde, l'évolution souhaitable de cette structure, en fonction d'un flux croissant d'animaux à engraisser, sera précisée.

I. L'ADAPTATION DES FLUX AUX STRUCTURES DE PRODUCTION

A. L'ajustement de l'offre à la demande

La coopérative qui sert de cadre à l'étude est un centre de relations agissant simultanément sur un ensemble de marchés. Confrontés d'une manière permanente avec les problèmes d'ajustement quantitatif et qualitatif de l'offre à la demande, les responsables de l'entreprise voient leurs possibilités d'action limitées par les contraintes que constituent, dans le court terme, tel type d'équipement, dans le long terme, les ressources financières qui conditionnent leur politique d'investissement. Cette situation conduit les responsables à mettre en place des structures capables d'absorber, dans les moins mauvaises conditions possibles, les distorsions entre l'offre et la demande, c'est-à-dire des structures permettant au niveau de la production, et en univers aléatoire, d'ajuster des flux d'animaux à des capacités d'installation. Cet ajustement est possible grace à l'existence d'une série de milieux-tampons dont les dimensions respectives varient, d'une période à l'autre, en fonction des couts de production des animaux, de leur temps d'attente dans une situation donnée, des coûts attachés à ces attentes, ainsi que de leurs prix au moment de la vente.

La connaissance des distributions de probabilités des principales variables et leur utilisation dans le cadre de la théorie des phénomènes d'attente permet d'établir les équations d'état qui définissent le „sous-système production" et, par conséquent, de déterminer la capacité minimale des installations nécessaires pour absorber un flux donné d'animaux. Cette capacité étant la résultante de deux paramètres, le nombre et la dimension des ateliers, un grand nombre de combinaisons permet de l'obtenir. L'optimisation de la fonction d'objectif nous donne, parmi toutes les solutions possibles, la solution la meilleure.

B. La répartition des flux d'animaux entre les différentes catégories d'ateliers

L'analyse statistique des flux a porté sur une période de trois ans. Elle couvre les exercices 1968–1969, 1969–1970, 1970–1971.

Le graphique 1 retrace l'évolution hebdomadaire des arrivées d'animaux à la coopérative au cours de l'exercice 1970–1971. L'étude de cette série a montré qu'elle obéissait à une loi de Poisson de même moyenne que la distribution observée.

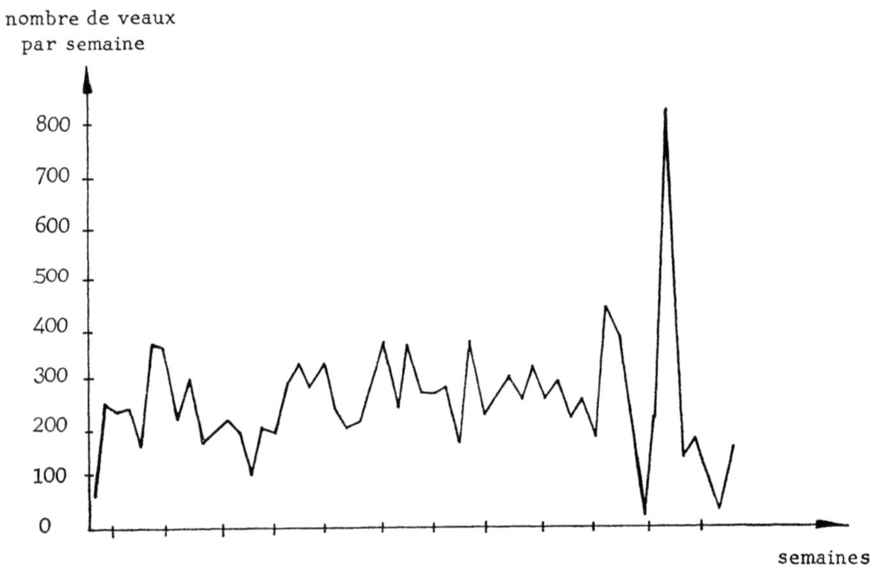

Graphique 1. Évolution hebdomadaire des arrivées d'animaux à la coopérative au cours de l'exercice 1970–1971.

Le graphique 2 représente la fonction de répartition complémentaire de la durée de présence des animaux dans les ateliers. La courbe ainsi obtenue caractérise une distribution d'Erlang.*

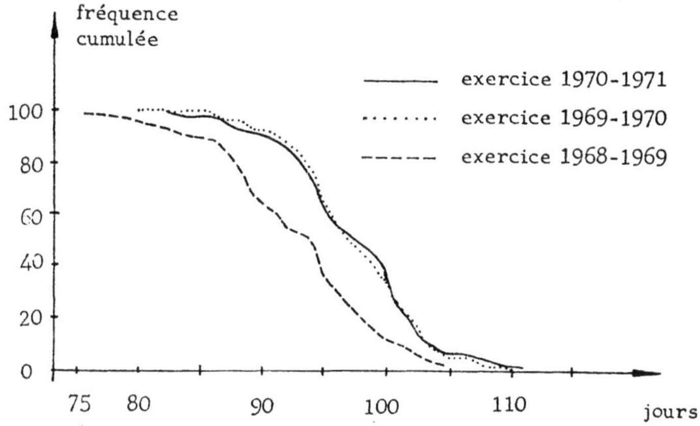

Graphique 2. Fonction de répartition des temps de présence des animaux dans les ateliers de production.

Les ateliers ont été regroupés en fonction de leur taux de service, c'est-à-dire en fonction du rapport de la capacité de l'atelier, exprimé en nombre de places, à la durée moyenne d'engraissement des animaux dans cet atelier au cours de la période considérée. En opérant de cette façon, on aboutit à l'établissement de 13 catégories d'ateliers. Par ailleurs, on a vérifié que les flux qui alimentent chaque catégorie d'ateliers sont poissonniens et que les durées d'engraissement peuvent être considérées comme obéissant à une loi exponentielle dans la mesure où l'attente des animaux que l'on met en place dans les ateliers est pratiquement nulle.

La demande de la coopérative, à un moment donné, se traduit par l'enlèvement des animaux. Cette demande correspond à des besoins qu'elle a exprimé antérieurement sous forme de commandes qui se sont concrétisées par la mise en place, chez les producteurs, des veaux dont l'engraissement lui fournira la viande dont elle a besoin. On peut exprimer cette situation en disant que la probabilité d'un délai de livraison, supérieure à la durée moyenne d'engraissement dans la catégorie correspondante d'ateliers, ne doit pas dépasser une valeur que l'on s'est fixée.

* Encore appelée loi γ_N. Cette fonction constitue une famille de distribution des temps de service qui couvre toute l'étendue des cas possibles entre le type exponentiel ($k = 1$) et celui où la durée d'engraissement est constante ($k = \infty$).

La fonction de répartition complémentaire de cette fonction est la suivante:

$$A_k(t) = e^{-\mu kt} \sum_{n=0}^{k-1} \frac{(\mu kt)^n}{n!} .$$

Cette probabilité (1) est fonction du nombre (S) d'ateliers de la catégorie considérée, du nombre moyen (μ) d'animaux engraissés par unité de temps dans cette catégorie, et du nombre moyen (λ) d'animaux livrés à la coopérative, c'est-à-dire préalablement commandés et mis en place dans les ateliers. La relation qui exprime la probabilité d'un temps d'attente (t) supérieure à la durée moyenne d'engraissement des animaux permet, au seuil de confiance que l'on s'est fixé, de calculer le nombre moyen d'animaux à mettre en place dans chaque catégorie d'ateliers.

(1) La probabilité d'une attente supérieure à la durée moyenne d'engraissement dans une catégorie donnée d'ateliers s'écrit:

$$p(>t) = e^{-S\mu t(1-\Psi/S)}p(>0)\,, \tag{1}$$

avec

$$p(>0) = \frac{S}{S!\,(1-\Psi/S)}\,p(0)\,, \tag{2}$$

et

$$p(0) = \frac{1}{\dfrac{\Psi^S}{S!\,(1-\Psi/S)} + \sum_{n=0}^{S-1} \dfrac{\Psi^n}{n!}}\,. \tag{3}$$

TABLEAU 1

Principales caractéristiques de la production et résultats obtenus pour une production moyenne journalière de 35 animaux

Catégorie d'ateliers	Nombre moyen de places disponibles par bande	Nombre d'ateliers	Production moyenne journalière (en nombre d'animaux)	Marge brute moyenne par place occupée
A	10	5	0,184	62,25
B	13	9	0,657	66,46
C	16	13	1,285	54,49
D	21	26	4,077	58,21
E	27	18	3,337	59,43
F	37	15	3,718	42,99
G	48	12	3,581	66,57
H	63	12	4,455	67,22
I	77	10	4,630	65,46
J	112	7	3,693	82,30
K	133	5	2,181	72,64
L	196	3	1,074	79,01
M	256	3	2,101	27,97
total		138	35	

La première relation exprime la probabilité d'une attente supérieure à une certaine valeur t, la seconde la probabilité qu'il y ait attente et la troisième la probabilité qu'il n'y ait pas d'animaux dans le système*

S est le nombre d'ateliers de la classe i (i = A, B, C, ..., M);

μ est le nombre moyen d'animaux engraissés par unité de temps dans un atelier de la classe i. $\mu = N/dN$ étant la capacité d'un atelier de la catégorie i et d la durée de présence (vide sanitaire compris) des animaux dans cette catégorie d'ateliers.

Ψ est l'intensité du trafic; c'est-à-dire la valeur du rapport λ/μ.

λ est le nombre moyen d'animaux livrés à la coopérative c'est-à-dire le nombre moyen d'animaux préalablement commandés et mis en place dans les ateliers.

Compte tenu de tous ces éléments, le tableau 1 donne pour l'exercice 1970–1971 et par catégorie d'ateliers, la répartition moyenne du flux aléatoire d'animaux à engraisser.

La répartition des animaux entre les différentes catégories d'ateliers n'est pas optimale, ou plus exactement, rien ne nous permet d'affirmer qu'elle le soit au regard d'un critère économique quelconque. Il s'agit, en l'occurence, de l'utilisation descriptive de la théorie des phénomènes d'attente pour rendre compte d'un processus d'approvisionnement, d'engraissement, et de vente. Par contre, dans la deuxième partie de l'étude, nous l'utiliserons à des fins normatives pour définir une évolution souhaitable à partir d'une structure de production donnée.

II. EVOLUTION DES STRUCTURES EN FONCTION DES FLUX

A. La fonction économique

Il s'agit de déterminer la fonction d'objectif du groupe constitué par l'ensemble des producteurs de veaux de boucherie adhérents à la coopérative. En général, l'agrégation des fonctions d'évaluation individuelles pose de difficiles problèmes. Ces difficultés résident moins dans le nombre des centres de décision que dans l'im-

* Les relations (1) at (2) peuvent encore s'écrire:

$$(1') \qquad p(>t) = \mathrm{e}^{-(S-C)t}\, p(>0)\,,$$

$$(2') \qquad p(>0) = \frac{\dfrac{C^S}{(S-C)(S-1)!}}{\displaystyle\sum_{n=0}^{S-1} \frac{C^n}{n!} + \frac{C^S}{(S-C)(S-1)!}}$$

dans lesquelles C est la densité des arrivées $C = S\eta$, η étant le taux d'occupation des ateliers de la catégorie i et S le nombre d'ateliers de cette catégorie. Dans ces deux relations l'unité de temps est la durée moyenne de présence des animaux dans les ateliers.

possibilité de trouver une fonction globale cohérente. Il est, en effet, évident que la stratégie qui paraît la meilleure à un membre de la „coalition" n'est pas nécessairement la meilleure pour un autre. Pour arbitrer entre les préférences des membres du groupe, il faut connaître les structures de préférence de ce groupe. Or „il existe plusieurs manières pour construire ou révéler cette structure: vote à la majorité absolue, vote à la majorité simple, pondération, règle de Pareto etc... Malheureusement, aucune d'elle ne garantit une structure de préférence collective, complète et transitive; le vote à la majorité permet de construire une structure d'évaluation complète mais souvent intransitive, la règle de dominance de Pareto, une structure transitive mais incomplète". Dans certains cas on peut après avoir sélectionné les solutions possibles, les comparer. Cette démarche qui assure la cohérence de l'ensemble étudié, implique que les différents centres de décision disposent d'une fonction numérique pour classer les solutions possibles. Existe-t-il une telle fonction dans une coopérative? Une coopérative est une organisation où, à côté d'un centre principal de décision, subsistent des centres plus ou moins autonomes. Les agriculteurs, adhérents à cette coopérative, se sont groupés pour trouver à leurs problèmes une solution meilleure que celle qu'ils auraient en agissant isolément. Cette attitude suppose qu'ils se sont mis d'accord, au préalable, sur le partage des avantages qu'ils tirent de cette coopération. Si cette condition est réalisée, il semble que l'on puisse admettre l'existence d'une fonction économique globale. Comme les producteurs apprécient l'efficacité de leur coopération aut travers des gains monétaires qu'elle leur permet d'obtenir, il a semblé réaliste de chercher à maximiser l'espérance mathématique du revenu global. De cette façon l'évaluation du système étudié est plus complète car elle porte sur l'ensemble des éléments qui le constituent.*

L'analyse statistique des marges brutes obtenues dans chacune des différentes catégories d'ateliers a montré que leurs distributions étaient gaussiennes. Pour tenir compte du caractère aléatoire des marges obtenues par les producteurs et pour apprécier les résultats économiques de l'augmentation prévue de la production, nous avons été amenés à réaliser des échantillons artificiels des lois de Gauss en utilisant la méthode de Monte Carlo.

* En toute rigueur il faudrait assortir le critère de l'espérance mathématique d'une contrainte fixant la probabilité de ruine de façon à transformer en utilité les bénéfices ou les pertes des producteurs. Cette probabilité n'a pu être évaluée et n'intervient donc pas dans les calculs. Ce qui, théoriquement, a pour conséquence de ne pas éliminer du domaine des structures possibles celles pour lesquelles la probabilité de ruine de certains producteurs n'est pas négligeable. En fait, la suite des calculs montre que les catégories d'ateliers qui sont parmi les premières a être éliminées sont celles pour lesquelles les écarts-types sont tels que la probabilité d'une marge negative pour chacune d'elles est élevée.

B. L'évolution optimale des structures

La fonction d'évaluation ayant été définie, nous sommes en mesure de déterminer l'évolution optimale des structures de production en fonction d'un flux aléatoire croissant d'animaux à engraisser. Pour cela, nous utiliserons la relation qui exprime la probabilité d'attente de la coopérative. Cette première série de calculs donne les solutions techniquement possibles, c'est-à-dire pour chaque catégorie, le nombre minimum d'ateliers nécessaires pour atteindre l'objectif de production que l'on s'est fixé. A titre d'exemple, nous donnons dans le tableau 2 le nombre d'ateliers de la dixième catégorie (J) pour différents niveaux de production, les valeurs du taux moyen de production journalière (λ) ainsi que le nombre de places occupées dans les ateliers.

Déterminer l'évolution optimale des structures de production revient à savoir si, pour atteindre les objectifs que l'on s'est fixés, le nombre des ateliers que l'on trouve initialement dans chacune des treize catégories va augmenter, et si oui de combien? se maintenir au niveau où il se trouve, ou diminuer jusqu'à s'annuler. Poser ainsi le problème revient à envisager toutes les combinaisons possibles de ces 13 catégories d'ateliers prises, successivement, une à une, deux à deux, trois à trois, etc... et ne retenir, chaque fois, que la combinaison qui maximise la fonction d'objectif.

Les résultats de ces calculs figurent sur le graphique 3 et appellent les observations suivantes:

1. Le modèle proposé donne des solutions optimales conditionnelles c'est à-dire qu'il indique ce qu'il faut faire lorsque, à partir d'une situation donnée, on se propose de supprimer, de maintenir ou de développer certaines catégories d'ateliers.

TABLEAU 2

Evolution de quelques paramètres des ateliers de la catégorie J

Nombre d'ateliers de la catégorie J	Taux moyen de production journalière λ	Nombre moyen de places occupées = nombre moyen d'animaux engraissés
2	0,579	65
3	1,198	135
4	2,007	226
5	2,726	307
6	3,575	402
⋮	⋮	⋮
38	34,248	3 853
39	35,237	3 964
40	36,215	4 074
41	37,204	4 185
42	38,192	4 297
⋮	⋮	⋮

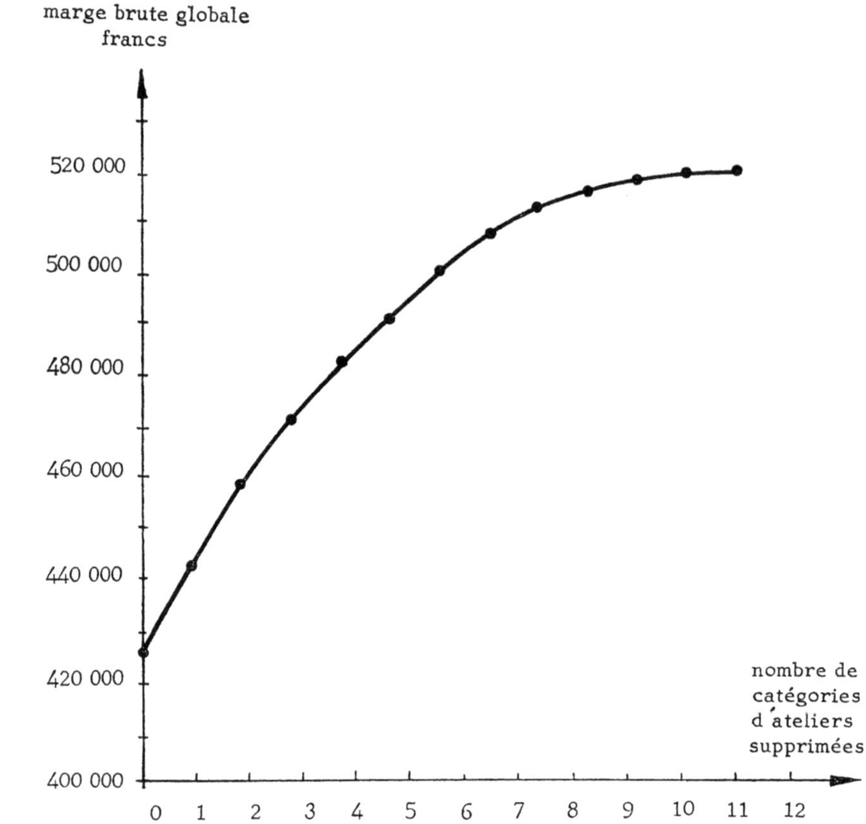

catégories d'ateliers maintenues

A	F	D	D	D	D	D	D	D	C	B	B	A
B	A	F	F	F	E	E	E	E	D	C	C	B
C	B	A	M	I	F	F	F	F	E	D	D	C
D	C	B	A	M	I	G	G	G	F	E	E	D
E	D	C	B	A	M	I	H	H	G	F	F	E
F	E	E	C	B	A	M	I	I	H	G	G	F
G	G	G	E	C	B	A	M	K	I	H	H	G
H	H	H	G	E	C	B	A	M	K	I	I	H
I	I	I	H	G	G	C	B	A	M	K	K	I
J	J	J	I	H	H	H	C	B	A	M	L	K
K	K	K	J	J	J	J	J	C	B	A	M	L
L	L	L	K	K	K	K	K	J	J	J	A	M
M	M	M	L	L	L	L	L	L	L	L	J	J

catégories d'ateliers supprimées

Graphique 3. Évolution de la marge brute globale compte tenu de la suppression de différentes catégories d'ateliers et pour une production moyenne journalière de 55 animaux.

2. La politique optimale conduit à ne retenir, à terme, que les ateliers de la dixième catégorie J. Ce résultat est logique dans la mesure où les ateliers correspondants sont ceux qui procurent la marge unitaire la plus élevée, l'écart-type de la distribution de ces marges étant, par ailleurs, le plus faible. Ce résultat ne signifie pas, dans le court terme, la disparition brutale de toutes les autres catégories, mais indique l'objectif vers lequel on doit tendre à plus ou moins longue échéance.

3. Le cheminement prévu pour atteindre cet objectif n'est pas conforme à ce que l'on pourrait penser a priori. C'est ainsi que la suppression d'une catégorie d'ateliers conduit à l'élimination de la sixième (F). Or, si l'on se réfère au tableau 1, on constate que ce n'est pas cette catégorie, mais la treizième (M), qui donne par veau, la marge la plus faible. Si l'on envisage la suppression de deux catégories c'est la quatrième (D) et la sixième (F) qui „sortent". La treizième (M) demeure. Elle n'est éliminée que si on supprime trois catégories. Lorsque l'on élimine quatre catégories d'ateliers, on observe qu'aux trois précédentes (D, F et M) s'ajoute la neuvième (I). Or, cette dernière assure une marge unitaire supérieure à celles que procurent certaines catégories d'ateliers qui sont conservées et dont les écarts-types des distributions des marges sont comparables. Bien entendu, il n'est pas réaliste d'éliminer des ateliers qui assurent un revenu correct à ceux qui les exploitent. Pratiquement, cela signifie que la restructuration entreprise doit s'arrêter lorsque ce type d'atelier est mis en cause. Cette attitude conduit à conserver toutes les catégories d'ateliers qui n'ont pas encore été éliminées par le modèle.

4. Le modèle permet de calculer la variable duale associée à la contrainte de capacité de chaque type d'ateliers.

CONCLUSION

Dans une coopérative, l'attitude des adhérents est caractérisée par un effort de concertation pour définir une politique leur assurant des avantages supérieurs à ceux qu'ils obtiendraient en agissant isolément. L'opposition des intérêts individuels ne disparaît pas pour autant. Aussi, pour assurer au groupe un minimum de cohésion est-il nécessaire de se mettre d'accord, au préalable sur le partage des bénéfices.

La recherche d'une politique commune peut être effectuée à l'aide du modèle proposé. Celui-ci explore des trajectoires possibles et donne des solutions optimales conditionnelles, c'est-à-dire qu'il indique, pour un objectif de production donné, la meilleure structure de production lorsqu'on envisage la suppression, le maintien ou le développement d'une ou de plusieurs catégories d'ateliers.

L'adoption des solutions qu'il propose conduit à envisager l'élimination des catégories dans lesquelles, chez certaines, la probabilité d'obtenir une marge négative est telle que le répartition du surplus collectif prévu par le règlement ne saurait assurer une quelconque rentabilité aux ateliers concernés.

L'élimination d'ateliers qui assurent, à leurs propriétaires, un revenu moyen relativement bon, n'est pas réaliste. L'information fournit par le modèle, à ce sujet, est, néanmoins, intéressante dans la mesure où elle permet de fixer, à un moment donné de l'évolution observée, une limite supérieure aux interventions qui ont pour objectif la mise en place d'une structure optimale de production.

Le problème essentiel se situe au niveau des ateliers dont le maintien, à un moment donné, est nécessaire à la réalisation de l'équilibre offre-demande. Dans ces ateliers, les résultats généralement obtenus par animal sont relativement satisfaisants, mais la dimension de certains d'entre eux est insuffisante pour assurer à leurs propriétaires un revenu socio-économique considéré comme minimum. Si les autres activités de ces agriculteurs ne leur permettent pas d'obtenir ce revenu, le problème se pose de savoir dans quelle mesure la coopérative doit assurer aux produits qu'ils fournissent un prix unitaire qui, à qualité égale, serait supérieur à celui qu'obtiendraient les agriculteurs possédant des ateliers de grande dimension. Les résultats que donne le modèle justifient une telle politique qui, d'ailleurs, en modifiant les données de base du problème, conduit à la définition d'une nouvelle structure optimale. On peut, de cette façon, envisager un système de prix différentiels et déterminer la structure de production qui, logiquement, en découle. Réciproquement, on peut à partir d'une structure de production considérée comme souhaitable à un moment donné, calculer le système de prix que cette structure implique.

RÉFÉRENCE BIBLIOGRAPHIQUES

C. Broussolle, R. Hovelaque: L'organisation d'un atelier industriel de production porcine en avenir aléatoire. Recherches d'Econ. et Sociol. rurales (1970), n° 3, 41–75.

C. Broussolle: L'adaptation des structures de production aux exigences des industriels. Economie Rurale (1971), n° 90, 61–67.

C. Broussolle: L'offre des produits animaux dans des ensembles agro-industriels. Coll. INRA de la Station d'Eco. Ru. de Rennes, oct. 1971, 89 p.

C. Broussolle: Evolution des structures de production et demande des produits animaux dans une coopérative. Annales d'Eco. et Sociol. rurales (1973), n° 1, 45–70.

C. Broussolle: L'influence des structures de production et celle des prix sur l'évolution des systèmes agroalimentaire (à paraître).

D. R. Cox, H. D. Miller: The theory of Stochastic Processes. Chapman and Hall 1965.

R. Cruon, A. Kaufmann: Les phénomènes d'attente. Dunod 1961.

A. Ja. Khintchine: Désserte d'un grand nombre d'usagers (théorie mathématique des services de masse). Dunod 1969.

S. C. Kolm: Le service des masses. Dunod 1971.

P. Le Gall: Les systèmes avec ou sans attente et les processus stochastiques. Dunod 1962.

Institut national de la Recherche agronomique

ON A CLASS OF BI- AND MULTIVARIATE DISTRIBUTIONS GENERATED BY MARGINAL WEIBULL DISTRIBUTIONS

JAN BUTKIEWICZ and ELŻBIETA HYS

WARSZAWA

In this paper we present a generalization of the Morgenstern system of bivariate distributions with one parameter in order to obtain a system of three- and n-variate distributions with specified marginal distributions. Then we consider those systems with Weibull marginal distributions.

Investigations of the strength of material caused W. Weibull to bring in the probability distribution called now his name into engineering applications. The Weibull distribution enables to characterize those features having nonnegative values like the strength or the fatigue of items. In general an item has many features described by a random vector with nonnegative components. So it may be useful to introduce a multivariate distribution taking into account Weibull distributions as its univariate marginal distributions. The construction of such bivariate distribution is easy achievable because of the Morgenstern system of bivariate distributions [2]

$$F(x_1, x_2) = F_1(x_1)\, F_2(x_2)\, [1 + \alpha(1 - F_1(x_1))\,(1 - F_2(x_2))] \tag{1}$$

with specified marginal distributions $F_i(x_i)$, $i = 1, 2$, and a parameter $\alpha \in \langle -1, 1 \rangle$.

A CLASS OF BIVARIATE DISTRIBUTIONS WITH WEIBULL MARGINAL DISTRIBUTIONS

The formula (1) by substituting

$$F_i(x_i) = 1 - \exp\left(-\left(\frac{x_i}{b_i}\right)^{\beta_i}\right), \quad i = 1, 2,$$

describes a class of bivariate distributions with Weibull marginal distributions

$$(2)\qquad F(x_1, x_2) = \left\{1 - \exp\left(-\left(\frac{x_1}{b_1}\right)^{\beta_1}\right)\right\}\left\{1 - \exp\left(-\left(\frac{x_2}{b_2}\right)^{\beta_2}\right)\right\} \times$$
$$\times \left\{1 + \alpha \exp\left(-\left(\frac{x_1}{b_1}\right)^{\beta_1} - \left(\frac{x_2}{b_2}\right)^{\beta_2}\right)\right\}$$

where $x_i > 0$, $\beta_i > 0$ (a curve shape parameter), $b_i > 0$ (a scale parameter), $i = 1, 2$, and a class parameter $\alpha \in \langle -1, 1\rangle$. From the formula (1) follows that probability densities of the Morgenstern system of distributions are of form

$$f(x_1, x_2) = f_1(x_1) f_2(x_2) \left[1 + \alpha(1 - 2F_1(x_1))(1 - 2F_2(x_2))\right]$$

where $f_i(x_i)$ denotes the density of marginal distribution $F_i(x_i)$, $i = 1, 2$. In a similar way from the formula (2) follows that the density functions of a class of bivariate distributions with Weibull marginal distributions are of form

$$(2a)\qquad f(x_1, x_2) = \frac{\beta_1}{b_1}\frac{\beta_2}{b_2}\left(\frac{x_1}{b_1}\right)^{\beta_1 - 1}\left(\frac{x_2}{b_2}\right)^{\beta_2 - 1}\left\{1 + \alpha\left(1 - 2\exp\left(-\left(\frac{x_1}{b_1}\right)^{\beta_1}\right)\right) \times\right.$$
$$\left.\times\left(1 - 2\exp\left(-\left(\frac{x_2}{b_2}\right)^{\beta_2}\right)\right)\right\}\exp\left(-\left(\frac{x_1}{b_1}\right)^{\beta_1} - \left(\frac{x_2}{b_2}\right)^{\beta_2}\right)$$

where

$$x_i > 0\,,\quad \beta_i > 0\,,\quad b_i > 0\,;\quad i = 1, 2\,,\quad \alpha \in \langle -1, 1\rangle\,.$$

The distribution functions (2) and the density functions (2a) are expressed by forms being symmetric in regard to variables; then it is sufficient to consider a conditional distribution of X_i, given $X_j = x_j$, $i \neq j$, $i, j = 1, 2$, if the joint density of X_1, X_2 is described by (2-2a). So the conditional density function of X_i, given $X_j = x_j$, $i \neq j$, $i, j = 1, 2$, is

$$f(x_i \mid x_j) = \left\{1 + \alpha\left(1 - 2\exp\left(-\left(\frac{x_i}{b_i}\right)^{\beta_1}\right)\right)\left(1 - 2\exp\left(-\left(\frac{x_j}{b_j}\right)^{\beta_2}\right)\right)\right\} \times$$
$$\times \frac{\beta_i}{b_i}\left(\frac{x_i}{b_i}\right)^{\beta_i - 1}\exp\left(-\left(\frac{x_i}{b_i}\right)^{\beta_i}\right).$$

The regression of X_i on X_j is expressible by a function

(3)

$$E(X_i \mid X_j = x_j) = b_i\Gamma\left(1 + \frac{1}{\beta_i}\right)\left\{1 + \alpha\left(1 - 2\exp\left(-\left(\frac{x_j}{b_j}\right)^{\beta_j}\right)\right)(1 - 2^{-1/\beta_i})\right\},$$

$i \neq j;\ i, j = 1, 2$. The conditional expectation $E(X_i \mid X_j = x_j)$ increases if $\alpha > 0$ (decreases if $\alpha < 0$) beginning from

$$b_i \Gamma\left(1 + \frac{1}{\beta_i}\right)[1 - \alpha(1 - 2^{-1/\beta_i})] \quad \text{to} \quad b_i \Gamma\left(1 + \frac{1}{\beta_i}\right)[1 + \alpha(1 - 2^{-1/\beta_i})]$$

provided that the variable x_j increases. Regarding (3) and denoting $E(X_i \mid X_j = x_j) = E_\alpha(X_i \mid x_j)$ it is easy to see a relationship that follows:

$$\tfrac{1}{2}[E_{-\alpha}(X_i \mid x_j) + E_\alpha(X_i \mid x_j)] = b_i \Gamma\left(1 + \frac{1}{\beta_i}\right), \quad i \neq j; \quad i, j = 1, 2.$$

This relationship indicates the symmetry of $E_\alpha(X_i \mid x_j)$, $E_{-\alpha}(X_i \mid x_j)$. The similar property have conditional moments of kth order

$$\tfrac{1}{2}[E_{-\alpha}(X_i^k \mid x_j) + E_\alpha(X_i^k \mid x_j)] = b_i^k \Gamma\left(1 + \frac{k}{\beta_i}\right).$$

The conditional variance of X_i, given $X_j = x_j$, is described by formula

$$\operatorname{Var}(X_i \mid x_j) = b_i^2 \Gamma\left(1 + \frac{2}{\beta_i}\right)\left\{1 + \alpha\left(1 - 2\exp\left(-\left(\frac{x_j}{b_j}\right)^{\beta_j}\right)\right)(1 - 2^{-2/\beta_i})\right\} -$$

$$- b_i^2 \Gamma^2\left(1 + \frac{1}{\beta_i}\right)\left\{1 + \alpha\left(1 - 2\exp\left(-\left(\frac{x_j}{b_j}\right)^{\beta_j}\right)\right)(1 - 2^{-1/\beta_i})\right\}^2,$$

$$i \neq j; \quad i, j = 1, 2.$$

It is easy to obtain

$$E(X_1^{u_1} X_2^{u_2}) = b_1^{u_1} b_2^{u_2} \Gamma\left(1 + \frac{u_1}{\beta_1}\right)\Gamma\left(1 + \frac{u_2}{\beta_2}\right)[1 + \alpha(1 - 2^{-u_1/\beta_1})(1 - 2^{-u_2/\beta_2})],$$

$$(u_1, u_2) \in R_+^2.$$

In particular, using this expression it is found the covariance between X_1 and X_2

$$\operatorname{Cov}(X_1, X_2) = \alpha b_1 b_2 \Gamma\left(1 + \frac{1}{\beta_1}\right)\Gamma\left(1 + \frac{1}{\beta_2}\right)(1 - 2^{-1/\beta_1})(1 - 2^{-1/\beta_2})$$

and the correlation coefficient

$$\varrho = \alpha \, . \, \frac{(1 - 2^{-1/\beta_1})(1 - 2^{-1/\beta_2})\Gamma\left(1 + \frac{1}{\beta_1}\right)\Gamma\left(1 + \frac{1}{\beta_2}\right)}{\left(\left[\Gamma\left(1 + \frac{2}{\beta_1}\right) - \Gamma^2\left(1 + \frac{1}{\beta_1}\right)\right]\left[\Gamma\left(1 + \frac{2}{\beta_2}\right) - \Gamma^2\left(1 + \frac{1}{\beta_2}\right)\right]\right)^{1/2}}.$$

We add that the least squares linear regression of X_i on X_j has as its equation

$$X_i = \alpha \cdot \frac{b_j \Gamma\left(1+\frac{1}{\beta_1}\right)\Gamma\left(1+\frac{1}{\beta_2}\right)(1-2^{-1/\beta_1})(1-2^{-1/\beta_2})}{b_i\left[\Gamma\left(1+\frac{2}{\beta_i}\right)-\Gamma^2\left(1+\frac{1}{\beta_i}\right)\right]} X_j +$$

$$+ b_j \Gamma\left(1+\frac{2}{\beta_i}\right)\left\{1-\alpha \cdot \frac{\Gamma^2\left(1+\frac{1}{\beta_i}\right)(1-2^{-1/\beta_1})(1-2^{-1/\beta_2})}{\Gamma\left(1+\frac{2}{\beta_i}\right)-\Gamma^2\left(1+\frac{1}{\beta_i}\right)}\right\}.$$

A CLASS OF THREEVARIATE DISTRIBUTIONS WITH WEIBULL MARGINAL DISTRIBUTIONS

The Morgenstern system of bivariate distributions with specified marginal distributions may be extended in case of three random variables in a such way

$$(4) \qquad F(x_1, x_2. x_3) = \left[1 + \sum_{1 \leqq r < s \leqq 3} \alpha_{rs} \bar{F}_r(x_r)\, \bar{F}_s(x_s) + \alpha_{123} \prod_{j=1}^{3} \bar{F}_j(x_j)\right] \prod_{j=1}^{3} F_j(x_j)$$

where $\bar{F}_i(x_i) = 1 - F_i(x_i)$ and $F_i(x_i)$ is a specified cumulative distribution function, $i = 1, 2, 3$, parameters $\alpha_{rs} \in \langle -1, 1 \rangle$, $1 \leqq r < s \leqq 3$.

The density functions describing the system of threevariate distributions with specified marginal distributions $F_i(x_i)$, $i = 1, 2, 3$, are of form

$$(4a) \qquad f(x_1, x_2, x_3) = \left[1 + \sum_{1 \leqq r < s \leqq 3} \alpha_{rs}(1 - 2F_r(x_r))(1 - 2F_s(x_s)) + \right.$$

$$\left. + \alpha_{123} \prod_{j=1}^{3} (1 - 2F_j(x_j))\right] \prod_{j=1}^{3} f_j(x_j).$$

The marginal bivariate distributions connected with distributions having cumulative distribution functions (4) or probability density functions (4a) are expressed by

$$F_{ij}(x_i, x_j) = F_i(x_i)\, F_j(x_j)\left[1 + \alpha_{ij} \bar{F}_i(x_i)\, \bar{F}_j(x_j)\right], \quad 1 \leqq i < j \leqq 3,$$

or

$$f_{ij}(x_i, x_j) = f_i(x_i) f_j(x_j)\left[1 + \alpha_{ij}(1 - 2F_i(x_i))(1 - 2F_j(x_j))\right], \quad 1 \leqq i < j \leqq 3.$$

A trivial case of independent random variables implies $\alpha_{ij} = 0$, $1 \leqq i < j \leqq 3$,

$\alpha_{1,2,3} = 0$. The conditional distributions connected with distributions (4a) may be described by conditional probability density functions

$$f(x_i \mid x_j, x_k) = f_i(x_i) \{1 + [1 + \alpha_{jk}(1 - 2 F_j(x_j))(1 - 2 F_k(x_k))]^{-1} \times$$

$$\times [(1 - 2 F_i(x_i)) (\alpha_{ij}(1 - 2 F_j(x_j) + \alpha_{ik}(1 - 2 F_k(x_k)) + \alpha_{123} \prod_{r=1}^{3} (1 - 2 F_r(x_r)]\}$$

where $i \neq j$, $i \neq k$, $j < k$; $i, j, k \in \{1, 2, 3\}$,

$$f(x_i, x_j \mid x_k) = f_i(x_i) f_j(x_j) \times$$

$$\times [1 + \sum_{1 \leqq r < s \leqq 3} \alpha_{rs}(1 - 2 F_r(x_r)) (1 - 2 F_s(x_s) + \alpha_{123} \prod_{r=1}^{3} (1 - 2 F_r(x_r))]$$

where $j > i$, $i \neq k$, $j \neq k$; $i, j, k \in \{1, 2, 3\}$.

Using the formulas (4-4a) in order to obtain a system of threevariate distributions with Weibull marginal distributions we assume

$$(5) \qquad F_i(x_i) = 1 - \exp(-x_i^{\beta_i}) \quad \text{or} \quad f_i(x_i) = \beta_i x_i^{\beta_i - 1} \exp(-x_i^{\beta_i}) ;$$

$$x_i > 0 , \quad \beta_i > 0 .$$

Omitting of scale parapeters b_i (the assumption $b_i = 1$) is made only to simplify expressions of cumulative distribution functions or probability density functions of a class of threevariate distributions with Weibull marginal distributions

$$(6) \qquad F(x_1, x_2, x_3) = [1 + \sum_{1 \leqq r < s \leqq 3} \alpha_{rs} \exp(-x_r^{\beta_r} - x_s^{\beta_s}) + \alpha_{123} \exp(-\sum_{n=1}^{3} x_r^{\beta_r})] \times$$

$$\times \prod_{j=1}^{3} (1 - \exp(-x_j^{\beta_j})) ,$$

$$(6a) \qquad f(x_1, x_2, x_3) = [1 + \sum_{1 \leqq r < s \leqq 3} \alpha_{rs}(1 - 2 \exp(-x_r^{\beta_r})) (1 - 2 \exp(-x_s^{\beta_s})) -$$

$$- \alpha_{123} \prod_{j=1}^{3} (1 - 2 \exp(-x_j^{\beta_j}))] \prod_{j=1}^{3} \beta_j x_j^{\beta_j - 1}(-x_j^{\beta_j}) ,$$

$$\alpha_{rs} \in \langle -1, 1 \rangle , \quad (x_1, x_2, x_3) \in R_+^3 .$$

The conditional probability density functions connected with the class of distributions (6a) have following forms

$$f(x_i \mid x_j, x_k) = \{1 + [(1 - 2 \exp(-x_i^{\beta_i})) (\alpha_{ij}(1 - 2 \exp(-x_j^{\beta_j})) +$$

$$+ \alpha_{ik}(1 - 2 \exp(-x_k^{\beta_k}))) - \alpha_{123} \prod_{r=1}^{3} (1 - 2 \exp(-x_r^{\beta_r}))] \times$$

$$\times [1 + \alpha_{jk}(1 - 2 \exp(-x_j^{\beta_j})) (1 - 2 \exp(-x_k^{\beta_k}))]^{-1}\} \beta_i x_i^{\beta_i - 1} \exp(-x_i^{\beta_i})$$

where $i \neq j,\ i \neq k,\ j < k;\ i, j, k \in \{1, 2, 3\}$,

$$f(x_i, x_j \mid x_k) = [1 + \sum_{1 \leq r < s \leq 3} \alpha_{rs}(1 - 2\exp(-x_r^{\beta_r}))(1 - 2\exp(-x_s^{\beta_0})) - $$
$$- \alpha_{123} \prod_{r=1}^{3} (1 - 2\exp(-x_r^{\beta_r}))] \prod_{\substack{r \neq k \\ r \in \{1,2,3\}}} \beta_r x_r^{\beta_r - 1} \exp(-x_r^{\beta_r})$$

where $i < j,\ i \neq k,\ j \neq k;\ i, j, k \in \{1, 2, 3\}$.

The regression of X_i on X_j, X_k is expressible by a function

$$E(X_i \mid x_j, x_k) = \{1 + [1 + \alpha_{jk}(1 - 2\exp(-x_j^{\beta_j}))(1 - 2\exp(-x_k^{\beta_k}))]^{-1} \times$$
$$\times (1 - 2^{1/\beta_i}) \times [\alpha_{ij}(1 - 2\exp(-x_j^{\beta_j})) +$$
$$+ \alpha_{ik}(1 - 2\exp(-x_k^{\beta_k})) - \alpha_{123} \prod_{r=1}^{3} (1 - 2\exp(-x_r^{\beta_r}))]\} \Gamma\left(1 + \frac{1}{\beta_i}\right),$$
$$i \neq j; \quad i \neq k; \quad j < k; \quad i, j, k \in \{1, 2, 3,\}.$$

Taking into account the following function

$$E(X_1^{u_1} X_2^{u_2} X_3^{u_3}) = [1 + \sum_{1 \leq i < j \leq 3} \alpha_{ij}(1 - 2^{-u_i/\beta_i})(1 - 2^{-u_j/\beta_j}) -$$
$$- \alpha_{123} \prod_{r=1}^{3} (1 - 2^{-u_r/\beta_r})] \prod_{r=1}^{3} \Gamma\left(1 + \frac{u_r}{\beta_r}\right), \quad (u_1, u_2, u_3) \in R_+^3,$$

we can obtain, for example, elements of the covariance matrix $\{\mathrm{Cov}(X_i, X_j)\}$; $(i, j = 1, 2, 3)$ namely

(7)

$$\mathrm{Cov}(X_i, X_j) = \begin{cases} \alpha_{ij} \Gamma\left(1 + \frac{1}{\beta_i}\right) \Gamma\left(1 + \frac{1}{\beta_j}\right)(1 - 2^{-1/\beta_i})(1 - 2^{-1/\beta_j}), & i \neq j, \\ \Gamma\left(1 + \frac{2}{\beta_i}\right) - \Gamma^2\left(1 + \frac{1}{\beta_i}\right), & i = j. \end{cases}$$

Making use of these elements it is easy to find that $\det\{\mathrm{Cov}(X_i, X_j)\} \neq 0$.

The elements of the correlation matrix $\{\varrho_{ij}\}$ are of form

$$\varrho_{ij} = \begin{cases} \alpha_{ij} \dfrac{(1 - 2^{-1/\beta_i})(1 - 2^{-1/\beta_j}) \Gamma\left(1 + \frac{1}{\beta_i}\right) \Gamma\left(1 + \frac{1}{\beta_j}\right)}{\left(\left[\Gamma\left(1 + \frac{2}{\beta_i}\right) - \Gamma^2\left(1 + \frac{1}{\beta_i}\right)\right]\left[\Gamma\left(1 + \frac{2}{\beta_j}\right) - \Gamma\left(1 + \frac{1}{\beta_j}\right)\right]\right)^{1/2}}; & i \neq j, \\ 1, \quad i = j. \end{cases}$$

A CLASS OF n-VARIATE DISTRIBUTIONS WITH WEIBULL MARGINAL DISTRIBUTIONS

We generalize the Morgenstern system of distributions with specified marginal distributions in the n-variate case using the following formula for the cumulative distribution function

$$F(x_1, \dots, x_n) = [1 + \sum_{s=2}^{n} \sum_{1 \leqq i_1 < \dots < i_s \leqq n} \alpha_{i_1 \dots i_s} \prod_{j=1}^{s} \bar{F}_{i_j}(x_{i_j})] \prod_{j=1}^{n} F_j(x_j) \tag{8}$$

where parameters $\alpha_{i_1 \dots i_s}$, $\underset{2 \leqq s \leqq n}{\text{Card}} \{\alpha_{i_1 \dots i_s}\} = 2^n - n - 1$, $\bar{F}_j(x_j) = 1 - F_j(x_j)$ and $F_j(x_j)$ denotes the cumulative distribution function of the specified distribution $j = 1, 2, \dots, n$. The probability density functions of this generalized Morgenstern system of distributions may be written

$$f(x_1, \dots, x_n) = [1 + \sum_{s=2}^{n} \sum_{1 \leqq i_1 < \dots < i_s \leqq n} \alpha_{i_1 \dots i_s} \prod_{j=1}^{s} (1 - 2 F_{i_j}(x_{i_j}))] \prod_{j=1}^{n} f_j(x_j) . \tag{8a}$$

The marginal distributions related to distributions given by (8a) have probability density functions of similar form

$$f(x_{k_1}, \dots, x_{k_m}) = [1 + \sum_{s=2}^{m} \sum_{k_1 \leqq r_1 < \dots < r_s \leqq k_m} \alpha_{r_1 \dots r_s} \prod_{j=1}^{s} (1 - 2 F_{r_j}(x_{r_j}))] \prod_{j=1}^{m} f_{k_j}(x_{k_j}) ,$$
$$2 \leqq m < n , \quad 1 \leqq k_j < k_{j+1} \leqq n .$$

Here it is easy to notice that parameters of (8-8a) must satisfy the following conditions

$$1 + \sum_{s=2}^{m} \sum_{k_1 \leqq r_1 < \dots < r_s \leqq k_m} \alpha_{r_1 \dots r_s} \prod_{j=1}^{s} \xi_{r_j} \geqq 0 \quad \text{where} \quad |\xi_{r_j}| \leqq 1$$

for $1 \leqq k_1 < k_m \leqq n$ and successively for $m = 2, 3, \dots, n$.

The conditional distributions connected with distributions (8a) have conditional density functions of form

$$f(x_{k_1}, \dots, x_{k_i} \mid x_{k_{i+1}}, \dots, x_{k_n}) = \tag{9}$$
$$= \{1 + [1 + \sum_{s=2}^{n-i} \sum_{k_{i+1} \leqq h_1 < \dots < h_s \leqq k_n} \alpha_{h_1 \dots h_s} \prod_{j=1}^{s} (1 - 2 F_{h_j}(x_{h_j}))]^{-1} \times$$
$$\times [\sum_{s=2}^{n} \sum_{\substack{k_1 \leqq r_1 < \dots < r_s \leqq k_n \\ (r_1, \dots, r_s) \neq (h_1, \dots, h_s) \\ s = 2, \dots, n-i}} \alpha_{r_1 \dots r_s} \prod_{j=1}^{s} (1 - 2 F_{r_j}(x_{r_j}))]\} \prod_{j=1}^{i} f_{k_j}(x_{k_j}) ,$$

$$k_j < k_{j+1} ; \quad j = 1, \dots, i-1 ; \quad k_r < k_{r+1} ; \quad r = i+1, \dots, n-1 ,$$
$$k_j \neq k_r .$$

The construction of a class of n-variable distributions with Weibull marginal distributions is made regarding (5) in (8) or (8a). Hence the such class of n-variable distributions with Weibull marginal distributions is described by cumulative distribution functions

(10) $$F(x_1, \ldots, x_n) =$$
$$= [1 + \sum_{s=2}^{n} \sum_{1 \leqq i_1 < \ldots < i_s \leqq n} \alpha_{i_1 \ldots i_s} \prod_{j=1}^{s} \exp(-x_{i_j}^{\beta_{i_j}})] \prod_{j=1}^{n} (1 - \exp(-x_j^{\beta_j}))$$

or by probability density functions

(10a) $$f(x_1, \ldots, x_n) =$$
$$= [1 + \sum_{s=2}^{n} \sum_{1 \leqq i_1 < \ldots < i_s \leqq n} (-1)^s \alpha_{i_1 \ldots i_s} \prod_{j=1}^{s} (1 - 2\exp(-x_{i_j}^{\beta_{i_j}}))] \prod_{j=1}^{n} \beta_j x_j^{\beta_j - 1} \exp(-x_j^{\beta_j}) .$$

The product moments of $X_1, \ldots, X_n$ having distribution (10a) may be obtained by using functions
$$E(X_1^{u_1} \ldots X_n^{u_n}) =$$
$$= [1 + \sum_{s=2}^{n} \sum_{1 \leqq i_1 < \ldots < i_s \leqq n} (-1)^s \alpha_{i_1 \ldots i_s} \prod_{j=1}^{s} (1 - 2^{-u_j/\beta_j})] \prod_{j=1}^{n} \Gamma\left(1 + \frac{u_j}{\beta_j}\right),$$
$$(u_1, \ldots, u_n) \in R_+^n .$$

The elements of the covariance matrix are like (7) and elements of the correlation matrix are like (7a).

The conditional probability density functions of the considered distributions follow from (5) and (9)
$$f(x_{k_1}, \ldots, x_{k_i} \mid x_{k_{i+1}}, \ldots, x_{k_n}) =$$
$$= \{1 + [1 + \sum_{s=2}^{n-i} \sum_{k_{i+1} \leqq h_1 < \ldots < h_s \leqq k_n} (-1)^s \alpha_{h_1 \ldots h_s} \prod_{j=1}^{s} (1 - 2\exp(-x_{h_j}^{\beta_{h_j}}))]^{-1} \times$$
$$\times [\sum_{s=2}^{n} \sum_{\substack{k_1 \leqq r_1 < \ldots < r_s \leqq k_n \\ (r_1, \ldots, r_s) \neq (h_1, \ldots, h_s) \\ s = 2, \ldots, n-i}} (-1)^s \alpha_{r_1 \ldots r_s} \prod_{j=1}^{s} (1 - 2\exp(-x_{r_j}^{\beta_{r_j}}))]\} \times$$
$$\times \prod_{j=1}^{i} \beta_{k_j} x_{k_j}^{\beta_{k_j} - 1} \exp(-x_{k_j}^{\beta_{k_j}}),$$
$$k_j < k_{j+1} ; \quad j = 1, \ldots, i-1 ; \quad k_r < k_{r+1} ; \quad r = i+1, \ldots, n-1 .$$

Hence the regression of X_{k_1} on $X_{k_2}, \ldots, X_{k_n}$ may be espresible by a function

$$E(X_{k_1} \mid x_{k_2}, \ldots, x_{k_n}) =$$

$$= \{1 + [1 + \sum_{s=2}^{n-1} \sum_{k_2 \leqq h_1 < \ldots < h_s \leqq k_n} (-1)^s \alpha_{h_1 \ldots h_s} \prod_{j=1}^{s} (1 - 2\exp(-x_{h_j}^{\beta_{h_j}}))]^{-1} \times$$

$$\times [\sum_{s=2}^{n} \sum_{\substack{k_1 \leqq r_1 < \ldots < r_s \leqq k_n \\ (r_1,\ldots,r_s) \neq (h_1,\ldots,h_s) \\ s=2,\ldots,n-1}} (-1)^s \alpha_{r_1 \ldots r_s} (1 - 2^{-1/\beta_{k_1}}) \prod_{\substack{j=1 \\ r_j \neq k_1}} (1 - 2\exp(-x_{r_j}^{\beta_{r_j}}))]\} \times$$

$$\times \Gamma\left(1 + \frac{1}{\beta_{k_1}}\right).$$

The estimation of parameters of considered class distributions is to be made regarding the parameters $\alpha_{i_1 \ldots i_s}$, $s = 2, \ldots, n$. The estimation of parameters connected with marginal univariate Weibull distribution is known, for exemple, in [1].

It is evident that the Morgenstern systems of n-variate distributions may be proposed as tool to construct multivariate distributions with specified marginal univariate distributions.

REFERENCES

[1] S. Firkowicz: Statystyczne badanie wyrobów. WNT Warszawa 1970.

[2] D. Morgenstern: Einfache Beispiele zweidimensionaler Verteilungen. Mittelangsblatt für Math. Statistik *8* (1956), 234—235.

[3] W. Weibull: A statistical theory of the strength of material. Ingeniörs Vetenskaps Akademiens Handligar, No 151, Stockholm 1939.

[4] W. Weibull: The phenomenon of rupture in solids. Ingeniörs Vetenskaps Akademiens Handligar No 153, Stockholm 1939.

Polish Academy of Sciences

and

Ministry of Sciences, Higher Education and Technology

Institute for Organization, Management and Control Sciences

ON THE GENERAL STOCHASTIC EPIDEMIC WITH IMMIGRATION IN LARGE POPULATIONS*

VINCENZO CAPASSO

BARI

We examine the general stochastic epidemic with immigration and an exponential distributed incubation period, for large (infinite) populations.

We prove a theorem of existence and uniqueness of a solution of our problem in the Banach space of summable sequences, using the methods of semigroup theory.

Finally, we indicate how the initial value problem for such a process can be approximated by a problem in a finite dimensional space.

1. INTRODUCTION

Many authors have contributed to the study of the general stochastic epidemic, which was originally proposed by Bartlett [3], (see for example the papers by Bailey [1], [2], Whittle [17], Foster [6], Kendall [12], Siskind [15], and Gani [7], [8], [9] among others).

In this paper I shall refer to a general stochastic epidemic for a large population (that I shall assume to be infinite). As in the model by Neymann-Scott [13], each infectious individual may communicate the disease to one or more individuals which, after the contagion enter an incubation period, after which themselves become infectives. The length of this incubation period is not fixed, but is an exponentially distributed random variable. Immigration of both infectives and infected individuals not yet infectives is allowed. At any time, removal of infectives is possible through recovery or death.

---

* Work performed under the auspices of the C.N.R. (Gruppo Nazionale per l'Analisi e le sue Applicazioni).

The equations of the problem shall be examined by using the theory of semigroups. By means of methods studied by Kato [10] and extensively applied by Belleni-Morante [4] to neutron theory, I shall prove existence and uniqueness of a solution in a suitable Banach space. Equations for expected values are also given. Finally I indicate how our problem can be approximated by a suitable problem in a finite-dimensional space.

The stochastic epidemic model considered here is that for which at time $t \geqq 0$ there are in circulation in a population of infinite size

(i) $0 \leqq n$ infectious individuals,
(ii) $0 \leqq m$ infected individuals subject to their incubation period and not yet infectious.

Let the probabilities of possible transitions in the time interval, $[t, t + \Delta t[$ be

(1.1) $\pi_l \, \Delta t$ = Prob {one infectious individual be removed through recovery or death} ;
(1.2) $\pi_f \, \Delta t$ = Prob {one infectious individual produces an infection} ;
(1.3) $\theta(s)$ = Prob {during an infection by one infectious individual s individuals are infected entering their incubation period} .

$$0 \leqq \theta(s) \leqq 1\,, \quad s = 1, 2, \ldots\,; \quad \sum_{s=1}^{\infty} \theta(s) = 1\,.$$

(1.4) $S \, \Delta t$ = Prob {immigration of one infectious individual} ;
(1.5) $S_1 \, \Delta t$ = Prob {immigration of one infected individual subject to his incubation period} ;
(1.6) $\lambda \, \Delta t$ = Prob {one infected becomes infectious} .

Under the hypothesis that each individual acts independently of each other, we get the following system of Kolmogorov (forward) differential equations for the probabilities $P(n, m; t)$ to have at time $t \geqq 0$, n infectives and m individuals subject to their incubation period:

$$\begin{aligned}(2)\qquad \frac{\mathrm{d}}{\mathrm{d}t} P(n, m; t) = {} & \pi_l\{(n + 1)\, P(n + 1, m; t) - n\, P(n, m; t)\} + \\ & + \pi_f n\{\sum_{s=1}^{m} \theta(s)\, P(n, m - s; t) - P(n, m; t)\} + \\ & + \lambda\{(m + 1)\, P(n - 1, m + 1; t) - m\, P(n, m; t)\} + \\ & + S\{P(n - 1, m; t) - P(n, m; t)\} + S_1\{P(n, m - 1; t) - P(n, m; t)\}\end{aligned}$$

where we let

$$P(n, m; t) \equiv 0 \quad \text{if} \quad n < 0 \quad \text{or} \quad m < 0\,.$$

System (2) must be supplemented with an initial condition of the form

$$(3)\qquad P(n, m; 0) = P_0(n, m) \quad (n, m) \in \mathbf{N} \times \mathbf{N}$$

where

$$\forall (n, m) \in \mathbf{N} \times \mathbf{N}: \quad 0 \leqq P_0(n, m) \leqq 1 \tag{4.1}$$

and

$$\sum_{n=0}^{\infty} \sum_{m=0}^{\infty} P_0(n, m) = 1 \,. \tag{4.2}$$

In particular, one might assume as initial conditions the following:

$$P(n, m; 0) = 1 \quad \text{if} \quad \begin{cases} n = k \in \mathbf{N}\,, \\ m = h \in \mathbf{N}\,, \end{cases}$$

$$= 0 \qquad \text{otherwise}$$

but we shall continue to work with the more general case (3).

2. THE INITIAL VALUE PROBLEM

Due to (4.2), we now introduce the Banach space X of all double sequences of real numbers

$$f = (f(r, s))_{(r,s) \in \mathbf{N} \times \mathbf{N}} \in \mathbf{R}^{\mathbf{N} \times \mathbf{N}}$$

such that

$$\|f\| = \sum_{r,s} |f(r, s)| < \infty \,. \tag{5}$$

In the following we shall denote by X_+ the set of all positively defined vectors $x \geqq 0$ of X.

Let us further define the following operators

$$[Af]_{n,m} = -(\pi_a n + \lambda m) f(n, m) + \pi_l (n + 1) f(n + 1, m) + \tag{6}$$

$$+ \pi_f n \sum_{s=1}^{m} \theta(s) f(n, m - s) + \lambda (m + 1) f(n - 1, m + 1)\,,$$

$$[Mf]_{n,m} = S f(n - 1, m) + S_1 f(n, m - 1)\,, \quad n > 0\,; \quad m > 0\,, \tag{7.1}$$

$$[Mf]_{0,m} = S_1 f(0, m - 1)\,, \qquad m > 0\,, \tag{7.2}$$

$$[Mf]_{n,0} = S f(n - 1, 0)\,, \qquad n > 0\,, \tag{7.3}$$

$$[Mf]_{0,0} = 0\,, \tag{7.4}$$

where

(8) $$\pi_a = \pi_l + \pi_f$$

and $[Af]_{n,m}$ denotes the $(n + 1, m + 1)$-th component of the transformed vector Af. The domain $\mathscr{D}(A)$ is assumed to be the following

(9) $$\mathscr{D}(A) = \{f \in X \mid \sum_{n,m} |[Af]_{n,m}| < +\infty\} .$$

From (7.1)–(7.4) it follows that we can assume

$$\mathscr{D}(M) = X .$$

Let us now introduce the following auxiliary operators; if

(10) $$\mathscr{D} = \{f \in X \mid \sum_{n,m} n|f(n, m)| < \infty, \ \sum_{n,m} m|f(n, m)| < \infty\}$$

we set

$$\forall f \in \mathscr{D}, \quad \forall (n, m) \in \mathbf{N} \times \mathbf{N} :$$

(11) $$[Hf]_{n,m} = (\pi_a n + \lambda m) f(n, m) ,$$

(12) $$[Kf]_{n,m} = \pi_l(n + 1) f(n + 1, m) +$$

$$+ \pi_f n \sum_{s=1}^{m} \theta(s) f(n, m - s) + \lambda(m + 1) f(n - 1, m + 1)$$

where we can assume

(13) $$\mathscr{D}(H) = \mathscr{D}(K) = \mathscr{D} .$$

For, given any $f \in \mathscr{D}$, we have $\forall p \in \mathbf{N}$, $\forall q \in \mathbf{N}$:

(14) $$\sum_{n=0}^{p} \sum_{m=0}^{q} n \sum_{s=1}^{m} \theta(s) |f(n, m - s)| = \sum_{n=0}^{p} \sum_{s=1}^{q} \sum_{m=s}^{q} n \, \theta(s) |f(n, m - s)| =$$

$$= \sum_{s=1}^{q} \theta(s) \sum_{n=0}^{p} \sum_{m=s}^{q} n|f(n, m - s)| \leq \sum_{n=0}^{\infty} \sum_{m=0}^{\infty} n|f(n, m)| < \infty$$

since $\sum_{s=1}^{\infty} \theta(s) = 1$. It also follows from (14) that, $f \in \mathscr{D}(A)$. Consequently, $\mathscr{D} \subset \mathscr{D}(A)$

and $-H + K \subset A$.

REMARK 1. Observe that definition (10) is justified by the fact that the mean number

$$\langle N(t)\rangle = E[N(t)] = \sum_{n=0}^{\infty} \sum_{m=0}^{\infty} n f(n, m) \tag{15}$$

of infectives, and the mean number of infected individuals under their incubation period

$$\langle M(t)\rangle = E[M(t)] = \sum_{n=0}^{\infty} \sum_{m=0}^{\infty} m f(n, m) \tag{16}$$

exist and are finite provided that, $f \in \mathscr{D}_+ \equiv \mathscr{D} \cap X_+$ (it should also be noted that it should be $0 \leqq f(n, m) \leqq 1$, for a correct probabilistic meaning), where we recall that

$$X_+ = \{f \in X \mid \forall (n, m) \in \mathbf{N} \times \mathbf{N} : f(n, m) \geqq 0\}\,. \tag{17}$$

Moreover, let $\mathscr{D}_0$ be the linear manifold spanned by the canonical base of X:

$$\mathscr{D}_0 = \{f \in X \mid f \text{ has only a finite number of nonzero components}\}\,. \tag{18}$$

$\mathscr{D}_0$ is dense in X and it is contained in $\mathscr{D}$. We define $A_0 = A_{\mathscr{D}_0}$.

Finally, we introduce operators

$$G_r = -H + rK\,, \quad \mathscr{D}(G_r) = \mathscr{D} \tag{19}$$

where r is a real parameter, $r \in [0, 1[$.

3. PROPERTIES OF H AND OF K

Along the lines suggested by Kato [10], we can prove the following lemmas and next Theorem 1.

LEMMA 1. *For every* $\alpha > 0$: $(\alpha I + H)^{-1} \in \mathscr{B}(X)$ [10, p. 149], $(\alpha I + H)^{-1} [X_+] \subset \subset X_+$, $\|(\alpha I + H)^{-1}\| \leqq 1/\alpha$; $H \in \mathscr{C}(X)$ [10], p. 164), $H[\mathscr{D}_+] \subset X_+$, $\mathscr{D}(H) = \mathscr{D} \supset \mathscr{D}_0$ *is dense in* X.

Lemma 1 follows directly from the definition (11) of H and from the explicit expression of $(\alpha I + H)^{-1}$; $\forall f \in X$, $\forall (n, m) \in \mathbf{N} \times \mathbf{N}$, $\forall \alpha > 0$:

$$[(\alpha I + H)^{-1} f]_{n,m} = (\alpha + \pi_a n + \lambda m)^{-1} f(n, m)\,.$$

LEMMA 2. $K[\mathscr{D}_+] \subset X_+$; $\forall f \in \mathscr{D} : \|Kf\| \leqq \|Hf\|$; $\forall f \in \mathscr{D}_+ : \|Kf\| = \|Hf\|$.
Which follows from the definition (12) of K and inequality (14).

LEMMA 3. $\forall r \in [0, 1[: G_r \in \mathscr{G}(1, 0)$ [11, Ch. 9]; *the generated semigroup* $Z_r(t) = \exp(tG_r)$ *is such that*: (i) $Z_r(t)[X_+] \subset X_+$ $\forall t \geqq 0$; (ii) $\forall r, r' \in [0, 1[$, $r \leqq r'$, $\forall f \in X_+$, $\forall (n, m) \in \mathbf{N} \times \mathbf{N}$: $[Z_r(t) f]_{n,m} \leqq [Z_{r'}(t) f]_{n,m}$.

Thus $\{Z_r(t)\}$ is an increasing function of r; as

$$\forall t \in [0, +\infty[, \ \forall r \in [0, 1[: \|Z_r(t)\| \leqq 1$$

it follows that

$$\forall f \in X, \quad \forall t \in [0, +\infty[: \exists Z(t) f = \lim_{r \to 1-} Z_r(t) f. \tag{20}$$

THEOREM 1. (a) $\{Z(t)\}$ *is a semigroup such that* $\forall t \in [0, +\infty[: \|Z(t)\| \leqq 1$; $Z(t)[X_+] \subset X_+$;

(b) *The limit in* (20) *holds uniformly in each finite interval of* t; *i.e., fixed a* $\bar{t} \in [0, +\infty[: \forall \varepsilon > 0 \, \exists \delta > 0 \ni' \forall t \in [0, \bar{t}[, \ 1 - \delta < r < 1 : \|Z(t) - Z_r(t)\| < \varepsilon$;

(c) *If* G *is the generator of* $Z(t)$, *then* $A_0 \subset -H + K \subset G \subset A$;

(d) *If there is a semigroup* $\{Z'(t); 0 \leqq t < +\infty\}$ *such that* $Z'(t)(X_+) \subset X_+$ *and its generator is an extension of* A_0, *then*

$$\forall t \in [0, +\infty[, \quad \forall f \in X_+, \quad \forall (n, m) \in \mathbf{N} \times \mathbf{N} : [Z(t) f]_{n,m} \leqq [Z'(t) f]_{n,m}.$$

Theorem 1 is a consequence of Lemma 3 (see [10]).

4. THE INITIAL VALUE PROBLEM

Due to the results of Theorem 1, the initial-value problem

$$\begin{cases} \dfrac{du}{dt} = [G - (S + S_1) I + M] u, & t \geqq 0, \\ \lim\limits_{t \to 0^+} u(t) = u_0 \in \mathscr{D}(G) \end{cases} \tag{21}$$

admits a unique solution of the form

$$u(t) = \exp[t(G - (S + S_1) I + M)] u_0, \quad t \geqq 0, \quad \forall u_0 \in \mathscr{D}(G) \tag{22}$$

where $\forall t \geqq 0$:

$$\exp\left[t(G-(S+S_1)I+M)\right][X_+] \subset X_+ \,; \quad \left\|\exp\left[t(G-(S+S_1)I+M)\right]\right\| \leqq 1 .$$

In fact, we have [10, p. 495]

$$\forall t \geqq 0, \quad \forall f \in \mathscr{D}(G) : \exp\left[t(G+M)\right] f = \sum_{j=0}^{\infty} T^{(j)}(t) f \tag{23}$$

where $\forall f \in \mathscr{D}(G)$, $\forall t \geqq 0$:

$$T^{(0)}(t) f = \exp(tG) f = Z(t) f , \tag{24.1}$$

$$T^{(j+1)}(t) f = \int_0^t Z(t-s)\, M\, T^{(j)}(s) f \, \mathrm{d}s, \quad j = 0, 1, 2, \ldots \tag{24.2}$$

and where, according to (7.1)–(7.4):

$$\forall f \in X : \|Mf\| \leqq (S+S_1)\|f\| \,; \quad M[X_+] \subset X_+ .$$

It follows, by (a) of Theorem 1, that

$$\forall t \geqq 0 : T^{(j)}(t)[X_+] \subset X_+ \,; \quad \|T^{(j)}(t) f\| \leqq \left[((S+S_1)\,t)^j/j!\right] \|f\| , \tag{25}$$

we can thus conclude that

$$\forall t \geqq 0 : \exp\left[t(G+M)\right][X_+] \subset X_+ \,; \quad \left\|\exp\left[t(G+M)\right]\right\| \leqq \exp\left[(S+S_1)\,t\right].$$

Hence,

$$\forall t \geqq 0 : \exp\left[t(G+M-(S+S_1)I)\right][X_+] \subset X_+ \,;$$

$$\left\|\exp\left[t(G+M-(S+S_1)I)\right]\right\| \leqq 1$$

as desired.

We conclude that, given $u_0 \in \mathscr{D}(G) \cap X_+$, the solution $u(t)$ of system (21) is such that

$$\forall t \geqq 0 : u(t) \in \mathscr{D}(G) \cap X_+ \,; \quad \|u(t)\| \leqq \|u_0\| .$$

REMARK 2. If we assume, in particular,

$$u_0 = \{P_0(0,0), P_0(1,0), P_0(0,1), \ldots\} \in \mathscr{D}_+ = \mathscr{D}(H) \cap X_+ = \mathscr{D}(-H+K) \cap X_+$$

with

$$\|u_0\| = \sum_{n,m} P_0(n,m) = 1$$

then relations (4.1), (4.2) are satisfied. In addition we have

$$E[N(0)] = \sum_{n,m} n\, P_0(n,m) < \infty \,; \quad E[M(0)] = \sum_{n,m} m\, P_0(n.\, m) < \infty . \tag{26}$$

It also follows that, at any $t \geqq 0$

$$u(t) = \{u(0, 0; t), u(0, 1; t), u(1, 0; t), u(1, 1; t), \ldots\} \in \mathscr{D}(G) \cap X_+ \subset \mathscr{D}(A) \cap X_+$$

since, according to (c) of Theorem 1, $G \subset A$. In this case we have

$$\forall t \geqq 0 : G(u(t)) = A(u(t))$$

and system (21) becomes just the "physical" system (2) + (3). This system thus admits a positive solution $u(t) \in X_+$, but for the time being we can only state that

$$\forall t \geqq 0 : \|u(t)\| = \sum_{n,m} u(n, m; t) \leqq 1 .$$

Of course, these statements remain valid if we take as initial conditions

$$u_0 = \{0, 0, \ldots, P_0(k, h) = 1, 0, 0, \ldots\}$$

because in this case $u_0 \in \mathscr{D}_0 \cap X_+ \subset \mathscr{D}(G) \cap X_+$.

5. NORM PRESERVATION

Kato ([10], Theorem 3. (ii)] states that:

THEOREM 2. *In order that $\{Z(t)\}$, the semigroup generated by G, be a transition semigroup, i.e.*

$$\forall t > 0, \quad \forall x \in X_+ : \|Z(t)\, x\| = \|x\| \tag{27}$$

it is necessary and sufficient that the following condition be satisfied.

(ii) *For every $\alpha > 0$, the characteristic equation*

$$A_0^* x^* = \alpha x^* \tag{28}$$

has no solution $x^* \neq 0$, where A_0^* is the adjoint of A_0 defined in X^*, the conjugate space of X.

In this statement "every $\alpha > 0$" may be replaced by "some $\alpha > 0$". When this condition is satisfied, $\{Z(t)\}$ is the only semigroup the infinitesimal generator of which is an extension of A_0.

The importance of relation (27) in our case will be obvious in few steps. Actually we will need that

$$\forall t \geqq 0, \quad \forall f \in X_+ : \|\exp[t(G + M - (S + S_1) I)] f\| = \|f\| . \tag{29}$$

Anyhow, let us begin with the proof of condition (ii) in Theorem 2.

Let X^* denote the space of all bounded linear forms on X defined by sequences

$$f^* = (f^*(n, m))_{(n,m)\in \mathbf{N}\times\mathbf{N}} \in \mathbf{R}^{\mathbf{N}\times\mathbf{N}}, \quad \|f^*\| = \sup_{n,m} |f^*(n, m)| < +\infty .$$

We need to define now operator A_0^*. Let then $f \in \mathscr{D}_0 \equiv \mathscr{D}(A_0)$. From the definition of $\mathscr{D}_0$:

$$\exists n_0 \in \mathbf{N}, \quad \exists m_0 \in \mathbf{N} \ni' \forall n > n_0, \quad \forall m > m_0 : f(n, m) = 0 .$$

We then find, if $f^* \in X^*$:

$$(f^*, A_0 f) = \sum_{n,m} f^*(n, m) [A_0 f]_{n,m} = \sum_{n=0}^{n_0} \sum_{m=0}^{m_0} g^*(n, m) f(n, m) \tag{30}$$

where

$$\begin{aligned} g^*(n, m) = {} & -(\pi_a n + \lambda m) f^*(n, m) + \pi_l n f^*(n-1, m) + \\ & + \pi_f n \sum_{s=1}^{\infty} \theta(s) f^*(n, m+s) + \lambda m f^*(n+1, m-1) . \end{aligned} \tag{31}$$

For g^* to belong to X^* we only need that $\sup_{n,m} |g^*(n, m)| < \infty$. Thus, if we set

$$\mathscr{D}(A_0^*) = \{ f^* \in X^* \mid g^* \in X^* \} \tag{32}$$

we may state that

$$\forall f^* \in \mathscr{D}(A_0^*) : [A_0^* f^*]_{n,m} = g^*(n, m) . \tag{33}$$

Finally we can verify (ii) in Theorem 2. Let $\alpha > 0$, and $f^* \in \mathscr{D}(A_0^*)$. Consider equation

$$(\alpha I - A_0^*) f^* = 0 . \tag{34}$$

If $(n, m) = (0, 0)$ we obtain

$$[(\alpha I - A_0^*) f]_{0,0} = 0$$

from which $\alpha f^*(0, 0) = 0$ and then $f^*(0, 0) = 0$. Let now $(n, m) \neq (0, 0)$. Equation (34) gives

(35)

$$\begin{aligned} f^*(n, m) = {} & \frac{\pi_l n}{\pi_a n + \lambda m + \alpha} f^*(n-1, m) + \frac{\pi_f n}{\pi_a n + \lambda m + \alpha} \sum_{s=1}^{\infty} \theta(s) f^*(n, m+s) + \\ & + \frac{\lambda m}{\pi_a n + \lambda m + \alpha} f^*(n+1, m-1) \end{aligned}$$

from which

$$(36)\qquad |\varphi^*(n, m)| \leqq \frac{\pi_l n}{\pi_a n + \lambda m + \alpha} |\varphi^*(n-1, m)| +$$

$$+ \frac{\pi_f n}{\pi_a n + \lambda m + \alpha} \sum_{s=1}^{\infty} \theta(s) |\varphi^*(n, m+s)| +$$

$$+ \frac{\pi_f}{\pi_a n + \lambda m + \alpha} \sum_{s=1}^{\infty} \theta(s)\, s |\varphi^*(n, m+s)| + \frac{\lambda m}{\pi_a n + \lambda m + \alpha} |\varphi^*(n+1, m-1)|$$

where we have set

$$\forall (n, m) \neq (0, 0) : \varphi^*(n, m) \equiv \frac{f^*(n, m)}{n + m}.$$

If f^* is a non-trivial solution of (34), φ^* will be such that

$$\lim_{n,m \to \infty} |\varphi^*(n, m)| = 0.$$

This implies

$$\exists (\bar{n}, \bar{m}) \in \mathbf{N} \times \mathbf{N} \ni' \mu = |\varphi^*(\bar{n}, \bar{m})| = \max_{n,m} |\varphi^*(n, m)|.$$

From (36) we thus obtain

$$(37)\qquad \pi \leqq \mu \frac{\pi_a \bar{n} + \lambda \bar{m} + \pi_f \langle s \rangle}{\pi_a \bar{n} + \lambda \bar{m} + \alpha}$$

where

$$(38)\qquad \langle s \rangle = \sum_{s=1}^{\infty} s\, \theta(s).$$

Relation (37) is impossible if $\alpha > \alpha_0 \equiv \pi_f \langle s \rangle$ unless $\mu = 0$, which implies

$$\forall (n, m) \neq (0, 0) : \varphi^*(n, m) = 0 \quad \text{and} \quad \forall (n, m) \neq (0, 0) : f^*(n, m) = 0.$$

We conclude that equation (34) admits only the trivial solution provided that $\alpha > \alpha_0$. Conclusions of Theorem 2 then hold; i.e.

$$(39)\qquad \forall f \in X_+, \quad \forall t \geqq 0 : \|\exp(tG) f\| = \|f\|$$

It also follows from relations (24.1), (24.2) that

$$\forall f \in X_+, \quad \forall t \geqq 0, \quad \forall j \in \mathbf{N} : \|T^{(j)}(t) f\| = [((S + S_1)\, t)^j / j!]\, \|f\|$$

and, consequently

$$\forall f \in X_+, \quad \forall t \geqq 0 : \|\exp[t(G + M)]f\| = \exp[(S + S_1)t]\,\|f\|$$

hence

(40) $$\forall f \in X_+, \quad \forall t \geqq 0 : \|\exp[t(G + M - (S + S_1)I)]f\| = \|f\|.$$

Relation (40) implies that

(41) $$\forall t \geqq 0, \quad \forall u_0 \in \mathscr{D}(G) \cap X_+ : \|u(t)\| = \|u_0\|.$$

REMARK 3. If u_0 is as in Remark 2, (41) implies that $u(t) = \exp[t(G + M - (S + S_1)I)]\,u_0$ is the physical solution of system (2) + (3). In other words:

$$u(t) = (u(0, 0; t), u(1, 0; t), u(0, 1; t), u(1, 1; t), \ldots) =$$
$$= (P(0, 0; t), P(1, 0; t), P(0, 1; t), P(1, 1; t), \ldots)$$

and

$$\|u(t)\| \equiv \sum_{n,m} P(n, m, t) = 1$$

in agreement with the probabilistic meaning of the $P(n, m; t)$'s.

6. THE EQUATIONS FOR EXPECTED VALUES

Using methods similar to those used until now, it is possible to prove in a rigorous way (by an extension of methods employed in [4]) that the expected values of $N(t)$, the number of infectives and $M(t)$, the number of individuals subject to their incubation period, exist at any $t \geqq 0$, and must satisfy the following equations:

(42) $$\frac{d}{dt}\langle N(t)\rangle = -\pi_l\langle N(t)\rangle + \lambda\langle M(t)\rangle + S,$$

(43) $$\frac{d}{dt}\langle M(t)\rangle = \pi_f\langle s\rangle\,\langle N(t)\rangle - \lambda\langle M(t)\rangle + S_1.$$

An analysis of stability for this system of equations gives the following condition of asymptotic stability [14]:

(44) $$\pi_f\langle s\rangle < \pi_l$$

which can be obtained by imposing that the eigenvalues of the matrix of coefficients all have negative real parts.

Relation (44) has an obvious physical meaning. An analogous argument may be used for second moments.

7. FINITE-DIMENSIONAL APPROXIMATION OF THE PROBLEM

It is possible to show [4] that the initial value problem (21) can be approximated by a suitable problem in a finite-dimensional space $X_{\nu,\mu}$. The space $X_{\nu,\mu}$ is defined as the set of all $(\nu + 1) \times (\mu + 1)$-tuples of real numbers

$$f^{(\nu,\mu)} = (f^{(\nu,\mu)}(n, m))_{0 \leqq n \leqq \nu; 0 \leqq m \leqq \mu}$$

with norm

$$\|f^{(\nu,\mu)}\| = \sum_{n=0}^{\nu} \sum_{m=0}^{\mu} |f^{(\nu,\mu)}(n, m)| . \tag{45}$$

An operator $Q^{(\nu,\mu)} \in \mathscr{B}(X, X_{\nu,\mu})$ is defined as follows:

$$Q^{(\nu,\mu)}f = g^{(\nu,\mu)} ; \quad g^{(\nu,\mu)} = (f(n, m))_{0 \leqq n \leqq \nu; 0 \leqq m \leqq \mu} \quad \text{if} \quad f \in \mathscr{D}(Q^{(\nu,\mu)}) = X . \tag{46}$$

It is easy to show that

$$\|Q^{(\nu,\mu)}\| \leqq 1 , \quad \lim_{\nu,\mu \to \infty} \|Q^{(\nu,\mu)}f\|_{\nu,\mu} = \|f\| \ \forall f \in X .$$

Hence, the sequence $\{X_{\nu,\mu}\}_{(\nu,\mu) \in N \times N}$ is a sequence of Banach spaces approximating X [5], [16] Finally, the following operations from $X_{\nu,\mu}$ to $X_{\nu,\mu}$ are defined:

$$\begin{aligned}[G_r^{(\nu,\mu)} f^{(\nu,\mu)}]_{n,m} &= -(\pi_a n + \lambda m) f^{(\nu,\mu)}(n, m) + \\ &+ r\{\pi_l(n + 1) f^{(\nu,\mu)}(n + 1, m) + \\ &+ \pi_f n \sum_{s=1}^{m} \theta(s) f^{(\nu,\mu)}(n, m - 1) + \lambda(m + 1) f^{(\nu,\mu)}(n - 1, m + 1)\}\end{aligned} \tag{47.1}$$

if $n \leqq \nu - 1;\ m \leqq \mu - 1$;

$$\begin{aligned}[G_r^{(\nu,\mu)} f^{(\nu,\mu)}]_{\nu,m} &= -(\pi_a \nu + \lambda m) f^{(\nu,\mu)}(\nu, m) + \\ &+ r\{\pi_f \nu \sum_{s=1}^{m} \theta(s) f^{(\nu,\mu)}(\nu, m - 1) + \lambda(m + 1) f^{(\nu,\mu)}(\nu - 1, m + 1)\}\end{aligned} \tag{47.2}$$

if $m \leqq \mu - 1$;

$$(47.3)\qquad [G_r^{(\nu,\mu)} f^{(\nu,\mu)}]_{n,\mu} = -(\pi_a n + \lambda\mu) f^{(\nu,\mu)}(n, \mu) +$$

$$+ r\{\pi_l(n+1) f^{(\nu,\mu)}(n+1, \mu) + \pi_f n \sum_{s=1}^{\mu} \theta(s) f^{(\nu,\mu)}(n, \mu - s)\}$$

if $n \leqq \nu - 1$;

$$(47.4)\qquad [G_r^{(\nu,\mu)} f^{(\nu,\mu)}]_{\nu,\mu} =$$

$$= -(\pi_a \nu + \lambda\mu) f^{(\nu,\mu)}(\nu, \mu) + r\{\pi_f \nu \sum_{s=1}^{\mu} \theta(s) f^{(\nu,\mu)}(\nu, \mu - s)\}$$

with $\mathscr{D}(G_r^{(\nu,\mu)}) = Q^{(\nu,\mu)}[\mathscr{D}]$ and $r \in [0, 1[$;

$$(47.5)\qquad [M^{(\nu,\mu)} f^{(\nu,\mu)}]_{n,m} = = S f^{(\nu,\mu)}(n-1, m) + S_1 f^{(\nu,\mu)}(n, m-1) \qquad \text{if } n > 0, \quad m > 0;$$

$$(47.6)\qquad [M^{(\nu,\mu)} f^{(\nu,\mu)}]_{n,0} = S f^{(\nu,\mu)}(n-1, 0) \qquad \text{if } n > 0;$$

$$(47.7)\qquad [M^{(\nu,\mu)} f^{(\nu,\mu)}]_{0,m} = S_1 f^{(\nu,\mu)}(0, m-1) \qquad \text{if } m > 0;$$

$$(47.8)\qquad [M^{(\nu,\mu)} f^{(\nu,\mu)}]_{0,0} = 0;$$

with $\mathscr{D}(M^{(\nu,\mu)}) = X_{\nu,\mu}$.

Consider the following initial-value problem

$$(48)\qquad \begin{cases} \dfrac{d}{dt} u^{(\nu,\mu)}(t) = [G_r^{(\nu,\mu)} + M^{(\nu,\mu)} - (S + S_1) I^{(\nu,\mu)}] u^{(\nu,\mu)}(t), \quad t > 0, \\ \lim\limits_{t \to 0+} \| u^{(\nu,\mu)}(t) - u_0^{(\nu,\mu)} \|_{\nu,\mu} = 0; \quad u_0^{(\nu,\mu)} \in \mathscr{D}(G_r^{(\nu,\mu)}). \end{cases}$$

It can be shown that, if r is close to one, and if we choose $u_0^{(\nu,\mu)} = Q^{(\nu,\mu)} u_0$, the solution $u(t)$ of problem (21) is such that

$$(49)\qquad \lim_{\nu,\mu \to \infty} \| Q^{(\nu,\mu)} u(t) - u^{(\nu,\mu)}(t) \| = 0$$

uniformly with respect to $t \in [0, \bar{t}]$.

And this has an obvious interpretation: the first $(\nu + 1) \times (\mu + 1)$ components of the matrix $u(t) \in X$ can be approximated by the corresponding components of $u^{(\nu,\mu)} \in X_{\nu,\mu}$ with an error not larger than an arbitrarily small $\varepsilon > 0$.

Acknowledgements. I am indebted to Prof. A. Belleni-Morante for helpful criticisms and suggestions.

REFERENCES

[1] N. T. J. Bailey: The total size of a general stochastic epidemic. Biometrika *40* (1953), 177 to 185.

[2] N. T. J. Bailey: The mathematical theory of epidemics. Griffin, London 1957.

[3] M. S. Bartlett: Some evolutionary stochastic processes. J. Roy. Statist. Soc. Ser. B *11* (1949), 211—229.

[4] A. Belleni-Morante: On Chapman-Kolmogorov equations for neutron population in a multiplying assembly. To be published.

[5] A. Belleni-Morante and G. Busoni: Multigroup neutron transport. J. Math. Phys. *13* (1972), 1146.

[6] F. G. Foster: A note on Bailey's and Whittle's treatment of a general stochastic epidemic. Biometrika *42* (1955), 123—125.

[7] J. Gani: On a partial differential equation of epidemic theory, I. Biometrika *52* (1965), 617—622.

[8] J. Gani: On a partial differential equation of epidemic theory, II. The model with immigration. Office of Naval Research technical report RM-124 at Michigan State University.

[9] J. Gani: On the general stochastic epidemic. In: Proceedings of the Fifth Berkeley Symposium on Mathematical Statistics and Probability, Berkeley and Los Angeles 1965—66. University of California Press, 1967, 4, 271—279.

[10] T. Kato: On the semi-groups generated by Kolmogoroff's differential equations. J. Math. Soc. Japan *6* (1954), 1, 1—15.

[11] T. Kato: Perturbation theory for linear operators. Springer, New York 1966.

[12] D. G. Kendall: Deterministic and stochastic epidemics in closed populations. In: Proceedings of the Third Berkeley Symposium on Mathematical Statistics and Probability, Berkeley and Los Angeles 1956. University of California Press, 1956, 4, 149—165.

[13] J. Neyman, E. L. Scott: A stochastic model of epidemics. In: Stochastic Models in Medicine and Biology (edited by J. Gurland). University of Wisconsin Press, Madison 1964, 45—83.

[14] E. Polak, E. Wong: Notes for a first course on linear systems. Van Nostrand, New York 1970.

[15] V. Siskind: A solution of the general stochastic epidemic. Biometrika *52* (1965), 613—616.

[16] H. F. Trotter: Pacific J. Math. *8* (1958), 887.

[17] P. Whittle: The outcome of a stochastic epidemic — A note on Bailey's paper. Biometrika *42* (1955), 116—122.

Universita' di Bari

Istituto di Analisi Matematica

ON THE AGE STRUCTURE OF CELL-SIZE DEPENDENT BRANCHING PROCESSES

PETER CLIFFORD

BRISTOL

It is perhaps tautological to say that the factor which determines the time of division of a bacterium is its physiological state. That the traditional age dependent branching process provides an inadequate model for growing bacteriological populations is not surprising. Stochastic models which involve a physiological state vector for each cell present complex analytical problems. The multitype processes make a discrete approximation to the continuously varying development of a cell. In earlier papers Clifford and Sudbury [3], [12] have studied models which involve a continuous measure of development, the cell-size. The models induce correlations between the life-spans of cells and the earlier work was devoted to the calculation of these correlations. In the present paper, the age structure of such processes is studied. The special case in which the size of cells grows exponentially is shown to be related to a suggestion made by Harris. Explicit results are obtained including the almost sure convergence of the age distribution.

INTRODUCTION

Experimentation has shown that the interdivision times of bacterial cells belonging to the same family are dependent unlike the interdivision times of the Bellman-Harris age-dependent branching process. Positive correlations are observed between sisters [7], [8], [10] and negative correlations are observed between mothers and daughters [8]. Harris [5] has indicated a way in which sister correlations can be introduced into an age-dependent process. This has been investigated by Crump and Mode [4]. The mathematical device is not accompanied by a phenomenological justification and since mothers and daughters remain independent, the model is unsatisfactory.

In 1932, Bayne-Jones and Adolph [1] suggested that the time of division was primarily determined by the size of the organism. The hypothesis was first put on a quantitative basis by Koch and Schaechter [6] and has been studied in more general forms by others [2], [3], [9], [12]. The consequences for correlations are that, roughly speaking, long-lived and therefore large cells will give rise on the average to large and therefore short-lived cells and vice versa.

These cell-size-dependent processes resemble the general branching processes of Harris [5] and the branching diffusion processes treated by Skorohod [11]. The essential feature is that the lifespan of a cell depends on its 'type' at birth and the 'type' of the offspring depends on the length of life of the dividing cell. For the case in which cell-size grows exponentially much of the analysis of mean functions has been carried out by Powell [9]. In particular a simplifying hypothesis has been introduced by Powell which enables correlations to be calculated readily [3], [12]. In this paper Powell's hypothesis is used to facilitate the derivation of theorems concerning the almost sure convergence of the age and size distributions in a cell-size-dependent branching process.

DEFINITIONS

We suppose cells grow continuously such that the rate of increase of size s with respect to time t is given by

$$\frac{\mathrm{d}s}{\mathrm{d}t} = p(s)$$

where p is a positive function of s alone. The probability that a cell of size s will divide in the next interval of time τ is $\lambda(s)\,\tau + o(\tau)$. We allow infinite values for λ and interpret these as corresponding to instantaneous division. (Koch and Schaecter consider $p(s) \propto s$ and λ arbitrary.) It is further assumed that bacteria divide exactly into half and there is no death. Powell's hypothesis is that there exist positive numbers a and b such that, $\lambda(s) = 0$, $s < a$ and $\lambda(s) = \infty$, $s \geqq b$ and that $2a \geqq b$. Without loss of generality it can be assumed that $a = \frac{1}{2}$, $b = 1$. The cell-size-dependent process begins with a single individual of size $\frac{1}{2}$ at time $t = 0$. It was shown in [3], [12] that such processes when sampled by generation result in appropriate correlations between related cells which can be varied by suitable choice of λ and p. $Z(t)$ is defined to be the number of cells present at time t. $Z(t, s, x)$ is the number of cells at time t whose size is less than s and whose age is less than x.

THE UNDERLYING AGE-DEPENDENT PROCESS

If a cell is born with size less than a half than its size at division S will necessarily be in the interval $(\frac{1}{2}, 1]$ and will be independent of its initial size. Furthermore

$$P(S > s) = \exp\left\{-\int_{1/2}^{s} \lambda(x)/p(x)\,\mathrm{d}x\right\}.$$

Thus one of the consequences of Powell's hypothesis is that the sizes at division of all cells are independent. Consider a particular division and look at the total time T taken for the mother to grow from size $\frac{1}{2}$ to size S plus the time taken for one of the daughters to grow from size $S/2$ back to size $\frac{1}{2}$.

$$T = \{P(S) - P(\tfrac{1}{2})\} + \{P(\tfrac{1}{2}) - P(S/2)\} = P(S) - P(S/2)$$

where

$$P(s) = \int_{1/4}^{s} \mathrm{d}x/p(x)\,.$$

Thus underlying the cell-size-dependent branching process there is an age-dependent process with independent life spans which are copies of T.

STATEMENT OF THEOREM

Denoting $P(S) - P(\frac{1}{2})$ by T_1 and letting g and G stand for the density and distribution function of T_1 and T with subscripts where appropriate we have the following theorem:

THEOREM. If $g \in L_p$ for some $p > 1$ then

$$\frac{Z(t, s, x)}{n_1 \mathrm{e}^{\alpha t}} \xrightarrow{\text{a. s.}} \beta\, A(s, x)\, W$$

for every $s, x \geqq 0$ where

$$\beta = 2\left[1 - \alpha \int_0^\infty \mathrm{e}^{-\alpha t}(1 - G_1(t))\,\mathrm{d}t\right]$$

and $A(s, x)$ is a bivariate distribution function. The random variable W and the constants n_1 and α correspond to the underlying age-dependent process. Thus

$$\phi(\theta) = E(\mathrm{e}^{-\theta W})$$

satisfies the equation

$$\phi(\theta) = \int_0^\infty \phi^2(\theta \mathrm{e}^{-\alpha u})\, g(u)\,\mathrm{d}u\,.$$

α is the positive root of

$$2\int_0^\infty \mathrm{e}^{-\alpha t}\, g(t)\,\mathrm{d}t = 1$$

and

$$n_1 = \left[4\alpha \int_0^\infty t\mathrm{e}^{-\alpha t}\, g(t)\,\mathrm{d}t\right]^{-1}.$$

The expression for $A(s, x)$ is a little unwieldy and we merely give the size distribution.
Thus

$$\beta A(s, \infty) = \beta - \int_{P(s)-P(1/2)}^{\infty} 2\alpha e^{-\alpha t}(1 - G(t))\, dt, \quad s \geqq \tfrac{1}{2},$$

$$= 4\alpha \int_0^{\infty} e^{-\alpha t}\{G_1(t) - G(t + P(\tfrac{1}{2}) - P(s/2))\}\, dt, \quad s < \tfrac{1}{2}.$$

The proof of this theorem is direct once the underlying age-dependent process has been identified.

CONCLUSION

The correlation induced in the interdivision times of sisters is seen to be the result of sisters having a common early childhood until reaching size $\frac{1}{2}$. This is related to the suggestion of Harris. Moreover mother-daughter correlation is induced by virtue of a negatively associated period towards the end of the mother's division cycle. It should be noted that more distantly-related cells have no correlation but that second order correlations could be induced by transforming the branching process yet again in a similar manner.

REFERENCES

[1] S. Bayne-Jones, E. F. Adolph: Growth in size of micro-organisms measured from motion pictures III. Bacterium coli. J. Cell. Comp. Physiol. *2* (1932), 329—348.

[2] G. I. Bell, E. C. Anderson: Cell growth and division I. Biophys. J. *7* (1967), 329—351.

[3] P. Clifford, A. W. Sudbury: The linear cell-size-dependent branching process. J. Appl. Prob. *9* (1972), 687—696.

[4] K. S. Crump, C. J. Mode: An age-dependent branching process with correlations among sister cells. J. Appl Prob. *6* (1969), 205—210.

[5] T. E. Harris: The Theory of Branching Processes. Springer, Berlin 1963.

[6] A. L. Koch, M. Schaechter: A model for statistics of the cell division process. J. Gen. Microbiol. *29* (1962), 435—454.

[7] H. E. Kubitschek: Normal distribution of cell generation rate. Expl. Cell. Res. *26* (1962), 439—451.

[8] E. O. Powell: Some features of the generation times of individual bacteria. Biometrika *42* (1955), 16—44.

[9] E. O. Powell. A note on Koch and Schaechter's hypothesis about growth and fission of bacteria. J. Gen. Microbiol. *37* (1964), 231—249.

[10] M. Schaechter J. P. Williamson, J. R. Hood, A. L. Koch: Growth, cell and nuclear division in some bacteria. J. Gen. Microbiol. *29* (1962), 421—434.

[11] A. V. Skorohod: Branching Diffusion Processes. Theory of Prob. and its Applic. *9* (1967), 492—497.

[12] A. W. Sudbury, P. Clifford: Some results for general cell-size-dependent branching processes. J. Appl. Prob. *10* (1973), 289—298.

University of Bristol

School of Mathematics

INFORMATION PATTERNS IN STOCHASTIC POINT PROCESS DIFFERENTIAL GAMES

ANTHONY EPHREMIDES

COLLEGE PARK

In stochastic control problems or in stochastic differential games the choice of an observation model is critical. In particular for control or game problems involving a random point process it is possible to consider the Fokker-Planck type of equation, and the process itself as the observation equation. In the context of the single-server queueing problem the state equation becomes $\dot{p}_t = -(\alpha_t + \beta_t) p_t + \alpha_t$, where p_t = prob. that the server is busy, α_t = the rate parameter of the arrival process, controlled by one player, β_t = the parameter of the service time, controlled by another player. This equation describes a bilinear system. Choosing a linear cost functional and a variety of associated point processes as observables, the conceptual problems of separation principles, existence of value, and saddle point solutions arise. The results obtained in this paper are shown to be applicable to the optimization of resource allocation systems that can be modeled as co-operative or competitive games. The martingale representation theorems play a central role in this discussion.

1. INTRODUCTION

A classical model of a dynamical system is given by

(1) $$\dot{x}_t = f_t(x_t, u_t, \omega)$$

where x_t denotes the state of the system, u_t denotes a control function, and $\omega \in \Omega$: the sample space of a probability space. A typical stochastic control problem is then formulated as follows: An observation model given by

$$y_t = g_t(x_t, u_t, \omega)$$

is assumed, and a cost function is chosen given by

$$K(u) = \mathrm{E} \int_0^T k(u_t, x_t)\, \mathrm{d}t$$

for a fixed T. Then a function u_t^* must be determined to minimize K. Usually one is interested in determining a feedback or closed-loop, causal control law as a function of the observations, that is of the form

$$u_t^* = u(y_s, s \leqq t).$$

A stochastic differential game can be described by a similar model with the difference that in Eq. (1) the function f_t depends on two control laws u_t and v_t, that is

$$\dot{x}_t = f_t(x_t, u_t, v_t, \omega)$$

and so do the observations function g_t and the cost function K. In simple cases, called zero sum games, the objective is to determine u_t^* and v_t^* as functions of $y_s, s \leqq t$ so that

$$K(u^*, v^*) = \min_u \max_v K(u, v) = \max_u \min_v K(u, v).$$

That is, a pair of strategies must be determined that realizes a saddle point of K. Usually it is assumed that the minimaximization and the maximinimization are carried out by different individuals who are called the players of the game. The u-player wants to maximize and the v-player wants to minimize. Of course there are generalized versions of this model that allow for different cost functions, larger numbers of players, and co-operative as well as conflict situations.

The so called *LQG* problem (Linear, Quadratic, Gaussian) has been solved for simple cases of additive random disturbances [1], [2]. In problems of estimation that use similar models, interest has developed recently in models where the observations process y_t is a jump point process like, for example, the counting process of Poisson distributed arrival times. Such models are useful in communications at optical frequencies [3], [4], in signal processing with impulsive (shot) noise, or jitter [5], in the biomedical applications [6], [7], and in queueing, resource allocation, and operations research problems [8], [9]. A major difficulty in dealing with such models lies in the fact that the point process models are not as analytically tractable as Gaussian process models. Recent representation results on jump processes in terms of stochastic integrals over martingales [10], [11] have enhanced the use of jump process models by providing some means of their mathematical representation and analysis.

2. POINT PROCESS GAMES

In the context of the previously described models it is possible to consider queueing optimization problems. For example consider a simple service station with one server whose service time γ_s, starting at time s, is a random variable distributed in an exponential-like fashion, that is

$$p\{\gamma_s \leqq t - s + \mathrm{d}t / \gamma_s > t - s\} = \beta_t \, \mathrm{d}t$$

where β_t obviously "measures" his service capabilities at time t. Assume that customers arrive randomly in time with Poisson distributions, that is

$$p\{\text{one arrival in } (t, t + \mathrm{d}t]\} = \alpha_t\, \mathrm{d}t$$

where α_t is the average intensity rate of arrivals at time t. It is reasonable to propose that, subject to some constraints, the quantities α_t and β_t should be chosen suitably to maximize the average flow of customers. If M_t denotes the total number of exiting customers in $(0, t]$, then the function K given by

$$K(\alpha, \beta) = \mathrm{E}[M_T] - c \int_0^T (\beta_t + \alpha_t)\, \mathrm{d}t$$

could be a cost function to be maximized by the choice of α_t, β_t. The second term is included to indicate a "fuel" cost for large β_t and α_t.

For the sake of studying a more interesting example, suppose that in the previous model the service station is in fact a border corridor that is defended by a sentry who engages in battle of duration γ_s intruders who arrive at time s. Then the process M_t is the number of penetrators of the border. Clearly the α-player would want to choose α_t in order to maximize $\mathrm{E}[M_T]$ and the β-player would want to select β_t to minimize $\mathrm{E}[M_T]$. Allowing for "fuel" costs one may now consider the conflict situation described by the following cost function

$$K(\alpha, \beta) = \mathrm{E}[M_T] + c \int_0^T (\beta_t - \alpha_t)\, \mathrm{d}t$$

where $c > 0$ is a constant, and where a saddle point of K must be determined.

It is clear that to solve this simple, hypothetical problem, one must decide on the observation and the state models, that is, in slightly different terminology, on the information patterns available to the players and on the underlying dynamical model that must fit a central underlying quantity.

Typically one may consider that the process M_t is continuously observed and that the state coincides with this process. Then since it may be argued that M_t is a Markov or a diffusion process, one has a dynamical equation of the form of Eq. (1) with which he can proceed. The difficulty lies in determining f_t, which, even in this simple problem, does not have a tractable form. For this reason, an alternative approach may be followed in modeling the dynamics. If p_t denotes the probability that the sentry (or server) is busy at time t, then from a simple analysis of the M/M/S problem of queueing theory it can be shown that

$$\dot{p}_t = -(\alpha_t + \beta_t)\, p_t + \alpha_t\,. \tag{2}$$

Eq. (2) has a very simple bi-linear form. It is therefore possible to consider p_t as the "state" x_t of the system and M_t as the "observation" process y_t. In this case the form of the observations equation g_t is not readily available. However the separation into a basic state equation of simple form and an observation process whose statistics are simply related to p_t can be of substantial help. For example, suppose that instead of observing M_t the players observe the equivalent "indicator" process I_t given by

$$I_t = \begin{cases} 1, & \text{if the server is busy at time } t, \\ 0, & \text{otherwise}. \end{cases}$$

Then clearly one has

$$E[I_t] = p_t$$

or

$$p_t = p[I_t = 1].$$

In the remainder of the paper a summary of results concerning this border violation problem are provided, first under the assumption of perfect information that is when $y_t = p_t$, and then under the assumption that $y_t = I_t$. The concepts and techniques that arise from the study of this sample problem can then be generalized to model and analyze more realistic, complex problems of similar nature.

3. PERFECT INFORMATION CASE

In this case it is assumed that both players can observe the state p_t at every $0 \leqq t \leqq T$. First, one verifies that the system satisfies trivially various sets of sufficient conditions for the existence of a saddle point, as for example the ones in Thm. 2.5.1 of [12], namely the existence of a unique solution to the state equation for any given controls α_t, β_t, the latter's continuity in the controls, and the linearity of the pay-off in α_t, β_t. That the saddle point consists of pure strategies will be concluded a posteriori, after finding and exhibiting such pure strategies. To find them, Pontryagin's maximum (minimum) principle will be used [13] in the standard control theoretic fashion by finding the $\max_\alpha [\min_\beta (K)]$, which, by the existence of value for the game, will be equal to $\min_\beta [\max_\alpha (K)]$. This approach has the advantage of supplying as a by-product the solution to the control problem of designing the defense β_t for known rate α_t (typical in simple models of air traffic control systems), by specifying the function $\beta_t(\alpha_t)$ in the intermediate stage of the maximin procedure.

Of course the analysis can be carried out in terms of cost-to-go from an arbitrary starting position (π, τ) in the playing space $[0, 1] \times [0, T]$. Thus prescriptions for optimal strategies will be supplied to each player in case the opponent deviates from optimal play and consequenty forces the state out of the optimal trajectory. Furthermore arbitrary boundary conditions for p at the beginning of the play must be allowed.

The Min part: the system is described by

$$\dot{p}_t = -(\alpha_t + \beta_t)\, p_t + \alpha_t,$$

$$\dot{q}_t = (\alpha_t + \beta_t)\, q_t - \alpha_t$$

where q_t is the co-state, and with boundary conditions

$$p_\tau = \pi\,, \quad q_T = 0\,.$$

The Hamiltonian is given by

$$H(p_t, q_t; \alpha_t, \beta_t) = (c - p_t q_t)\,\beta_t + (p_t + q_t - p_t q_t - c)\,\alpha_t\,.$$

The H-minimal control will be

$$\beta_t = \begin{cases} 0\,, & p_t q_t < c\,, \\ \text{indeterminate}\,, & p_t q_t = c\,, \\ \beta_M\,, & p_t q_t > c\,. \end{cases}$$

Already it should be noted that for $c \geqq 1$ the overall game assumes the trivial solution $\alpha_t \equiv \beta_t \equiv 0$, since it is then more expensive to both players to play the game. So it will be assumed that $0 < c < 1$.

The solution can be best illustrated graphically. As it turns out the relative magnitude of β_M and α_t is critical to the form of the solution. Thus one can distinguish the case of the so called weak defense, that is when

$$\beta_M \leqq \min_{0 \leqq t \leqq t_1} \alpha_t \frac{1 - \sqrt{c}}{\sqrt{c}}$$

and the case of strong defense, that is

$$\beta_M \geqq \max_{0 \leqq t \leqq t_1} \alpha_t \frac{1 - \sqrt{c}}{\sqrt{c}}\,.$$

For the case that

$$\min_{0 \leqq t \leqq t_1} \alpha_t \frac{1 - \sqrt{c}}{\sqrt{c}} < \beta_M < \max_{0 \leqq t \leqq t_1} \alpha_t \frac{1 - \sqrt{c}}{\sqrt{c}}$$

which represents an intermediate situation, the answer can be obtained in all but a portion of the playing space, as will be explained later. However in the maximization portion of the maximin process this difficulty will be circumvented. There are two critical instants t_1, and t_2 defined by

(3)
$$\int_{t_1}^{T} \alpha_t \, dt = \ln \frac{1}{1 - \sqrt{c}}\,,$$

(4)
$$\int_{t_2}^{T} \alpha_t \, dt = \ln \frac{1}{1 - c}\,.$$

Eq. (3) implies that $q_{t_1} = \sqrt{c}$, and Eq. (4) that $q_{t_2} = c$ while the function ψ_t denotes the locus of the points at which $p_t q_t = \sqrt{c}$, with $\psi_t = p_t$.

A function θ_t representing the locus of the points at which $p_t q_t = c$ with $\theta_t = p_t$, can be defined parametrically by

(5)
$$c = \left\{\psi_s - \int_t^s \alpha_\xi \exp\left[-\int_\xi^s (\alpha_\eta + \beta_M)\,d\eta\right] d\xi\right\} \left\{\frac{c}{\psi_s} + \int_t^s \alpha_\xi \exp\left[\int_\xi^s (\alpha_\eta + \beta_M)\,d\eta\right] d\xi\right\}$$

and

(6) $$\theta_t = \psi_s - \int_t^s \alpha_\xi \exp\left[-\int_\xi^s (\alpha_\eta + \beta_M)\,\mathrm{d}\eta\right]\mathrm{d}\xi\,, \quad t_1 \leqq s < t_2\,, \quad t \leqq t_1\,.$$

This function need not exist in the intermediate defense case because the above equations may not have solutions. The fact that we do not know the conditions for existence of solutions in the intermediate case accounts for the incomplete specification of the solution.

The Max part: there are three regimes of solutions for β_t in the min part, namely $\beta = 0$, $\beta =$ indeterminate, and $\beta = \beta_M$. For each regime Pontryagin's principle is used to obtain that the H-maximal control is given by

$$\alpha_t = \begin{cases} 0\,, & p_t + q_t - p_t q_t < c \\ \text{indeterminate}\,, & p_t + q_t - p_t q_t = c\,, \\ \alpha_M\,, & p_t + q_t - p_t q_t > c \end{cases}$$

for the regimes corresponding to $\beta = 0$ and $\beta = \beta_M$, and by

$$\alpha_t = \begin{cases} 0\,, & p_t + q_t - (1/\sqrt{c})\, p_t q_t < c\,, \\ \text{indeterminate}\,, & p_t + q_t - (1/\sqrt{c})\, p_t q_t = c\,, \\ \alpha_M\,, & p_t + q_t - (1/\sqrt{c})\, p_t q_t > c \end{cases}$$

for the regime $\beta =$ indeterminate. In all cases it is argued that α_t cannot be indeterminate. Therefore always α_t is either α_M or 0, for the cases of weak and strong defenses. Whether $\alpha_t = 0$, or $\alpha_t = \alpha_M$ depends on the location in the playing space subregions as defined by the graphs of the functions ψ_t, θ_t, and the lines $p_t = \sqrt{c}$, $p_t = c$. The intermediate case does not arise, since in the region of possible ambiguity we have $\alpha_t = \alpha_M$, so that the defense can be determined to be either weak or strong. There is also a dispersal surface given by

$$\delta_t = 1 - \frac{\alpha_M(1 - c)(T - t)}{1 - \exp\left[-\alpha_M(T - t)\right]}$$

and below which $\alpha_t = 0$. Thus the closed-loop, feedback strategies are completely specified for both players, in the perfect information case. The solution to equation (5) is given by

$$t = s - \frac{1}{\alpha_M + \beta_M} \ln \frac{\dfrac{\alpha_M}{\alpha_M + \beta_M} - \dfrac{c}{\psi_s}}{\dfrac{\alpha_M}{\alpha_M + \beta_M} - \psi_s}\,, \quad t_1 \leqq s \leqq t < t_2$$

when

$$\beta_M < \alpha_M \frac{1 - \sqrt{c}}{\sqrt{c}}\,.$$

4. IMPERFECT INFORMATION CASE

The quantity p_t that was considered to be the "state" in the formulation of the problem is clearly unobservable and one can only compute it from the differential equation (2) if α_t and β_t are completely known. This not being the case it is clear that an additional observation equation must be considered. Since a solution for the perfect information case has been obtained, one would hope that a separation theorem [14], [15] might hold, that is

$$\alpha_t = \alpha(\hat{p}_t), \quad \beta_t = \beta(\hat{p}_t)$$

where $\alpha(\cdot)$, $\beta(\cdot)$ are the saddle point solution derived before, and $\hat{p}_t$ the max likelihood estimate of p_t given y_s, $s \leqq t$.

Showing a separation theorem has not been accomplished. Thus, as a suboptimal attempt, it was decided to try the min part solution assuming such a separation. Reasonable approximate solutions would then be obtained for resource allocation problems (like airport facilities control). Towards this end it was shown that $\hat{p}_t$ can be obtained by just solving Eq. (2) causally, with $\hat{\alpha}_t$ substituted for α_t, and where $\hat{\alpha}_t$ is the max likelihood estimate of α_t given y_s, $s \leqq t$. This result was proven rigorously for $\alpha_t \equiv$ unknown constant and heuristically for arbitrary α_t. Methods for obtaining $\hat{\alpha}_t$ are numerous [16], [17] and, since Eq. (2) is relatively simple, the method was tried in the computer with results comparing favorably to the perfect information case.

The imperfect information case has only been investigated imperfectly, but indications are that it is amenable to further analysis, particularly by the use of the Markov property and the martingale representation. In the case that $y_t = I_t$ the existence of a saddle point can be shown.

5. GENERALIZATIONS

Dimension: by considering many sentries one arrives at a matrix equation with the state vector being $\bar{p}(t)$ where $p_i(t)$ is the probability that i sentries are engaged at time t. The method of analysis generalizes straight-forwardly, although analytical solutions in closed form are hard to obtain. Similar extensions can be accomplished when more than one corridor is considered. The form of the system is then nontrivially bilinear and recent results [18] in studying such systems through their equivalent simple, almost non-dynamical, linear systems can be used to infer properties of the trajectories. The generalization becomes more interesting when switching of sentries from corridor to corridor is allowed. Then one gets an improved model for resource allocation problems.

Assignment: when switching is allowed the model can be constructed as follows. Consider 2 corridors and 1 sentry. Let

$$p\left[\frac{\text{of sentry being transferred to corridor 2 in } (t, t + \mathrm{d}t]}{\text{given he is in corridor 1 and idle at } t}\right] = \delta_t^1 \, \mathrm{d}t$$

and let the probability of the opposite switch be $\delta_t^2 \, \mathrm{d}t$. The attack rates for the two corridors are α_t^1, α_t^2 and the parameter of the sentry's capability is β_t as before. Let $p_t^0, p_t^1, p_t^2, p_t^3$ represent the probabilities of the four possibilities that the sentry is in corridor 1 and idle, or in 1 and busy, or in 2 and idle, or in 2 and busy at time t respectively. Then allowing for switches only during idle periods and denoting by $\bar{p}_t$ the vector of the above probabilities the system is described by

$$\dot{\bar{p}}_t = \begin{bmatrix} -(\alpha_t^1 + \delta_t^1) & \beta_t & \delta_t^2 & 0 \\ \alpha_t^1 & -\beta_t & 0 & 0 \\ \delta_t^1 & 0 & -(\alpha_t^2 + \delta_t^2) & \beta_t \\ 0 & 0 & \alpha_t^2 & -\beta_t \end{bmatrix} \cdot \bar{p}_t$$

with pay-off (cost) function given by

$$K = \int_0^T c(\beta_t + \delta_t^1 + \delta_t^2 - \alpha_t^1 - \alpha_t^2) \, \mathrm{d}t + \mathrm{E}[M_T] .$$

It is of course assumed that the two attack processes are mutually independent and that the switching intensities are independent of the service times.

The above system is non-trivially bilinear, and similar analysis as before is still applicable. More elementary treatments of restricted cases have been already done [19]. They involve simplifying the assignment strategy to allow only one switch during the course of the game. Then the optimum delta rule is to find a max likelihood decision time t_c and then make a MAP decision on the two attack processes.

Of course the general problem that involves many corridors, several sentries and includes assignment functions has a large dimensionality and therefore, although properties of the solutions can be infered analytically, the actual solutions would be rather obtained computationally.

Finally the possibility of sentry liquidation can be incorporated in the previous model by defining an additional intensity h_t such that $h_t \, \mathrm{d}t$ represents the probability of elimination in $(t, t + \mathrm{d}t]$ conditioned appropriately on the sentry's state at time t.

REFERENCES

[1] R. D. Behn, Y. C. Ho: On a class of linear stochastic differential games. IEEE Transactions on Automatic Control *14* (1969), 3, 227—240.

[2] I. B. Rhodes, D. G. Luenberger: Nondeterministic differential games with constrained state estimators. IEEE Transactions on Automatic Control *14* (1969), 5, 476—481.

[3] I. Bar David: Communications under the Poisson regime. IEEE Transactions on Information Theory *13* (1969), 1, 31—37.

[4] R. S. Kennedy: Communication through optical scattering channels. Proceedings of the IEEE *58* (1970), 10, 1651—1665.

[5] F. J. Beutler, O. A. Z. Leneman: On the statistics of random pulse processes. Information and Control *18* (1971), 4, 326—341.

[6] D. L. Snyder: Information processing for observed jump processes. Information and Control *22* (1973), 1, 69—78.

[7] A. S. French, A. V. Holden: Alias free sampling of neuronal spike trains. Kybernetik *8* (1971), 2, 165—171.

[8] P. A. W. Lewis: Stochastic point processes. Wiley, New York 1972.

[9] B. V. Gnedenko, I. N. Kovalenko: Introduction to queueing theory. Israel Program for Scientific Translations 1968.

[10] R. Boel, P. Varaiya, E. Wong: Martingales on jump processes: representation results and applications. University of California, Berkeley 1973.

[11] A. Segall: A martingale approach to modeling, estimation, and detection. Stanford University, Stanford 1973.

[12] A. Friedman: Differential games. Wiley-Interscience, New York 1971.

[13] M. Athans, P. Falb: Optimal control. McGraw-Hill, New York 1966.

[14] W. M. Wonham: On the separation theorem of stochastic control. SIAM Journal on Control *6* (1968), 3, 312—326.

[15] D. L. Snyder, I. B. Rhodes: Filtering and control performance bounds with implications on asymoptic separation. Automatica *8* (1972), 747—753.

[16] D. L. Snyder: Filtering and detection for doubly stochastic poisson processes. IEEE Transactions on Information Theory *16* (1972), 1, 91—102.

[17] I. Rubin: Regular point processes and their detection. IEEE Transactions on Information Theory *16* (1972), 5, 547—557.

[18] R. W. Brockett: System theory on group manifolds and coset spaces. SIAM Journal on Control *10* (1972), 265—281.

[19] A. Ephremides: The analysis of the border violation problem as a point process game. In: Proceedings of Eleventh Annual Allerton Conference on Circuit and System Theory, Allerton 1973, 505—514.

University of Maryland

Electrical Engineering Department

ZUR SIMULATION STETIGER STOCHASTISCHER WIRTSCHAFTMODELLE

LUDWIG FAHRMEIR, HELMUT BEECK

MÜNCHEN

Many economic models are now constructed in a form in which time is considered a continuous variable. This paper deals with methods to simulate stochastic dynamic models which may be described by diffusion processes or by stochastic differential equations. It will be shown that numerical results obtained by these methods on digital and hybrid computers converge weakly to the exact solution of the given model.

1. EINFÜHRUNG

Zufällig gestörte dynamische Systeme mit kontinuierlichem Zustands- und Parameterraum traten zuerst bei technischen Problemstellungen auf. Sie gewinnen jedoch auch in der Wirtschaftstheorie zunehmend an Bedeutung, da sie geeignete Modelle für viele Bereiche [3], [7], [9], [10], [11], [12], [13], [14], [15], [16] liefern.

Wir gehen von einem einfachen ökonomischen Modell aus [7]. Sei $x(t)$ die Preisabweichung eines Preises von einem Gleichgewichtspunkt zum Zeitpunkt t. Die Veränderung der Preisabweichung setze sich zusammen aus der Trendbewegung des Marktes und aus unerklärten (zufälligen) Störgrößen im Marktablauf. Dann kann man für den Zuwachs $\mathrm{d}x(t)$ ansetzen

$$\text{(1)} \qquad \mathrm{d}x(t) = a(t, x(t))\,\mathrm{d}t + b(t, x(t))\,\mathrm{d}w(t)\,, \quad x(0) = x_0\,.$$

Dabei ist $w(t)$, $t \geqq 0$, ein Wiener-Prozeß auf einem Wahrscheinlichkeitsraum $(\Omega, \mathscr{A}, p)$, $a(t, x)$ charakterisiert die Trendbewegung und $b(t, x)$ die zufälligen Eigenschaften des Marktes im Zustand x zur Zeit t. (1) ist zu verstehen als stochastische Differentialgleichung (SDGL) im Sinne von Ito ([1], [6], [7]) und darf daher nicht als gewöhnliche DGL für die Pfade $x(t, \omega)$, $\omega \in \Omega$, bei gegebenem Pfad $w(t, \omega)$ interpretiert werden!

Andere Modelle werden nicht durch SDGLen, sondern durch Diffusionsprozesse beschrieben [9], [13], [14]. Unter gewissen Bedingungen sind jedoch Lösungen von SDGLen und Diffusionsprozesse statistisch äquivalent:

(1) sei nun eine Vektor-DGL, d. h. $x(t)$, $a(t, x)$ seien d-dim. Vektoren, $w(t)$ ein m-dim. Wiener-Prozeß mit unabhängigen Komponenten und $b(t, x)$ eine $d \times m$-Matrix. Für $a(t, x)$ und $b(t, x)$ gelte:

Es existiere ein $K > 0$ mit

(2a) $$|a(t, x) - a(t, y)| + |b(t, x) - b(t, y) \leqq K|x - y| ,$$

(2b) $$|a(t, x)|^2 + |b(t, x)|^2 \leqq K^2(1 + |x|^2) .$$

Sind a und b darüberhinaus stetig in t, so ist $x(t)$ ein d-dim. Diffusionsprozeß mit Driftvektor $a(t, x)$ und Diffusionsmatrix $d(t, x) = b(t, x) . b(t, x)'$. Ist umgekehrt ein Diffusionsprozeß vorgegeben, so kann man immer $b(t, x)$ so wählen, daß $d = b . b'$ gilt. Daher unterscheiden wir im folgenden nicht mehr zwischen Lösungen von SDGLen und Diffusionsprozessen [1].

Im nächsten Abschnitt definieren wir nun numerisch realisierbare Prozesse, die Diffusionsprozesse approximieren und so zur Simulation verwendet werden können.

2. DEFINITION DER APPROXIMATIVEN PROZESSE

Sei $[0, T]$ das betrachtete Interval und

$$0 = t_0 < t_1 < \dots < t_k < t_{k+1} < \dots < t_n = T$$

eine Zerlegung dieses Intervalls. Wie definieren den Prozeß $x_n(t)$ durch

(3) $$x_n(0) = x_0 ,$$

$$x_n(t_{k+1}) = x(t_k) + a(t_k, x_n(t_k))(t_{k+1} - t_k) + b(t_k, x_n(t_k))(w(t_{k+1}) - w(t_k))$$

für die Zerlegungspunkte und durch lineare Interpolation dazwischen. Die Diskretisierung (2) entspricht Vorwärtsdifferenzen, also einem expliziten Verfahren. Wählen wir allgemeiner gemischte Differenzen, so wird $x_n(t)$ erklärt für $0 \leqq \lambda \leqq 1$ durch

(4) $$x_0(0) = x_0 ,$$

$$x_n(t_{k+1}) = x_n(t_k) + a(\lambda t_{k+1} + (1 - \lambda) t_k, \lambda x(t_{k+1}) +$$
$$+ (1 - \lambda) x(t_k))(t_{k+1} - t_k) + b(\lambda t_{k+1} + (1 - \lambda) t_k, \lambda x(t_{k+1}) +$$
$$+ (1 - \lambda) x(t_k))(w(t_{k+1}) - w(t_k))$$

für die Zerlegungspunkte und durch lineare Interpolation dazwischen. Für $\lambda = 0$ erhält man wieder (2). Für $\lambda \neq 0$ erhält man implizite Verfahren. Auch Zwischenformen aus (2) und (3) sind denkbar, etwa

$$(5) \qquad x_n(t_{k+1}) = x_n(t_k) + a(t_k, x(t_k))(t_{k+1} - t_k) + \\ + b(t_k, \lambda x(t_{k+1}) + (1 - \lambda) x(t_k))(w(t_{k+1}) - w(t_k)).$$

Es wird sich im Abschnitt 3 zeigen, daß (4) und (5) bei gleichem λ denselben Diffusionsprozeß approximieren. Kritisch ist nur welche „Zwischenstelle" für das zweite Argument von $b(\cdot, \cdot)$ eingesetzt wird. Die Zuwächse $w(t_{k+1}) - w(t_k)$ können am Digitalrechner erzeugt werden: die Komponenten $w_i(t_{k+1}) - w_i(t_k)$ sind nach Definition des Wiener-Prozesses alle $(0, (t_{k+1} - t_k)^2)$ – normalverteilt, d. h. man erhält sie aus normalverteilten Zufallszahlen, die von einem Zufallszahlengenerator erzeugt werden.

Um Diffusionsprozesse am Analogteil eines Hybridrechners zu simulieren, löst man folgendes gewöhnliche DGL-System für die Pfade $x_n(t, \omega)$:

$$(6) \qquad \frac{dx_n(t)}{dt} = a(t, x_n(t)) + b(t, x_n(t))\, r^n(t), \quad x_n(0) = x_0 .$$

Dabei ist $r^n(t) = (r_1^n(t), \ldots, r_m^n(t))'$ stückweise stetiges Gaußsches Rauschen mit unabhängigen Komponenten, $E_i^n(t) = 0$, und mit „deltaähnlichen" Kovarianzfunktionen $K_i^n(t)$, die für $n \to \infty$ gegen die Deltafunktion $\delta(t)$ konvergieren („deltaähnlich" ist in [4], [5] streng gefaßt, n kann als physikalische Kenngröße aufgefaßt werden). Am Rechner steht $r^n(t)$ etwa als physikalischer Rauschgenerator zur Verfügung.

3. SCHWACHE KONVERGENZ DER APPROXIMATIVEN PROZESSE GEGEN DIFFUSIONSPROZESSE

Für manche Anwendungen (z. B. Austrittsprobleme wie in Abschnitt 4) ist es bequemer statt des Intervalls $[0, T]$ das Intervall $[0, \infty)$ zu betrachten. Entsprechend erstreckt sich dann die Zerlegung über $[0, \infty)$. Die folgenden Ergebnisse enthalten jedoch alle den Spezialfall $[0, T]$.

Die Pfade $x(t, \omega)$, $t \geqq 0$, von Diffusionsprozessen sind stetig, ebenso die Pfade $x_n(t, \omega)$, $t \geqq 0$, der approximativen Prozesse aus Abschnitt 2, beides also Elemente des Funktionsraumes $C[0, \infty)$. Dieser Raum läßt sich so metrisieren ([4]), daß die Topologie der kompakten Konvergenz (= gleichmäßige Konvergenz auf allen beschränkten Teilmengen) induziert wird. $\mathscr{B}_C$ sei die σ – Algebra der Borelschen Teilmengen.

Die (meßbaren) Abbildungen x bzw. x_n: $(\Omega, \mathscr{A}, p) \to (C, \mathscr{B}_C)$ die $\omega \in \Omega$ die Pfade $x(t, \omega)$ bzw. $x_n(t, \omega)$, $t \geqq 0$, zuordnen, erzeugen auf $(C, \mathscr{B}_C)$ die Bildmaße P bzw. P_n von p. P bzw. P_n heißt Verteilung von $x(t)$ bzw. $x_n(t)$.

Wir sagen $x_n(t)$ *konvergiert schwach gegen* $x(t)$, wenn die Verteilungen schwach konvergieren, d. h. wenn gilt $P_n \Rightarrow P$ ([2]).

Satz 1. *$a(t, x)$, $b(t, x)$ erfüllen in jedem Intervall $[0, i]$, $i = 1, 2, \ldots$, die Bedingungen (2a, b) und seien dort beschränkt. Dann konvergiert für* $\sup (t_{k+1} - t_k) \to \to 0$ *der durch (3) definierte Näherungsprozeß $x_n(t)$ schwach gegen die Lösung $x(t)$ von (1).*

Beweis. Es genügt Satz 1 für $d = m = 1$ zu beweisen. Hinreichend für die schwache Konvergenz ist [4]:

a) die endlich-dimensionalen Verteilungen von $x_n(t)$ konvergieren schwach gegen die der Lösung $x(t)$, und

b) es existieren Konstante $a_i, b_i, c_i > 0$, so daß für alle $i = 1, 2, \ldots$ und $r, s \in \in [0, i]$

$$E\|x_n(r) - x_n(s)\|^{a_i} \leqq c_i |r - s|^{1+b_i} \tag{7}$$

gilt. Da nach [6] (S. 218–220), $x_n(t)$ im quadratischen Mittel gegen $x(t)$ konvergiert und daraus die Konvergenz der endlich-dimensionalen Verteilungsfunktionen folgt, ist nur noch b) zu zeigen.

Sei $r, s \in [0, i]$, o.B.d.A $r > s$, und t_i der nächste von s links gelegene Zerlegungspunkt, t_j der nächste von r links gelegene Zerlegungspunkt. Aus (3) folgt dann

$$\begin{aligned} x_n(t_{i+1}) - x_n(s) &= a(t_i, x_n(t_i))\,(t_{i+1} - s) + \\ &+ \left(\frac{t_{i+1} - s}{t_{i+1} - t_i}\right)^{1/2} b(t_i, x_n(t_i))\,(w(t_{i+1}) - w(s))\,, \end{aligned} \tag{8a}$$

$$\begin{aligned} x_n(r) - x_n(t_j) &= a(t_j, x_n(t_j))\,(r - t_j) + \\ &+ \left(\frac{r - t_j}{t_{j+1} - t_j}\right)^{1/2} b(t_j, x_n(t_j))\,(w(r) - w(t_j)) \end{aligned} \tag{8b}$$

wenn man beachtet, dass z. B. in (8a)

$$\left(\frac{t_{i+1} - s}{t_{i+1} - t_i}\right)^{1/2} (w(t_{i+1}) - w(t_i))$$

verteilt ist wie $w(t_{i+1}) - w(s)$ und in die Abschätzung (7) nur die Verteilungen eingehen.

Im folgenden setzen wir

$$s_0 := s\,,\ s_1 := t_{i+1}, \ldots, s_{l-1} := t_j\,,\quad s_l := r$$

(wenn zwischen r und s l Zerlegungspunkte liegen) und

$$a_0 := a(t_i, x_n(t_i))\,,\quad b_0 := \left(\frac{t_{i+1} - s}{t_{i+1} - t_i}\right)^{1/2} b(t_i, x_n(t_i))\,,$$

$$b_{l-1} := \left(\frac{r - t_j}{t_{j+1} - t_j}\right)^{1/2} b(t_j, x_n(t_j))$$

ansonsten

$$a_k := a(s_k, x_n(s_k))\,,\quad b_k := b(s_k, x_n(s_k))\,,\quad \Delta w_k := w(s_{k+1}) - w(s_k)\,.$$

Dann gilt

$$x_n(r) - x_n(s) = \sum_{k=0}^{l-1} (x_n(s_{k+1}) - x_n(s_k)) =$$

$$= \sum_{k=0}^{l-1} a_k(s_{k+1} - s_k) + \sum_{k=0}^{l-1} b_k\, \Delta w_k = S_1 + S_2$$

und

$$\mathscr{E}|x_n(r) - x_n(s)|^4 \leqq 8(\mathscr{E}S_1^4 + \mathscr{E}S_2^4)\,,$$

$$\mathscr{E}S_1^4 = \mathscr{E}\Big(\sum_{k=0}^{l-1} a_k(s_{k+1} - s_k)\Big)^4 \leqq K_i^4\Big(\sum_{k=0}^{l-1}(s_{k+1} - s_k)\Big)^4 = K_i^4(r - s)^4 \leqq$$

$$\leqq K_i^2 \,.\, i^2(r - s)^2$$

da in $[0, i]$ nach Voraussetzung $|a| \leqq K_i$.

$$\mathscr{E}S_2^4 = \mathscr{E}\Big(\sum_{k=0}^{l-1} b_k\, \Delta w_k)^4\Big) + 6\sum_{k=0}^{l-1} \mathscr{E}\Big(\sum_{i=0}^{k-1} b_i\, \Delta w_i\Big)^2 b_k^2\, \Delta^2 w_k = S_{21} + S_{22}\,,$$

$$S_{21} = 3\sum_{k=0}^{l-1} \mathscr{E}b_k^4(s_{k+1} - s_k)^2 \leqq 3K_i^4(r - s)^2\,,$$

$$S_{22} = 6\sum_{k-0}^{l-1} \mathscr{E}\Big(\sum_{i=0}^{k-1} b_i^2\, \Delta^2 w_i + \sum_{i \neq j} b_i \,.\, b_j\, \Delta w_i\, \Delta w_j\Big)\, b_k^2\, \Delta^2 w_k \leqq$$

$$\leqq 6\sum_{k=0}^{l-1} K_i^4(s_k - s_0)\,(s_{k+1} - s_k) \leqq 6K_i^4(r - s)^2\,.$$

Wir erhalten also

$$\mathscr{E}|x_n(r) - x_n(s)|^4 \leqq 8(K_i^4 i^2(r - s)^2 + 3K_i^4(r - s)^2 + 6K_i^4(r - s)^2) = c_i(r - s)^2\,.$$

Damit ist die Ungleichung (7) und Satz 1 bewiesen.

Satz 2. *Zusätzlich zu den Voraussetzungen in Satz 1 sei $b(t, x)$ einmal stetig differenzierbar und die Ableitung beschränkt in $[0, i]$ für alle $i = 1, 2, \ldots$. Dann konvergiert der durch (4) oder (5) erklärte Prozeß $x_n(t)$ für* $\sup(t_{k+1} - t_k) \to 0$ *schwach gegen die Lösung $x(t)$ der folgenden* SDGL:

$$(9) \qquad dx(t) = [a(t, x(t)) + \lambda b_x(t, x(t)) * b(t, x(t))]\, dt + b(t, x(t))\, dw(t)$$

wobei die i-te Komponente von $b_x * b$ erklärt ist durch

$$(10) \qquad (b_x * b)_i := \sum_{j=1}^{m} \sum_{k=1}^{d} (\partial b_{ij}/\partial x_k)\, b_{kj}\,, \quad 1 \leqq i \leqq d\,.$$

Beweis. Wie beim Beweis zu Satz 1 sind a) und b) zu zeigen. Für $\lambda = 0.5$ konvergiert nach [6] $x_n(t)$ im quadratischen Mittel gegen $x(t)$, der Beweis läßt sich leicht auf $\lambda \neq 0$ verallgemeinern, so daß daraus wie vorhin a) folgt. Für b) entwickelt man zunächst

$$b(\bar{\lambda}, \lambda x_n(t_{k+1}) + (1 - \lambda)\, x_n(t_k)) =$$

$$= b(\bar{\lambda}, x_n(t_k)) + \lambda \frac{db(\bar{\lambda}, \bar{x})}{dx} (x_n(t_{k+1}) - x_n(t_k))$$

wobei $\bar{\lambda} = \lambda t_{k+1} + (1 - \lambda)\, t_k$ und $\bar{x}$ eine Stelle zwischen $x_n(t_k)$ und $\lambda x_n(t_{k+1}) + (1 - \lambda)\, x_n(t_k)$ ist. Damit gilt nach (4)

$$x_n(t_{k+1}) - x_n(t_k) = a(\cdot, \cdot)\,(t_{k+1} - t_k) + b(\lambda, x(t_k))\, \Delta w_k +$$

$$+ \lambda \frac{db(\cdot, \cdot)}{dx} \Delta w_k(x_n(t_{k+1}) - x_n(t_k))\,,$$

$$x_n(t_{k+1}) - x_n(t_k) = [a(\cdot, \cdot)\,(t_{k+1} - t_k) + b(\cdot, \cdot)\, \Delta w_k]\,.$$

$$.\left[1 + \lambda \frac{db}{dx} \Delta w_k + 0(\Delta w_k)^2\right].$$

Nach dieser Umformung schätzt man wieder ähnlich wie beim Beweis zu Satz 1 ab.

Satz 3. *$a(t, x)$, $b(t, x)$ seien in jedem Intervall $[0, i]$, $i = 1, 2, \ldots$, einmal bzw. zweimal stetig differenzierbar und samt diesen Ableitungen beschränkt. Dann konvergiert für $n \to \infty$ der durch (6) definierte Prozeß $x_n(t)$ gegen die Lösung $x(t)$ der* SDGL (9) *mit* $\lambda = 0.5$.

Beweis. Siehe [4].

Bemerkung. Zum Begriff der schwachen Konvergenz bei stochastischen Prozessen muß auf die Literatur verwiesen werden. Für die Anwendung wichtig ist, daß aus $P_n \Rightarrow P$ die Konvergenz der endlich-dimensionalen Verteilungen von $x_n(t)$ gegen die entsprechenden endlich-dimensionalen Verteilungen von $x(t)$ folgt. Der Vorteil der stärkeren Aussage $P_n \Rightarrow P$ wird sich bei Satz 4 und Satz 5 in Abschnitt 4 erweisen. Die Sätze 1, 2, 3 gelten in entsprechender Weise für das Intervall $[0, T]$.

Als *Faustregel* ergibt sich:

Ist eine stochastische Differentialgleichung mit den Koeffizienten $\bar{a}(t, x)$, $\bar{b}(t, x)$ vorgegeben, so bestehen folgende Möglichkeiten sie am *Digitalrechner* zu simulieren:

a) mittels der Vorwärtsdifferenzen nach (3) mit

$$a(t, x) := \bar{a}(t, x), \quad b(t, x) := \bar{b}(t, x);$$

b) mittels gemischter Differenzen $(0 < \lambda < 1)$ oder Rückwärtsdifferenzen $(\lambda = 1)$ nach (4) oder (5) mit

$$a(t, x) := \bar{a}(t, x) - \lambda \bar{b}_x(t, x) * \bar{b}(t, x), \quad b(t, x) := \bar{b}(t, x).$$

Wesentlich dabei ist die *Beachtung der i.a. nicht verschwindenden trendbeeinflussenden Terme* $- \lambda \bar{b}_x * \bar{b}$.

Am *Hybridrechner* ist die SDGL zu simulieren durch (6) mit

$$a(t, x) := \bar{a}(t, x) - \tfrac{1}{2}\bar{b}_x(t, x) * \bar{b}(t, x), \quad b(t, x) := \bar{b}(t, x).$$

Wesentlich ist wiederum die *Beachtung der trendbeeinflussenden Terme* $-\frac{1}{2}\bar{b}_x * \bar{b}$.

4. ANWENDUNGEN

Wir gehen von stetigen stochastischen Wirtschaftsmodellen aus, die mathematisch als Diffusionsprozeß $x(t)$ beschreibbar sind. Dabei kann $x(t)$ auf Grund theoretischer Überlegungen durch Angabe von Drift- und Diffusionskoeffizienten (oder Übergangswahrscheinlichkeiten) oder durch eine SDGL (1) definiert sein.

Ein Beispiel für den ersten Fall ist das Modell von ([9], S. 47ff) für das Kaufverhalten eines Konsumenten der sich zum Zeitpunkt t zwischen zwei Marken einer Ware entscheidet. Bezeichnet $x(t)$ die Wahrscheinlichkeit, daß sich der Konsument zum Zeitpunkt t für die erste Marke entscheidet, so ist nach diesem Modell $x(t)$ ein Diffusionsprozeß mit Driftkoeffizient $a(t, x) = a - (a + b)x$ und Diffusionskoeffizient $b(t, x) = c(x - x^2)$. Die Konstanten a, b, c werden dabei empirisch bestimmt.

Den zweiten Fall wollen wir am Beispiel eines geschlossenen dynamischen Leontief-Modells erläutern.

$x_i(t)$, $i = 1, \ldots, n$ sei die Produktion der Industrie i (Industrie n sind die Haushalte). Für $x(t) = (x_1(t), \ldots, x_n(t))'$ gibt dann Leontief (1953) ein System von linearen Differentialgleichungen an:

$$\frac{\mathrm{d}x(t)}{\mathrm{d}t} = A\, x(t)\,. \tag{11}$$

A gibt dabei die Beziehungen zwischen den einzelnen Industrien wieder. Wenn nun die Elemente von A durch eine zufällige Störung überlagert sind, geht (11) über in eine lineare SDGL [3].

Im folgenden nehmen wir nun an, daß der Diffusionsprozeß $x(t)$, der das wirtschaftliche Modell beschreibt, in der Gestalt (1) mit gegebenen Koeffizienten $a(t, x)$ und $b(t, x)$ – die auch empirisch gewonnen sein können – vorliegt. Mit Hilfe der Sätze 1, 2, 3 bzw. der Faustregel läßt sich eine Modellsimulation durchführen. Oft interessieren Aussagen über das Verhalten von gewissen Funktionalen des Prozesses $x(t)$ (etwa max $x(t)$, erstes Über- oder Unterschreiten von Schranken u. ä.).

Dazu seien die Abbildungen x bzw. x_n wie oben erklärt. Weiter sei eine Abbildung Y

$$Y : (C, \mathscr{B}_C) \to (R^n, \mathscr{B}_n) \tag{12}$$

gegeben. Hintereinanderschalten von x bzw. x_n und Y liefert die Abbildung

$$\begin{aligned} Y \circ x\ &: (\Omega, \mathscr{A}, p) \xrightarrow{x} (C, \mathscr{B}_C, P) \xrightarrow{Y} (R^n, \mathscr{B}_n)\,, \\ Y \circ x_n &: (\Omega, \mathscr{A}, p) \xrightarrow{x_n} (C, \mathscr{B}_C, P) \xrightarrow{Y} (R^n, \mathscr{B}_n)\,. \end{aligned} \tag{13}$$

Ist Y zusätzlich meßbar, so sind auch $Y \circ x$, $Y \circ x_n$ meßbar und somit $y = Y \circ x$ und $y_n = Y \circ x_n$ Zufallsvariable auf $(\Omega, \mathscr{A}, p)$ mit Werten $y(\omega)$ bzw. $y_n(\omega)$ in $(R^n, \mathscr{B}_n)$.

Für ihre Verteilungsfunktionen $F(x)$ bzw. $F_n(x)$ gilt nun folgender

SATZ 4. *Ist $Y : (C, \mathscr{B}_C) \to (R^n, \mathscr{B}_n)$ meßbar und P – fast überall stetig und sind außerdem die Voraussetzungen zu Satz 1, 2 oder 3 erfüllt, so konvergiert für $n \to \infty$ $F_n(x)$ schwach gegen $F(x)$.*

Beweis. Direkte Anwendung von [2], Theorem 5.1 auf die Sätze 1, 2, 3.

ANWENDUNG: Wollen wir etwa die Wahrscheinlichkeit $p(y \leqq x)$ für ein solches Funktional y des Prozesses $x(t)$ vorhersagen, so nähern wir zunächst $p(y \leqq x)$ durch $p(y_n \leqq x)$ an, indem wir $x(t)$ durch $x_n(t)$ approximieren. Durch wiederholte numerische Simulation von $x_n(t)$ kann dann $p(y_n \leqq x)$ durch die relative Häufigkeit angenähert werden. Man gelangt so zu einem Monte-Carlo-Verfahren zur näherungsweisen Berechnung von $p(y \leqq x)$.

BEISPIEL: In der kollektiven Risikotheorie [15] heißen die durch (1) definierten Prozesse mit $d = m = 1$ Risikoprozesse 2. Art. Man interessiert sich für den zufälligen Zeitpunkt τ, zu dem zum erstenmal das Ereignis $x(t) < 0$ eintritt, wenn das Startkapital $x(0) = x_0$ positiv ist. Die Wahrscheinlichkeit $p(\tau \leqq t_0)$, daß der „Ruin" $x(t) < 0$ bis zum Zeitpunkt t_0 eingetreten ist, heißt „Ruinwahrscheinlichkeit". τ wird charakterisiert durch

$$(14) \qquad x(t) \geqq 0 \quad \text{für} \quad t \leqq \tau \quad \text{und} \quad \min_{\tau \leqq t \leqq \tau + \varepsilon} x(t) < 0 \quad \text{für alle} \quad \varepsilon > 0\,.$$

Die analytische Berechnung von $p(\tau \leqq t_0)$ bereitet i.a. große Schwierigkeiten. Wir können jedoch Satz 4 und die daran anschließende Bemerkung anwenden. Sei τ_n der Zeitpunkt, wann zum erstenmal $x_n(t) < 0$ eintritt. Dann gilt

SATZ 5. *Sind die Voraussetzungen zu Satz* 1, 2 *oder* 3 *erfüllt, so konvergiert für* $n \to \infty$ *die Ruinwahrscheinlichkeit* $p(\tau_n \leqq t_0)$ *des Prozesses* $x_n(t)$ *gegen die Ruinwahrscheinlichkeit* $p(\tau \leqq t_0)$ *von* $x(t)$.

Beweis. Siehe [3].

Satz 5 führt so zu einer *Monte-Carlo-Methode zur Berechnung der Ruinwahrscheinlichkeiten.*

Eine andere Anwendungsmöglichkeit besteht darin, die Sätze 1, 2, 3 zu *Monte-Carlo-Untersuchungen von Filter- oder Prognosealgorithmen* heranzuziehen. Dazu sei des ökonomische Modell wieder durch (1) beschrieben. Die interessierende Größe $x(t)$ sei aber nich selbst beobachtbar, sondern nur eine „gestörte" Größe $z(t)$:

$$(15) \qquad \mathrm{d}z(t) = c(t, x(t))\,\mathrm{d}t + \mathrm{d}v(t)\,;$$

$v(t)$ ist wieder ein Wiener-Prozeß.

(15) kann auch in diskretisierter Form vorliegen, was einer Beobachtung zu diskreten Zeitpunkten entspricht. Gesucht ist beidemale eine erwartungstreue Schätzfunktion minimaler Varianz für $x(t)$ beim Filterproblem oder $x(t + h)$, $h > 0$, beim Prognoseproblem [7], [8]. Reale ökonomische Daten $z(t)$ sind zum Testen von Lösungsalgorithmen für beide Probleme meist zu wenig bekannt. Deshalb testet man solche Lösungsalgorithmen zweckmäßigerweise durch Monte-Carlo-Untersuchungen, wobei einfach $x(t)$ und $z(t)$ am Computer in der hier beschriebenen Art und Weise simuliert werden.

LITERATUR

[1] L. ARNOLD: Stochastische Differentialgleichungen. R. Oldenbourg, München 1973.
[2] P. BILLINGSLEY: Convergence of Probability Measures. Wiley, New York 1968.
[3] L. FAHRMEIR: Prognoseprobleme bei zufällig gestörten dynamischer Systemen. In: Proc. in Op. Research 3. Physica, Würzburg 1974, 289—298.
[4] L. FAHRMEIR: Schwache Konvergenz von Wahrscheinlichkeitsmaßen und ein hybrides Monte-Carlo-Verfahren für lineare Randwertaufgaben. Diss. TU München 1972.
[5] L. FAHRMEIR: Schwache Konvergenz gegen Diffusionsprozesse. ZAMM *54* (1974), 4, T245.
[6] R. Z. HASMINSKII: Stabilität von Differentialgleichungssystemen bei zufälligen Störungen der Parameter (russ.). Nauka, Moskau 1969.
[7] H. HAUPTMANN: Schätz- und Kontrolltheorie in stetigen dynamischen Wirtschaftsmodellen. Springer Verlag, Lecture Notes in Economics and Mathematical Systems, Berlin 1971.
[8] A. JAZWINSKI: Stochastic Processes and Filtering Theory. Academic Press, London 1970.
[9] D. B. MONTGOMERY, G. L. URBAN: Applications of Management Science in Marketing. Prentice-Hall, London 1970.
[10] P. C. B. PHILLIPS: The problem of identification in finite parameter continuous time models. J. of Econometrics *1* (1973) 4, 351—362.
[11] E. A. ROBINSON: Stochastic Differential Theory of Price. Econometrica *27* (1959).
[12] S. S. SENGUPTA: Operations Research in Seller's Competition. Wiley, New York 1968.
[13] G. TINTNER, W. POLLAN: Ein logarithmisch-normaler Diffusionsprozeß mit Anwendungen auf die wirtschaftliche Entwicklung Österreichs. Leipzig: Wiss. Zeitschr. der Karl-Marx-Univ. *17* (1965).
[14] G. TINTNER, J. SENGUPTA: Stochastic Economics. Academic Press, New York 1972.
[15] K. H. WOLFF: Versicherungsmathematik. Springer Verlag, Wien 1970.
[16] C. R. WYMER: Econometric Estimation of Stochastic Differential Equation Systems. Econometrica *40* (1972), 3, 565—578.

TECHNISCHE UNIVERSITÄT MÜNCHEN

INSTITUT FÜR ANGEWANDTE MATHEMATIK

POISSONSCHE ZUFÄLLIGE PUNKTFOLGEN UND ORTSABHÄNGIGE VERDÜNNUNGEN

KARL-HEINZ FICHTNER

JENA

In der vorliegenden Arbeit wird ein Konvergenzsatz für unabhängige Überlagerungen verdünnter zufälliger Punktfolgen bewiesen. Aus diesem Satz werden dann ein „Gesetz der großen Zahlen“ für zufällige Maße und, in Verallgemeinerung eines Resultates von Rényi [5] und Srivastava [6], eine Charakterisierung der Poissonschen zufälligen Punktfolgen abgeleitet.

1. GRUNDBEGRIFFE UND RESULTATE

1.1. Zufällige Maße

Es seien E ein vollständiger, separabler, metrischer Raum und $\mathfrak{L}$ die σ-Algebra der Borelmengen aus E. Mit $\mathfrak{A}$ bezeichnen wir das System aller beschränkten Mengen aus $\mathfrak{L}$ und mit M die Menge aller Maße μ auf dem meßbaren Raum $[E, \mathfrak{L}]$, die auf $\mathfrak{A}$ endlich sind. $\mathfrak{M}$ möge die kleinste σ-Algebra auf M sein, bezüglich der alle Abbildungen der Gestalt $\zeta_A(\mu) = \mu(A)$ $(A \in \mathfrak{A})$ meßbar sind. Unter einem zufälligen Maß (z.M.) auf E verstehen wir ein Verteilungsgesetz Q auf $[M, \mathfrak{M}]$. Das durch

$$\Lambda_Q(A) = \int \mu(A)\, Q(\mathrm{d}\mu)\,; \quad \forall A \in \mathfrak{A}$$

charakterisierte Maß auf $[E, \mathfrak{L}]$ heißt Intensitätsmaß von Q. Ist $\Lambda_Q \in M$, so sagen wir, daß Q von endlicher Intensität ist. Weiter bezeichnen wir mit $\mathop{*}\limits_{i=1}^{n} {}^iQ$ die Faltung der z.M. ${}^1Q, \ldots, {}^nQ$.

Sei U die Menge aller nichtnegativen, beschränkten meßbaren und finiten Funktionen auf E. Jedes z.M. Q ist eindeutig charakterisiert durch sein Laplacesches Funktional L_Q, definiert durch

$$L_Q(u) = \int \exp\left(-\int u(x)\, \mu(\mathrm{d}x)\right) Q(\mathrm{d}\mu)\,; \quad \forall u \in U\,.$$

Eine Folge z.M. $(Q_n)_{n\in N}$ (N – Menge der natürlichen Zahlen) konvergiert schwach gegen ein z.M. Q, wenn gilt:

$$L_{Q_n}(u) \xrightarrow[n\to\infty]{} L_Q(u)\,; \quad \forall u \in U\,.$$

1.2. Verdünnung zufälliger Punktfolgen

Es sei $\overline{M}$ die Menge aller ganzzahligen Maße aus M. Ein z.M. P heißt zufällige Punktfolge (z.Pf.), wenn $P(\overline{M}) = 1$ ist. P läßt sich, entsprechend jedem Element aus $\overline{M}$, deuten als ein Verteilungsgesetz, das die Lage eines Ensembles von Punkten oder Teilchen im Raum E beschreibt.

Eine meßbare Abbildung von E in das Intervall $[0, 1]$ nennen wir *Verdünnung*. Man kann nämlich aus P mit Hilfe von q eine neue z.Pf. P_q ableiten, indem man für jeden Punkt (in $x \in E$) unabhängig von allen anderen (mit Wahrscheinlichkeit $q(x)$) auswürfelt, ob er an seinem Ort verbleibt oder aus E entfernt wird.

1.3. Charakterisierung Poissonscher z.Pf.

Für eine Verdünnung q und jedes $\mu \in M$ bezeichnen wir mit $q \circ \mu$ das durch

$$(q \circ \mu)(A) = \int_A q(x)\,\mu\,\mathrm{d}(x)\,; \quad \forall A \in \mathfrak{A}$$

charakterisierte Maß. Ist $P^{(\mu)}$ die Poissonsche z.Pf. mit dem Intensitätsmaß μ, so gilt:

$$P_q^{(\mu)} = P^{(q\circ\mu)}\,. \tag{1}$$

Weiter erhält man für eine Folge Poissonscher z.Pf. $P^{(\mu_1)}, \ldots, P^{(\mu_s)}$

$$\mathop{*}_{i=1}^{s} P^{(\mu_i)} = P \exp\Big[\Big(\sum_{i=1}^{s} \mu_i\Big)\Big]\,. \tag{2}$$

Ist $q_1, \ldots, q_s$ eine Folge von Verdünnungen mit $\sum q_i(x) = 1$ für alle $x \in E$, dann muß deshalb $P^{(\mu)} = \mathop{*}_{i=1}^{s} P_{q_i}^{(\mu)}$ sein. Diese Eigenschaft ist charakteristisch für die Poissonschen z.Pf. Wir erhalten nämlich:

Satz 1. *Es seien P eine z.Pf. auf E und $q_1, \ldots, q_s$ eine Folge von Verdünnungen mit den Eigenschaften*

$$\max_{1\leq i\leq s} \sup_{x\in A} q_i(x) < 1\,; \quad \forall A \in \mathfrak{A}\,,$$

$$\sum_{i=1}^{s} q_i(x) = 1\,, \qquad \forall x \in E\,.$$

P ist genau dann Poissonsch, wenn $P = \mathop{*}_{i=1}^{s} P_{q_i}$ ist.

1.4. Schwache Konvergenz gegen Poissonsche z.Pf.

Für eine z.Pf. P auf E und eine Folge $A_1, \ldots, A_m \in \mathfrak{A}$ bezeichnen wir mit $P^{A_1,\ldots,A_m}$ das Verteilungsgesetz des zufälligen Vektors $\zeta_{A_1}, \ldots, \zeta_{A_m}$ über dem Wahrscheinlichkeitsfeld $[M, \mathfrak{M}, P]$. Eine Folge z.Pf. $({}^nP)_{n\in N}$ konvergiert schwach gegen eine z.Pf. P, wenn für alle Folgen $A_1, \ldots, A_m$ paarweise disjunkter Mengen aus $\mathfrak{A}$ die Konvergenz der Variationsabstände

$$\left\| {}^nP^{A_1,\ldots,A_m} - P^{A_1,\ldots,A_m} \right\| \xrightarrow[n\to\infty]{} 0$$

statt hat.

Jedem $n \in N$ sei nun eine endliche Folge von Verdünnungen $q_{n,1}, \ldots, q_{n,s_n}$ zugeordnet. Die Folge $(q_{n,1}, \ldots, q_{n,s_n})_{n\in N}$ heißt *Verdünnungsschema. Infinitesemal* nennen wir das Verdünnungsschema, wenn für alle $A \in \mathfrak{A}$ die Konvergenz

$$\max_{1\leq i\leq s_n} \sup_{x\in A} q_{n,i}(x) \xrightarrow[n\to\infty]{} 0$$

statt hat und die Folge $(\sup_{x\in A} \sum_{i=1}^{s_n} q_{n,i}(x))_{n\in N}$ beschränkt ist. Wir erhalten folgenden Satz:

SATZ 2. *Es seien P eine z.Pf. endlicher Intensität, μ ein Maß auf $[E, \mathfrak{L}]$ und $(q_{n,1}, \ldots, q_{n,s_n})_{n\in N}$ ein infinitesemales Verdünnungsschema. Findet für alle $A \in \mathfrak{A}$ die Konvergenz*

$$\sum_{i=1}^{s_n} \int_A q_{ni}(x)\, \Lambda_P(\mathrm{d}x) \xrightarrow[n\to\infty]{} \mu(A)$$

statt, dann konvergiert die Folge z.Pf. $(\mathop{*}_{i=1}^{s_n} P_{q_{n,i}})_{n\in N}$ schwach gegen die Poissonsche z.Pf. $P^{(\mu)}$.

Von Interesse ist der folgende Spezialfall von Satz 2.

COROLLAR ZU SATZ 2. *Es seien P eine z.Pf. endlicher Intensität, h eine reelle Funktion auf E und $(q_{n,1}, \ldots, q_{n,s_n})$ ein infinitesemales Verdünnungsschema. Findet für alle $x \in E$ die Konvergenz*

$$\sum_{i=1}^{s_n} q_{n,i}(x) \xrightarrow[n\to\infty]{} h(x)$$

statt, dann konvergiert die Folge $(\mathop{*}_{i=1}^{s_n} P_{q_{n,i}})_{n\in N}$ schwach gegen die Poissonsche z.Pf. $P^{(h\circ\Lambda_P)}$.

In Verbindung mit Satz 2 erhebt sich die Frage, ob auch für z.Pf. P, die nicht endlicher Intensität sind, die Folge $(\mathop{*}_{i=1}^{s_n} P_{q_{n,i}})_{n\in N}$ schwach konvergiert, wenn $(q_{n,1}, \ldots$ $\ldots, q_{n,s_n})_{n\in N}$ ein infinitesemales Verdünnungsschema ist. Falls für alle $n \in N$, alle

$i = 1, \ldots, s_n$ und alle $x \in E$ $q_{n,i}(x) = 0$ ist, erhalten wir trivialerweise die Konvergenz. Für den Fall, daß das infinitesemale Verdünnungsschema *positiv*, das heißt für alle $A \in \mathfrak{A}$

$$\liminf_{n \to \infty} \inf_{x \in A} \sum_{i=1}^{s_n} q_{n,i}(x) > 0$$

ist, erhalten wir jedoch:

SATZ 3. *Es seien P eine z.Pf. auf E und* $(q_{n,1}, \ldots, q_{n,s_n})_{n \in N}$ *ein positives infinitesemales Verdünnungsschema.*

Ist P nicht von endlicher Intensität, dann existiert ein $A \in \mathfrak{A}$, *so daß für alle* $k = 0, 1, 2, \ldots$ *die Konvergenz*

$$(\mathop{*}_{i=1}^{s_n} P_{q_{n,i}})^A(k) \xrightarrow[n \to \infty]{} 0$$

statt hat.

Mit Hilfe von Satz 3 und des Corollars zu Satz 2 gelangen wir nun leicht zu einer charakteristischen Eigenschaft der z.Pf. endlicher Intensität auf E.

SATZ 4. *Es seien* $(q_{n,1}, \ldots, q_{n,s_n})_{n \in N}$ *ein positives infinitesemales Verdünnungsschema, so daß für alle* $x \in E$ *die Folge* $(\sum_{i=1}^{s_n} q_{n,i}(x))_{n \in N}$ *konvergiert.*

Eine z.Pf. P auf E ist genau dann von endlicher Intensität, wenn die Folge $(\mathop{*}_{i=1}^{s_n} P_{q_{n,i}})_{n \in N}$ *schwach konvergiert.*

1.5. Ein Gesetz der großen Zahlen für zufällige Maße

Es sei q eine Verdünnung. Durch

$$T_q(\mu) = q \circ \mu \; ; \quad \forall \mu \in M$$

ist dann eine meßbare Abbildung von M in M definiert, die jedes z.M. Q in ein neues z.M. $Q_{[q]}$ überführt. Mit Hilfe der doppelt stochastischen Poissonschen z.Pf. kann man nun aus den Sätzen 2 und 4 die folgende Beziehung ableiten.

SATZ 5. *Es seien Q ein z.M. auf E und* $(q_{n,1}, \ldots, q_{n,s_n})_{n \in N}$ *ein infinitesemales Verdünnungsschema mit*

$$\sum_{i=1}^{s_n} q_{n,i}(x) = 1 \; ; \quad \forall n \in N \, , \quad \forall x \in E \, .$$

a) *Q ist genau dann von endlicher Intensität, wenn die Folge* $(\mathop{*}\limits_{i=1}^{s_n} Q_{[q_{n,i}]})_{n\in N}$ *schwach gegen ein z.M. konvergiert.*

b) *Ist Q von endlicher Intensität, so konvergiert die Folge* $(\mathop{*}\limits_{i=1}^{s_n} Q_{[q_{n,i}]})_{n\in N}$ *schwach gegen* δ_{Λ_Q}.

2. BEWEISE

2.1. Beweis von Satz 2

Da nach Voraussetzung für alle $A \in \mathfrak{A}$ die Konvergenz

$$\sum_{i=1}^{s_n} \int_A q_{n,i}(x)\, \Lambda_P(\mathrm{d}x) \xrightarrow[n\to\infty]{} \mu(A)$$

statt hat, konvergiert die Folge Poissonscher z.Pf. $(P^{(\Sigma q_{n,i}\circ \Lambda_P)})_{n\in N}$ schwach gegen $P^{(\mu)}$. Andererseits ist wegen (1) und (2) für alle $n \in N$

$$P \exp\left[\left(\sum_{i=1}^{s_n} q_{n,i}\circ \Lambda_P\right)\right] = \mathop{*}\limits_{i=1}^{s_n} P_{q_{n,i}}^{(\Lambda_P)} .$$

Es genügt deshalb zum Beweis von Satz 2 zu zeigen, daß für alle Folgen $A_1, \ldots, A_m$ paarweise disjunkter Mengen aus $\mathfrak{A}$ die Konvergenz

$$\text{(3)} \qquad \left\|\left(\mathop{*}\limits_{i=1}^{s_n} P_{q_{n,i}}\right)^{A_1,\ldots,A_m} - \left(\mathop{*}\limits_{i=1}^{s_n} P_{q_{n,i}}^{(\Lambda_P)}\right)^{A_1,\ldots,A_m}\right\| \xrightarrow[n\to\infty]{} 0$$

statt hat. Dazu seien für jedes $n \in N$ alle $i = 1. \ldots, s_n$ und alle $j = 1, \ldots, m$ die Folge $A_{j,1}^{i,n}, \ldots, A_{j,u(i,j,n)}^{i,n} \in \mathfrak{A}$ eine Zerlegung der Menge A_j und $a_{j,1}^{i,n}, \ldots, a_{j,u(i,j,n)}^{i,n}$ eine Folge nichtnegativer, reeller Zahlen mit:

$$\text{(4)} \qquad q_{ni}(x) - \left(\frac{1}{n}\right)^i \leqq a_{j,k}^{i,n} \leqq q_{ni}(x)\,; \quad \forall x \in A_{j,k}^{i,n} .$$

Weiter definieren wir für alle $n \in N$ und alle $i = 1, \ldots, s_n$ eine Verdünnung $\bar{q}_{n,i}$ durch:

$$\bar{q}_{n,i}(x) =_{\mathrm{Def}} \sum_{j=1}^{m} \sum_{k=1}^{u(i,j,n)} a_{j,k}^{i,n} K_{A_{j,k}^{i,n}}(x)\,; \quad \forall x \in E .$$

Dabei bezeichnet für jedes $B \in \mathfrak{L}$ K_B die Indikatorfunktion der Menge B. Setzen wir der Kürze halber $\Theta^{i,n} = A_{1,1}^{i,n}, \ldots, A_{m,u(i,m,n)}^{i,n}$, dann erhalten wir für alle $n \in N$ folgende Ungleichung:

(5)
$$\|(\mathop{*}_{i=1}^{s_n} P_{q_{n,i}})^{A_1,\ldots,A_m} - (\mathop{*}_{i=1}^{s_n} P_{q_{n,i}}^{(\Lambda_P)})^{A_1,\ldots,A_m}\| \leqq$$
$$\leqq \sum_{i=1}^{s_n} (\|P_{q_{n,i}}^{\Theta^{i,n}} - P_{\bar{q}_{n,i}}^{\Theta^{i,n}}\| + \|P_{\bar{q}_{n,i}}^{\Theta^{i,n}} - P_{\bar{q}_{n,i}}^{(\Lambda_P)\Theta^{i,n}}\| + \|P_{\bar{q}_{n,i}}^{(\Lambda_P)\Theta^{i,n}} - P_{\bar{q}_{n,i}}^{(\Lambda_P)\Theta^{i,n}}\|).$$

In [2] wird gezeigt, daß für zwei Verdünnungen q_1, q_2, eine z.Pf. $\bar{P}$ und jede Zerlegung $B_1, \ldots, B_r \in \mathfrak{L}$ der Menge $A = \bigcup_{i=1}^{m} A_i$ gilt:

$$\|\bar{P}_{q_1}^{B_1,\ldots,B_r} - \bar{P}_{q_2}^{B_1,\ldots,B_r}\| \leqq 2\,\Lambda_P(A)\,.\sup_{x\in A} |q_1(x) - q_2(x)|\,.$$

Aus (4) können wir schließen, daß stets

$$\sup_{x\in A} |\bar{q}_{n,i}(x) - q_{n,i}(x)| \leqq \left(\frac{1}{n}\right)^i$$

ist. Wir erhalten deshalb die folgende Abschätzung

(6)
$$\sum_{i=1}^{s_n} (\|P_{q_{n,i}}^{\Theta^{i,n}} - P_{\bar{q}_{n,i}}^{\Theta^{i,n}}\| + \|P_{\bar{q}_{n,i}}^{(\Lambda_P)\Theta^{i,n}} - P_{q_{n,i}}^{(\Lambda_P)\Theta^{i,n}}\|) \leqq 4\,\Lambda_P(A) \sum_{i=1}^{s_n} \left(\frac{1}{n}\right)^i .$$

Ist weiter q eine Verdünnung der Gestalt

$$q(x) = \sum_{i=1}^{r} a_i K_{B_i}(x)\,; \quad \forall x \in E$$

mit $\Lambda_P(A)\,.\max_{1\leqq i\leqq r} a_i \leqq 1$, so kann man zeigen, (siehe [2]), daß gilt:

$$\|P_q^{B_1,\ldots,B_r} - P^{(\Lambda_P)B_1,\ldots,B_r}\| \leqq 2\,.\sum_{i=1}^{r} a_i \left(\frac{\partial f}{\partial z_i}(1,\ldots,1) - \frac{\partial f}{\partial z_i}(1 - a_1, \ldots, 1 - a_r)\right) +$$
$$+ \Lambda_P(B_i)\,(1 - \exp(-\sum_{j=1}^{r} a_j\,\Lambda_P(B_j)))\,.$$

Dabei bezeichnet f die erzeugende Funktion des Verteilungsgesetzes $P^{B_1,\ldots,B_m}$. Wir erhalten deshalb unter Berücksichtigung von (4) für alle $n \in N$ mit $\Lambda_P(A)\,.\max_{1\leqq i\leqq s_n} \sup_{x\in A}\,.$ $.\,q_{ni}(x) \leqq 1$ die folgende Abschätzung:

(7)
$$\sum_{i=1}^{s_n} \|P_{\bar{q}_{n,i}}^{\Theta^{i,n}} - P_{\bar{q}_{i,n}}^{(\Lambda_P)\Theta^{i,n}}\| \leqq$$
$$\leqq 2(\sup_{x\in A} \sum_{i=1}^{s_n} q_{n,i}(x))\,(\sum_{k=1}^{\infty} k(1 - (1 - \max_{1\leqq i\leqq s_n} \sup_{x\in A} q_{n,i}(x))^{k-1})\,P^A(k) +$$
$$+ \Lambda_P(A)\,(1 - \exp(-\Lambda_P(A) \max_{i\leqq i\leqq s_n} \sup_{x\in A} q_{n,i}(x)))\,.$$

Die Ergebnisse unter (5), (6) und (7) zusammengefaßt, ergibt nun für alle $n \in N$ mit $\Lambda_P(A) \cdot \max_{1 \leqq i \leqq s_n} \sup_{x \in A} q_{n,i}(x) \leqq 1$

(8)
$$\left\|\left(\mathop{*}_{i=1}^{s_n} P_{q_{n,i}}^{A_1,\ldots,A_m}\right) - \left(\mathop{*}_{i=1}^{s_n} P_{q_{n,i}}^{(\Lambda_P)A_1,\ldots,A_m}\right)\right\| \leqq 4\,\Lambda_P(A) \sum_{i=1}^{\infty} \left(\frac{1}{n}\right)^i +$$
$$+ 2\left(\sup_{x \in A} \sum_{i=1}^{s_n} q_{n,i}(x)\right) \left(\sum_{k=1}^{\infty} k\left(1 - \left(1 - \max_{1 \leqq i \leqq s_n} \sup_{x \in A} q_{n,i}(x)\right)^{k-1}\right) P^A(k) +\right.$$
$$\left. + \Lambda_P(A)\left(1 - \exp\left(-\Lambda_P(A) \max_{1 \leqq i \leqq s_n} \sup_{x \in A} q_{n,i}(x)\right)\right)\right).$$

Wie man leicht sieht, findet die Konvergenz

(9)
$$\sum_{i=1}^{\infty} \left(\frac{1}{n}\right)^i \xrightarrow[n \to \infty]{} 0$$

statt. Wegen

$$\lim_{n \to \infty} \max_{1 \leqq i \leqq s_n} \sup_{x \in A} q_{n,i}(x) = 0$$

und da P von endlicher Intensität ist, erhalten wir:

(10)
$$\lim_{n \to \infty} \sum_{k=1}^{\infty} k\left(1 - \left(1 - \max_{1 \leqq i \leqq s_n} \sup_{x \in A} q_{n,i}(x)\right)^{k-1}\right) P^A(k) = 0\,,$$

(11)
$$\lim_{n \to \infty} \left(1 - \exp\left(-\Lambda_P(A) \max_{1 \leqq i \leqq s_n} \sup_{x \in A} q_{n,i}(x)\right)\right) = 0\,.$$

Nach Voraussetzung war die Folge $\left(\sup_{x \in A} \sum_{i=1}^{s_n} q_{n,i}(x)\right)_{n \in N}$ beschränkt. Wir können deshalb aus (8), (9), (10) und (11) auf die Gültigkeit von (3) schließen. Satz 2 ist damit bewiesen.

2.2. Beweis von Satz 3

Nach Voraussetzung ist die z.Pf. P nicht von endlicher Intensität. Es existiert dann ein $A \in \mathfrak{A}$, so daß $\Lambda_P(A) = \infty$ ist. Ausgehend von P konstruieren wir nun zu jedem $m \in N$ eine z.Pf. ^{m}P auf die folgende Weise: Befinden sich in A mehr als m Punkte, so werden alle Punkte aus E entfernt. Befinden sich in A höchstens m Punkte, so werden nur die Punkte außerhalb von A entfernt. Bezeichnen wir für jedes Verteilungsgesetz G auf $(N \cup \{0\})^s$ mit $F(G)$ die erzeugende Funktion von G, dann gilt folgende Beziehung:

Hilfssatz. *Sind q_1 und q_2 Verdünnungen mit $q_1(x) \leqq q_2(x)$ für alle $x \in E$, dann gilt für alle $m \in N$ und alle $z \in [0, 1]$:*

$$F({}^mP_{q_1}^A)(z) \leqq F(P_{q_2}^A)(z)\,.$$

Beweis. Wir wollen ohne Beschränkung der Allgemeinheit annehmen, daß q_2 und damit auch q_1 nur auf A von Null verschieden ist. Für alle $n \in N$ und $j = 1, 2$ existiert dann eine Verdünnung q_j^n der Gestalt

$$q_j^n(x) = \sum_{i=1}^{u(n)} a_i^{n,j} K_{A_i^n}(x) ; \quad \forall x \in E$$

so daß stets $q_1^n(x) \leqq q_2^n(x)$ ist und die Konvergenz

$$\sup_{x \in E} |q_j^n(x) - q_j(x)| \xrightarrow[n \to \infty]{} 0 \tag{12}$$

statt hat. Die Folge $A_1^n, \ldots, A_{u(n)}^n \in \mathfrak{A}$ möge darüber hinaus eine Zerlegung der Menge A sein. Wegen

$$F({}^m P_{q_j^n}^A)(z) = F({}^m P_{q_j^n}^{A_1^n, \ldots, A_{u(n)}^n})(z, \ldots, z)$$

und

$$F({}^m P_{q_j^n}^{A_1^n, \ldots, A_{u(n)}^n})(z_1, \ldots, z_{u(n)}) =$$
$$= F({}^m P^{A_1, \ldots, A_m})(1 + a_1^{n,j}(z_1 - 1), \ldots, 1 + a_{u(n)}^{n,j}(z_{u(n)} - 1))$$

können wir nun zeigen, daß für alle $z \in [0, 1]$ und alle $k, m, n \in N$ mit $k \geqq m$ gilt:

$$F({}^m P_{q_1^n}^A)(z) \geqq F({}^k P_{q_2^n}^A)(z) . \tag{13}$$

Wie in [2] gezeigt wird, ist für jede z.Pf. $\bar{P}$ und zwei Verdünnungen $\bar{q}_1$ und $\bar{q}_2$

$$\|\bar{P}_{\bar{q}_1}^A - \bar{P}_{\bar{q}_2}^A\| \leqq 2\, \Lambda_{\bar{P}}(A) \sup_{x \in A} |\bar{q}_1(x) - \bar{q}_2(x)| .$$

Da für jedes $m \in N$ ${}^m P$ von endlicher Intensität ist, können wir aus (12) und (13) schließen, daß für alle $z \in [0, 1]$ und alle $m \leqq k$ gilt:

$$F({}^m P_{q_1}^A)(z) \geqq F({}^k P_{q_2}^A)(z) . \tag{14}$$

Wie man leicht überprüft, ist

$$\lim_{k \to \infty} \|{}^k P^A - P^A\| = 0 .$$

Wegen

$$\|{}^k P_{q_2}^A - P_{q_2}^A\| \leqq \|{}^k P^A - P^A\|$$

findet deshalb auch die Konvergenz

$$\|{}^k P_{q_2}^A - P_{q_2}^A\| \xrightarrow[k \to \infty]{} 0$$

statt. Damit gilt auch

$$F(P_{q_2}^A)(z) = \lim_{k \to \infty} F({}^k P_{q_2}^A)(z) ; \quad \forall z \in [0, 1] .$$

In Verbindung mit (14) ergibt das jedoch die Behauptung des Hilfssatzes. □

Wir zeigen nun zum Beweis von Satz 3, daß für alle $z \in [0, 1)$ die Konvergenz

(15) $$F\big(\big(\mathop{*}_{i=1}^{s_n} P_{q_{n,i}}\big)^A\big)(z) \xrightarrow[n\to\infty]{} 0$$

statt hat. Dazu sei $(\bar{q}_{n,1}, \ldots, \bar{q}_{n,s_n})_{n\in N}$ ein Verdünnungsschema, so daß für alle $n \in N$

$$\inf_{m\geqq n} \sum_{i=1}^{s_m} q_{m,i}(x) = \sum_{i=1}^{s_n} \bar{q}_{n,i}(x)\,; \quad \forall x \in E$$

und für alle $i = 1, \ldots, s_n$ $\bar{q}_{n,i}$ durch $q_{n,i}$ beschränkt ist. Unter Verwendung des Hilfssatzes erhalten wir für alle $m, n \in N$ und alle $z \in [0, 1]$:

(16) $$F\big(\big(\mathop{*}_{i=1}^{s_n} P_{q_{n,i}}\big)^A\big)(z) \leqq F\big(\big(\mathop{*}_{i=1}^{s_n} {}^mP_{\bar{q}_{n,i}}\big)^A\big)(z)\,.$$

Setzen wir

$$h(x) = \lim_{n\to\infty} \sum_{i=1}^{s_n} \bar{q}_{n,i}(x)\,; \quad \forall x \in E\,,$$

$$b = \varliminf_{n\to\infty} \inf_{x\in A} \sum_{i=1}^{s_n} q_{n,i}(x)$$

dann folgt aus dem Corollar zu Satz 2, daß für alle $m \in N$ die Folge $\big(\mathop{*}_{i=1}^{s_n} {}^mP_{\bar{q}_{n,i}}\big)_{n\in N}$ schwach gegen die Poissonsche z.Pf. $P^{(h\circ \Lambda_{mP})}$ konvergiert. Unter Verwendung von (16) erhalten wir deshalb

$$\varlimsup_{n\to\infty} F\big(\big(\mathop{*}_{i=1}^{s_n} P_{q_{ni}}\big)^A\big)(z) \leqq e^{b\Lambda_{mP}(A)(1-z)}$$

für alle $m \in N$ und alle $z \in [0, 1)$. Weil nach Voraussetzung $b > 0$ und $\lim_{m\to\infty} \Lambda_{mP}(A) = \infty$ ist, können wir daraus auf die Gültigkeit von (15) schließen. Satz 3 ist damit bewiesen.

2.3. Beweis von Satz 1

Wir müssen noch zeigen, daß eine z.Pf. P Poissonsch ist, wenn sie sich in der Gestalt

(17) $$P = \mathop{*}_{i=1}^{s} P_{q_i}$$

darstellen läßt. Für eine Folge z.Pf. ${}^1P, \ldots, {}^sP$ und eine Verdünnung q gilt stets:

$$\big(\mathop{*}_{i=1}^{s} {}^iP\big)_q = \mathop{*}_{i=1}^{s} ({}^iP)_q\,.$$

Weiter gilt für jede z.Pf. P und zwei Verdünnungen $\bar{q}_1, \bar{q}_2$ $(P_{\bar{q}_1})_{\bar{q}_2} = P_{(\bar{q}_1 \cdot \bar{q}_2)}$. Wegen

$$\max_{1 \leq i \leq s} \sup_{x \in A} q_i(x) < 1 ; \quad \forall A \in \mathfrak{A}$$

und $\sum_{i=1}^{s} q_i(x) = 1$ für alle $x \in E$ können wir aus (17) auf die Existenz eines infinitesemalen Verdünnungsschemas $(q_{n,1}, \ldots, q_{n,s_n})_{n \in N}$ mit $\sum_{i=1}^{s_n} q_{ni}(x) = 1$ schließen, so daß für alle $n \in N$ gilt:

$$P = \mathop{*}_{i=1}^{s_n} P_{q_{n,i}} . \tag{18}$$

Wegen Satz 4 muß P deshalb endlicher Intensität sein und damit nach Aussage des Corollar zu Satz 2 die Folge $(\mathop{*}_{i=1}^{s_n} P_{q_{n,i}})_{n \in N}$ schwach gegen die Poissonsche z.Pf. $P^{(\Lambda_P)}$ konvergieren. In Verbindung mit (18) bedeutet das jedoch, daß P selbst Poissonsch ist. Satz 1 ist damit bewiesen.

2.4. Beweis von Satz 5

Für eine z.Pf. $\bar{P} = \int P^{(\mu)} \bar{Q}(\mathrm{d}\mu)$ ist (vergleiche [4]):

$$L_{\bar{P}}(u) = L_{\bar{Q}}(1 - \exp(-u)) ; \quad \forall u \in U .$$

Weiter hat das Laplacesche Funktional einer Poissonschen z.Pf. die Gestalt:

$$L_{P}(\lambda)(u) = \exp\left(- \int (1 - \exp(-u(x))) \right) \lambda(\mathrm{d}x) ; \quad \forall u \in U .$$

Ist $({}^nP)_{n \in N}$ eine Folge z.Pf. mit ${}^nP = \int P^{(\mu)}\, {}^nQ(\mathrm{d}\mu)$, so erhalten wir deshalb folgende Beziehungen:

2.4.1. *Konvergiert die Folge z.M.* $({}^nQ)_{n \in N}$ *schwach gegen* ${}^1Q$, *so konvergiert die Folge z.Pf.* $({}^nP)_{n \in N}$ *schwach gegen* ${}^1P$.

2.4.2. *Konvergiert die Folge* $({}^nP)_{n \in N}$ *schwach gegen die Poissonsche z.Pf.* $P^{(\lambda)}$, *so konvergiert die Folge* $({}^nQ)_{n \in N}$ *schwach gegen* δ_λ.

Sei nun $P = \int P^{(\mu)} Q(\mathrm{d}\mu)$. Wie man leicht überprüft, ist $\Lambda_P = \Lambda_Q$ und für alle $n \in N$

$$\mathop{*}_{i=1}^{s_n} P_{q_{n,i}} = \int P^{(\mu)} (\mathop{*}_{i=1}^{s_n} Q_{[q_{n,i}]})\, \mathrm{d}\mu) .$$

Ist Q von endlicher Intensität, so können wir deshalb unter Verwendung von 2.4.2 aus dem Corollar zu Satz 2 schließen, daß die Folge $(\mathop{*}_{i=1}^{s_n}{}_{[q_{n,i}]})_{n \in N}$ schwach gegen δ_{Λ_Q}

konvergiert. Konvergiert umgekehrt die Folge $(\overset{s_n}{\underset{i=1}{*}} Q_{[q_{n,i}]})_{n\in N}$ schwach gegen ein z.M., so muß nach 2.4.1 auch die Folge $(\overset{s_n}{\underset{i=1}{*}} P_{q_{n,i}})_{n\in N}$ schwach konvergieren und deshalb nach Satz 4 P von endlicher Intensität sein. Daraus ergibt sich schließlich der Beweis von Satz 5.

LITERATUR

[1] K. H. Fichtner: Charakterisierung Poissonscher zufälliger Punktfolgen und infinitesemale Verdünnungsschemata. Math. Nachr. (im Druck).

[2] K. H. Fichtner: Schwache Konvergenz von unabhängigen Überlagerungen verdünnter zufälliger Punktfolgen. Math. Nachr. (im Druck).

[3] J. Kerstan, K. Matthes, J. Mecke: Unbegrenzt teilbare Punktprozesse. Akademieverlag, Berlin (im Druck).

[4] J. Mecke: Eine charakteristische Eigenschaft der doppelt stochastischen Poissonschen Prozesse. Z. Wahrscheinlichkeitstheorie verw. Gerb. *11* (1968), 74–81.

[5] A. Rényi: On two mathematical models of traffic on a divided highway. J. Appl. Prob. *1* (1964), 311–320.

[6] R. C. Srivastava: On a characterization of the Poisson Process. J. Appl. Prob. *8* (1971), 615–616.

Friedrich-Schiller-Universität

Sektion Mathematik

DIE ANZAHL DER KREUZUNGSPUNKTE EINER FUNKTION MIT EINEM GAUSSSCHEN PROZESS

WERNER FIEGER

KARLSRUHE

Ist $\{X(t) : t \in [0, 1]\}$ ein über dem Zeitintervall $[0, 1]$ definierter stochastischer Prozeß mit stetigen Pfaden, so läßt sich der Erwartungswert der Anzahl $U(0; X(\cdot))$ der Null-Niveau-Überkreuzungen als Burkill-Unterteilungsintegral berechnen

$$\mathrm{E}U(0; X(\cdot)) = \int_0^1 q_1$$

mit $q_1((t, t']) := p(X(t) \,.\, X(t') < 0)$, sofern $p(X(t) = 0) = 0$ für jedes $t \in [0, 1]$ gilt. Der Erwartungswert $\mathrm{E}U(0; X(\cdot))$ existiert genau dann (mit einem endlichen Wert), wenn dieses Burkill-Unterteilungsintegral einen endlichen Wert hat. Dieses Ergebnis bleibt auch ohne die Voraussetzung der Stetigkeit der Pfade für separable stochastische Prozesse richtig, sofern man die Anzahl der Null-Niveau-Überkreuzungen entsprechend definiert.

Ist $\gamma(t)$ eine auf $[0, 1]$ stetige Funktion und $\{X(t) : t \in [0, 1]\}$ ein stochastischer Prozeß mit stetigen Pfaden und mit $p(X(t) = \gamma(t)) = 0$ für jedes $t \in [0, 1]$, so ist entsprechend der Erwartungswert der Anzahl $U(\gamma(\cdot); X(\cdot))$ der Überkreuzungen des Prozesses $\{X(t) : t \in [0, 1]\}$ mit der Funktion $\gamma \mid [0, 1]$ gleich dem Burkill-Unterteilungsintegral $\int_{(0,1]} q$ der Intervallfunktion

$$q((t, t']) := p([X(t) - \gamma(t)] \,.\, [X(t') - \gamma(t')] < 0)\,.$$

Ist $\{X(t) : t \in [0, 1]\}$ ein separabler Gaußscher Prozeß mit $\mathrm{E}\, X(t) \equiv 0$, $\mathrm{var}\, X(t) \equiv 1$ und mit der Kovarianzfunktion $r(t, t') = \mathrm{E}\, X(t) \,.\, X(t')$, so hat der Erwartungswert $\mathrm{E}\; U(\gamma(\cdot); X(\cdot))$ genau dann einen endlichen Wert, wenn $\gamma(t)$ auf $[0, 1]$ von beschränkter Schwankung ist und $\int_0^1 \sqrt{(1 - r)}$ mit endlichem Wert existiert [2]. Leadbetter und Cryer [5] haben für einen Gaußschen Prozeß $\{Y(t) : t \in [0, 1]\}$ den Erwartungswert $\mathrm{E}\; U(0, Y(\cdot))$ angegeben unter der zusätzlichen Voraussetzung, daß $m(t) = \mathrm{E}\; Y(t)$ einmal stetig differenzierbar ist und die zweite gemischte partielle Ableitung von $R(t, t') = \operatorname{cov}(Y(t), Y(t'))$ stetig ist (vgl. Spezialfall 3 in § 2). Weitere Arbeiten über die Anzahl der Kreuzungspunkte, insbesondere der Niveaukreuzungspunkte, sind in [4] zusammengestellt.

Im folgenden (Sätze (2.1) und (2.4)) berechnen wir den Erwartungswert E $U(\gamma(\cdot), X(\cdot))$ der Anzahl der Kreuzungspunkte der Funktion $\gamma(t)$ mit dem Gaußschen Prozeß $\{X(t) : t \in [0, 1]\}$ für den Fall, daß $\gamma(t)$ und $s_2(t) := \int_0^t \sqrt{(1-r)}$ stetig von t abhängen. In § 1 sind die Aussagen, die über das Burkill-Unterteilungsintegral benötigt werden, zusammengestellt.

1. EINIGE AUSSAGEN ÜBER DAS BURKILL-UNTERTEILUNGSINTEGRAL

$F(t, t')$ sei eine reelle, für $0 \leqq t < t' \leqq 1$ definierte Funktion. Üblicherweise nennt man die „Intervallfunktion" $F(t, t')$ über $[a, b]$ (mit $0 \leqq a < b \leqq 1$) Burkill-unterteilungsintegrabel (kurz: B-integrabel) mit dem Integralwert $\alpha = \int_a^b F$, wenn zu jedem $\varepsilon > 0$ eine Zerlegung $Z(\varepsilon)$ von $[a, b]$ existiert derart, daß für jede Zerlegung $Z : a = t_0 < t_1 < \ldots < t_n = b$, die alle Teilungspunkte von $Z(\varepsilon)$ (und eventuell weitere Punkte) als Teilungspunkte enthält (abgekürzt $Z \geqq Z(\varepsilon)$),

$$\left| \sum_{\nu=1}^{n} F(t_{\nu-1}, t_\nu) - \alpha \right| < \varepsilon$$

gilt.

Wir betrachten allgemeiner Funktionen $F(\bar{t}; t, t')$, definiert für $0 \leqq t < \bar{t} \leqq t' \leqq$ $\leqq 1$; $F(\bar{t}; t, t')$ hängt also von dem Intervall $(t, t']$ und dem „Zwischenpunkt" $\bar{t} \in (t, t']$ ab. Entsprechend betrachten wir die um Zwischenpunkte $\bar{t}_1, \ldots, \bar{t}_n$ erweiterten Zerlegungen $\tilde{Z} = \{t_0, t_1, \ldots, t_n; \bar{t}_1, \ldots, \bar{t}_n\}$ von $[a, b] : a = t_0 < t_1 < \ldots < t_n =$ $= b$, $\bar{t}_\nu \in (t_{\nu-1}, t_\nu]$ für $\nu = 1, \ldots, n$.

(1.1) DEFINITION. *Wir nennen die erweiterte Intervallfunktion $F(\bar{t}; t, t')$ über $[a, b]$ B-integrabel mit dem Integralwert $\alpha = \int_a^b F$, wenn es zu jedem $\varepsilon > 0$ eine Zerlegung $Z(\varepsilon)$ von $[a, b]$ gibt derart, daß für jede erweiterte Zerlegung $\tilde{Z} =$ $= \{t_0, t_1, \ldots, t_n; \bar{t}_1, \ldots, \bar{t}_n\}$ von $[a, b]$, die alle Teilungspunkte von $Z(\varepsilon)$ als Teilungspunkte enthält,*

$$\left| \sum_{\nu=1}^{n} F(\bar{t}_\nu; t_{\nu-1}, t_\nu) - \alpha \right| < \varepsilon$$

gilt.

Es sei $s(t)$ eine über $[0, 1]$ monoton nicht fallende, stetige Funktion. Gibt es zu jedem $\varepsilon > 0$ eine Zerlegung $Z(\varepsilon) : 0 = \tau_0 < \tau_1 < \ldots < \tau_n = 1$ von $[0, 1]$ und ein

$\delta(\varepsilon) > 0$ derart, daß für beliebige $0 \leqq t_1 \leqq t'_1 \leqq t_2 < t'_2 < \ldots < t_r < t'_r \leqq 1$ mit $\sum_{\eta=1}^{r} (s(t'_\varrho) - s(t_\varrho)) < \delta(\varepsilon)$ und mit der Eigenschaft, daß jeder der Teilungspunkte $\tau_1, \ldots, \tau_n$ nicht innerhalb der Intervalle $(t_1, t'_1), \ldots, (t_r, t'_r)$ liegt, stets $\left|\sum_{\varrho=1}^{r} F(t_\varrho, t'_\varrho)\right| < \varepsilon$ gilt, so nennen wir die Intervallfunktion F bezüglich der Funktion s totalstetig.

Wir betrachten nun zwei über $[0, 1]$ B-integrable Intervallfunktionen $F_1(t, t')$, $F_2(t, t')$ mit $0 \leqq F_2(t, t') \leqq c_0$ für beliebige t, t' mit $0 \leqq t < t' \leqq 1$. Weiter sei $g(t, u, v)$ eine für $t \in [0, 1]$, $u \in \boldsymbol{R}$, $v \in [0, c_0]$ definierte Funktion mit $g(t, 0, 0) = 0$ und mit der Eigenschaft, daß $h(t, u, v) := \lim_{\eta \to 0} g(t, \eta u, \eta v)/\eta$ existiert für jedes $t \in [0, 1]$, $u \in \boldsymbol{R}$, $v \in [0, +\infty)$. $h(t, u, v)$ ist bei festem t positiv homogen von (u, v) abhängig (d.h. es gilt für beliebiges $\lambda \geqq 0$ stets $h(t, \lambda u, \lambda v) = \lambda \,.\, h(t, u, v)$). $s(t)$ sei eine monoton nicht fallende stetige Funktion, $\sigma \mid \mathfrak{B}(0, 1)$ das über der Gesamtheit der Borelschen Teilmengen von $(0, 1]$ definierte, eindeutig bestimmte endliche Maß mit $\sigma((a, b]) = s(b) - s(a)$. Wir setzen weiter voraus, daß F_1, F_2 totalstetig bezüglich s sind. Dies hat zur Folge, daß es ein endliches Maß $\Psi_2 \mid \mathfrak{B}(0, 1)$ und ein endliches signiertes Maß

$$\Psi_1 \mid \mathfrak{B}(0, 1) \quad \text{gibt mit} \quad \Psi_1((a, b]) = \int_a^b F_1, \quad \Psi_2((a, b]) = \int_a^b F_2 \,,$$

und daß diese beiden signierten Maße $\Psi_1 \mid \mathfrak{L}(0, 1)$, $\Psi_2 \mid \mathfrak{B}(0, 1)$ totalstetig (in dem üblicherweise für Maße erklärten Sinn) bezüglich $\sigma \mid \mathfrak{B}(0, 1)$ sind. Dann können wir mit zwei geeignet gewählten Baire'schen Funktionen $\psi_1(t)$, $\psi_2(t)$ die signierten Maße Ψ_1, Ψ_2 darstellen als Lebesgue-Stieltjes-Integrale $\Psi_\nu(B) = \int_B \psi_\nu(t)\, d\sigma$ für $B \in \mathfrak{B}(0, 1)$, $\nu = 1, 2$; ohne Beschränkung der Allgemeinheit können wir $\psi_2(t) \geqq 0$ für jedes $t \in [0, 1]$ voraussetzen.

Wir definieren nun $G(\bar{t}; t, t') := g(\bar{t}, F_1(t, t'), F_2(t, t'))$ und interessieren uns für weitere Bedingungen, die die Existenz von $\int_0^1 G$ sicherstellen.

(1.2) Satz. *Hängt $h(t, u, v)$ stetig von (t, u, v) ab, und gilt*

$$\frac{g(t, \eta u, \eta v)}{\eta} \xrightarrow[\eta \downarrow 0]{} h(t, u, v)$$

gleichmäßig auf $\{(t, u, v) : 0 \leqq t \leqq 1,\ v \geqq 0,\ u^2 + v^2 = 1\}$, so ist G über $[0, 1]$ B-integrabel und der Wert von $\int_0^1 G$ läßt sich auch als Lebesgue-Stieltjes-Integral berechnen:

$$\int_0^1 G = \int_{(0,1]} h(t, \psi_1(t), \psi_2(t)) \,.\, d\sigma \,.$$

Zum Beweis vergleiche man [3].

2. DER ERWARTUNGSWERT VON $U(\gamma(\cdot), X(\cdot))$

Wir gehen in diesem Paragraphen von einem Gaußschen Prozeß $\{X(t) : t \in [0, 1]\}$ mit stetigen Pfaden und mit der Kovarianzfunktion $r(t, t')$ aus; es sei $\mathrm{E}\, X(t) \equiv 0$ und $\mathrm{var}\, X(t) \equiv 1$ auf $[0, 1]$. $s_1(t)$ sei die Totalvariation der stetigen Funktion γ, die über $[0, 1]$ von beschränkter Schwankung vorausgesetzt wird:

$$s_1(t) = \int_0^t \tilde{\gamma} \quad \text{mit} \quad \tilde{\gamma}(t, t') := |\gamma(t) - \gamma(t')| ;$$

weiter setzen wir voraus, daß $s_2(t) = \int_0^t \sqrt{(1 - r)}$ ebenfalls stetig ist. Wir setzen $F_1(t, t') := \gamma(t') - \gamma(t)$ und $F_2(t, t') := [1 - r(t, t')]^{1/2}$. Die Totalstetigkeit von F_1 bezüglich $s(t) := s_1(t) + s_2(t)$ folgt aus $|F_1(t, t')| \leqq s_1(t') - s_1(t) \leqq s(t') - s(t)$; daß F_2 bezüglich $s_2(t)$ und damit auch bezüglich $s(t)$ totalstetig ist, kann man leicht aus [1, Lemma (1.2)] herleiten. Bezeichnen wir wieder mit $\sigma \mid \mathfrak{B}(0, 1)$ das durch $\sigma((a, b]) = s(b) - s(a)$ eindeutig festgelegte Maß, so ist mit geeigneten Baire'schen Funktionen $\psi_1(t), \psi_2(t)$ (mit $\psi_2(t) \geqq 0$ auf $[0, 1]$)

$$\gamma(\tau') - \gamma(\tau) = \int_\tau^{\tau'} F_1 = \int_{(\tau,\tau']} \psi_1(t) \, . \, \mathrm{d}\sigma$$

(und damit

$$s_1(\tau) = \int_{(0,\tau]} |\psi_1(t)| \, . \, \mathrm{d}\sigma \;)$$

und

$$s_2(\tau') - s_2(\tau) = \int_\tau^{\tau'} F_2 = \int_{(\tau,\tau']} \psi_2(t) \, . \, \mathrm{d}\sigma \, ;$$

hieraus folgt

$$|\psi_1(t)| + \psi_2(t) = 1 \quad \text{für } \sigma\text{-fast alle} \quad t \in [0, 1] \, .$$

Zur Abkürzung bezeichnen wir mit $\varphi(x) = 1/\sqrt{(2\pi)} \, . \exp\{-x^2/2\}$ die Dichte und mit $\Phi(x) = \int_{-\infty}^x \varphi(y) \, . \, \mathrm{d}y$ die Verteilungsfunktion der Normalverteilung. Damit definieren wir

$$h(\gamma, u, v) := \begin{cases} \varphi(\gamma) \, . \, |u| \, , & \text{falls} \quad v = 0 \, , \\ \varphi(\gamma) \, . \left\{ 2 \, . \, (\sqrt{2}) \, . \, v \, . \, \varphi\left(\dfrac{u}{v \, . \sqrt{2}}\right) + u \, . \left(2 \, . \, \Phi\left(\dfrac{u}{v \, . \sqrt{2}}\right) - 1 \right) \right\} , & \text{falls} \quad v > 0 \, . \end{cases}$$

Dann gilt

(2.1) Satz.

$$\mathrm{E}\, U(\gamma(\cdot), X(\cdot)) = \int_{(0,1]} h(\gamma(t), \psi_1(t), \psi_2(t)) \, . \, \mathrm{d}\sigma \, .$$

Zum Beweis haben wir für die Intervallfunktion

$$q(t, t') = p([X(t) - \gamma(t)] \,.\, [X(t') - \gamma(t')] < 0)$$

die Beziehung

$$\int_0^1 q = \int_{(0,1]} h(\gamma(t), \psi_1(t), \psi_2(t)) \,.\, d\sigma$$

nachzuweisen.

Mit G, G_1 bezeichnen wir im folgenden zwei unabhängige normalverteilte Zufallsgrößen mit Erwartungswert 0 und Varianz 1. γ_0 wählen wir so, daß $|\gamma(t)| \leqq \gamma_0$ auf $[0, 1]$. Wir definieren

$$C_0 := \{(\gamma, u, v) : \gamma,\ u \in \boldsymbol{R},\ v \in [0, \sqrt{2}]\} \,,$$

$$C_1 := \{(\gamma, u, v) : |\gamma| \leqq \gamma_0,\ u \in [-1, 1],\ v > 0,\ u^2 + v^2 = 1\}$$

und weiter für $(\gamma, u, v) \in C_0$

$$g(\gamma, u, v) := p([(1 - v^2) \,.\, G_1 + v \,.\, \sqrt{(2 - v^2)} \,.\, G - \gamma + u] \,.\, [G_1 - \gamma] < 0) \,.$$

Für $(\gamma, u, v) \in C_0$ und $\eta \in [0, v^{-1} \,.\, \sqrt{2})$ setzen wir

$$e_1(\gamma, u, v, \eta, z) := p(v \,.\, \sqrt{(2 - \eta^2 v^2)} \,.\, G > -u + \eta v^2 \gamma - (1 - \eta^2 v^2)\, z) \,,$$

$$e_2(\gamma, u, v, \eta, z) := p(v \,.\, \sqrt{(2 - \eta^2 v^2)} \,.\, G < -u + \eta v^2 \gamma - (1 - \eta^2 v^2)\, z) \,,$$

$$k(\gamma, u, v, \eta) := \int_{-\infty}^{0} e_1(\gamma, u, v, \eta, z) \,.\, dz + \int_0^{+\infty} e_2(\gamma, u, v, \eta, z) \,.\, dz$$

und

$$r(\gamma, u, v, \eta) := \int_{-\infty}^{0} \{\varphi(\gamma + \eta z) - \varphi(\gamma)\} \,.\, e_1(\gamma, u, v, \eta, z) \,.\, dz +$$

$$+ \int_0^{+\infty} \{\varphi(\gamma + \eta z) - \varphi(\gamma)\} \,.\, e_2(\gamma, u, v, \eta, z) \,.\, dz \,.$$

Da $X(t'), X(t)$ verteilt sind wie G_1, $r(t, t') \,.\, G_1 + [1 - r^2(t, t')]^{1/2} \,.\, G$ $(=\{1 - [1 - r(t, t')]\} \,.\, G_1 + [1 - r(t, t')]^{1/2} \,.\, \{2 - [1 - r(t, t')]\}^{1/2} \,.\, G)$, ist

$$q(t, t') = g(\gamma(t'), \gamma(t') - \gamma(t), [1 - r(t, t')]^{1/2}) =$$

$$= g(\gamma(t'), F_1(t, t'), F_2(t, t')) \,.$$

Bei der Berechnung von $\int_0^1 q$ stützen wir uns auf Satz (1.2). Zunächst beweisen wir einige Lemmata.

(2.2) Lemma.

$$\varphi(\gamma) \,.\, k(\gamma, u, v, \eta) \xrightarrow[\eta \downarrow 0]{} h(\gamma, u, v) \text{ \textit{gleichmäßig auf} } C_1 \,.$$

Beweis. 1. Für $v > 0$ erhält man durch partielle Integration

$$k(\gamma, u, v, \eta) = \frac{2v \,.\, \sqrt{(2 - \eta^2 v^2)}}{1 - \eta^2 v^2} \,.\, \varphi\left(\frac{-u + \eta v^2 \gamma}{v \,.\, \sqrt{(2 - \eta^2 v^2)}}\right) +$$

$$+ \frac{-u + \eta v^2 \gamma}{1 - \eta^2 v^2} \left\{2\Phi\left(\frac{-u + \eta v^2 \gamma}{v \,.\, \sqrt{(2 - \eta^2 v^2)}}\right) - 1\right\}.$$

2. Zu $\varepsilon > 0$ können wir $\beta(\varepsilon) > 0$ mit $\varphi(\beta) < \varepsilon/2$ für jedes $\beta \geqq \beta(\varepsilon)$ bestimmen. Für $(\gamma, u, v) \in C_1$ mit $|u/v| \geqq 4\,\beta(\varepsilon)$ und für $\eta \leqq \eta(\varepsilon) := \min\{1, \beta(\varepsilon)/\gamma_0\}$ gilt

$$\left|\frac{-u + \eta v^2 \gamma}{v \,.\, \sqrt{(2 - \eta^2 v^2)}}\right| \geqq \beta(\varepsilon)$$

und somit

$$\left|\varphi\left(\frac{-u + \eta v^2 \gamma}{v \,.\, \sqrt{(2 - \eta^2 v^2)}}\right) - \varphi\left(\frac{-u}{v \,.\, \sqrt{2}}\right)\right| < \varepsilon\,.$$

Da für $(\gamma, u, v) \in C_1$ mit $|u/v| \leqq 4\,\beta(\varepsilon)$ gleichmäßig

$$\frac{-u + \eta v^2 \gamma}{v \,.\, \sqrt{(2 - \eta^2 v^2)}} \xrightarrow[\eta \downarrow 0]{} -\frac{u}{v \,.\, \sqrt{2}}$$

und wegen der gleichmäßigen Stetigkeit von $\varphi(x)$ folglich im gleichen Bereich gleichmäßig

$$\varphi\left(\frac{-u + \eta v^2 \gamma}{v \,.\, \sqrt{(2 - \eta^2 v^2)}}\right) \xrightarrow[\eta \downarrow 0]{} \varphi\left(-\frac{u}{v \,.\, \sqrt{2}}\right)$$

gilt, erhalten wir die letzte Konvergenzaussage auch gleichmäßig in C_1. Da gleichmäßig für $(\gamma, u, v) \in C_1$

$$\frac{2v \,.\, \sqrt{(2 - \eta^2 v^2)}}{1 - \eta^2 v^2} \xrightarrow[\eta \downarrow 0]{} 2v \,.\, \sqrt{2}$$

gilt, strebt der erste Summand von (1) gleichmäßig in C_1 gegen $\varphi(-u/(v \,.\, \sqrt{2})) \,.\, 2v\,\sqrt{2}$.

3. Ähnlich beweist man die gleichmäßige Konvergenz des zweiten Summanden von (1) gegen $-u \,.\, \{2\Phi(-u/(v \,.\, \sqrt{2})) - 1\}$. □

(2.3) Lemma.

$$r(\gamma, u, v, \eta) \xrightarrow[\eta \downarrow 0]{} 0 \quad \textit{gleichmäßig auf } C_1\,.$$

Beweis. 1. Es sei $b \in [\frac{1}{2}, +\infty)$, $\alpha \geqq 0$ und $c \in \boldsymbol{R}$. Für x mit $|x| \geqq 2$ gilt

$$\varphi_b(x) := \frac{b}{\sqrt{(2\pi)}} \,.\, \exp\left\{-\frac{b^2 x^2}{2}\right\} \leqq \varphi_{1/2}(x)\,.$$

Ist $\alpha - |c/b| \geqq 2$, so erhält man leicht

$$0 < \int_{-\infty}^{-\alpha} p(G > -bz + c) \, . \, dz \leqq \int_{\alpha - |c/b|}^{+\infty} x \, . \, \varphi_{1/2}(x) \, . \, dx$$

und

$$0 < \int_{\alpha}^{+\infty} p(G < -bz + c) \, . \, dz \leqq \int_{\alpha - |c/b|}^{+\infty} x \, . \, \varphi_{1/2}(x) \, . \, dx \, .$$

2. Da für $(\gamma, u, v) \in C_1$ und $0 \leqq \eta \leqq \min\{\frac{1}{2}, 1/2\gamma_0\}$ die Abschätzungen

$$\frac{1}{2} < b := \frac{1 - \eta^2 v^2}{v \, . \, \sqrt{(2 - \eta^2 v^2)}} < +\infty \, ,$$

und mit

$$c := \frac{-u + \eta v^2 \gamma}{v \, . \, \sqrt{(2 - \eta^2 v^2)}} \quad \text{auch} \quad \left|\frac{c}{b}\right| \leqq 2$$

gelten, erhalten wir unter Berücksichtigung von Beweisteil 1

$$\left| \int_{-\infty}^{-\alpha} [\varphi(\gamma + \eta z) - \varphi(\gamma)] \, . \, e_1(\gamma, u, v, \eta, z) \, . \, dz + \right.$$

$$\left. + \int_{\alpha}^{+\infty} [\varphi(\gamma + \eta z) - \varphi(\gamma)] \, . \, e_2(\gamma, u, v, \eta, z) \, . \, dz \right| \leqq 4 \, . \int_{\alpha - 2}^{+\infty} x \, . \, \varphi_{1/2}(x) \, . \, dx \, .$$

3. Wählt man zu vorgegebenem $\varepsilon > 0$ die Zahl $\alpha \geqq 4$ so, daß

$$\int_{\alpha - 2}^{+\infty} x \, . \, \varphi_{1/2}(x) \, . \, dx < \frac{\varepsilon}{8} \, ,$$

und weiter $\eta_0 \in (0, \min\{\frac{1}{2}, 1/2\gamma_0\}]$ so, daß

$$|\varphi(\gamma + \eta z) - \varphi(\gamma)| \leqq \frac{\varepsilon}{4\alpha}$$

für $\gamma \in [-\gamma_0, \gamma_0]$, $\eta \in [0, \eta_0]$ und $z \in [-\alpha, \alpha]$ gilt, so erhalten wir wegen $|e_\nu(\gamma, u, v, \eta, z)| \leqq 1$ sofort $|r(\gamma, u, v, \eta)| < \varepsilon$; es gilt also

$$r(\gamma, u, v, \eta) \xrightarrow[\eta \downarrow 0]{} 0$$

gleichmäßig in C_1. □

Beweis zu Satz (2.1). Nach den Ausführungen vor Lemma (2.2) genügt es wegen der Stetigkeit von $h(\gamma, u, v)$,

$$\frac{1}{\eta} \, . \, g(\gamma(t), \eta u, \eta v) \xrightarrow[\eta \downarrow 0]{} h(\gamma(t), u, v)$$

gleichmäßig für $t \in [0, 1]$, $u \in [-1, +1]$, $v \geqq 0$ mit $u^2 + v^2 = 1$ zu beweisen.

Benutzt man die Substitution $z = (x - \gamma)/\eta$, so erhält man

$$\frac{1}{\eta} \cdot g(\gamma, \eta u, \eta v) =$$

$$= \frac{1}{\eta} \cdot \int_{-\infty}^{\gamma} p((1 - \eta^2 v^2) \cdot x + \eta v \cdot (2 - \eta^2 v^2)^{1/2} \cdot G - \gamma + \eta u > 0) \cdot \varphi(x) \cdot \mathrm{d}x +$$

$$+ \frac{1}{\eta} \cdot \int_{\gamma}^{+\infty} p((1 - \eta^2 v^2)\, x + \eta v \cdot (2 - \eta^2 v^2)^{1/2} \cdot G - \gamma + \eta u < 0) \cdot \varphi(x) \cdot \mathrm{d}x =$$

$$= \int_{-\infty}^{0} p((1 - \eta^2 v^2) \cdot \eta z - \eta^2 v^2 \gamma + \eta v \cdot (2 - \eta^2 v^2)^{1/2} \cdot G + \eta u > 0) \cdot \varphi(\gamma + \eta z) \cdot \mathrm{d}z +$$

$$+ \int_{0}^{+\infty} p((1 - \eta^2 v^2) \cdot \eta z - \eta^2 v^2 \gamma + \eta v (2 - \eta^2 v^2)^{1/2} \cdot G + \eta u < 0) \cdot \varphi(\gamma + \eta z) \cdot \mathrm{d}z =$$

$$= \int_{-\infty}^{0} \varphi(\gamma + \eta z) \cdot e_1(\gamma, u, v, \eta, z) \cdot \mathrm{d}z + \int_{0}^{+\infty} \varphi(\gamma + \eta z) \cdot e_2(\gamma, u, v, \eta, z) \cdot \mathrm{d}z =$$

$$= \varphi(\gamma) \cdot k(\gamma, u, v, \eta) + r(\gamma, u, v, \eta) .$$

Aus Lemma (2.2) und Lemma (2.3) erhalten wir daher unmittelbar

$$\frac{1}{\eta}\, g(\gamma, \eta u, \eta v) \xrightarrow[\eta \downarrow 0]{} h(\gamma, u, v)$$

gleichmäßig auf C_1.

Weiter gilt für $u = 1$, $v = 0$

$$\frac{1}{\eta}\, g(\gamma, \eta, 0) = \frac{1}{\eta} \cdot p([G_1 - \gamma + \eta] \cdot [G_1 - \gamma] < 0) =$$

$$= \frac{1}{\eta} \cdot \int_{\gamma - \eta}^{\gamma} \varphi(x) \cdot \mathrm{d}x \xrightarrow[\eta \downarrow 0]{} \varphi(\gamma)$$

gleichmäßig für $\gamma \in [-\gamma_0, \gamma_0]$ und für $u = -1$, $v = 0$ ebenso

$$\frac{1}{\eta}\, g(\gamma, -\eta, 0) = \frac{1}{\eta} \cdot p([G_1 - \gamma - \eta] \cdot [G_1 - \gamma] < 0) =$$

$$= \frac{1}{\eta} \cdot \int_{\gamma}^{\gamma + \eta} \varphi(x) \cdot \mathrm{d}x \xrightarrow[\eta \downarrow 0]{} \varphi(\gamma)$$

gleichmäßig für $\gamma \in [-\gamma_0, \gamma_0]$.

Wegen $|\gamma(t)| \leqq \gamma_0$ für $t \in [0, 1]$ erhalten wir daher insgesamt

$$\frac{1}{\eta} \cdot g(\gamma(t), \eta u, \eta v) \xrightarrow[\eta \downarrow 0]{} h(\gamma(t), u, v)$$

gleichmäßig auf $\{(t, u, v) : t \in [0, 1], u \in R, v \geqq 0 \text{ mit } u^2 + v^2 = 1\}$. □

Für die Anwendungen ist folgender Satz nützlich.

(2.4) SATZ. *Ist* $\tilde{\sigma} \,|\, \mathfrak{B}(0, 1)$ *ein atomfreies Maß, und gelten mit geeigneten Baireschen Funktionen* $\tilde{\psi}_1(t), \tilde{\psi}_2(t)$ *die Beziehungen*

$$\gamma(t) = \int_{(0,t]} \tilde{\psi}_1(\tau) \cdot d\tilde{\sigma},$$

$$s_2(t) = \int_{(0,t]} \tilde{\psi}_2(\tau) \cdot d\tilde{\sigma},$$

so ist

$$\mathrm{E}\, U(\gamma(\cdot), X(\cdot)) = \int_{(0,1]} h(\gamma(t), \tilde{\psi}_1(t), \tilde{\psi}_2(t)) \cdot d\tilde{\sigma}.$$

Der Beweis von Satz (2.4) ergibt sich sofort aus der Tatsache, daß σ bezüglich $\tilde{\sigma}$ totalstetig ist mit der Dichte

$$\frac{d\sigma}{d\tilde{\sigma}}(t) = |\tilde{\psi}_1(t)| + \tilde{\psi}_2(t)$$

und folglich

$$\psi_1(t) \cdot (|\tilde{\psi}_1(t)| + \tilde{\psi}_2(t)) = \tilde{\psi}_1(t)$$

und

$$\psi_2(t) \cdot (|\tilde{\psi}_1(t)| + \tilde{\psi}_2(t)) = \tilde{\psi}_2(t)$$

für $\tilde{\sigma}$-fast alle $t \in [0, 1]$ gilt; denn es gilt dann

$$\int_{(0,1]} h(\gamma(t), \tilde{\psi}_1(t), \tilde{\psi}_2(t)) \cdot d\tilde{\sigma} =$$

$$= \int_{(0,1]} h(\gamma(t), \psi_1(t), \psi_2(t)) \cdot (|\tilde{\psi}_1(t)| + \tilde{\psi}_2(t)) \cdot d\tilde{\sigma} =$$

$$= \int_{(0,1]} h(\gamma(t), \psi_1(t), \psi_2(t)) \cdot d\sigma = \mathrm{E}\, U(\gamma(\cdot), X(\cdot)).$$ □

Wir betrachten nun einige Spezialfälle.

1. Ist $\gamma(t) \equiv \gamma_1$ auf $[0, 1]$, so erhalten wir also $\gamma'(t) \equiv 0$ und folglich nach Satz (2.1)

$$\mathrm{E}\, U(\gamma_1, X(\cdot)) = \frac{\sqrt{2}}{\pi} \cdot \exp\left\{-\frac{\gamma_1^2}{2}\right\} \cdot \int_0^1 \psi_2(t) \cdot \mathrm{d}\sigma$$

und, da hier $\psi_1(t) \equiv 0$, $\psi_2(t) \equiv 1$, $s(t) = \int_0^t \sqrt{(1 - r)}$, mit

$$\mathrm{E}\, U(\gamma_1, X(\cdot)) = \frac{\sqrt{2}}{\pi} \cdot \exp\left\{-\frac{\gamma_1^2}{2}\right\} \cdot \int_0^1 \sqrt{(1 - r)}$$

das bekannte Ergebnis [2].

2. Ist $r(t, t') \equiv 1$ auf $[0, 1] \times [0, 1]$, so haben wir also mit einer geeigneten Gaußschen Zufallsgröße G mit $E\, G = 0$, $\operatorname{var} G = 1$ für jedes $t \in [0, 1]$ stets $X(t) = G$. Wir setzen weiter voraus, daß $\gamma(t)$ stetig differenzierbar ist. Bei festem $G = x_0$ ist $U(\gamma(\cdot), X(\cdot)) = U(\gamma(\cdot), x_0)$ gleich der Anzahl der Überkreuzungen des Niveaus x_0 durch die Funktion γ. Benutzt man die Funktion

$$\delta_n(y) := \begin{cases} \dfrac{n}{2} & \text{für} \quad y \in \left[-\dfrac{1}{n}, \dfrac{1}{n}\right] \\ 0 & \text{sonst}, \end{cases}$$

so erhält man

$$U(\gamma(\cdot), x_0) = \lim_{n\to\infty} \int_0^1 \delta_n(\gamma(t) - x_0) \cdot |\gamma'(t)| \cdot \mathrm{d}t\,.$$

Geht man davon aus, daß man Grenzwertbildung und Integration in ihrer Reihenfolge vertauschen kann, so ergibt sich

$$\begin{aligned} \mathrm{E}\, U(\gamma(\cdot), X(\cdot)) &= \mathrm{E}\left(\lim_{n\to\infty} \int_0^1 \delta_n(\gamma(t) - G) \cdot |\gamma'(t)| \cdot \mathrm{d}t\right) = \\ &= \lim_{n\to\infty} \int_0^1 \mathrm{E}(\delta_n(\gamma(t) - G)) \cdot |\gamma'(t)| \cdot \mathrm{d}t = \\ &= \int_0^1 |\gamma'(t)| \cdot \lim_{n\to\infty} \mathrm{E}(\delta_n(\gamma(t) - G)) \cdot \mathrm{d}t = \\ &= \int_0^1 |\gamma'(t)| \cdot \varphi(\gamma(t)) \cdot \mathrm{d}t\,, \end{aligned}$$

da

$$\lim_{n\to\infty} \mathrm{E}(\delta_n(\gamma(t) - G)) = \lim_{n\to\infty} \frac{n}{2} \cdot p\left(\gamma(t) - \frac{1}{n} \leqq G \leqq \gamma(t) + \frac{1}{n}\right) = \varphi(\gamma(t))$$

gilt. Dieses Ergebnis stimmt überein mit dem Ergebnis, das man in diesem Spezialfall unmittelbar aus Satz (2.4) erhält.

3. Es sei $\{Y(t) : t \in [0, 1]\}$ ein Gaußscher Prozeß mit stetigen Pfaden, $m(t) := = \mathrm{E}\, Y(t)$ und $R(t, t') := \operatorname{cov}(Y(t), Y(t'))$. Wir setzen $R(t, t) > 0$ für jedes $t \in [0, 1]$ voraus. Die Anzahl $U(0; Y(\cdot))$ der Null-Niveau-Überkreuzungen des Prozesses $\{Y(t) : t \in [0, 1]\}$ ist gleich der Anzahl $U(\gamma(\cdot), X(\cdot))$ der Überkreuzungen des Prozesses $X(t) := [R(t, t)]^{-1/2} . (Y(t) - m(t))$ mit der Funktion $\gamma(t) := [R(t, t)]^{-1/2} . m(t)$.

Setzen wir voraus, daß $m(t)$ stetig und einmal differenzierbar mit Lebesgue-integrablem $m'(t)$ ist und alle partiellen Ableitungen zweiter Ordnung von $R(t, t')$ stetig sind, so können wir $\mathrm{E}\, U(0, Y(\cdot))$ mit Hilfe von Satz (2.4) berechnen.

Es gilt

$$\gamma'(t) = [R(t, t)]^{-1/2} . m'(t) - [R(t, t)]^{-3/2} . m(t) . R_2'(t, t)$$

(wegen $R_1'(t, t) = R_2'(t, t)$; wir bezeichnen mit $R_\nu'(t, t')$ die erste partielle Ableitung nach der ν-ten Variablen, entsprechende Bezeichnungen benutzen wir für die partiellen Ableitungen zweiter Ordnung). Für die Kovarianzfunktion $r(t, t') = [R(t, t) . . R(t', t')]^{-1/2} . R(t, t')$ des stochastischen Prozesses $\{X(t) : t \in [0, 1]\}$ erhalten wir

$$r_2'(t, t) = 0$$

und

$$r_{22}''(t, t) = -\frac{R_{12}''(t, t)}{R(t, t)} + \frac{(R_2'(t, t))^2}{(R(t, t))^2},$$

und somit

$$r(t, t + h) = r(t, t) + h . r_2'(t, t) + \frac{h^2}{2} . r_{22}''(t, t) + o(h^2) =$$

$$= 1 + \frac{h^2}{2} . (R(t, t))^{-2} . \{(R_2'(t, t))^2 - R(t, t) . R_{12}''(t, t)\} + o(h^2) .$$

Wegen der (gleichmäßigen) Stetigkeit von $R_{12}''(t, t')$ in $[0, 1] \times [0, 1]$ gilt für das Restglied $o(h^2)$ auch

$$\frac{1}{h^2} . o(h^2) \xrightarrow[h \downarrow 0]{} 0$$

gleichmäßig für $t \in [0, 1]$. Wir erhalten daher

$$s_2(t) = \int_0^t \sqrt{(1 - r)} = \frac{1}{\sqrt{2}} . \int_0^t (R(\tau, \tau))^{-1} . \sqrt{\{R(\tau, \tau) . R_{12}''(\tau, \tau) - (R_2'(\tau, \tau))^2\}} . d\tau .$$

Setzen wir schließlich noch voraus, daß $s_2(t)$ auf $[0, 1]$ streng monoton wachsend ist, so erhalten wir als unmittelbare Anwendung von Satz (2.4)

$$\mathrm{E}\, U(0, Y(\cdot)) = \int_0^1 H(t) . dt$$

mit

$$H(t) := \varphi([R(t,t)]^{-1/2} . m(t)) . \frac{\{R(t,t) . R''_{12}(t,t) - [R'_2(t,t)]^2\}^{1/2}}{R(t,t)} .$$

$$. \{2\varphi(A(t)) - A(t) . (2\Phi(-A(t)) - 1)\},$$

wobei zur Abkürzung

$$A(t) := \frac{m'(t) . R(t,t) - m(t) . R'_2(t,t)}{\sqrt{\{R(t,t)^2 . R''_{12}(t,t) - R(t,t) . [R'_2(t,t)]^2\}}}$$

gesetzt ist. Das gleiche Ergebnis haben auf einem anderen Wege Leadbetter und Cryer [5] erhalten.

LITERATUR

[1] W. Fieger: Die Transformation von Burkill-Unterteilungsintegralen. Math. Ann. *183* (1969), 115–129.

[2] W. Fieger: Die Anzahl der γ-Niveau-Kreuzungspunkte von stochastischen Prozessen. Z. Wahrscheinlichkeitstheorie *18* (1971), 227–260.

[3] W. Fieger: Zur Burkill-Integrabilität einer Funktion B-integrabler Intervallfunktionen. In Vorbereitung.

[4] M. R. Leadbetter: Point processes generated by level crossings. In: Stochastic point processes. Ed. P.A.W. Lewis. Wiley, New York 1972, 436–467.

[5] M. R. Leadbetter, J. D. Cryer: On the mean number of curve crossings by non stationary normal processes. Ann. Math. Stat. *36* (1965), 509–516.

Universität Karlsruhe

Institut für Mathematische Statistik

ON THE STABILITY OF SPATIALLY HOMOGENEOUS CLUSTER FIELDS

KLAUS FLEISCHMANN

BERLIN

This paper deals with a clustering model which is described in detail by J. Kerstan, K. Matthes and J. Mecke ([2], chap. 6).

Let D be a point process (p.p.) in Euclidean space R^s, its intensity measure ϱ_D being a probability law, i.e. the mean number of points with respect to D is equal to one. Let $P \neq \delta_o$ be a stationary p.p. in R^s with finite intensity. Each point of P is clustered independently of all other points of P and spatially homogeneously according to the cluster distribution D. After superposition of all clusters the clustered p.p. $P_{[D]}$ is obtained. D is called *stable*, if there is a stationary p.p. $P \neq \delta_o$ with finite intensity and which is cluster invariant with respect to D, i.e. P is a solution of the equation $P_{[D]} = P$. For the stability of D *necessary* conditions were deduced (partly under additional assumptions) by Liemant and Matthes (cf. [3], [2, § 6.7 and § 6.9]).

Let $a(D)$ be the probability law of the total number of points with respect to D. In the following we neglect the trivial case $a(D) = \delta_1$. A *sufficient* condition formulated only in terms of ϱ_D and $a(D)$ has been found by Liemant [3]. He has shown that D is stable for $s \geq 3$, if the second moment of $a(D)$ is finite and the characteristic function $\hat{\varrho}_D$ of ϱ_D has property

$$\text{(A)} \qquad \limsup_{\|t\| \to +\infty} |\hat{\varrho}_D(t)| < 1 .$$

The purpose of the present paper is to show that assumption (A) may be neglected (cf. theorem 1.1.), if we only presuppose that ϱ_D is s-dimensional, i.e. that the following condition is satisfied:

(B) all linear submanifold U of R^s with $\varrho_D(U) = 1$ are s-dimensional.

By the finiteness of the second moment of $a(D)$ the spectral theory of p.p. can be used in the proofs, and the constructed cluster invariant p.p. has finite second moments.

1. NOTATIONS AND RESULT

Let M be the set of all integer valued measures Φ on the σ-algebra $[R^s, \mathfrak{R}^s]$ of the Borel sets of Euclidean space R^s, for which $\Phi(B)$ is finite for compact $B \subset R^s$. Let $\mathfrak{M}$ be the smallest σ-algebra of subsets of M so that the real functions

$$\Phi \to \Phi(B) \quad (\Phi \in M)$$

are measurable for compact $B \subset R^s$. The translation by x $(x \in R^s)$ is defined by

$$T_x\Phi = \int \Phi(\mathrm{d}y)\, \delta_{y-x}(\cdot) \quad (\Phi \in M).$$

By setting

$$[D]_{(x)} = D(T_{-x}\chi \in (\cdot)) \quad (x \in R^s)$$

a spatially homogeneous cluster field in R^s is obtained from a probability law D on $[M, \mathfrak{M}]$. If the infinite convolution

$$\mathop{*}_{x \in R^s, \Phi(\{x\}) > 0} (([D]_{(x)})^{\Phi(\{x\})})$$

exists for a $\Phi \in M$, it will be denoted by $[D]_{(\Phi)}$. For a p.p. P in R^s, that is a probability law P on $[M, \mathfrak{M}]$, we then obtain the clustered p.p. $P_{[D]} = \int P(\mathrm{d}\Phi)\, [D]_{(\Phi)}$, if this integral makes sense. P is called *cluster invariant* (with respect to D), if $P_{[D]}$ exists and is equal to P. For the existence of a cluster invariant stationary $P \neq \delta_o$ with finite intensity it is necessary, that the mean number of points with respect to D is equal to one, i.e. that the intensity measure ϱ_D of D is a probability law. Let $\boldsymbol{D}$ be the set of all these cluster distributions D.

$D \in \boldsymbol{D}$ is called *stable*, if there is a stationary cluster invariant p.p. P with intensity i_p, $0 < i_p < +\infty$. If P in addition has finite second moments, i.e. for all compact $B \subset R^s$ the integrals $\int P(\mathrm{d}\Phi)\, \Phi^2(B)$ are finite, D is called *stable of second order*.

We say that ϱ_D is *s-dimensional*, if each linear sumanifold U of R^s with $\varrho_D(U) = 1$ has dimension s. Now we will formulate the result of this paper:

1.1. THEOREM. *Let be $s \geqq 3$, $D \in \boldsymbol{D}$, $\int D(\mathrm{d}\chi)\, \chi^2(R^s) < +\infty$ and let ϱ_D be s-dimensional. Then the cluster distribution D is stable of second order.*

2. THE SPECTRAL MEASURE OF CLUSTERED P.P.

In this paragraph we will compile some facts from the spectral theory of p.p. in R^s (cp. [1, chap. III; 3]).

A measure σ on $[R^s, \mathfrak{R}^s]$ is called a *tempered measure* if the integral $\int \sigma(\mathrm{d}t)\,.\,. 1 + \|t\|^2)^{-p}$ converges for some positive p.

Let the real function φ on R^s be equal to zero outside a bounded domain, and let all partial derivatives $D^\alpha\varphi$ of any order exist and be continuous. Let K be the linear space of all these φ. A sequence $\varphi_1, \varphi_2, \ldots$ from K is called a zero-sequence, if all φ_n vanish outside some common bounded domain and $D^\alpha\varphi_1, D^\alpha\varphi_2, \ldots$ uniformly converges to 0 for any α. Each $\varphi \in K$ will be associated with the Fourier-transform $\hat{\varphi}$:

$$\hat{\varphi}(t) = \int \mu(\mathrm{d}x)\, \varphi(x)\, \mathrm{e}^{\mathrm{i}(t,x)} \quad (t \in R^s)$$

(μ = Lebesgue measures in R^s). φ, ψ being elements of K, $\Phi \in M$ and P a p.p. in R^s we denote

$$\langle \varphi, \Phi \rangle = \int \Phi(\mathrm{d}x)\, \varphi(x), \quad M_P\varphi = \int P(\mathrm{d}\Phi)\, \langle \varphi, \Phi \rangle,$$

$$B_P(\varphi, \psi) = \int P(\mathrm{d}\Phi)\, \langle \varphi, \Phi \rangle \langle \psi, \Phi \rangle$$

if the last two integrals exist. If the expression

$$C_P(\varphi, \psi) = B_P(\varphi, \psi) - M_P\varphi M_P\psi \quad (\varphi, \psi \in K)$$

makes sense, C_P is called the *correlation functional* of P. By a theorem of Bochner-Schwartz [1, theorem II.3.3] we now obtain the following.

2.1. *For each stationary p.p. P in R^s with finite second moments there exists the correlation functional C_P having a representation*

$$C_P(\varphi, \psi) = \int \sigma_P(\mathrm{d}t)\, \hat{\varphi}(t)\overline{\hat{\psi}(t)} \quad (\varphi, \psi \in K).$$

In this case σ_P is a uniquely defined tempered measure in R^s, the spectral measure of P.

Let $D \in \boldsymbol{D}$ with $\int D(\mathrm{d}\chi)\, \chi^2(R^s) < +\infty$. If we denote the cube $[0, 1)^s$ by E it holds

$$[D]_{(\Phi)}(\mathrm{d}\chi)\, \chi^2(E) \leqq \int \Phi(\mathrm{d}x) \int [D]_{(x)}(\mathrm{d}\chi)\, \chi^2(E) + \left(\int \Phi(\mathrm{d}x) \int [D]_{(x)}(\mathrm{d}\chi)\, \chi(E)\right)^2$$

for $\Phi \in M$ (provided that $[D]_{(\Phi)}$ exists). We have

$$\left(\int \Phi(\mathrm{d}x) \int [D]_{(x)}\, (\mathrm{d}\chi)\, \chi(E)\right)^2 \leqq \int \varrho_D(\mathrm{d}x)\, \Phi^2(E - x)$$

and thus we obtain

$$\int P_{[D]}(\mathrm{d}\Phi)\,\Phi^2(E) \leqq i_P \int \mu(\mathrm{d}x) \int D(\mathrm{d}\chi)\,\chi^2(E - x) + \int P(\mathrm{d}\Phi) \int \varrho_D(\mathrm{d}x)\,\Phi^2(E - x) \leqq$$

$$\leqq i_P\, 2^{2s} \int D(\mathrm{d}\chi)\,\chi^2(R^s) + \int P(\mathrm{d}\Phi)\,\Phi^2(E) < +\infty\,.$$

Thus for such D the stationary p.p. $P_{[D]}$ also has second moments if P has second moments.

We denote the characteristic function of ϱ_D by $\hat{\varrho}_D$,

$$\hat{\varrho}_D(t) = \int \varrho_D(\mathrm{d}x)\,\mathrm{e}^{\mathrm{i}(t,x)} \quad (t \in R^s)\,,$$

and

$$\dot{B}_D(t) = \int D(\mathrm{d}\chi) \left| \int \chi(\mathrm{d}x)\,\mathrm{e}^{\mathrm{i}(t,x)} \right|^2 \quad (t \in R^s)\,.$$

$\dot{B}_D$ being bounded by $\int D(\mathrm{d}\chi)\,\chi^2(R^s)$ and $\dot{B}_D - |\hat{\varrho}_D(t)|^2 \geqq 0$ being fulfilled. The following holds (cf. [3, theorem 4.4])

2.2. *Let the stationary p.p. P have second moments and let $\int D(\mathrm{d}\chi)\,\chi^2(R^s) < +\infty$ hold for $D \in \boldsymbol{D}$. Then the spectral measure of $P_{[D]}$ has the form*

$$\sigma_{P_{[D]}}(\mathrm{d}t) = (2\pi)^{-s}\, i_P \mu(\mathrm{d}t)\,(\dot{B}_D(t) - |\hat{\varrho}_D(t)|^2) + \sigma_P(\mathrm{d}t)\,|\hat{\varrho}_D(t)|^2\,.$$

Let the integrals

$$\int \mu(\mathrm{d}t)\,(1 - |\hat{\varrho}_D(t)|^2)^{-1}\,(\dot{B}_D(t) - |\hat{\varrho}_D(t)|^2)\,|\hat{\varphi}(t)|^2 \quad (\varphi \in K)$$

converge. Then

$$\sup_{n=1,2,\ldots} \int \mu(\mathrm{d}t) \sum_{j=0}^{n-1} |\hat{\varrho}_D(t)|^{2j}\,(\dot{B}_D(t) - |\hat{\varrho}_D(t)|^2)\,|\hat{\varphi}(t)|^2 < +\infty \quad (\varphi \in K)\,.$$

From the stationary Poisson p.p. P_μ of intensity 1 the p.p. $P_n = (\ldots((P_\mu)_{[D]})\ldots)_{[D]}$ is obtained after n-fold clustering. According to 2.2 for its correlation functional C_{P_n} we have

$$C_{P_n}(\varphi, \varphi) = (2\pi)^{-s} \left[\int \mu(\mathrm{d}t) \sum_{j=0}^{n-1} |\hat{\varrho}_D(t)|^{2j}\,(\dot{B}_D(t) - |\hat{\varrho}_D(t)|^2)\,|\hat{\varphi}(t)|^2 + \right.$$

$$\left. + \int \mu(\mathrm{d}t)\,|\hat{\varrho}_D(t)|^{2n}\,|\hat{\varphi}(t)|^2 \right] \quad (\varphi \in K,\ n = 1, 2, \ldots)$$

and we have

$$\sup_{n=1,2,\ldots} C_{P_n}(\varphi, \varphi) < +\infty \quad (\varphi \in K).$$

So because of $i_{P_n} = 1$, $n = 1, 2, \ldots$,

$$\sup_{n=1,2,\ldots} \int P_n(\mathrm{d}\Phi)\, \Phi^2(E) < +\infty$$

is valid and accordingly

$$\inf_{n=1,2,\ldots} P_n(\Phi(E) > 0) > 0.$$

Thus any accumulation point (in the sense of weak topology) of the sequence $n^{-1} \sum_{j=1}^{n} P_j$ has positive intensity, so that there is a cluster invariant stationary nontrivial p.p. with finite second moments (cf. [2, p. 384]). So one obtains (cf. [3, theorem 5.4]).

2.3. *If the measure*

$$\mu(\mathrm{d}t)\,(1 - |\hat{\varrho}_D(t)|^2)^{-1}\,(\dot{B}(t) - |\hat{\varrho}_D(t)|^2)$$

is tempered for $D \in \boldsymbol{D}$ *with* $\int D(\mathrm{d}\chi)\, \chi^2(R^s) < +\infty$, *then* D *is stable of second order.*

3. PROOF OF THE THEOREM

1. We assume that the constants r, R satisfy the condition $0 < r < R < +\infty$. By choosing a sufficiently small positive ε we can obtain that for each given real x with $r < |x| < R$ in each semiclosed interval

$$Y_j = [(2j-1)\,\varepsilon,\ (2j+1)\,\varepsilon) \quad (j \in \Gamma = \{0, \pm 1, \pm 2, \ldots\})$$

with suitably chosen real $c_{j,x}$ the function f_x,

$$f_x(t) = 2\,|\sin xt| \quad (t \in R^1), \tag{1}$$

satisfies the inequality

$$f_x(t) \geqq r|t - c_{j,x}| \quad (t \in Y_j,\ j \in \Gamma)$$

and

$$\inf_{t \in Y_j} |t - c_{j,x}| = 0 \quad (r \leqq |x| \leqq R,\ j \in \Gamma)$$

holds.

2. For each $x \in R^s$ the function f_x,

$$f_x(t) = 2|\sin(x, t)| \quad (t \in R^s),$$

is constant on each hyperplane $\{t : (x, t) = \text{const.}\}$, but in the direction $e_x = x\|x\|^{-1}$ it behaves similar to (1). Thus in analogy to the first step of proof we may choose an $\varepsilon > 0$ so, that in each cube

$$Y_j = \mathop{\times}_{i=1}^{s} [(2j_i - 1)\varepsilon, (2j_i + 1)\varepsilon) \quad (j = [j_1, \ldots, j_s] \in \Gamma^s)$$

with suitably chosen real $c_{j,x}$ the inequality

$$f_x(t) \geqq r|(e_x, t) - c_{j,x}| \quad (t \in Y_j,\ j \in \Gamma^s)$$

with

(2) $$\inf_{t \in Y_j} |(e_x, t) - c_{j,x}| = 0$$

is valid for all x with $r \leqq \|x\| \leqq R$. In this case the $c_{j,x}$ may be chosen in such a way, that the mappings

$$x \to c_{j,x} \quad (r \leqq \|x\| \leqq R)$$

are measurable. From (2) it follows

(3) $$|c_{j,x}| \leqq \sup_{t \in Y_j} \|t\| \quad (r \leqq \|x\| \leqq R,\ j \in \Gamma^s).$$

3. Let an s-dimensional probability law λ in R^s satisfy $\lambda(r \leqq \|x\| \leqq R) = 1$. In this case using the second step of proof the function f,

$$f(t) = 4 \int \lambda(\mathrm{d}x) \sin^2(x, t) \quad (t \in R^s),$$

can be estimated as follows:

$$f(t) \geqq r^2 \int \lambda(\mathrm{d}x) |(e_x, t) - c_{j,x}|^2 \quad (t \in Y_j,\ j \in \Gamma^s).$$

4. For $j \in \Gamma^s$ we interpret the non-negative function f^j,

(4)
$$f^j(t) = \int \lambda(\mathrm{d}x) |(e_x, t) - c_{j,x}|^2 =$$
$$= \int \lambda(\mathrm{d}x) (e_x, t)^2 - 2 \int \lambda(\mathrm{d}x) (e_x, t)\, c_{j,x} + \int \lambda(\mathrm{d}x)\, c_{j,x}^2 \quad (t \in R^s),$$

as an elliptic paraboloid in R^{s+1}, because λ was assumed to be s-dimensional (The last two integrals exist because of (3).) The axis of this paraboloid is orthogonal to R^s. The shape of the paraboloid is completely determined by the first term in (4), which is independent of j. The abscissa of the vertex of the paraboloid will be denoted by c_j $(c_j \in R^s)$:

$$\min_{t \in R^s} f^j(t) = f^j(c_j).$$

The paraboloid can be enveloped in a rotational paraboloid, so that the two axes and the vertices are identical and the shape of the rotational paraboloid is independent of j. In other words, for f^j there exists an estimation

$$f^j(t) \geqq \alpha \|t - c_j\|^2 + f^j(c_j) \geqq$$

$$\geqq \alpha \|t - c_j\|^2 \quad (t \in R^s,\ j \in \Gamma^s)$$

for some positive α independent of j.

5. The last two steps of proof result in the inequality

$$f(t) \geqq \beta \|t - c_j\|^2 \quad (t \in Y_j,\ j \in \Gamma^s),$$

β being a positive constant. For $s \geqq 3$ the integrals

$$I_j = \int_{Y_j} \mu(\mathrm{d}t)\, \|t - c_j\|^{-2} = \int_{Y_j - c_j} \mu(\mathrm{d}t)\, \|t\|^{-2} \quad (j \in \Gamma^s)$$

converge and

$$I_j \leqq \sup_{x \in R^s} \int_{Y_0 + x} \mu(\mathrm{d}t)\, \|t\|^{-2} = \int_{Y_0} \mu(\mathrm{d}t)\, \|t\|^{-2}$$

holds. Therefore one has

$$\sup_{j \in \Gamma^s} \int_{Y_j} \mu(\mathrm{d}t) \left(\int \lambda(\mathrm{d}x) \sin^2(x, t) \right)^{-1} < +\infty \quad (s \geqq 3).$$

6. Let ν be an s-dimensional probability law in R^s. There exist positive constants r, R, so that even $\nu((\cdot) \mid r \leqq \|x\| \leqq R)$ is s-dimensional. So according to the previous steps of proof we find an "equidistant" decomposition $(Y_j)_{j \in \Gamma^s}$ of R^s for which

$$\sup_{j \in \Gamma^s} \int_{Y_j} \mu(\mathrm{d}t) \left(\int \nu(\mathrm{d}x) \sin^2(x, t) \right)^{-1} < +\infty \quad (s \geqq 3)$$

is valid.

7. Let the assumptions of theorem 1.1. be satisfied. Then the probability law ν on R^s corresponding to the characteristic function $|\hat{\varrho}_D|^2$ is s-dimensional and

$$1 - |\varrho_D(t)|^2 = 1 - \int \nu(\mathrm{d}x) \cos(x, t) =$$

$$= 2 \int \nu(\mathrm{d}x) \sin^2 (2^{-1}x, t) \quad (t \in R^s)$$

holds. Because of

$$\dot{B}_D(t) - |\hat{\varrho}_D(t)|^2 \leqq \int D(\mathrm{d}\chi)\, \chi^2(R^s) \quad (t \in R^s)$$

we get by means of the sixth step of proof: There exists an "equidistant" decomposition $(Y_j)_{j \in \Gamma^s}$ of R^s with

$$\sup_{j \in \Gamma^s} \int_{Y_j} \mu(\mathrm{d}t)\,(1 - |\hat{\varrho}_D(t)|^2)^{-1}\,(\dot{B}_D(t) - |\hat{\varrho}_D(t)|^2) < +\infty .$$

Hence the measure

$$\sigma(\mathrm{d}t) = \mu(\mathrm{d}t)\,(1 - |\hat{\varrho}_D(t)|^2)^{-1}\,(\dot{B}_D(t) - |\hat{\varrho}_D(t)|^2)$$

is tempered and according to 2.3. the cluster distribution D is stable of second order. So the proof is completed.

REFERENCES

[1] И. М. Гельфанд, Н. Я. Виленкин: Обобщенные функции, выпуск 4. Физматгиз, Москва 1961.

[2] J. Kerstan, K. Matthes, J. Mecke: Unbegrenzt teilbare Punktprozesse. Akademie-Verlag, Berlin 1974.

[3] A. Liemant: Invariante zufällige Punktfolgen. Wiss. Z. Friedrich-Schiller-Univ. Jena, math.-naturw. R. *18* (1969), 361—372.

Akademie der Wissenschaften der DDR

Zentralinstitut für Mathematik und Mechanik

A MARKOV CHAIN FOR TYPE COUNTS IN VOCABULARY SUBGROUPS

J. Gani

Sheffield and Canberra

Literary texts consist of words from two groups, each subdivided into several subgroups. A continuous time Markov chain model is formulated for the text-writing process, and its forward Kolmogorov equations derived. The p.g.f. of the process is obtained in an explicit form, and relevant expectations of type counts are found from it.

1. INTRODUCTION

In any English literary text, word types fall into two groups: those in the very large Group I consist of nouns, adjectives, verbs and adverbs, while the small Group II contains pronouns, prepositions, conjunctions, auxiliary verbs, articles and interjections. Studies carried out by Yule [4] on the total number of words listed in the 1933 *Shorter Oxford English Dictionary* led to estimates of approximately 100,000 for Group I and 500 for Group II. Readers interested in a more detailed account of problems of vocabulary, are referred to Chapter 5 of Williams [3].

When estimating an author's total potential vocabulary it proves useful to compare his type count, that is the number of words used in the text for the first time, with his token count or total number of words. A simple discrete time Markov chain model was proposed by Brainerd [1] to study this relationship in the works of Shakespeare. The only distinction between words made in this, was essentially that between words in Groups I and II; Gani [2] reconsidered the model in continuous time, and suggested some extensions to it.

The main purpose of the present paper is to examine one such extension: a general Markov chain model in continuous time for type and token counts, which will take note of the 4 Group I, and the 6 Group II subgroups. It will be directly apparent that its structure is similar to that for a sampling model with replacement from a finite population consisting of two groups, each with a number of subgroups; the type count is then analogous to the count of new items in the sample, while the token count is equivalent to the total number of items sampled.

2. FORMULATION OF THE MARKOV CHAIN MODEL

Let us consider a text-writing process in continuous time $t \geqq 0$ such that the *type* counts for nouns, adjectives, verbs and adverbs in Group I are respectively

$$0 \leqq X_1(t) \leqq M_1, \quad 0 \leqq X_2(t) \leqq M_2, \quad 0 \leqq X_3(t) \leqq M_3, \quad 0 \leqq X_4(t) \leqq M_4.$$

Here M_1, M_2, M_3, M_4 are the total number of such words contained in the author's vocabulary; it is assumed for simplicity that a word is written instantaneously at the moment when it is first penned. For the less important words of Group II, we shall not record the type counts; let

$$0 \leqq Y_1(t) \leqq y_1, \ 0 \leqq Y_2(t) \leqq y_2, \ \ldots, \ 0 \leqq Y_6(t) \leqq y_6,$$

be the *token* counts for pronouns, prepositions, conjunctions, auxiliary verbs, articles and interjections respectively; $y_1 + y_2 + \ldots + y_6$ will then be the total number of Group II words written up to time $t \geqq 0$. Let $Y(t)$ denote the total token count up to time $t \geqq 0$; we shall assume that it forms a Poisson process with parameter λ. There is then a probability $0 < p < 1$ that a word belongs to Group I, and $q = 1 - p$ that it belongs to Group II.

We now postulate the following continuous time Markov chain model for the process. Precisely as in Brainerd [1], it is assumed that the token counts for each of the Group II subgroups are distributed with the respective probabilities $\beta_1, \ldots, \beta_6$; $\sum_{k=1}^{6} \beta_k = 1$. More exactly, if in the interval $(t, t + \delta t)$ a word from subgroup k $(k = 1, \ldots, 6)$ of Group II occurs, then

$$\begin{aligned}(2.1)\quad & \Pr\{Y(t + \delta t) = n + 1;\ Y_i(t + \delta t) = y_i + \delta_{ik};\ X_j(t + \delta t) = x_j \mid Y(t) = n; \\ & Y_i(t) = y_i;\ X_j(t) = x_j\} = \lambda q \beta_k \,\delta t + o(\delta t) \quad (i = 1, \ldots, 6;\ j = 1, \ldots, 4),\end{aligned}$$

where, from now on, δ_{ik} (and later δ_{jl}) is the Kronecker delta.

For words in Group I, however, it is assumed that the probability that a new type from subgroup l $(l = 1, \ldots, 4)$ occurs, will be proportional to the number $M_l - x_l$ of unused words in this subgroup, so that

$$\begin{aligned}(2.2)\quad & \Pr\{Y(t + \delta t) = n + 1;\ Y_i(t + \delta t) = y_i;\ X_j(t + \delta t) = x_j + \delta_{jl} \mid Y(t) = n; \\ & Y_i(t) = y_i;\ X_j(t) = x_j\} = \lambda p\{\alpha_l(M_l - x_l)\,\delta t + o(\delta t)\} \\ & (i = 1, \ldots, 6;\ j = 1, \ldots, 4),\end{aligned}$$

where the $\alpha_l > 0$ are intensity parameters for the occurrence of unused subgroup l words. The probability that no new type occurs in Group I for a further token from this group is clearly

$$\text{(2.3)}\quad \Pr\{Y(t+\delta t) = n+1;\ Y_i(t+\delta t) = y_i;\ X_j(t+\delta t) = x_j \mid Y(t) = n;$$
$$Y_i(t) = y_i;\ X_j(t) = x_j\} = \lambda p\{1 - \sum_{l=1}^{4} \alpha_l (M_l - x_l)\}\,\delta t + o(\delta t)$$
$$(i = 1, \ldots, 6;\ j = 1, \ldots, 4)\,.$$

The process described is a continuous time Markov chain. The forward Kolmogorov equations for its probabilities $P(n; y_i; x_j; t) = \Pr\{Y(t) = n;\ Y_i(t) = y_i;$ $X_j(t) = x_j \mid Y(0) = 1;\ Y_i(0);\ X_j(0)\}$ $(i = 1, \ldots, 6;\ j = 1, \ldots, 4)$ are readily found to be

$$\text{(2.4)}\quad \frac{\mathrm{d}P(n; y_i; x_j; t)}{\mathrm{d}t} = -\lambda P + \lambda q \sum_{k=1}^{6} \beta_k P(n-1; y_i - \delta_{ik}; x_j; t) +$$
$$+ \lambda p \sum_{l=1}^{4} \alpha_l (M_l - x_l + 1)\, P(n-1; y_i; x_j - \delta_{jl}; t) +$$
$$+ \lambda p\{1 - \sum_{l=1}^{4} \alpha_l (M_l - x_l)\}\, P(n-1; y_i; x_j; t)$$
$$(i = 1, \ldots, 6;\ j = 1, \ldots, 4)\,,$$

for appropriate values of $0 \leqq x_j \leqq n - \sum_{i=1}^{6} y_i$, $\sum_{i=1}^{6} y_i \leqq n$. If we denote the joint generating function of the process by

$$\Phi(u; w_i; v_j; t) = \sum_{n=1}^{\infty} u^n \sum_{y_i} w_1^{y_1} \ldots w_6^{y_6} \sum_{x_j} v_1^{x_1} \ldots v_4^{x_4}\, P(n; y_i; x_j; t)$$
$$0 \leqq |u|,\ |w_i|,\ |v_j| \leqq 1\,,$$

then this satisfies the first order partial differential equation

$$\text{(2.5)}\quad \frac{\partial \Phi}{\partial t} = -\lambda \Phi + \lambda\{u(q \sum_{i=1}^{6} \beta_i w_i + p[1 - \sum_{j=1}^{4} \alpha_j M_j (1 - v_j)])\}\, \Phi +$$
$$+ \lambda p u \left\{\sum_{j=1}^{4} \alpha_j v_j (1 - v_j) \frac{\partial \Phi}{\partial v_j}\right\}.$$

We shall set the initial condition as

$$\text{(2.6)}\quad \Phi(u; w_i; v_j; 0) = u(q \sum_{i=1}^{6} \beta_i w_i + p \sum_{j=1}^{4} \alpha_j' M_j v_j)\,,$$

where $\alpha'_j = \alpha_j / \sum_{j=1}^{4} \alpha_j M_j$, so that $p\alpha'_j M_j$ is the probability that the first word belongs to subgroup j of Group I and $q\beta_i$ that it belongs to subgroup i of Group II. We now proceed to find the solution of (2.5).

3. EXPLICIT SOLUTION OF THE P.G.F.

Using the standard Lagrange method for solving (2.5), we see that the auxiliary equations are

$$\text{(3.1)} \qquad \frac{\mathrm{d}t}{1} = -\frac{\mathrm{d}v_j}{\lambda p u \alpha_j v_j(1-v_j)} = \frac{\mathrm{d}w_i}{0} = \frac{\mathrm{d}u}{0} =$$

$$= \frac{\mathrm{d}\Phi}{\lambda\Phi\{u[q\sum_i \beta_i w_i + p(1-\sum_j \alpha_j M_j(1-v_j))] - 1\}}\,.$$

From these, we obtain

$$\text{(3.2)} \qquad u = a\,, \quad w_i = b_i \quad (i = 1, \ldots, 6)\,,$$

and

$$\text{(3.3)} \qquad \left(\frac{v_j - 1}{v_j}\right) \exp(-\lambda p u \alpha_j t) = c_j \quad (j = 1, \ldots, 4)\,,$$

where a, b_i, c_j denote constants.

Finally

$$\text{(3.4)} \qquad \frac{\mathrm{d}\Phi}{\Phi} = \lambda\,\mathrm{d}t\{u[q\sum_i \beta_i w_i + p(1-\sum_j \alpha_j M_j(1-v_j))] - 1\} =$$

$$= \lambda\,\mathrm{d}t\left\{u\left[q\sum_i \beta_i w_i + p\left(1 - \sum_j \alpha_j M_j \frac{c_j \exp(\lambda p u \alpha_j t)}{1 - c_j \exp(\lambda p u \alpha_j t)}\right)\right] - 1\right\},$$

as a result of (3.3). Integrating this equation, with u, w_i held constant because of (3.2), we find

$$\text{(3.5)} \qquad \Phi \exp[-\lambda t(uq\sum_i \beta_i w_i + up - 1)] \prod_j [1 - c_j \exp(\lambda p u \alpha_j t)]\, M_j = d\,,$$

or

$$\text{(3.6)} \qquad \Phi \exp[-\lambda t(uq\sum_i \beta_i w_i + up - 1)] \prod_j v_j^{-M_j} = d\,,$$

where d is also a constant.

Thus we can write

(3.7)

$$\Phi \exp\left[-\lambda t(uq\sum_i \beta_i w_i + up - 1)\right] \prod_j v_j^{-M_j} = \Psi\left[u; w_i; \frac{v_j - 1}{v_j}\exp(-\lambda pu\alpha_j t)\right],$$

expressing Φ in terms of an arbitrary function Ψ; here Φ is subject to the initial condition (2.6). It follows from (3.7) that for $t = 0$,

$$u(q\sum_i \beta_i w_i + p\sum_j \alpha_j' M_j v_j)\prod_j v_j^{-M_j} = \Psi\left(u; w_i; \frac{v_j - 1}{v_j}\right).$$

Setting $z_j = (v_j - 1)/v_j$, so that $v_j = (1 - z_j)^{-1}$, the function Ψ is defined as

(3.8)
$$\Psi(u; w_i; z_j) = u\left(q\sum_i \beta_i w_i + p\sum_j \frac{\alpha_j' M_j}{1 - z_j}\right)\prod_j (1 - z_j)^{M_j}.$$

We finally find from (3.7) that

(3.9)
$$\Phi(u; w_i; v_j; t) =$$

$$= \exp\left[\lambda t(uq\sum_i \beta_i w_i + up - 1)\right]\prod_j v_j^{M_j}\Psi\left[u; w_i; \frac{v_j - 1}{v_j}\exp(-\lambda pu\alpha_j t)\right] =$$

$$= \exp\left[\lambda t(uq\sum_i \beta_i w_i + up - 1)\right] \times$$

$$\times u\left[q\sum_i \beta_i w_i + p\sum_j \frac{\alpha_j' M_j v_j}{\exp(-\lambda pu\alpha_j t) + v_j[1 - \exp(-\lambda pu\alpha_j t)]}\right] \times$$

$$\times \prod_j \{\exp(-\lambda pu\alpha_j t) + v_j[1 - \exp(-\lambda pu\alpha_j t)]\}^{M_j} =$$

$$= \exp\left[\lambda t(uq\sum_i \beta_i w_i + up - 1)\right] \times$$

$$\times u\{q\sum_i \beta_i w_i \prod_j \{\exp(-\lambda pu\alpha_j t) + v_j[1 - \exp(-\lambda pu\alpha_j t)]\}^{M_j} +$$

$$+ p\sum_j \alpha_j' M_j v_j \{\exp(-\lambda pu\alpha_j t) + v_j[1 - \exp(-\lambda pu\alpha_j t)]\}^{M_j - 1} \times$$

$$\times \prod_{l \neq j}\{\exp(-\lambda pu\alpha_l t) + v_l[1 - \exp(-\lambda pu\alpha_l t)]\}^{M_l}\} =$$

$$= \sum_{n=1}^{\infty} u^n G_n(w_i; v_j; t).$$

This provides a fully explicit expression for the p.g.f. Φ, from which we can evaluate relevant mean type counts.

4. EXPECTATIONS OF TYPE COUNTS

The quantities of greatest interest in studies of an author's vocabulary are the expectations of the Group I word type counts. We can readily evaluate these for any given token count $Y(t) = n$ as

$$(4.1) \qquad E\{X_j(t) \mid Y(t) = n\} = \frac{\partial G_n(1;\, 1;\, t)}{\partial v_j} \{G_n(1; 1; t)\}^{-1} =$$

$$= M_j\{1 - (1 - p\alpha_j')(1 - p\alpha_j)^{n-1}\} \quad (j = 1, \ldots, 4).$$

We see that as the number of tokens $Y(t) = n$ increases, the expectation of types approaches geometrically the total number M_j of words in subgroup j. It is also easily seen that the Group II word token counts for a given $Y(t) = n$ are

$$(4.2) \qquad E\{Y_i(t) \mid Y(t) = n\} = \frac{\partial G_n(1;\, 1;\, t)}{\partial w_i} \{G_n(1; 1; t)\}^{-1} = nq\beta_i\,.$$

The effect of subdividing Group I words into their 4 subgroups is to modify the size of the expected type count. We note that if Group I formed a single undivided pool of $M = \sum_{j=1}^{4} M_j$ types from which the type count at $t \geqq 0$ was $X(t)$, then

$$(4.3) \qquad E\{X(t) \mid Y(t) = n\} = M\{1 - (1 - pM^{-1})(1 - p\alpha)^{n-1}\},$$

where α is the single parameter governing the occurrence of new words in the text; it would be reasonable to assume that $\alpha M = \sum_{j=1}^{4} \alpha_j M_j$.

Let us now consider the sum of the expectations of the 4 Group I subgroups

$$(4.4) \qquad \sum_{j=1}^{4} M_j\{1 - (1 - p\alpha_j')(1 - p\alpha_j)^{n-1}\} = M - \sum_{j=1}^{4} M_j(1 - p\alpha_j')(1 - p\alpha_j)^{n-1}\,.$$

Since $\alpha_j' = \alpha_j / \sum_{j=1}^{4} \alpha_j M_j$, then $\alpha_j' M_j = \alpha_j M_j / \sum_{j=1}^{4} \alpha_j M_j$ so that (4.4) is

$$M - \sum_{j=1}^{4} M_j \left(1 - \frac{p\alpha_j}{\sum_j \alpha_j M_j}\right)(1 - p\alpha_j)^{n-1}\,.$$

The type count will be decreased or increased depending on whether

$$(4.5) \qquad \sum_{j=1}^{4} M_j \left(1 - \frac{p\alpha_j}{\alpha M}\right)(1 - p\alpha_j)^{n-1} < M(1 - pM^{-1})(1 - p\alpha)^{n-1}\,,$$

or the reverse inequality holds. The inequality (4.5) holds if

$$\sum_{j=1}^{4} M_j \left(1 - \frac{p\alpha_j}{\alpha M}\right) \{(1 - p\alpha_j)^{n-1} - (1 - p\alpha)^{n-1}\} < 0\,,$$

$$(4.6) \qquad \sum_{j=1}^{4} M_j \left(1 - \frac{p\alpha_j}{\alpha M}\right) p(\alpha - \alpha_j) \sum_{r=0}^{n-2} (1 - p\alpha_j)^{n-2-r} (1 - p\alpha)^r < 0\,,$$

$$\sum_{j=1}^{4} M_j p(\alpha - \alpha_j)\, h_j < 0\,.$$

Here, we have written for the set of positive weights h_j,

$$h_j = \left(1 - \frac{p\alpha_j}{\alpha M}\right) \sum_{r=0}^{n-2} (1 - p\alpha_j)^{n-2-r} (1 - p\alpha)^r \quad (j = 1, \ldots, 4)\,.$$

In some numerical studies, the details of which are given in Gani (1975), it is shown that the effect of subdivision into subgroups is to reduce the size of the expected type count, so that the inequality (4.6) effectively holds.

The variances of $X_j(t)$ given $Y(t)$ can also be obtained; it is readily seen that these are zero at $t = 0$ and also as $t \to \infty$. If Group I is treated as a single word pool, with $\alpha = M^{-1}$, the variance of $X(t)$ is a maximum for values of the taken count around $n = \frac{5}{4}M$. The value of the standard deviation is, however, relatively small compared with the expectation itself, and can be neglected for all practical purposes.

The model considered is a fairly simple but realistic one, except for the assumption of independence of individual words, since these are usually grouped in linguistic phrases. It would be interesting to study it numerically in greater detail and possibly carry out some Monte Carlo simulations. It would also prove useful to refine the model in order to achieve greater accord with available data; in studies of Shakespeare's vocabulary the numerical results obtained, while improving on previous ones, are still not adequate and leave many unanswered questions.

8

REFERENCES

[1] B. BRAINERD: On the relation between types and tokens in literary text. J. Appl. Prob. *9* (1972), 507—518.
[2] J. GANI: Stochastic models for type counts in a literary text. In: Perspectives in Probability and Statistics, Papers in Honour of M. S. Bartlett. Academic Press, London 1975, 313—323.
[3] C. B. WILLIAMS: Style and Vocabulary: Numerical Studies. Griffin, London 1970.
[4] G. D. YULE: The Statistical Study cf Literary Vocabulary. Cambridge U.P. 1944.

UNIVERSITY OF SHEFFIELD

DEPARTMENT OF PROBABILITY & STATISTICS

and

CSIRO

DIVISION OF MATHEMATICS AND STATISTICS

ON THE REALIZATION OF STOCHASTIC PROCESSES BY PROBABILITY DISTRIBUTIONS IN FUNCTION SPACES

PETER GÄNSSLER

BOCHUM

1. INTRODUCTION

Denoting by $\mathbf{R}[\mathbf{N}]$ the set of real [natural] numbers and by I the unit interval $[0, 1]$ let $x = (x_t)_{t\in I}$ be an $\mathbf{R}$-valued stochastic process which may be considered as a p-distribution (= probability distribution) Q_x defined on the σ-field $\mathscr{B}_I$ spanned by the field $\mathscr{Z}_I$ of cylindersets in $\mathbf{R}^I$, Denoting by $Q_{x_{t_1}\dots x_{t_n}}$, $t_i \in I$, $n \in \mathbf{N}$, the finite dimensional distributions pertaining to x, the question about additional conditions on the $Q_{x_{t_1}\dots x_{t_n}}$ ensuring the existence of a p-distributions μ on the trace σ-field $\mathscr{B}_I^X = \mathscr{B}_I \cap X$ of nice function spaces $X \subset \mathbf{R}^I$, with finite dimensional marginal distributions $\mu_{\{t_1,\dots,t_n\}}$ coinciding with $Q_{x_{t_1}\dots x_{t_n}}$, has a long history and one knows about many sufficient conditions under which this question has an affirmative answer; in that case we will say that x *can be realized in* $(X, \mathscr{B}_I^X)$.

Early in 1937, Kolmogorov showed (cf. [6] VII. Theorem 3.1) that the condition

$$E[|x_{t'} - x_t|^\gamma] \leqq \beta|t' - t|^\alpha, \quad t', t \in I, \tag{1.1}$$

where β and γ are strictly positive constants and α is a constant greater than 1, is sufficient for x to be realizable in the space $(C, \mathscr{B}_I^C)$; here $C = C(I)$ is the space of all continuous functions on I and (cf. [6] VII, Theorem 2.1) $\mathscr{B}_I^C$ coincides with $\mathscr{B}(C)$, the Borel σ-field in C pertaining to the topology of uniform convergence on I.

Concerning the subspace C of $\mathbf{R}^I$ similar sufficient conditions can be found in [1], Theorem 12.4, as well as (for separable processes) in [5], III.5; Billingsley showed that the condition

$$P\{|x_{t'} - x_t| \geqq \varepsilon\} \leqq \varepsilon^{-\gamma}|F(t') - F(t)|^\alpha, \quad t', t \in I, \tag{1.2}$$

with $\gamma \geqq 0$, $\alpha > 1$, $\varepsilon > 0$ arbitrary, and $F : I \to \mathbf{R}$ monotone non-decreasing and continuous, is also sufficient for x to be realizable in $(C, \mathscr{B}(C))$.

Another important example of a nice space $X \subset \mathbf{R}^I$ is the so-called Skorohod-space $X = D = D(I)$ of all $f \in \mathbf{R}^I$ which are right-continuous and have left-hand limits (cf. [1], Chapter 3). Also in that case we know about sufficient conditions for a stochastic process $x = (x_t)_{t \in I}$ to be realizable in $(D, \mathscr{B}_I^D)$, where (cf. [1], Theorem 14.5) $\mathscr{B}_I^D$ is the same as $\mathscr{B}(D)$, the Borel σ-field in D pertaining to the Skorohod topology. Billingsley ([1], Theorem 15.7) showed that the following two conditions

(1.3) $$P\{|x_{t'} - x_t| \geqq \varepsilon\,, \; |x_{t''} - x_t| \geqq \varepsilon\} \leqq \varepsilon^{-\gamma}(F(t'') - F(t'))^{\alpha}\,, \quad 0 \leqq t' \leqq t \leqq t'' \leqq 1\,,$$

with $\gamma \geqq 0$, $\alpha > 1$, $\varepsilon > 0$ arbitrary, $F : I \to \mathbf{R}$ monotone non-decreasing and continuous, and

(1.4) For any $\varepsilon > 0$ and $t \in [0, 1[$:

$$\lim_{h \to 0} Q_{x_t x_{t+h}}(\{(\beta_1, \beta_2) : |\beta_1 - \beta_2| \geqq \varepsilon\}) = 0\,,$$

are sufficient for x to be realizable in $(D, \mathscr{B}(D))$.

Now, if one asks for *necessary and sufficient* conditions for x to be realizable in $(C, \mathscr{B}(C))$ or in $(D, \mathscr{B}(D))$, the only answer known to the author is a criterion presented in ([4], Theorem 3) which was first established by Doob in [2] and which is concerned with $X = C$. The proof given in [4] is along the lines of Kolmogorov's construction of p-measures in $\mathbf{R}^I$.

In the same way the present paper gives a general criterion for the realizability of stochastic processes $x = (x_t)_{t \in I}$ as p-distributions on certain function spaces $X \subset \mathbf{R}^I$ (Theorem 2.1) which yields the Doob-Mann-Criterion and a similar criterion for $X = D$ (based on a theorem of Heidemann [3] (Theorem 3.1)) as special cases. It follows in a rather elementary way that the necessary and sufficient conditions found for both cases are implied by (1.2) and (1.3)–(1.4), respectively (and not vice versa, as it can be easily shown by examples).

Furthermore this provides us with a new proof of the results cited above. It should be emphasized that this is achieved without using the theory of weak convergence of p-measures (Prohorov's theorem). Therefore we think that the presentation given here may also serve to be used within an elementary course on probability theory.

2. THE MAIN RESULT

With the notation introduced in 1. let μ be a p-measure on $\mathscr{B}_I^X$ with finite dimensional marginal distributions $\mu_{\{t_1, \ldots, t_n\}}$ defined by

(2.1) $$\mu_{\{t_1, \ldots, t_n\}}(B) = \mu(\pi^{-1}_{\{t_1, \ldots, t_n\}}(B) \cap X)\,, \quad B \in \mathscr{B}_n\,, \quad t_1 < t_2 < \ldots < t_n\,, \quad t_i \in I\,, \quad n \in \mathbf{N}\,,$$

where $\mathscr{B}_n$ denotes the Borel σ-field in $\mathbf{R}^n$, and $\pi_{\{t_1,\ldots,t_n\}} : \mathbf{R}^I \to \mathbf{R}^n$ is defined by $\pi_{\{t_1,\ldots,t_n\}}(f) = (f(t_1), \ldots, f(t_n))$ for $f \in \mathbf{R}^I$.

It is easy to see that every μ on $\mathscr{B}_I^X$ is uniquely determined by its finite dimensional marginal distributions.

DEFINITION 2.1. If $x = (x_t)_{t\in I}$ is an $\mathbf{R}$-valued stochastic process with finite dimensional distributions $Q_{x_{t_1}\ldots x_{t_n}}$, $t_i \in I$, $n \in \mathbf{N}$, we say that x *can be realized in* $(X, \mathscr{B}_I^X)$ with $X \subset \mathbf{R}^I$, iff there exists a p-measure μ on $\mathscr{B}_I^X$ so that the finite dimensional marginal distributions $\mu_{\{t_1,\ldots,t_n\}}$ coincide with $Q_{x_{t_1},\ldots,x_{t_n}}$, $t_i \in I$, $n \in \mathbf{N}$.

In the following we shall confine ourselves to function spaces $X \subset \mathbf{R}^I$ fulfilling the following regularity conditions. Examples where these conditions are fulfilled will be given in section 3.

REGULARITY CONDITIONS. Let $x = (x_t)_{t\in I}$ be a $\mathbf{R}$-valued stochastic process on some p-space $(\Omega, \mathscr{F}, P)$ and for $T \subset I$ let $\mathscr{F}_T$ denote the σ-field spanned by $\{x_t : t \in T\}$. Let $X \subset \mathbf{R}^I$ be a function space with the following properties (2.1)–(2.3):

(2.1) There exists an "X-*function*" w^X, i.e. a mapping $(f, \delta, T) \to w_f^X(\delta, T) \in \mathbf{R}_+$, defined for $f \in \mathbf{R}^I$, $\delta > 0$, $T \subset I$, in such a way that

(a) $w_f^X(\delta, T)$ is monotone decreasing in δ (for all $f \in \mathbf{R}^I$, $T \subset I$),
$w_f^X(\delta, S_1) \leqq w_f^X(\delta, S_2)$ if $S_1 \subset S_2 \subset I$, S_2 finite,
$w_f^X(\delta, S) \leqq w_f^X(\delta, I)$ if $S \subset I$, S finite,
$w_f^X(\delta, S_n) \uparrow w_f^X(\delta, T)$ for finite $S_n \uparrow T \subset I$;

(b) $w_f^X(\delta, T) = w_g^X(\delta, T)$ whenever $f, g \in \mathbf{R}^I$ coincide on $T \subset I$.
Furthermore, for $S = \{t_1, t_2, \ldots, t_n\}$ (with $t_1 < t_2 < \ldots < t_n$, $t_i \in I$) there exists a continuous function $\Phi_{\delta,S}^X : \mathbf{R}^n \to \mathbf{R}$ such that for all $f \in \mathbf{R}^I$, $\delta > 0$: : $w_f^X(\delta, S) = \Phi_{\delta,S}^X(f(t_1), \ldots, f(t_n))$;

(c) $f \in X \Leftrightarrow \lim_{\delta\to 0} w_f^X(\delta, I) = 0$.

(2.2) For any $t_1 < t_2 < \ldots < t_m$ $(t_i \in I,\ m \in \mathbf{N})$ and any $(a_1, a_2, \ldots, a_m) \in \mathbf{R}^m$ there exists $f \in X$ with $f(t_i) = a_i$, $i = 1, 2, \ldots, m$.

(2.3) If T is a dense subset of I and if $\psi \in \mathbf{R}^T$ fulfils $\lim_{\delta\to 0} w_\psi^X(\delta, T) = 0$, then there exists $\tilde{\psi} \in X$ coinciding with ψ on T (i.e. $rest_T\, \tilde{\psi} = \psi$). (We remark that $w_\psi^X(\delta, T)$ is defined by $w_\psi^X(\delta, T) = w_f^X(\delta, T)$ for some $f \in \mathbf{R}^I$ with $rest_T\, f = \psi$, this definition being unique according to (2.1) (b).)

DEFINITION 2.2. Let $x = (x_t)_{t\in I}$ be a $\mathbf{R}$-valued stochastic process on some p-space $(\Omega, \mathscr{F}, P)$ and $X \subset \mathbf{R}^I$ be a function space with the properties (2.1)–(2.3). We shall say that x *fulfils Condition* (X) iff

(2.4) For any $\varepsilon, \eta > 0$ there exists $\delta = \delta(\varepsilon, \eta) > 0$ such that for all finite $S \subset I$
$P(\{\omega \in \Omega : w^X_{x(\omega)}(\delta, S) \leqq \varepsilon\}) \geqq 1 - \eta$.

We remark that according to (2.1) (b) property (2.4) is a condition on the finite dimensional distributions of the stochastic process x; ($\omega \mapsto w^X_{x(\omega)}(\delta, S)$ is $\mathscr{F}_S$-measurable for all finite $S \subset I$). Now we can state our main result:

THEOREM 2.1. Let $x = (x_t)_{t \in I}$ be an **R**-valued stochastic process on some p-space $(\Omega, \mathscr{F}, P)$ and $X \subset \mathbf{R}^I$ be a function space with the properties (2.1)–(2.3). Then the following two assertions are equivalent:

(i) x can be realized in $(X, \mathscr{B}^X_I)$,

(ii) x fulfils Condition (X), i.e. (2.4) holds.

Proof. (ii) ⇒ (i): According to Lemma 4.1 (i) we obtain that

$$\mu_0(\{f \in X : (f(t_1), \ldots, f(t_n)) \in B\}) = Q_{x_{t_1} \ldots x_{t_n}}(B), \quad B \in \mathscr{B}_n, \quad \{t_1, \ldots, t_n\} \subset I,$$

defines a p-content on $\mathscr{Z}_I \cap X$. We shall show:

(a) μ_0 is σ-additive on $\mathscr{Z}_I \cap X$, i.e. for any $A_j \in \mathscr{Z}_I \cap X$, $j \in \mathbf{N}$, with $A_j \downarrow \emptyset$ we have $\lim\limits_{j \to \infty} \mu_0(A_j) = 0$.

This will be proved by showing that

(b) whenever $A_j \in \mathscr{Z}_I \cap X$, $j \in \mathbf{N}$, with $A_j \downarrow$ and $\lim\limits_{j \to \infty} \mu_0(A_j) =: L > 0$, then $\bigcap\limits_{j \in \mathbf{N}} A_j \neq \emptyset$.

According to Lemma 4.1 (ii) we may assume without loss of generality that $A_j = \pi^{-1}_{\{t_1, \ldots, t_{n_j}\}}(K_j) \cap X$, K_j compact in $\mathbf{R}^{n_j}$, $n_1 < n_2 < \ldots$, with $T := \bigcup\limits_{j \in \mathbf{N}} \{t_1, \ldots, t_{n_j}\}$ dense in I. Since by assumption x fulfils Condition (X), for any $\varepsilon = \eta = L2^{-m}$, $m \in \mathbf{N}$, there exists $\delta_m = \delta(L2^{-m}, L2^{-m}) > 0$ such that for all $j \in \mathbf{N}$

$$\mu_0(B^{(m)}_j) = P(\{w^X_{x(\cdot)}(\delta_m, \{t_1, \ldots, t_{n_j}\}) \leqq L2^{-m}\}) \geqq 1 - L2^{-m},$$

where

$$B^{(m)}_j = \{f \in X : w^X_f(\delta_m, \{t_1, \ldots, t_{n_j}\}) \leqq L2^{-m}\} \in \mathscr{Z}_I \cap X$$

(cf. (2.1) (b)). By (2.1) (a) we may assume w.l.o.g. that $\delta_m \downarrow 0$.

If we put $A^{(m)}_j = A^{(m-1)}_j \cap B^{(m)}_j$, $j \in \mathbf{N}$, $m \geqq 2$, $A^{(1)}_j = A_j$, $j \in \mathbf{N}$, then (since $A_j \downarrow$ and $B^{(m)}_j \downarrow$ for all $m \in \mathbf{N}$) $A^{(m)}_j \downarrow$ for all $m \in \mathbf{N}$ and $A^{(m)}_j = \pi^{-1}_{\{t_1, \ldots, t_{n_j}\}}(K^{(m)}_j)$ with $K^{(m)}_j \subset \mathbf{R}^{n_j}$; here $K^{(m)}_j$ turns out to be compact for all $j \in \mathbf{N}$, since $K^{(m)}_j$ occurs as an intersection of a closed set (cf. (2.1) (b)) with $K^{(m-1)}_j$ (where $K^{(1)}_1 = K_1$ is compact).

Next we claim that for every $m \in \mathbf{N}$

(c) $$\mu_0(A^{(m)}_j) \geqq L/2 + L2^{-m} \quad \text{for all} \quad j \in \mathbf{N}.$$

(b) implies that $\mu_0(A_j^{(1)}) = \mu_0(A_j) \geqq L$, i.e. (c) holds for $m = 1$; assuming $\mu_0(A_j^{(m-1)}) \geqq L/2 + L2^{-(m-1)}$ for all $j \in \mathbf{N}$ it follows from $\mu_0(B_j^{(m)}) \geqq 1 - L2^{-m}$, $j \in \mathbf{N}$, that $\mu_0(A_j^{(m)}) = \mu_0(A_j^{(m-1)} \cap B_j^{(m)}) \geqq L/2 + L2^{-m}$; hence (c) holds for all $m \in \mathbf{N}$.

Now, considering the diagonal sequence $(A_p^{(p)})_{p\in\mathbf{N}}$, we have $A_p^{(p)} \subset A_p^{(p-1)} \subset \subset A_{p-1}^{(p-1)}$, i.e. $A_p^{(p)} \downarrow$; furthermore, $A_p^{(p)} = \pi_{\{t_1,\ldots,t_{n_p}\}}(K_p^{(p)}) \cap X$ with $K_p^{(p)}$ compact in $\mathbf{R}^{n_p}$ and $\lim_{p\to\infty} \mu_0(A_p^{(p)}) \geqq L/2 > 0$.

Hence there exists for any $p \in \mathbf{N}$ $f_p \in A_p^{(p)}$ and (since $\{f_p(t) : p \in \mathbf{N}\}$ is bounded for all $t \in T$) using the diagonal procedure we find a subsequence $(\bar{f}_p)_{p\in\mathbf{N}}$ with $\lim_{p\to\infty} \bar{f}_p(t_i) = \psi(t_i)$ for all $t_i \in T$; as $K_j^{(j)}$ is closed we have $(\psi(t_1), \ldots, \psi(t_{n_j})) \in K_j^{(j)}$ and therefore $A_j \supset A_j^{(j)} \supset \{f \in X : (f(t_1), \ldots, f(t_{n_j})) = (\psi(t_1), \ldots, \psi(t_{n_j}))\} = C_j$ for all $j \in \mathbf{N}$. Therefore it suffices to show that $\bigcap_{j\in\mathbf{N}} C_j \neq \emptyset$, i.e. that

(d) there exists $\tilde{\psi} \in X$ such that $rest_T\, \tilde{\psi} = \psi$.

According to (2.3) (d) holds if we show that

(e) $$\lim_{\delta\to 0} w_\psi^X(\delta, T) = 0 .$$

But (e) follows immediately from the definition of ψ: Since for all $p \in \mathbf{N}$ and $m \leqq p$ $\bar{f}_r \in A_p^{(p)} \subset A_p^{(m)}$ and therefore $\bar{f}_r \in B_p^{(m)}$ for all but a finite number of $r \in \mathbf{N}$, we obtain according to (2.1) (b) that $w_\psi^X(\delta_m, \{t_1, \ldots, t_{n_p}\}) \leqq L2^{-m}$ for all m and p with $m \leqq p$. Hence it follows from (2.1) (a) that $w_\psi^X(\delta_m, T) \leqq L2^{-m}$ for all $m \in \mathbf{N}$, i.e. $\lim_{\delta\to 0} w_\psi^X(\delta, T) = 0$.

This proves (b), i.e. μ_0 is σ-additive on $\mathscr{Z}_I \cap X$ and therefore μ_0 can be uniquely extended to a p-measure μ on $\sigma(\mathscr{Z}_I \cap X) = \sigma(\mathscr{Z}_I) \cap X = \mathscr{B}_I \cap X = \mathscr{B}_I^X$, where by the construction $\mu_{\{t_1,\ldots,t_n\}} = Q_{x_{t_1},\ldots,x_{t_n}}$, $t_i \in I$, $n \in \mathbf{N}$; hence x can be realized in $(X, \mathscr{B}_I^X)$.

(i) ⇒ (ii): For $\varepsilon, \delta > 0$ let $S(\varepsilon, \delta) := \{f \in X : w_f^X(\delta, I) > \varepsilon\}$. (2.1) (a) and (2.1) (c) imply that for $\delta_n \downarrow 0$ $S(\varepsilon, \delta_n) \downarrow \emptyset$ and therefore $\lim_{n\to\infty} \mu_*(S(\varepsilon, \delta_n)) = 0$, where μ_* denotes the inner measure pertaining to μ on $\mathscr{B}_I^X$. Therefore it follows that for any $\varepsilon, \eta > 0$ there exists $\delta = \delta(\varepsilon, \eta) > 0$ such that $\mu_*(\{f \in X : w_f^X(\delta, I) > \varepsilon\}) \leqq \eta$; hence for any finite $S \subset I$ it follows from (2.1) (a) that $\mu(\{f \in X : w_f^X(\delta, S) > \varepsilon\}) = P(\{\omega \in \Omega : : w_{x(\omega)}^X(\delta, S) > \varepsilon\}) \leqq \eta$, i.e. (2.4) holds. □

3. APPLICATIONS

As mentioned in the introduction we will face our attention to the function spaces $C = C(I)$ of all continuous functions on I and to the Skorohod space $D = D(I)$. Before stating the results for D we consider the third space $\hat{D} = \hat{D}(I)$ considered in [3] which is defined in the following way.

DEFINITION 3.1. $\hat{D} = \hat{D}(I) := \{f \in \mathbf{R}^I : f \text{ fulfils } (3.1)-(3.3)\}$, with

$$\text{For all} \quad t \in [0, 1[\; f(t+0) = \lim_{s \downarrow t} f(s) \quad \text{exists};\tag{3.1}$$

$$\text{For all} \quad t \in]0, 1]\; f(t-0) = \lim_{s \uparrow t} f(s) \quad \text{exists};\tag{3.2}$$

$$\text{For all} \quad t \in]0, 1[\; f(t) = f(t+0) \quad \text{or} \quad f(t) = f(t-0).\tag{3.3}$$

We remind that $D = D(I) = \{f \in \mathbf{R}^I : f \text{ fulfils } (3.1), (3.2) \text{ and } (3.4)\}$, with

$$\text{For all} \quad t \in [0, 1[\; f(t) = f(t+0).\tag{3.4}$$

Furthermore, for the function spaces $X = C$ and $X = \hat{D}$ we will consider the following X-functions w^C and $w^{\hat{D}}$, respectively:

$$w_f^C(\delta, T) = \sup_{\substack{|t-t'|<\delta \\ t,t' \in T}} |f(t) - f(t')|, \quad \text{for} \quad f \in \mathbf{R}^I, \quad \delta > 0, \quad T \subset I;\tag{3.5}$$

$$w_f^{\hat{D}}(\delta, T) = \sup_{\substack{t-\delta<t_1 \leq t < t_2 < t+\delta \\ t_1, t, t_2 \in T}} \min\left(|f(t_1) - f(t)|, \; |f(t) - f(t_2)|\right), \quad \text{for} \quad f \in \mathbf{R}^I, \quad \delta > 0, \quad T \subset I,\tag{3.6}$$

(where we make the convention that $\sup \emptyset := 0$).

Now, as to $X = C$, using w^C as defined in (3.5) it follows at once that (2.1)–(2.3) are fulfilled, hence theorem 2.1 yields the Doob-Mann-Criterion ([4], Theorem 3). Furthermore, using Theorem 12.2 in [1] it can be shown in a rather elementary way that (1.2) implies that $x = (x_t)_{t \in I}$ fulfils Condition (C) and is therefore realizable in $(C, \mathscr{B}(C))$.

As to $X = \hat{D}$ using $w^{\hat{D}}$ as defined in (3.6) it follows from Lemma 4.2 and Lemma 4.3 proved in [3] that (2.1)–(2.3) are fulfilled:

(2.1) (a) and (b) are obvious and (2.1) (c) is the content of Lemma 4.3. Also (2.2) holds and (2.3) follows from the following considerations. If T is dense in I and if $\psi \in \mathbf{R}^T$ fulfils $\lim_{\delta \to 0} w_\psi^{\hat{D}}(\delta, T) = 0$ then it follows from Lemma 4.2 that for any $t_0 \in [0, 1[$ (for any $t_0 \in]0, 1]$) there exist the right-hand (left-hand) limits $\lim_{T \ni t \downarrow t_0} \psi(t)$

$(\lim_{T \ni t \uparrow t_0} \psi(t))$. Hence we may define $\tilde{\psi}(t_0) = \lim_{T \ni t \downarrow t_0} \psi(t)$ for $t_0 \in [0, 1[\cap (I \setminus T)$, $\tilde{\psi}(t) = \psi(t)$ for all $t \in T$, and $\tilde{\psi}(1) = \lim_{T \ni t \uparrow 1} \psi(t)$ if $1 \notin T$. It follows that $w_{\tilde{\psi}}^{\hat{D}}(\delta, I) = w_{\tilde{\psi}}^{\hat{D}}(\delta, T)$ for every $\delta > 0$ and therefore $\lim_{\delta \to 0} w_{\tilde{\psi}}^{\hat{D}}(\delta, I) = 0$, i.e. $\tilde{\psi} \in \hat{D}$ according to Lemma 4.3.

As $rest_T\, \tilde{\psi} = \psi$ by the construction, (2.3) is also fulfilled, and Theorem 2.1 yields the following result of Heidemann:

THEOREM 3.1. Let $x = (x_t)_{t \in I}$ be a **R**-valued stochastic process on some p-space $(\Omega, \mathscr{F}, P)$. Then the following two assertions are equivalent:

(i) x can be realized in $(\hat{D}, \mathscr{B}_I^{\hat{D}})$,

(ii) x fulfils Condition $(\hat{D})$, i.e. for any $\varepsilon, \eta > 0$ there exists $\delta = \delta(\varepsilon, \eta) > 0$ such that for all finite $S \subset I$

$$P(\{\omega \in \Omega : w_{x(\omega)}^{\hat{D}}(\delta, S) \leqq \varepsilon\}) \geqq 1 - \eta\,.$$

Furthermore, using Theorem 12.5 in [1] it can be shown again in a rather elementary way that (1.3) implies that $x = (x_t)_{t \in I}$ fulfils Condition $(\hat{D})$ and is therefore realizable in $(\hat{D}, \mathscr{B}_I^{\hat{D}})$. For the next result, concerned with $X = D$, it turns out that (1.4) is essential. Before stating this result the following example due to Heidemann shows that

(3.7) (i) (or (ii)) in Theorem 3.1 does not imply (1.4) and

(3.8) (1.4) does not imply (i) (or (ii)) in Theorem 3.1 .

EXAMPLE 3.1. Let $x = (x_t)_{t \in I}$ be defined on some p-space $(\Omega, \mathscr{F}, P)$ by

$$x_t(\omega) = \begin{cases} 1 & \text{if } \; 0 \leqq t \leqq 2^{-1} \\ 0 & \text{if } \; 2^{-1} < t \leqq 1 \end{cases}, \quad \omega \in \Omega\,, \quad t \in I\,.$$

Then x fulfils Condition $(\hat{D})$ but for $t = 2^{-1}$, $0 < \varepsilon < 1$ and $2^{-1} \geqq h > 0$ we obtain $Q_{x_t x_{t+h}}(\{(\beta_1, \beta_2) : |\beta_1 - \beta_2| \geqq \varepsilon\}) = 1$; this proves (3.7). As to (3.8) let $x = (x_t)_{t \in I}$ be defined by

$$x_t(\omega) = \begin{cases} (-1)^n & \text{if } \; t \in [1 - 2^{-(n-1)}, 1 - 2^{-n}[\,, \\ 0 & \text{if } \; t = 1\,, \end{cases} \quad n \in \mathbf{N}\,, \quad \omega \in \Omega\,, \quad t \in I\,.$$

Then (1.4) holds true but x does not fulfil Condition $(\hat{D})$: Given $\varepsilon_0 \in]0, 2[$, $\eta_0 = 2^{-1}$ and $\delta > 0$ choose $n_0 \in \mathbf{N}$ so that $2^{-n_0} < \delta$; then $P(\{\omega \in \Omega : w_{x(\omega)}^{\hat{D}}(\delta, S) \leqq \varepsilon_0\}) = P(\emptyset) = 0$ for $S_0 = \{1 - 2^{-n_0}, 1 - 2^{-(n_0+1)}, 1 - 2^{-(n_0+2)}\}$, i.e. (ii) of Theorem 3.1 is not fulfilled and this proves (3.8).

Now, on the basis of theorem 3.1 we could prove the following theorem:

THEOREM 3.2. A stochastic process $x = (x_t)_{t \in I}$ can be realized in the Skorohod space $(D, \mathscr{B}_I^D) = (D, \mathscr{B}(D))$ if and only if (1.4) and (i) (or (ii)) of theorem 3.1 are fulfilled.

Proof. Assume that x can be realized in $(D, \mathscr{B}(D))$; as $D \subset \hat{D}$ $\mu(\hat{D} \cap A) := := \mu(A \cap D)$, $A \in \mathscr{B}_I$ defines a p-measure $\hat{\mu}$ on $\mathscr{B}_I^{\hat{D}}$ whose finite dimensional marginal distributions are the same as those pertaining to μ. Hence x can be realized in $(\hat{D}, \mathscr{B}_I^{\hat{D}})$ and therefore (i) (or (ii)) if theorem 3.1 is fulfilled. It remains to show that for any $\varepsilon > 0$ and $t \in [0, 1[$ $\lim_{h \downarrow 0} \mu(\{f \in D : |f(t) - f(t + h)| \geqq \varepsilon\}) = 0$. Let $h_n \downarrow 0$, then by (3.4) $\bigcap_{m \geqq n} \{f \in D : |f(t) - f(t + h_m)| < \varepsilon\} \uparrow D$ and therefore

$$\lim_{n \to \infty} \mu(\bigcap_{m \geqq n} \{f \in D : |f(t) - f(t + h_m)| < \varepsilon\}) = \mu(D) = 1\,,$$

hence

$$\lim_{n \to \infty} \mu(\{f \in D : |f(t) - f(t + h_n)| \geqq \varepsilon\}) = 0\,.$$

To prove the opposite direction, assume x fulfils (1.4) and (i) (or (ii)) of Theorem 3.1. As $X = D$ fulfils (2.2) it follows from Lemma 4.1 (i) that $\mu_0(\pi^{-1}_{\{t_1,\ldots,t_n\}}(B) \cap D) = = Q_{x_{t_1}\ldots x_{t_n}}(B)$, $B \in \mathscr{B}_n$, $t_i \in I$, defines a p-content on $\mathscr{Z}_I \cap D$. It remains to show that μ_0 is σ-additive on $\mathscr{Z}_I \cap D$, i.e. for any $A_j \in \mathscr{Z}_I \cap D$, $j \in \mathbf{N}$, with $A_j \downarrow \emptyset$ we have $\lim_{j \to \infty} \mu_0(A_j) = 0$. As in the proof of Theorem 3.1 let $A_j = = \pi^{-1}_{\{t_1,\ldots,t_{n_j}\}}(B_j) \cap D$, $B_j \in \mathscr{B}_{n_j}$, $n_1 < n_2 < \ldots$, and $T = \bigcup_{j \in \mathbf{N}} \{t_1, \ldots, t_{n_j}\}$ dense in I; furthermore, let $Z_j := \pi^{-1}_{\{t_1,\ldots,t_{n_j}\}}(B_j)$.

Let us assume to the contrary that $\lim_{j \to \infty} \mu_0(A_j) = L > 0$. According to (ii) it follows from theorem 3.1 that there exists a p-measure ν on $\mathscr{B}_I^{\hat{D}}$ such that $\nu_{\{t_1,\ldots,t_n\}} = = Q_{x_{t_1}\ldots x_{t_n}}$ for all $\{t_1, \ldots, t_n\} \subset I$; hence $\mu_0(A_j) = \nu(A_j')$ for $A_j' = Z_j \cap \hat{D}$, $j \in \mathbf{N}$. Let $t_j \in T$, $t_j \neq 1$, and $N_{t_j} := \{f \in \hat{D} : f(t_j) \neq f(t_j + 0)\}$. We shall show below that

$$(+) \qquad N_{t_j} \in \mathscr{B}_I^{\hat{D}}$$

and

$$(++) \qquad \nu(N_{t_j}) = 0\,.$$

Therefore $M_j = \hat{D} \setminus \bigcup_{k=1}^{j} N_{t_k} \in \mathscr{B}_I^{\hat{D}}$ and $\nu(M_j) = 1$, hence $\mu_0(A_j) = \nu(A_j') = = \nu(A_j' \cap M_j)$ for all $j \in \mathbf{N}$ and therefore $\lim_{j \to \infty} \nu(A_j' \cap M_j) = L > 0$. Since ν is σ-additive on $\mathscr{B}_I^{\hat{D}}$ and $A_j' \cap M_j \downarrow$ it follows that there exists $\psi \in \bigcap_{j \in \mathbf{N}} A_j' \cap \bigcap_{j \in \mathbf{N}} M_j$, i.e. $\psi \in \hat{D}$, $\psi(t_j) = \psi(t_j + 0)$ for all $t_j \in T \setminus \{1\}$ and $\psi \in \bigcap_{j \in \mathbf{N}} Z_j$. $\psi \in \hat{D}$ implies by (3.1) that for all $t \in [0, 1]$ $\psi(t + 0)$ exists. Put $\tilde{\psi}(t) = \psi(t + 0)$ for $t \in [0, 1[$ and $\tilde{\psi}(1) = = \psi(1)$; then $\tilde{\psi} \in D$, $\tilde{\psi}(t_j) = \psi(t_j)$ for all $t_j \in T$ and therefore $\tilde{\psi} \in (\bigcap_{j \in \mathbf{N}} Z_j) \cap D = = \bigcap_{j \in \mathbf{N}} A_j$, is a contradiction. It remains to prove (+) and (++). Let $t_j \in T \setminus \{1\}$; by (1.4) for every $\varepsilon > 0$ and $m \in \mathbf{N}$ there exists $h_m(\varepsilon) > 0$ such that $Q_{x_{t_j} x_{t_j + h_m(\varepsilon)}}(\{(\beta_1, \beta_2) : : |\beta_1 - \beta_2| \geqq \varepsilon\}) < 2^{-m}$, where we may assume w.l.o.g. that $h_m(\varepsilon) \downarrow 0$ if $m \to \infty$. Now, $N_{t_j} = \bigcup_{0 < \varepsilon \in \mathbf{Q}} N_{t_j}(\varepsilon)$ with $N_{t_j}(\varepsilon) = \bigcap_{n \in \mathbf{N}} \bigcup_{m \geqq n} \{f \in \hat{D} : |f(t_j) - f(t_j + h_m(\varepsilon))| \geqq \varepsilon\}$. ($\mathbf{Q}$ is the set of rational numbers), hence $(+)$ follows at once and $(++)$ follows from

$$\nu(N_{t_j}(\varepsilon)) = \lim_{n \to \infty} \nu\big(\bigcup_{m \geqq n} \{f \in \mathbf{R}^I : |f(t_j) - f(t_j + h_m(\varepsilon))| \geqq \varepsilon\} \cap \hat{D}\big) \leqq$$

$$\leqq \lim_{n \to \infty} \sum_{m=n}^{\infty} Q_{x_{t_j} x_{t_j + h_m(\varepsilon)}}(\{(\beta_1, \beta_2) : |\beta_1 - \beta_2) \geqq \varepsilon\}) \leqq \lim_{n \to \infty} \sum_{m=n}^{\infty} 2^{-m} = 0 . \qquad \square$$

4. AUXILIARY RESULTS

For the sake of completeness let us collect here some Lemmata which were needed within the proofs of some of the results in the preceeding sections. They are well known or rather easy to prove; hence the proof will be left to the reader.

LEMMA 4.1. Let $X \subset \mathbf{R}^I$ be a function space with the property (2.2) and let $x = (x_t)_{t \in I}$ be an $\mathbf{R}$-valued stochastic process with finite dimensional distributions $Q_{x_{t_1} \dots x_{t_n}}$, $t_i \in I$, $n \in \mathbf{N}$; then

(i) $$\mu_0(\pi^{-1}_{\{t_1, \dots, t_n\}}(B) \cap X) = Q_{x_{t_1} \dots x_{t_n}}(B), \quad B \in \mathscr{B}_n,$$

defines a p-content on $\mathscr{Z}_I \cap X$;

(ii) If $A_j = \pi^{-1}_{\{t_1, \dots, t_{n_j}\}}(B_j) \cap X$, $j \in \mathbf{N}$, is a monotone decreasing sequence of sets in $\mathscr{Z}_I \cap X$ with $B_j \in \mathscr{B}_{n_j}$, $n_1 < n_2 < \dots$, and $\lim_{j \to \infty} \mu_0(A_j) > 0$, then there exist compact $K_j \subset \mathbf{R}^{n_j}$ such that $A'_j = \pi^{-1}_{\{t_1, \dots, t_{n_j}\}}(K_j) \cap X$, $j \in \mathbf{N}$, is monotone decreasing, $A'_j \subset A_j$ for all $j \in \mathbf{N}$ and $\lim_{j \to \infty} \mu_0(A'_j) > 0$.

LEMMA 4.2. Let T be a dense subset of I and $\psi \in \mathbf{R}^T$; then (with $w^{\hat{D}}_{\psi}$ as defined in (3.6)) $\lim_{\delta \to 0} w^{\hat{D}}_{\psi}(\delta, T) = 0$ implies that for any $t \in [0, 1[$ (for any $t \in]0, 1]$) there exist the right-hand (left-hand) limits $\lim_{T \ni s \downarrow t} \psi(s) (\lim_{T \ni s \uparrow t} \psi(s))$.

LEMMA 4.3. $f \in \hat{D}$ if and only if $\lim_{\delta \to 0} w^{\hat{D}}_f(\delta, I) = 0$.

ACKNOWLEDGEMENT. The author wishes to thank Mr. D. Heidemann and Mr. W. Hummitzsch for some critical remarks and for their careful reading of an earlier version of the manuscript.

REFERENCES

[1] P. Billingsley: Convergence of probability measures. John Wiley and Sons, New York 1968.
[2] J. L. Doob: Probability in function spaces. Bull. Am. Math. Soc. *53* (1947), 15–30.
[3] D. Heidemann: Wahrscheinlichkeitsmaße in Funktionenräumen. Diplomarbeit RUB 1973 (unpublished).
[4] H. B. Mann: On the realization of stochastic processes by probability distributions in function spaces. Sankhya *11* (1951), 3–8.
[5] J. Neveu: Mathematical foundations of the calculus of probability. Holden-Day, Inc., San Francisco 1965.
[6] K. R. Parthasarathy: Probability measures on metric spaces. Academic Press, New York and London 1967.

Ruhr-University
Mathematical Institute

TESTS OF GOODNESS-OF-FIT FOR STATIONARY PROCESSES

Theo Gasser

Zürich

Goodness-of-fit tests as e.g. the χ^2-test and the Kolmogorov-Smirnov test assume independent variables. A sampling study indicates, how the distribution under the hypothesis and the power of the X^2-statistic are affected, when introducing certain correlation patterns. To improve the power, an inverse filtering procedure is suggested for processes which can be well represented by an autoregressive-moving-average model. For a second model, consisting of a Gaussian or non-Gaussian signal plus Gaussian noise, sample skewness and kurtosis are introduced and their asymptotical properties are derived.

1. INTRODUCTION

All the classical goodness-of-fit tests work with independent and identically distributed random variables (i.i.d. variables). We want to drop the assumption of independence, while the second assumption is analogously generalized to stationarity. The very simple hypothesis of independence cannot just be abandoned, but has to be replaced by some other model. The first model consists of processes generated by a mixed autoregressive-moving-average scheme (ARMA) of order (P, Q):

Model I:

$$X_t = \sum_{k=1}^{P} A_k X_{t-k} + \sum_{k=1}^{Q} B_k Z_{t-k} + Z_t$$

with: Z_t a process of i.i.d. variables;

A_k satisfy (for stationarity):

$$1 - A_1 z - \dots - A_P z^P = 0 \quad \text{has all roots outside the unit circle ;}$$

B_k satisfy (for invertibility):

$$1 - B_1 z - \ldots - B_Q z^Q = 0 \quad \text{has all roots outside the unit circle.}$$

The second part is devoted to a different model:

MODEL II:

$$X_t = S_t + Z_t$$

S_t: Gaussian or non-Gaussian signal;

Z_t: Gaussian noise.

One important difference between the models is the following: Model II allows that the process – or at least a signal part – is concentrated in some frequency bands, which is not possible with model I. Electroencephalographic (EEG) analysis gave the motivation for this study, and there model I is in many cases of doubtful value.

Physiological hypotheses led to the question, if the EEG is Gaussian. The amplitude distribution was usually tested with the χ^2-test, in few cases with the Kolmogorov-Smirnov test; examples are: Saunders [4], Elul [3], Weiss [7]. The possible fallacies by the strongly correlated patterns in EEG-analysis were ignored; in many cases additional correlation was introduced by digitizing the continuous record at too high frequency.

2. DISTRIBUTION AND TAIL PROBABILITIES OF X^2 UNDER THE HYPOTHESIS

For n i.i.d. observations $\{X_i : i = 1, \ldots, n\}$ we form K groups with probabilities p_i^0 under the null-distribution. The test statistic

$$X^2 = \sum_{i=1}^{K} \left(\frac{n_i - np_i^0}{np_i^0} \right)^2$$

is asymptotically distributed as χ^2_{K-S-1}, if S parameters are determined as multinomial maximum likelihood estimators.

To study the distribution of X^2 under the hypothesis, ARMA-realizations of length $n = 200$ are generated:

$$X_t = \sum_{k=1}^{P} A_k X_{t-k} + \sum_{k=1}^{Q} B_k Z_{t-k} + Z_t \, .$$

The white noise distribution is taken as Gaussian for simplicity (invariance to linear transformations). The number of samples in the Monte Carlo study is $N = 1000$. The number of classes is $K = 10$, of equal probability size under the hypothesis. The parameters mean and variance are estimated by maximum likelihood estimators, so that X^2 is no longer asymptotically distributed as χ^2_{K-S-1} (Chernoff and Lehmann [1]). (The effect decreases very fast with increasing K.) Table 1 gives the number of rejections at the 10%, 5% and 1% significance level.

The following conclusions are substantiated by table 2, and additional computed examples: For low to moderate correlation (processes 2 and 6), the distribution of X^2 does not change significantly. Strongly correlated processes show severe deviations,

TABLE 1

Distribution of X^2 under the hypothesis

No	Parameters	Rejects at 10%	Rejects at 5%	Rejects at 1%
1	$P = 0,\ Q = 0$: White Noise	116	62	10
2	$P = 1,\ Q = 0$: $A_1 = 0.3$	95	46	10
3	$P = 1,\ Q = 0$: $A_1 = 0.75$	171	98	21
4	$P = 1,\ Q = 0$: $A_1 = 0.9$	358	274	136
5	$P = 1,\ Q = 0$: $A_1 = -0.9$	164	79	17
6	$P = 2,\ Q = 0$: $A_1 = 0.4,\ A_2 = -0.5$	109	48	11
7	$P = 2,\ Q = 0$: $A_1 = 1.1,\ A_2 = -0.85$	125	64	17
	Thinned Sequence:			
3	$P = 1,\ Q = 0$: $A_1 = 0.75$	117	61	12
4	$P = 1,\ Q = 0$: $A_1 = 0.9$	117	64	5
	Ideal expected value	100	50	10

TABLE 2

Power of Normal against Uniform G

ARMA-Parameters	Rejections at 10% level	Rejections at 5% level	Rejections at 1% level
$P = 0,\ Q = 0$: White Noise	881	746	448
$P = 1,\ Q = 0$: $A_1 = 0.6$	158	91	27
Inversely Filtered	860	741	459
$P = 2,\ Q = 0$: $A_1 = 1.1,\ A_2 = -0.85$	143	68	11
Inversely Filtered	863	723	438

if they are of low frequency type (process 4), but much less so if they are of high frequency type (process 5).

A thinning procedure reduces correlation and leads to an improvement in the approximation to the χ^2-distribution (see last two rows of table 1). For a realization $\{X_i : i = 1, \ldots, n\}$ of a process X_k with spectrum $f(\nu)$ and correlation function $r(t)$, one should thin the n observations to n_{mod} in the following way:

$$n_{\mathrm{mod}} = \frac{\left(\int_{-1/2}^{1/2} f(\nu)\,\mathrm{d}\nu\right)^2}{\int_{-1/2}^{1/2} f^2(\nu)\,\mathrm{d}\nu} = \frac{n}{\int_{-\infty}^{+\infty} r^2(t)\,\mathrm{d}t}.$$

3. POWER

Due to a central limit type argument, a goodness-of-fit test applied to a linear process X_t

$$X_t = \sum_{k=-\infty}^{\infty} a_k Z_{t-k}$$

may drastically lose power, compared to a test for the i.i.d. process Z_t with distribution G. We need again a sampling experiment to judge the extent of breakdown of power. With the hypothesis that X_t is normal, I choose the following three alternatives:

1. G uniform (shorter tails than normal),
2. G double exponential (longer tails than normal),
3. G distributed as χ_4^2 (asymmetry).

From tables 2, 3 and 4, and additional computed material, the following can be concluded: The breakdown of power is more severe for symmetric alternatives, it is usually more accentuated for stronger correlated patterns, and it is much more severe for high frequency patterns than for low frequency patterns.

For model I, i.e. for ARMA-type processes, there is a direct and constructive way out of this difficulty:

(i) Estimate the parameters A_k, B_k of the ARMA-scheme.

(ii) The estimates $\hat{A}_k$, $\hat{B}_k$ are used to inversely filter the series X_t, which results in a pseudo-i.i.d. series $\hat{Z}_t$.

(iii) Apply the χ^2-test to $\hat{Z}_t$.

The rows marked "Inversely Filtered" in tables 2–4 show how successful this proposal is.

4. SKEWNESS AND KURTOSIS AS TESTS OF NORMALITY

EEG-analysts who have tested their data for normality, have preferably used the χ^2-test, in few cases the Kolmogorov-Smirnov test. Shapiro, Wilk and Chen [5] have demonstrated, that these two statistics have poor power as a test for normality compared to other statistics, as e.g. skewness and kurtosis. These two statistics can also be generalized to the correlated case.

The 3rd and 4th order sample cumulants $\hat{k}_3$, $\hat{k}_4$ of a realization $\{X_k : k = 1, \ldots, n\}$ from a stationary process X_t are defined in the usual way:

$$\hat{k}_3 = n^{-1}s_3 - 3n^{-2}s_2s_1 + 2n^{-3}s_1^3\,,$$

$$\hat{k}_4 = n^{-1}s_4 - 4n^{-2}s_3s_1 - 3n^{-2}s_2^2 + 12n^{-3}s_2s_1^2 - 6n^{-4}s_1^4$$

with

$$s_L = \sum_{k=1}^{n} X_k^L\,.$$

TABLE 3

Power of Normal against Double Exponential G

ARMA-Parameters	Rejections at 10% level	Rejections at 5% level	Rejections at 1% level
$P = 0$, $Q = 0$: White Noise	924	864	691
$P = 1$, $Q = 0$: $A_1 = 0.6$	382	285	119
Inversely Filtered	905	847	666
$P = 2$, $Q = 0$: $A_1 = 0.4$, $A_2 = -0.5$	310	229	85
Inversely Filtered	904	843	650

TABLE 4

Power of Normal against $G \sim \chi_4^2$

ARMA-Parameters	Rejections at 10% level	Rejections at 5% level	Rejections at 1% level
$P = 0$, $Q = 0$: White noise	999	999	993
$P = 1$, $Q = 0$: $A_1 = -0.6$	489	356	169
Inversely Filtered	999	999	991
$P = 2$, $Q = 0$: $A_1 = 0.4$, $A_2 = -0.5$	665	546	304
Inversely Filtered	998	995	983

THEOREM. We assume that the stationary Gaussian process X_t, $t = 0, \pm 1, \pm 2, \ldots$ with integrated spectrum $F(\lambda)$ and covariance function R_k fulfills:

$$\sum_{k=0}^{\infty} |R_k| < \infty .$$

Then the following holds:

A. $n^{1/2}\hat{k}_3$ and $n^{1/2}\hat{k}_4$ are asymptotically normal with zero mean and finite variance.

B.
$$\operatorname{Var}(\hat{k}_3) = 6n^{-1} \sum_{k=-\infty}^{\infty} R_k^3 + 0(n^{-1}),$$

$$\operatorname{Var}(\hat{k}_4) = 24n^{-1} \sum_{k=-\infty}^{\infty} R_k^4 + 0(n^{-1}).$$

Assertion A could be proved with a general mixing condition for X_t. Using a theorem by Sun [6], we can obtain the much simpler assumption for the R_k for our particular case. The details of the proof, as well as the cumbersome calculations for assertion B, are omitted. Sample skewness $\hat{\beta}_1$ and sample kurtosis $\hat{\beta}_2$ are obtained from $\hat{k}_3$, $\hat{k}_4$ by normalizing:

$$\hat{\beta}_1 = \frac{\hat{k}_3}{\hat{\sigma}^{3/2}}, \quad \hat{\beta}_2 = \frac{\hat{k}_4}{\hat{\sigma}^4}.$$

COROLLARY. $n^{1/2}\hat{\beta}_1$ and $n^{1/2}\hat{\beta}_2$ are asymptotically normally distributed with the following variances:

$$\operatorname{Var}(\hat{\beta}_1) = 6n^{-1} \sum_{k=-\infty}^{\infty} r_k^3 + 0(n^{-1}),$$

$$\operatorname{Var}(\hat{\beta}_2) = 24n^{-1} \sum_{k=-\infty}^{\infty} r_k^4 + 0(n^{-1})$$

(r_k = correlation function of X_t).

The proof is simple and is again omitted.

This theorem has to be supplemented by a discussion of the asymptotical properties of the statistics $\hat{k}_3$, $\hat{k}_4$ under the alternative. With $c_k(t_1, \ldots, t_{k-1})$ we designate the k-th order cumulant of $X(0)\,X(t_1), \ldots, X(t_{k-1})$.

THEOREM. Let us assume that X_t is a strictly stationary process with finite moments of arbitrary order. The statistics $\hat{k}_3$ and $\hat{k}_4$ are asymptotically unbiased and consistent estimates of $c_3(t, t)$ resp. $c_4(t, t, t)$ if

$$\sum_{t_1=-\infty}^{\infty} \cdots \sum_{t_{k-1}=-\infty}^{\infty} |c_k(t_1, \ldots, t_{k-1})| < \infty , \quad k = 2, 3, \ldots ,$$

and they are asymptotically normally distributed if

$$\sum_{t_1=-\infty}^{\infty} \cdots \sum_{t_{k-1}=-\infty}^{\infty} |t_j c_k(t_1, \ldots, t_{k-1})| < \infty, \qquad k = 2, 3, \ldots, \quad j = 1, \ldots, k-1 .$$

The asymptotical unbiasedness and the consistency can be proved by straightforward calculations. The asymptotical normality is based on the fact that cumulants of $\hat{k}_3$ and $\hat{k}_4$ higher than order 2 vanish asymptotically. In classical statistics, i.e. in the i.i.d. case, a moment condition would be sufficient; for stochastic processes, we need in addition the above assumption about asymptotical independence.

For applications, the two series

$$\sum_{k=-\infty}^{\infty} r_k^3, \quad \sum_{k=-\infty}^{\infty} r_k^4$$

have to be truncated to a finite order and the correlation function r_k has to be replaced by the respective estimate.

Skewness $\hat{\beta}_1$ and kurtosis $\hat{\beta}_2$ should be used for testing the Gaussianity of the signal S_t in model II. The application to EEG-samples revealed a frequent violation of the assumed normality of the amplitude distribution (Dumermuth, Gasser and Strehl-Marquardt [2]). If one wants to apply this method to model II for a hypothesis different from the normal one, one has to transform the data according to the transformation leading from the hypothetical distribution to the normal one.

In many situations of practical interest, one of the two models suggested here will be adequate. For correlated data, there will however be no universal recipes; we should use any apriori information and physical insight available to arrive at a reasonable model.

ACKNOWLEDGEMENT. I would like to thank Prof. P. J. Huber for stimulating discussions on this topic.

REFERENCES

[1] H. CHERNOFF, E. L. LEHMANN: The use of maximum likelihood estimates in χ^2-tests for goodness of fit. Ann. Math. Stat. *23* (1954), 579–586.

[2] G. DUMERMUTH, TH. GASSER, E. MARQUARDT-STREHL: Amplitude distributions fo normal EEG. To be submitted to: Electroenceph. clim. Neurophysiol.

[3] R. ELUL: Gaussian behavior of the electroencephalogram during performance of mental task. Science *164* (1969), 328–331.

[4] M. G. SAUNDERS: Amplitude probability density studies on alpha and alpha-like patterns. Electroenceph. clin. Neurophysiol. *15* (1963), 761–767.

[5] S. S. SHAPIRO, M. B. WILK, H. J. CHEN: Comparative study of various tests for normality. J. Am. Stat. Ass, *63* (1968), 1343–1372.

[6] T. C. SUN: Some further results on cental limit theorems for non-linear functions of a normal process. J. of Math. & Mech. *14*, (1965), 71–85.

[7] M. H WEISS: Non-Gaussian properties of the EEG during sleep. Electroenceph. clin. Neurophysiol. *34* (1973), 200–202.

EIDGENÖSSISCHE TECHNISCHE HOCHSCHULE ZÜRICH

FACHGRUPPE FÜR STATISTIK

and

UNIVERSITÄTS-KINDERKLINIK ZÜRICH

CYBERNETIC MODELLING OF QUEUEING & STORAGE SYSTEMS

A. Ghosal

New Delhi

Great advances have been made in the field of development of appropriate probability models for queueing and storage systems and derivation of their analytical solutions during the last 25 years. Nevertheless many of the problems in the control and design of water reservoirs inventory systems and queueing networks which an Operational Research worker or a practical statistician faces are not amenable to these methods. In most of these complex problems one has to apply simulation to derive approximate solutions.

This paper highlights the cybernetic approach to the study of complex systems, and emphasises on the development of isomorphic models. Through a theorem on restrictive isomorphism, attempt has been made to search for a simpler system which is isomorphic with the parent system with respect to a (or a few) desired property (or properties). The corresponding statistical problem is to develop approximate probability distributions for complex systems.

1. INTRODUCTION

Many of the problems in Statistics & Probability generate from real life problems; in the initial stages the mathematical treatment, the design of the study etc. are made by keeping in mind the practical problems in view, but after a certain time the progress of work is channeled into two directions – one concerning mostly with the development of theoretical treatment, which may be remotely connected with the problem in view, and the other direction involving too much of a pragmatic approach neglecting theoretical titbits. The work on the stochastic treatment of queueing & storage systems has also reached such a stage. Great advances have been made in the field of development of probability models and their analytical solution [13], [15], nevertheless many of the problems in the control and design of water reservoirs, inventory systems and queueing networks are not amenable to these methods. The work on probability models hinge mostly on the derivation of exact analytical solutions for determination of waiting time distribution or the distribution of the content (either of a reservoir or inventory system), given certain specified forms of the input and output distribution to the system. In course of practical work, the present author came across many situations in which the tailormade solutions could not be utilised; but useful contributions could be done for the user of the results by resorting to approximate methods. This paper highlights cybernetic approach of defining a system, however, complex it is and of getting approximate solution through numerical methods (see also [5], [6]).

The basic theme of the paper is to explore whether a complex system can be defined by an approximate equivalent simple system. This property has been referred to as restrictive isomorphism in which two systems are isomorphic to each other in respect of a particular property. For example, there may be a network queue comprising several queues in series and/or in parallel and a unit has passed through the entire network for the completion of a desired job; service times at various points may have complex relations among themselves and also with the inter-arrival time. Such a problem is commonly observed in many production systems, viz. steel plant, a fertiliser plant or in fact, any complex plant, but analytical solution for such a system does not exist except for trivial assumptions.

2. UNIFIED MODEL

2.1 The *u*-filter

The analogy between the stochastic model in queueing and storage systems has long been recognised [2], [5], [13], [18]. How a single model describes the single server queueing system, a simpler reservoir system or a simple inventory system has been elucidated by Ghosal ([5], Table 1.1, p. 3). The basic single server queueing model is given by

$$Z_{n+1} = \min [Z_n + X_n - Y_n, k]^+ \tag{2.1}$$

where Z_n is the waiting time of the nth customer (storage level at the beginning of the nth time period), X_n the service time (amount of input in the nth period), Y_t the inter-arrival time between nth and $(n + 1)$st customers (release at the end of the nth period), k the capacity of the dam (k may be finite or infinite).

In the equation (2.1) above we find that the waiting time of the nth customer (or the storage level at the nth time period) is obtained from the knowledge of u $(=X - Y)$. If u is identically distributed independent random variables (i.d.i.r.v.), then from the knowledge of these random noises for the past, we can determine the waiting time of the storage level (Z), both for the finite and infinite systems. This is called the *u-filter*. In the case of infinite system, the Z_n can be expressed as supremum of the sum of *u*'s,

$$Z_n = \sup_{1<r<n} \sum_{j=1}^{r} u_j . \tag{2.2}$$

When a stationary probability distribution of Z_t exists, we can express it, both for discrete and continuous cases, in terms of either an integral equation or a simultaneous equation. In the language of control theory, we get Z as an output from a system subject to white noise u and a set R representing constraints like capacity restriction etc.:

$$f : (r, R) \to Z . \tag{2.3}$$

The equation (2.3) can be expressed in terms of flow diagrams for models with various complications and feedback links. An interesting observation was made by Ghosal [3] that though a single server queueing system was analogous to a simple dam or an inventory system, the same did not hold true for a system of two queues in series and two dams in series where in both cases the output from the first system is an input in the second. Nevertheless, the concept of *u*-filter, in terms of random noises, holds true for both such systems.

We can, thus, develop models for complicated systems by taking into consideration random elements at each stage and constraints, like the capacity constraints. In most of such complicated cases we are not in a position to solve analytically for the probability distribution of Z_t at various phases, except for trivival assumptions like exponential distributions for service and inter-arrival times at each stage, etc. Unfortunately, however, in practical work in operational research, in most cases such complicated models are come across, and they do not lend exact analytical solutions. A number of approximations are resorted to by the worker in doing simulation exercises on various assumptions. In a number of practical situations met an important question arises, viz; whether the error in assuming a simplified model is large, if the error is within reasonable limits, should we assume a simplified model for a complex situation?

3. ISOMORPHIC SYSTEMS

With a view to exploring the possibility of developing simplified models for complex situations, an investigation into the isomorphic systems in cybernetic modelling was made. Let a system be defined by $\{X, R\}$ where $X = (x_1, \ldots, x_n)$ is a set of elements and the set R denotes the behaviour characteristics of the system [10]. The set R may include constraints, rules of operation on the set X, and various control phenophena for a system defined by X, R; we get an output Y where $f:\{X, R\} \rightarrow Y$. A completely deterministic system gives and output with probability one, while in a stochastic system in while either X or R or both are stochastic, we get an output Y_j with probability π_j ($j = 1, 2, \ldots$) [5], [7].

3.1 Definitions

A model $\{X, R\}$ may be isomorphic with the actual system either in its entirety or with respect to a few characteristics. We give below the definitions of strict isomorphism and restricted isomorphism as below [7], [8].

DEFINITION 3.1 (Strict Isomorphism). Let the two systems S_1 and S_2 be represented respectively by $\{X_1, R_1\}$ and $\{X_2, R_2\}$, where X_i denotes the set of elements and R_i characteristic of the system S_i ($i = 1, 2$). Then S_1 and S_2 are strictly isomorphic if (a) is possible to assign to every element of X_1 uniquely an element of X_2 and vice-versa; (b) the elements of characteristic R_1 can be uniquely assigned to R_2 and vice-versa. Then we state

$$S_1 \text{ i } S_2 \,. \tag{3.1}$$

DEFINITION 3.2 (Restricted Isomorphism). Restricted isomorphism between two systems S_1 and S_2 implies either of the following:

(i) the systems S_1 and S_2 are isomorphic with respect to a set of properties P only: this is defined by

$$(S_1 \text{ i } S_2)_P \,; \tag{3.2}$$

(ii) the systems S_1 and S_2 are isomorphic only with respect to output functions under given maps:

$$f_1(X_1, R_1) = f_2(X_2, R_2) \,. \tag{3.3}$$

The definition of restricted isomorphism as above is due to the present author [7]. In a larger number of problems in operational research and statistics there is a scope for getting solutions through a system which is isomorphic with the actual system in the restricted sense mentioned above.

3.2 Application

The idea of restricted isomorphism in stochastic systems is elucidated below. Suppose there is a storage (or queueing) system in which inputs come from various sources, their probability distribution functions (p.d.f.'s) are known, mutual relations among them defined above, constraints (e.g. capacity), release mechanism (deterministic or stochastic) all come under the set R defined above. Let us call this system as S_1. We may be interested in the d.f. of the storage level Z_1, given by $F_1(x) = = \text{prob}(Z \leqq x)$, under stationary condition. If the system is such that we cannot obtain an exact analytical solution for $F_1(x)$, we may look for a simpler system S_2 given by $\{X_2, R_2\}$ such that its level (Z_2) follows a d.f. $F_2(x)$. Then the system S_2 is isomorphic to the actual system if (given k)

$$F_1(x) = F_2(x), \quad x \in (0, k). \tag{3.4}$$

Thus isomorphism is defined in terms of equivalence in d.f. In cases where the congruent relation (3.4) does not exist (as it is in most of the practical cases), we accept S_1 as an equivalent system if we fail to reject the hypothesis $F_1 = F_2$.

For such a problem, therefore, our approach would be as follows: (i) to determine through numerical methods (simulation) the d.f. of the actual system; (ii) to search for a simpler system whose parameters can be determined in terms of parameters of the actual complex system; (iii) finally to test whether the cumulative d.f. of the simpler system differs significantly from the actual system.

3.3 Tests

There are a numer of test for testing the hypothesis H_0 $(F_1 = F_2)$. A test which appeals to practicians to test whether the non-overlapping area between F_1 and F_2 (A) is significantly different from zero: A is given by

$$A = \int_0^\infty |(F_1(x) - F_2(x)| \, \mathrm{d}x \tag{3.5}$$

(A may be assumed to follow truncated $\mathrm{N}(0, \sigma)$ on the range $[0, \infty)$.

To the knowledge of the author, a general test for H_0 $(A = 0)$ does not exist, though it may not be difficult to construct such a test to see whether M/M/1 queueing system with different parameters are same or not. (However, a graphical representation of A sometimes gives a good practical hunch – though such an approach is not acceptable by statisticians.)

Resorting to ordered samples from the actual and assumed distributions appear to be a reasonable approach. We may apply Smirnov's test [17],

$$W^2 = \int_0^\infty \{S_n(x) - F_2(x)\}^2 \, \mathrm{d}F_2(x) \tag{3.6}$$

where $S_n(x) = r/n$ if $x_{(r)} \leqq x \leqq x_{(r+1)}$, the ordered sample being $(x_{(1)}, \ldots, x_{(n)})$). We known $\mathrm{E}(W^2) = 1/6n$, $\mathrm{Var}(W^2) = (4n - 3)/180n^3$.

Kolmogorov's criterion [11], given by

$$(3.7) \qquad D_n = \sup_x |S_n(x) - F_2(x)|$$

has better properties, and detailed tables are available for calculation (see Kendall and Stuart [8]).

Ultimately the problem is one in fiducial inference on the set of parameters $\boldsymbol{\theta}(\theta_1, \theta_2)$ of the distribution $F_2(x)$. We look for the region $\boldsymbol{\theta} \in \Omega$ within which variation of $\boldsymbol{\theta}$ is permissible for given probability level. The final step is to expres $\boldsymbol{\theta}$ in terms of the parameters of actual system.

4. SEARCH FOR SIMPLE ISOMORPHIC SYSTEM

The search for simpler systems is treated in this section with respect to network queues. The same can be extended to the case of network dams.

4.1 Network Queues

We come across network of queues in many practical situations, mostly in processing plants. For example, the finished product may be processed in a number of stages – at each stage there is a queue due to the processing plant being pre-occupied. We describe the situation obtained in queues in series in which a customer (unit) enters the first queue, on getting its service there moves to the second queue, etc. so that waiting is involved at each stage. Finally he leaves the system after getting his service from the last queue in series. This network of queues is termed GI/G/1 → → ·/G/1 → ... in the literature (Ghosal [5]).

Let the inter-arrival time in the first queue be Y_n (between nth and $(n + 1)$th customers), $X_n^{(i)}$ the service time of the nth customer in the ith queue, $Z_n^{(i)}$ the waiting time of the nth customer in the ith queue $(i = 1, 2, \ldots)$; then if the capacity is unrestricted we have

$$(4.1) \qquad Z_{(n+1)}^{(1)} = [Z_n^{(1)} + X_n^{(1)} - Y_n^{(1)}]^+ , \quad Z^{(1)} \in [0, \infty) ,$$

$$(4.2) \qquad Z_{n+1}^{(2)} = \begin{cases} [Z_n^{(2)} + X_n^{(2)} - X_{n+1}^{(1)}]^+ & \text{if } Z_{n+1}^{(1)} > 0 , \\ [Z_n^{(2)} + X_n^{(2)} - X_{n+1}^{(1)} + V_n]^+ & \text{if } Z_{n+1}^{(1)} = 0 , \end{cases}$$

where

$$V_n = Z_n^{(1)} + X_n^{(1)} - Y_n^{(1)} = -I_n ,$$

where I_n is the idle time of the server after departure of the nth customer in the first queue. The solution for $F^{(i)}(x)$, the d.f. in the ith ($i \geqq 2$) has been obtained analytically only for a limited class of problems. For example $F^{(i)}(x)$ in an M/M/1 → ·/M/1 → → ... system is the same as the d.f. in a simple M/M/1 system ($i = 2, 3, \ldots$) (see Reich [16]); $F^{(2)}(x)$ in M/E_2/1 → ·/M/1 behaves asymptotically as a hyper-exponential distribution [4]. In a number of practical studies done by the author and his associates it has been observed through numerical studies that in many complex situations the waiting time behaved approximately like a hyper-exponential distribution – even in a few cases like an exponential distribution

$$\left(F^{(i)}(x) = 1 - \sum_{j=i}^{m} c_j \, . \exp\left(-a_j x\right); \quad m = 1 \quad \text{for M/M/1}\right).$$

4.2 Bounds for Waiting time

In practical situations it is useful to known the upper and lower bounds of waiting time. Studies in this direction have been done for GI/G/1 by Kingman [9], Marshall [12]. We shall give here the upper bound for the waiting time in two queues in series, GI/G/1 → ·/G/1. Results are briefly described below.

If we define $R_n^{(i)}$ as the idle time of the nth customer in the ith queue ($i = 1, 2, \ldots$), then we get for the first queue [12]

$$Z_{n+1}^{(1)} - R_n^{(1)} = Z_n^{(1)} + u_n^{(1)} \tag{4.3}$$

where $R_n^{(i)} = I_n^{(i)}$ (idle time) if $Z_{n+1}^{(i)} \leqq 0$ and 0 otherwise, $u_n^{(i)} = X_n^{(i)} - Y_n$.

For the second queue we get

$$Z_{n+1}^{(2)} - R_n^{(2)} = Z_n^{(2)} + X_n^{(2)} - X_{n+1}^{(1)} - R_n^{(1)} . \tag{4.4}$$

On squaring both sides of (4.4) and taking expectations, we get assuming independence between $Z^{(i)}$ and $u^{(i)}$ ($E(Z^{(i)}u^{(i)}) \sim E(Z^{(i)})\,E(u^{(i)})$),

$$E(X_n^{(2)})^2 = \{E(u_n^{(2)}\}^2 + 2\sum_{i=1}^{2} E(Z_n^{(i)})\,E(u_n^{(i)}) . \tag{4.5}$$

Since $E(X_n^{(2)}) = -E(u_n^{(2)})$ under stationary condition,

$$\operatorname{Var}(X_n^{(2)}) = \operatorname{Var}(u_n^{(2)}) + 2\sum_{i=1}^{2} E(Z_n^{(i)})\,E(u_n^{(i)}) . \tag{4.6}$$

Since Var (·) is non-negative, we get (leaving suffixes)

$$\operatorname{Var}(u^{(2)}) + 2\sum_{i=1}^{2} E(Z^{(i)})\,E(u^{(i)}) > 0 . \tag{4.6}$$

For stable queues, we get $E(u^{(i)}) < 0$. From Marshall [12] we get bounds of the first queue

$$l \leqq E(Z^{(1)}) \leqq -\mathrm{Var}\,(u^{(1)})/2\,E(u^{(1)})\,, \tag{4.7}$$

where l is the solution of

$$x = \int_{-x}^{\infty} G(u)\,du\,, \tag{4.8}$$

where $G(u) = \mathrm{prob}\,(u^{(1)} \geqq u)$. Hence we get the upper bound of $E(Z^{(2)})$ as

$$E(Z^{(2)}) \leqq -\{\mathrm{Var}\,(u^{(2)}) + 2l\,E(u^{(1)})\}/2\,E(u^{(2)})\,. \tag{4.9}$$

Thus given the first two moments of $u^{(1)}$, $u^{(2)}$, we get the upper bound of $E(Z^{(2)})$ numerically. The lower bound of $E(Z^{(2)})$ can also be determined approximately on similar lines as followed by Marshall [12]. We determine bounds for $E(Z^{(j)})$ for subsequent queues in series, by writing j for 2 in (4.6). Detailed results on this study will be reported by the author in a subsequent communication.

4.3 Determination of simpler systems

An exercise done by the author and his associates in a number of simulation studies has been to study whether the waiting time distribution in second or subsequent queue could be ascribed to an approximately equivalent single queue system. For example, suppose the equivalent system had been an M/M/1 queue in which inter-arrival time and service time followed exponential distributions with parameters λ and μ respectively, then we would get

$$E(Z) = \varrho/(\mu - \lambda) \tag{4.10}$$

and

$$H(x) = \varrho \exp - (\mu - \lambda)\,x\,, \quad x \in (0, \infty)\,, \tag{4.11}$$

where $H(x) = \mathrm{prob}\,(Z \geqq x)$, $\varrho = \lambda/\mu < 1$.

Comparison of (4.9) and (4.10) gives a suggestion as to the region within which to look for values of parameters λ and μ purely from the knowledge of $u^{(1)}$ and $u^{(2)}$ processes. If we simulate the waiting time distribution for the jth queue ($j \geqq 2$), then through graphical process by plotting $H(x)$ (actual) against $H(x)$ we would estimate parameters λ and μ. Finally, tests described in Sec. 3 would have to be applied to examine validity of hypothesis.

In a simulated experiment done for a processing plant in a three-stage queueing system, in which at no stage it was an M/M/1 queue, the waiting time distribution in the last queue could however be predicted through a M/M/1 with $\lambda = 0.11$ and $\mu = 0.41$.

4.4 Network Dams

The problem of two or more dams in series in which the release or overflow from one dam goes to the next dam is also amenable to similar modelling technique. The level in the second dam in series getting input $X_t^{(2)} + \min(Y_t^{(1)}, Z_t^{(1)} + X_t^{(1)})$ where $Y_t^{(1)}$ may be a const. (say m_1) is given by ([5], p. 73 eqn. (3.26)]

$$(4.12) \qquad Z_{t+1}^{(2)} = \begin{cases} [Z_t^{(2)} + X_t^{(2)} + Y_t^{(1)} - Y_t^{(2)}] & \text{if } Z_t^{(1)} > 0\,, \\ [Z_t^{(2)} + X_t^{(2)} + Z_{t-1}^{(1)} + X_{t-1}^{(1)} - Y_t^{(2)}]^+ & \text{if } Z_t^{(1)} = 0\,. \end{cases}$$

Inequalities for this system will be reported in a subsequent communication.

5. REMARKS

There are a number of other problems of cybernetic modelling come across in queueing and storage problems. For example, the server may adapt himself to the changing rate or arrival of customers; in a dam we may come across the problem of adaptive release schemes (see Ghosal [5]). These are not described here; there is scope for applying control-theoretic approach in these problems.

Work has been done on state-dependent queues (Bhat [1]); though not much work on situations in which parameters "learn" has been done. This paper highlights mainly the problem of determining isomorphic (or equivalent) systems, and deriving approximate solutions thereform. Coming years will see further exploration of approximate distributions and their solutions. There is scope for work of development of equivalent systems for simulation work; the work of Moran [14] in deriving equivalent bivariate normal distribution for bivariate gamma distributions shows great promise of application in queueing and storage systems. In a general way, dual systems for finite capacity systems form a special class of isomorphic systems; interesting solutions for the parent system can be obtained from the dual ([5], Chapter 2).

REFERENCES

[1] U. Naryan Bhat: Elements of Applied Stochastic Processes. John Wiley, New York 1972.

[2] J. Gani: Problems in the probability theory of storage systems. J. Roy. Stat. Sec. B *19* (1957), 181—206.

[3] A. Ghosal: Contributions to the theory of dams. Unpublished D. Phil thesis, Calcutta Univ. 1962.

[4] A. Ghosal: Queues in series. J. Roy. Stat. Soc. B *24* (1962), 359—363.

[5] A. GHOSAL: Some aspects of queueing and storage systems. Lecture Notes in Oper. Res. and Math. Systems Vol. 23. Springer Verlag, Heidelberg 1970.
[6] A. GHOSAL: Cybernetic queueing systems. Cahiers du Centre d. Res. Oper. *13* (1971), 188—202.
[7] A. GHOSAL: Some problems in applied cybernetics. SCIMA *2* (1973), 37—52.
[8] M. G. KENDALL, A. STUART: Advanced theory of Statistics. Vol. 2. Charles Griffin, London 1967.
[9] J. F. C. KINGMAN: Some inequalities in Queueing. Biometrika *49* (1962), 315—324.
[10] J. KLIR, M. VALACH: Cybernetic Modelling. Illiffe Books, London 1967.
[11] A. KOLMOGOROV: Sulla determinazione empirica di uria legge di distribuzione. G. Ist. Ital. Attuari *4* (1933), 83.
[12] K. T. MARSHALL: Some inequalities in Queueing. Opns. Res. *16* (1968), 651—665.
[13] P. A. P. MORAN: The theory of storage. Methuens, London 1958.
[14] P. A. P. MORAN: Statistical inference with bivariate gamma distribution. Biometrika *56* (1969), 627—634.
[15] N. U. PRABHU: Queues and Inventory. John Wiley, New York 1965.
[16] E. REICH: Waiting times when queues are in tandem. Ann. Math. Stat. *28* (1975), 768—773.
[17] N. V. SMIRNOV: Sur la distribution de w^2. C.R. Acad. Sci. Paris *202* (1936), 449.
[18] W. L. SMITH: On the distribution of queueing times. Proc. Camb. Phil. Soc. *49* (1953), 449—461.

COUNCIL OF SCIENTIFIC AND INDUSTRIAL RESEARCH

Postscript. During the last two years more work has been done by the author and his associates. Some of these results are given in the working paper "Isomorphic Queues and Related Problems" (G 1/76, Graduate Center of Management, Baruch College, City University of New York, 1976).

SCHEDULING, QUEUES AND DYNAMIC ALLOCATION INDICES

JOHN GITTINS and PETER NASH

OXFORD CAMBRIDGE

A theorem is stated giving conditions under which the optimal allocation of a scarce resource between different tasks of uncertain characteristics is expressible by a priority index. When this can be done the problem is much simpler. The paper aims to give an overview of the kind of situations in which such a result does or does not hold, and the mathematical treatment is informal. Applications are described to priority queues, chemical research, search procedures, and the two-armed bandit problem.

1. INTRODUCTION

A *total pre-ordering* between the members $a, b, \ldots$ of a set C may be defined to be a transitive and reflexive relation with the property that, for any $a, b \in C$, either $a \succ b$, or $b \succ a$. For all physically interesting total pre-orderings a corresponding scale function $f : C \to R$ exists with the property that $f(a) \geqq f(b) \Leftrightarrow a \succ b$. The relation may then conveniently be defined by reference to the scale f. Greenwich Mean Time and temperature are familiar examples of such scales.

In the field of decision-making under uncertainty the important relation is that of preference. A typical member of C now consists of an uncertain situation in which one of a set of mutually exclusive events must occur. Von Neumann and Morgenstern [26] and Savage [21] have shown that the assumption that preference constitutes a total pre-ordering, and obeys some other extremely plausible conditions, implies the existence of scales both of utility and of probability, and that the utility of an uncertain situation may be found by taking expectations over the different possible events. Schlaifer [22] goes on to show in detail how these results may be used to analyse, and greatly reduce, the problems of decision under conditions of uncertainty.

Thus there is plenty of evidence in different fields, including that of decision-making under uncertainty, that relations of total pre-ordering are a big help in the description and analysis of complex systems. With this thought in mind let us now consider the following problems.

PROBLEM 1. *Job i $(i = 1, 2, \ldots)$ takes a time t_i to complete, and costs c_i for every time unit for which its completion is postponed. Only one job can be done at a time. In what order should the jobs be done?*

PROBLEM 2. *At the beginning of any particular day there are x_i $(i = 1, 2)$ tons of gold in mine i. Working mine i is successful with fixed probability p_i, in which case the proportion of gold extracted in a day's mining is r_i. If mining on a particular day is unsuccessful, no gold is extracted and the mining programme comes to and end. On each day just one of the mines may be worked. With given initial amounts of gold in the two mines, how should the mine to be worked on each successive day be chosen? The aim is to maximise the total expected amount of gold extracted.*

It is easy to check that the answer to problem 1 is to do the jobs in order of decreasing values of the ratio c_i/t_i. For problem 2, Bellman [4] has shown that on any particular day the mine to work is the one for which $p_i r_i x_i/(1 - p_i)$ is larger. Thus in both cases there is a total pre-ordering amongst jobs or mines, with all possible parameter values, which may be used to decide which job to do, or which mine to work, next.

In both problems effort is available at a given rate, and the question is how it should be allocated between different competing tasks. When, as in these cases, the answer is given by a scale function defining a total pre-ordering of priority for the various possible tasks, the value taken by the scale function for a particular task will be termed the *dynamic allocation index* (DAI) for that task. The word 'dynamic' is in recognition of the fact that the index for a task may change as effort is expended on it.

The purpose of this paper is to describe some of the conditions under which dynamic allocation indices might be expected to exist, to state a theorem formalizing these ideas, and then to exhibit situations where the theorem may be applied.

2. DISCUSSION OF PROBLEMS 1 AND 2

Problem 1 is reminiscent of the well-known Knapsack Problem (e.g. see Greenberg [13]), in which item i is worth c_i and has weight t_i. In filling the knapsack, items for which c_i/t_i is large obviously have first priority once again. If the maximum total weight $T \gg \max t_i$ the solution to the problem, roughly speaking, is to start filling the knapsack with items with the largest values of c_i/t_i, and then, when only a small amount of additional weight is allowed, to consider as well the desirability of packing items in so as not to waste any of the weight allowance. It is, of course, this second feature of an optimal policy which makes the problem interesting.

Thus the Knapsack Problem is not trivial, and in particular no DAI exists, because of the conjunction of a finite maximum total weight and the discrete nature of the individual items. It is thus natural to ask what happens to the solution of problem 1 if a limit is set to the total time available, for example by stipulating that some other means of doing the jobs will become available at time T, and will finish them all in negligible time. The answer, not surprisingly, is that as T approaches we should, so to speak, bear in mind that a partially completed job means the time spent working on it has been wasted, and try as far as possible to avoid this happening. Once again the packing phenomenon, this time of jobs into the remaining time, means the optimal solution is messy, and there is no DAI. More generally, we can say that whenever the effort allocation problem is constrained by a finite time horizon some careful packing may be required, and this may well mean that no DAI exists.

A finite time horizon is an example of an overall time dependence. It amounts to the assertion that whatever we do after a certain time, it counts for nothing. As we shall see, other, less extreme, ways of discounting the future may also mean that there is no DAI. One might suppose that the way to cope with a problem with an overall time dependence is to make the DAI for a task dependent on time, as well as on the parameters of the task itself. This, however, does not work in general, as can be seen by thinking once again of the Knapsack Problem. If we are not to waste the weight allowance we must choose a combination of items whose total weight is close to T. Whether this is possible with a particular item included obviously depends on the other items which are available, as well as on the amount of unappropriated weight allowance. Thus there is no scale of priority defined with respect to individual items. Similar reasoning applies to problem 1.

Problem 1 has the property that the DAI for job i, c_i/t_i, where now t_i is interpreted as the remaining time needed for completion, increases as work proceeds on job i. In problem 2 the converse is true, since the amount of gold left in a mine can only decrease. This being so it is not surprising that, as Bellman notes, the DAI, $p_i r_i x_i/(1 - p_i)$, depends only on the expected gain and loss from the next day's work on mine i. If there was a possibility that, say, the tenth day's work on mine 1 might produce more gold than the first day, it might be worth working mine 1 even if mine 2 is likely to produce higher initial yields. Since there is no such possibility, the allocation is determined simply by the prospects of immediate gain or loss, and we find that there is a DAI expressed in these terms. This might be termed the 'eat, drink and be merry for tomorrow we die' strategy. As we shall see, this is fairly generally optimal for tasks whose prospective returns are bound to deteriorate, and this is a situation, therefore, which favours the existence of an optimal strategy expressible in terms of DAI's.

However, for problem 2, DAI's only exist in a restricted sense since, as Bellman shows, the result fails to extend to more than two mines. In problem 2 there is a linkage between the two mines, in the sense that all work must stop when a day's work has been unsuccessful. Our theorem, stated in the next section, rules out linkages between the different tasks. Its failure to apply to problem 2 shows that, in general, this independence is necessary.

In summary, then, we have reason to expect that some form of time homogeneity, or stationarity, tasks with deteriorating prospects of payoff, and absence of linkages between tasks, are all characteristics which favour the existence of DAI's.

3. A DAI THEOREM

To state the theorem we shall use the terminology of Markov decision processes. Consider a system defined on a general state space Θ, of which x denotes a typical member. The set of decision times Z^+ consists of the positive integers, t denoting a typical member. The *control set-function* U defined on $\Theta \times Z^+$ gives the set $U(x, t)$ of available controls at the point (x, t), u being a typical member of $U(x, t)$. $P(x, u, t)$ is a probability measure defined on some σ-field of subsets of Θ. It determines the state of the system at time $t + 1$, given that at time t the system was in state x and control u was applied. The set of probability measures $P(x, u, t)$ is the law of motion for the system. $x(t)$ denotes the state of the system at time t for a particular trajectory of the process. The reward associated with a transition from x at t to y at $t + 1$ when control u is applied at time t is $R(x, y, u, t)$, a real number.

We shall call such a process a bandit process if the following conditions hold. (a) $U(x, t)$ is independent of t (so we shall write simply $U(x)$) and consists either of the two controls c (continue) and w (wait), or simply of c alone. (b) $P(x, c, t)$ is independent of t. (c) If w is applied at (x, t), then with probability one the system goes to $(x, t + 1)$. (d) $\exists a$ $(0 < a < 1)$ such that $R(x, y, c, t) = a^t R(x, y, c, 0)$ $\forall x, y, t$. (e) $R(x, x, w, t) = 0$ $\forall x, t$.

Thus the control w simply delays the evolution of the process for one time unit, and subsequent rewards are discounted by the factor a. We shall abbreviate the notation by writing $R(x, y, c, t) = a^t R(x, y)$.

The members of a family $\mathscr{F}$ of bandit processes are said to be *alternative* under the following conditions. (f) The processes take place simultaneously. (g) Every process starts, at $t = 0$, from a state in which w is allowed. (h) For any t, c must be applied to just one process and w to all the others. (i) The factor a is the same for every member of $\mathscr{F}$ which may, however, have different state-spaces and laws of motion.

A *strategy* for the family $\mathscr{F}$ is a function from the product of Z^+ and of the state-spaces for the individual processes onto the set of alternative bandit processes. It prescribes, for every state of $\mathscr{F}$, the process to which c is to be applied. The bandit processes correspond to the tasks which we considered earlier, and our object, as before, is to find the strategy which maximises the total expected gain (or minimises total expected loss). It turns out that, provided the total expected gain is bounded, the following theorem holds.

THEOREM. (i) *The unique optimal strategy is given by a stationary dynamic allocation index* $v(x)$ (*in general a different function for each bandit process*).

(ii) *For a given bandit process,* $v(x)$ *may be defined as the unique function satisfying the equation*

$$v(x) = E\left[\sum_{t=0}^{t^*-1} a^t R(x(t), x(t+1)) + a^{t^*} v(x)\right]. \tag{1}$$

Here $x(0) = x$, c is applied for all t, the expectation is taken over all possible trajectories, and t^* is the random time such that

$$t^* = \min_{t>0} \{t \mid v(x(t)) \leqq v(x)\},$$

$$t^* = \infty \text{ if there is no such } t.$$

For the case of just two alternative bandit processes part (i) of the theorem remains true if the discount rate a is different for the two processes.

A weaker version of part (i) of the theorem, which we shall call the DAI theorem, was proved by Gittins and Jones [11], and the proof of part (i) is almost identical.

Once the existence of a DAI has been established its scale may be calibrated by reference to standard processes for which $\Theta = \{\emptyset\}$, and $R(\emptyset, \emptyset, c, t) = a^t R(\emptyset, \emptyset) = a^t_\mu$. Given any bandit process B in state x, consider the family of alternative processes which consists just of process B and a standard process D. There is just one value $\mu(x)$ of μ for which it is optimal to apply c to either of the two processes. The function $v(x)$ defined as $\mu(x)/(1 - a)$ must satisfy equation (1). Nash [18] has shown that the solution is unique.

Any monotonic function of $v(x)$ will, of course, do equally well as a DAI function. The solution of (1), however, has the distinction of being straightforward to calculate. An initial approximation to $v(x)$ may be substituted into the right hand side of (1), which may be evaluated to give a second approximation, and so on. The procedure always converges.

Now suppose that for some value $x = x(0)$, $U(x) = \{c, w\}$ and $E[R(x, x(1)) \mid x(0) = x] > E[R(x(t), x(t+1)) \mid x(t)]$, for all $t > 0$ and for all trajectories starting from x. A bandit process B with this property will be described as *deteriorating* from x. From the above characterisation of $v(x)$ as the constant rate of return from a standard process which it would be equally desirable to continue at $t = 0$, it is clear that $t^* = 1$ and, from (1), $v(x) = E[R(x, x(1)) \mid x(0) = x]/(1 - a)$. Monotonic transformations of DAI functions are allowed, so the DAI theorem has the corollary:

COROLLARY 1. *The dynamic allocation index of a deteriorating bandit process may be defined as the expected reward resulting from applying c for one time unit.*

This result may also be regarded as a corollary of theorem 3.3 by Chow et al. [7], which deals with a stopping problem in what they call the monotone case. Its particular interest here is that it confirms that, when the prospects of payoff are deteriorating in the way specified, $(1 - a)\, v(x)$ has a simple form. This version of the DAI is, in fact, independent of the discount factor a, therefore so is the optimal strategy. A strategy which is optimal for any discount factor clearly remains optimal for a weighted average of discount factors. Thus, if every bandit process is deteriorating, the DAI theorem continues to hold if a^t is replaced in condition (d), and in (1), by $\int_0^1 a^t\, dF(a)$ for some probability distribution $F(a)$, a representation which includes a large class of non-negative decreasing functions of t.

The DAI theorem has been stated for a discrete time variable. A similar result holds in continuous time, as may be shown by letting the intervals between successive occasions when the control may be changed tend to zero. For the continuous time version of corollary 1 the words 'one time unit' are replaced by 'an infinitesimal time interval'.

4. APPLICATIONS OF THE DAI THEOREM

1. Problem 1

Each job may be regarded as a bandit process, whose state $x = (s, I)$, s being the time so far spent on the job, and I taking the values 1 or 0 according as the job has so far been completed or not. Non-zero losses occur only when I changes from 0 to 1. If this is at time t for job i, the loss is $c_i t$.

The DAI theorem holds for rewards which, if they occur at time t, are discounted by the factor a^t. However, if $a = 1 - \varepsilon$, then $a^t = 1 - \varepsilon t + o(\varepsilon)$ as $\varepsilon \to 0$. Thus for values of a just less than one the rewards associated with a bandit process consist of a contribution which is independent of time, negative contributions proportional to the time at which they occur, and terms of smaller order of magnitude. The time-independent contributions are independent of the strategy chosen, so we have a reward structure which, apart from terms of order $o(\varepsilon)$, is of the desired form, if we replace the loss $c_i t$ by a reward $c_i a^t/(1 - a)$. It is easy to see that $(1 - a)\, v_i(a, 0) \to$ $\to c_i/t_i$ as $a \to 1$, where $v_i(a, s)$ is the solution of equation (1) for job i when $x =$ $= (s, 0)$. It follows that c_i/t_i may be taken as the DAI for job i in the original problem.

This we already knew from more elementary considerations. But having shown that the problem can be dealt with by means of the DAI theorem we can immediately obtain some more general results.

Instead of the known service time t_i, let the service time for job i have the probability distribution F_i, with $f_i(t) = \mathrm{d}F_i(t)/\mathrm{d}t$, $P_i(t) = 1 - F_i(t)$, and completion rate $\varrho_i(t) = f_i(t)/P_i(t)$.

In the discussion of the deterministic version of the problem it was implicitly assumed that $U_i(s, 1) = \{c, w\}\ \forall s > t_i$. It was unnecessary to specify $U_i(s, 0)$ because we should clearly not want to stop a partially completed job even if it were allowed. We now maintain the same assumption for $U_i(s, 1)$, and for $U_i(s, 0)$ simply assume that whether w is included or not depends only on the expired service time s. This new assumption seems to be implied by our notation, but it could easily be avoided by defining a richer state-space.

Taking limits as before now gives

$$v_i^*(s) = \lim_{a\to 1}(1 - a)\, v_i(a, s) = \sup C_i \frac{\int_0^y f_i(s + t)\,\mathrm{d}t}{\int_0^y P_i(s + t)\,\mathrm{d}t}, \qquad (2)$$

where the supremum is taken over all positive values of y for which $w \in U_i(s + y, 0)$. Once again, $v_i^*(s)$ is a valid DAI. This result was first proved, using different methods, by Sevcik [24], for the case when w is always allowed.

It includes some previously known results as special cases. For example, if the expected remaining processing time for job i never exceeds its initial value, $v_i^*(s) = c_i\, \varrho_i(s)$, which leads to a time-sharing optimal strategy if there is more than one job of this type. We may think of a time-sharing strategy as the limit, as the basic time unit tends to zero, of a strategy which applies c cometimes to one job and sometimes to another.

The main scope for practical applications of these results seems to be in production scheduling and scheduling of jobs on computers.

2. Pharmaceutical Research

Now let the jobs be pharmaceutical research projects, which it is reasonable to think of as terminating instantaneously when a viable drug has been discovered. Since the life of such a project is measured in years, exponential rather than linear discounting of future rewards is more appropriate. The research director's problem is how to allocate his research chemists between the different computing projects. Insofar as chemists are interchangeable between projects the more general model for problem 1 is applicable, $c_i a^t$ now being the reward if project i is completed at time t. For this application there is not much point in disallowing w, and (1) gives

$$v_i(a, s) = \sup_{y>0} C_i \frac{\int_0^y f_i(s + t)\, a^t\, dt}{\int_0^y P_i(s + t)\, a^t\, dt}.$$

This problem has been discussed elsewhere by the present authors, [9], [18], [19].

3. Multi-Armed Bandit

Bellman [3] gave a Bayesian formulation of the two-armed bandit problem with no memory limitation and an infinite time horizon. Our multi-armed bandit is the obvious generalisation.

There are n machines, $1, 2, \ldots, n$, which may be used in any order for an infinite sequence of trials. Each trial results in either a success or a failure. A success at the t'th trial means a reward of a^t and a failure means a gain of 0. For machine i ($i = 1, 2, \ldots, n$) the probability of a success at any trial is p_i. These probabilities are not known, though there are independent prior probability distributions for each of them before the sequence of trials starts. The problem is to determine the strategy which maximises the total expected payoff.

Bandit processes were so named because the different machines, or arms, of the multi-armed bandit may be regarded as a family of such processes. The state x_i of arm i is the current posterior distribution for p_i. For a given prior distribution this is defined by the two jointly sufficient statistics n_i, the number of trials so far on arm i, and r_i, the number of them which have been successful.

For this problem equation (1) has no obvious solution, though it may, as already mentioned, be used as the basis of an iterative calculation of the DAI function $v_i(x)$. As the numerical work by Yakowitz [27] shows, this is well within the capabilities of modern computers. For a given prior distribution, $v_i(n_i, r_i)$ may easily be calculated for all n_i and r_i, since for moderately large n_i and r_i the mean of the posterior distribution is a close underestimate.

4. A Multi-Population Search Problem

There are n populations each of which has an infinite number of members. The members of each population have scores, with different unknown distributions of scores for each population. The object is to find by sequential sampling a member of any of the populations with a score above some target value, after taking as small a total sample as possible.

Like the n arms of a multi-armed bandit, the n populations may be regarded as a family of bandit processes. DAI functions again may be calculated iteratively using equation (1). Details of such a calculation are given by Jones [17], and Gittins [10] describes the relevance of this model to the search for new drugs.

5. SUPERPROCESSES AND A DAI FOR QUEUES

We now describe an extension of the theory to cover the case when new bandit processes join the original family, which we now refer to as the *queue*, in a Bernoulli arrival stream.

Consider a set $\{L_i : i = 1, 2, \ldots, N\}$ of classes of bandit processes. Each class L_i consists of an infinite number of copies of a bandit process i, with state space Θ_i, law of motion P_i, reward function R_i, and discount factor a. At time $t = 0$ the queue contains a finite set of processes $\{ij \in L_i : i = 1, 2, \ldots, N;\ j = 1, 2, \ldots, n_i\}$. New arrivals occur in an independent Bernoulli stream for each class. The probability of an arrival at time t belonging to class L_i is p_i. Define $\Theta = \bigcup_{i=1}^{N} \Theta_i$, and $\hat{\Theta}$ to consist of all finite non-empty sequences of members of Θ, so that $\Theta \subset \hat{\Theta}$. A *strategy* is now a function $S : \hat{\Theta} \to \{1, 2, \ldots, N\} \times Z^+$, and specifies, for each state x of the queue, the process to which the control c is applied.

A *superprocess* (Q, S) is a queue Q together with a strategy S. A superprocess itself satisfies the conditions for a bandit process if we make the stipulation that the arrival stream is switched off whenever the control w is applied, and switched on again when c is applied. When c is applied to the superprocess as a whole, the strategy S picks out the individual bandit process to be continued.

The notion of a superprocess means that a queue may be looked at in a number of different ways. At any time t the set of bandit processes in the queue may be partitioned into groups with one or more members. Such a group constitutes a superprocess if any arrivals which occur when c is applied to a member of the group are themselves assigned to the group, and provided there is a strategy associated with the group.

We now proceed to show how this idea may be used to define a DAI for a family of bandit processes with Bernoulli arrivals.

Consider what we might term the maximal partition of the queue at time zero, by which we mean that every group contains just one of the bandit processes initially present. With each such group associate any strategy S_0 (the same strategy for each group), thereby transforming them into superprocesses. Thus the DAI theorem may be applied, and there is a unique DAI function ν_1 defined on $\hat{\Theta}$ which is the solution to equation (1).

$\Theta \subset \hat{\Theta}$, so the restriction ν_1 of $\hat{\nu}_1$ to Θ defines a DAI function $\nu_{1,i}(x_i)$ for each class L_i of the original bandit processes. Thus ν_1 defines a strategy S_1 for the queue, now regarded as a collection of the original bandit processes rather than of superprocesses. Our procedure may now be repeated, using S_1 in place of S_0, to give another DAI function ν_2 and corresponding strategy S_2, and so on.

It may be shown, Nash [18], that this iterative procedure, which is very similar to what Howard [15] calls policy iteration, always converges to a unique DAI function and associated strategy S, and that S is the unique optimal strategy. We shall call such a DAI function self-consistent, and summarise this discussion by stating a further corollary of the DAI theorem.

COROLLARY 2. *The DAI theorem remains true when arrivals occur in a Bernoulli process. In part* (ii) *the word 'self-consistent' is now inserted before the word 'solution'.*

For corollary 2 the expectation in equation (1), and the definition of t^*, are with respect to realisations of superprocesses. This makes for increased complexity, but a computed iterative solution is in principle possible as before.

A continuous time version of corollary 2, with Poisson arrival streams, follows in similar fashion from the continuous time DAI theorem.

The problem with random (i.e. Bernoulli or Poisson) arrivals may be shown to have the following general properties.

(i) If $\nu_i(x_i) \geqq \nu_j(o_j)$, where o_j denotes the state of a member of L_j on arrival, then L_j can be ignored in calculating $\nu_i(x_i)$.

(ii) $\nu_i(x_i) \geqq \nu_{o,i}(x_i)$, where the subscript o denotes the value of the DAI with no arrivals, with equality if and only if $\nu_{o,i}(x_i) \geqq \nu_j(o_j)$, $\forall j$.

(iii) If every bandit process is deteriorating from its arrival state, the DAI with arrivals is a monotonic function of the DAI with no arrivals, so that in this sense corollary 1 still holds.

The case of problem 1 with general service time distributions, Poisson arrivals and linear losses, has a particularly neat solution. As before, letting a tend to 1 gives the required linear loss structure, though now this leads to an infinite total loss, and our criterion is the minimisation of the expected cost per unit time. This procedure is familiar in the theory of Markov decision processes (e.g., see Hordijk [14]).

It turns out, by examining equation (1), that in this case the DAI with arrivals is a monotonic function of the DAI with no arrivals (see Nash [18]). Thus the general solution (2) continues to hold when there are Poisson arrivals. This result was conjectured for the case when w is always allowed by Sevcik [24]. Its validity in particular cases had previously been proved by several authors, including Aczel [1], Avi-Itzkak et al. [2], Bronstein [5], Bronstein and Veklerov [6], Cox and Smith [8], Jaiswal [16], Rykov and Lemberg [20], Schrage [23], Sevcik [24], and Simonovits [25]. Note that the superprocess idea allows us, for example, to forbid pre-emption except when an arrival occurs, a case which has received some attention.

6. THE MULTI-SERVER CASE

The optimality of a DAI strategy is less common when there is more than one server. For example, in the simple generalisation of the original version of problem 1 to more than one server c_i must be independent of i, and for the deteriorating case identical completion rates, or some other similarly strong condition, is required. These results are due to Glazebrook [12].

REFERENCES

[1] M. Aczel: The effect of introducing priorities. Operations Research *8* (1960), 730—733.

[2] B. Avi-Itzkak, I. Brosh, P. Naor: On discretionary priority queueing. Z. für Angewandte Mathematik und Mechanik *44* (1964), 235—242.

[3] R. E. Bellman: A problem in the sequential design of experiments. Sankhya *16* (1956). 221—229.

[4] R. E. Bellman: Dynamic Programming. Oxford University Press, London 1957.

[5] E. O. Bronstein: On multiple level service with one server. Tech. Kibernetika (1969), 5, 59—64.

[6] E. O. Bronstein, E. B. Veklerov: On one type of service discipline. In: Mass service in information transmission systems. Nauka, Moscow 1969, 54—59.

[7] Y. S. Chow, H. Robbins, D. Siegmund: Great expectations. Houghton Mifflin, Geneva, 1971.

[8] D. R. Cox, W. L. Smith: Queues. Methuen, London 1961.

[9] J. C. Gittins: Optimal resource allocation in chemical research. Advances in Applied Probability *1* (1969), 238—270.

[10] J. C. Gittins: How many eggs in a basket. R & *D* Management *3* (1973), 2, 73—81.

[11] J. C. Gittins, D. M. Jones: A dynamic allocation index for the sequential design of experiments. Progress in Statistics (editor J. Gani). North Holland, Amsterodam, 1974, 241—266.

[12] K. D. Glazebrook: Stochastic Scheduling, Ph.D. disertation, Cambridge University Engineering Department 1976.

[13] H. Greenberg: An algorithm for the computation of knapsack functions. J. Math. Analysis and its Applications *26* (1969).

[14] A. Hordijk: Dynamic programming and Markov potential theory. Math. Centrum, Amsterdam 1974.

[15] R. A. Howard: Dynamic programming and Markov processes. Wiley, New York 1960.

[16] N. K. Jaiswal: Priority queues. Academic Press, London 1968.

[17] D. M. Jones: A sequential method for industrial chemical research. M. Sc. thesis, University College of Wales, Aberystwyth 1970.

[18] P. Nash: Optimal allocation of resources between research projects, Ph. D. thesis, Cambridge University, England 1973.

[19] P. Nash, J. C. Gittins: A Hamiltonian approach to optimal stochastic resource allocation. Adrances in Applied Probability *9* (1977).

[20] V. V. Rykov, E. Y. Lemberg: On optimal dynamic priority systems for mass service. Tech. Kibernetika (1967), 2, 25—34.

[21] L. J. SAVAGE: The Foundations of Statistics. Wiley, New York 1954.
[22] R. SCHLAIFER: Analysis of decisions under uncertainty. McGraw Hill, New York 1967.
[23] L. E. SCHRAGE: A proof of the optimality of the shortest remaining processing time discipline. Operations Research *16* (1968), 687—689.
[24] K. C. SEVCIK: The use of service time distributions in scheduling. Technical report CSRG-14 University of Toronto 1972.
[25] A. SIMONOVITS: Direct comparison of different priority queueing disciplines. Studia Scientiarum Math. Hungarica (to appear).
[26] J. VON NEUMANN, O. MORGENSTERN: Theory of games and economic behaviour. Second edition. Priceton University Press, Oxford 1953.
[27] S. J. YAKOWITZ: Mathematics of adaptive control processes. Elsevier, New York 1969.

UNIVERSITY OF OXFORD and UNIVERSITY OF CAMBRIDGE
MATHEMATICAL INSTITUTE — ENGINEERING DEPARTMENT

RANDOM GRAPH THEOREMS

GEOFFREY GRIMMETT

OXFORD

INTRODUCTION

In an earlier paper [3] McDiarmid and I studied the chromatic number of a random graph on n vertices in which each edge is present with a prescribed probability p independently of the presence or absence of the other edges. Such random graphs differ from those which Erdös and Rényi considered in their paper [1] on the evolution of random graphs, for they were interested in graphs on n vertices with $N(n)$ edges where N is a prescribed function and each N-subset of the set of possible edges occurs with equal probability. They noted that there is no essential difference between the two approaches.

Let Ω be the set of all undirected graphs on the vertex set $\{1, 2, 3, \ldots\}$ which have neither loops nor multiple edges. For each integer n and $\omega \in \Omega$ let ω_n be the finite graph derived from ω by deleting all except the first n vertices of ω. To each such ω_n we assign the probability

$$\mathbf{P}(\omega_n) = p^e q^{\binom{n}{2}-e}$$

where e is the number of edges of ω_n, and $p = 1 - q$ satisfies $0 < p < 1$. It is clear that the function $\mathbf{P}$ may be extended to a probability measure on the space $(\Omega, \mathscr{F})$, where $\mathscr{F}$ is the σ-field of subsets of Ω generated by the finite-dimensional sets $\{\omega \in \Omega : \omega_n = G\}$ as G ranges over all graphs on the vertices $\{1, 2, \ldots, n\}$ and n takes all possible values.

In [3] we studied the number $\Gamma_n(\omega)$ of colours which a particular graph colouring algorithm uses to colour the vertices of ω_n. This algorithm colours the vertices in numerical order as follows. Suppose that $\{c_1, c_2, c_3, \ldots\}$ are the available colours. Colour vertex 1 with c_1, and proceed to colour the remaining vertices in increasing numerical order using c_j to colour vertex i $(1 < i \leqq n)$ where j is the least positive integer such that no vertex already coloured c_j is joined to i. We showed that

$$\lim_{n\to\infty} \left(\Gamma_n \frac{\log n}{n}\right) = \log 1/q$$

with probability one and in any mean. Let $\sigma_n(\omega)$ be the number of vertices of ω_n coloured with c_1. Then

$$\lim_{n\to\infty} \frac{\sigma_n}{\log n} = (\log 1/q)^{-1}$$

with probability one and in any mean. In this paper I prove a stronger statement about the distribution of σ_n for large n which implies that the central moments of σ_n are bounded uniformly in n. In general it seems to be very difficult to describe the limiting distribution of the invariants of ω_n when they are suitably normalized. It is only possible for the variables $\{\sigma_n\}$ because they form a Markov chain.

In the second section of this paper I present a generating function for the rank polynomial of ω_n and deduce the expected chromatic polynomial of ω_n. The rank polynomials may also be used to find the expected number of components of a random graph.

THE LIMITING DISTRIBUTION OF THE CHAIN $\{\sigma_n\}$

The sequence $\{\sigma_n\}$ is a Markov chain with transition probabilities

$$\mathbf{P}(\sigma_{n+1} = j \mid \sigma_n = i) = \begin{cases} q^i & \text{if } j = i + 1\,, \\ 1 - q^i & \text{if } j = i\,, \end{cases}$$

and the initial distribution $\sigma_1 = 1$. The distribution of σ_n may be calculated formally, but I am interested only in its limiting distribution as $n \to \infty$.

THEOREM. *The chain $\{\sigma_n\}$ satisfies*

(1) $$\limsup_{n\to\infty} F(\sigma_n - \log_r n) = F(Y)\,,$$

(2) $$\liminf_{n\to\infty} F(\sigma_n - \log_r n) = F(Y + 1)$$

where $F(X)$ is the distribution function of X, Y is a random variable with specified distribution, and $qr = 1$.

The sequence $\sigma_n - \log_r n$ $(n \geqq 1)$ converges almost nowhere because in any infinite time interval it has infinitely many jump discontinuities of size arbitrarily close to one with probability one. In this sense the result of the theorem is the best possible.

The theorem holds also for the limiting distribution of the continuous time birth process $X(t)$ with $X(0) = 0$ and parameters $\lambda_n = q^n$. Indeed a similar result holds for birth processes whose parameters are such that

$$\left\{\frac{\lambda_{n+1}}{\lambda_n} : n = 0, 1, 2, \ldots\right\}$$

is a monotone convergent sequence whose limit λ satisfies $0 < \lambda < 1$.

The variable Y in (1) and (2) is given by

$$Y = \begin{cases} \frac{1}{2}\log_r (r^2 - 1) - \log_r Z & \text{if } \ Z > 0\,, \\ 0 & \text{if } \ Z = 0 \end{cases} \tag{3}$$

where

$$Z = \sum_{j=0}^{\infty} Z_j$$

is the sum of independent exponential random variables with expectations $q^j(1 - q^2)^{1/2}$. It is easy to show that Y has finite moments, and we deduce that

$$\limsup_{n\to\infty} \mathbf{E}\left|\sigma_n - \log_r n\right|^i < \infty$$

for any integer i.

Let

$$T_n = \inf\{j : \sigma_j = n\}$$

be the time of arrival of the nth point, and let

$$U_n = \frac{T_n}{v(n)}$$

where $v(n)^2$ is the variance of T_n. The proof of the theorem follows from the next lemma.

LEMMA. *The variable U_n converges to Z in distribution.*

Proof. Let $T_0 = 0$ and let

$$X_j = T_{j+1} - T_j\,, \quad j \geqq 0\,, \tag{4}$$

be the interarrival times of the chain. Then the X_j are independent geometric random variables distributed as

$$P(X_j = i) = q^j(1 - q^j)^{i-1}\,, \quad i \geqq 1\,. \tag{5}$$

Now, let $s \geqq 0$ and let φ_n be given by

$$\varphi_n(s) = \mathbf{E}(e^{-sU_n}).$$

Then by (4)

$$\varphi_n(s) = \prod_{j=0}^{n-1} \left(1 - r^j \left(1 - \exp\left(\frac{s}{v(n)}\right)\right)\right)^{-1}$$

where by (5)

$$v(n)^2 = \operatorname{var}(T_n) = \sum_{j=0}^{n-1} (r^{2j} - r^j) = \frac{r^{2n} - 1}{r^2 - 1} - \frac{r^n - 1}{r - 1}. \tag{6}$$

For any $m \geqq 0$ and $n \geqq m + 2$ we have that

$$\varphi_n(s) = Q_{m,n} \prod_{j=0}^{m} \left(1 - r^{n-1-j} \left(1 - \exp\left(\frac{s}{v(n)}\right)\right)\right)^{-1} \tag{7}$$

where

$$Q_{m,n} = \prod_{j=m+1}^{n-1} \left(1 - r^{n-1-j} \left(1 - \exp\left(\frac{s}{v(n)}\right)\right)\right)^{-1}.$$

But

$$1 \geqq Q_{m,n} \geqq \exp\left(-r^n \left(1 - \exp\left(\frac{s}{v(n)}\right)\right) \sum_{j=m+1}^{n-1} q^{j+1}\right)$$

since $1 + x \leqq e^x$, and

$$r^{n-1-j} \left(1 - \exp\left(\frac{s}{v(n)}\right)\right) \to -sq^j(1 - q^2)^{1/2}$$

as $n \to \infty$ by (6). Thus, letting $n \to \infty$ and $m \to \infty$ in (7) we deduce that

$$\varphi_n(s) \to \prod_{j=0}^{\infty} (1 + sq^j(1 - q^2)^{1/2})^{-1} = \mathbf{E}(e^{-sZ})$$

and so $U_n \to Z$ in distribution.

Proof of the theorem. We may extend the domain of the function v from the positive integers to the real numbers exceeding one by

$$v(x) = \left(\frac{r^{2x} - 1}{r^2 - 1} - \frac{r^x - 1}{r - 1}\right)^{1/2}.$$

Let $[x]$ denote the integer part of x. Then

$$\mathbf{P}(\sigma_n \geqq u + \log_r n) = \mathbf{P}(T_{[u + \log_r n]} \leqq n).$$

Writing $s = \log_r n$ we have that

$$\mathbf{P}(\sigma_n \geqq u + \log_r n) = \mathbf{P}\left(U_{[u+s]} \leqq \frac{r^s}{v([u+s])}\right).$$

But, for all large s

$$v(u+s-1) \leqq v([u+s]) \leqq v(u+s)$$

and so

$$\mathbf{P}(\sigma_n \geqq u + \log_r n) \geqq \mathbf{P}\left(U_{[u+s]} \leqq \frac{r^s}{v(u+s)}\right) \to \mathbf{P}(Z \leqq q^u(r^2-1)^{1/2})$$

as $s \to \infty$. Similarly

$$\mathbf{P}(\sigma_n \geqq u + \log_r n) \leqq \mathbf{P}\left(U_{[u+s]} \leqq \frac{r^s}{v(u+s-1)}\right) \to \mathbf{P}(Z \leqq q^{u-1}(r^2-1)^{1/2})$$

as $s \to \infty$. Defining Y by (3) gives

$$\limsup_{n\to\infty} F(\sigma_n - \log_r n) \leqq F(Y),$$

$$\liminf_{n\to\infty} F(\sigma_n - \log_r n) \geqq F(Y+1)$$

where $F(X)$ is the distribution function of the random variable X, and it is easy to see that equality holds in these equations.

THE RANK POLYNOMIAL OF ω_n

The rank polynomial of a graph G is defined by

$$W(x, y) = \left(\frac{x}{y}\right)^n \sum x^{-k} y^{e+k}$$

where the summation is over all subsets of the edge set of G, e and k are the numbers of edges and connected components of the resulting subgraph and n is the number of vertices of G. The chromatic polynomial of G is given by the relation

(8) $$f(\lambda) = \lambda^n W(-\lambda^{-1}, -1).$$

Let $W_n(x, y)$ be the expected value of the rank polynomial of ω_n, and let $N(n, k, e)$ be the number of graphs on n labelled vertices with e edges and k connected components. Then

$$W_n(x, y) = \sum_{e,k} x^{n-k} y^{e-n+k} N(n, k, e)\, p^e$$

since each (n, k, e) graph is a subgraph of a random graph with n vertices with probability p^e. Thus

$$W_n(x, y) = \left(\frac{x}{y}\right)^n G_n(y/x, py) \tag{9}$$

where

$$G_n(y, z) = \sum_{e,k} y^k z^e N(n, k, e).$$

It is not hard to show that

$$G(x, y, z) = \sum_n \frac{x^n}{n!} G_n(y, z) = \left(\sum_{i=0}^{\infty} \frac{x^i}{i!}(1 + z)^{i(i-1)/2}\right)^y \tag{10}$$

by a method similar to that used by Gilbert [2] who counted connected labelled graphs. So the exponential generating function

$$W(x, y; t) = \sum_n \frac{t^n}{n!} W_n(x, y)$$

is given by

$$W(x, y; t) = G(xt/y, y/x, py) \tag{11}$$

where G is given by (10). Setting $p = 1$ in this relation (11) gives a generating function for the rank polynomials of the complete graphs, and provides an alternative derivation of the result proved by Tutte [4].

Let $f_n(\lambda)$ be the expected number of λ-colourings of ω_n. Then by (8) and (9)

$$f_n(\lambda) = G_n(\lambda, -p)$$

and

$$f(\lambda; t) = \sum_n \frac{t^n}{n!} f_n(\lambda) = G(t, \lambda, -p) = \left(\sum_{i=0}^{\infty} \frac{t^i}{i!} q^{\binom{i}{2}}\right)^\lambda$$

by (10). This may be derived directly.

Generating functions for the expected value, and higher moments, of the number $k_n(\omega)$ of connected components of ω_n may be derived from the rank polynomial generating function. For, let

$$M_n(s) = \mathbf{E}(s^{k_n})$$

be the moment generating function of k_n. Then

$$M_n(s) = \sum p^e q^{\binom{n}{2} - e} s^k$$

where the summation is over all edge subsets of the complete graph on $\{1, 2, \ldots, n\}$, and e and k are the numbers of edges and connected components in the resulting subgraph. Thus

$$K(s, t) = \sum_n \frac{t^n}{n!} q^{-\binom{n}{2}} M_n(s)$$

is given by

$$K(s, t) = \sum_{n,k,e} \frac{t^n}{n!} s^k \left(\frac{p}{q}\right)^e N(n, k, e) = G(t, s, p/q) = (F(t))^s$$

where

$$F(t) = \sum_i \frac{t^i}{i!} q^{-\binom{i}{2}} .$$

This provides generating functions for the moments of k_n by differentiation. For example

$$\sum_n \frac{t^n}{n!} q^{-\binom{n}{2}} \mathrm{E}(k_n) = F(t) \log_c F(t) .$$

REFERENCES

[1] P. Erdös, A. Rényi: On the evolution of random graphs. Publ. Math. Inst. Hung. Acad. Sci. *5A* (1960), 17—61.

[2] E. N. Gilbert: Enumeration of labelled graphs. Can. J. Math. *8* (1956), 405—411.

[3] G. R. Grimmett, C. J. H. McDiarmid: On colouring random graphs. Math. Proc. Camb. Phil. Soc. *77* (1975), 313—324.

[4] W. T. Tutte: On dichromatic polynomials. J. Comb. Th. *2* (1967), 301—320.

University of Oxford

Mathematical Institute

РАСПРЕДЕЛЕНИЕ АБСОЛЮТНОГО МАКСИМУМА ПУАССОНОВСКИХ И ВИНЕРОВСКИХ ПРОЦЕССОВ НА ЦЕПИ МАРКОВА

Д. В. Гусак

Киев

В сообщении излагаются результаты по изучению распределения абсолютного максимума для пуассоновских и винеровских процессов, определенных на цепи Маркова. Для пуассоновских процессов, заданных на цепи Маркова, обобщается результат Королюка [1] по определению потенциала для обыкновенных пуассоновских процессов. С помощью этого обобщения находится представление для распределения абсолютного максимума, удобное при изучении его асимптотического поведения на ∞. Для винеровских процессов на цепи Маркова распределение абсолютного максимума определяется соответствующим предельным переходом в соотношениях, установленных в [2] для двойного интегрального преобразования распределения максимума на конечном интервале.

1.

Пусть $\{x_t, t \geqq 0\}$ — эргодическая цепь Маркова с конечным числом состояний $k = 1, 2, \ldots, n$ и матрицей переходных вероятностей

$$(1)\qquad \boldsymbol{P}(t) = \|\mathscr{P}\{x_t = r \mid x_0 = k\}\| = \exp\{t\boldsymbol{Q}\},$$
$$\boldsymbol{Q} = \boldsymbol{N}[\boldsymbol{P} - \boldsymbol{I}], \quad \boldsymbol{N} = \|\delta_{kr}\nu_k\| \quad (k = 1, 2, \ldots, n),$$

$0 < \nu_k < \infty$ — параметр показательного распределения времени τ_k сидения цепи x_t в состоянии k, $\boldsymbol{P}$ — матрица преходных вероятностей вложенной цепи $z_n = x_{\sigma_n + 0}$, σ_n — моменты изменения состояний цепи x_t.

Рассмотрим последовательность обобщенных пуассоновских процессов $\xi_k(t)$ ($t \geqq 0$, $\xi_k(0) = 0$, $k = 1, 2, \ldots, n$) с производящими функциями (п.ф.)

$$M e^{-\alpha\xi_k(t)} = \exp\{t\psi_k(\alpha)\}, \quad \psi_k(\alpha) = \alpha + \int_0^\infty (e^{-\alpha x} - 1)\, d\Pi_k(x).$$

Обозначим

$$\Psi(\alpha) = \|\delta_{kr}\, \psi_k(\alpha)\|$$

и рассмотрим процесс $\xi(t)$ ($t \geqq 0, \xi(0) = 0$), определяемый на цепи x_t с помощью п.ф.

$$\Phi_t(\alpha) = \|M(e^{-\alpha\xi(t)}, x_t = r \mid x_0 = k)\| = \exp\{t(\Psi(\alpha) + Q)\}. \tag{2}$$

Наряду с $\xi(t)$ будем рассматривать винеровский процесс $W(t)$ ($t \geqq 0$, $W(0) = 0$), определеяемый на цепи x_t с помощью п.ф.

$$\|M(e^{-\alpha W(t)}, \quad x_t = r \mid x_0 = k)\| = \exp\left\{t\left(Q + \frac{\alpha^2}{2}B^2 - \alpha A\right)\right\}, \tag{3}$$

$$B^2 = \|\delta_{kr} b_k^2\|, \quad b_k^2 \geqq 0; \quad A = \|\delta_{kr} a_k\|.$$

Пусть θ_s ($s > 0$) случайная величина с распределением $\mathscr{P}\{\theta_s > t\} = \exp\{-st\}$ ($t \geqq 0$),

$$P_s \equiv \|\mathscr{P}\{x_{\theta_s} = r \mid x_0 = k\}\| = s(sI - Q)^{-1}.$$

Непосредственным обращением матрицы $sI - Q$ нетрудно представить P_s в виде

$$P_s = s\,\Delta^{-1}(s)\,A(s), \quad \Delta(s) = |sI - Q|$$

$A(s)$ – соответствующая матрица алгебраических дополнений к элементам матрицы $sI - Q$. Из эргодичности цепи x_t следует существование стационарного распределения, определяемого соотношением

$$P^\infty = \lim_{s\to 0} P_s = \lim_{s\to 0} s\,\Delta^{-1}(s)\,A(0),$$

$$P^\infty = \|p_{kr}^\infty\|, \quad p_{kr}^\infty = \varrho_r > 0,$$

где $\lim_{s\to 0} s^{-1}\Delta(s) = (q^*)^{-1}$, $0 < q^* < \infty$ и $A(0)$ – матрица с положительными элементами: $A(0) = q^* P^\infty$.

Предположим, что выполняется условие

$$\tilde{\Pi}(-\alpha_0) = \int_0^\infty e^{\alpha_0 x}\, d\Pi(x) < \infty \quad (\alpha_0 > 0, \quad \Pi(x) = \|\delta_{kr}\,\Pi_k(x)\|). \tag{4}$$

Из результатов Миллера [3] следует, что максимальное собственное число $\lambda_1(\alpha)$ (перроновский корень) матрицы $\Psi(\alpha) + Q$ обладает свойством выпуклости при $\alpha > -\alpha_0$, причем $\lambda_1(0) = 0$ и возможны три случая:

a) $\lambda_1(\alpha_1) = \lambda_1(0) = 0$ ($\alpha_1 = -\varrho < 0$),

в) $\lambda_1(0) = \lambda_1'(0) = 0$ ($\alpha_1 = 0$),

c) $\lambda_1(\alpha_1) = \lambda_1(0) = 0$ ($\alpha_1 = \varrho > 0$).

Мы ограничимся случаем a). В этом случае $\lambda = 0$ является перроновским корнем уравнения $|\boldsymbol{\Psi}(-\varrho) + \boldsymbol{Q} - \lambda I| = 0$. Заметим, что

$$\Delta(\lambda, \alpha) = |\boldsymbol{\Psi}(\alpha) + \boldsymbol{Q} - \lambda I| = \prod_{k=1}^{n} (\lambda_k(\alpha) - \lambda)\,,$$

$$\Delta(\lambda, 0) = -\lambda \prod_{k=2}^{n} (\lambda_k(0) - \lambda)\,,$$

$$-\prod_{k=2} \lambda_k(0) = \lim_{\lambda \to 0} \lambda^{-1} \Delta(\lambda, 0) = q^*\,,$$

$$\Delta(\lambda, -\varrho) = -\lambda \prod_{k=2}^{n} (\lambda_k(-\varrho) - \lambda)\,,$$

$$-\prod_{k=2}^{n} \lambda_k(-\varrho) = \lim_{\lambda \to 0} \lambda^{-1} \Delta(\lambda, -\varrho)\,,$$

$$\lim_{\alpha \to 0} \alpha^{-1} \Delta(0, \alpha) = \lim_{\alpha \to 0} \alpha^{-1} \lambda_1(\alpha) \prod_{k=2}^{n} \lambda_k(0) = -q^* \lambda_1'(0)\,,$$

$$\lim_{\alpha \to 0} \alpha(\boldsymbol{\Psi}(\alpha) + \boldsymbol{Q})^{-1} = -\boldsymbol{A}(0) \lim_{\alpha \to 0} \alpha\, \Delta^{-1}(0, \alpha) = \frac{A(0)}{q^* \lambda_1'(0)} = \frac{1}{\lambda_1'(0)} \boldsymbol{P}^{\infty}\,.$$

Из выпуклости $\lambda_1(\alpha)$ и соотношения a) следует, что $\lambda_1'(0) > 0$. Воспользовавшись результатом из [4] по обращению возмущенных матриц, $\lambda_1'(0)$ можно выразить через усредненный по стационарному распределению сдвиг процесса $\xi(t)$

(5) $$\lambda_1'(0) = -\sum_{k=1}^{n} \varrho_k m_k = -m_0 \quad (m_k = \boldsymbol{M}\xi_k(1))\,.$$

Таким образом, при выполнении условия a) усредненный сдвиг m_0 процесса $\xi(t)$ отрицательный.

Из предположения о том, что перроновский корень $\lambda_1(\alpha)$ удовлетворяет условию a), следует, что существует предел $\boldsymbol{C}_\varrho = \lim_{\alpha \to -\rho} (\alpha + \varrho)(\boldsymbol{\Psi}(\alpha) + \boldsymbol{Q})^{-1}$, причем $\boldsymbol{C}_\varrho = \boldsymbol{D}'\boldsymbol{P}^{\infty}\boldsymbol{D}''$, где $\boldsymbol{D}'$ и $\boldsymbol{D}''$ – диагональные матрицы с положительными элементами. В дальнейшем по аналогии с обозначением m_0 в (5) все усредненные по стационарному распределению характеристики будут отмечаться нулевыми индексами. Например:

$$d_0' = \sum_{k=1}^{n} \varrho_k d_k'\,, \quad \tilde{\Pi}_0(0) = \sum_{k=1}^{n} \varrho_k \tilde{\Pi}_k(0)\,, \quad \tilde{\Pi}_0'(-\varrho) = \sum_{k=1}^{n} \varrho_k \tilde{\Pi}_k'(-\varrho)\,.$$

Рассмотрим процесс $\xi(t)$ $(t \geqq 0)$ с п.ф. (2) и обозначим

(6) $$\xi^+ = \sup_{0 \leqq t < \infty} \xi(t)\,.$$

Пусть ζ_k – показательное время между скачками процесса $\xi_k(t)$ с параметром $0 < \lambda_k < \infty$, ξ_k – скачки процесса $\xi_k(t)$, $F_k(x) = \mathscr{P}\{\xi_k < x\}$. Из соотношения

$$F_{kr}^{+}(x) \equiv \lim_{t\to\infty} \mathscr{P}\{\sup_{0\leqq u\leqq t} \xi(u) < x,\ x_t = r \mid x_0 = k\} =$$
$$= \mathscr{P}\{\xi^{+} < x,\ x_\infty = r \mid x_0 = k\} =$$
$$= \mathscr{P}\{\xi_k - \zeta_k < x,\ \xi^{+} < x + \zeta_k - \xi_k,\ \zeta_k < \tau_k,\ x_\infty = r \mid x_0 = k\} +$$
$$+ \sum_{j\neq k} p_{kj}\mathscr{P}\{\xi^{+} < x + \tau_k,\ \tau_k < \zeta_k,\ x_\infty = r \mid x_0 = j\}$$

нетрудно вывести интегро-дифференциальное уравнение, которое в матричном виде запишется так:

(7)
$$\boldsymbol{L}\boldsymbol{F}^{+}(x) \equiv \frac{\mathrm{d}\boldsymbol{F}^{+}(x)}{\mathrm{d}x} + \int_0^\infty \mathrm{d}\Pi(z)\,[\boldsymbol{F}^{+}(x-z) - \boldsymbol{F}^{+}(x)] + \boldsymbol{Q}\boldsymbol{F}^{+}(x) = 0 \quad (x > 0)\,,$$
$$\boldsymbol{F}^{+}(x) \equiv \|F_{kr}^{+}(x)\| = 0\ (x < 0)\,, \quad \mathrm{d}\boldsymbol{\Pi}(z) = \|\delta_{kr}\lambda_k\,\mathrm{d}F_k(z)\|\,.$$

Прежде чем найти решение задачи (7) с неизвестным начальным условием (т.к. $\boldsymbol{F}^{+}(+0)$ – неопределено), рассмотрим задачу с заданным начальным условием

(8)
$$\boldsymbol{L}\boldsymbol{R}(x) = \boldsymbol{0}\ (x > 0)\,,$$
$$\boldsymbol{R}(x) = \boldsymbol{0}\ (x < 0)\,, \quad \boldsymbol{R}(+0) = \boldsymbol{P}^\infty\,.$$

Матричное решение $\boldsymbol{R}(x)$, определяемое задачей (8) играет ту же роль, что и скалярная функция $R(x)$ в [1] с единичным начальным условием. Исходя из свойств матричной функции $\boldsymbol{\Psi}(\alpha) + \boldsymbol{Q}$ по аналогии с результатом для скалярного случая в [1] будем искать $\boldsymbol{R}(x)$ в виде

(9)
$$\boldsymbol{R}(x) = \boldsymbol{P}^\infty \boldsymbol{X}_0 + \boldsymbol{C}_\varrho \boldsymbol{Y}_0 \mathrm{e}^{-\rho x} + \boldsymbol{V}(x)\,,$$

где диагональные матрицы $\boldsymbol{X}_0$, $\boldsymbol{Y}_0$ и убывающая быстрее $\mathrm{e}^{-\rho x}$ $(x \to \infty)$ матричная функция $\boldsymbol{V}(x)$ будут определены в ходе доказательства следующего утверждения.

Теорема 1. Если для $\xi(t)$ $(t \geqq 0)$ с п.ф. (2) выполнены условия (4) и а), то

(10)
$$\boldsymbol{R}(x) = (1 + \tilde{\Pi}_0(0))^{-1}\,\boldsymbol{P}^\infty - \frac{\mathrm{e}^{-\rho x}}{\varrho \tilde{\Pi}_0'(-\varrho)\, d_0'}\,\boldsymbol{C}_\varrho (\boldsymbol{D}'')^{-1} + \boldsymbol{V}(x)\,,$$

(11)
$$\boldsymbol{F}^{+}(x) = \boldsymbol{P}^\infty - \frac{1 + \tilde{\Pi}_0(0)}{\varrho \tilde{\Pi}_0'(-\varrho)\, d_0'}\,\mathrm{e}^{-\rho x}\,\boldsymbol{C}_\rho (\boldsymbol{D}'')^{-1} + (1 + \tilde{\Pi}_0(0))\,\boldsymbol{V}(x)\,,$$

где $V(x)$ — убывающая быстрее $e^{-\rho x}$ $(x \to \infty)$ функция, преобразование Лапласа которой удовлетворяет уравнению

$$(12) \qquad (\Psi(\alpha) + Q)\,\tilde{V}(\alpha) = \Phi(\alpha),$$

$$\Phi(\alpha) = P^{\infty} - (I + \tilde{\Pi}(\alpha))\,[P^{\infty}X_0 + C_\varrho Y_0] +$$

$$+ \varrho(\alpha + \varrho)^{-1}\,[\tilde{\Pi}(\alpha) - \tilde{\Pi}(-\varrho)]\,C_\varrho Y_0 =$$

$$= P^{\infty} - (\Psi(\alpha) + Q)^{-1}\,(\alpha^{-1}P^{\infty}X_0 + (\alpha + \varrho)^{-1}\,C_\varrho Y_0),$$

$$P^{\infty}X_0 = (1 + \tilde{\Pi}_0(0))^{-1}\,P^{\infty}, \quad C_\varrho Y_0 = \frac{-1}{\varrho d_0' \tilde{\Pi}_0'(-\varrho)}\,C_\varrho(D'')^{-1}.$$

Доказательство. Подставляя (3) в уравнение (8) и учитывая, что

$$(13) \qquad (Q + \Psi(-\varrho))\,C_\varrho = 0$$

для $V(x)$ получаем интегро-дифференциальное уравнение

$$LV(x) = 0 \ (x > 0) \ (V(x) = -P^{\infty}X_0 - C_\varrho Y_0 e^{-\rho x},\ x < 0),$$

которое после интегрирования приводит к интегральному уравнению

$$(14) \qquad V(x) + \int_0^x \Pi(y)\,V(x - y)\,dy - Q\int_x^{\infty} V(y)\,dy =$$

$$= \int_x^{\infty} \Pi(y)\,[P^{\infty}X_0 + C_\varrho Y_0\,e^{-\rho(x-y)}]\,dy,$$

$$(V(0) = QV(0) + \tilde{\Pi}(0)\,P^{\infty}X_0 + \tilde{\Pi}(-\varrho)\,C_\varrho Y_0).$$

Применяя к (14) преобразование Лапласа, получаем уравнение (12), правая часть которого имеет вид

$$\Phi(\alpha) = P^{\infty}(I - X_0) - C_\varrho Y_0 - \tilde{\Pi}(\alpha)\,[P^{\infty}X_0 + C_\varrho Y_0] +$$

$$+ \varrho(\alpha + \varrho)^{-1}\,[\tilde{\Pi}(\alpha) - \tilde{\Pi}(-\varrho)]\,C_\varrho Y_0.$$

Для определения X_0 и Y_0 воспользуемся следующими условиями, которые должны выполняться для $\tilde{V}(\alpha)$:

1. $\lim_{\alpha \to 0} \alpha\,\tilde{V}(\alpha) = \lim_{\alpha \to 0} P^{\infty}\,\Phi(\alpha) = 0,$
2. $\lim_{\alpha \to -\rho} (\alpha + \varrho)\,V(\alpha) = \lim_{\alpha \to -\rho} C_\varrho\,\Phi(\alpha) = 0.$

Заметим, что из (10) следуют соотношения

$$(15) \qquad P^{\infty}(I + \tilde{\Pi}(-\varrho))\,C_\varrho = 0, \quad 1 + \tilde{\Pi}_0(-\varrho) = 0.$$

В силу условия 1)

$$P^{\infty}[I - (I + \tilde{\Pi}(-\varrho))\, C_{\varrho} Y_0 - (1 + \tilde{\Pi}_0(0))\, P^{\infty} X_0] = 0\,.$$

Учитывая (15), отсюда находим

$$P^{\infty} X_0 = (1 + \tilde{\Pi}_0(0))^{-1}\, P^{\infty}\,. \tag{16}$$

Условие 2) приводит к соотношению

$$C_{\varrho}[P^{\infty} - (I + \tilde{\Pi}(-\varrho))\,(P^{\infty} X_0 + C_{\varrho} Y_0) + \varrho \tilde{\Pi}'(-\varrho)\, C_{\varrho} Y_0] = 0\,,$$

которое с учетом (15) принимает вид

$$C_{\varrho} P^{\infty} + \varrho C_{\varrho} \tilde{\Pi}'(-\varrho)\, C_{\varrho} Y_0 = 0\,.$$

Отсюда находим

$$C_{\varrho} Y_0 = -\frac{1}{\varrho d_0' \, \tilde{\Pi}_0'(-\varrho)}\, C_{\varrho}(D'')^{-1}\,. \tag{17}$$

В скалярном случае (когда управляемая цепь Маркова имеет единственное состояние и рассматриваемый процесс является обычным пуассоновским процессом) соотношения (16) и (17) определяют постоянные величины, через которые выражается константа Крамера.

Воспользовавшись соотношением((15) и представлением

$$I + \tilde{\Pi}(\alpha) = \alpha^{-1}(\Psi(\alpha) - Q)$$

можно получить несколько упрощенное выражение для $\Phi(\alpha)$

$$\Phi(\alpha) = P^{\infty} - \alpha^{-1}(\Psi(\alpha) + Q)^{-1}\, [P^{\infty} X_0 + \alpha(\alpha + \varrho)^{-1}\, C_{\varrho} Y_0]\,.$$

2.

Рассмотрим винеровский процесс $\{W(t), t \geqq 0\}$, определяемый на цепи Маркова x_t при помощи п.ф. (3). В [2] рассматривается случай, когда B^2 – невырожденная диагональная матрица. Здесь ограничение невырожденности снимается и предполагается, что цепь x_t кроме эргодичности удовлетворяет условию обратимости (см. [5] гл. IY)

$$PR^{-1} = R^{-1} P^{\mathrm{T}} \quad (R = \|\delta_{kr} \varrho_k\|,\ P^{\mathrm{T}} = \|p_{kr}^{\mathrm{T}}\|,\ p_{kr}^{\mathrm{T}} = p_{rk})\,. \tag{19}$$

Из этого условия следует, что производящая матрица Q цепи x_t симметризуема и матрица

$$K(s) = (sI - Q)\, C \quad (C = NR^{-1})$$

симметрична и при $s > 0$ положительно определена.

Введем обозначения

$$D^2 = \left\| \tfrac{1}{2} b_k^2 \nu_k \varrho_k^{-1} \delta(b_k^2 > 0) + \delta(b_k^2 = 0) \right\| ,$$

$$I_\pm = \left\| e_k^\pm \delta_{kr} \right\| .$$

Элементы $e_k^\pm$ определяются следующим образом:

если $b_k^2 > 0$, то $e_k^\pm = 1$;

если $b_k^2 = a_k = 0$, то $e_k^\pm = 0$;

если $b_k^2 = 0$, $a_k < 0$, то $e_k^+ = 0$, $e_k^- = 1$;

если $b_k^2 = 0$, $a_k > 0$, то $e_k^+ = 1$, $e_k^- = 0$.

Имеет место следующее утверждение, обобщающее лемму 2 из [2].

Лемма. Если выполнено условие (19), то при $\operatorname{Im} \alpha = 0$ и $s \geqq 0$ справедливо разложение

$$K(s) + \frac{\alpha^2}{2} B^2 C - i\alpha A C = (X_-(s) + i\alpha I_-)\, D^2 (X_+(s) - i\alpha I_+) , \tag{20}$$

обратные величины множителей которого аналитичны соответственно в областях $\mp \operatorname{Im} \alpha > 0$ при $s > 0$.

Следует заметит, что при $s > 0$ $K(s) = X_-(s)\, D^2 X_+(s)$. Поэтому существуют обратные матрицы $K^{-1}(s)$ и $X_\pm^{-1}(s)$,

$$sK^{-1}(s) = C^{-1} P_s , \quad Z_+(s) = X_+^{-1}(s)\, I_+ , \quad Z_-(s) = I_-\, X_-^{-1}(s)$$

и (20) можно переписать в виде

$$K(s) + \frac{\alpha^2}{2} B^2 C - i\alpha A C = (I + i\alpha I_-\, X_-^{-1}(s))\, K(s)\, (I - i\alpha X_+^{-1}(s)\, I_+) =$$

$$= (I + i\alpha\, Z_-(s))\, K(s)\, (I - i\alpha\, Z_-(s)) .$$

Матрицы $Z_\pm(s)$ удовлетворяют уравнениям

$$Z_-(s)\, K(s)\, Z_+(s) = B^2 ,$$

$$K(s)\, Z_+(s) - Z_-(s)\, K(s) = A .$$

Обозначим

$$W^+(t) = \sup_{0 \leqq u \leqq t} W(u) , \quad W^+ = \sup_{0 \leqq t < \infty} W(t) .$$

Теорема 2. Если выполнено условие (19), то

(21) $$Me^{i\alpha W^+(\theta_s)} \equiv \|M(e^{i\alpha W^+(\theta_s)}, x_{\theta_s} = r \mid x_0 = k)\| = $$
$$= C(I - i\alpha X_+^{-1}(s) I_+)^{-1} C^{-1} P_s .$$

Пр этом условие невырождаемости $X_+(s)$ при $s \to 0$

(22) $$\lim_{s \to 0} X_+(s) = X_+ , \quad |X_+| \neq 0 ,$$

является необходимым и достаточным для того, чтобы абсолютный максимум был с вероятностью 1 ограничен и

(23) $$Me^{-uW^+} = C(I + uX_+^{-1} I_+) C^{-1} P^\infty .$$

При этом в невырожденном случае $B^2 > 0$ (т.е., когда при всех k $b_k^2 > 0$) и в случае $B^2 \geqq 0$, когда при некоторых k $b_k^2 > 0$, а при остальных k' $b_{k'}^2 = 0$, но $a_{k'} > 0$,

$$I_+ = I , \quad \mathscr{P}\{W^+ = 0\} = 0 .$$

В остальных случаях I_+ – вырождена и

(24) $$\mathscr{P}\{W^+ = 0\} = \lim_{u \to \infty} Me^{-uW^+} = \lim_{\varepsilon \to 0} \varepsilon C(\varepsilon I + X_+^{-1} I_+) CP^\infty \neq 0 .$$

Доказательство 1-ой части теоремы аналогично доказательству соответствующего соотношения теоремы в [2]. При осуществлении доказательства используется факторизационное разложение (20) и уравнение, которому удовлетворяет совместное распределение $\{W(t), W^+(t)\}$

$$P_z(t, x) = \|\mathscr{P}\{W(t) < z + x, W^+(t) < z, x_t = r \mid x_0 = k\}\| \quad (z > 0) ;$$

(25) $$\frac{\partial P_z}{\partial t} = \frac{1}{2} \frac{\partial^2 P_z}{\partial x^2} B^2 - \frac{\partial P_z}{\partial x} A + P_z Q \quad (x > 0) ,$$

$$P_z(t, x) = P_z(t, 0) \quad (x \geqq 0) ; \qquad P_z(0, x) = \delta(-z \leqq x) I \quad (x \leqq 0) .$$

Остальные соотношения теоремы 2 устанавливаются предельным переходом ($s \to 0$).

Заметим, что $X_\pm = \lim_{s \to 0} X_\pm(s)$ не могут быть одновременно невырожденными, так как при $s \to 0$ из соотношения

$$K(s) = X_-(s) D^2 X_+(s)$$

следует, что

$$-QC = X_- D^2 X_+ ,$$

где в силу вырожденности Q, хотя бы одна из матриц $X_\pm$ вырожена. Если $a_0 = \sum \varrho_k a_k = 0$, обе матрицы $X_\pm$ вырождены. Если $a_0 < 0$, X_+ – невырождена.

3.

В случае непрерывного достижения положительного уровня процессом $\xi(t)$ с п.ф. $\Phi_t(\alpha) = \exp\{t\Psi(\alpha)\}$, где

$$\Psi(\alpha) = i\alpha A - \frac{\alpha^2}{2} B^2 + \int_{-\infty}^{0} (e^{i\alpha x} - 1)\, d\Pi(x)\,, \tag{26}$$

($a_k > 0$ для тех k, при которых $b_k^2 = 0$) из результатов работы [6] следует, что распределение ξ^+ выражается с помощью нормированной положительной части распределения

$$\int_0^\infty \mathscr{P}\{\xi(t) > x\}\, dt = \left\| \int_0^\infty \mathscr{P}\{\xi(t) > x,\ x_t = r \mid x_0 = k\}\, dt \right\| \quad (x \geqq 0)\,.$$

Теорема 3. Если для процесса $\xi(t)$, определяемого на цепи x_t с помощью п.ф. (26) выполнено условие

$$\int_0^\infty \mathscr{P}\{\xi(t) > 0\}\, dt < \infty\,, \tag{27}$$

то при $x \geqq 0$

$$\mathscr{P}\{\xi^+ > x\} = \int_0^\infty \mathscr{P}\{\xi(t) > x\}\, dt \left[\int_0^\infty \mathscr{P}\{\xi(t) > 0\}\, dt\right]^{-1} \boldsymbol{P}^\infty\,. \tag{28}$$

ЛИТЕРАТУРА

[1] В. С. Королюк: Граничные задачи для сложного пуассоновского процесса. Теория вероят. и ее примен. (1974), № 1, 3—14.

[2] Д. В. Гусак: Распределение экстремумов винеровских процессов со сдвигом, управляемых конечной цепью Маркова. In: Trans. Sixth Prague Conf. 1971. Academia, Prague 1973, 207—212.

[3] H. D. Miller: A convexity property in the theory of random variables defined on a finite Markov chain. Ann. Math. Statist, *32* (1961), 1260—1270.

[4] В. С. Королюк, А. Ф. Турбин: Про один метод доведення граничных теорем для деяких функцiоналiв вiд напiвмарковских процесiв. УМЖ (1972), № 2, 233—238.

[5] Дж. Кемени, Дж. Снелл: Конечные цепи Маркова. Изд. Наука, Москва 1970.

[6] D. V. Gusak: On the continuous passage through a fixed level of a homogeneous process with independent increments on a Markov chain. Lecture Notes in Mathematics, v. 330. Springer-Verlag, Berlin—Heidelberg—New York 1973, 95—103.

Академия наук УССР

Институт математики

SOME ASPECTS OF AUTOMATIC SYSTEMS OF STATISTICAL INFERENCE

TOMÁŠ HAVRÁNEK

PRAGUE

If we consider automatic methods of data processing based on automatic generating (by computer) of questions about data and reduction (by computer) of obtained information by the help of deduction, then, if these methods have statistical background, some questions of statistical nature arise. Usually these methods enables us test on data large systems of hypotheses; the questions are then tests. In such situation we have to consider the problem of preserving characteristics of tests if deduction is used and problems of simultaneous inference (error rates). Such problems for some situations are formulated and solved in the presented paper.

1. INTRODUCTION

1.1. A class of interesting methods of automatic data analysis was investigated by logicians in latest years (see [5], [6] and [7]); this class of methods can be described by a theory called ALIOS (Automatic Listing of Important Observational Statements; see [8]). Concepts, this theory deals which, are: a set of observational sentences (say *Sent*), i.e. sentences which can be stated about data and truthfulness of which depends only on data, and a set of possible data samples (say $\mathscr{M}$; elements from $\mathscr{M}$ are called models). The truth-value of a $\varphi \in Sent$ given a model $M \in \mathscr{M}$ is denoted by $\|\varphi\|_M$ and must be effectively computable; we denote $Tr(M) = \{\varphi \in Sent;$ φ *true in* $M\}$.

On Sent deduction rules are considered; such a rule I must be sound, i.e. if φ is an immediate consequence of $\varphi_1, \ldots, \varphi_n$ w.r.t. to I then $\varphi_1, \ldots, \varphi_n \in Tr(M)$ implies $\varphi \in Tr(M)$ (for each $M \in \mathscr{M}$). We write $(\varphi_1, \ldots, \varphi_n)/\varphi \in I$ or $\frac{e}{\varphi} \in I$ for "φ is an immediate consequence of $\varphi_1, \ldots, \varphi_n$ ($e = \{\varphi_1, \ldots, \varphi_n\}$)". Note that if I is sound then $\frac{e}{\varphi} \in I$ implies $\{M; e \subseteq Tr(M)\} \subseteq \{M; \varphi \in Tr(M)\}$.

An *research problem* (r-problem) is given by a set $RQ \subseteq Sent$ of relevant questions (sentences for which we want to know whether they are true or not) and by a deduction rule I (which formalizes the research worker's ability to see immediate consequence of his statements). The mentioned methods (based on the usage of computers) are constructed for situations in which RQ is a large set of questions. Then the set in which we are interested (given data M) is $RQ \cap Tr(M)$ (the set of relevant truths). This set can be also large. To solve the r-problem $\langle RQ, I\rangle$ means to construct automatically on the base of the data M a *representation* of relevant truths; i.e. a set $X \subseteq Sent$ such that $X \subseteq Tr(M)$ and $RQ \cap Tr(M) \subseteq I(X)$. (In words: X is a set of true sentences and all relevant truths are immediate consequences of the sentences in X.) The aim is to obtain X as small as possible.

A systematic theory of such procedures is developed in the prepared publication [9]. Practical usage of various methods is described in [14].

1.2. The construction of the set RQ of relevant questions can be motivated statistically. Namely, some theoretical sentences of statistical type (e.g. some statements about distributions) can be associated with some sentences from *Sent*. We restrict ourselves to the case in which a pair $\Phi = \langle \Phi^+, \Phi^- \rangle$ of a hypothesis and an alternative corresponds to each $\varphi \in RQ$ in such a way that whenever φ is true (in the data we have at disposal) and $\langle \Phi^+, \Phi^- \rangle$ corresponds to φ then we *infer* Φ^- (reject the hypothesis). We suppose that the sentences Φ^+, Φ^- are sentences of a formal calculus. Let ι be the (one-one) mapping associating with each $\varphi \in RQ$ the corresponding pair $\langle \Phi^+, \Phi^- \rangle$ and put $TQ = \iota(RQ)$ (theoretical questions).

On the other hand we suppose that to each $\alpha \in (0, 0.5]$ and each $\Phi \in TQ$ we have a $\varphi(\alpha) \in Sent$ such that $\varphi(\alpha)$ is a test of Φ on the level α, i.e. $P(\varphi(\alpha) \in Tr(M) \mid \Phi^+) \leqq \alpha$. These tests satisfy the following monotonicity condition: for each $M \in \mathcal{M}$ if $\varphi(\alpha) \in$ $\in Tr(M)$ then $\varphi(\alpha') \in Tr(M)$ for each $\alpha' \geqq \alpha$. (Our convention is: if $\varphi(\alpha)$ is a test for Φ on the level α then $\varphi(\alpha')$ is a test for the same Φ on the level α'.) Consequently an RQ containing a test of Φ on a given level can be constructed. If it cannot lead to misunderstanding we shall write φ instead of $\varphi(\alpha)$.

1.3. Example. We shall give examples of reasonable candidates for TQ and RQ:

a) Hájek, Havel and Chytil [5] (see also [8] and [13]): Consider binary quantities $F_1, \ldots, F_k$; $\Phi_{\varkappa,\delta} = \langle \neg(\varkappa \Rightarrow_p \delta), \varkappa \Rightarrow_p \delta \rangle$, where $\varkappa \Rightarrow_p \delta$ means $P(\delta/\varkappa) \geqq p$ (p is given number); $\varkappa$ is an elementary conjunction and δ an elementary disjuntion from $F_1, \ldots, F_k$). TQ is then $\{\Phi_{\varkappa,\delta}\}_{\varkappa,\delta}$. The corresponding $\varphi(\alpha)$ (symbolically $\varkappa \rightarrow_{p,\alpha} \delta$) is true if the lower bound of one sided $(1 - \alpha)$ confidence interval (given by the frequencies of $\varkappa$ and δ in a data M) is greater then p.

b) Hájek [6], [8]: $2 \times 2 \times 2 \times \ldots \times 2$ contingency tables are considered (binary quantities as above, k e.g. equals to 30). $\Phi_1 = \langle F_1 \nsim F_2, F_1 \sim F_2 \rangle$, $\Phi_2 = \langle F_1 \nsim F_2$, $F_1 \sim \neg F_2 \rangle \ldots$ $\Phi_j = \langle F_1 \nsim F_2 \& F_3, F_1 \sim F_2 \& F_3 \rangle, \ldots$; generally $\Phi = \langle \varkappa \nsim \delta$, $\varkappa \sim \delta \rangle$, where $\varkappa$, δ are elementary conjuctions from $F_1, \ldots, F_k$; $\varkappa \nsim \delta$ is independence hypothesis (i.e. for 2 × 2 × 2 case $F_1 \sim F_2$ means $p_{ij.} = p_{i..}p_{.j.}$, $F_1 \sim F_2 \& F_3$ means $p_{111} = p_{1..}p_{.11}$) and $\varkappa \sim \delta$ is the alternative of positive association. Fisher exact test is used.

c) A complete system of hypotheses is considered by Victor [11]: Consider general three-dimensional contingency tables; hypotheses are as in (b) and, moreover, $\Phi_0^+ : p_{ijk} = p_{i..}p_{.j.}p_{..k}$ and hypotheses of conditioned independence are also considered. The usual χ^2-tests are used.

d) Bendová, Havránek [12]: F_1, F_2 are names of real quantities and $\Phi_i = \langle \Phi_0^+, \Phi_i^- \rangle$, where Φ_0^+ is independence hypothesis and Φ_i^- are particular regression alternatives. Rank tests of the Wilcoxon type are used.

1.4. For RQ-sets in the previous examples we have e.g. following deduction rules: For (a) $(\varkappa \& \varkappa' \rightarrow_{p,\alpha} \delta)/(\varkappa \rightarrow_{p,\alpha} \delta \vee \neg \varkappa')$. For (b) $(\varkappa \sim_\alpha \delta, \varkappa \ll \varkappa')/(\varkappa' \sim_\alpha \delta)$, where $\varkappa$ is a subconjunction of $\varkappa'$. Here $\varkappa \sim_\alpha \delta$ is true if the exact Fisher test accepts the alternative of positive association on the level α; $\varkappa \ll \varkappa'$ is true if $n(\varkappa \& \delta) \leqq n(\varkappa' \& \delta)$, $n(\varkappa \& \neg\delta) \geqq n(\varkappa' \& \neg\delta)$, $n(\neg k \& \delta) \geqq n(\neg\varkappa' \& \delta)$ and $n(\neg\varkappa \& \neg\delta) \leqq n(\neg\varkappa' \& \neg\delta)$ (n means the frequency in the given data). Note that some stronger but more complicated deduction rules can be formulated. We see that it is useful to consider some sentences from $Sent - RQ$ (auxiliary questions; AQ). A solution $X \subseteq RQ$ is called a *direct* solution.

There are two classes of questions of statistical nature: 1. questions of local interpretation; i.e. of preservation of properties of particular tests φ in such procedures; 2. questions of global interpretation, ie. of interpretation (or significance) of the whole solution X and/or the set of relevant truths $I(X) \cap RQ$.

In the present paper are these questions briefly sketched and some answers for the simpliest situations are proposed.

2. LOCAL INTERPRETATION

2.1. We shall suppose that the function $\|\varphi\|$ is measurable for each $\varphi \in Sent$ (remember that $\|\varphi\|$ is effectively computable indicator of a set in $\mathcal{M}$). We consider probability distributions on $\mathcal{M}$ (more precisely on $\langle \mathcal{M}, \mathcal{A} \rangle$, where $\mathcal{A}$ is a σ-algebra of subsets of $\mathcal{M}$). Hypotheses and alternatives are then conditions to such distributions ($P(A \mid \mathcal{H})$ is then to be read "the probability of an event under a distribution satisfying $\mathcal{H}$").

2.2. From our definition of a solution we see that if for some $\varphi \in RQ$ and $M \in \mathcal{M}$ we have $\varphi \in Tr(M)$ then either $\varphi \in X$ or there is an $e \subseteq X \subseteq RQ \cap AQ$ such that $\frac{e}{\varphi} \in I$ (X is a solution for given M).

We use a procedure $\mathscr{X}$ giving for each $M \in \mathscr{M}$ a solution X_M instead of testing (evaluating) all φ from RQ. Consider the probability of the rejection of Φ^+ for a $\Phi \in TQ$. Let $\varphi = \iota^{-1}(\Phi)$. The following assertion can be stated.

2.3. THEOREM. For any probability distribution on $\mathscr{M}$ and each $\varphi \in RQ$ we have

$$P(\varphi \in Tr(M)) = P(\varphi \in X_M \cup \bigcup_{\frac{e}{\varphi} \in I} e \subseteq X_M).$$

Proof. Use the fact that $\{M; \varphi \in Tr(M)\} = \{M; \varphi \in X_M\} \cup \bigcup_{\frac{e}{\varphi} \in I} \{M; e \subseteq X_M\}$.

The inclusion $\supseteq$ follows from the soundness of I (see 1.1) and $\subseteq$ follows from the above mentioned basic property of solutions.

2.4. PROPOSITION. Consider a test $\varphi \in RQ$ (on a given level). Then the testing of $\Phi = \langle \Phi^+, \Phi^- \rangle$ (if $\iota(\varphi) = \Phi$) with the use of the procedure $\mathscr{X}$ has the same probabilities of the both first and second kind errors as φ.

2.5. We can consider all $\varphi \in RQ$ to be tests on a given level α $(RQ(\alpha))$, then the obtained solution is called the *solution on the level* α $(X(\alpha))$. Suppose now that if $\frac{e}{\varphi} \in I$ then $e = \{\psi\}$ and $\varphi, \psi \in RQ(\alpha)$ for some α. Such deduction rule is called *invariant* if the following holds: if for some

$$\alpha \in (0, 0.5] \quad \frac{\psi(\alpha)}{\varphi(\alpha)} \in I$$

then for each

$$\alpha \in (0, 0.5] \quad \frac{\psi(\alpha)}{\varphi(\alpha)} \in I.$$

2.6. THEOREM. Let an r-problem $\mathscr{P} = \langle RQ(\alpha), I \rangle$ be given. Let I be invariant and let $X_M(\alpha)$ be a solution of $\mathscr{P}$. Let $0 < \alpha' \leqq \alpha$. Then $X'_M = \{\varphi(\alpha') \in Tr(M); \varphi(\alpha) \in X_M(\alpha)\}$ is a solution of $\langle RQ(\alpha'), I \rangle$.

Proof. Remember the assumptions from 1.2; if $\varphi(\alpha) \in RQ(\alpha)$ then $\varphi(\alpha')$ is a test of the same Φ. If $\psi(\alpha') \in X'_M$ then $\psi(\alpha)/\varphi(\alpha) \in I$ iff $\psi(\alpha')/\varphi(\alpha') \in I$ and $I(\psi(\alpha')) \subseteq \subseteq Tr(M)$; on the other hand if a $\varphi(\alpha') \in Tr(M)$ then $\varphi(\alpha) \in Tr(M)$ and so $\varphi(\alpha) \in \in I(X_M(\alpha))$. □

2.7. REMARK. We can consider a more general case: $e = \{\psi(\alpha)\} \cup e_1$, where $e_1 \subseteq AQ$. Then $X'_M = Y_M \cup (AQ \cap X_M)$, where $Y_M = \{\varphi(\alpha') \in Tr(M); \varphi(\alpha) \in X_M \cap \cap RQ\}$. An example of such deduction rule is (b) from 1.4.

2.8. Let $\varphi(\alpha) \in X_M$ for some $M \in \mathscr{M}$. We consider the critical level $\alpha_0 = \alpha(\varphi, M) = \inf\{\alpha; \varphi(\alpha) \in Tr(M)\}$. If I is invariant then we know that $\alpha(\varphi', M) \leqq \alpha_0$ for each $\varphi' \in I(\varphi(\alpha) \cup (AQ \cap X_M)) \cap RQ$. The procedures described in [6] and [7] (see also [14]) give in fact (1) a solution X_M on a given level and (2) the critical levels for sentences of $X_M \cap RQ$.

3. GLOBAL INTERPRETATION

3.1. Consider a set RQ and corresponding $TQ = \iota(RQ)$; these sets are supposed to be finite. Let card $(TQ) = r$. Recall that TQ is a set of pairs of sentences of the form $\langle \Phi^+, \Phi^- \rangle$. Say that a sentence Ψ *belongs* to $\langle \Phi^+, \Phi^- \rangle$ if Ψ is either the sentence Φ^+ or the sentence Φ^-. A set Z of sentences is a *component* of TQ if (1) for each $\Psi \in Z$ there is a $\Phi \in TQ$ such that Ψ belongs to Φ and (2) the conjunction $\&Z$ of all members of Z is satisfiable (not contradictory).

REMARK. (1) It follows from our finitness assumption that each component is contained in a maximal component. (2) It follows from the condition (2) above that there is no $\langle \Phi^+, \Phi^- \rangle \in TQ$ such that both Φ^+ and Φ^- is in Z.

We assume, in accordance with methods of [5]–[9], [11]–[13], that $\mathscr{H}_0 = \{\Phi^+; \Phi \in TQ\}$ is a maximal component. As a measure of quality of a given solution X we can use the following probabilities: (1) $P(\{M; I(X) \cap RQ \cap Tr(M) = \emptyset\} \mid \& \& \mathscr{H}_0)$ and (2) $P(\{M; I(X) \cap RQ \cap D \subseteq Tr(M)\} \mid \& \mathscr{H})$, where $\mathscr{H}$ is a maximal component, $\mathscr{H} \neq \mathscr{H}_0$ and $D = \{\iota^{-1}(\Phi);\ \Phi^- \in \mathscr{H}\}$. We shall now investigate (1) – the probability of the global first kind error. We can see immediately the following:

3.2. LEMMA. Let α be a number $(0 < \alpha)$. The condition $P(\iota^{-1}(\Phi) \in Tr(M) \mid \Phi^+) \leqq \alpha$ for each $\Phi^+ \in \mathscr{H}_0$ is necesary for $(1) \leqq \alpha$ for any solution of a given r-problem.

3.3. Let φ be a sentence from $RQ \cap Tr(N)$ for a given data N. The sentence $\varphi^N_{\text{crit}} = \varphi(\alpha')$, where $\alpha' = \alpha(\varphi, N)$ is called the *critical strengthening* of φ (similarly for a set $Y \subseteq RQ \cap Tr(N) : Y^N_{\text{crit}} = \{\varphi^N_{\text{crit}}; \varphi \in Y\}$). We can now consider instead the given solution X his strengthening $X^N_{\text{crit}} = (X \cap RQ)^N_{\text{crit}} \cup (X - RQ)$. If I is invariant note that: (1) $\iota(I(X) \cap RQ) = \iota(I(X^N_{\text{crit}}) \cap RQ)$; we make the same inferences from $I(X) \cap RQ$ as from $I(X^N_{\text{crit}}) \cap RQ$. (2) All tests from $I(X^N_{\text{crit}}) \cap RQ$ are on level less then $\alpha_{\max} = \max\{\alpha(\varphi, N);\ \varphi \in RQ \cap X\}$.

3.4. LEMMA. Let X be a solution (obtained from given data). Then: (1) If each $\varphi \in X$ is on the level less or equal to α/k, where $k = \text{card}\,(I(X) \cap RQ)$, then the probability (1) is less or equal to α. (2) Suppose I to be invariant. Let, for each $\varphi \in I(X) \cap RQ$, $\alpha_\varphi = \min\{\alpha(\varphi, N);\ \psi \in X \cap RQ,\ \varphi \in I(\psi \cup e)$ for some $e \subseteq X\}$. Then the condition $\sum_{\varphi \in I(X) \cap RQ} \alpha_\varphi \leqq \alpha$ is sufficient to be $(1) \leqq \alpha$ for X^N_{crit}.

Proof. The both assertions follows from the fact that $\{M; I(X) \cap RQ \cap Tr(M) = \emptyset\} = \bigcup_{\varphi \in I(X) \cap RQ} \{M; \varphi \in Tr(M)\}$ (for (2) we substitute X^N_{crit} for X). □

3.5. We can consider the global qualities of solutions in connection with error rates used in multiple comparisons problems (see e.g. [1] or [10]). The first error rate, which we are going to consider, is the inferencewise error rate: i.e.r. = number of errorneous inferences/number of inferences. An inference is here a usage of the rule $\frac{\varphi}{\Phi^-}$. This inference is errorneous if Φ^+ is true.

3.6. THEOREM. Consider an r-problem $\mathscr{P} = \langle RQ, I \rangle$. Let I be invariant and let X be a solution of $\mathscr{P}$ (given data N). Then for X^N_{crit} and each maximal component

$$E(\text{i.e.r.} \mid \& \mathscr{H}) \leqq \alpha_{\max} .$$

Proof. Remember that $\alpha_{\max} = \max \{\alpha(\varphi, N); \varphi \in X \cap RQ\}$. Let $I(X^N_{\text{crit}}) \cap RQ = \{\varphi_1, \ldots, \varphi_k\}$ and let $\Phi_1, \ldots, \Phi_k$ be the corresponding theoretical pairs ($\Phi_i = \langle \Phi_i^+, \Phi_i^- \rangle$). Assuming $\& \mathscr{H}$ we have i.e.r. $= k^{-1}(\sum_{\Phi_i^+ \in \mathscr{H}} \|\varphi_i\|)$. ($\|\varphi_i\|$ is the function assigning to each M the value $\|\varphi_i\|_M$). So i.e.r. attends maximum under the assumption $\mathscr{H}_0$. Then $E(\text{i.e.r.} \mid \& \mathscr{H}_0) = E(k^{-1} \sum_{i=1}^{k} \|\varphi_i\| \mid \& \mathscr{H}_0) \leqq k^{-1} \sum_{i=1}^{k} P(\|\varphi_i\| = 1 \mid \Phi_i^+) \leqq \leqq k^{-1} k \alpha_{\max} = \alpha_{\max}$.

3.7. If our aim is to obtain as far as possible complete information on the given data and if we should not make a (reliable) conclusion depending directly on the simultaneous correctness of the set $RQ \cap Tr(M)$, the usage of i.e.r. is appropriate (cf. Cox in [2]). The original aim of the above mentioned methods was to give such a complete information and a possibility of choice of some interesting hypotheses for the further testing.

3.8. The most rigorous error rate is the following first experimentwise error rate:
e.e.r.$_1$ = number of errorneous inferences/number of experiments. We have an $M \in \mathscr{M}$, i.e. the output of one experiment (in a sense). Then under condition $\mathscr{H}_0$ is e.e.r.$_1 = \sum_{i=1}^{k} \|\varphi_i\|$.

3.9. THEOREM. Under assumptions of 3.5 if $\alpha_\varphi \leqq \alpha/k$ for each $\varphi \in I(X) \cap RQ$ then for X^N_{crit} and each maximal component $\mathscr{H}$ it is $E(\text{e.e.r.}_1 \mid \& \mathscr{H}) \leqq \alpha$.

Proof is analogous to 3.5. Use Lemma 3.3.

3.10. The present concept of error rate corresponds to R. A. Fisher's concept of error rate in multiple comparisons tests (B [4]). If we have a solution X we see immediately $I(X)$. We can choose a set of interesting statements $S \subseteq I(X) \cap RQ$ and infer from them. From the point of error rates it is the same as if we tested only

$\{\Phi;\ \iota^{-1}(\Phi) \in S\}$. If each $\varphi \in S$ is on the level $\leqq \alpha/k$ then $E(\text{e.e.r.}_1 | \& \mathscr{H}) \leqq \alpha$ (where e.e.r.$_1$ is restricted to S). We can use also the following consideration: let $X{\downarrow}S$ be a minimal subset of X such that $S \subseteq I(X{\downarrow}S) \cap RQ$. Then we aply Theorem 3.9 to $(X{\downarrow}S)^N_{\text{crit}}$. (Lemma 3.4 can be also used.)

3.11. Considering the second experimentwise error rate e.e.r.$_2$ = number of experiments with one or more errorneous inferences/number of experiments, the inner structure of the set TQ can be used (see the paper [3] by Gabriel).

We suppose now that we have in addition to an invariant rule I an inference rule I_T on theoretical sentences such that if $\dfrac{\Phi^-}{\Psi^-} \in I_T$ then $\iota^{-1}(\Phi)(\alpha)/\iota^{-1}(\Psi)(\alpha) \in I$; let this rule be sound and complete in the following sense: if $\dfrac{\Phi^-}{\Psi^-} \in I_T$ then Ψ^+ logically implies $\Phi^+(\Psi^+ \Rightarrow \Phi^+)$ and if $\Psi^+ \Rightarrow \Phi^+$ then Ψ^- is I_T-provable from Φ^-. Consider a direct solution X of a given r-problem (given data N). Denote $TTr = \{\Phi^-;\ \Phi \in \iota(X)\}$. Let φ_0 be an element of $I(X) \cap RQ$ and $\Phi_0 = \iota(\varphi_0)$. The system of hypotheses (pairs) $\mathscr{F} = \{\Phi;\ \Phi_0^-$ is I_T-provable from Φ^- and $\Phi^- \in I_T(\Psi^-)$ for a $\Psi^- \in TTr\}$ is called the system based on Φ_0. Note that if $\mathscr{F}$ is based on Φ_0 then $\Phi_0^+ \Rightarrow \Phi^+$ for each $\Phi \in \mathscr{F}$.

3.12. THEOREM. Let X be a direct solution of a given r-problem (and given data N). Choose a $\varphi_0 \in I(X) \cap RQ$. Let $\mathscr{F}$ is the system based on Φ_0. Then: (1) The system $\{\langle \Phi^+, \iota^{-1}(\Phi)\rangle\}_{\Phi \in \mathscr{F}}$ is coherent; (2) if we denote $\alpha^0 = \min\{\alpha(\varphi, N);\ \Phi_0^- \in I_T(\iota(\varphi)^-)$ and $\varphi \in X\}$ then $\alpha(\varphi_0, N) \leqq \alpha^0$. (3) If we denote $\alpha^1 = \max\{\alpha(\varphi, N);\ \Phi_0^- \in I_T(\iota(\varphi)^-)$ and $\varphi \in X\}$ and $X^0 = \{\varphi^N_{\text{crit}};\ \varphi \in X$ and $\Phi_0^- \in I_T(\iota(\varphi)^-)\}$ then the probability that $I(X^0) \cap RQ$ rejects one or more true hypotheses from $\mathscr{F}$ is less or equal to α^1 under condition Φ_0^+.

Proof. (1) A system $\mathscr{S} = \{\Phi^+, \iota^{-1}(\Phi)\}$ is coherent (Gabriel [3]) if the following holds: If $\Phi^+ \Rightarrow \Psi^+$ then, for each α and $M \in \mathscr{M}$, $\iota^{-1}(\Psi)(\alpha) \in Tr(M)$ implies $\iota^{-1}(\Phi)(\alpha) \in Tr(M)$; use the completeness of I_T, the relation between I_T and I and the soundness of I.

(2) can be seen immediately.

(3) From (1) we see that, for each α and each $M \in \mathscr{M}$, $\iota^{-1}(\Phi)(\alpha) \in Tr(M)$ implies $\iota^{-1}(\Phi_0)(\alpha) \in Tr(M)$ (for each Φ such that $\Phi^- \in TTr$); in particular for α^1. Consequently, if we recall the definition of $\mathscr{F}$ it is

$$\{M;\ \iota^{-1}(\Phi_0)(\alpha^1) \in Tr(M)\} \subseteq \bigcup_{\varphi} \{M;\ \varphi(\alpha^1) \in Tr(M) \text{ and } \iota(\varphi) \in \iota(I(X^0) \cap RQ) \cap \mathscr{F}\}. \qquad \square$$

3.13. REMARK. (1) Note that the soudness of I_T as supposed in 3.10 is a sufficient condition for (3) from Theorem 3.14. We need not the completeness of I_T. (2) If we define $\mathscr{F}^o = \{\iota(\varphi); \varphi_0$ is I-provable from φ and $\varphi \in I(\psi) \cap RQ$ for a $\psi \in X\}$ (note that $\mathscr{F}^0 \supseteq \mathscr{F}$), assume satisfiability of $\& \mathscr{F}^+$ (where $\mathscr{F}^+ = \{\Phi^+; \Phi \in \mathscr{F}\}$ and replace, in the point (3) of Theorem 3.12, $\mathscr{F}$ by $\mathscr{F}^0$ and condition Φ_0^+ by $\& \mathscr{F}^+$, then the assertion holds without any assumptions about I_T. (3) In the practical treatment of solutions it is very helpfull to use both the above results and results of 3.10.

3.14. Example. $\Phi^+ : \neg(\varkappa \Rightarrow_p \delta)$, $\Phi^- : \varkappa \Rightarrow_p \delta$; $I_T = \left\{\dfrac{\varkappa_1 \& \varkappa_2 \Rightarrow_p \delta}{\varkappa_1 \Rightarrow_p \delta \vee \neg \varkappa_2}\right\}$. If $\varkappa \Rightarrow_p \delta \in$ $\in TTr$ we choose $\Phi_0^+ : \neg(\emptyset \Rightarrow_p \delta \vee \neg\varkappa_2)$. It is clear that Φ_0^- can be infered from different sentences (by I_T). On the other hand the inference rule I_T corresponding to $\dfrac{\varkappa \sim_\alpha \delta, \varkappa \preccurlyeq \varkappa'}{\varkappa' \sim_\alpha \delta}$ is not sound and cannot be used; we can use 3.13 (2); consider $I(\Psi \cup e)$, where $e \subseteq AQ$.

4. APPENDIX

4.1. The investigated procedures are consistent in a sense to be described. Suppose that $M = \langle |M|, f_1, \ldots, f_k \rangle$, where $|M|$ is a finite set of objects and $f_1, \ldots, f_k$ are tables of values of measured quantities on M. Suppose that if $\{|M|_n\}$ is a sequence of distinct sets then each sequence of random variables measured on different sets from $\{|M|_n\}$ is independent. Let now X_n be a sequence of obtained solutions and $(\text{i.e.r.})_n$ the corresponding error rates under $\mathscr{H}_0$. Let $E((\text{i.e.r.})_n | \& \mathscr{H}_0) \leqq \alpha_n$. Let $\alpha_n \leqq \alpha$. The number of inferences from the each model M_j is a finite number $n_j \leqq$ $\leqq$ const. Then $\operatorname{Var}(\text{i.e.r.})_j \leqq C$ for each j.

4.2. Theorem. Consider global error rate $gl(\text{i.e.r.})_n = (\sum_{j=1}^n m_j) / \sum_{j=1}^n n_j$, where m_j is the number of errorneous inferences from M_j. Then $P(\lim_{n \to \infty} gl(\text{i.e.r.})_n \leqq \alpha | \& \mathscr{H}_0) = 1$.

Proof. We have $(\sum^n m_j)/\sum^n n_j = (\sum^n n_j(m_j/n_j))/\sum^n n_j = \sum^n n_j Y_j / \sum^n n_j$. By Kolmogorov's inequality we obtain $P(\max_{1 \leqq k \leqq n} ((\sum^k n_j Y_j)/\sum^k n_j - (\sum^k n_j \alpha_j)/\sum n_j| \geqq \varepsilon) \leqq$ $\leqq C \sum^n n_j^2 (\varepsilon \sum^n n_j)^{-2}$. We know that $\lim (\sum^n n_j^2 / (\sum^n n_j)^2) = 0$ and $\lim ((\sum^n n_j \alpha_j)/\sum^n n_j) \leqq$ $\leqq \alpha$.

Analogous statements can be proved for the other error rates.

Acknowledgement. I wish to thank Dr. P. Hájek, CSc., for his helpfull and valuable comments and support of my work, and to thank Ing. Dr. M. Chytil, CSc., for many stimulating discussions on GUHA methods.

REFERENCES

[1] L. N. Balaam, W. T. Federer: Error rate basis. Technometrics *7* (1965), 260—262.
[2] D. R. Cox: A remark on multiple comparisons methods. Technometrics *7* (1965), 223—224.
[3] K. R. Gabriel: Simultaneous test procedures — some theory of multiple comparisons. Annals of Math. Statisti. *40* (1969), 224—250.
[4] R. A. Fisher: The design of experiments. Oliver and Boyd, London 1936.
[5] P. Hájek, I. Havel, M. Chytil: The GUHA method of automatic hypotheses determination, Computing *1* (1966), 293—308.
[6] P. Hájek: Problém obecného pojetí metody GUHA (The question of a general notion of the GUHA method). Kybernetika 4 (1968), 505—515.
[7] P. Hájek, K. Bendová, Z. Renc: The GUHA method and the three valued logic. Kybernetika *7* (1971), 421—435.
[8] P. Hájek: Automatic listing of important observational statements I—III. Kybernetika *9* (1973), 187—205 and *10* (1974), 95—124.
[9] P. Hájek, T. Havránek: Mechanized hypothesis formation (a monography; in print.)
[10] R. O'Neil, G. B. Wetheril: The present state of multiple comparisons methods. JRSS B *33* (1971), 218—250.
[11] N. Victor: Zur Klassifizierung mehrdimensionaler Konfidenctafeln. Biometrics *28* (1972), 427—441.
[12] K. Bendová, T. Havránek: Výběr regresních vztahů (Choosing of regression models). Sdělení MSBÚ ČSAV *2* (1) (1973), 1—19.
[13] T. Havránek: The statistical modification and interpretation of GUHA method. Kybernetika 7 (1971), 13—21.
[14] P. Hájek, I. Havel, T. Havránek, M. Chytil, J. Rauch and Z. Renc: Metoda GUHA — skripta (The GUHA method — mimeographed), Dům techniky ČVTS, České Budějovice 1975.

Czechoslovak Academy of Sciences

Mathematical Centre of Biology

RANDOM MEASURES OF GLEASON TYPE

EWA HENSZ and RYSZARD JAJTE

ŁÓDŹ

The aim of this paper is to give a spectral representation for random measures of Gleason type. Random Gleason measures may be treated as a non-commutative version of stochastic processes with independent increments. Theorems 1 and 2 give the representation of such processes in terms of S-operator families. In particular, the operator-valued measure N (Theorem 2) plays the same role as the Lévy-Khinchine spectral measure in the theory of infinitely divisible distributions and it is a very convenient tool in consideration concerning convergence and compactness of sequences of random Gleason measures.

We shall consider a separable Hilbert space H, $\dim H \geqq 3$, with the inner product $(\ ,\ )$. Let us denote by $\mathfrak{N}$ the lattice of all orthogonal projectors acting in H.

A non-negative function μ defined on $\mathfrak{N}$ and satisfying the condition $\mu(\sum_{j=1}^{\infty} E_j) = \sum_{j=1}^{\infty} \mu(E_j)$ for every sequence $E_1, E_2, \ldots$ of mutually orthogonal operators from $\mathfrak{N}$ is called a *Gleason measure.*

The Gleason theorem [1] states that every such measure on $\mathfrak{N}$ is of the form $\mu(E) = \operatorname{tr} ME$, $E \in \mathfrak{N}$, where M is an S-operator, i.e. M is non-negative, symmetric and with a finite trace. Moreover, M is uniquely determined by μ.

Let $(\Omega, \mathscr{F}, P)$ be a probability space and let $L_0(\Omega, \mathscr{F}, P)$ denote a space of random variables on $(\Omega, \mathscr{F}, P)$ with the topology of convergence in probability P.

By the *random Gleason measure* we mean a stochastic process $\xi : \mathfrak{N} \to L_0(\Omega, \mathscr{F}, P)$ such that for every sequence $E_1, E_2, \ldots$ of mutually orthogonal projectors the random variables $\xi(E_1), \xi(E_2), \ldots$ are independent and the equality

$$\xi(\sum_{j=1}^{\infty} E_j) = \sum_{j=1}^{\infty} \xi(E_j) \tag{1}$$

holds. In the sequel we shall always assume that all the random variables $\xi(E)$, $E \in \mathfrak{N}$, are symmetric and infinitely divisible. Such situation corresponds to the considering of non-atomic usual measures in the classical (commutative) case.

Denote by $\hat{\xi}(E, *)$ the characteristic function (Fourier transform) of $\xi(E)$, i.e. $\hat{\xi}(E, t) = \mathbf{E} \exp(it\, \xi(E))$, $t \in \mathbf{R}$.

Recall that a family M_t, $t \in T$, of S-operators is said to be compact if the following conditions are fulfilled:

$$(2) \qquad 1^\circ \qquad \sup_{t \in T} \operatorname{tr} M_t < \infty$$

$$2^\circ \qquad \lim_{n \to \infty} \sup_{t \in T} \sum_{k=n+1}^{\infty} (M_t e_k, e_k) = 0$$

for some orthonormal basis $(e_k)_{k=1}^{\infty}$ in H.

THEOREM 1. If ξ is a random Gleason measure then there exists a one-parameter family M_t, $t \in \mathbf{R}$, of S-operators, such that

$$(3) \qquad \hat{\xi}(E, t) = \exp(-\operatorname{tr} M_t E), \quad E \in \mathfrak{N},$$

and, moreover, the family M_t satisfies the following conditions:

$$(\alpha) \qquad M_0 = 0, \quad M_{-t} = M_t, \quad t \in \mathbf{R};$$

(β) M_t, $t \in \mathbf{R}$, is locally compact i.e. for every $T > 0$ the family M_t, $|t| \leqq T$, is compact;

$$(\gamma) \qquad \lim_{t \to 0} (M_t x, x) = 0 \quad \text{for} \quad x \in H;$$

(δ) the quadratic form $(M_t x, x)$ is negative definite (in wide sense), i.e. for each sequence of complex numbers $\alpha_1, \ldots, \alpha_n$, $n = 1, 2, \ldots$, the condition $\sum_{j=1}^{n} \alpha_j = 0$ implies

$$\sum_{j,k=1}^{n} \alpha_j \bar{\alpha}_k (M_{t_j - t_k} x, x) \leqq 0$$

for, $t_j \in \mathbf{R}$, $j = 1, \ldots, n$ and $x \in H$.

Conversely, if the family M_t. $t \in \mathbf{R}$, of S-operators satisfies the conditions (α) ... (δ) then there exists a random Gleason measure ξ such that (3) holds.

Proof. Let ξ be a random Gleason measure and let us consider a sequence $E_1, E_2, \ldots$ of mutually orthogonal projectors. Denoting $E = \sum_{k=1}^{\infty} E_k$ and using the independence of $\xi(E_k)$ we obtain

$$(4) \qquad \hat{\xi}(E, t) = \lim_{n \to \infty} \prod_{k=1}^{n} \hat{\xi}(E_k, t)$$

for all $t \in \mathbf{R}$. Notice that $0 \leqq \hat{\xi}(E, t) \leqq 1$ ([4], p. 297). Putting

$$(5) \qquad \psi(E, t) = -\ln \hat{\xi}(E, t) \quad \text{for} \quad E \in \mathfrak{N}$$

we have $\psi(E, t) \geqq 0$ and by (4) the equality

$$\psi(E, t) = \sum_{k=1}^{\infty} \psi(E_k, t)$$

holds. Therefore, for each $t \in \mathbf{R}$, $\psi(\cdot, t)$ is a Gleason measure and by Gleason theorem it can be written in form

$$(6) \qquad \psi(E, t) = \operatorname{tr} M_t E, \quad t \in \mathbf{R}, \quad E \in \mathfrak{N},$$

where M_t is an S-operator acting in H. Thus, by (5) and (6), we obtain (3). We shall show that M_t, $t \in \mathbf{R}$, satisfies $(\alpha) \dots (\delta)$. In fact, $\hat{\xi}(E, 0) = 1$, so $\operatorname{tr} M_0 E = 0$ for $E \in \mathfrak{N}$. Putting $E = I$ we obtain $M_0 = 0$. Further, by the distribution symmetry of $\xi(E)$ we have $\hat{\xi}(E, -t) = \hat{\xi}(E, t)$. Hence $\operatorname{tr} M_t E = \operatorname{tr} M_{-t} E$ for any $E \in \mathfrak{N}$ and $M_t = M_{-t}$. To show (β) notice that for given $T > 0$ there exists $\delta > 0$ such that $\inf_{|t| \leqq T} \hat{\xi}(I, t) > \delta > 0$. Thus, by (3), we have $\sup_{|t| \leqq T} \operatorname{tr} M_t < -\ln \delta < \infty$.

Let now $(e_k)_{k=1}^{\infty}$ be an orthonormal basis in H. Denote by $\bar{e}_k$ an operator projecting the subspace spanned by e_k. Putting $E_n = \sum_{k=n+1}^{\infty} \bar{e}_k$ we obtain

$$(7) \qquad \operatorname{tr} M_t E_n = \sum_{k=n+1}^{\infty} (M_t e_k, e_k).$$

On the other hand, $\xi(E_n) = \sum_{k=n+1}^{\infty} \xi(\bar{e}_k) \to 0$ in probability as $n \to \infty$. Hence

$$\lim_{n \to \infty} \inf_{|t| \leqq T} \hat{\xi}(E_n, t) = 1$$

and, further, by virtue of (3)

$$(8) \qquad \lim_{n \to \infty} \sup_{|t| \leqq T} \operatorname{tr} M_t E_n = 0.$$

Taking into account (7) and (8) we complete the proof of (β).

The condition (γ) is obtained from the continuity of $\hat{\xi}(E, *)$. The negative definiteness of quadratic forms $(M_t x, x)$ follows from theorem of Johansen [3].

Conversely, suppose that a family M_t, $t \in \mathbf{R}$, of S-operators satisfying the conditions $(\alpha) \dots (\delta)$, is given. We can put

$$(9) \qquad \varphi_E(t) = \exp(-\operatorname{tr} M_t E), \quad E \in \mathfrak{N}.$$

By the theorem of Johansen and by the assumptions $(\alpha) \dots (\delta)$ φ_E is a characteristic function of some symmetric infinitely divisible distribution. Now, for every finite sequence $E_1, \ldots, E_k$ we can form the probability distributions $Q_{E_1,\ldots,E_k}$ in $\mathbf{R}^k$ putting

$$\hat{Q}_{E_1,\ldots,E_k}(t_1, \ldots, t_k) = \prod_{j=1}^{k} \varphi_{E_j}(t_j) = \exp\left(-\operatorname{tr} \sum_{j=1}^{k} M_{t_j} E_j\right) \tag{10}$$

as the characteristic function. It is easy to verify that the family of distributions $Q_{E_1,\ldots,E_k}$ is consistent.

By the well-known theorem of Kolmogorov there exists a stochastic process $\xi(E)$, $E \in \mathfrak{N}$ over some $(\Omega, \mathscr{F}, P)$ such that the joint distribution of random vector $[\xi(E_1), \xi(E_2), \ldots, \xi(E_k)]$ is $Q_{E_1,\ldots,E_k}$. In particular, we have (3).

Let $E_1, E_2, \ldots$ be a sequence of mutually orthogonal projectors. From the form of the characteristic function (10) we conclude that $\xi(E_1), \xi(E_2), \ldots$ are independent, by (3) we have $\xi(E_1) + \ldots + \xi(E_n) = \xi(\sum_{k=1}^{n} E_k)$. Denoting $E = \sum_{k=1}^{\infty} E_k$ we obtain for every $t \in \mathbf{R}$

$$\hat{\xi}\left(\sum_{k=1}^{n} E_k, t\right) = \exp\left(-\operatorname{tr} M_t \sum_{k=1}^{n} E_k\right) \to \exp\left(-\operatorname{tr} M_t E\right) = \hat{\xi}(E, t)\,.$$

Hence

$$\sum_{k=1}^{n} \xi(E_k) \to \xi(E) \quad \text{in} \quad \mathrm{L}_0(\Omega, \mathscr{F}, P)\,.$$

Thus ξ is a random Gleason measure satisfying the equality (3). □

Let N be a function defined on the borelian σ-field $\mathscr{B}$ in $\mathbf{R}^+$ taking its values in the space of all S-operators.

N is said to be an S-operator-valued spectral measure if for every sequence $B_1, B_2, \ldots$ of disjoint Borel sets in $\mathbf{R}^+$ we have

$$\sum_{j=1}^{n} N(B_j) \to N\left(\bigcup_{j=1}^{\infty} B_j\right) \quad (\text{as } n \to \infty)$$

in the sense of S-operator convergence i.e. for all $x \in H$ we have

$$\sum_{j=1}^{n} (N(B_j)\, x, x) \to \left(N\left(\bigcup_{j=1}^{\infty} B_j\right) x, x\right) \quad \text{as} \quad n \to \infty$$

and the family $\sum_{j=1}^{n} N(B_j)$, $n = 1, \ldots,$ of S-operators is compact.

THEOREM 2. If ξ is a random Gleason measure then

$$\hat{\xi}(E, t) = \exp\left\{-\int_0^\infty (1 - \cos \lambda t)\,(1 + \lambda^2)\,\lambda^{-2} \operatorname{tr} N(\mathrm{d}\lambda)\, E\right\} \tag{11}$$

where N is an S-operator-valued spectral measure defined on the Borel σ-field $\mathscr{B}$, $E \in \mathfrak{N}$.

Conversely, if N is an S-operator-valued spectral measure then there exists a random Gleason measure ξ such that (11) holds.

Proof. Let ξ be a random Gleason measure. By the formula of Lévy-Khinchine we can write

$$\hat{\xi}(E, t) = \exp\left\{-\int_0^\infty (1 - \cos \lambda t)(1 + \lambda^2)\lambda^{-2}\,\mu(E, d\lambda)\right\}, \tag{12}$$

$$t \in \mathbf{R}, \quad E \in \mathfrak{N},$$

where $\mu(E, *)$ is a non-negative finite Borel measure in $\mathbf{R}^+$. We shall first prove that for every $B \in \mathscr{B}$ $\mu(*, B)$ is a Gleason measure on $\mathfrak{N}$. Indeed, for a sequence $E_1, E_2, \ldots$ of mutually orthogonal projectors we can verify, using (12), that for $n = 1, 2, \ldots$

$$\exp\left\{-\int_0^\infty (1 - \cos \lambda t)(1 + \lambda^2)\lambda^{-2}\sum_{k=1}^n \mu(E_k, d\lambda)\right\} =$$

$$= \exp\left\{-\int_0^\infty (1 - \cos \lambda t)(1 + \lambda^2)\lambda^{-2}\,\mu\Big(\sum_{k=1}^n E_k, d\lambda\Big)\right\}.$$

Hence, by the uniqueness of the Lévy-Khinchine representation we obtain

$$\sum_{k=1}^n \mu(E_k, *) = \mu\Big(\sum_{k=1}^n E_k, *\Big), \quad n = 1, 2, \ldots \tag{13}$$

On the other hand, the relation (1) implies, by virtue of Gnedenko theorem ([4], p. 300), the weak convergence of measures $\mu(\sum_{k=1}^n E_k, *)$ to the measure $\mu(E, *)$ where $E = \sum_{k=1}^\infty E_k$. In particular, using (13) we obtain that $\sum_{k=1}^n \mu(E_k, \mathbf{R}^+)$ converges to $\mu(E, \mathbf{R}^+)$ as $n \to \infty$. Thus the series $\sum_{k=1}^\infty \mu(E_k, B)$ is convergent uniformly with respect to $B \in \mathscr{B}$. We can define the measure ν putting $\nu(B) = \sum_{k=1}^\infty \mu(E_k, B)$, $K \in \mathscr{B}$.

By the uniqueness of the weak limit we have $\nu = \mu(E, *)$, so the condition $\sum_{k=1}^\infty \mu(E_k, B) = \mu(E, B)$ holds for $B \in \mathscr{B}$.

Using the Gleason theorem we can write $\mu(E, B)$ as $\operatorname{tr} N(B) E$ for $B \in \mathscr{B}$, $E \in \mathfrak{N}$, where $N(B)$ is an S-operator. Therefore, by (12) we obtain the representation (11). It remains to prove that N is an S-operator-valued spectral measure. Let $(B_j)_{j=1}^\infty$ be a sequence of disjoint Borel sets in $\mathbf{R}^+$ and $B = \bigcup_{j=1}^\infty B_j$. For $E \in \mathfrak{N}$ $\mu(E, *)$ is a measure, so we have in terms of traces

$$\operatorname{tr} N(B) E = \lim_{n\to\infty} \sum_{j=1}^n \sum_{l=1}^\infty (N(B_j) E e_l, e_l) \tag{14}$$

where $(e_l)_{l=1}^{\infty}$ is an orthonormal basis. Let $(\eta_k)_{k=1}^{\infty}$ be an arbitrary zero-one sequence. Putting in (14) $E = \sum_{k=1}^{\infty} \eta_k \bar{e}_k$ we obtain

$$\text{(15)} \qquad \operatorname{tr} N(B)\, E = \lim_{n\to\infty} \sum_{l=1}^{\infty} \eta_l \sum_{j=1}^{n} (N(B_j)\, e_l, e_l)\,.$$

Now, we notice that the limit of the form $\lim_{n\to\infty} \sum_{l=1}^{\infty} a_{nl}\eta_l$, where $a_{nl} = \sum_{j=1}^{n} (N(B_j)\, e_l, e_l)$, $n, l = 1, 2, \ldots$, exists for every zero-one sequence $(\eta_k)_{k=1}^{\infty}$. By Schur theorem ([2] or [5]) we obtain

1° $$\lim_{n\to\infty} a_{nl} = d_l\,, \quad l = 1, 2, \ldots$$

2° $$\sup_n \sum_{l=1}^{\infty} |a_{nl}| < \infty\,,$$

3° the series $\sum_{l=1}^{\infty} |a_{nl}|$ converges uniformly with respect to n.

It is now easy to verify that the condition 1° implies the convergence of quadratic forms $\sum_{j=1}^{n} (N(B_j)\, x, x)$ to $(N(B)\, x, x)$, $x \in H$, and the conditions 2° and 3° imply the compactness of the sequence $\sum_{j=1}^{n} N(B_j)$. Thus N is an S-operator-valued spectral measure.

Conversely, let an S-operator-valued spectral measure N be given. For an arbitrary $E \in \mathfrak{N}$ $\operatorname{tr} N(*)\, E$ is a finite nonnegative borelian measure on $\mathbf{R}^+$. Let us put

$$\text{(16)} \qquad \mu(E, t) = \int_0^{\infty} (1 - \cos \lambda t)\,(1 + \lambda^2)\, \lambda^{-2} \operatorname{tr} N(\mathrm{d}\lambda)\, E\,, \quad t \in \mathbf{R}\,.$$

By the theorem of Lévy-Khinchine $-$ $\mu(E, *)$ is the logarithm of some symmetric infinitely divisible characteristic function. On the other hand, we can show that for $t \in \mathbf{R}$ $\mu(*, t)$ is a Gleason measure. Indeed, if $E_1, E_2, \ldots$ is a sequence of suitable projectors and $E = \sum_{k=1}^{\infty} E_k$ then we have

$$\sum_{k=1}^{n} \operatorname{tr} N(B)\, E_k = \operatorname{tr} N(B) \sum_{k=1}^{n} E_k \to \operatorname{tr} N(B)\, E$$

as $n \to \infty$ for all Borel sets B in $\mathbf{R}^+$. Thus the sequence $\sum_{k=1}^{n} \operatorname{tr} N(*)\, E_k$ of measures is weakly convergent to $\operatorname{tr} N(*)\, E$. Hence

$$\sum_{k=1}^{n} \mu(E_k, t) = \mu\Big(\sum_{k=1}^{n} E_k, t\Big)$$

converges to $\mu(E, t)$ for all t.

Now, by the Gleason theorem we can write $\mu(E, t)$ in the form $\mathrm{tr}\, M_t E$, where M_t is an S-operator. Keeping in mind that $-\mathrm{tr}\, M_{(*)}E$ is the logarithm of symmetric infinitely divisible characteristic function and using the theorem of Johansen we verify that the family M_t, $t \in \mathbf{R}$, satisfies the conditions $(\alpha) \ldots (\delta)$ of Theorem 1. Therefore, there exists a random Gleason measure ξ such that $\hat{\xi}(E, t) = \exp(-\mathrm{tr}\, M_t E)$. By (16) we conclude that $\hat{\xi}$ has the required representation (11). □

REFERENCES

[1] A. M. GLEASON: Measures on the Closed Subspaces of a Hilbert Space. J. Math. and Mechanics *6* (1957), 885—894.

[2] G. H. HARDY: Divergent series. Oxford 1959.

[3] S. JOHANSEN: An Application of Extreme Point Methods to the Representation of Infinitely Divisible Distributions Z. Wahrscheinlichkeitstheorie *5* (1966), 304—316.

[4] M. LOÈVE: Probability Theory. Second edition, Wiley, New York 1957.

[5] I. SCHUR: Über lineare Transformationen in der Theorie der unendlichen Reichen. Journ. für Reine und Angewandte Math. *151* (1921), 79—111.

ŁÓDŹ UNIVERSITY

MATHEMATICAL INSTITUTE

FUNCTIONALS OF BROWNIAN MOTION

TAKEYUKI HIDA

NAGOYA

We discuss stochastic processes which are quadratic functionals of Brownian motion, and show the importance of the role of martingales being expressed by the double stochastic integral in Itô's sense. A typical example of such martingales has been dealt with by G. Kallianpur and the author, while here is presented another kind of martingales.

1. INTRODUCTION

We shall deal with the stochastic processes which are formed from the Brownian motion $B(t, \omega)$, $t \geqq 0$. Specifically, we shall be interested in the processes expressed as quadratic functionals of the Brownian motion. A typical example of such a process has been discussed by G. Kallianpur and the author [4], where they discuss the square of a multiple Markov Gaussian process expressing it as a sum of martingales, thereby they give an explicit algorithm to obtain the nonlinear predictor.

We shall present, in this paper, another kind of stochastic processes which are viewed as quadratic functionals of the Brownian motion. Namely, we shall discuss those processes obtained by the stochastic integral in Itô's sense, where the integrand is taken to be a Gaussian martingale.

The significance of our approach may be illustrated from two view points. One is concerned with the role of a martingale in the investigation of stochastic processes which are living in the space of multiple Wiener integrals of degree more than one, in particular of degree two. The other is connected with the so-called infinite dimensional analysis. If the integrand is a multiple Wiener integral, then the stochastic integral increases its degree by one. One may ask if there is a converse operation. We may answer in several ways, however, if we take the time propagation into account, our approach seems to give a plausible answer. The operation $\mathrm{d}/\mathrm{d}B(t)$, formally writing, is exactly what we need.

We restrict our attention mostly to quadratic functionals of a Brownian motion, however the author has hopes that may of our results can be extended to general nonlinear functionals.

2. MARTINGALES IN $\mathscr{H}_2$

Let $\mathscr{H}_n$ be the collection of the multiple Wiener integrals, due to N. Wiener and K. Itô, of degree n, and let $\mathscr{B}_t$ be the σ-field generated by the $B(s, \omega)$, $s \leqq t$. Any martingale $\{X(t), \mathscr{B}_t, t \geqq 0\}$ such that $X(t) \in \mathscr{H}_n$ for every t and that $E\,X(t) \equiv 0$ can be expressed in the form

$$X(t) = \int_0^t \int_0^t \cdots \int_0^t F(u_1, u_2, \ldots, u_n)\, \mathrm{d}B(u_1)\, \mathrm{d}B(u_2) \ldots \mathrm{d}B(u_n)\,, \tag{1}$$

where $F(u_1, u_2, \ldots, u_n)$ is a symmetric, locally square integrable function on R^n.

For $n = 2$, $F(u_1, u_2)$ $(= F(u_2, u_1))$ may be thought of as an integral kernel. We can therefore appeal to the general theory of the integral operator to obtain the eigenfunction expansion of $F(u_1, u_2)$. Being associated with this expansion, $X(t)$ itself can be developed into a sum of independent random variables, each of which is the product of the square of a standard Gaussian random variable and the eigenvalue of F.

Let $X(t)$, $t \geqq 0$, be an N-ple Markov Gaussian process in the restricted sense (see [1]) with the proper canonical representation

$$X(t) = \sum_{i=1}^{N} f_i(t) \int_0^t g_i(u)\, \mathrm{d}B(u)\,, \tag{2}$$

where the f_i's form a fundamental system of solutions of an N-th order ordinary differential equation $L_t f = 0$ associated with $X(t)$, while so do the g_i's for $L_u^* g = 0$, L_u^* being the adjoint operator for L_t. In order to investigate the best nonlinear predictor for the process

$$Y(t) = X(t)^2 - E(X(t)^2)\,, \quad t \geqq 0\,,$$

we have found in [4] that the martingales

$$M_{ij}(t) = \int_0^t \int_0^t g_i(u)\, g_j(v)\, \mathrm{d}B(u)\, \mathrm{d}B(v) = \tag{3}$$

$$= \tfrac{1}{2} \int_0^t \int_0^t \{g_i(u)\, g_j(v) + g_j(u)\, g_i(v)\}\, \mathrm{d}B(u)\, \mathrm{d}B(v)$$

play an important role. We note that the $Y(t)$ as well as the $M_{ij}(t)$'s lives in the space $\mathscr{H}_2$ and that the kernel $\frac{1}{2}\{g_i(u)\,g_j(v) + g_j(u)\,g_i(v)\}$, which is a symmetric $L^2(R^2)$-function (see [3]) associated with $M_{ij}(t)$, is degenerated, namely it is a Goursat kernel of order 2.

Quite contrary to a Goursat kernel we now use an $L^2(R^2)$-kernel of the form

$$F(u, v) = g(u \vee v)\,h(u \wedge v)\,, \quad g, h \in \mathscr{C}^1([0, \infty))\,, \tag{4}$$

to define

$$M(t) = \int_0^t \int_0^t F(u, v)\,\mathrm{d}B(u)\,\mathrm{d}B(v) = \tag{5}$$

$$= 2\int_0^t g(u)\left\{\int_0^u h(v)\,\mathrm{d}B(v)\right\}\mathrm{d}B(u)\,, \quad t \geqq 0\,.$$

With this notation we have

THEOREM 1. i) $(M(t), \mathscr{B}_t, t \geqq 0)$ *is a martingale.*

ii) *For fixed* t, $M(t)$ *is expressed as a linear combination of infinitely many independent random variables, each of which is the square of a standard Gaussian variable.*

iii) *The characteristic function* $\chi(z)$, $z \in R$, *of* $M(t)$ *is given by*

$$\chi(z) = \delta(-2\mathrm{i}z; F)^{-1/2}\,,$$

where $\delta(\cdot\,; F)$ *is the modified Fredholm determinant of* F *given by* (4).

Proof. The assertion i) is obvious. For the proofs of ii) and iii) we refer to [3]. An additional consideration which we need is that the eingevalues of the $F(u, v)$ can be obtained by observing the eigenvalues of the following differential equation of Sturm-Liouville type:

$$\left(\frac{g(u)}{h'(u)}\,\eta'(u)\right)' + \lambda\,\eta'(u) = 0\,. \tag{6}$$

□

Given such a martingale $M(t)$, we can easily form a random measure $\mathscr{M}$ in a usual manner:

$$\mathscr{M} = \left\{M(A);\ A \text{ Borel}, \iint\limits_{u\in A, u \geqq v} F(u, v)^2\,\mathrm{d}u\,\mathrm{d}v < \infty\right\}$$

with the relation $M(A) = \iint\limits_{u\in A, u\geqq v} F(u, v)\,\mathrm{d}B(u)\,\mathrm{d}B(v)$. We are now ready to discuss a stochastic process living in $\mathscr{H}_2$ given by

$$Z(t) = \int_0^t G(t, u)\,\mathrm{d}M(u)\,.$$

In fact, we can think of the canonical representation by the same way as in the Gaussian case (cf. [1]). In addition, we can discuss the probability distribution of $Z(t)$ with the help of Theorem 1.

3. STOCHASTIC DIFFERENTIAL EQUATIONS WITH A MARTINGALE COEFFICIENT

This section is devoted to an observation from the point of view of the infinite dimensional analysis. Originally we are inspired by a biological phenomena which is expressed mathematically by a stochastic differential equation

$$\mathrm{d}X(t) = a\,X(t) + M(t)\,\mathrm{d}B(t)\,, \tag{7}$$

where $M(t)$ is a continuous martingale relative to the σ-field $\mathscr{B}_t$ generated by the $B(s)$, $s \leqq t$, (note that $\mathrm{d}B(t)$ stands for the idealized internal "fluctuation" of a microbe). If we put an initial condition $X(0) = 0$, the solution of (8) is expressible as

$$X(t) = \int_0^t \mathrm{e}^{a(t-u)}\,M(u)\,\mathrm{d}B(u)\,. \tag{9}$$

For our purpose we may assume $a = 0$, so that we now consider the equation

$$\begin{aligned} \mathrm{d}X(t) &= M(t)\,\mathrm{d}B(t)\,, \\ X(0) &= 0\,. \end{aligned} \tag{10}$$

Further we restrict our attention to the case where $M(t) \in \mathscr{H}_n$ for some n. The solution $X(t)$ of (10) is again a martingale which is in $\mathscr{H}_{n+1}$. The following assertion is easily proved (for example see [2] § 10).

THEOREM 2. *Suppose that a symmetric* $L^2(R^n)$*-function* $G(u_1, u_2, \ldots, u_n)\,.$ $\cdot\,\chi_{[0,t]^n}(u_1, u_2, \ldots, u_n)$ *is a kernel of the integral representation of* $M(t)$. *Then*

$$F(u_1, u_2, \ldots, u_{n+1}) = \int_0^{u_1} G(u_2, \ldots, u_{n+1}) \otimes \chi_{\mathrm{d}u_1}(u_1)\,, \quad t \geqq u_1 \geqq u_2 \ldots \geqq u_{n+1}\,, \tag{11}$$

is associated with $X(t)$ *the solution of* (10), *where* $\chi_{\mathrm{d}u}$ *is the indicator function of the interval* $(u, u + \mathrm{d}u]$, $\mathrm{d}u > 0$.

Thus we have determined an explicit correspondence between martingales $M(t)$ in $\mathscr{H}_n$ and $X(t)$ in $\mathscr{H}_{n+1}$ in terms of the kernel. We now think of $M(t)$ is the derivative of $X(t)$ with respect to $B(t)$; formally

$$\mathrm{d}X(t)/\mathrm{d}B(t) = M(t)\,. \tag{12}$$

It is noted the operation $d/dB(t)$ reduces the degree of multiple Wiener integral by one. If, in particular, we consider the case $n = 1$, then $M(t)$ is nothing but an additive Gaussian process, expressible as, say $\int_0^t h(v)\,dB(v)$, and $X(t)$ is expressed in the form

$$X(t) = \int_0^t \int_0^u h(v)\,dB(v)\,dB(u)$$

which is a process discussed in § 2. With the formal expression (12) in mind, we can discuss a stochastic process given by the equation (10). Details will be reported in a separate paper.

REFERENCES

[1] T. HIDA: Canonical Representations of Gaussian Processes and their Applications. Memoires College of Sci., University of Kyoto, A, *33* Math. (1960), 109—155.

[2] T. HIDA: Stationary Stochastic Processes. Princeton University Press, Princeton. N. J. 1970.

[3] T. HIDA: Quadratic Functionals of Brownian Motion. J. of Multivariate Analysis *1* (1971), 58—69.

[4] G. KALLIANPUR, T. HIDA: The square of a Gaussian Markov Process and Nonlinear Prediction J. of Multivariate Analysis *5* (1975), 451—461.

[5] P. LÉVY: Processus Stochastiques et Mouvement Brownien. Deuxième Édition. Gauthier-Villars, Paris 1965.

NAGOYA UNIVERSITY

DEPARTMENT OF MATHEMATICS

AN IMPROVEMENT OF J. F. SHAPIRO'S TURNPIKE THEOREM FOR THE HORIZON OF FINITE STAGE DISCRETE DYNAMIC PROGRAMS

KARL HINDERER and GERHARD HÜBNER

KARLSRUHE and HAMBURG

The potential usefulness of the well-known turnpike theorem in [12] is for practical purposes rather limited, since numerical evaluation of the estimate given in [12] for the turnpike horizon N^* gives poor results, unless the discount factor β is rather small. We contribute to the problem in the following directions:

(i) Considerably improved estimates for N^* are derived, including cases where $\beta \geqq 1$.

(ii) For dynamic programs with absorbing sets further improvement is derived.

(iii) Some numerical examples are given, including a rather detailed analysis of Howard's toymaker example for arbitrary values of β.

1. INTRODUCTION

We are concerned with stationary, finite-stage, discrete dynamic programs $(S, A, D, p, r, \beta, V^0)$, the components of which have the following meaning:

S and A are the finite *state space* and *action space*, respectively.

$D(s)$ is the *set of actions* available when the system is in state s, and $D := \{(s, a) \in \in S \times A: a \in D(s)\}$.

$p(s, a, t)$ is the (conditional) probability that the system moves to state t when it has been in state s one stage before and when action a is taken.

$r(s, a)$ is the *reward* one obtains during a single stage when the system is in state s and action a has been taken.

$\beta \in (0, \infty)$ is the *discount factor*.

$V^0(s)$ is the *terminal reward* one obtains when the system is at the end in state s.

In addition, we shall use the following terminology:

F is the set of (one-stage) *decision functions*, i.e. functions from S into A such that $f(s) \in D(s)$ for all states s.

N is the *horizon*, i.e. the number of stages.

Δ_N is the (finite) set of deterministic Markovian *N-stage policies* $\pi = (f_0, f_1, \ldots$ $\ldots, f_{N-1})$, $f_\nu \in F$.

$V^N(s)$ is the *maximal (conditional) expected reward* for the N-stage model when s is the initial state.

Formulations and proofs simplify considerably by the use of the following three monotone operators L: $\boldsymbol{R}^S \to \boldsymbol{R}^D$, $U : \boldsymbol{R}^S \to \boldsymbol{R}^S$ and $L_f : \boldsymbol{R}^S \to \boldsymbol{R}^S$ defined by

$$(1.1) \qquad (Lv)(s, a) := r(s, a) + \beta \sum_{t \in S} p(s, a, t)\, v(t)\,,$$

$$(1.2) \qquad (Uv)(s) := \max_{a \in D(s)} Lv(s, a)\,,$$

$$(1.3) \qquad (L_f v)(s) := Lv(s, f(s))\,, \quad f \in F\,.$$

A decision function f will be called a *maximizer* of the map $H : D \to \boldsymbol{R}$, iff

$$(1.4) \qquad H(s, f(s)) = \max_{a \in D(s)} H(s, a) \quad \text{for all} \quad s \in S\,.$$

Lack of space does not allow to give a detailed description of the probabilistic model underlying our dynamic program; it may be found e.g. in [3] and [11].

The general approach for the determination of an optimal N-stage policy and of the so-called *value function* V^N consists in the *method of backward induction*, which may be formulated by the following two well-known theorems:

(A_N) *The value function V^N may be computed recursively by the value iteration $V^{n+1} = UV^n$, $0 \leqq n < N$.*

(B_N) *If φ_n is a maximizer of LV^{n-1} for $1 \leqq n \leqq N$, then the policy $(\varphi_N, \varphi_{N-1}, \ldots$ $\ldots, \varphi_1)$ is optimal.*

It follows that φ_n is a maximizer of LV^{n-1} iff φ_n is an "optimal initial decision function" for the n-stage problem.

Now we are ready to formulate the problem of the present paper. When applying the method of backward induction to numerical problems, one often is led after a few stages to the conjecture that the set of maximizers of LV^n is independent of n for sufficiently large n; or, more specifically, that there exists some decision function f satisfying the following.

Condition (T): *f is the unique maximizer of LV^n for all $n \geqq k$.*

If we are able to show that f in fact satisfies condition (T), then we have made a big step forward concerning the solutions of the dynamic programs with arbitrary horizon $N > k$ (provided k is not too 'large'): We have to apply value iteration only k times in order to get V^k and an optimal policy $\pi \in \Delta_k$; but then $(f, f, \ldots, f, \pi) \in \Delta_N$

is optimal for the N-stage model, and the value iteration for the computation of V^N simplifies to the linear recurrence

$$V^{n+1} = L_f V^n, \quad k \leqq n < N. \tag{1.5}$$

One possible approach for proving (T) is the so-called method of *decision exclusion* (when applied to finite-stage dynamic programs); c.f. [10] and [9]. It has the drawback that it does not tell us in advance whether and at what stage k the successive exclusion of non-optimal actions will finally filter out a decision function satisfying (T).

Another method for proving (T) has been developped by Shapiro [12], and this will be improved in the subsequent sections. As in [12] we generalize the problem slightly. Let us denote by M_n the (non-empty) set of maximizers of LV^{n-1}, $n \in N$. Given $M \subset F$, the number

$$N^*(M) := \inf\{k \in N : M_n \subset M \text{ for all } n \geqq k\}, \tag{1.6}$$

(which may also be infinite) is called the *turnpike horizon* of M. We are especially interested in finding decision functions f, for which the turnpike horizon

$$N^*(f) = \inf\{k \in N : f \text{ is the unique maximizer of } LV^{n-1} \text{ for all } n \geqq k\} \tag{1.7}$$

is finite and yields reasonable upper estimates.

2. FINITENESS OF TURNPIKE HORIZONS AND UPPER ESTIMATES

LEMMA 2.1. *If (V^n) converges to a real-valued function V, then V is a fixed point of U, and $\overline{\lim}_n M_n \subset M$, where M is the set of maximizers of LV.*

Proof. Since S is finite, LV^n converges to LV. Now it is easy to see, that the finiteness of $D(s)$ implies

$$\max_{a \in D(s)} Lv_n(s, a) \to \max_{a \in D(s)} Lv(s, a) \quad (n \to \infty)$$

for any sequence of functions $v_n : S \to R$ converging to v. Therefore

$$V^{n+1}(s) = U\, V^n(s) = \max L V^n(s, a) \to \max L V(s, a) = U\, V(s),$$

hence $V = UV$. Next, if $g \in \overline{\lim}\, M_n$, then there exist positive integers n_k such that $g \in M_{n_k}$, i.e.

$$L_g V^{n_k - 1} = U V^{n_k - 1} = V^{n_k}, \quad k \in N.$$

Now $k \to \infty$ yields $L_g V = V = UV$, hence $g \in M$. □

THEOREM 2.2. *If* (V^n) *converges to a real-valued function* V, *then the set of maximizers of* LV *has a finite turnpike horizon.*

The special case $\beta < 1$ was proved by Shapiro [12]; cp. also ([2], p. 51). Theorem 2.2 follows from lemma 2.1 easily along the lines drawn in [12]. We remark that theorem 2.2 remains valid when V is extended real-valued and when U is applicable to V, in particular, when $r \geqq 0$ and $V^0 \geqq 0$.

It will be possible to prove turnpike theorems even when (V^n) does not converge, by replacing $\lim V^n$ by an arbitrary fixed point V of U (if it exists). In that case denote by M the set of maximizers of LV. If $M \neq F$, then there are pairs (s, a) for which $V(s) - LV(s, a)$ is positive. Put then

$$(2.1) \qquad \varrho := \text{least positive of the numbers } V(s) - LV(s, a)\,, \quad (s, a) \in D\,.$$

If $M = F$, then trivially $N^*(M) = 1$.

An important role will play the number

$$(2.2) \qquad \alpha := \max_{(s,a),(s',a') \in D} \left[1 - \sum_{t \in S} \min\left(p(s, a, t),\ p(s', a', t)\right)\right],$$

which, as shown by Hübner in [9], may provide for a useful contraction property of the dynamic program.

In the next theorem, the span of a function $v : S \to \boldsymbol{R}$ is defined by $\operatorname{sp} v := := \max v - \min v$, and $\inf \emptyset := \infty$, as usual.

THEOREM 2.3. *Let* V *be a fixed point of* U, *and let* M *be the set of maximizers of* LV. *Then, if* $M \neq F$,

$$(2.3) \quad N^*(M) \leqq n_0 := \inf\{k \in \boldsymbol{N} : \alpha\beta \operatorname{sp}(V - V^{n-1}) < \varrho \text{ for all } n \geqq k\}\,.$$

Proof. For any $v, w \in \boldsymbol{R}^S$ a slight generalization of theorem 1.3 in [9] yields for any $(s, a) \in D$

$$(2.4) \qquad \left|Uv(s) - Lv(s, a) - (Uw(s) - Lw(s, a))\right| \leqq \alpha\beta \operatorname{sp}(v - w)\,.$$

Fix some $n \geqq n_0$ and some $f \notin M$. Then it suffices to show $f \notin M_n$. Now $V - L_f V$ attains its (positive) supremum at some point i, hence $V(i) - L_f V(i) \geqq \varrho$. Now we get by the definition of n_0

$$\begin{aligned} UV^{n-1}(i) - L_f V^{n-1}(i) &\geqq UV(i) - L_f V(i) - \alpha\beta \operatorname{sp}(V - V^{n-1}) \\ &\geqq \varrho - \alpha\beta \operatorname{sp}(V - V^{n-1}) > 0\,. \end{aligned}$$

Therefore $f \notin M_n$. □

REMARKS. 1. Theorem 2.3 and its proof remain valid if we replace V by some solution of the equation $v + c = Uv$, where $c \in \boldsymbol{R}$ is arbitrary, and if we put $\varrho :=$ least positive of the numbers $UV(s) - LV(s, a)$, $(s, a) \in D$. This extension is useful for the case $\beta = 1$, $\alpha < 1$: From $\alpha < 1$ one can conclude, that each of the stochastic matrices $(p(s, f(s), t))$, $f \in F$, is completely ergodic in the sense of ([8], p. 60). But then, as is well-known (cf. e.g. ([11], p. 28ff.)), the optimal one-stage average reward $c(s)$ is independent of s, and there exists a solution V of $v + c = Uv$.

2. A proof similar to that one of theorem 2.3 yields the following variant: If we have available estimates $c_n \leqq V - V^n \leqq d_n$, $n \in N_0$, then

$$N^*(M) \leqq \inf\{k \in N : d_n - \beta c_{n-1} < \varrho \text{ for all } n \geqq k\}\,. \tag{2.5}$$

For an efficient application of theorem 2.3 we need estimates for $\operatorname{sp}(V - V^{n-1})$. If we insert the rough estimate (using the sup-norm $\|\cdot\|$)

$$\alpha\beta \operatorname{sp}(V - V^{n-1}) \leqq 2\beta\|V - V^{n-1}\| \leqq 2\beta^n(1-\beta)^{-1}\,\|V^1 - V^0\|\,, \quad \beta < 1\,, \tag{2.6}$$

into (2.3), we get for $\beta < 1$ and $M \neq F$ the estimate

$$N^*(M) \leqq n_1 := 1 + \left[\left(\frac{\log \dfrac{\varrho(1-\beta)}{2\|V^1 - V^0\|}}{\log \beta}\right)^+\right] < \infty\,. \tag{2.7}$$

(Here we use $x/0 := \infty$ for $x > 0$; $\log \infty := \infty$; $x^+ := \max(x, 0)$; $[x] :=$ largest integer not greater than x.) For the special case $V^0 \equiv 0$, (2.7) is equivalent to the estimate given by Shapiro [12]. It can be considerably improved (cf. section 4), by applying to (2.3) the estimate

$$\operatorname{sp}(Uv - Uw) \leqq \alpha\beta \operatorname{sp}(v - w)\,, \quad v, w \in \mathbf{R}^S\,, \tag{2.8}$$

derived by Hübner ([9], theorem 1.1). Thus we get

THEOREM 2.4. *Let V be a fixed point of U, and let M be the set of maximizers of LV. If $M \neq F$ and $\alpha\beta < 1$, then*

$$N^*(M) \leqq n_2 := 1 + \left[\left(\frac{\log \dfrac{\varrho}{\operatorname{sp}(V - V^0)}}{\log(\alpha\beta)}\right)^+\right] < \infty\,. \tag{2.9}$$

Again by the application of (2.8) one may generalize (2.9) to

$$N^*(M) \leqq n_2(k) := k + \left[\left(\frac{\log \dfrac{\varrho}{\operatorname{sp}(V - V^{k-1})}}{\log(\alpha\beta)}\right)^+\right], \quad k \in N\,, \tag{2.10}$$

which might be useful if some stages in the value iteration have been already computed.

3. DYNAMIC PROGRAMS WITH ABSORBING SETS

Some of the results in section 2 can be improved when there exists a non-empty absorbing set in the sense of the following

DEFINITION. *A set $G \subset S$ is called absorbing (for the dynamic program under consideration) if for any $s \in G$ and any $a \in D(s)$ we have $r(s, a) = 0$, $\sum_{t \in G} p(s, a, t) = 1$, and $V^0(s) = 0$.* (Dynamic programs with non-empty absorbing sets are also called 'terminating processes' (c.f. [11]) or 'first-passage time problems' (c.f. [5]).) Obviously there exists a maximal absorbing set G^*, for which we define an operator $T: \mathbf{R}^S \to \mathbf{R}^S$ by

$$(3.1) \qquad (Tv)(s) := \max_{a \in D(s)} \sum_{t \in S - G^*} p(s, a, t)\, v(t), \quad s \in S.$$

Results for dynamic programs with absorbing sets are given in [7] and [13], from which we cite the following ones: Let $\beta < \beta^* := \sup_{m \in N} \|T^m 1\|^{-1/m}$ (which is equivalent to the condition that $\beta^m \|T^m 1\| < 1$ for some $m \in N$). Then

(i) there exists $V := \lim_n V^n$, and V is the unique fixed point of U which vanishes on G^*;

(ii) V may be computed by policy iteration.

It follows from (i) and from theorem 2.2, that for $\beta < \beta^*$ the set of maximizers of LV has a finite turnpike horizon. Moreover, an analogue to (2.7) could be derived, but we turn immediately to an analogue of theorem 2.4.

THEOREM 3.1. *Let us assume that $\beta^m \|T^m 1\| < 1$, that V is the unique fixed point of U which vanishes on G^*, and that the set M of maximizers of LV is different from F. Then*

$$(3.2) \qquad N^*(M) \leqq n_3 := 1 + \left[\left(\frac{\log \dfrac{\varrho \|T^m 1\|^{1+1/m}}{\alpha \operatorname{sp}(V - V^0)}}{\log (\beta \|T^m 1\|^{1/m})} \right)^+ \right].$$

Proof. It can be shown that

$$(3.3) \qquad \operatorname{sp}(V - V^n) \leqq \beta^n \|T^m 1\|^{n/m - 1} \operatorname{sp}(V - V^0).$$

Now (3.2) follows from theorem 2.3 by easy computation. □

4. EXAMPLES

Example 1

Examples given in [4] and [1] show, that in the case $\beta = 1$ there need not exist a decision function with finite turnpike horizon; the same is true for some $\beta < 1$ in a deterministic example given in [12]. As noted by Rieder, the stochastic example in [4] may be modified in the following way, in order to exhibit the above mentioned phenomenon also for some values of $\beta < 1$. Put $S := A := \{1, 2\}$, $D(1) := A$, $D(2) := \{1\}$, $V^0 \equiv 0$, $0.75 < \beta \leqq 1$. p and r are given by table 1, where $\lambda := (4\beta)^{-1}$. Using the fact, that $d_n := LV^n(1, 2) - LV^n(1, 1) = 0.2 - (V^n(1) - V^n(2))/4$, one can show by induction on the number of stages that $M_{2n+1} = \{g\}$, $M_{2n+2} = \{f\}$, $n \in N_0$, where $f :\equiv 1$ and $g(1) := 2$, $g(2) := 1$.

TABLE 1

(s, a)	$p(s, a, 1)$	$p(s, a, 2)$	$r(s, a)$
(1, 1)	2λ	$1 - 2\lambda$	1
(1, 2)	λ	$1 - \lambda$	1.2
(2, 1)	3λ	$1 - 3\lambda$	0

Example 2

Here we give a rather detailed analysis of the finite stage version of Howards toymaker example, definied by $S := A := D(s) := \{1, 2\}$, p and r as in table 2, $\beta \in (0, \infty)$, $V^0 \equiv 0$.

(i) At first we determine the fixed points V_β of U_β and the set M_β of maximizers of LV_β. Elementary computation shows: If V_β is a fixed point of U_β, then

$$LV_\beta(s, 1) \gtreqless LV_\beta(s, 2) \quad \text{for} \quad s \in S \Leftrightarrow 3\beta(V_\beta(1) - V_\beta(2)) \lesseqgtr 20\,. \tag{4.1}$$

TABLE 2

(s, a)	$p(s, a, 1)$	$p(s, a, 2)$	$r(s, a)$
(1, 1)	0.5	0.5	6
(1, 2)	0.8	0.2	4
(2, 1)	0.4	0.6	−3
(2, 2)	0.7	0.3	−5

In particular, only $f :\equiv 1$ and $g :\equiv 2$ can be maximizers of LV_β. If f or g is a maximizer, then $V_\beta = L_f V_\beta$ or $V_\beta = L_g V_\beta$, respectively. It is easy to see that each of these two systems of linear equations has a solution iff $\beta \notin \{1, 10\}$, and that then the unique solutions are

$$(4.2) \qquad v_f = 3w \begin{pmatrix} 20 - 17\beta \\ 13\beta - 10 \end{pmatrix}, \quad v_g = 2w \begin{pmatrix} 20 - 11\beta \\ 34\beta - 25 \end{pmatrix},$$

where $w := ((1 - \beta)(10 - \beta))^{-1}$. From (4.1) and (4.2) we conclude: (α) U_β has a (unique) fixed point iff $\beta \notin \{1, 10\}$; (β) v_f is the unique fixed point of U_β and f is the unique maximizer of LV iff $3\beta(v_f(1) - v_f(2)) < 20$, which is equivalent to $\beta < \beta_0 := 20/29$. Analysing v_g in the same way, we get as final result:

(a) $\beta \in \{1, 10\} \Rightarrow U_\beta$ has no fixed point,

(b) $\beta \notin \{1, 10\} \Rightarrow U_\beta$ has a unique fixed point V_β,

(c) $\beta \in (0, \beta_0) \Rightarrow V_\beta = v_f, \quad M = \{f\}$,

(d) $\beta = \beta_0 \Rightarrow V_\beta = v_f = v_g, \quad M = F$,

(e) $\beta \in (\beta_0, 1) \cup (1, 10) \cup (10, \infty) \Rightarrow V_\beta = v_g, \quad M = \{g\}$.

In addition, if $\beta = 1$, then $c := 2$ is the maximal average reward, $V := \begin{pmatrix} 10 \\ 0 \end{pmatrix}$ is a solution of $v + c = Uv$, and g is the unique maximizer of LV; c.f. ([8], p. 41).

(ii) Next we are going to evaluate the estimates n_1 and n_2 for the turnpike horizon of M. One easily computes $\alpha = 0.4$, hence M has finite turnpike horizon for all $\beta \in (0, 2.5)$ by theorem 2.4 and remark 1. For the computation of ϱ one observes that, if e.g. $M = \{f\}$, then $\varrho = \min(v_f - L_g v_f)$ and $LV(s, 1) - LV(s, 2) = 2 - 0.3\beta(v_f(1) - v_f(2))$. It follows that $\varrho(\beta) = |(20 - 29\beta)(10 - \beta)^{-1}|$. Since $\|r\| = 6$, we get for Shapiro's bound

$$(4.3) \qquad n_1(\beta) = 1 + \left[\left(\frac{\log \dfrac{(1 - \beta)\,|20 - 29\beta|}{12(10 - \beta)}}{\log \beta} \right)^+ \right], \quad \beta \in (0, 1);$$

in particular $n_1(\beta) \to \infty$ $(\beta \uparrow 1)$. Some values of $n_1(\beta)$ are given in table 3. On the other hand, since $\operatorname{sp} V_\beta = |V_\beta(1) - V_\beta(2)| = 90/|10 - \beta|$, our estimate n_2 from theorem 2.4 and remark 1 is

$$(4.4) \qquad n_2(\beta) = 1 + \left[\left(\frac{\log \dfrac{|20 - 29\beta|}{90}}{\log (0.4\beta)} \right)^+ \right], \quad \beta \in (0, 2.5);$$

in particular $n_2(\beta) = 3$ in a neighborhood of $\beta = 1$. Some values of $n_2(\beta)$ are given in table 3.

(iii) Next we show how to obtain in our simple problem the exact value of $N^*(M)$ for $\beta \in (0, \infty)$, which is useful for checking the quality of the estimates n_1 and n_2. From

$$Lv(s, 1) - Lv(s, 2) = 2 - 0.3\beta(v(1) - v(2)) =: 2 - 0.3\beta\, z(v),$$

we conclude that $f \in M^{n+1}$ iff $z(V^n) \leqq 20/3\beta$ and that $g \in M^{n+1}$ iff $z(V^n) \geqq 20/3\beta$. Moreover, $z(V^n) = 9 + 0.1\beta\, z(V^{n-1}) = 9 \sum_{m=0}^{n-1} (0.1\beta)^m$. Now one shows easily: If $\beta \geqq 10$, then $z(V^n) \geqq 20/3\beta$ for $n \geqq 1$, hence $N^*(g) = 2$. If $\beta < 10$, then $z(V^n)$ converges non-decreasingly to $9/(1 - 0.1\beta)$, which is not larger than $20/3\beta$ iff $\beta \leqq \beta_0$. Therefore, if $0 < \beta \leqq \beta_0$, then $z(V^n) < 20/3\beta$ for $n \geqq 0$, hence $N^*(f) = 1$. Finally, if $\beta_0 < \beta < 10$, then $N^*(g) = k$ iff $z(V^{k-2}) \leqq 20/3\beta < z(V^{k-1})$, i.e. iff $\beta \in (\beta_k, \beta_{k-1}\rangle$, where β_k is the least positive root of the equation $270\,(0.1\beta)^k - 29\beta + 20 = 0$. Thus one gets $\beta_1 = 10$, $\beta_2 = 20/27 = 0.7407$, $\beta_3 = 0.6928$, and also $\beta_n \to \beta_0 = 20/29 = 0.689655$.

From table 3 we realize that for the economically relevant values of β, namely in a neighborhood of $\beta = 1$, the estimate n_2 is very good, whereas n_1 either is poor ($\beta < 1$) or is even not defined ($\beta \geqq 1$).

TABLE 3

β	0+	0.2	0.4	0.6	0.68	0.689	0.6897	0.69	0.7	0.9
n_1	1	2	4	10	19	27	34	29	9	50
n_2	1	1	2	3	5	7	9	8	5	3
N^*	1	1	1	1	1	1	5	4	3	2
β	0.95	0.98	0.99	1	1.1	1.5	2	2.4	2.5−	≧2.5
n_1	111	320	704	—	—	—	—	—	—	—
n_2	3	3	3	3	3	3	4	15	∞	—
N^*	2	2	2	2	2	2	2	2	2	2

Example 3

(Optimal lot size problem; cf. ([14], p. 114) and [6]).

A customer orders $x := 10$ pieces of a certain product, let us say some sensitive electronic system. At each day the firm may produce a batch of either one or two pieces. There is probability $p := 0.9$ that a produced piece turns out to be faultless, independent of the quality of the other pieces. The production costs per unit are $c := 1$ and the set-up costs for one batch are $b := 5$. There is no discounting, i.e. $\beta = 1$. After N days all 10 pieces have to be delivered, otherwise the firm must pay penalty costs of $d := 10$ for every piece not yet produced. Finally, excess production is of no value.

We take $S := \{z \in N_0 : z \leqq x\}$, where s_n denotes the number of pieces still to be produced at the beginning of the n-th day, $n \in N_0$. Moreover $A := \{1, 2\}$, where a_n denotes the size of the batch used at the n-th day. Obviously $G^* := \{0\}$ is the largest absorbing set, and without loss of generality we take $D(0) := \{1\}$ and $D(s) := A$ for $s \neq 0$. One can show that $T^m 1(s)$ is the maximal probability that the infinite stage process is not absorbed by the set G^* during the first m stages. From this one can conclude that $\|T^m 1\| = \sum_{k=0}^{9} \binom{m}{k} 0.9^k \, 0.1^{m-k}$. Since $\|T^m 1\| < 1$ for $m \geqq 10$, we know that $\beta^* > 1$, and therefore theorem 3.1 is applicable. For the evaluation of the estimate n_3 we need $W := -V$, which may be computed by recursion in state space according to $W(0) = 0$ and

(4.5)

$$W(s) = \min_{a \in \{1,2\}} \left\{ \left[5 + a + 0.1^a \sum_{v=1}^{\min(s-1,a)} \binom{a}{v} 9^v W(s-v) \right] / (1 - 0.1^a) \right\}, \quad s > 0 .$$

From (4.5) one can show that $M = \{f\}$, where $f(s) := 1$ for $s = 0,1$, and $f(s) := 2$ for $s > 1$, and that $\mathrm{sp}\,(V - V^0) = 59.79$ and $\varrho = 0.4$. Some values of the estimate $n_3(m)$ in (3.2) are given in table 4.

The authors are grateful to Dr. G. Quelle for his help in preparing an early version of the present paper.

TABLE 4

m	10	11	12	13	14	15	16...19	20	21
$n_3(m)$	128	59	41	34	30	29	28	29	29

REFERENCES

[1] J. BATHER: Optimal decision procedures for finite Markov chains. Part I: Examples. Adv. Appl. Pr. *5* (1973), 328–339.
[2] M. J. BECKMANN: Dynamic programming of economic decisions. Springer, Berlin 1968.
[3] D. BLACKWELL: Discrete dynamic programming. Ann. Math. Stat. *33* (1962), 719–726.
[4] B. W. BROWN: On the iterative method of dynamic programming on a finite space discrete time Markov process. Ann. Math. Stat. *36* (1965), 1279–1285.
[5] C. DERMAN: Finite state Markovian decision processes. Academic Press, New York 1970.
[6] K. HINDERER: On the application of dynamic programming to economic lot size problems. Oper. Res. Verfahren *X* (1971), 57–75.
[7] K. HINDERER: Neuere Resultate in der Stochastischen Dynamischen Optimierung. Z. Angew. Math. Mech. *55* (1975), T16–T26.
[8] R. HOWARD: Dynamic programming and Markov processes. Technology Press/Wiley, New York 1960.
[9] G. HÜBNER: Improved procedures for eliminating suboptimal actions in Markov programming by the use of contraction properties. This volume 257–263.
[10] J. MACQUEEN: A test for suboptimal actions in Markovian decision problems. Oper. Res. *15* (1967), 559–561.
[11] H. MINE, S. OSAKI: Markovian decision processes. Elsevier, New York 1970.
[12] J. F. SHAPIRO: Turnpike planning horizons for a Markovian decision model. Management Sci. *14* (1968), 292–300.
[13] A. F. VEINOTT: Discrete dynamic programming with sensitive discount optimality criteria. Ann. Math. Stat. *40* (1969), 1635–1660.
[14] D. J. WHITE: Dynamic programming. Oliver and Boyd, Edinburgh 1969.

UNIVERSITÄT KARLSRUHE

INSTITUT FÜR MATHEMATISCHE STATISTIK

UNIVERSITÄT HAMBURG

INSTITUT FÜR MATHEMATISCHE STOCHASTIK

IMPROVED PROCEDURES FOR ELIMINATING SUBOPTIMAL ACTIONS IN MARKOV PROGRAMMING BY THE USE OF CONTRACTION PROPERTIES

GERHARD HÜBNER

HAMBURG

In applying the method of value iteration one frequently observes that the relative values for the n-th stage converge very rapidly with increasing n, whereas the absolute values converge slowly (discounting factor β near one) or even diverge ($\beta \geqq 1$). This fact is used by MacQueen [10], [11] and others to give good bounds for the value of the infinite horizon problem and, in addition, for the elimination of suboptimal actions in the early stages. This elimination can be improved by the use of an upper bound δ for the convergence rate. In case of $\delta < \beta$ the improvement has two effects: it reduces computing time and it allows the application to the finite horizon case with some $\beta \geqq 1$.

1. INTRODUCTION AND RESULTS

Let us consider a finite Markov decision problem consisting of a finite set of states S, a finite set of actions A, a set $D \subset S \times A$, the i-sections D_i of which (assumed to be non-empty) being the set of admissible actions in state i, a transition law $P = (p_{ij}^a, (i, a) \in D, j \in S)$ with $\sum_j p_{ij}^a = 1$ for all $(i, a) \in D$, an immediate expected reward r_i^a, $(i, a) \in D$, a discounting factor β, for which we only assume $0 < \beta < \infty$, and a final payment $V_0 : S \to \boldsymbol{R}$.

For the maximal expected n-stage reward V_n we have the following recursion, called value iteration:

$$(1.1) \qquad V_n(i) = \max_{a \in D_i} (r_i^a + \beta \sum_{j \in S} p_{ij}^a V_{n-1}(j)), \quad n \in \boldsymbol{N}, \quad i \in S.$$

Using the operators $L : \boldsymbol{R}^S \to \boldsymbol{R}^D$, $Lx(i, a) := r_i^a + \beta \sum_j p_{ij}^a x(j)$ and $U : \boldsymbol{R}^S \to \boldsymbol{R}^S$, $Ux(i) := \max_{a \in D_i} Lx(i, a)$ the recursion (1.1) reduces to $V_n = UV_{n-1}$, $n \in \boldsymbol{N}$.

By the relations

$$\text{(1.2)} \qquad \max(V_{n+1} - V_n) \leqq \beta \max(V_n - V_{n-1}), \quad \min(V_{n+1} - V_n) \geqq \beta \min(V_n - V_{n-1})$$

(the maximum of a vector being the maximum of its components) one obtains (cp. MacQueen [10], Porteus [14])

$$\text{(1.3)} \qquad V'_{n-1,N} \leqq V'_{n,N} \leqq V_N \leqq V''_{n,N} \leqq V''_{n-1,N}, \; N \geqq n,$$

$$V'_{n,N} := V_n + \sum_{\nu=1}^{N-n} \beta^\nu \min(V_n - V_{n-1}),$$

$$V''_{n,N} := V_n + \sum_{\nu=1}^{N-n} \beta^\nu \max(V_n - V_{n-1}).$$

The special importance of these bounds lies in the fact that in many practical problems the differences $V''_{n,N} - V'_{n,N}$ converge very rapidly to zero with increasing n, $n \leqq N$, N being large and fixed. An estimate of the rate of convergence is given by the following theorem. (The proofs of 1.1 and 1.2 will be given in section 2.)

1.1. THEOREM. *Define the span of a vector by* $\operatorname{sp} x := \max x - \min x$. *Then we have*

$$\text{(1.4)} \qquad \operatorname{sp}(V_{n+1} - V_n) \leqq \delta \operatorname{sp}(V_n - V_{n-1}), \quad n \in N$$

with

$$\delta := \beta \max_{(i,a),(i',a') \in D} \left(1 - \sum_j \min(p^a_{ij}, p^{a'}_{i'j})\right)$$

and

$$\delta \leqq \delta' := \beta\left(1 - \sum_j \min_{(i,a) \in D} p^a_{ij}\right) \leqq \beta.$$

In papers of MacQueen [11], Porteus [14], Grinold [2], Hastings, Mello [4] and others, treating only the case $\beta < 1$ and $N = \infty$, bounds of the type (1.3) are used to eliminate suboptimal actions in early stages of the value iteration (1.1). Let us call an action $a \in D_i$ *suboptimal for n and i* ($n \in N$, $i \in S$), if

$$\text{(1.5)} \qquad LV_{n-1}(i, a) < V_n(i)$$

holds. We therefore are interested in the differences $UV_{n-1}(i) - LV_{n-1}(i, a)$ and in their behaviour as n increases.

1.2. LEMMA. *For* $(i, a) \in D$ *we have*

$$\text{(1.6)} \qquad UV_n(i) - LV_n(i, a) \geqq UV_{n-1}(i) - LV_{n-1}(i, a) - \delta_{iaa^*} \operatorname{sp}(V_n - V_{n-1})$$

with

$$\delta_{iaa'} := \beta\left(1 - \sum_j \min(p^a_{ij}, p^{a'}_{ij})\right), \quad a' \in D_i,$$

a^* defined by $UV_{n-1}(i) = LV_{n-1}(i, a^*)$, and

$$\delta_{iaa'} \leqq \delta_{ia} := \max_{a' \in D_i} \delta_{iaa'} \leqq \delta_i := \max_{a \in D_i} \delta_{ia} \leqq \delta \leqq \delta' \leqq \beta\,,$$

$$\delta_i \leqq \delta'_i := \beta\bigl(1 - \sum_j \min_{a \in D_i} p^a_{ij}\bigr) \leqq \delta' \leqq \beta\,.$$

Together with (1.4) we obtain by induction

1.3 Theorem. *For $(i, a) \in D$ and $N \geqq n$ the following inequality holds:*

$$UV_N(i) - LV_N(i, a) \geqq \tag{1.7}$$

$$\geqq UV_{n-1}(i) - LV_{n-1}(i, a) - \delta_{ia} \sum_{\nu=0}^{N-n} \delta^\nu \operatorname{sp}\left(V_n - V_{n-1}\right).$$

This theorem is used as a criterion for suboptimality of actions in the following obvious way:

1.4 Corollary. (a) *The action a is suboptimal for i and N' with $n - 1 \leqq N' \leqq N$, if*

$$UV_{n-1}(i) - LV_{n-1}(i, a) > \delta_{ia} \sum_{\nu=0}^{N-n} \delta^\nu \operatorname{sp}\left(V_n - V_{n-1}\right). \tag{1.8}$$

(b) *In case of $\delta < 1$ action a is suboptimal for i and all $N \geqq n - 1$, if*

$$UV_{n-1}(i) - LV_{n-1}(i, a) > \delta_{ia}(1 - \delta)^{-1} \operatorname{sp}\left(V_n - V_{n-1}\right). \tag{1.9}$$

(c) *If in addition V_N converges to a finite vector V, it follows $V(i) > LV(i, a)$, if (1.9) holds.*

We obtain the result of MacQueen [11] as special case of 1.4(c), if we replace δ_{ia} and δ by β.

2. CONTRACTION PROPERTIES OF STOCHASTIC MATRICES AND APPLICATION TO THE DECISION MODEL

Let $A = (a_{ij})$ be a stochastic $k \times m$-matrix, i.e. a matrix with non-negative elements and $\sum_j a_{ij} = 1$ for $1 \leqq i \leqq k$. We easily see that the span of a vector defined in 1.1 as $\operatorname{sp} x := \max x - \min x$ is a semi-norm, i.e. $\operatorname{sp} x \geqq 0$, $\operatorname{sp}(x + y) \leqq \operatorname{sp} x + \operatorname{sp} y$, $\operatorname{sp}(\lambda x) = |\lambda| \operatorname{sp} x$, and has the additional property $\operatorname{sp}(x + \lambda \mathbf{1}) = \operatorname{sp} x$, $x \in \mathbf{R}^d$, $\mathbf{1}$ being a vector whose components are all 1. This leads to the following fundamental result:

2.1. THEOREM. *Define* $\gamma_{il} := 1 - \sum_j \min(a_{ij}, a_{lj})$, $i, l \in \{1, \ldots, k\}$, $\gamma_A := \max_{i,l} \gamma_{il}$ *and* $\gamma'_A := 1 - \sum_j \min_i a_{ij}$. *It follows*

$$\gamma_{il} \leqq \gamma_A \leqq \gamma'_A \leqq 1, \tag{2.1}$$

$$\sum_j (a_{ij} - a_{lj})\, x_j \leqq \gamma_{il} \operatorname{sp} x \tag{2.2}$$

and

$$\operatorname{sp} Ax \leqq \gamma_A \operatorname{sp} x. \tag{2.3}$$

Proof. The inequalities (2.1) are trivial. Let $i, l \in \{1, \ldots, k\}$ and $b_j := \min(a_{ij}, a_{lj})$. Then we have $\sum_j a_{ij}x_j - \sum_j a_{lj}x_j = \sum_j (a_{ij} - b_j)\, x_j - \sum_j (a_{lj} - b_j)\, x_j \leqq \sum_j (a_{ij} - b_j) \max x - \sum_j (a_{lj} - b_j) \min x = (1 - \sum_j b_j)(\max x - \min x) = \gamma_{il} \operatorname{sp} x$ and therefore $\operatorname{sp} Ax = \max_i \sum_j a_{ij}x_j - \min_l \sum_j a_{lj}x_j = \max_{i,l} \sum_j (a_{ij} - a_{lj})\, x_j \leqq \gamma_A \operatorname{sp} x$. □

2.2. COROLLARY. *Let* A_1, A_2 *be two stochastic* $k \times m$*-matrices. Let* A_1/A_2 *be the superposition of* A_1 *and* A_2 *(*$2k \times m$*-matrix). Then we have* $\max A_1x - \min A_2x \leqq \gamma_{A_1/A_2} \operatorname{sp} x$.

Proof. It is $\max A_1x - \min A_2x \leqq \max (A_1/A_2)\, x - \min (A_1/A_2)\, x \leqq \gamma_{A_1/A_2} \operatorname{sp} x$. □

Weaker forms of 2.1 are found in the theory of finite Markov chains (see e.g. Kemeny, Snell [9], p. 69) with $\gamma'''_A := 1 - 2 \min_{i,j} a_{ij}$ and in White [20], where the bound $\gamma''_A := 1 - \max_j \min_i a_{ij}$ is used. One easily realizes that $\gamma'_A \leqq \gamma''_A$ and $\gamma'_A \leqq \gamma'''_A$ $(m \geqq 2)$. An additional application of $\gamma_A, \gamma'_A, \gamma''_A$ is found in the so called Markov theorem for inhomogeneous Markov chains (see Hajnal [3], Mott [13], Sarymsakov [16] and Seneta [18], p. 104), where a Markov chain with transition matrices P_i proves to be weak ergodic, if $\sum_i (1 - \gamma_{P_i})$ diverges. A lower bound to γ_A is found in the absolute value of the second largest latent root of A, which, too, plays a role in the asymptotic analysis of Markov chains (cp. Morton [12] and e.g. Seneta [18], p. 7).

2.3. LEMMA. *In inequality* (2.3) *the value* γ_A *is the best possible, i.e. for each* A *there exists a vector* $y \in \mathbf{R}^m$ *with* $\operatorname{sp} Ay = \gamma_A \operatorname{sp} y$.

Proof. Assume $\gamma_A = \gamma_{kl} \neq 0$ and set $y_j := 1$ if $a_{kj} > a_{lj}$, else $y_j := 0$. From $a_{k\cdot} \neq a_{l\cdot}$ and $\sum_j (a_{kj} - a_{lj}) = 0$ follows $\operatorname{sp} y = 1$ and $\operatorname{sp} Ay \geqq \sum_j a_{kj}y_j - \sum_j a_{lj}y_j = \sum_j (a_{kj} - a_{lj})^+ = \sum_j (a_{kj} - \min(a_{kj}, a_{lj})) = \gamma_{kl}$. □

Let us now return to our decision model (see section 1). A mapping $f: S \to A$ with $f(i) \in D_i$ is called *admissible decision function*. We define $L^f: \boldsymbol{R}^S \to \boldsymbol{R}^S$ by $L^f x(i) := Lx(i, f(i))$, and similarly $p_{ij}^f := p_{ij}^{f(i)}$, $P^f := (p_{ij}^f, i, j \in S)$. Let f_n be one of the decision functions with $L^{f_n} V_{n-1} = U V_{n-1}$, $n \in N$. It follows

$$\max(V_{n+1} - V_n) = \max(L^{f_{n+1}} V_n - L^{f_n} V_{n-1}) \leqq$$
$$\leqq \max(L^{f_{n+1}} V_n - L^{f_{n+1}} V_{n-1}) = \max(\beta P^{f_{n+1}}(V_n - V_{n-1}))$$

and similarly

$$\min(V_{n+1} - V_n) \geqq \min(\beta P^{f_n}(V_n - V_{n-1}))$$

from which we easily conclude (1.2) and

$$\operatorname{sp}(V_{n+1} - V_n) \leqq \beta(\max P^{f_{n+1}}(V_n - V_{n-1}) - \min P^{f_n}(V_n - V_{n-1})).$$

This leads together with 2.2 and $\beta\gamma_A \leqq \delta$, $A := P^{f_{n+1}}/P^{f_n}$, to $\operatorname{sp}(V_{n+1} - V_n) \leqq \leqq \delta \operatorname{sp}(V_n - V_{n-1})$ (1.4). It remains to prove (1.6). We take $(i, a) \in D$ and set $a^* := f_n(i)$. Then we have $UV_n(i) - LV_n(i, a) \geqq L^{f_n}(V_{n-1} + (V_n - V_{n-1}))(i) - L(V_{n-1} + (V_n - V_{n-1}))(i, a) = UV_{n-1}(i) - LV_{n-1}(i, a) + \beta \sum_j (p_{ij}^{a^*} - p_{ij}^a)(V_n - V_{n-1})(j) \geqq UV_{n-1}(i) - LV_{n-1}(i, a) - \delta_{iaa^*} \operatorname{sp}(V_n - V_{n-1})$ where the last inequality follows from (2.2) and the definitions of $\beta\gamma_{il}$ and δ_{iaa^*}.

3. CONCLUSIONS

Looking at the definitions of $\delta, \delta', \delta_{ia}, \delta_i, \delta_i'$ (see 1.1, 1.2) and the necessary computing effort it seems best to use only the (weaker) bound δ' instead of δ and δ_{ia}. (Calculation of δ' will generally take less time than one step of value iteration.) In case of $\delta' \geqq 1$, however, it is necessary to compute δ in order to apply (1.9), if possible at all. In certain circumstances it may be efficient to use δ_i' or even δ_{ia}, especially if there are only a few actions available at each stage.

If, after having computed δ' and possibly δ, δ_{ia} or δ_i', we have $\delta' < 1$ or $\delta < 1$, the next step is to compute V_n by (1.1), $n = 1, 2, \ldots$, and to test suboptimality of actions at each stage by (1.9), replacing δ and δ_{ia} by δ', δ_i' if necessary. Oftenly by this procedure we find some stage n_0, at which for each $i \in S$ only one action remains. Otherwise we choose some arbitrary stage n_0 and set $f_n = f_{n_0}$ for $n \geqq n_0$. In both cases we obtain

(i) a sequence of optimal (in the second case: nearly optimal)* decision functions f_n, $n \in N$, being constant at least from n_0 onward,

(ii) the exact sequence of values V_n, $1 \leqq n \leqq n_0$, and

(iii) bounds for V_n, $n \geqq n_0$, by (1.3).

---

* Cp. Hinderer [5].

If we only are interested in a problem with finite but large horizon N, we may use the same procedure with the suboptimality test 1.4(a), even in the case of $\delta \geqq 1$.

In general the bounds (1.3) diverge as N increases, n being fixed. Only if $\beta < 1$ (and in certain models with absorbing structure, see Hinderer [5] and Hinderer, Hübner [6]) the differences $V''_{n,N} - V'_{n,N}$ remain bounded, and may therefore be reduced to an arbitrarily small value by continuing value iteration sufficiently far (cp. (1.4)). If however $\beta \geqq 1$, one may use $w := \lim_{N\to\infty} \beta^{-N}(V_{N+1} - V_N)$ as a valuation of the optimal process, a kind of average return per period (cp. Dirickx [1]), which at least in case of $\delta < \beta$ is finite and constant and may be approximated by $\min \beta^{-n}(V_{n+1} - V_n) \leqq w \leqq \max \beta^{-n}(V_{n+1} - V_n)$ (cp. (1.2), (1.4)).

We may vary the described procedures by using the results of Hastings, Mello [4] giving some computational ease, and by omitting the suboptimality test at some stages. It is possible, too, to adapt the values of $\delta, \delta', \delta'_i$, and δ_{ia} to the varying size of the matrix P, caused by the elimination of suboptimal actions, but this seems to be efficient only in very special cases.

To give numerical values we look at some examples found in Howard [7]: In "the toymaker's problem" we have $\delta = \delta' = 0.4\beta$, $\delta_{ia} \equiv 0.3\beta$, in the "taxicab operation" example one obtains $\delta = \delta' = 0.875\beta$ and $0.5\beta \leqq \delta_{ia} \leqq 0.875\beta$. For the "baseball" and "automobil replacement" problems, however, there is the poor result $\delta = \delta' = \beta$.

Another application of 1.1 is an improvement of the "turnpike theorem" of Shapiro [19], see Hinderer, Hübner [6].

The results of the present paper may be generalized to the case of infinite sets S and A. An application to semi-Markov decision processes (cp. Jewell [8], Porteus [14] and Schellhaas [17]) at first glance seems not to be possible.

I wish to give my thanks to Prof. K. Hinderer who stimulated my work on this topic by his own work and by his valuable advice.

REFERENCES

[1] Y. M. I. DIRICKX: Deterministic discrete dynamic programming with discount factor greater than one: structure of optimal policies. Man. Sci. *20* (1973), 32—43.

[2] R. C. GRINOLD: Elimination of suboptimal actions in Markov decision problems. Operations Res. *21* (1973), 848—851.

[3] J. HAJNAL: Weak ergodicity in non-homogeneous Markov chains. Proc. Cambr. Phil. Soc. *54* (1958), 233—246.

[4] N. A. J. HASTINGS, J. C. M. MELLO: Tests for suboptimal actions in discounted Markov programming. Man. Sci. *19* (1973), 1019—1022.

[5] K. HINDERER: Estimates for finite state dynamic programs. J. Math. Anal. Appl. *55* (1976), 207—238.

[6] K. HINDERER, G. HÜBNER: An improvement of J. F. Shapiro's turnpike theorem for the horizon of finite stage discrete dynamic programs. This volume 245—255.

[7] R. A. HOWARD: Dynamic programming and Markov processes. Wiley, New York 1960.

[8] W. S. JEWELL: Markov renewal programming I + II. Operations Res. *11* (1963), 938—971.

[9] J. G. KEMENY, J. L. SNELL: Finite Markov chains. Van Nostrand, Princeton, N. J. 1960.

[10] J. MACQUEEN: A modified dynamic programming method for Markovian decision problems. J. Math. Anal. Appl. *14* (1966), 38—43.

[11] J. MACQUEEN: A test for suboptimal actions in Markovian decision problems. Operations Res. *15* (1967), 559—561.

[12] TH. E. MORTON: On the asymptotic convergence rate of cost differences for Markovian decision processes. Operations Res. *19* (1971), 244—248.

[13] J. L. MOTT: Conditions for the ergodicity of non-homogeneous finite Markov chains. Proc. Roy. Soc. Edinburgh Sec. A *64* (1951), 369—380.

[14] E. L. PORTEUS: Some bounds for discounted sequential decision processes. Man. Sci. *18* (1971), 7—11.

[15] D. REETZ: A decision exclusion algorithm for a class of Markovian decision processes. Zeitschr. Operations Res. *20* (1976), 125—131.

[16] T. A. SARYMSAKOV: On inhomogeneous Markov chains. Doklady A.N.S.S.S.R. *120* (1958), 465—467.

[17] H. SCHELLHAAS: Zur Extrapolation in Markoffschen Entscheidungsmodellen mit Diskontierung. Zeitschr. Operations Res. *18* (1974), 91—104.

[18] E. SENETA: Non-negative matrices. Allen & Unwin, London 1973.

[19] J. F. SHAPIRO: Turnpike planning horizons for a Markovian decision model. Man. Sci. *14* (1968), 292—300.

[20] D. J. WHITE: Dynamic programming, Markov chains and the method of successive approximations. J. Math. Anal. Appl. *6* (1963), 373—376.

UNIVERSITÄT HAMBURG

INSTITUT FÜR MATHEMATISCHE STOCHASTIK

ON MARKOV POLICIES IN CONTINUOUS TIME DISCOUNTED DYNAMIC PROGRAMMING

ADAM IDZIK

WARSZAWA

We consider the problem of discounted dynamic programming with a continuous time parameter (CDP) when the Markov policies are used. We give an axiomatization of such discounted CDP. We also give necessary and sufficient conditions for the existence of an optimal policy. Analogously to the discrete case we formulate improvement's theorems and a theorem on the existence of a (p, ε)-optimal policy in a class of semi-Markov policies.

1. INTRODUCTION

During the last several years, problems of dynamic programming with a discrete time parameter (DDP) have been investigated by many authors (for example [1], [2], [3], [4], [5], [7], [8], [11]). Also many authors have investigated problems of continuous time dynamic programming (CDP) (for example [5], [6], [10]). Howard [5] and Miller [10] have restricted themselves to finite state and action spaces. Kakumanu [6] has studied the case with countable state and action spaces. The papers [5], [6], [10] have been based on the differential structure of the Markov decision processes.

In this paper we extend some Blackwell's ideas contained in papers [1] and [3] to CDP. We consider only the discounted case where the Markov policies are used. We give an axiomatization of such discounted CDP and in some sense its characterization. We formulate also the necessary and sufficient conditions for the existence of an optimal policy. Analogously to respective theorems in [3] we formulate improvement's theorems and a theorem on the existence of (p, ε)-optimal policy in a class of semi-Markov policies.

2. PRELIMINARY DEFINITIONS

Let T denote the set of nonnegative real numbers (the set of time – parameters). Elements of T are called moments (we denote them by letters t, k). The set of earlier moments than the moment t (i.e. the set $\{k : k \in T,\ k < t\}$) we denote by $T \mid t$, and the set of non-earlier moments than the moment t (i.e. the set $\{k : k \in T,\ t \leqq k\}$) we denote by $t \mid T$.

Let X and D be arbitrary metric spaces. We call them the *set of states* and the *set of available decisions* at each state, respectively. The elements of X we denote by x, y, z and the elements of D by d. Maps between sets we denote by small letters of the Greek alphabet (excluding β). And thus, for example, a *strategy* is an arbitrary Borel measurable map $\sigma : X \to D$. Further, we shall write simply a measurable map instead of a Borel measurable map. A *policy* is an arbitrary measurable map $\pi : X \times T \to D$. Thus, for a fixed t $\pi(\cdot, t)$ is the strategy. By $\pi \mid t$, respectively $t \mid \pi$, we denote a policy π restricted to the set $T \mid t$, respectively $t \mid T$.

For fixed policies π_1, π_2, maps defined as follows:

2.1. $$^t\pi_2(x, k) := \pi_2(x, k - t) \quad \text{for} \quad x \in X, \quad t \in T, \quad k \in t \mid T,$$

2.2. $$\pi_1^t(x, k) := \pi_2(x, t + k) \quad \text{for} \quad x \in X, \quad \text{and} \quad t, k \in T,$$

2.3. $$(\pi_1 \mid t, t \mid \pi_2)(x, k) := \begin{cases} \pi_1(x, k) & \text{for} \quad k < t, \\ \pi_2(x, k) & \text{for} \quad t \leqq k, \end{cases} \quad x \in X,$$

2.4. $$(\pi_1 \mid t, {}^t\pi_2)(x, k) := \begin{cases} \pi_1(x, k) & \text{for} \quad k < t, \\ \pi_2(x, k - t) & \text{for} \quad t \leqq k, \end{cases} \quad x \in X,$$

are measurable. Thus, the maps defined by 2.2–2.4 are policies. A policy π is *t-periodic* if $\pi = \pi^t$; we denote it by $\pi^{(t)}$. A policy π is *stationary* if $\pi(x, t_1) = \pi(x, t_2)$ for $x \in X$ and $t_1, t_2 \in T$.

3. THE DISCOUNTED CDP WHEN THE MARKOV POLICIES ARE USED

By $M(X)$ we denote the Banach space of all bounded measurable real functions on X with the norm $\|v\| = \sup\{|v(x)| : x \in X\}$. Let us assign to every policy π and every moment $t \in T$, a linear nonnegative operator on $M(X)$, depending only on $\pi \mid t$. We shall write then $P_{\pi|t} : M(X) \to M(X)$. For convenience we shall write $P_{x,\pi|t}(v)$

instead of $[P_{\pi|t}(v)](x)$ and $P_{\pi|t}(B)$, where B is a Borel subset of X, instead of $P_{\pi|t}(\chi_B)$ with χ_B being the characteristic function of B. Thus, for instance, $P_{x,\pi|t}(B)$ is meaningfull. The norm of an operator $P_{\pi|t}$ is defined in the usual way as $\|P_{\pi|t}\| = \sup\{\|P_{\pi|t}(v)\| : \|v\| = 1\}$. And by o we denote the superposition. Thus, $[P_{\pi|t} \circ P_{\bar{\pi}|k}](v) = P_{\pi|t}[P_{\bar{\pi}|k}(v)]$.

We formulate the following axioms:

I. $$\|P_{\pi|t}\| \leqq 1 \quad \text{for all } \pi \text{ and } t\,.$$

II. $$P_{\pi|t} \circ P_{\pi_1{}^t|k-t} = P_{(\pi|t,t|\pi_1)|k}\,, \quad P_{\pi|0} = I$$

for all π, π_1, t and k; $t \leqq k$
(the generalized Chapman-Kolmogorov's equation).

III. For fixed x and $\pi \mid t$ $P_{x,\pi|t}(\cdot)$ is a Borel measure.

IV. For a fixed Borel set $B \subset X$ and a fixed policy π $P_{x,\pi|t}(B)$ is a measurable function of (x, t).

From the axiom IV follows (see [9], p. 15).

3.1. For a fixed policy π and for each function $v(x, t) \in M(X \times T)$, the function

$$\mu(y, t) = \int_X v(x, t)\, P_{y,\pi|t}(\mathrm{d}x) \in M(X \times T)\,.$$

In this paper the axiom IV will be used in the form 3.1 only.

Now, let us imagine a system which may change states. Let it be in a state $x \in X$ at the moment $t = 0$. We may control this system using a respective policy. We assume that our system moves in a probabilistic way, and at every moment t probability distribution defined on Borel subsets of X is given. Namely $P_{x,\pi|t}(B)$: the probability of coming from the initial state x to one of the states of the set $B \subset X$ at the moment t, if we use the policy π.

Furthermore, we assume that our system gives us a payoff (which may be negative). A payoff we define in the following way:

Let $\varrho : X \times D \to R$ be a bounded, measurable function i.e. $\varrho \in M(X \times D)$.

LEMMA 1. The function ϱ_π defined by $\varrho_\pi(x, t) := \varrho(x, \pi(x, t))$ is a measurable function.

Proof. Because a map $\lambda : X \times T \to X$ defined by $\lambda(x, t) := x$ is measurable and a map $\varphi : X \times T \to X \times D$ defined by $\varphi(x, t) := (x, \pi(x, t))$ is also measurable, then $\varrho_\pi = \varrho \circ \varphi$ is measurable as a superposition of measurable maps.

The function ϱ_π is a *density of payoff* for a policy π. Let a number β^t $(0 < \beta < 1)$ denote a worth of a unit of the payoff at the moment t (the number β is the discount factor). A *payoff function of* an *initial state* x and a *policy* $\pi \mid t$ (in other words:

an expected payoff if our system starts from a state x and we use a policy π to a moment t) is the function

$$(3.2) \qquad v(x, \pi \mid t) := \int_{T|t} \beta^k \int_X \varrho(y, \pi(y, k))\, P_{x,\pi|k}(d, y)\, dk\,,$$

where the first integral is Lebesgue's integral.

For a fixed policy $\pi \mid t$, from 3.1, $v(x, \pi \mid t) \in M(X)$. A function $v(x, \pi) := := v(x, \pi \mid \infty) := \lim_{t\to\infty} v(x, \pi \mid t)$ is a *payoff function of* an *initial state* x and *policy* π. Because for every policy $\pi \mid t$, $|v(x, \pi \mid t)| \leqq K(\beta)$, where $K(\beta)$ is a constant, then $v(x, \pi)$ is well-defined and $v(x, \pi) \in M(X)$. Further, we shall use the notation $v(\pi \mid t)(x) := v(x, \pi \mid t)$.

Summing up, a discounted CDP problem, when the Markov policies are used, is defined by four objects: X, D, ϱ and a collection of operators $P_{\pi|t}$ satisfying the axioms I–IV.

REMARK 1. The definitions given above we may formulate restricting ourselves to some subset of all policies. For instance, policies piece-wise stationary or policies piece-wise continuous in time. However, this subset must be closed with respect to the operations 2.2–2.4.

4. THE CHARACTERIZATION OF SUCH A DISCOUNTED CDP

Let us observe that

LEMMA 2. Every operator $P_{\pi|t}$ may be presented in an integral form, namely

$$4.0. \qquad P_{x,\pi|t}(v) = \int_X v(y)\, P_{x,\pi|t}(dy)\,, \qquad v \in M(X)\,.$$

Proof. If v is a simple function, then 4.0 is obvious. Because each function $v \in M(X)$ is the limit of a uniformly convergent sequence of simple functions, then the equation 4.0 is true for all function $v \in M(X)$.

Now, we prove the following

THEOREM 1.

$$v(x, \pi) = v(x, \pi \mid t) + \beta^t \int_X v(y, \pi^t)\, P_{x,\pi|t}(dy) = v(x, \pi \mid t) + \beta^t P_{x,\pi|t}(v(\pi^t))$$

for each $x \in X$ and every policy π.

Proof. It is sufficient to show, that

$$\int_{t|T} \beta^k \int_X \varrho(y, \pi(y, k))\, P_{x,\pi|k}(\mathrm{d}y)\, \mathrm{d}k = \beta^t \int_X v(y, \pi^t)\, P_{x,\pi|t}(\mathrm{d}y).$$

But for $k, t \in T$ and $t \leqq k$ it follows from axiom II that

$$\int_X \varrho(y, \pi(y, k))\, P_{x,\pi|k}(\mathrm{d}y) = \int_X \left[\int_X \varrho(y, \pi(y, k))\, P_{z,\pi|k-t}(\mathrm{d}y)\right] P_{x,\pi|t}(\mathrm{d}z).$$

Then

$$\int_{t|T} \beta^k \int_X \varrho(y, \pi(y, k))\, P_{x,\pi|k}(\mathrm{d}y)\, \mathrm{d}k =$$

$$= \beta^t \int_{t|T} \beta^{k-t} \int_X \left[\int_X \varrho(y, \pi^t(y, k - t))\, P_{z,\pi^t|k-t}(\mathrm{d}y)\right] P_{x,\pi|t}(\mathrm{d}z)\, \mathrm{d}k =$$

$$= \beta^t \int_T \beta^k \int_X \left[\int_X \varrho(y, \pi^t(y, t))\, P_{z,\pi^t|k}(\mathrm{d}y)\right] P_{x,\pi|t}(\mathrm{d}z)\, \mathrm{d}k =$$

$$= \beta^t \int_X \int_T \beta^k \left[\int_X \varrho(y, \pi^t(y, k))\, P_{z,\pi^t|k}(\mathrm{d}y)\right] \mathrm{d}k\, P_{x,\pi|t}(\mathrm{d}z) =$$

$$= \beta^t \int_X v(z, \pi^t)\, P_{x,\pi|t}(\mathrm{d}z).$$

The last equality follows from 3.1 and Fubini's theorem.

Now, we shall give necessary and sufficient conditions for the existence of an optimal policy. At the beginning let us formulate a few definitions. For $\omega_1, \omega_2 \in M(X)$ we write $\omega_1 \leqq \omega_2$ if $\omega_1(x) \leqq \omega_2(x)$ for each $x \in X$, and $\omega_1 < \omega_2$ if $\omega_1 \leqq \omega_2$ and $\omega_1 \neq \omega_2$. For two policies π_1 and π_2 we write $\pi_1 \leqq \pi_2$ if $v(\pi_1) \leqq v(\pi_2)$, and $\pi_1 < \pi_2$ if $v(\pi_1) < v(\pi_2)$. We say, that $\bar{\pi}$ is an *optimal policy* if $\pi \leqq \bar{\pi}$ for every policy π. Let us denote $v(x, \bar{\pi}) = \mu(x)$ for the optimal policy $\bar{\pi}$.

THEOREM 2 (a necessary condition). If $\bar{\pi}$ is the optimal policy, then at every moment t the optimality equation

4.1. $$\mu(x) = \sup_{\pi|t} \left[v(x, \pi \mid t) + \beta^t P_{x,\pi|t}(\mu)\right]$$

and

4.2. $$\pi \leqq (\pi \mid t, {}^t\bar{\pi})$$

are fulfilled.

Proof. From Theorem 1 follows, that

$$\mu(x) = v(x, \bar{\pi} \mid t) + \beta^t \int_X v(y, \bar{\pi}^t)\, P_{x,\bar{\pi}|t}(\mathrm{d}y) \leqq$$

$$\leqq \sup_{\pi|t} \left[v(x, \pi \mid t) + \beta^t \int_X v(y, \bar{\pi}^t)\, P_{x,\pi|t}(\mathrm{d}y) \right] \leqq$$

$$\leqq \sup_{\pi|t} \left[v(x, \pi \mid t) + \beta^t \int_X v(y, \bar{\pi})\, P_{x,\pi|t}(\mathrm{d}y) \right] =$$

$$= \sup_{\pi|t} \left[v(x, \pi \mid t) + \beta^t \int_X \mu(y)\, P_{x,\pi|t}(\mathrm{d}y) \right].$$

The last inequality is true because $\bar{\pi}$ is optimal. From the optimality of $\bar{\pi}$ and from Theorem 1 we have

$$\mu(x) \geqq v(x, (\pi \mid t, {}^t\bar{\pi})) = v(x, \pi \mid t) + \beta^t \int_X v(y, \bar{\pi})\, P_{x,\pi|t}(\mathrm{d}y) =$$

$$= v(x, \pi \mid t) + \beta^t \int_X \mu(y)\, P_{x,\pi|t}(\mathrm{d}y)$$

for every $\pi \mid t$. And thus

$$\mu(x) \geqq \sup_{\pi|t} \left[v(x, \pi \mid t) + \beta^t \int_X \mu(y)\, P_{x,\pi|t}(\mathrm{d}y) \right].$$

Then 4.1 follows from two inequalities obtained above.

Now, let us suppose that there exist such x, t and π that $v(x, (\pi \mid t, {}^t\bar{\pi}) < v(x, \pi)$. Using Theorem 1 we get

$$\int_X v(y, \bar{\pi})\, P_{x,\pi|t}(\mathrm{d}y) < \int_X v(y, \pi^t)\, P_{x,\pi|t}(\mathrm{d}y).$$

But $\bar{\pi}$ is optimal i.e. $v(y, \bar{\pi}) \geqq v(y, \pi^t)$ for each $y \in X$, and thus

$$\int_X v(y, \bar{\pi})\, P_{x,\pi|t}(\mathrm{d}y) \geqq \int_X v(y, \pi^t)\, P_{x,\pi|t}(\mathrm{d}y)$$

what contradicts the inequality obtained above.

A theorem in some sense reversed is true.

THEOREM 3 (a sufficient condition). If at some moment t the payoff function of a policy $\bar{\pi} : v(x, \bar{\pi}) = \mu(x)$ fulfils the conditions 4.1 and 4.2, then $\bar{\pi}$ is optimal.

Proof. Let us fix x. Obviously for $t = 0$ the theorem is true (follows from 4.2). Let us assume that $t > 0$. From 4.1 follows that

$$v(x, \bar{\pi}) = \sup_{\pi/t} \left[v(x, \pi \mid t) + \beta^t \int_X v(y, \bar{\pi})\, P_{x,\pi|t}(\mathrm{d}y) \right] \geqq$$

$$\geqq v(x, \pi \mid t) + \beta^t \int_X v(y, \bar{\pi})\, P_{x,\pi|t}(\mathrm{d}y) = v(x, (\pi \mid t, {}^t\bar{\pi}))$$

for every policy π. Thus by 4.2 we have

$$v(x, \bar{\pi}) \geqq v(x, (\pi \mid t, {}^t\bar{\pi})) \geqq v(x, \pi)$$

for every policy π; what ends our proof.

Now, for every policy $\pi \mid t$ let us define an operator $L_{\pi|t} : M(X) \to M(X)$ as follows

4.3. $$L_{\pi|t}(\omega) = v(\pi \mid t) + \beta^t P_{\pi|t}(\omega), \quad \omega \in M(X).$$

The operator $L_{\pi|t}$ has the following interpretation: $L_{\pi|t}(\omega)(x)$ is an expected payoff as a function of an initial state x, if we are starting from x using a policy π to a moment t and the payoff is holding with the density ϱ_π to the moment t, and if we shall yet in a state y, we obtain a reward $\omega(y)$.

PROPERTIES (of $L_{\pi|t}$ operator):

4.4. $L_{\pi|t}$ is monotone, i.e. $\mu \leqq \nu$ implies $L_{\pi|t}(\mu) \leqq L_{\pi|t}(\nu)$.

4.5. For any constant $c \geqq 0$ $L_{\pi|t}(\mu + c) \leqq L_{\pi|t}(\mu) + \beta^t c$.

4.6. For any policy $\bar{\pi}$ $L_{\pi|t}(v(\bar{\pi})) = v((\pi \mid t, {}^t\bar{\pi}))$.

Proof of this properties of $L_{\pi|t}$ is immediate.

We prove now one condition more for the optimality of a policy.

THEOREM 4 (see Lemma 2 in [7]). If $(\pi \mid t_0, {}^{t_0}\bar{\pi}) \leqq \bar{\pi}$ for some $t_0 > 0$ and all π, then $\bar{\pi}$ is optimal.

Proof. Let us fix $\tilde{\pi}$. From our assumption we have $L_{\tilde{\pi}|t_0}(v(\bar{\pi})) = v(\bar{\pi})$. By monotonicity of $L_{\pi|t_0}$ it is easy to obtain, using a few times the property 4.6, that $(\tilde{\pi} \mid nt_0, {}^{nt_0}\bar{\pi}) \leqq \bar{\pi}$ for $n = 1, 2, 3, \ldots$ Tending with n to infinity we obtain $v(\tilde{\pi}) \leqq v(\bar{\pi})$ for every $\tilde{\pi}$.

COROLLARY. A policy $\bar{\pi}$ is optimal if and only if its payoff function fulfils the optimality equation 4.1 at some moment $t_0 > 0$.

Proof. If $\bar{\pi}$ is optimal, then by Theorem 2 it fulfils the optimality equation at every moment t_0. On the other hand if its payoff function $v(x, \bar{\pi}) = \mu(x)$ fulfils 4.1 at the moment $t_0 > 0$, then

$$v(x, \bar{\pi}) = \sup_{\pi | t_0} \left[v(x, \pi \mid t_0) + \beta^{t_0} P_{x,\pi | t_0}(v(\bar{\pi})) \right] \geqq$$
$$\geqq v(x, \pi \mid t_0) + \beta^{t_0} P_{x,\pi | t_0}(v(\bar{\pi})) = v(x, (\pi \mid t_0, {}^{t_0}\bar{\pi}))$$

for every policy π, i.e. $(\pi \mid t_0, {}^{t_0}\bar{\pi}) \leqq \bar{\pi}$. And by Theorem 4 $\bar{\pi}$ is optimal.

5. IMPROVEMENT'S THEOREMS

At the begining we prove

THEOREM 5 (Howard improvement). If $\pi < (\bar{\pi} \mid t_0, {}^{t_0}\pi)$, then $\pi < \bar{\pi}^{(t_0)}$, where $\bar{\pi}^{(t_0)}$ is t_0-periodic policy such that $\bar{\pi}^{(t_0)} \mid t_0 = \bar{\pi} \mid t_0$ $(t_0 > 0)$.

Proof. From the assumption we have $L_{\bar{\pi}|t_0}(v(\pi)) \geqq v(\pi)$. By monotonicity of $L_{\bar{\pi}|t_0}$ it is easy to show, that $(\bar{\pi} \mid t_0, {}^{t_0}\pi) = (\bar{\pi}^{(t_0)} \mid nt_0, {}^{nt_0}\pi)$ for every $n \geqq 1$. Tending with n to infinity we obtain $\pi < (\bar{\pi} \mid t_0, {}^{t_0}\pi) \leqq \bar{\pi}^{(t_0)}$.

Now, we shall generalize a definition of the policy. Let a policy π and a moment $t > 0$ will be fixed in this paragraph. We say, that a *generalized policy* in time-interval $[jt, (j + 1)t)$ $\pi_{\gamma_j} : X \times X \times \{jt \mid T \mid (j + 1)t\} \to D$ is *(π, t)-generated* if there is a partition γ_j of X into Borel sets X_{ij} $(i = 0, 1, 2, \ldots)$ such that $X = \bigcup_{i=0}^{\infty} X_{ij}$ and $\pi_{\gamma_j}(x, \cdot, \cdot) = \pi^{it}(\cdot, \cdot)$ for a process starting from an initial state $x \in X_{ij}$ at a moment jt $(j = 0, 1, 2, \ldots)$. Let us denote $\bar{\gamma} = (\gamma_0, \gamma_1, \gamma_2, \ldots)$. We say, that a *generalized policy* $\pi_\gamma : X \times X \times T \to D$ is *(π, t)-generated* if there is a sequence of generalized policies π_{γ_j} $(j = 0, 1, 2, \ldots)$ (π, t)-generated defined above and $\pi_{\bar{\gamma}} \mid \{jt \mid T \mid (j + 1)t\} = \pi_{\gamma_j}$. Further, we shall write a policy $\pi_{\gamma_j}(\pi_{\bar{\gamma}})$ instead of a generalized policy $\pi_{\gamma_j}(\pi_{\bar{\gamma}})$, because the above notation of it will inform us what kind of policy we use.

Let π_{γ_0} be (π, t)-generated policy. Analogously to 4.3 we define an operator $L_{\pi_{\gamma_0}} : M(X) \to M(X)$ as follows

5.1. $$L_{\pi_{\gamma_0}}(\omega)(x) := v(x, \pi_{\gamma_0}) + \beta^t P_{\pi_{\gamma_0}}(\omega)(x) :=$$
$$:= v(x, \pi^{it} \mid t) + \beta^t P_{x,\pi^{it}|t}(\omega) \quad \text{for} \quad x \in X_{i0}, \quad \omega \in M(X).$$

From 5.1 we have $v(\pi_{\gamma_0}) = L_{\pi_{\gamma_0}}(0)$. If $\pi_{\bar{\gamma}}$ $[\bar{\gamma} = (\gamma_0, \gamma_1, \ldots)]$ is a (π, t)-generated policy then

5.2. $$v(\pi_{\bar{\gamma}} \mid nt; \mu) := L_{\pi_{\gamma_0}} \circ L_{\pi_{\gamma_1}} \circ \ldots \circ L_{\pi_{\gamma_{n-1}}}(\mu)$$

is an *expected payoff* for *policy* $\pi_{\bar{\gamma}}$ to a *moment nt with a final reward* μ $(n = 1, 2, 3, \ldots)$. For a (π, t)-generated policy $\pi_{\bar{\gamma}}$ we define the payoff function by

5.3. $$v(\pi_{\bar{\gamma}}) := \lim_{n \to \infty} v(\pi_{\bar{\gamma}} \mid nt; 0)$$

$v(\pi_{\bar{\gamma}})$ is well-defined function of $M(X)$.

It is easy to imagine a (π, t)-generated t-periodic policy $(: \pi_{\bar{\gamma}}^{(t)}$ or $: \pi_{\gamma 0}^{(t)})$. Evidently we must have $X_{ij} = X_{ik}$ $(i, j, k = 0, 1, 2, \ldots)$ and $\bar{\gamma} = (\gamma_0, \gamma_0, \gamma_0, \ldots)$. The process starts from a state $x \in X_{i0}$ and we use a policy $\pi^{it} \mid t$ up to the moment t. If the process will be in a state $y \in X_{j0}$ at the moment t we use a policy $\pi^{jt} \mid t$ and so on. From 5.3 we have

5.4. $$v(\pi_{\gamma 0}^{(t)}) = \lim_{n \to \infty} L_{\pi_{\gamma 0}}^{n}(0) .$$

An operator $L_{\pi_{\gamma 0}}$ is a contraction, so by the fixed point Banach theorem we have

5.5. $$v(\pi_{\gamma 0}^{(t)}) = L_{\pi_{\gamma 0}}(v(\pi_{\gamma 0}^{(t)})) .$$

REMARK 2. a) A (π, t)- generated policy $\pi_{\bar{\gamma}}$ we may interpret as an instructed policy at moments jt $(j = 0, 1, 2, \ldots)$ by the policy π. If we are in some state at a moment jt we choose a decision (depending on a partition $\bar{\gamma}$) which a policy we should use up to the moment $(j + 1)\, t$ $(j = 0, 1, 2, \ldots)$. A chosen policy depends on a behaviour of the policy π in time-intervals $[kt, (k + 1)\, t)$ $(k = 0, 1, 2, \ldots)$.

b) If π is stationary, then for every t (π, t)-generated policies π_{γ_j} and $\pi_{\bar{\gamma}}$ are also stationary and equal to $jt \, |\pi| \, (j + 1)\, t$ and π, respectively.

Now, we formulate the second improvement theorem.

THEOREM 6 (Eaton-Zadeh improvement). For two policies $\pi_{\gamma_{10}} : X \times X \times T \mid t \to D$ and $\pi_{\gamma_{20}} : X \times X \times T \mid t \to D$ (π_1, t)- and (π_2, t)-generated, respectively, we define a (π_3, t)-generated policy

$$\pi_{\gamma_{30}}(x, \cdot, \cdot) := \begin{cases} \pi_{\gamma_{10}}(x, \cdot, \cdot) & \text{for} \quad x \in X_1 = \{x : v(x, \pi_{\gamma_{20}}^{(t)}) \leqq v(x, \pi_{\gamma_{10}}^{(t)})\} \\ \pi_{\gamma_{20}}(x, \cdot, \cdot) & \text{for} \quad x \in X_2 = \{x : v(x, \pi_{\gamma_{10}}^{(t)}) < v(x, \pi_{\gamma_{20}}^{(t)})\} \end{cases}$$

where $\gamma_{30} = \{X_1 \cap Y_i, \ X_2 \cap Z_j\}$ $i, j = 0, 1, 2, \ldots$ $(\gamma_{10} = \{Y_i\}_{i=0,1,\ldots}, \ \gamma_{20} = \{Z_j\}_{j=0,1,\ldots})$ and a policy π_3 is defined (not uniquely) by the policies π_1 and π_2 (for example: $\pi_3^{2it} \mid t = \pi_1^{it} \mid t$ and $\pi_3^{(2i+1)t} \mid t = \pi_2^{it} \mid t$, $i = 0, 1, 2, \ldots$).

Then for every $x \in X$

$$v(x, \pi_{\gamma_{30}}^{(t)}) \geqq \max\,[v(x, \pi_{\gamma_{10}}^{(t)}), v(x, \pi_{\gamma_{20}}^{(t)})] .$$

Proof (is similar to the proof given by A. Maitra in the discrete time case, see [3]). For every $\omega \in M(X)$.

$$L_{\pi_{\gamma 30}}(\omega) = L_{\pi_{\gamma 10}}(\omega) \quad \text{on the set} \quad X_1$$

and

$$L_{\pi_{\gamma 30}}(\omega) = L_{\pi_{\gamma 20}}(\omega) \quad \text{on the set} \quad X_2$$

Putting $\bar{\omega} = \max\left[v(\pi_{\gamma 10}^{(t)}), v(\pi_{\gamma 20}^{(t)})\right]$ we obtain

$$L_{\pi_{\gamma 30}}(\bar{\omega}) = L_{\pi_{\gamma 10}}(\bar{\omega}) \geqq L_{\pi_{\gamma 10}}(v(\pi_{\gamma 10}^{(t)})) = v(\pi_{\gamma 10}^{(t)}) = \bar{\omega} \quad \text{on the set} \quad X_1 .$$

The last equality is true by 5.5. And analogously

$$L_{\pi_{\gamma 30}}(\bar{\omega}) \geqq \bar{\omega} \quad \text{on the set} \quad X_2 .$$

Thus

$$L_{\pi_{\gamma 30}}(\bar{\omega}) \geqq \bar{\omega} \quad \text{on whole} \quad X ,$$

so that

$$v(\pi_{\gamma 30}^{(t)}) \geqq \bar{\omega} .$$

6. SEMI-MARKOV POLICIES; (p, ε)-OPTIMALITY

A policy $\pi_{\mathrm{SM}} : X \times X \times T \to D$ is *semi-Markov* (*SM-policy*) if there is a partition γ_0 of X into Borel sets $X_0, X_1, \ldots$ such that $X = \bigcup_{i=0}^{\infty} X_i$ and if there is a sequence of policies $\pi_0, \pi_1, \pi_2, \ldots$ such that $\pi_{\mathrm{SM}}(x, \cdot, \cdot) = \pi_i(\cdot, \cdot)$ for the process starting from a state $x \in X_i$. A payoff function of SM-policy π_{SM} is defined as

6.1. $$v(x, \pi_{\mathrm{SM}}) := v(x, \pi_i) \quad \text{for} \quad x \in X_i .$$

It is easy to observe that SM-policy is a limit case of a (π, t)-generated policy $(t = \infty)$. Every policy is of course SM-policy.

Let $P(X)$ denote the set of all probability measures (simply: probabilities) defined over the Borel subsets of X. For any $p \in P(X)$, and any $\varepsilon > 0$ $\bar{\pi}$ is *(p, ε)-optimal* if $p\{v(\bar{\pi}) + \varepsilon < v(\pi)\} = 0$ for every policy π. Of course, if $\bar{\pi}$ is (p, ε)-optimal for every $p \in P(X)$ and every $\varepsilon > 0$, then is optimal.

THEOREM 7. For any $p \in P(X)$ and any $\varepsilon > 0$ there is a (p, ε)-optimal SM-policy in the class of all SM-policies.

Proof is based on the same method as the proof of Theorem 1 in [3] and we omit it.

REMARK 3. Using our notations and similar methods as in [3] it is easy to prove an analogical theorem to Theorem 6 in [3].

ACKNOWLEDGMENT. I am very grateful to Professor J. Łoś and Dr. M. Wycech-Łoś for their valuable advices and many helpful suggestions during the preparation of this paper.

REFERENCES

[1] D. Blackwell: On the functional equation of dynamic programming. J. Math. Anal. Appl. *2* (1961), 273—276.

[2] D. Blackwell: Discrete dynamic programming. The Annals of Math. Statist. *33* (1962), 719—726.

[3] D. Blackwell: Discounted dynamic programming. The Annals of Math. Statist. *36* (1965), 226—235.

[4] K. Hinderer: Foundations of Non-stationary Dynamic Programming. Springer-Verlag, Berlin—Heidelberg 1970.

[5] R. A. Howard: Dynamic Programming and Markov Processes. Wiley, New York 1960.

[6] P. Kakumanu: Continuously discounted Markov decision model with countable state and action space. The Annals of Math. Statist. *42* (1971), 919—926.

[7] A. Maitra: Dynamic programming for countable state systems. Sankhyá Ser. A *27* (1965), 241—248.

[8] A. Maitra: Discounted dynamic programming on compact metric spaces. Sankhyá Ser. A *30* (1968), 211—216.

[9] P. A. Meyer: Probability and Potentials. Blaisdell Publishing Company: Waltham, Massachusetts—Toronto—London 1966.

[10] B. L. Miller: Finite state continuous time Markov decision processes with a finite planning horizon. SIAM J. Control *6* (1968), 266—280.

[11] R. E. Strauch: Negative dynamic programming. The Annals of Math. Statist. *37* (1966), 871—890.

Polish Academy of Sciences

Computation Centre

OSCILLATIONS OF MONOTONE AMPLITUDE AND APPLICATION

J. P. IMHOF

GENEVA

A duality result of Lindley expressing probabilities for the position after n steps of a random walk restricted to a horizontal strip as absorption probabilities for unrestricted walk is proved by introducing the notion of oscillations of monotone amplitude. Additional information is obtained as well as extension to arbitrary restricting boundary.

1. INTRODUCTION

Let S be the walk determined by the sums $S_j = \sum_1^j X_k$ of randon variables $X_1, \ldots, X_n$, $j = 0, 1, \ldots, n$, with $S_0 = 0$. Setting up impenetrable barriers at levels 0 and $a > 0$ produces restricted walk W given by

$$W_0 = 0\,, \quad W_j = \min\{a, (W_{j-1} + X_j)^+\}\,, \quad j = 1, \ldots, n\,.$$

The reverse of S is the walk $T = \{T_0, T_1, \ldots, T_n\}$, $T_0 = 0$, $T_j - T_{j-1} = X_{n-j+1}$, $j = 1, \ldots, n$. For given $y \in [0, a]$, call T^y the walk resulting from absorption of T in $]y, \infty[$ and in $]-\infty, y - a]$, defined by introducing

$$\alpha = \begin{cases} \min\{j : 0 \leqq j \leqq n,\ T_j \notin]y - a, y]\} & \text{if there is such } j\,, \\ \infty\,, \quad \text{otherwise} \end{cases}$$

and setting for $k = 0, 1, \ldots, n$,

$$T_k^y = \begin{cases} T_k & \text{if} \quad k < \alpha\,, \\ T_\alpha & \text{if} \quad k \geqq \alpha\,. \end{cases}$$

A classical result due to Lindley [1] expresses in probabilistic terms, when the X_j are independent identically distributed, the duality relation

$$\{W_n > y\} = \{T_n^y > y\}, \quad 0 \leqq y \leqq a. \tag{1}$$

A recent proof of (1) has been given by Phatarford, Speed and Walker [3] based on the explicit representation, justified by induction,

$$W_n = \max_{1 \leqq j \leqq n} \min \{a, (T_1 + a)^+, \ldots, (T_{j-1} + a)^+, T_j^+\}. \tag{2}$$

We show that an expression equivalent to this rather hermetic formula follows from simple geometric considerations involving successive oscillations of decreasing amplitude of S, and the corresponding oscillations of increasing amplitude of T. (1) can then be subdivided into sharper equations taking into account which barrier W visits last. This geometric approach carries over with only minor adaptations to the case of a walk restricted by arbitrary boundary, and a corresponding absorbed reverse walk.

2. OSCILLATIONS

Quite obviously, $W_j = 0$ whenever j is a (non-strict) downladder index of S, i.e. when $S_j \leqq S_k$ for $0 \leqq k < j$. If therefore i_0 = "index of last minimum of 0, $S_1, \ldots, S_n$", $W_{i_0} = 0$ and W_n depends only on the behavior of S from i_0 on. Relevant are in fact the successive oscillations of decreasing ampitude of S. For $s = 0, 1, \ldots, n$, let

$$u_s = \max \{k : S_k = \max_{s \leqq j \leqq n} S_j\}, \quad l_s = \max \{k : S_k = \min_{s \leqq j \leqq n} S_j\}, \tag{3}$$

and write $i_0 = l_0$, $i_1 = u_{i_0}$, $i_2 = l_{i_1}$, $i_3 = u_{i_2}, \ldots$ and so forth alternatively until reaching a last $i_\lambda = n$. The r-th oscillation of decreasing amplitude of S extends from i_{r-1} to i_r. Thus we count as first oscillation the one starting upwards at i_0, even when $u_0 < l_0$ in which case it is preceeded by a maximal downwards oscillation which is irrelevant for our purposes. Notice that $\lambda = 0 \Leftrightarrow S_n = \min\{0, S_1, \ldots, S_n\}$. When $\lambda > 0$, odd numbered oscillations go upwards. The amplitudes of successive oscillations are $A_r = |S_{i_r} - S_{i_{r-1}}|$, $r = 1, \ldots, \lambda$ and

$$S_{i_1} > S_{i_3} > \ldots \geqq S_n \geqq \ldots > S_{i_2} > S_{i_0}; \quad A_1 > A_2 > \ldots > A_\lambda > 0. \tag{4}$$

Let

$$\mu = \begin{cases} \max\{r = 1, \ldots, \lambda : A_r \geqq a\} & \text{if } \lambda > 0 \text{ and } A_1 \geqq a, \\ 0 & \text{otherwise}. \end{cases}$$

LEMMA. $W_{i_0} = 0$, $W_{i_1} = a, \dots, W_{i_\mu} = 0$ or a according as μ is even or odd. Consequently

$$(5) \qquad W_n = \begin{cases} S_n - S_{i_\mu} & \text{when } \mu \text{ is even}, \\ a + S_n - S_{i_\mu} & \text{when } \mu \text{ is odd}. \end{cases}$$

This hardly requires formal proof: the first μ oscillations of S having amplitude $\geqq a$, one checks easily that each of them sends W across from one barrier to the other (and W may make intermediate back-and-forth trips while the oscillation of S progresses). If any oscillations are left, they have amplitude $<a$ so that, from i_μ on, W "runs parallel to S" without reaching the barriers any more, and (5) results.

For the reverse walk T, let $j_r = n - i_{\lambda-r}$ (thus $j_0 = 0$), $B_r = A_{\lambda-r}$ and $\nu = \lambda - \mu$. The r-th oscillation (of increasing amplitude) of T extends from j_{r-1} to j_r and has amplitude B_r. (4) becomes, if $\lambda > 0$:

$$T_{j_\lambda} > T_{j_{\lambda-2}} > \dots \geqq 0 \geqq \dots > T_{j_{\lambda-3}} > T_{j_{\lambda-1}};$$

$$0 < B_1 < \dots < B_\nu < a \leqq B_{\nu+1} < \dots < B_\lambda .$$

Notice that one could define $j_1, j_2, \dots, \lambda$ and ν directly on T, rather than via S. The numbering of oscillations would then stop in all cases at the index j_λ of the first maximum of T. We remark that "μ even" can occur when $\lambda > 0$ and is then equivalent to $T_{j_\nu} > 0$, and occurs when $\lambda = 0$ in which case $T_{j_\nu} = T_0 = 0$. Similarly "μ odd" can occur when $\nu > 0$ and means then $T_{j_\nu} < 0$, and occurs when $\nu = 0$ in which case again $T_{j_\nu} = 0$. Thus (5) can be written as

$$(6) \quad W_n = \begin{cases} T_{j_\nu} \text{ if } \lambda - \nu \text{ is even, i.e. if } T_{j_\nu} > 0 \text{ or } \lambda = 0, \\ a + T_{j_\nu} \text{ if } \lambda - \nu \text{ is odd, i.e. if } T_{j_\nu} < 0 \text{ or } \nu = 0 \text{ and } \lambda \text{ is odd}. \end{cases}$$

The successive oscillations of T can be thought of as generating a doubly monotone saw edge with teeth pointing up, while those of $T + a$ make a parallel saw edge with imbricated teeth pointing down. (6) says that W_n is achieved at the extremity of the last tooth of T (of $T + a$) which is not lower (not higher) than the tip of the next imbricated tooth of $T + a$ (of T). This is conveniently expressed through the extremal sequences Y and Z defined by

$$Y_j = \min\{T_k + a : 0 \leqq k \leqq j\}, \quad Z_j = \max\{T_k : 0 \leqq k \leqq j\}, \quad j = 0, 1, \dots, n .$$

THEOREM.

$$(7) \qquad W_n = \max\{\min(Y_j, Z_j) : 0 \leqq j \leqq n\} .$$

Proof. As long as oscillations have amplitude $<a$, Y remains above Z, but $Y_j \leqq Z_j$ occurs at the latest when $j = j_{\nu+1}$. Specifically, if $\lambda - \nu$ is even and $\nu < \lambda$, the $(\nu + 1)$-st oscillation of T is downwards so $Z_{j_\nu} = Z_{j_\nu+1} = \dots = Z_{j_{\nu+1}}$ while $Y_{j_\nu} > Z_{j_\nu} \geqq Y_{j_{\nu+1}}$. $\min(Y_j, Z_j)$ attains therefore a maximum equal to $Z_{j_\nu} = T_{j_\nu}$. This is true also if $\nu = \lambda > 0$, while the case $\lambda = 0$, i.e. $T_0 = \max\{T_0, \ldots, T_n\}$ gives directly $\max \min\{Y_j, Z_j\} = 0 = W_n$. If on the contrary $\lambda - \nu$ is odd, the $(\nu+1)$-st oscillation of T is upwards so $Y_{j_\nu} = \ldots = Y_{j_{\nu+1}}$ while $Z_{j_\nu} < Y_{j_\nu} \leqq Z_{j_{\nu+1}}$. The maximum of $\min(Y_j, Z_j)$ is then $Y_{j_\nu} = a + T_{j_\nu}$. Thus (6) implies (7).

The equivalence of (7) and (2) is easy to check. It is clear also that (7) implies (1), because $W_n > y$ occurs if and only if there is a first $\alpha \leqq n$ where $Y_\alpha > y$ and $Z_\alpha > y$. This α necessarily occurs during an upwards oscillation of T, so that T becomes $> y$ before $T + a$ becomes $\leqq y$, or equivalently $T_n^y > y$.

The duality (1) can be broken down into several subcomponents. Leaving aside the cases $W_n = 0$ or a, suppose now $0 < y < a$ and let $b = a - y$, $\eta = i_\mu$, $\varrho = j_\nu = n - \eta$. Thus η is the time of last visit of W to either 0 or a. Let the hitting time of $]-\infty, -b]$ by T be

$$\tau = \begin{cases} \min\{j = 0, 1, \ldots, n : T_j \leqq -b\} & \text{if some } \ T_j \leqq -b\,, \\ \infty \quad \text{otherwise}\,. \end{cases}$$

Property. For $0 < y < a$,

$$(8) \qquad \{W_\eta = 0,\ 0 < W_n \leqq y,\ \max_{\eta < j \leqq n} (W_j - W_n) \geqq b\} = \{0 < T_\varrho \leqq y,\ \tau < \varrho\}\,,$$

$$(9) \qquad \{W_\eta = 0,\ 0 < W_n \leqq y,\ \max_{\eta < j \leqq n} (W_j - W_n) < b\} = \{0 < T_\varrho \leqq y,\ \tau > \varrho\}\,,$$

$$(10) \qquad \{W_\eta = a,\ 0 < W_n \leqq y\} = \{T_\varrho \leqq -b\}\,.$$

Proof. The statement relative to W in (8) is equivalent, in terms of S, to

$$(11) \qquad \{0 \leqq \mu < \lambda,\ \mu \text{ is even},\ y - S_n + S_\eta \geqq 0 \text{ and } S_{i_{\mu+1}} - S_n \geqq b\}\,.$$

In fact, μ is even $\Leftrightarrow W_\eta = 0$, in which case $0 < W_n \Leftrightarrow \mu < \lambda$ and then, for $j > \eta$, $W_j = S_j - S_\eta$ while $W_j - W_n = S_j - S_n$ has its maximum at $j = i_{\mu+1}$, which justifies (11). When expressed in terms of T, this yields the right-hand member of (8) because μ even $< \lambda \Leftrightarrow T_\varrho > 0$, and then $Z_\varrho = T_\varrho = S_n - S_\eta \leqq y$ while $\tau < \varrho \Leftrightarrow \tau \leqq j_{\nu-1} \Leftrightarrow T_{j_{\nu-1}} = S_n - S_{i_{\mu+1}} \leqq -b$. The proof of (9) is the same, and as for (10) each member equals $\{\mu < \lambda,\ \mu \text{ is odd},\ S_\eta - S_n \geqq b\}$.

Putting together (8) and (9) gives, because $\tau \neq \varrho$

$$\{W_\eta = 0,\ 0 < W_n \leqq y\} = \{0 < T_\varrho \leqq y\}\,.$$

3. GENERAL BOUNDARY

The fact that we have considered paths obtained through successive summations is irrelevant in the above proof of (1). Thus one could substitute to sums, values defined by a relation of type $S_{n+1} = f(S_n, X_{n+1})$, assuming f monotone in the first argument so that T is well defined, a circumstance investigated by Loynes [2]. For brevity, we stick to sums and indicate how (1) extends to the case of arbitrary boundary, when definitions are adapted.

Suppose the boundary is $h = (f, g)$, where $f > g$ are two functions defined on $\{0, 1. \ldots, n\}$. Assume $g_0 = 0$, let $f_0 = a > 0$ the restricted walk W^h resulting from S when h is impenetrable is given by

$$W_0^h = 0\,, \quad W_j^h = \min\{f_j, \max(W_{j-1}^h + X_j, g_j)\}\,.$$

Oscillations of S must now be defined in accordance with the lower-g and upper-f profiles. To this end let $S^f = S - f + a$, $S^g = S - g$ and define as in (3)

$$u_s = \max\{k : S_k^f = \max_{s \leqq j \leqq n} S_j^f\}\,, \quad l_s = \max\{k : S_k^g = \min_{s \leqq j \leqq n} S_j^g\}\,.$$

Starting at $i_0 = l_0$, ends-of-oscillations are defined like previously as $i_1 = u_{i_0}, \ldots$ until $i_\lambda = n$. The h-amplitude of the r-th oscillation is

$$A_r^h = \begin{cases} S_{i_{r-1}}^f - S_{i_r}^g & \text{for } r \text{ even}\,, \\ S_{i_r}^f - S_{i_{r-1}}^g & \text{for } r \text{ odd}\,. \end{cases}$$

The relations (4) become, if $f_n - g_n = b$,

$$S_{i_1}^f > S_{i_3}^f > \ldots \geqq S_n^f = S_n^g + a - b\,, \quad S_n^g \geqq \ldots > S_{i_2}^g > S_{i_0}^g\,,$$
$$A_1^h > A_2^h > \ldots > A_\lambda^h > a - b\,.$$

Thus, oscillations of negative h-amplitude can occur when $b > a$. The μ-th oscillation is defined as before, but in terms of h-amplitudes, and (5) becomes

$$W_n^h = \begin{cases} S_n - S_{i_\mu}^g & \text{if } \mu \text{ is even}\,, \\ a + S_n - S_{i_\mu}^f & \text{if } \mu \text{ is odd}\,. \end{cases}$$

Writing f', g' for the functions with values $f_j' = f_{n-j}$, $g_j' = g_{n-j}$, we define with the reverse walk T the extremal sequences

$$Y_j^{f'} = \min\{T_k + f_k' : 0 \leqq k \leqq j\}\,, \quad Z_j^{g'} = \max\{T_k + g_k' : 0 \leqq k \leqq j\}\,.$$

Then formula (7) translates, with completely similar proof, into

$$W_n^h = \max\{\min(Y_j^{f'}, Z_j^{g'}) : 0 \leqq j \leqq n\}\,.$$

If $g_n \leqq y \leqq f_n$, $W_n^h > y$ means like previously that $Z_j^{g'} > y$ occurs before $Y_j^{f'} \geqq y$, i.e. $T_j > y - g_j'$ occurs before $T_j \leqq y - f_j'$. Introducing the walk $T^{h,y}$ resulting from absorption of T at values $> y - g'$ or $\leqq y - f'$, with the convention that $T^{h,y}$ is given the value $+\infty(-\infty)$ as soon as absorbed in the upper (lower) region, one concludes that (1) extends to

$$\{W_n^h > y\} = \{T_n^{h,y} = +\infty\}\,.$$

4. FINAL COMMENT

When the summands $X_1, \dots, X_n$ are exchangeable with symmetric, continuous law, the number of oscillations of increasing amplitude and the number of the subsequent ones of decreasing amplitude of S have an invariant joint law. In particular, the total number of oscillations has the same law as the number of records in the sequence $0, X_1, \dots, X_n$, so that Rényi's basic limit results in [4] apply. Numbers of oscillations will be investigated elsewhere.

REFERENCES

[1] D. V. LINDLEY: The theory of queues with a single server. Proc. Camb. phil. Soc. *48* (1952), 277—289.

[2] R. M. LOYNES: On a property of the random walks describing simple queues and dams. J.R. Statist. Soc. B *27* (1965), 125—129.

[3] R. M. PHATARFOD, T. P. SPEED and M. WALKER: A note on random walks. J. Appl. Prob *8* (1971), 198—201.

[4] A. RÉNYI: Théorie des éléments saillants d'une suite d'observations. In: Colloquium on Combinatorial Methods in Probability Theory, Aarhus (1962), 104—115.

UNIVERSITÉ DE GENÈVE

SECTION DE MATHÈMATIQUES

RECENT DEVELOPMENTS IN ERGODIC THEORY

KONRAD JACOBS

ERLANGEN

The purpose of this paper is to give a survey over some major fields of current research in ergodic theory and to depict some cluster of problems and results in more detail. For other surveys see e.g. Rokhlin [1], [2], [5], Oxtoby [1], Halmos [1], [2], Jacobs [2], Ornstein [7], [8], Sinai [9], Friedman-Ornstein [2]. I didn't aim at completeness at all and apologize to those authors who miss their papers in the bibliography, in particular Akcoglu, Sucheston and Jones who have done important work on iterates of operators recently.

Let $\boldsymbol{R}$ be the real line and $S = (S_t)_{t \in \boldsymbol{R}}$ a one-parameter group of bijections of a set Ω. For every $\omega \in \Omega$ we call $S\omega = \{S_t\omega \mid t \in \boldsymbol{R}\}$ the *orbit* of ω under (the action of) S. The situation thus described is the basic framework of ergodic theory which investigates the global behavior of orbits. The couple (Ω, S) si called a *flow*; ergodic theory is the orbit theory of flows.

This basic scheme is often varied:

1. If we replace $\boldsymbol{R}$ by the set $\boldsymbol{Z}$ of all integers, our one-parameter group boils down to a single bijection $S_1 = T$ and its iterates $\{T^n\omega \mid n \in \boldsymbol{Z}\}$ is now called the *orbit* of ω under T. We call this the *time discrete case* in contrast to the original *time continuous situation*. Let us mention a natural relation between them: If T is a bijection on Ω and f a strictly positive real function on Ω, we get a flow on the ordinate set $\Omega_f = \{(\omega, t) \mid \omega \in \Omega, 0 \leqq t < f(\omega)\}$ of f most naturally: a point (ω, t) moves up with uniform speed 1 until it reaches the "ceiling" $f(\omega)$ and then falls down into $(T\omega, 0)$ etc. The flow (Ω_f, S) thus obtained is called the *flow built under the function* f (for T).

2. We may consider $\boldsymbol{R}^n$ instead of $\boldsymbol{R}$ and study the action of a n-parameter group on Ω; or we may consider $\boldsymbol{Z}^\nu$ instead of $\boldsymbol{Z}$ and study power products of ν commuting bijections $T_1, \ldots, T_\nu$ on Ω. We call these situations *multi-parameter* cases.

There is always a natural *isomorphy problem*. We display it in the time discrete case of a single bijection. Let T be a bijection of Ω and T' a bijection of Ω'. A bijection φ of Ω onto Ω' is called an *isomorphy* (for T, T') if $T' \circ \varphi = \varphi \circ T$ holds. Couples (Ω, T) fall now into isomorphy classes and we have the standard *isomorphy problem* of describing all classes and finding nice representatives.

Our basic framework will always be implemented with *additional structures*, and $S = (S_t)_{t \in \boldsymbol{R}}$ etc. is then assumed to consist of automorphisms of this structure. Likewise, attention is restricted to isomorphies with respect to the given structures which effectuates a refinement of the isomorphy classification. We mention the following cases:

a) *Measure-theoretical structure*: A σ-field $\mathscr{B}$ is given in Ω and the S_t resp. T are assumed to be $\mathscr{B}$-measurable as well as their inverses. In the time continuous case we consider also the natural σ-field $\mathscr{A}$ in $\boldsymbol{R}$ and assume the measurability of the map $(\omega, t) \to S_t\omega$ with respect to the product σ-field $\mathscr{B} \times \mathscr{A}$ in $\Omega \times \boldsymbol{R}$ and $\mathscr{B}$ in Ω. – Attention is focused on *invariant normalized* measures m on $\mathscr{B}$ (i.e. in Ω) : $m(S_tF) = m(F)$ resp. $m(T^{-1}F) = m(F)$ $(F \in \mathscr{B})$ is required along with $m(\Omega) = 1$. A quadruple $(\Omega, \mathscr{B}, m, (S_t)_{t \in \boldsymbol{R}})$ is then called a *(measure-theoretical) flow*, and a quadruple $(\Omega, \mathscr{B}, m, T)$ is then called a *(measure-theoretical) dynamical system*. We don't go into the problem of existence of invariant measures here; they exist in a broad variety of natural examples. The oldest example is a Hamiltonian flow in $\boldsymbol{R}^{6N}$ which preserves Lebesgue measure, according to Liouville's theorem. Quite in the opposite corner we encounter one of the simplest examples: Bernoulli schemes. Let A be finite, say $A = \{1, \ldots, a\}$ and $p = (p_1, \ldots, p_a)$ a probability vector, representing a normalized measure p on the σ-field $\mathscr{P}(A)$. Consider the compact metric space

$$\Omega = A^{\boldsymbol{Z}} = \{\omega = (\ldots, \omega_{-1}, \omega_0, \omega_1, \ldots) \mid \omega_t \in A \ (t \in \boldsymbol{Z})\}.$$

The shift

$$T: (\ldots, \omega_{-1}, \omega_0, \omega_1, \ldots) \to (\ldots, \omega_0, \omega_1, \omega_2, \ldots)$$

is a homeomprohism of Ω. The natural σ-field $\mathscr{B}$ of Ω coincides with the product σ-field $\mathscr{P}(A)^{\boldsymbol{Z}}$ and the product measure $m = p^{\boldsymbol{Z}}$ is T-invariant. $(\Omega, \mathscr{B}, m, T)$ is called the *Bernoulli-scheme* based on p. If p is equidistribution, i.e. $p = (1/a, \ldots, 1/a)$, the Bernoulli scheme is shortly called a-shift. In any case, $\Omega = A^{\boldsymbol{Z}}$ is also called *a-shift space.*

If we deal with measure-theoretical ergodic theory, isomorphies are supposed to send one invariant measure into the other; nullsets are dropped and added ad libitum. One of the simplest examples of isomorphy is the natural one between 2-shift and the so-called bakers transformation. – *Measure-theoretical ergodic theory* investigates those properties of (orbits of) dynamical systems which can be formulated in a purely measure-theoretical way and hence are invariant under measure-theoretical isomorphy. A typical example is G. D. Birkhoff's classical individual ergodic theorem which in the time discrete case states that

$$\lim_{n \to \infty} \frac{1}{n} \sum_{k=0}^{n-1} f(T^k \omega)$$

exists, is finite and constant along orbits m – a.e. in Ω, for every dynamical system $(\Omega, \mathscr{B}, m, T)$ and every $f \in L^1_m$. Taking $f = 1_F$ for some $F \in \mathscr{B}$, we obtain the existence (m-a.s.) of mean sojourn times. Clearly this is a statement on the global behaviour of orbits. Another bunch of examples is provided by the classical concepts of ergodicity and mixing. A dynamical system $(\Omega, \mathscr{B}, m, T)$ with $m(\Omega) = 1$ is called

1. *ergodic* if $E, F \in \mathscr{B}$, $m(E)$, $m(F) > 0$ implies $m(E \cap T^t F) > 0$ for some $t \in Z$. Clearly this means some "mixing" behaviour and reflects a global behaviour of orbits: almost all of them go almost everywhere;

2. *strongly mixing* if

$$\lim_{t \to \infty} m(E \cap T^t F) = m(E)\, m(F) \quad (E, F \in \mathscr{B})\,;$$

3. *mixing of degree r* if

$$\lim_{n \to \infty} m(T^{t_{n1}} E_1 \ldots T^{t_{nr}} E_r) = m(E_1) \ldots m(E_r) \quad (E_1, \ldots, E_r \in \mathscr{B})$$

for all r-tuples of sequences $t_{n1}, \ldots, t_{nr}$ $(n = 1, 2, \ldots)$ such that

$$\lim_n |t_{ni} - t_{nj}| = \infty \quad (i \neq j)\,.$$

Nearly every concrete example of a dynamical system is ergodic. Strong mixing amounts to mixing of degree 2 and implies ergodicity. No example of a dynamical system which is mixing of degree r but not of degree $r + 1$ is known for any $r = 2, 3, \ldots$ For relations of ergodicity and mixing to spectral theory see § 1.

The isomorphy problem in the measure-theoretical situation has brought forth, among others, two major of isomorphy invariants: *spectral type* and *entropy*. We will deal with the spectral type in § 1, but entropy will be discussed in this introduction, preceded by a sketch of *generator theory*. We stick to the time discrete case for the sake of simplicity.

Let $(\Omega, \mathscr{B}, m, T)$ be a measure-theoretical dynamical system. A finite partition $\{A_1, \ldots, A_a\} = \alpha \subseteq \mathscr{B}$ is called a *generator* of length a if its translates $T^t\alpha (t \in Z)$ generate $\mathscr{B}$ mod m. In this case the mapping $\Omega \ni \omega \to (\ldots, \omega_{-1}, \omega_0, \omega_1, \ldots) \in A^Z$ defined by $\omega_t = j \in A = \{1, \ldots, a\}$ iff $T^t\omega \in A_j$; sends m into a shift invariant normalized measure m' in A^Z and, under mild regularity assumptions on $(\Omega, \mathscr{B}, m)$, constitutes a measure-theoretical isomorphism of $(\Omega, \mathscr{B}, m, T)$ with the dynamical system arising from A^Z, m' and shift. Conversely, an isomorphism of $(\Omega, \mathscr{B}, m, T)$ with a dynamical system living in a-shift space immediately yields a generator of length a. It is just the inverse image of the partition of A^Z according to the component no. 0, under the given isomorphism. Let us sum up these facts by saying: a generator of length a makes the system *live in a*-shift space. The measure m' in A^Z is Bernoulli iff the partitions $T^t\alpha$ $(t \in Z)$ are *independent* in the usual probabilistic sense; in that case way say that α is *Bernoulli*. Thus isomorphy to a Bernoulli scheme essentially amounts to the existence of a *Bernoulli generator*.

Rokhlin [3], [4] was the first to show that every dynamical system with a countably generated σ-field (plus some mild regularity conditions) can live in a shift space based on a countable alphabet, which amounts to the existence of a *countable generator*. The question of existence of a *finite generator* has much to do with entropy, as we shall see.

If α is a generator for $(\Omega, \mathscr{B}, m, T)$, then the σ-field $\mathscr{B}_0$ generated by $\bigcup_{t=-\infty}^{0} T^t\alpha$ has the properties:

1. $$T\mathscr{B}_0 \supseteq \mathscr{B}_0 .$$

2. $\bigcup_{t=1}^{\infty} T^t\mathscr{B}_0$ generates $\mathscr{B}$ mod m.

If α is Bernoulli, we have even

3. $$\bigcap_{t=-\infty}^{0} T^t\mathscr{B}_0 = \{\emptyset, \Omega\} \bmod m .$$

If $(\Omega, \mathscr{B}, m, T)$ allows of a σ-field $\mathscr{B}_0 \subseteq \mathscr{B}$ with the properties 1.–3., we call it a *K-system*. Ornstein [4] has shown that there are K-systems which don't allow of a Bernoulli generator (not even a countable one), i.e. are not isomorphic to Bernoulli schemes (see also Ornstein-Shields [1] and furthermore Ornstein [5], [6], [7]). The property of being a K-System clearly is an isomorphy invariant.

Let $(\Omega, \mathscr{B}, m)$ be any normalized measure space. If $\alpha, \beta \subseteq \mathscr{B}$ are partitions of Ω, $\alpha \vee \beta$ denotes their common refinement. If $\alpha = \{A_1, \ldots, A_a\}$, we call

$$H(\alpha) = -\sum_{k=1}^{a} m(A_k) \log m(A_k)$$

the entropy of α (with the convention $0 \log 0 = 0$). We have the general estimate $0 \leqq H(\alpha) \leqq \log a$. If $(\Omega, \mathscr{B}, m, T)$ is a dynamical system, the limit

$$h(\alpha) = \lim_{t\to\infty} \frac{1}{t} H(\alpha \vee T\alpha \vee \ldots \vee T^{t-1}\alpha)$$

exists and is called the *mean entropy* of α. Again we have $0 \leqq h(\alpha) \leqq \log a$. The extended real number

$$h(\Omega, \mathscr{B}, m, T) = h(T) = \sup_{\alpha} h(\alpha)$$

is called the *entropy* of $(\Omega, \mathscr{B}, m, T)$; it is clearly an isomorphy invariant. If α is a generator, then $h(\alpha) = h(T)$, which implies $0 \leqq h(T) \leqq \log a$ for the length a of any such generator. In particular, a finite generator cannot exist if $h(T) = \infty$. If α is a Bernoulli generator, we have $h(T) = h(\alpha) = H(\alpha)$. This latter observation was the historical beginning of entropy theory for dynamical systems (Kolmogorov [1], [2], Sinai [1], [2], see e.g. Parry [4], Jacobs [1], [2], Rokhlin [2], [5]). The problem of finite generators was solved by Krieger [3] in 1969 (see also Krieger [6], Denker [3], [4]): If $(\Omega, \mathscr{B}, m, T)$ is ergodic and $h(\Omega, \mathscr{B}, m, T) = h(T) < \infty$, then there exists a generator α of any finite length a such that $h(T) < \log a$.

Kolmogorov's evaluation $-\sum_{k=1}^{a} p_k \log p_k$ of the entropy of a Bernoulli scheme displayed a whole continuum of nonisomorphic Bernoulli schemes. In particular, the a-shifts $(a = 2, 3, \ldots)$ are pairwise nonisomorphic. It is a natural question whether two Bernoulli schemes with equal entropy are isomorphic. After preparatory work by Meshalkin [1], Blum-Hanson [1], [2], Jacobs [1], Keane [1], Sinai [5], Zsidó [1] and Adler-Weiss [1], the problem was solved in the positive by Ornstein [1], [2] (see also Smorodinski [1], Friedman-Ornstein [1], [2]). In § 2 we report on subsequent discoveries of Bernoulli generators. i.e. isomorphies to Bernoulli schemes in numerous other cases.

Analogous definitions are to be made in the time continuous case. The ergodicity of a certain system corresponding to an ideal gas is the classical ergodic hypothesis (see Sinai [4], [7]) on which we report in § 5.

b) *Topological structure*: A topology is given in Ω and the S_t resp. T are assumed to be homeomorphisms. In the time continuous case we consider also the natural topology on $\boldsymbol{R}$ and assume that the mapping $(\omega, t) \to S_t\omega$ is continuous with respect to the product topology in $\Omega \times \boldsymbol{R}$ and the given topology in Ω. The discipline which works within this frame is called *topological dynamics*. – Attention is focused on the case where Ω is compact metric. A non-empty closed invariant set is called *minimal* (invariant) if it is minimal with respect to these three properties. Compactness plus Zorn's lemma implies the existence of minimal subsets of arbitrary nonempty closed invariant sets such as orbit closures. Minimal orbit closures can be characterized by an almost periodicity property of the points under consideration. A point Ω is called *nonwandering* under the homeomorphism T of Ω if for any neighborhood U of ω there is a $t \neq 0$ in $\boldsymbol{Z}$ such that $U \cap T^t U \neq \emptyset$. The points of a minimal set are always nonwandering. Certain developments in differentiable dynamics (see c)) have created particular interest in the set of all nonwandering points (Smale [1], [2], [3], Bowen [1], [2], [3], [4], Sigmund [1], [2], [3], [4]). We report on some related results in § 6 and § 7. We mention in passing that topological dynamics has undergone an ample evolution under the stimulus of a paper of Furstenberg [1] on so-called *distal* homeomorphisms (see e.g. Ellis [1]). In a very classical situation, namely flows in the plane, an amazing classification theory has been built by Beck [1] and his collaborators.

A particularly strong link between topological dynamics and measure theoretical ergodic theory is provided by the theory of *strictly ergodic sets* on which we report in § 3.

Typical phenomena of topological dynamics assume nearly combinatorial features if we study them in the case of shift space $A^{\boldsymbol{Z}}$ for a finite alphabet A plus shift homeomorphism. The corresponding discipline is called *symbolic dynamics* (see e.g. Gottschalk-Hedlund [1], Hedlund [1]) and is touched upon in § 3.

c) *Differentiable structure*: Ω is assumed to be a smooth manifold and the S_t resp. T are assumed to be diffeomorphisms. In the time continuous case we make even some differentiability assumptions on the manifold mapping $(\omega, t) \to S_t\omega$ of $\Omega \times \boldsymbol{R} \to \Omega$. The theory which works in this frame is called *differentiable dynamics*; it includes classical mechanics as a special case (see Abraham-Marsden [1], Arnold-Avez [1], Sternberg [1]). We report on a few of its aspects in § 6. The theory must be considerably enlarged in order to include the non-smooth phenomenon of collision of particles and models for the classical gas (see Sinai [4], [7], Bunimovic [1], [2], [3], Kubo [1], [2], [3]) on which we report in § 5. A cross-relation between symbolic and differentiable dynamics is discussed in § 7.

1. SPECTRAL THEORY

There are two aspects under which dynamical systems have been linked with general spectral theory.

1. The spectral type of a dynamical system

Let U be a unitary transformation in the Hilbert space H. The couple (H, U) is then called a *unitary element*. Hilbert space isomorphies classify the unitary elements into *(unitary) spectral types*. Properties of the spectral type of (H, U) are called *spectral properties* of (H, U). Koopman observed [1] that with every dynamical system $(\Omega, \mathscr{B}, m, T)$ a unitary element (L^2_m, U_T) is associated by $(U_T f)(\omega) = f(T\omega)$ (m-a.e., $f \in L^2_m$). It is called the *unitary element of* $(\Omega, \mathscr{B}, m, T)$ and it was this association which enabled v. Neumann [1] 1931 to prove the mean ergodic theorem for dynamical systems (for the first time and) by purely spectral theoretical methods. The spectral type of (L^2_m, U_T) is also called the *spectral type of the dynamical system* $(\Omega, \mathscr{B}, m, T)$ and its spectral properties are also called *spectral properties* of the latter. Obviously isomorphic dynamical systems have the same spectral type. The converse is true if $(\Omega, \mathscr{B}, m, T)$ has the spectral property of having pure point spectrum, plus some regularity conditions (v. Neumann [2], see e.g. Jacobs [1]). It is not true for Bernoulli shifts: they have, if nontrivial, all the same spectral type (to be called countable Lebesgue spectrum below) but are nonisomorphic if their entropies differ. Several important properties of dynamical systems turn out to be spectral, at least if we assume $0 < m(\Omega) < \infty$ (and hence w.l.o.g. $m(\Omega) = 1$):

1. $(\Omega, \mathscr{B}, m, T)$ is *ergodic* iff $H_0 = \{h \mid h \in L_m^2, U_T h = h\}$ is one-dimensional (the presence of the constants in L_m^2, which is due to our assumption $m(\Omega) < \infty$, implies that H_0 is always at least one-dimensional), a spectral property which is usually shortly stated as the simplicity of the (always present) eigenvalue 1.
2. $(\Omega, \mathscr{B}, m, T)$ is *strongly mixing* iff H_0 is one-dimensional and $\lim_{n\to\infty} (U_T^n f, g) = 0$ $(f, g \in H_0^\perp)$.
3. A certain property which is between ergodicity and strong mixture can be defined conveniently in spectral terms: $(\Omega, \mathscr{B}, m, T)$ is called *weakly mixing* if H_0 is of dimension one and there is no eigenvalue other than 1, i.e. we have continuous spectrum in the orthogonal complement of H_0.

For examples of weakly but not strongly mixing dynamical systems see Maruyama [1], Kakutani [2].

Let us list some more spectral properties of dynamical systems:

$(\Omega, \mathscr{B}, m, T)$ is said to have *simple spectrum* if H_0 is one-dimensional and there is some $f \in H_0^\perp$ such that $H_0^\perp$ is the closed lirear span of $\{U_T^n f \mid n \text{ integer}\}$.

$(\Omega, \mathscr{B}, m, T)$ is said to have *Lebesgue spectrum of power p* if H_0 is one dimensional and there is a set M of power p and an orthonormal base $(f_{jk})_{j\in M, k \text{ integer}}$ of $H_0^\perp$ such that $U_T f_{jk} = f_{j,k+1}$ ($j \in M$, k integer). It can be shown that this implies mixing of all degrees.

If $(\Omega, \mathscr{B}, m, T)$ is isomorphic to a Bernoulli scheme, then it has countable Lebesgue spectrum (see e.g. Jacobs [1]). But also K-systems have countable Lebesgue spectrum (see e.g. Jacobs [1]), and some of them are not isomorphic to Bernoulli schemes (Ornstein [4], Ornstein-Shields [1]).

Today the spectral type of many concrete examples of dynamical systems has completely been cleared up (for an exception see Keane [2]). We have already mentioned the "Bernoulli" case. Another interesting case is constituted by Gaussian $(\Omega, \mathscr{B}, m, T)$ with a given spectral measure μ (on $<-\frac{1}{2}, \frac{1}{2})$) and its associated covariance function $\hat{\mu}$. Recall that $(\Omega, \mathscr{B}, m, T)$ is called Gaussian if $\Omega = \boldsymbol{R}^{\boldsymbol{Z}}$ and the finite-dimensional projections send m into Gaussian (i.e. normal) probability distributions. If φ_t denotes the t-th component mapping $(t \in \boldsymbol{Z})$, then

$$\hat{\mu}(s) = \int \varphi_t \varphi_{t+s} \, \mathrm{d}m = \int_{1/2}^{1/2} \mathrm{e}^{-2\pi i s u} \mu(\mathrm{d}u) \quad (s, t \in \boldsymbol{Z}).$$

Here many spectral properties of $(\Omega, \mathscr{B}, m, T)$ can be read directly from properties of μ or $\hat{\mu}$.

a) First of all, many Gaussian Systems are isomorphic to Bernoulli schemes and hence have countable Lebesgue spectrum. Among these are all those for which μ is equivalent to Lebesgue measure (Newton [1], Smorodinski [2]), in particular those with a trivial tail field (Urbanik [1]), but even all those which are K-systems (Vershik [1], without proof).

b) A Gaussian $(\Omega, \mathscr{B}, m, T)$ is ergodic iff it is weakly mixing of all degrees iff μ has no point masses (Maruyama [1], Grenander [1], Leonov [1]).
c) A Gaussian $(\Omega, \mathscr{B}, m, T)$ is strongly mixing iff it is strongly mixing of all degrees iff $\lim_{n\to\infty} \hat{\mu}(n) = 0$ (Itô [1], [2], [3], Maruyama [1], Leonov [1], Totoki [1], Dugdale [1]).
d) A Gaussian $(\Omega, \mathscr{B}, m, T)$ is a K-system iff it has countable Lebesgue spectrum iff $\mu \neq 0$ and μ is absolutely continuous with respect to Lebesgue measure.
e) There is a Gaussian $(\Omega, \mathscr{B}, m, T)$ which is weakly but strongly mixing (Maruyama [1], Newton [1]).
f) There is a Gaussian $(\Omega, \mathscr{B}, m, T)$ which has simple continuous spectrum (Girsanov [1], Newton [1]).
g) There is a Gaussian $(\Omega, \mathscr{B}, m, T)$ which is mixing of all degrees but does not have Lebesgue spectrum (Maruyama [1], Dugdale [1], Totoki [1], Newton [1]).

For a neat unified treatment of these points see Gundel [1]. Kushnirenko [1] has characterized pure point spectrum by the vanishing of all "sequence entropies".

Katok [1] has characterized pure point spectrum as well as given an evaluation of the entropy by the so-called *speed of cyclic approximation.*

In spite of the indicated ample material there seems to be no example of a dynamical system with simple Lebesgue spectrum.

Even if a special question like the one just mentioned allows no answer yet, it is still meaningful to tackle the general question: Which are the spectral types that can be spectral types of dynamical systems $(\Omega, \mathscr{B}, m, T)$?

Let us restrict attention to the case $m(\Omega) = 1$. Then it is clear that a unitary element (H, U) is (Hilbert) isomorphic to (L^2_m, U_T) only if H contains elements which correspond to indicator functions of sets in $\mathscr{B}$; in particular there must be an element $e \in H$ which corresponds to the constant $1 \in L^2_m$. Trying to reformulate properties of indicator functions in a purely spectral fashion, we arrive at

DEFINITION 1.1. Let (H, U) be a unitary element. A subset E of H with a distinguished member e is called a *D-family* if it has the following properties:

1. $UE = E$ and H is the closed linear span of E.

2. $$(f, g) \geqq 0 \quad (f, g \in E).$$

3. $$\|f\|^2 = (f, f) = (f, e) \quad (f \in E).$$

4. There is a constant $C > 0$ such that every $f, g \in E$ there is a $h \in E$ such that
$$(h, e) = (h, f) = (h, g) \geqq C(f, g).$$

5. If $f, g \in E$, $(f, 1) = (f, g)$ then $g - f \in E$.

6. $$\|e\| = 1, \quad Ue = e.$$

THEOREM 1.2 (Denker [2]). A unitary element (H, U) is (Hilbert) isomorphic to the unitary element (L_m^2, U_T) of some dynamical system $(\Omega, \mathscr{B}, m, T)$ with $m(\Omega) = 1$ iff there is a D-family in H.

The Hilbert isomorphism constructed for the "if" part of this theorem sends the D-family E into a subfamily of $\{1_F \mid F \in \mathscr{B}\}$ and e into the constant 1. Denker's result continuous earlier work by Choksi [1], [2].

2. The v. Neumann algebra associated with a nonsingular automorphism

Although there are many examples of dynamical systems $(\Omega, \mathscr{B}, m, T)$ in the sense of invariant measure, we know that among all nonsingular i.e. nullset preserving transformations T of a measure space $(\Omega, \mathscr{B}, m)$ (say, normalized to $m(\Omega) = 1$) those that admit an invariant σ-finite measure equivalent to m are the exception and not the rule in the sense of a (Baire) category statement (Ionescu-Tulcea [1]). It is thus natural to try to build up a large part of ergodic theory with the assumption of an invariant measure replaced by the weaker assumption that T be nonsingular, and this has e.g. been done for the individual ergodic theorem by Hurewicz [1] and Dowker [1]; also it is clear that the concept of ergodicity needs for its formulation only the concept of a nullset and hence can be set up straightforwardly for non-singular transformations.

Let T be any nonsingular transformation of the normalized measure space $(\Omega, \mathscr{B}, m)$, i.e., to be precise, a bijection of Ω which sends $\mathscr{B}$ into $\mathscr{B}$ and the m-nullsets onto the m-nullsets. Let $(\tilde{\Omega}, \tilde{\mathscr{B}}, \tilde{m})$ be the product measure space of $(\Omega, \mathscr{B}, m)$ and the integers $\boldsymbol{Z} = \{i = 0, \pm 1, \ldots\}$ endowed with counting measure. We are going to construct a v. Neumann algebra of operators (i.e. a weakly operator closed algebra which contains the adjoint of each of its members) in the Hilbert space $L_{\tilde{m}}^2$, and we construct it as the smallest v. Neumann algebra containing certain two families of operators derived from T. We will now describe these two families.

In order to describe the first family, let us consider measurable mappings π of Ω into the countable set of all permutations of $\boldsymbol{Z}$ which move only finitely many elements of $\boldsymbol{Z}$. If there is a nonsingular transformation S of $(\Omega, \mathscr{B}, m)$ such that

$$ST^i\omega = T^{\pi(w)i}\omega \quad (m\text{-a.e., } i \in \boldsymbol{Z}),$$

then π and S determine each other mod m and we may write $\pi(S, \omega)$ instead of $\pi(\omega)$. Now our first family consists of all operators of the form $\beta(S)$ in $L_{\tilde{m}}^2$ where $(\beta(S)\tilde{f})\,(\omega, i) = \tilde{f}(\omega, \pi(S, \omega)^{-1}\, i)$ $(\tilde{f} \in L_{\tilde{m}}^2, \omega \in \Omega, i \in \boldsymbol{Z})$.

The second family consists of all operators of the form $\beta(a)$ in $L_{\tilde{m}}^2$ where $a \in L_m^\infty$ and $(\beta(a)\tilde{f})\,(\omega, i) = a(T^i\omega)\tilde{f}(\omega, i)$ $(\tilde{f} \in L_{\tilde{m}}^2, \omega \in \Omega, i \in \boldsymbol{Z})$.

The smallest v. Neumann algebra containing these two families is denoted by $\tilde{W}(T)$. Our problem is to show how spectral properties of $\tilde{W}(T)$ depend upon ergodic properties of T.

The concept of ergodicity is defined for nonsingular transformations T of $(\Omega, \mathscr{B}, m)$ in an obvious fashion.

THEOREM 1.3 (Murray – v. Neumann [1], v. Neumann [3]. If T is ergodic then $\tilde{W}(T)$ is a factor, i.e. its center consists exactly of all constant multiples of the identity operator.

Let us now restrict attention to ergodic transformations T and determine the type of the arising factors $\tilde{W}(T)$ in the Murray – v. Neumann classification scheme. We impose the regularity condition that $(\Omega, \mathscr{B}, m)$ be a Lebesgue space.

THEOREM 1.4. (Dye [1], Krieger [1], [2], [4], [5], [8]).

1. For all nonsingular ergodic T which admit finite invariant measures equivalent to m the v. Neumann algebra $\tilde{W}(T)$ is the same up to isomorphy and is a factor of type II_1 (i.e. – admits finite traces separating the elements of the algebra).
2. For all nonsingular ergodic T which admit an infinite σ-finite measure equivalent to m the v. Neumann algebra $\tilde{W}(T)$ is the same up to isomorphy and is a factor of type II_∞ (i.e. admits no finite lower weakly continuous trace $\neq 0$).
3. For all nonsingular ergodic T which admit no invariant (finite or σ-finite) measure equivalent to m the v. Neumann algebra $\tilde{W}(T)$ is a factor of type III. Varying T in this class we obtain an infinity of pairwise nonisomorphic algebras $\tilde{W}(T)$.

The obvious question whether the isomorphy or non-isomorphy of the $\tilde{W}(T)$ can be directly read from the corresponding nonsingular transformations T is extremely difficult to answer.

DEFINITION 1.5. Let $(\Omega, \mathscr{B}, m)$ and $(\Omega', \mathscr{B}', m')$ be normalized Lebesgue spaces and T resp. T' nonsingular transformations in Ω resp. Ω'. A bimeasurable bijection of m-almost all of Ω onto m'-almost all of Ω' which sends the nullsets of m onto the nullsets of m' and T-orbits onto T'-orbits is called a *weak equivalence* between T and T'.

Dye [1] proved that two nonsingular transformations which admit a finite equivalent invariant measure are always weakly equivalent. The definitive result in this direction was obtained by Krieger in 1974:

THEOREM 1.6 (Krieger [8]). Two nonsingular ergodic transformations T, T' are weakly equivalent iff the corresponding v. Neumann algebras $\tilde{W}(T)$, $\tilde{W}(T')$ are isomorphic.

2. THE UBIQUITY OF THE BERNOULLI PROPERTY

Simplifying to a certain extent one can say that the basic model of *probability theory* is the *Bernoulli model*: the space Ω of all bisequences from a finite or possibly countable alphabet A endowed with a product measure. The essential feature of this model is the independence of the component variables; often one assumes stationarity, i.e. the assumption that all components have the same distribution. Assumptions of this simplicity entail theorems which make up the typical picture of classical probability theory: the law of large numbers, the arcsin law, the central limit theorem, the loglog law.

A Bernoulli model with equidistributed components is a measure-theoretical *Bernoulli scheme*, but it owes its incorporation into ergodic theory only to the present stage of generality of the latter, and not to its origins in mechanics. It is an achievement of the ingeniously simplifying human mind to pass over from a mechanical system like a tossed coin or die to a model like a Bernoulli scheme over an alphabet of two or six letters. It is a basic question whether this passage can be justified in view of properties of mechanical systems revealing a "hidden Bernoulli structure".

One method of justification consists in a proof of measure-theoretical isomorphy to Bernoulli schemes of mechanics-born dynamical systems. In spite of the fact that the isomorphies are only measure-theoretical ones and hence not directly exploitable for the design of physical experiments, they strongly support the belief that random is a legal citizen in mechanics.

Measure-theoretical isomorphy theory as it stands today thanks to the achievements of Kolmogorov [1], [2], Sinai [1], [2], [3], [5], [6], Zsidó [1], Meshalkin [1], Blum-Hanson [1], [2], Keane [1], Adler-Weiss [1], Katznelson [1] and in particular Ornstein [1], [2], [3], [4], Friedman-Ornstein [1], [2], Smorodinski [1], Gallavotti-Ornstein [1] started with the question when two Bernoulli schemes are isomorphic, and gave the answer: iff their entropies are equal (Ornstein [1], [2], before Ornstein [1], Adler-Weiss [1] had shown that in a special case, ergodic torus automorphisms are isomorphic iff their entropies are equal). The general proof that every ergodic torus automorphism is isomorphic to a Bernoulli scheme was given by Katznelson [1], but this was only the beginning of a whole series of discoveries of isomorphies of dynamical systems to Bernoulli schemes. To mention the newest achievement first, Miles-Thomas [1] have shown in 1974 that every ergodic continuous automorphism of a compact metric group is isomorphic to a Bernoulli scheme, thus generalizing the result of Katznelson [1] and bringing the attempt of Aoki-Totoki [1] to a positive end. Friedman-Ornstein [1] proved that mixing Markov chains with finite state space are isomorphic to Bernoulli schemes, including thereby a class of measure-preserving diffeomorphisms investigated by Sinai [6]. Gaussian processes with a spectral measure equivalent to Lebesgue measure are isomorphic to Bernoulli schemes (Newton [1], Smorodinski [2]). G. Ishitani (unpublished) has proved that nearly every piecewise linear transformation of the unit interval leads to an invariant measure such that the corresponding dynamical system is measure-theoretically isomorphic to a Bernoulli scheme. Joint work of S. Itô, I. Shiokawa, M. Mori, G. Ishitani, U. Mitsou, Y. Takahashi and Y. Ogata (unpublished symposium notes, Koyoto 1973) has shown the same result for β-transformations of the unit interval, i.e. transformations of the form $T_\beta : x \to \beta x \bmod 1$ (see also Takahashi [1]). The existence of a finite T_β-invariant measure μ_β absolutely continuous with respect to Lebesgue measure, and the proof that the corresponding dynamical system has entropy $\log \beta$ goes already back to Rényi [1] (see also Parry [2]). The same japanese authors have also proved a central limit theorem for sequences f, $T_\beta f$, $T_\beta^2 f, \ldots$ with suitable $f \in L_{\mu_\beta}^{2+\delta}$ by a skillful application of Ibragimov-Linnik's [1] central limit theorem. They have carried over some of these results to a similar but somewhat more general situation (unpublished).

3. STRICT ERGODICITY

Let Ω be the shift space of all bisequences over a finite alphabet A. A closed shift invariant subset Ω_0 of Ω is also called a *subshift over A*. It is called *strictly ergodic* iff it is minimal invariant and for every $f \in C(\Omega)$ the averages $(1/n)(f + f_0 T + \ldots + \ldots + f_0 T^{n-1})$ tend to a constant limit uniformly on Ω_0. It is easily seen that the latter condition is equivalent to the unique existence of a shift invariant normalized measure m_0 in Ω_0 (Oxtoby [1]); the constant limit in question is nothing but $\int f \, dm_0$ and m_0 is necessarily ergodic. Thus to every strictly ergodic subshift a measure theoretical dynamical system is associated in a natural and unique fashion. It is natural to ask whether every measure-theoretical ergodic dynamical system (with a normalized measure) is measure-theoretically isomorphic to a strictly ergodic one, or, as we shall say, whether it can live in a strictly ergodic set.

There are two obvious necessary conditions: the entropy of the given measure-theoretical system must not exceed the log of the alphabet length $|A|$; should it be equal to $\log |A|$, then it can live in Ω only as the Bernoulli (=product) measure corresponding to the equidistribution over A; but this measure cannot be carried by a strictly ergodic subshift, as can be immediately seen. Thus we are bound to restrict ourselves to the case of finite entropy strictly majorized by $\log |A|$. But in this case we can use Krieger's generator theorem (Krieger [3], [6], Denker [3], [4]) in order to make live our system in a subshift and our problem comes down to the question whether this subshift can be further modified into a strictly ergodic one, the modification implying an isomorphy in the measure-theoretical sense. Krieger [7] proved that this can actually be done (see also Denker [1]).

The whole theory was initiated by a fundamental paper of Jewett [1]. It should be mentioned that in the case of infinite entropy an analogous result can be obtained with the help of a countable alphabet A (Krieger [7], Denker [1]).

It is obvious to ask whether this time-discrete theory can be paralleled by a theory for the time-continuous case. The basic problem here is the choice of a suitable time-continuous analogous to the shift space. If we interprete bisequences of symbols from a finite alphabet as possible outputs of discrete physical measuring apparatus, it is natural to propose, for the time-continuous case, a time translation invariant space of functions on the real time axis which can be outputs of time-continuous physical measuring apparatus. One such space is

$$\mathrm{Lip}_1 = \{\omega \mid \omega : \boldsymbol{R} \to \langle 0, 1\rangle,\ |\omega(s) - \omega(t)| \leqq |s - t|\ (s, t \in \boldsymbol{R})\}\,.$$

It is compact metric in the compact uniform topology and the time translations induce a continuous-time flow of homeomorphisms on Lip_1 which we call the *shift flow*. Now it should be clear what a *strictly ergodic subshift* of Lip_1 is.

After a first breakthrough in Jacobs [4] (on the footsteps of Jewett [1]), Denker-Eberlein [1] achieved the final result: Every ergodic time-continuous flow in the measure-theoretical sense which is representable as a flow built under a function, i.e. which is "reasonable", can live on a strictly ergodic subshift of Lip_1. As in the time-discrete case, a generator theorem (Krengel [1], [2]; another generator theorem by Eberlein [2] turned out to be unnecessarily sharp for this purpose) plays a decisive intermediary rôle.

Our sketch has laid emphasis on definitive and general results about strict ergodicity so far. The historical origin of the whole theory was, however, the question raised by participants of the V. Berkeley Symposium in 1965 whether positive entropy is compatible with strict ergodicity at all. Seemingly, strict ergodicity was so close to determinism that one could quite well surmise that it would imply entropy 0 (as it had turned out e.g. in the case of pure point spectrum). Hahn-Katznelson [1] gave, however, a construction of a strictly ergodic subshift with positive entropy. Their device was to imitate a Bernoulli shift as much as possible within the limits of strict ergodicity. The construction was rather analytic in style. It was natural to ask what degree of variety of measure-theoretical properties of dynamical systems could be achieved by *machinal* (*combinatorial*) constructions of strictly ergodic subshifts over a finite alphabet. A machinal approach was obviously possible since machines can print out bisequences of symbols, and one single bisequence determines a subshift as its orbit closure, the subshifts strict ergodicity being readable from simple properties of the bisequence. It is thus reasonable to speak of strictly ergodic bisequences with positive entropy etc. The machinal approach is well motivated by the recent development in algorithmic mathematics.

In his Berkeley paper [1], Kakutani had exhibited spectral properties of a class of strictly ergodic bisequences which were constructed machinally along a combinatorial idea of M. Morse (Morse [1], Hedlund-Morse [1], Gottschalk-Hedlund

[2]). Keane [2] generalized Kakutani's results. His algorithms produced strictly ergodic bisequences with arbitrary "rational" point spectrum and also a nonempty continuous spectrum. The entropy turned out to be 0, and the same was the case for the so-called Toeplitz bisequences (Jacobs-Keane [1], Neveu [1], Eberlein [1]; see also Coven-Keane [1]). It was Grillenberger [1], [2] who first gave a machinal construction for strictly ergodic bisequences whose entropy comes even arbitrarily close to the log of the alphabet length, and even of K-systems. It is hard to read from the final version of his papers [1], [2] that he had originally imitated the computation of entropy by the speed of cyclic approximation as given by Katok [1].

4. STOCHASTIC FIELDS AND ERGODIC THEORY

Let $T = Z^d$ be the d-dimensional lattice. If $A \neq \emptyset$ is a finite set, then $\Omega = A^T = \{\omega = (\omega_s)_{s\in T} \mid \omega_s \in A \ (s \in T)\}$ is a compact metric space. For every $t \in T$, the mapping $\theta_t : \Omega \to \Omega$ defined by $(\theta_t\omega)_s = \omega_{s+t}$ $(\omega \in \Omega,\ s \in T)$ is a homeomorphism of Ω and $\theta_s \circ \theta_t = \theta_{s+t}$ $(s, t \in T)$. Let $I = \{m \mid m$ is a normalized measure on Ω, $\theta_t m = m \ (t \in T)\}$ be the convex weakly compact set of all normalized measures that are invariant under all θ_t $(t \in T)$. Wiener's extension of F. D. Birkhoff's ergodic theorem states, for every $m \in I$ and $f \in L^1_m$, the existence and finiteness of

$$\lim_{n\to\infty} \frac{1}{|T_n|} \sum_{t\in T_n} \theta_t f$$

m-a.e., where $T_1 \subseteq T_2 \subseteq \ldots$ are finite subsets of T increasing to T suitably. I is in fact a simplex and its extremal points are called *ergodic* measures. They can be characterized by the fact that they are trivial on the measurable subsets of Ω which are invariant under all θ_t $(t \in T)$. Ergodicity of m is implied by triviality on the tail field $\mathscr{F}_\infty = \bigcap_{V\subseteq T \text{ finite}} \mathscr{F}_{T-V}$, where generally $\mathscr{F}_S$ denotes the σ-field generated by the component mappings belonging to $S \subseteq T$. A finite partition $\alpha \subseteq \mathscr{F}_T$ of Ω is called *Bernoulli* for $m \in I$ if the transported partitions $\theta_t\alpha$ are independent and generate $\mathscr{F}$ mod m. We call m *Bernoulli* if such a partition exists. This sketch outlines the general framework of ergodic theory in *multishift space* Ω.

Gibbsian statistical mechanics can be subsumed under this framework. Let $P = (P_V)_{\emptyset \neq V \subseteq T \text{ finite}}$ be a family of stochastic kernels from Ω to Ω such that

1. $P_V(\omega, .)$ is carried by the finite set $\{\eta \mid \eta \in \Omega,\ \eta_t = \omega_t \ (t \notin V)\}$.
2. $P_V(\cdot, F)$ is a $\mathscr{F}_{T\setminus V}$-measurable function for every $F \in \mathscr{F}_T$.
3. $P_W P_V = P_W$ where the product of two kernels is defined as usual for any finite $\emptyset \neq V \subseteq W \subseteq T$. P is then called a *specification*.

A probability measure m in Ω is called a *stochastic field with specification P* if for every bounded measurable real function $\mathscr{F}$ on Ω

$$E_m(f \mid \mathscr{F}_{T\setminus V}) = P_V(\cdot, f) \quad m\text{-a.e.}$$

where $E_m(\cdot \mid \mathscr{F}_{T\setminus V})$ denotes the conditional expectation with respect to $\mathscr{F}_{T\setminus V}$ for m and $P_V(\cdot, f)$ is defined as usual. The set $C(P)$ of all stochastic fields specification P is nonempty, convex and compact metric if, as we shall henceforth assume, P is *continuous* in the sense that $P(\cdot, f) \in C(\Omega)$ $(f \in C(\Omega))$. In fact every weak limit of a sequence $P_{V_n}(\omega_n, \cdot)$ with finite $V_n \nearrow T$ is in $C(P)$ and the totality of these limits form a compact subset $C_\infty(P)$ of $C(P)$ which we call the *Martin boundary* of P. Let us assume an abstract notation for the Martin boundary and denote, for every $\delta \in C_\infty(P)$, by m_δ the probability measure on $\mathscr{F}_T$ which δ is.

The classical martingale theorem leads, for every $m \in C(P)$, to the m-a.e. existence of

$$(1) \qquad \varrho(\omega) = \lim P_{V_n}(\omega, \cdot) \in C_\infty(P) .$$

It is easy to show

$$P_{\varrho(\omega)}(\varrho(\cdot) = \varrho(\omega)) = 1 \quad m\text{-a.e.}$$

We call

$$\Delta = \{\delta \mid \delta \in C_\infty(P),\ m_\delta(\varrho(\cdot) = \delta) = 1\}$$

the *essential part* of $C_\infty(P)$. It is a Borel subset of the compact metric $C_\infty(P)$ and hence bears a natural σ-field $\mathscr{D}$. (1) can be reviewed as $\varrho(\omega) \in \Delta$ m-a.e. and we can consider ϱ as a measurable map from Ω to Δ (after suitable redefinition on a set which is null for all $m \in C(P)$). It is in fact $\mathscr{F}_\infty$-measurable. This map sends every $m \in C(P)$ into a probability distribution P_m on Δ whose barycenter in the sense of Choquet is just m: $m(f) = \int_\Delta m_\delta(f)\, P_m(\mathrm{d}\delta)$ $(f \in C(\Omega))$. Thus we can identify $C(P)$ with the simplex of all probability distributions on Δ. The extremal points of $C(P)$ correspond to the one-point masses on Δ; this allows to characterize the extremal points of $C(P)$ by their triviality on $\mathscr{F}_\infty$. This is a result of Preston [1] and Georgii [2]. We follow the presentation of Föllmer [2] here.

Let us now make the additional assumption that P is (translation) *invariant*, i.e.

$$P_{V_n+t}(\omega, F) = P_{V_n}(\theta_t\omega, \theta_t F) \quad (\omega \in \Omega,\ F \in \mathscr{F}_T,\ t \in T) .$$

In this case the set $C_0(P) = C(P) \cap I$ of all invariant stochastic fields with specification P is nonempty convex and compact. The θ_t $(t \in T)$ induce a group of measurable mappings in Δ and Wiener's ergodic theorem leads to an identification of the extremal points of $C_0(P)$ with the one-point masses on a certain $\Delta_0 \subseteq C_0(P)$. In particular, the extremal points of $C_0(P)$ turn out to be *ergodic* (Georgii [2], Preston [1], Föllmer [2]).

The interest in all this machinery stems from the following classical case. Let, for every finite $\emptyset \neq V \subseteq T$, $\Omega_V = A^V$ and $\omega \to \omega_V$ the natural projection from Ω to Ω_V.

Assume that for every such V a real function u_V is given on Ω_V. Fot the sake of simplicity we assume that $u_V = 0$ if the euclidean diameter of V exceeds a certain number $r > 0$. We call $u = (u_V)_{\emptyset \neq V \subseteq T \text{ finite}}$ a *potential of range r*. Put

$$U_V(\omega) = \sum_{\emptyset \neq W \subseteq T \text{ finite}, W \cap V = \emptyset} u_W(\omega_W) \quad (\emptyset \neq V \subseteq T \text{ finite}, \ \omega \in \Omega).$$

Then

$$P_V(\omega, \{\eta\}) = \frac{1}{Z_V(\omega)} e^{-u_V(\eta)} \ (\omega, \eta \in \Omega, \ \omega_{T\setminus V} = \eta_{T\setminus V})$$

with

$$Z_V(\omega) = \sum_{\tau_{T\setminus V} = \omega_{T\setminus V}} e^{-u_V(\tau)}$$

(this sum is finite!) defines a specification P. It is called the *Gibbs specification with potential* $u = (u_V)_{\emptyset \neq V \subseteq T \text{ finite}}$. It has the following *Markov property of order r*:

If $\delta_r V$ denotes the set of all lattice points $\notin V$ whose euclidean distance from some point of V is $\leqq r$, then $P_V(\cdot, F)$ is $\mathscr{F}_{\delta_r V}$-measurable. It can be shown that every specification with this property is a Gibbs specification with a suitable potential of range r (Averintzev [1], [2], Spitzer [1], [3], Sullivan [1], Grimmett [1], Georgii [3]). In the case $r = 1$ we speak of *next-neighbor interaction* or the *Ising model* in statistical mechanics. The Markov property of order r implies the continuity of P, and if $u_{V+t}(\omega_{V+t}) = u_V((\theta_t\omega)_V)$ ($\emptyset \neq V \subseteq T$ finite, $\omega \in \Omega$, $t \in T$), then P is invariant and our whole previous theory applies.

There are cases of Gibbs specifications based on a next neighbor potential where $C(P) = C_\infty(P) = C_0(P)$ consists of a single point. In a special case, Ledrappier [1] has shown that this single invariant probability distribution is Bernoulli. There are also cases where $C_\infty(P)$, and even Δ_0 consists of more than one point. One then speaks of *phase transition*, and again it is known only in special cases that the extremal points in Δ_0 are Bernoulli (Gallavotti [1], di Liberto-Gallavotti-Rosso [1]).

For a survey over the field the reader should consult Dobrushin [1], [2], [3], Georgii [1], [3], Kac [1], Spitzer [2], [3]. For its combinatorial aspects see Montroll [1], Zimmermann [1]. An application to economical theory was made by Föllmer [1].

5. THE ERGODIC HYPOTHESIS

In its early days, ergodic theory was concerned with the mechanical system which corresponds to a classical gas: A large number N of elastic balls in an elastic box. This system doesn't belong to mechanics in the usual sense, because it is not smooth. The reflections of particles play an essential rôle and destroy smoothness thoroughly. It is nevertheless possible to define a flow in a phase space representing the system and to pose the question of invariant measure and ergodicity. The famous *ergodic hypothesis* is nothing but the positive answer to the latter question. It has always been assumed as true by physicists, but a mathematical proof was announced only in 1963 by Sinai [4]. In fact Sinai announced that the system is even a *K-flow*. The details of the proof are still to appear, but an important simplified system is at present rather well known: the Sinai billiards (Sinai [7]).

Consider the 2-dimensional flat torus $\boldsymbol{R}^2$ mod 1 and cut out a finite number of disjoint strictly convex open areas with disjoint closures and smooth boundaries. Let Q (closed) connected be the remaining part of the torus and $M = Q \times S^1$ the space of all unit line elements over Q. Imagine a single point moving in Q with speed 1 and reflection at the boundary δQ of Q according to the classical law. If we replace M by M' by identifying line elements over the boundary δQ of Q which are connected by the classical reflection law, we may represent our motion by a flow $(S_t)_{t\in\boldsymbol{R}}$ on M'. This flow can be represented as flow built under a function on the base space M'_0 which arises from $\delta Q \times S^1$ by abovementioned identification. The function is nothing but the time which elapses until the next reflection. Lebesgue measure m on M' is left invariant by $(S_t)_{t\in\boldsymbol{R}}$, and it is easy to find a finite invariant measure m_0 for the transformation T on the base space M'_0.

Sinai proved that $(S_t)_{t\in\boldsymbol{R}}$ (with m) is a K-flow and T forms a K-system with m_0. Various implements to this result (focusing parts admitted in the boundary, elastic reflections, central limit theorem) have been found by Bunimovic [1], [2], Kubo [1] (see also the detailed survey Kubo [2]), but the most striking new result is due to Gallavotti-Ornstein [1]: The measure-theoretical dynamical system given by T and m_0 is measure-theoretically isomorphic to a Bernoulli scheme with finite entropy, and the measure-theoretical flow given by $(S_t)_{t\in\boldsymbol{R}}$ and m is a Bernoulli flow.

Although the isomorphy implied here is only a measure-theoretical one and hence doesn't allow immediate physical conclusions, it is perhaps the utmost contribution of ergodic theory known at present to the question why mechanical systems can show a probabilistic behaviour. It can be expected that the system representing the ideal gas has the same Bernoulli properties.

Now what are the steps which must be done in order to pass over from the billiard described here to the classical gas? 1. We must replace the torus of dimension 2 by a box of dimension 3 with reflecting walls. 2. We must replace the fixed "holes" and our moving point by a system of moving balls. There is some hope that the basic ideas are already displayed in the billard case and these two steps, however intricate, pose mainly technical problems.

6. QUALITATIVE MECHANICS

Modern mechanics considers smooth flows $(S_t)_{t\in R}$ on a compact smooth manifold M and the banachable separable space X^∞ of the associated vector fields f. A *generic property* of flows on M is a property fulfilled by all vector fields from a residual subset of X^∞, i.e. a countable intersection of everywhere dense open sets.

Properties of interest concern density of orbits, in particular of periodic ones, the set $\Omega(f)$ of all *nonwandering points* etc. Here a nonwandering point x is one for which every neighborhood U of x fulfills $U \cap S_t\, U \neq \emptyset$ for $|t|$ arbitrary large.

The whole theory can be set up under relaxed differentiability conditions and we get the banachable space X^r of all C^r vector fields on M.

If the dimension of M is even, we can consider a skew symmetric C^∞ 2-form ω on M and define a vector field f to be ω-symplectic if the associated flow leaves ω invariant. Inserting an ω-symplectic vector field into, say, the first argument of ω, we get a closed 1-form $\bar{\omega}(f)$. If $\bar{\omega}(f) = \mathrm{d}H$ for some function H, then f is called ω-Hamiltonian. Let X^r_ω be the banachable space of all C^r ω-Hamiltonian vector fields.

For nearly a decade the so-called *Closing Lemma* for banachable spaces of vector fields on a smooth compact manifold M has played a central rôle in attempts to prove the generic character of certain properties of flows (see e.g. Pugh [1], [2].) The statement of this lemma is, roughly speaking, that for every nonwandering point x of a given vector field f in that space, we can find a neighboring f' in the same space, such that a periodic point for f' is near x. It was only in 1974 that Pugh and Robinson [1] proved the closing lemma for a variety of spaces of vector fields which was wide enough to cover cases of classical interest. One of their most striking applications concern the space X^1_ω: generically in this space, the closed orbits of a Hamiltonian C^1 vector field f are dense in compact energy surfaces (i.e. submanifolds of the form $H = \text{const}$ where $\mathrm{d}H = \bar{\omega}(f)$). This proves an old conjecture of Poincaré [1].

7. EMBEDDING OF SUBSHIFTS INTO DIFFEOMORPHISMS

Ergodic theory in its early days, about 100 years ago, was, at least partly, intended to be a global theory of Hamiltonian flows, with special emphasis on mean value and equidistribution problems. At this stage, formulated in the laguage of today, the subject of the theory were smooth flows of diffeormorphisms on manifold which left a certain 2-form invariant. Mathematicians recognized an abstract pattern in the problem and reformulated it in terms of topological or measure theoretical dynamical system. The more general the definitions became, the larger became the scope of examples. Kronecker's and Weyl's [1] rotations of the circle were still diffeomorphisms or smooth flows of such. Probability theory contributed the example of shifts and subshifts whose basic space was a totally disconnected compact metric space. It looked as if there would be never any connection of these examples with the domain of diffeomorpisms and hence with the origins of ergodic theory.

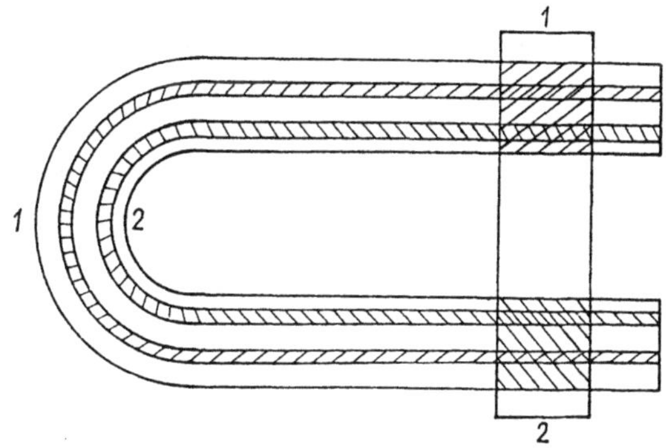

Fig. 1.

It was therefore quite surprising to learn that a shift space can be identified with the set of nonwandering points (i.e. the points with the property that each of their neighborhoods intersects itself after transformation with a suitable power of the diffeomorphism) of a certain diffeomorphism of a finite plane disc in such a fashion that the shift becomes nothing but a restriction of the latter. This identification was achieved by Smale's [1] horseshoe construction (for a detailed presentation see Nitecki [1]) whose visualization in the case of 2-shift immediately justifies its name.

Let us shortly explain it with the help of Fig. 1. The diffeomorphism acts in such a fashion that it flattens the vertical bar into a horizontal one, bends the latter into the shape of a horseshoe and puts this figure into a position indicated by Fig. 1 to the effect that the edges 1, 2 of the bar correspond to the edges 1, 2 of the horseshoe, respectively. The picture shows how the shaded areas of the vertical bar are transformed by the diffeomorphism. Its inverse produces vertical striplets instead of horizontal ones. Iterating this process in both directions we get "grills" of striplets of "higher orders". The intersection of all these grills is a totally disconnected subset of the plane and it is not hard to identify it with 2-shift space such that the restriction of the diffeomorphism to it becomes the shift.

Similar, but more complicated constructions apply to shifts with more than 2 symbols. Smale's construction endows the diffeomorphism with an additional important property usually called "Axiom A".

In the theory of dynamical systems, both in a topological and in a measure theoretical framework, the concept of a *subshift of finite type* has played a rôle of increasing importance since it was introduced by Parry [1] in 1964 (see e.g. Bowen [1], [2], [3], [4], Grillenberger [1], [2]). If A is a finite alphabet and W any family of A-words of a fixed length r, then the set Ω_W of all bisequences whose sections of length r belong to W is a compact shift invariant subset of A-shift space. The restriction of shift to Ω_W is called the subshift of finite type r determined by W. It is natural to ask whether every subshift of finite type can be identified with an axiom A diffeomorphism of a ball in some R^n such that the subshift becomes, say, an isolated component of the set of all nonwandering points of the latter.

In his paper [1], Williams sketched a construction which embeds a given subshift of finite type into an axiom A diffeomorphism on a 3-dimensional manifold. Independently, and at the same time, Kollmer [1] extended methods of Smale [1] and Nitecki [1] obtaining an embedding of a given subshift of finite type into a subset Ω of an arbitrary manifold M of dimension $\geqq 3$ such that, for a suitable diffeomorphism T of M

1. T is hyperbolic on M(this is part of axiom A),
2. Ω is isolated in the sense that every point in $M \setminus \Omega$ eventually leaves every open neighborhood of Ω either in the past or the future.

BIBLIOGRAPHY

[1] R. ABRAHAM, J. E. MARSDEN: Foundations of mechanics. XXXVIII + 296 pp., Benjamin, New York—Amsterdam 1967.

[1] R. ADLER, B. WEISS: Entropy, a complete metric invariant for automorphisms of the torus. Proc. Nat. Acad. Sci. USA *57* (1967), 1573—1576.

[1] N. AOKI, H. TOTOKI: Ergodic automorphisms of T^∞ are Bernoulli transformations, Publ. RIMS Kyoto *10* (1975), 535—544.

[1] V. I. ARNOLD, A. AVEZ: Ergodic problems of classical mechanics. X + 286 pp., Benjamin, New York—Amsterdam 1968.

[1] M. B. AVERINTZEV: On a means of description of random fields with discrete arguments. Prob. pered. inf. *6* (1970), 100—108.

[2] M. B. AVERINTZEV: Description of Markovian random fields by Gibbsian conditional probabilities. Teor. ver. prim. *17* (1972), 21—35; engl. transl. in Th. of Prob. Appl. *17* (1972), 20—33.

[1] A. BECK: Continuous flows in the plane. Grundl. d. math. Wiss. vol. 201. Springerverlag, Berlin-Heidelberg-New York, X + 462 pp. (1974).

[1] J. R. BLUM, D. L. HANSON: On the isomorphism problem for Bernoulli schemes. Bull. AMS *69* (1963), 221—223.

[2] J. R. BLUM, D. L. HANSON: On the isomorphism problem for Bernoulli schemes. 20 pp., preprint 1962.

[1] R. BOWEN: Topological entropy and Axiom A. In: Global Analysis, Proc. Symp. Pure Math. AMS vol. XIV (1970).

[2] R. BOWEN: Markov partitions and minimal sets for Axiom A diffeormorphisms. Amer. J. Math. *92* (1970), 907—918.

[3] R. BOWEN: Markov partitions for Axiom A diffeormorphisms. Amer. J. Math. *92* (1970), 725—747.
[4] R. BOWEN: Periodic points and measures for Axiom A diffeomorphisms. Trans. AMS *154* (1971), 377—397.
[1] L. A. BUNIMOVIC: A central limit theorem for scattering billards. Dokl. Akad. Nauk SSSR *204* (1972), 778—781.
[2] L. A. BUNIMOVIC: On ergodic properties of nearly scattering billards. Dokl. Akad. Nauk SSSR *211* (1973), 1024—1026.
[1] L. A. BUNIMOVIC, JA. G. SINAI: On the fundamental theorem of the theory of scattering billards. Mat. Sbor. *20* (1973), 415—431.
[1] J. R. CHOKSI: Unitary operators induced by measurable transformations. J. Math. Mech. *16* (1966), 83—100.
[2] J. R. CHOKSI: Unitary operators induced by measurable transformations. J. Math. Mech. *17* (1968), 785—801.
[1] E. M. COVEN, M. KEANE: The structure of substitution minimal sets. TAMS *162* (1971), 89—102.
[1] M. DENKER: On strict ergodicity. Math. Z. *134* (1973), 231—253.
[2] M. DENKER: Induced Measure-preserving transformations. 17 pp., preprint 1974.
[3] M. DENKER: Finite generators for ergodic measure-preserving transformations. Z. Wahrsch. verw. Geb. *29* (1974), 45—55.
[4] M. DENKER: Einige Bemerkungen zu Erzeugersätzen. Z. Wahrsch. verw. Geb. *29* (1974), 57—64.
[1] M. DENKER, E. EBERLEIN: Ergodic flows are strictly ergodic. Advances in Math. *13* (1974), 437—473.
[1] F. DI LIBERTO, G. GALLAVOTTI, L. ROSSO: Markov processes, Bernoulli schemes, and Ising model. Commun. Math. Phys. *33* (1973), 259—282.
[1] R. L. DOBRUSHIN: Existence of phase transitions in models of lattice gas. In.: Proc. V. Berkeley Symp. Math. Stat. Prob. III, (1967), 73—87.
[2] R. L. DOBRUSHIN: Description of a random field by means of conditional probabilities and its regularity conditions. Teor. ver. prim. *13* (1968), 197—224; engl. transl. in Th. Prob. Appl. *13* (1968), 197—224.
[3] R. L. DOBRUSHIN: Investigation of Gibbs states in 3-dimensional lattice systems. Teor. ver. prim. *18* (1973), 261—279.
[1] Y. N. DOWKER: A new proof for the general ergodic theorem, Ann. Sci. Math. Szeged *12*, Part B (1950), 162—166.
[1] J. K. DUGDALE: An example of an automorphism which is mixing of all degrees but which is not a Kolmogorov automorphism. Proc. Camb. Phil. Soc. *62* (1966), 687—691.
[1] H. A. DYE: On groups of measure preserving transformations I. Amer. J. Math. *85* (1959), 119—159.
[1] E. EBERLEIN: Toeplitzfolgen und Gruppentranslationen. Archiv d. Math. *22* (1971), 291—301.
[2] E. EBERLEIN: Einbettung von Strömungen in Funktionsräume durch Erzeuger vom endlichen Typ. Z. f. Wahrsch. verw. Geb. *27* (1973), 277—281.
[3] E. EBERLEIN: A generator theorem for flows, Bol. Soc. Brasil Mat. *5* (1974), 45—50.
[1] R. ELLIS: Lectures on topological dynamics. XVI + 211 pp., Benjamin, New York 1969.
[1] H. FÖLLMER: Random economies with many interacting agents. J. Math. Econ. *1* (1974), 51—62.
[2] H. FÖLLMER: Phase transitions and Martin boundary. 12 pp., preprint 1974.
[1] N. FRIEDMAN, D. ORNSTEIN: On isomorphism of weak Bernoulli transformations. Advances in Math. *5* (1971), 365—394.
[1] H. FURSTENBERG: The structure of distal flows. Amer. J. Math. *85* (1963), 477—515.

[1] G. GALLAVOTTI: Ising model and Bernoulli schemes in one dimension. Commun. Math. Phys. *32* (1973), 183–190.
[1] G. GALLAVOTTI, D. ORNSTEIN: Billards and Bernoulli schemes, Comm. Math. Phys. *38* (1974), 83–101.
[1] H. O. GEORGII: Phasenübergang 1. Art bei Gittergasmodellen. IX + 167 S., Lecture Notes in Physics.
[2] H. O. GEORGII: Two remarks on equilibrium states. Comm. Math. Phys. *32* (1973), 107–118.
[3] H. O. GEORGII: Stochastische Felder und ihre Anwendung auf Interaktionssysteme. IV + 141 S., Vorlesungsausarbeitung, Heidelberg 1974.
[1] I. V. GIRSANOV: On spectra of dynamical systems generated by stationary Gaussian processes. Dokl. Akad. Nauk SSSR *119* (1958), 851–853.
[1] W. H. GOTTSCHALK, G. HEDLUND: Topological dynamics. AMS publ. no. 36, 2nd edition Providence (AMS).
[2] W. H. GOTTSCHALK, G. HEDLUND: A characterization of the Morse minimal set. Proc. AMS *15* (1964), 70–74.
[1] U. GRENANDER: Stochastic processes and statistical inference. Archiv Mat. *1* (1952), 195–277.
[1] CHR. GRILLENBERGER: Constructions of strictly ergodic systems, I. Given entropy. Z. Wahrsch. verw. Geb. *25* (1973), 323–334.
[2] CHR. GRILLENBERGER: Constructions of strictly ergodic systems. II. K-systems. Z. Wahrsch. verw. Geb. *25* (1973), 335–342.
[1] G. R. GRIMMETT: A theorem about random fields. Bull. London Math. Soc. *5* (1973), 81–84.
[1] H. GUNDEL: Mischungseigenschaften Gauß'scher Automorphismen. Diplomarbeit, Göttingen 1973.
[1] F. HAHN, Y. KATZNELSON: On the entropy of uniquely ergodic transformations. Trans. AMS *126* (1967), 335–360.
[1] P. R. HALMOS: Lectures in Ergodic Theory. Math. Soc. Japan, 1956.
[2] P. R. HALMOS: Recent progress in ergodic theory. Bull. AMS *67* (1961), 70–80.
[1] G. HANSEL, J. P. RAOULT: Ergodicité, uniformité et ergodicité unique. Ind. Univ. Math. J. *23* (1973), 221–237.
[1] G. A. HEDLUND: Endomorphisms and automorphisms of the shift dynamical system. Math. Syst. Th. *3* (1969), 320–375.
[1] G. A. HEDLUND, M. MORSE: Unending chess, symbolic dynamics and a problem in semi-groups. Duke Math. J. *11* (1944), 1–7.
[1] W. HUREWICZ: Ergodic theorem without invariant measure. Ann. of Math. *45* (1944), 192–206.
[1] I. A. IBRAGIMOV, YU. V. LINNIK: Independent and stationary sequences of random variables. Wolters-Noordhoff, Groningen 1971.
[1] A. IONESCU-TULCEA: On the category of certain classes of transformations in ergodic theory. TAMS *114*, (1965), 261–279.
[1] K. ITÔ: On the ergodicity of a certain stationary process. Proc. Imp. Acad. Tokyo *20* (1944), 54–55.
[2] K. ITÔ: A cinematic theory of turbulence. Proc. Imp. Acad. Tokyo *20* (1944), 120–122.
[3] K. ITÔ: Complex multiple Wiener integral. Jap. J. Math. *22* (1952), 63–66.
[1] K. JACOBS: Lecture Notes in Ergodic Theory. 2 vol., Aarhus 1963.
[2] K. JACOBS: Einige neuere Ergebnisse der Ergodentheorie. Jber. DMV *67* (1965), 143–182.
[3] K. JACOBS: Combinatorial constructions in ergodic theory. In: Proc. Int. Conf. Funct. Anal. Tokyo (1969), 398–399.
[4] K. JACOBS: Lipschitz functions and the prevalence of strict ergodicity for continuous-time flows. Lect. Notes Math. vol. *160*. 87–124, Springerverlag Berlin–Heidelberg–New York 1970.

[1] K. JACOBS, M. KEANE: 0-1-sequences of Toeplitz type. Z. Wahrsch. verw. Geb. *13* (1969), 123—131.
[1] R. JEWETT: The prevalence of uniquely ergodic systems. J. Math. Mech. *19* (1970), 717—729.
[1] M. KAC: Lectures on mathematical aspects of phase transitions. In: Stud. in Math. Phys. (ed. Barut), Reidel, Dordrecht 1973, 51—79.
[1] S. KAKUTANI: Ergodic theory of shift transformations. In: Proc. V. Berkeley Symp. Stat. Prob. II, (1967), 405—414.
[2] S. KAKUTANI: Examples of ergodic measure preserving transformations which are weakly mixing but not strongly mixing. Lecture Notes in Math. vol. 318. Springerverlag Berlin—Heidelberg—New York 1973, 143—149.
[1] A. B. KATOK: Entropy and the approximation of dynamical systems by periodic transformations. Funkt. Analiz *1* (1967), 75—85.
[1] Y. KATZNELSON: Ergodic automorphisms of T^n are Bernoulli shifts. Israel J. Math. *10* (1971), 186—195.
[1] M. KEANE: Bernoulli-Schemata und Isomorphie. Diplomarbeit Göttingen 1965.
[2] M. KEANE: Generalized Morse sequences. Z. Wahrsch. verw. Geb. *10* (1968), 335—353.
[3] M. KEANE: Strongly mixing g-sequences. Invent. Math. *16* (1972), 309—324.
[1] H. KOLLMER: Über die Einbettung von Teilshifts in differenzierbare dynamische Systeme. Diplomarbeit Erlangen 1973.
[1] A. N. KOLMOGOROV: A new metric invariant of transitive dynamical systems and automorphisms of Lebesgue spaces. Dokl. Akad. Nauk SSSR *119* (1958), 861—864.
[2] A. N. KOLMOGOROV: On the entropy per time unit as a metric invariant of automorphisms. Dokl. Akad. SSSR *124* (1959), 754—753.
[1] B. O. KOOPMAN: Hamiltonian systems and linear transformations in Hilbert space. Proc. Nat. Acad. Sci. USA *17* (1931), 315—318.
[1] U. KRENGEL: On certain analogous difficulties in the investigation of flows in a probability space and of transformations in an infinite measure space. In: Proc. of a Symp. Fund. Anal. Monterey (ed. C. O. Wilke) Acad. Press, New York 1969, 75—91.
[2] U. KRENGEL: K-flows are forward deterministic, backward completely nondeterministic stationary point processes. J. Math. Anal. Appl. *35*, (1971) 611—620.
[3] U. KRENGEL: Recent results on generators in ergodic theory. In: Trans. VI. Prague Conf. 1971. Academia, Prague 1973, 465—482.
[1] W. KRIEGER: On non-singular transformations of a measure space I. Z. f. Wahrsch. verw. Geb. *11* (1969), 83—97.
[2] W. KRIEGER: On non-singular transformations of a measure space II. Z. f. Wahrsch. verw. Geb. *11* (1969), 98—119.
[3] W. KRIEGER: On entropy and generators of measure preserving transformations. TAMS *149* (1970), 453—464.
[4] W. KRIEGER: On constructing non-isomorphic hyperfinite factors of type III. J. funct. Anal. *6* (1970), 97—109.
[5] W. KRIEGER: On the Araki-Woods asymptotic ratio set and non-singular transformations of a measure space. Lect. Notes in Math. 160. Springerverlag, Berlin—Heidelberg—New York 1970, 158—177.
[6] W. KRIEGER: On generators in exhaustive σ-algebras of ergodic measurepreserving transformations. Z. f. Wahrsch. verw. Geb. *20* (1971), 75—82.
[7] W. KRIEGER: On unique ergodicity. In: Proc. VI. Berkeley Symp. Math. Stat. Prob. vol. II. UC Press, Berkeley—Los Angeles 1972, 327—345.
[8] W. KRIEGER: On ergodic flows and the isomorphism of factors. Math. Ann. *223* (1976), 19—70.

[1] I. KUBO: Ergodicity of the dynamical system of a particle on a downain with irregular walls. In: Proc. II. Japan—USSR Symp. Prob. Lect. Notes Math. vol. 330, Springerverlag, Berlin—Heidelberg—New York 1973, 287—295.

[2] I. KUBO: Ergodic properties of billards in the plane mod 1 (tentative title). Lecture Notes in Mathematics. Springerverlag, Berlin—Heidelberg—New York (to appear).

[1] A. G. KUSHNIRENKO: An upper estimate for the entropy of a classical dynamical system. Dokl. Akad. Nauk SSSR *161* (1965), 37—38.

[1] F. LEDRAPPIER: Messures d'èquilibre sur un reseau. Commun. Math. Phys. *33* (1973), 119—128.

[1] V. P. LEONOV: The use of the characteristic functional and semi-invariants in the ergodic theory of stationary processes. Sov. Math. *1* (1960), 878—881.

[1] G. MARUYAMA: The harmonic analysis of stationary stochastic processes. Mem. Fac. Sci. Kyushu Univ., Ser. A, *4* (1949), 45—106.

[1] L. D. MESHALKIN: A case of isomorphy of Bernoulli schemes. Dokl. Akad. Nauk SSSR *128* (1959), 41—44.

[1] G. MILES, R. K. THOMAS: The breakdown of automorphisms of compact topological groups. 29 pp., preprint 1974 (plus 2 more papers to follow).

[1] E. W. MONTROLL: Lattice statistics. In; Applied Comb. Math. New York 1964.

[1] M. MORSE: Recurrent geodesics on a surface of negative curvature. TAMS *22* (1921), 84—100.

[1] F. J. MURRAY, J. v. NEUMANN: On rings of operators. Ann. of Math. *37* (1936), 116—229.

[1] J. v. NEUMANN: Proof of the quasi-ergodic hypothesis. Proc. Nat. Acad. Sci. USA *18* (1932), 263—266.

[2] J. v. NEUMANN: Zur Operatorenmethode in der klassischen Mechanik. Ann. of Math. *33* (1932), 587—642.

[3] J. v. NEUMANN: On rings of operators III. Ann. of Math. *41* (1940), 94—161.

[1] J. NEVEU: Sur les suites de Toeplitz. Z. Wahrsch. verw. Geb. *13* (1969), 132—134.

[1] D. NEWTON: On Gaussian processes with simple spectrum. Z. Wahrsch. verw. Geb. *5* (1966), 207—209.

[1] Z. NITECKI: Differentiable Dynamics. XVIII + 282 pp., MIT Press, Cambridge—London 1971.

[1] D. ORNSTEIN: Bernoulli shifts with the same entropy are isomorphic. Advances in Math. *4* (1970), 337—352.

[2] D. ORNSTEIN: Two Bernoulli shifts with infinite entropy are isomorphic. Advances in Math. *5* (1971), 339—348.

[3] D. ORNSTEIN: Factors of Bernoulli shifts are Bernoulli shifts. Advances in Math. *5* (1971), 349—364.

[4] D. ORNSTEIN: A Kolmogorov automorphism that is not a Bernoulli shift. Advances in Math. *10* (1973), 49—62.

[5] D. ORNSTEIN: A K automorphism with no square root and Pinsker's conjecture. Advances in Math. *10* (1973), 89—102.

[6] D. ORNSTEIN: A mixing transformation for which Pinsker's conjecture fails. Advances in Math *10* (1973), 103—123.

[7] D. ORNSTEIN: An application of ergodic theory to probability theory. Ann. of Prob. *1* (1973), 43—65.

[8] D. ORNSTEIN: Ergodic theory, randomness, and dynamical systems. 141 pp., Yale Univ. Press, New Haven—London 1974.

[1] D. ORNSTEIN, P. SHIELDS: An uncountable family of K-automorphisms. Advances in Math. *10* (1973), 63—88.

[1] J. C. OXTOBY: Ergodic sets. Bull. AMS *58* (1952), 116—136.

[1] W. PARRY: Intrinsic Markov chains. TAMS *112* (1964), 55—66.

[2] W. PARRY: On Rohlin's formula for entropy. Acta Math. Acad. Sci. Hung. *15* (1964), 107—113.

[3] W. PARRY: Generators and strong generators in ergodic theory. Bull. AMS *72* (1966), 294—296.

[4] W. PARRY: Entropy and generators in ergodic theory. XII + 124 pp., Benjamin, New York—Amsterdam 1969.

[1] H. POINCARÉ: Les méthodes nouvelles de la mécanique cèleste, t. I. Paris 1892; Dover, New York 1957.

[1] C. J. PRESTON: Random fields. Lect. N. Math. *534* (1976).

[1] C. PUGH: The closing lemma. Amer. J. Math. *89* (1967), 956—1009.

[2] C. PUGH: An improved closing lemma and a general density theorem. Amer. J. Math. *89* (1967), 1010—1021.

[1] C. PUGH, and R. C. ROBINSON: The C^1 closing lemma, including Hamiltonians. 149 pp., preprint 1974.

[1] A. RÉNYI: Representations for real numbers and their ergodic properties. Acta Math. Acad. Sci. Hung. *8* (1957), 477—493.

[1] V. A. ROKHLIN: Selected topics in the metric theory of dynamical systems. Uspehi Mat. Nauk *4* (1949), 57—128; engl. transl. in Transl. AMS *49* (1966), 171—240.

[2] V. A. ROKHLIN: New progress in the theory of transformations with invariant measure. Uspehi Mat. Nauk *15* (1960), 1—26; engl. transl. in Russian Math. Surveys *15* (1960), 1—22.

[3] V. A. ROKHLIN: Generators in ergodic theory. Vestnik Leningr. Univ. Mat. Mech. Astron. (1963), 26—32.

[4] V. A. ROKHLIN: Generators in ergodic theory II. Vestnik Leningr. Univ. Mat. Mech. Astron. (1965), 68—72.

[5] V. A. ROKHLIN: Lectures on the entropy theory of transformations with invariant measure. Uspehi Mat. Nauk *22* (1967), 3—56; english translation in Russian Math. Surveys *22*, (1967), 1—52.

[1] K. SIGMUND: Generic properties of invariant measures for Axiom A—diffeomorphisms. Invent. Math. *11* (1970), 99—109.

[2] K. SIGMUND: Ergodic averages for Axiom A diffeomorphisms. Z. f. Wahrsch. verw. Geb. *20* (1971), 319—324.

[3] K. SIGMUND: On the space of invariant measures for hyperbolic flows. Amer. J. Math. *94* (1972), 31—37.

[4] K. SIGMUND: On mixing measures for Axiom A diffeomorphisms. Proc. AMS *36* (1972), 497—504.

[1] JA. G. SINAI: On the concept of entropy of a dynamical system. Dokl. Akad. Nauk SSSR *124* (1959), 768—771.

[2] JA. G. SINAI: On flows with finite entropy. Dokl. Akad. Nauk SSSR *125* (1959), 1200—1202.

[3] JA. G. SINAI: Weak isomorphy of transformations with invariant measure. Dokl. Akad. Nauk SSSR *147* (1962), 797—800; engl. transl. in Soviet Math. *3* (1962), 1725—1729.

[4] JA. G. SINAI: On the foundations of the ergodic hypothesis for a dynamical system of statistical mechanics. Dokl. Akad. Nauk SSSR *153* (1963), 1261—1264; engl. translation in Soviet Math. *4* (1963), 1818—1822.

[5] JA. G. SINAI: Weak isomorphism of transformations with invariant measure. Mat. Sb. *63* (1964), 23—42; engl. transl. in Transl. AMS, Ser. 2, *57* (1966), 123—143.

[6] JA. G. SINAI: Markovian partitions and U-diffeomorphisms. Funct. Analiz. *2* (1968), 64—89.

[7] JA. G. SINAI: Dynamical systems with elastic reflexion. Uspehi Mat. Nauk *25* (1970), 141—192.

[8] JA. G. SINAI: Gibbs measures in ergodic theory. Usp. Mat. Nauk *27* (1972), 21—64; english translation in Russian Math. Surveys *27* (1972), 21—69.

[9] JA. G. SINAI: Ergodic theory. In: The Boltzmann Equation (ed. Cohen-Thirring). Springer-verlag; Wien—New York 1973, 575—608.
[1] S. SMALE: Diffeomorphisms with many periodic points. In: Differential and Combinatorial Topology, Princeton 1965.
[2] S. SMALE: Differentiable dynamical systems. Bull. AMS *73* (1967), 747—817.
[3] S. SMALE: Ω-stability theorem. In: Global Analysis, Proc. of Symp. in Pure Math. AMS vol. XIV, (1970), 289—298.
[1] M. SMORODINSKI: Ergodic theory entropy. Lecture Notes in Math. 214. Springerverlag, Berlin—Heidelberg—New York, 1971.
[2] M. SMORODINSKI: A partition on a Bernoulli shift which is not weakly Bernoulli. Math. Syst. Th. *5* (1971), 201—203.
[1] F. SPITZER: Markov random fields and Gibbs ensembles. Amer. Math. Monthly *78* (1971), 142—154.
[2] F. SPITZER: Interaction of Markov processes. Advances in Math. *5* (1970), 246—290.
[3] F. SPITZER: Random fields and interacting particle systems. Lecture Notes Math. Assoc. Amer. 1971.
[1] S. STERNBERG: Celestial Mechanics I, II. XXIV + 158 pp., XVIII + 304 pp., Benjamin, New York 1969.
[1] W. G. SULLIVAN: Comm. Math. Phys. *33* (1973), 61—74.
[1] Y. TAKAHASHI: Isomorphisms of β-automorphisms to Markov automorphisms. Osaka J. Math. *10* (1973), 175—184.
[1] H. TOTOKI: The mixing property of Gaussian flows. Mem. Fac.-Sci., Kyushu Univ., Ser. A, *18* (1964), 136—139.
[1] K. URBANIK: Lectures on prediction theory. Lecture Notes in Mathematics vol. 44. Springer-verlag, Berlin—Heidelberg—New York, 1967.
[1] A. M. VERSHIK: Spectral and metric isomorphism of some normal dynamical systems. Sov. Math. *3* (1962), 693—696.
[1] H. WEYL: Über die Gleichverteilung von Zahlen mod. Eins. Math. Ann. *77* (1916), 313—352.
[1] R. F. WILLIAMS: Classification of subshifts of finite type. Ann. of Math. *98* (1973), 120—153.
[1] H. ZIMMERMANN: Hantelkonfigurationen auf Gittern und Berechnung der Zustandssumme für das planare Ising-Modell. Diplomarbeit, Erlangen 1972.
[1] L. ZSIDÓ: Sperante conditionate si teorema lui Sinai. Studii si Cercetari Mat. *20* (1968), 1045—1111.

UNIVERSITÄT ERLANGEN-NÜRNBERG

MATEMATISCHES INSTITUT

A NOTE ON THE GENERALIZED GAMMA DISTRIBUTION: THE STATISTIC $U = (X - Y)/(X + Y)$

Miloš Jílek

Prague

Let X_1, X_2 be two independent identically distributed random variables with the generalized gamma distribution. The form of the distribution of the statistic $U = (X_1 - X_2)/(X_1 + X_2)$ (as well as of some related statistics) is found, and, for some interesting particular cases of the generalized gamma distribution (gamma, Helmert, Maxwell, Weibull, exponential, Rayleigh), probability densities, distribution functions and moments are given.

1. INTRODUCTION

Under experimental conditions, there exist some situations in which only two parallel measurements may be done. When analyzing such pairs of experimental data, some functions of them are used (e.g. their sum, difference, product, ratio, etc.), and it may be useful to know probability distributions of appropriate statistics.

When investigating probability distributions of some functions of two normal distributed random variables X and Y, Agard [1] studied, among others, the distribution of the statistic

$$U = (X - Y)/(X + Y)\,, \tag{1}$$

which is important e.g. in studies on precision of laboratory methods, in quality control, in standardization of methods, etc.

Besides the normal distribution, however, there are some other distributions useful in treatment of experimental data; one of them is the generalized gamma distribution and its particular cases. Some functions of two generalized gamma distributed random variables were studied by Śródka [8] and Jakuszenkow [6], products and ratios of several independent generalized gamma distributed random variables were studied by Mathai [7]; however, the statistic U is not mentioned there. The aim of the present paper is to give some results concerning the probability distribution of the statistic U (and some related statistics) for the generalized gamma distribution as well as for its important particular cases.

2. NOTATION

Throughout this paper, the following notation will be used:

$$L(t) = \frac{(1-t)^\alpha}{(1+t)^\alpha + (1-t)^\alpha}, \quad |t| < 1;$$

$$A(t) = 2\left\{\frac{\alpha}{\beta^{p/\alpha}\,\Gamma(p/\alpha)}\right\}^2 \frac{(1-t)^{p-1}}{(1+t)^{p+1}}, \quad |t| < 1;$$

$$C(t) = \frac{1}{\beta}\left\{1 + \left(\frac{1-t}{1+t}\right)^\alpha\right\}, \quad |t| < 1;$$

$$M_{rs} = \int_0^1 \frac{t^s}{(1+t^2)^r}\,dt$$

(integrals of this form may be easily solved – cf., e.g., [3], formula (2.147.2)); $I_x(\mu, \nu)$ denotes the x-value of the incomplete beta function with parameters μ, ν

3. DISTRIBUTION OF U

THEOREM. Let X and Y be two independent identically distributed random variables with the probability density

$$f(t) = \begin{cases} \dfrac{\alpha}{\beta^{p/\alpha}\,\Gamma(p/\alpha)}\, t^{p-1} \exp(-t^\alpha/\beta), & \text{if} \quad t > 0, \\ 0 \quad \text{otherwise}, \end{cases} \tag{2}$$

where $\alpha, \beta, p > 0$.

Then the distribution of the statistic U given by (1) has the probability density

$$f_U(t) = \begin{cases} \dfrac{2\alpha}{B(p/\alpha, p/\alpha)} \dfrac{(1-t^2)^{p-1}}{[(1+t)^\alpha + (1-t)^\alpha]^{2p/\alpha}}, & \text{if} \quad -1 < t < 1, \\ 0 \quad \text{otherwise}, \end{cases} \tag{3}$$

and the distribution function

$$(4) \qquad F_U(t) = \begin{cases} 0, & \text{if } t \leqq -1, \\ I_{1-L(t)}(p/\alpha, p/\alpha), & \text{if } -1 < t < 1, \\ 1, & \text{if } t \geqq 1. \end{cases}$$

Proof. Joint probability density of (X, Y) has the form

$$f(x, y) = \begin{cases} \dfrac{\alpha^2}{\beta^{2p/\alpha}\{\Gamma(p/\alpha)\}^2} (xy)^{p-1} \exp\{-(x^\alpha + y^\alpha)/\beta\}, & \text{if } 0 < x, y < \infty, \\ 0 \text{ otherwise}; \end{cases}$$

setting

$$t = (x - y)/(x + y), \quad v = x,$$

we get

$$f(t, v) = \begin{cases} A(t)\, v^{2p-1} \exp\{-C(t)\, v^\alpha\}, & \text{if } -1 < t < 1, \quad 0 < v < \infty, \\ 0 \text{ otherwise}. \end{cases}$$

Hence, for $t \in (-1, 1)$,

$$f_U(t) = A(t) \int_0^\infty v^{2p-1} \exp\{-C(t)\, v^\alpha\}\, \mathrm{d}v = \frac{A(t)\, \Gamma(2p/\alpha)}{\alpha[C(t)]^{2p/\alpha}},$$

and after substituting in and rearrangement of the last expression we get (3). Then the distribution function has the form, for $t \in (-1, 1)$,

$$F_U(t) = \int_{-1}^t \frac{2\alpha}{B(p/\alpha, p/\alpha)} \frac{(1 - u^2)^{p-1}}{[(1 + u)^\alpha + (1 - u)^\alpha]^{2p/\alpha}}\, du,$$

and after substitution

$$y = \frac{(1 - u)^\alpha}{(1 + u)^\alpha + (1 - u)^\alpha}$$

we get that

$$F_U(t) = \frac{1}{B(p/\alpha, p/\alpha)} \int_{L(t)}^1 y^{p/\alpha - 1}(1 - y)^{p/\alpha - 1}\, \mathrm{d}y = 1 - I_{L(t)}(p/\alpha, p/\alpha),$$

and, therefore, $F_U(t)$ fulfils (4).

NOTE 1. Let us note that the distribution of U does not depend on the parameter β.

NOTE 2. There are some other statistics related to U and interesting from the view of applicability; e.g. $W = |U|$, $V = W\sqrt{2}$ (variation coefficient), $Z = 2W$ (ratio of the absolute difference and the average of two parallel measurements) etc. A survey of their distribution functions is given in Table 1 (distribution functions are equal to 0 for $t < \inf J$, and to 1 for $t > \sup J$).

NOTE 3. All odd moments of the statistic U are equal to zero, whereas all moments of other statistics mentioned above are non-zero. It seems useful, therefore, to study in detail some of statistics V, W, and Z; the simplest of them, W, was chosen, and probability densities, distribution functions and moments of W for some interesting particular cases of the generalized gamma diıribution are given in the next section. Table 2 gives simple relations between them and densities, distribution functions and moments of statistics U, V, and Z.

NOTE 4. Gamma distribution may be generalized in different ways, and, therefore, different forms of the generalized gamma distribution may be such obtained. However many of them – e.g. the form considered by Bain and Weeks [2] and Hager, Bain and Antle [4] – may be easily transformed into (2).

TABLE 1

Statistics U, V, W and Z, and their distribution functions

Statistic	Range J	Distribution function
$U = \dfrac{X - Y}{X + Y}$	$(-1, 1)$	$1 - I_{L(t)}(p/\alpha, p/\alpha)$
$V = W\sqrt{2}$	$\langle 0, \sqrt{2})$	$1 - 2\, I_{L(t/\sqrt{2})}(p/\alpha, p/\alpha)$
$W = \lvert U \rvert$	$\langle 0, 1)$	$1 - 2\, I_{L(t)}(p/\alpha, p/\alpha)$
$Z = 2\,W$	$\langle 0, 2)$	$1 - 2\, I_{L(t/2)}(p/\alpha, p/\alpha)$

3. PARTICULAR CASES

Some important distributions are particular cases of the generalized gamma distribution:

(I) $\alpha = 1$ .... *gamma distribution*;
(II) $\alpha = 2$ .... *Helmert distribution* ([5], formula (2.7));
or *seminormal distribution* [9];
(III) $\alpha = 2$, $p = 3$ ... *Maxwell distribution*;
(IV) $\alpha = p$ .... *Weibull distribution*;
(V) $\alpha = p = 1$ ...... *exponential distribution*;
(VI) $\alpha = p = 2$ ...... *Rayleigh distribution*.

Table 2

Relations between probability densities, distribution functions and moments of W and other statistics (U, V, and Z)

Statistic	Density	Distribution function	k-th moment
U	$\frac{1}{2} f_W(\lvert t\rvert)$, if $0 < \lvert t\rvert < 1$ $f_W(0)$, if $t = 0$ 0 otherwise	$\frac{1}{2}[1 - F_W(-t)]$, if $t < 0$ $\frac{1}{2}[1 + F_W(t)]$, if $t \geqq 0$	$\frac{1}{2}[1 + (-1)^k]\, E(W^k) = \begin{cases} E(W^k) \text{ for even } k \\ 0 \text{ for odd } k \end{cases}$
V	$\frac{1}{\sqrt{2}} f_W\left(\frac{t}{\sqrt{2}}\right)$, if $0 < t < \sqrt{2}$ $f_W(0)$, if $t = 0$ 0 otherwise	$F_W(t/\sqrt{2})$	$2^{-k/2}\, E(W^k)$
Z	$\frac{1}{2} f_W(t/2)$, if $0 < t < 2$ $f_W(0)$, if $t = 0$ 0 otherwise	$F_W(t/2)$	$2^{-k}\, E(W^k)$

The following scheme shows which distributions are particular cases of another ones (names of distributions in such pairs are connected with a bar; particular case is on left, more general case on the right hand side of such pair):

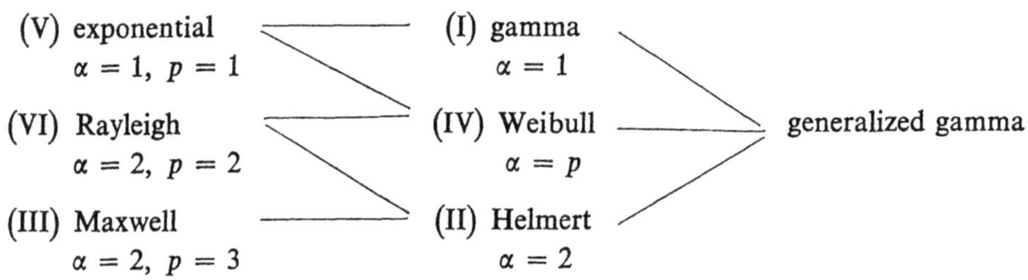

When studying properties of statistics in these particular cases, it is possible to use the general form of the distribution of U given by Theorem; however, it may be useful to know their particular form. Table 3 gives formulae of probability densities, distribution functions and moments of the statistic W for particular cases mentioned above (they are different, in some cases, from formulae obtainable immediately from the Theorem).

NOTES TO THE TABLE 3

(I) *Gamma distribution.* The form of the density given in the table follows from the Theorem and from the fact that $2^{2x-1}\, B(x, x) = B(1/2, x)$ (see, e.g., ([3], formula 8.384.4). When integrating this density (with the substitution $z = t^2$) we obtain distribution function in the form given in the table. — Moments μ_k may be written in another form as

$$\frac{1}{F(k/2,\, p;\; p + (k+1)/2;\; 1)},$$

where F denotes the hypergeometrical function (see [3], formula 9.122.1).

(II) *Helmert distribution.* Formulae given in table follow directly from the Theorem; formula for moments is valid for integer p only.

(III) *Maxwell distribution.* It is the particular case of the Helmert distribution; the form of the distribution function given in the table is obtained by the integration of the density ([3], formulae 2.141, 2.147, 2.148).

(IV) *Weibull distribution.* Form of the density follows from the Theorem, after integration we get the tabled form of the distribution function. Moments (in general case) are to be solved by numerical integration.

(V) *Exponential distribution.* It follows immediately from the Theorem, that U (as well as W) is distributed rectangularly.

(VI) *Rayleigh distribution.* Formulae follow immediately from the Theorem.

TABLE 3

Probability density, distribution function and moments of the statistic W in particular cases I— VI

Case	Density $f_W(t)$, $t \in (0, 1)$ *)	Distribution function $F_W(t)$, $t \in \langle 0, 1)$	Moments $E(W^k)$
I gamma	$\dfrac{2(1-t^2)^{p-1}}{B(\frac{1}{2}, p)}$	$I_{t^2}(\frac{1}{2}, p)$	$\dfrac{B\left(\frac{k+1}{2}, p\right)}{B(\frac{1}{2}, p)}$
II Helmert	$\dfrac{4(1-t^2)^{p-1}}{B(\frac{1}{2}, p/2)\,(1+t^2)^p}$	$1 - 2I_{(1+t)^2/2(1+t^2)}(p/2, p/2)$	$\dfrac{4}{B(p/2, p/2)} \sum_{j=0}^{p-1} (-1)^j \binom{p-1}{j} M_{p,2j+k}$ **)
III Maxwell	$\dfrac{8}{\pi} \dfrac{(1-t^2)^2}{(1+t^2)^3}$	$\dfrac{4}{\pi} \left\{ \dfrac{t(1-t^2)}{(1+t^2)^2} + \operatorname{arctg} t \right\}$	$\dfrac{8}{\pi} \{M_{3,k} - 2\,M_{3,k+2} + M_{3,k+4}\}$
IV Weibull	$\dfrac{4p(1-t^2)^{p-1}}{[(1+t)^p + (1-t)^p]^2}$	$\dfrac{(1+t)^p - (1-t)^p}{(1+t)^p + (1-t)^p}$	$4p \displaystyle\int_0^1 \dfrac{t^k(1-t^2)^{p-1}}{[(1+t)^p + (1-t)^p]^2}\,\mathrm{d}t$
V exponential	1	t	$(k+1)^{-1}$
VI Rayleigh	$2\,\dfrac{1-t^2}{(1+t^2)^2}$	$\dfrac{2t}{1+t^2}$	$2(M_{2,k} - M_{2,k+2})$

* $f_W(0) = \frac{1}{2}$ (appropriate term with $t = 0$)
** for integer p

REFERENCES

[1] J. Agard: Mélange de deux populations normales et étude de quelques fonctions $f(x, y)$ de variables normales x, y. Rev. Statist. appliquée *9* (1961), 4, 53—70.

[2] L. J. Bain and D. L. Weeks: Tolerance limits for the generalized gamma distribution. J. amer. statist. Assoc. *60* (1965), 1142—1152.

[3] И. С. Градштейн и И. М. Рыжик: Таблицы интегралов, сумм, рядов и произведений, 4-е изд. Гос. изд. физико-матем. лит., Москва 1962.

[4] H. W. Hager, L. J. Bain and C. E. Antle: Reliability estimation for the generalized gamma distribution and robustness of the Weibull model. Technometrics *13* (1971), 547—557.

[5] F. A. Haight: Index to the distributions of mathematical statistics. J. Res. nat. Bur. Stand. *65 B* (1961), 23—60.

[6] H. Jakuszenkow: On properties of the generalized gamma distribution. Demonstratio mathematica *7* (1974), 1, 13—22.

[7] A. M. Mathai: Products and ratios of generalized gamma variates. Skandinavisk Aktuarietidskrift (1972), 193—198.

[8] T. Śródka: On the distribution of product and ratio of powers of two independent random variables with the generalized gamma, Maxwell and Weibull distribution. Prace matematyczne *11* (1967), 77—85.

[9] J. F. Steffensen: On the seminormal distribution. Skandinavisk Aktuarietidskrift *20* (1937), 60—86.

Czechoslovak Academy of Sciences

Institute of Microbiology

ON THE BOUNDEDNESS OF STABLE PROCESSES

MAREK KANTER

MONTREAL

1. INTRODUCTION

By a stable process we mean a process $X = (X(t) \mid t \in T)$ such that all linear combinations of the random variables $X(t)$ have a stable distrition as defined in [5, p. 166].

The main theorem of this paper is that for any bounded stable process $X = (X(t) \mid t \in T)$ of index α, $E(\|X\|^\beta) < \infty$ for all $0 < \beta < \alpha$ where $\|X\|$ denotes the random variable $\sup_{t \in T} |X(t)|$. This theorem is best possible except in the case $\alpha = 2$, the case of Gaussian processes, when in fact $\|X\|$ has finite moment generating function as shown in [6], [11], and [16].

The theme of this paper is that the study of stable processes (and Gaussian processes in particular) can be facilitated by an axiomatic approach which depends only on the properties of stability and is therefore "dimension free". Accordingly, any theorem about stable random variables is not satisfactory unless it can be generalized directly to stable processes. It turns out that Feller's proof ([5], p. 215) of the fact any stable random variable of index α has moments of all order less than α can in fact be directly generalized as above. However since this was not apparent to us for several years we feel it is worthwhile to write down explicitly how to generalize Feller's proof. Furthermore our methods have some interesting consequences for stationary stable processes. For instance, we give an astonishingly simple proof of Belayev's theorem that any stationary Gaussian process whose spectral measure has a continuous component has unbounded sample functions, and generalize this theorem to stationary stable processes.

Previous work most closely related to this paper is in [4] and [7]. In [7] a similar axiomatic point of view was taken with respect to the study of stochastic integrals on stable processes. In [4] we proved the following theorem: if X is a stable process and S is a measurable linear subspace of R^T then $P(S) = 0$ or 1. (The theorems in [4] are more general and apply to competition measurable subspaces S, but the above is all we need.) This theorem is applied here to show that either $\|X\| < \infty$ with prob. 1 or $\|X\| = \infty$ with prob. 1.

An intriguing avenue of research we do not explore in this paper is to characterize those stable processes X for which $\|X\| < \infty$ a.s. By symmetrization we can assume X is symmetric stable, say of index α. In that case there exists a "spectral representation" i.e. a mapping $t \to f_t$ from T into $L^\alpha[0, 1]$ such that for all $t_1, \ldots, t_n$ the random variables $X(t_1), \ldots, X(t_n)$ have the same distribution as the stochastic integrals

$$\int_0^1 f_{t_1}(s)\,\mathrm{d}Y(s), \ldots, \int_0^1 f_{t_n}(s)\,\mathrm{d}Y(s).$$

(Here $Y(s)$ stands for a symmetric stable process of index α with homogeneous increments. For more on this see [8] or [9].) It follows that our problem is to characterize those subsets of $L^\alpha[0, 1]$ which are the image under the above map of the index set T of an almost surely bounded process $X = (X(t) \mid t \in T)$. Surely such a characterization can be found in terms of "metric entropy" or "ε-entropy" as has already been done for Gaussian processes [3, p. 68].

We end this introduction with some definitions and notation. T always stands for the index set of some stochastic process $X = (X(t) \mid t \in T)$ defined on some given probability space $(\Omega, \mathscr{B}, P)$. S stands for some linear subspace of R^T which is assigned outer measure 1 by the process X. $\mathscr{A}$ will always stand for the σ-field subsets of S generated by coordinate functionals. We shall describe the above situation simply by saying that X is a random vector on $(S, \mathscr{A})$. It is easy to check that if X and Y are stochastic processes with jointly defined finite dimensional distributions then, if X and Y are both random vectors on $(S, \mathscr{A})$, one can conclude that $aX + bY$ is also a random vector on $(S, \mathscr{A})$ for any $a, b \in R$. (All one needs to check is that the map $(x, y) \to ax + by$ from $(S \times S, \mathscr{A} \times \mathscr{A})$ into $(S, \mathscr{A})$ is measurable.)

DEFINITION 1.1. A random vector X on $(S, \mathscr{A})$ is stable of index $\alpha \in (0, 2]$ if for every n there exist $x_n \in R^T$ such that

$$X \sim \sum_1^n X_i/n^{1/\alpha} - x_n \tag{1.1}$$

where $X_1, \ldots, X_n$ are independent random vectors on $(S, \mathscr{A})$ all of which induce the same probability measure on $(S, \mathscr{A})$ as X does. ($X \sim Y$ means the random vectors X and Y induce the same probability measure on $(S, \mathscr{A})$.) X is strictly stable of index α if x_n can be set equal to 0. X is symmetric stable of index α if it satisfies (1.1) and is symmetric. If X is symmetric stable then it is clearly strictly stable.

Throughout the rest of this paper we shall assume the existence of a norm $\|\cdot\|$ on S which satisfies the following assumptions.

(A) For all $s > 0$, $\{x \mid \|x\| > s\} \in \mathscr{A}$.

(B) For every $\varepsilon > 0$, there exists an $\mathscr{A} \times \mathscr{A}$ measurable real valued function $h_x(z)$ defined for (x, z) in $S \times S$, and called an ε-approximate support for $\|\cdot\|$, such that

$h_x(z)$ is linear in z for fixed x and $\|x + z\| + \varepsilon \geqq \|x\| + h_x(z)$ for all (x, z) in $S \times S$. When $\varepsilon = 0$ a function $h_x(z)$ satisfying (1) and (2) is called a support function for the norm $\|\cdot\|$ at the point x. (See [15], p. 108 for a general proof of the existence of the support for any norm $\|\cdot\|$. The proof there, making use of the axiom of choice, does not prove measurability of the support in (x, z).)

If S is the space of all bounded functions defined on some countable set T with supremum norm $\|\cdot\|$ it is clear that $\|\cdot\|$ is $\mathscr{A}$-measurable. (However not all open sets need be $\mathscr{A}$-measurable since $(S, \|\cdot\|)$ is not separable.) If $\varepsilon > 0$ we define $h_x(z)$ as follows. Let $\{t_1, \ldots, t_n, \ldots\}$ be an enumeration of T. For $x \neq 0$, let j_0 be the least j with $x_{t_j} \neq 0$ and $|x_{t_j}| \geqq \|x\| - \varepsilon$. For $z = (z(t_n) \mid n \geqq 1)$ we define $h_x(z) = (x_{j_0}/|x_{j_0}|)\, z_{j_0}$. Clearly $h_x(z)$ is $\mathscr{A} \times \mathscr{A}$ measurable, and $\|x\| + h_x(z) - \varepsilon \leqq |x_{t_{j_0}}| + (x_{j_0}/|x_{j_0}|)\, z_{j_0} = (x_{j_0}/|x_{j_0}|)(x_{j_0} + z_{j_0}) \leqq \|x + z\|$. For $x = 0$, define $h_x(z) = 0$.

2. THE MAIN THEOREM

LEMMA 2.1. Let (Z, X) be a random vector on $(S \times S, \mathscr{A} \times \mathscr{A})$ such that the distribution of (Z, X) and $(-Z, X)$ are the same. Then

$$P[\|Z + X\| \geqq \|X\|] \geqq \tfrac{1}{2}.$$

Proof. Note first that the map $(z, x) \to \|z + x\|$ is $\mathscr{A} \times \mathscr{A}$ measurable, being the composition of measurable functions. Now let $\varepsilon > 0$. We shall prove that $P[\|Z + X\| \geqq \|X\| - \varepsilon] \geqq \frac{1}{2}$; letting $\varepsilon \to 0$ we get the required conclusion.

By assumptions A and B of the introduction we know there exists an ε-approximate support $h_x(z)$ for the norm $\|\cdot\|$. It follows that $P[\|Z + X\| \geqq \|X\| - \varepsilon] \geqq P[h_X(Z) \geqq 0]$. Now $h_X(Z) + h_X(-Z) = 0$ while $h_X(Z)$ is distributed like $h_X(-Z)$ by the hypotheses of the theorem. We conclude that $h_X(Z)$ has a symmetric distribution so $P[h_X(Z) \geqq 0] \geqq \frac{1}{2}$. Q.E.D.

LEMMA 2.2. Let $X_1, \ldots, X_n$ be symmetric independent random vectors in $(S, \mathscr{A})$. Then for $t > 0$

$$P[\|\sum_1^n X_i\| \geqq t] \geqq \tfrac{1}{2}(1 - \prod_1^n P[\|X_i\| \leqq t]). \tag{2.1}$$

Proof. Let τ be the least j for which $\|X_j\| = \max_{1 \leqq i \leqq n} \|X_i\|$ and let $Z = \sum_1^n X_i - X_\tau$. We see that the distribution of (Z, X_τ) and $(-Z, X_\tau)$ are the same (because the distribution of $(X_1, \ldots, X_{\tau-1}, X_\tau, X_{\tau+1}, \ldots, X_n)$ is the same as the distribution of $(X_1, \ldots, X_{\tau-1}, -X_\tau, X_{\tau+1}, \ldots, X_n)$.) This remains true if we condition by the event $[\|X_\tau\| > t]$. We now compute

$$P[\|\sum_1^n X_i\| \geqq t] = P[\|X_\tau + Z\| \geqq t] \geqq$$

$$\geqq P[\|X_\tau\| > t]\, P[\|Z + X_\tau\| \geqq \|X_\tau\| \mid \|X_\tau\| > t] \geqq$$

$$\geqq P[\|X_\tau\| > t]\,(\tfrac{1}{2}) = \tfrac{1}{2}(1 - \prod_1^n P[\|X_i\| \leqq t])$$

where the third line follows from Lemma 2.1. Q.E.D.

THEOREM 2.1. Let X be a symmetric stable random vector on $(S, \mathscr{A})$ of index $\alpha \in (0, 2]$. Then for any $\beta \in (0, \alpha)$ we have $E(\|X\|^\beta) < \infty$.

Proof. Let $X_1, \ldots, X_n$ in Lemma 2 be independent and distributed like X. Then, letting $X_1' = x_i/n^{1/\alpha}$, we have

$$P[\|\sum_1^n X_i'\| \geqq t] \geqq 1/2(1 - \prod_1^n P[\|X_i'\| \leqq t])\,. \tag{2.2}$$

We now use the definition of stability as well as the elementary inequality $\prod_1^n(1 - y_i) \leqq \leqq e^{-\Sigma y_i}$ valid for $y_i \in [0, 1]$ to rewrite (2.2) as

$$P[\|X\| \geqq t] \geqq 1/2(1 - e^{-nP\|X\| \geqq n^{1/\alpha}t})\,. \tag{2.3}$$

If we set $t = n^\varepsilon$ with $\varepsilon > 0$ we conclude that $n\, P[\|X\| \geqq n^{\varepsilon + 1/\alpha}]$ is bounded since $\lim_{t \to \infty} P[\|X\| \geqq t] = 0$ by hypothesis. This easily implies that $E(\|X\|^\beta) < \infty$. (See [10], p. 407 for the concluding steps of this proof.) Q.E.D.

REMARK. To prove Theorem 2.1 we could have used an inequality of the form derived in LeCam [12], instead of (2.1) above. However the inequalities in [12] seem to be not as elementary in nature as (2.1) being based on a multidimensional extension of Kolmogorov's concentration function inequality.

COROLLARY 2.1. Let X be a stable random vector on $(S, \mathscr{A})$ of index $\alpha \in (0, 2]$. Then for any $\beta \in (0, \alpha)$ we have $E(\|X\|^\beta) < \infty$.

Proof. If X' is independent of X and distributed like X then we have $E(\|X - X'\|^\beta) < \infty$ by Theorem 1.1. Applying Fubini's theorem we conclude the existence of $x' \in S$ for which $E\|X - x'\|^\beta < \infty$. We then note that

$$\|X\|^\beta \leqq (\|X - x'\| + \|x'\|)^\beta \leqq K_\beta(\|X - x'\|^\beta + \|x'\|^\beta)$$

and we are done. ($K_\beta = \max(1, 2^{\beta-1})$.) Q.E.D.

Corollary 2.2. Let $(X(t) \mid t \in T')$ be a stable process of index $\alpha \in (0, 2]$, and assume the process is separable. If $\sup_{t \in T'} |X(t)| < \infty$ a.s. then for any $\beta \in (0, \alpha)$ we have $E((\sup_{t \in T'} |X(t)|)^\beta) < \infty$.

Proof. By separability (see [13], p. 505) we can choose a sequence of points $t_n \in T'$ such that $\sup_{t \in T'} |X(t)| = \sup_n |X(t_n)|$ a.s. Let S be the linear space of bounded real valued functions defined on $T = \{t_n \mid n \geqq 1\}$ equipped with the supremum norm. Then Corollary 2.1 applies to $X = (X(t_n) \mid n \geqq 1)$ and $(S, \mathscr{A})$. Q.E.D.

3. APPLICATIONS TO STATIONARY PROCESSES

In this section we shall generalize a result of Belayev [1] on stationary Gaussian processes. Let $(Y_n \mid n = 0, \pm 1, \pm 2, \ldots)$ be an integer indexed sequence of independent, identically distributed symmetric stable random variables of index $\alpha \in (0, 2]$. If $(c_n \mid n = 0, \pm 1, \pm 2, \ldots)$ is an integer indexed sequence of real numbers with $\sum |c_n|^\alpha < \infty$ then we know by [7] that the moving average process

$$X_n = \sum_{k=-\infty}^{+\infty} c_{n-k} Y_k \tag{3.1}$$

is defined a.s. and is stationary. By [2, p. 158] the process X_n is in fact ergodic. (The argument there works for any moving average process such as (3.1) though it is given only for stationary Gaussian processes $X(t)$ with $\lim_{t \to \infty} E(X(t)\, X(0)) = 0$.)

Theorem 3.1. Let $X_n \neq k$ a.s. be a stationary ergodic process which is stable of index α. Then $P[\sup_n |X_n| = \infty] = 1$.

Proof. Assume that $P[\sup_n |X_n| = \infty] < 1$. Then by [4] we know that $\sup_n |X_n| < \infty$ a.s., hence $X = (X_n \mid n = 0, \pm 1, \ldots)$ is a stable random vector of index α on $(S, \mathscr{A})$ where S is the linear space of bounded integer indexed sequences of real numbers with sup norm $\| \ \|$. We now consider the stationary process $Z_n = \sup_{k \leqq n} |X_k|$. Since Z_n is non-decreasing and stationary we conclude the existence of a random variable Z such that $Z = Z_n$ a.s. for all n. Since $Z = \lim_{n \to -\infty} \sup_{k \leqq n} |X_k|$ we conclude that Z is measurable w.r. to the shift invariant sub σ-field generated by X_n hence Z is constant since X_n is ergodic.

We now get a contradiction: Since $X_n \neq k$ a.s. we know by [5], p. 280 that $E|X_n| = \infty$. Howerer this contradicts $|X_n| \leqq Z$. Q. E. D.

The above theorem holds without any change in proof if the stationary process X is indexed by a continuous time parameter t instead of the discrete time parameter n.

COROLLARY 3.1. Suppose $X = (X_n \mid n = 0, \pm 1, \pm 2, \ldots)$ is a stationary Gaussian process such that $E(X_n X_m) = \int_0^2 e^{is(n-m)}\, dF(s)$ where the spectral measure dF is not atomic. Then $P[\sup_n |X_n| = \infty] = 1$.

Proof. The hypotheses of the theorem show that we can write $X = Y + Z$ where Y and Z are independent stationary Gaussian processes such that Y has continuous spectral measure (i.e. the continuous part of dF) while Z has atomic spectral measure (i.e. the atomic part of dF). By [2] we know Y is ergodic so using Theorem 3.1 we conclude $P[\sup_n |Y_n| = \infty] = 1$.

Suppose now that $p[\sup |Y_n + Z_n| = \infty] < 1$. By [4] we conclude then that $P[\sup_n |Y_n + Z_n| = \infty] = 0$. Then by Theorem 2.1 we conclude that $E(\|Y + Z\|) <$ $< \infty$, hence by Fubini's theorem we have $E(\|Y\|) < \infty$. Of course this is a contradiction since $\|Y\| = \infty$ a.s. Q.E.D.

Belayev's theorem [1] is stated in terms of stationary Gaussian processes with continuous time parameter. Again there is no trouble in adapting the above proof to that case, first taking a separable modification of the process as in [13], p. 505, then proceeding as in Corollary 2.2.

We end by stating a generalization of Corollary 3.1 for stable process.

COROLLARY 3.2. Suppose $X_n = Y_n + Z_n$; $n = 0, \pm 1, \ldots$ where Y_n and Z_n are independent stationary stable process of index α such that Y_n is ergodic, $Y_n \neq k$ a.s. Then $P[\sup_n |X_n| = \infty] = 1$ a.s.

Proof. Exactly as for Corollary 3.1. Q.E.D.

REFERENCES

[1] Yu. K. Belayev: On the unboundedness of the sample functions of Gaussian processes. Theor. Probability Appl. *3* (1958), 327—329.

[2] H. Cramer and M. R. Leadbetter: Stationary and Related Stochastic Processes. Wiley, New York 1967.

[3] R. M. Dudley: Sample functions of the Gaussian process. Annals of Probability *1* (1973), 66—103.

[4] R. M. Dudley and M. Kanter: Zero-one laws for stable measures. Proceedings Amer. Math. Soc. *45* (1944), 1—8.

[5] W. Feller: Introduction to Probability Theory and its Applications 2. Wiley, New York 1966.

[6] X. Fernique: Integrabilite des vecteurs gaussiens. C. R. Acad. Sci. Paris *258*, (1970), 6058—6060.

[7] M. Kanter: Linear sample spaces and stable processes. Journal of Functional Analysis *9* (1972), 441—459.

[8] M. Kanter: A representation theorem for L^P spaces. Proceedings Amer. Math. Soc. *31* (1972), 472—474.

[9] M. Kanter: On the spectral representation for symmetric stable random variables. Z. Wahrscheinlichkeitstheorie and Verw. Gebiete *23* (1972), 1—6.

[10] M. Kanter: Stable laws and the imbedding of L^P spaces. Amer. Math. Monthly *80* (1973), 403—407.

[11] H. J. Landau and L. A. Shepp: On the supremum of a Gaussian process. Sankhya Ser A *32* (1971), 369—378.

[12] Lucien, Le Cam: An inequality on the concentration of sums of random variables. Ann. Math. Stat. (1962), 826.

[13] M. Loeve: Probability Theory. 3rd ed. Van Nostrand, Princeton 1963.

[14] G. Maruyama: The harmonic analysis of stationary stochastic processes. Mem. Fac. Sci. Kyusyu Univ. *A 4*, (1949), 45—106.

[15] A. W. Roberts and D. E. Varberg: Convex Functions. Academic Press, New York (1973), p. 108.

[16] A. V. Skorokhod: A note on Gaussian measures in a Banach space. Theory of Probability and its Applications *15*, p. 508.

Sir George Williams University

Department of Mathematics

ON THE ESTIMATION OF THE CHANGE POINT IN THE AUTOREGRESSIVE SEQUENCE

NERUTĖ KLIGIENĖ

VILNIUS

1. INTRODUCTION

This paper deals with the change detection problem in an observed time series. We come across such problems in the recognition of a signal in the noise and technical or medical diagnosis. There are many techniques for detecting the change point, most of which assume [1], [2] the given observation to be statistically independent random variables. Another widely used assumption is that the process observed before a change point should be independent of the process after the change. For example, if these processes are normal with known autocovariance functions, the problem has been solved in [3]. In this paper we shall do without the restrictive assumptions mentioned above because we deal with the estimation of a change in parameters of a stochastic difference equation, namely, an autoregression model.

2. STATEMENT OF THE PROBLEM

Let $\{X_t^{(\nu)},\ t = 0, \pm 1, \ldots\}$, $\nu = 1, 2$, be a stationary autoregressive sequence of random variables $X_t^{(\nu)}$, $\mathscr{E}X_t^{(\nu)} = 0$, defined by the stochastic difference equation of order p_ν:

(1)
$$X_t^{(\nu)} + a_1^{(\nu)} X_{t-1}^{(\nu)} + \ldots + a_{p_\nu}^{(\nu)} X_{t-p_\nu}^{(\nu)} = b_\nu V_t\,,$$

where V_t's are independently distributed, each according to $\mathcal{N}(0, 1)$ and the roots $z_j^{(\nu)}$, $j = 1, 2, \ldots, p_\nu$ of the polynomical equation $\sum_{i=0}^{p_\nu} a_i^{(\nu)} z^{p_\nu - i} = 0$, $a_0^{(\nu)} \equiv 1$, associated with (1) are less than 1 in absolute value: $|z_j^{(\nu)}| < 1$. The joint distribution of any set $X_1^{(\nu)}, X_2^{(\nu)}, \ldots, X_N^{(\nu)}$, $N > p_\nu$, is normal and determined by its $p_\nu \times p_\nu$ covariance matrix $\boldsymbol{K}^{(\nu)} = \{k_\tau^{(\nu)} = \mathscr{E}X_t^{(\nu)}X_{t+\tau}^{(\nu)},\ \tau = 0, 1, \ldots, p_\nu - 1\}$. Assume either $\boldsymbol{K}^{(\nu)}$ or the parameters $\boldsymbol{a}^{(\nu)} = (a_1^{(\nu)}, a_2^{(\nu)}, \ldots, a_{p_\nu}^{(\nu)})$, b_ν to be known.

Let us consider the non-stationary sequence $\{X_t,\ t = 0, \pm 1, \ldots\}$, $\mathscr{E}X_t = 0$, defined by

$$\text{(2)} \qquad X_t + a_1^{(1)}X_{t-1} + \ldots + a_{p_1}^{(1)}X_{t-p_1} = b_1 V_t\,, \quad t = \ldots, u_1 - 1, u_1\,;$$

$$\text{(3)} \qquad X_t + a_1^{(2)}X_{t-1} + \ldots + a_{p_2}^{(2)}X_{t-p_2} = b_2 V_t\,, \quad t = u_1 + 1, u_1 + 2, \ldots,$$

where a sudden change of parameters occurs at an unknown point u_1 that is independent of V_t's. The maximum likelihood (MLH) estimate $\hat{u}_1$ of the change point u_1 will be calculated from the given set $x_1, x_2, \ldots, x_N$, $N > p_\nu$, of the observations on the sequence $\{X_t\}$ defined by (2), (3). The main purpose is to investigate the accuracy of the available estimate $\hat{u}_1$ when N is large.

Let $f_X(x_1, x_2, \ldots, x_N \mid u_1 = u)$ be a conditional probability density function of the random vector $X = (X_1, X_2, \ldots, X_N)$ given the parameters $\boldsymbol{a}^{(1)}$, b_1 have changed to $\boldsymbol{a}^{(2)}$, b_2 at the point $u_1 = u$ where u is any value from the set $(p_1 + 1, p_1 + 2, \ldots \ldots, N - 1)$. According to equations (2) and (3) $f_X(x_1, x_2, \ldots, x_N \mid u_1 = u)$ is normal and logarithm of a likelihood (LLH) function $L_u = \ln f_{\boldsymbol{X}}(x_1, x_2, \ldots, x_N \mid u_1 = u)$ may be written as follows

$$\text{(4)} \quad L_u = C_1 + u \ln\left|\frac{b_2}{b_1}\right| - \tfrac{1}{2}\Big[\sum_{t=p_1+1}^{u} (x_t + a_1^{(1)}x_{t-1} + \ldots + a_{p_1}^{(1)}x_{t-p_1})^2/b_1^2 +$$
$$+ \sum_{t=u+1}^{N} (x_t + a_1^{(2)}x_{t-1} + \ldots + a_{p_2}^{(2)}x_{t-p_2})^2/b_2^2\Big] = C_2 + u \ln\left|\frac{b_2}{b_1}\right| -$$
$$- \tfrac{1}{2} \sum_{t=p_1+1}^{u} \Big[(x_t + a_1^{(1)}x_{t-1} + \ldots + a_{p_1}^{(1)}x_{t-p_1})^2/b_1^2 -$$
$$- (x_t + a_1^{(2)}x_{t-1} + \ldots + a_{p_2}^{(2)}x_{t-p_2})^2/b_2^2\Big]\,,$$

where constants $C_1 C_2$ indicate a part of L_u which is independent of an unknown parameter u.

In order to get MLH estimate $\hat{u}_1$ of the change point u_1 we must maximize the expression (4) over the set of admissible values u:

$$\text{(5)} \qquad L_{\hat{u}_1} = \max_{u = p_1+1, \ldots, N-1} L_u\,.$$

The absence of an explicit expression of MLH estimate $\hat{u}_1$ does not permit us to investigate its asymptotic properties directly and one has to consider some related statistic. We shall study a statistic of the LLH function to obtain an asymptotic distribution of its maximal value which will enable us to test hypotheses about the true value of the change point.

3. INVESTIGATION OF THE LLH STATISTIC

Let $L_u(X)$ be a statistic defined by the expression (4) where the observed values $x_1, \ldots, x_N$ are replaced by the random vector X. We can represent $L_u(X)$ as the sum

$$L_u(X) = C_2 + \sum_{t=p_1+1}^{u} Q_t,$$

where

$$Q_t = \ln \left|\frac{b_2}{b_1}\right| - \tfrac{1}{2}[(X_t + a_1^{(1)}X_{t-1} + \ldots + a_{p_1}^{(1)}X_{t-p_1})^2/b_1^2 - - (X_t + a_1^{(2)}X_{t-1} + \ldots + a_{p_2}^{(2)}X_{t-p_2})^2/b_2^2]. \tag{6}$$

Lemma 1. For any parameters $\boldsymbol{a}^{(\nu)}$, b_ν, $\nu = 1, 2$, of the autoregressive sequences $\{X_t^{(\nu)}, t = 0, \pm 1, \ldots\}$ (1), the inequality

$$\sum_{i,j=0}^{p_\nu} a_i^{(\mu)} a_j^{(\mu)} k_{i-j}^{(\nu)} > b_\nu^2, \quad \nu, \mu = 1, 2; \quad \nu \neq \mu, \tag{7}$$

holds.

Proof. It is known [4] that the linear predictor $\hat{x}_t^{(\nu)} = -\sum_{i=1}^{p_\nu} a_i^{(\nu)} x_{t-i}^{(\nu)}$ of $X_t^{(\nu)}$ (1) based on given values $x_{t-1}^{(\nu)}, \ldots, x_{t-p_\nu}^{(\nu)}$ has minimum mean square error

$$\mathscr{E}(X_t^{(\nu)} - \hat{x}_t^{(\nu)})^2 = \mathscr{E} b_\nu^2 V_t^2 = b_\nu^2.$$

On the other hand it is true

$$\mathscr{E} b_\nu^2 V_t^2 = \mathscr{E}(X_t^{(\nu)} + a_1^{(\nu)} X_{t-1}^{(\nu)} + \ldots + a_{p_\nu}^{(\nu)} X_{t-p_\nu}^{(\nu)})^2 = \sum_{i,j=0}^{p_\nu} a_i^{(\nu)} a_j^{(\nu)} k_{i-j}^{(\nu)}. \tag{8}$$

If we use any other value $\boldsymbol{a}^{(\mu)}$, $\mu \neq \nu$, instead of $\boldsymbol{a}^{(\nu)}$ in the predictor $\hat{x}_t^{(\nu)}$, the mean square error $\sum_{i,j=0}^{p_\mu} a_i^{(\mu)} a_j^{(\mu)} k_{i-j}^{(\nu)}$ would exceed b_ν^2. Note that

$$\sigma_{\mu\nu} > \left(\frac{b_\nu}{b_\mu}\right)^2, \quad \text{where} \quad \sigma_{\mu\nu} = b_\mu^{-2} \sum_{i,j=0}^{p_\mu} a_i^{(\mu)} a_j^{(\mu)} k_{i-j}^{(\nu)} \tag{9}$$

is another form of (7) that will be also used there. The lemma is proved.

Consider the random variables Q_t, $t = p_1 + 1, \ldots, N - 1$. The expressions (2), (3), (6) lead us to derive the conditional expectations m_1 and m_{2u} of the random variables Q_u given $u \leqq u_1$ or $u > u_1$:

(10)
$$m_1 = \mathscr{E}\{Q_u \mid u \leqq u_1\} = \ln\left|\frac{b_2}{b_1}\right| - \tfrac{1}{2}\left(1 - b_2^{-2} \sum_{i,j=0}^{p_2} a_i^{(2)} a_j^{(2)} k_{i-j}^{(1)}\right) = q - \tfrac{1}{2}(1 - \sigma_{21}),$$

(11)
$$m_{2u} = \mathscr{E}\{Q_u \mid u > u_1\} = q + \tfrac{1}{2}\left(1 - b_1^{-2} \sum_{i,j=0}^{p_1} a_i^{(1)} a_j^{(1)} k_{u-i,u-j}\right),$$

where $q = \ln|b_2/b_1|$ and $k_{s,t} = \mathscr{E}\{X_s X_t \mid s, t > u_1\}$ is the covariance function of non-stationary sequence after the change. It is not difficult [5] to justify the fact

(12)
$$\lim_{u \to \infty} k_{u-i,u-j} = k_{i-j}^{(2)}.$$

Then

(13)
$$m_2 = \lim_{u \to \infty} m_{2u} = q + \tfrac{1}{2}(1 - \sigma_{12}).$$

Lemma 2. The conditional expectations (10), (13) of Q_u's have the opposite signs: $m_1 > 0$, $m_2 < 0$.

The proof is omited. It is based on inequality (9) and power series expansions of functions included in (10), (13).

Setting to derive a conditional density of the random variable Q_u given either $u \leqq u_1$ or $u > u_1$ it is necessary to prove the following lemma.

Lemma 3. If a joint distribution of random variables (ξ_1, ξ_2) is nonsingular normal with $\mathscr{E}\xi_\nu = 0$, $\nu = 1, 2$, and the covariance matrix

(14)
$$\boldsymbol{B} = \begin{pmatrix} B_{11} & B_{12} \\ B_{21} & B_{22} \end{pmatrix}, \quad B_{\nu\mu} = \mathscr{E}\xi_\nu \xi_\mu, \quad \nu, \mu = 1, 2,$$

then the probability density function $f_\chi(x)$ of a random variable $\chi = \xi_1^2 - \xi_2^2$ is

(15)
$$f_\chi(x) = \frac{\exp\left(-\dfrac{D}{4|\boldsymbol{B}|}x\right)}{2\pi|\boldsymbol{B}|^{1/2}} K_0\left[\frac{(D^2 + 4|\boldsymbol{B}|)^{1/2}}{4|\boldsymbol{B}|}|x|\right],$$

where $|\boldsymbol{B}|$ is a determinant of the matrix $\boldsymbol{B}$ (14), $-D = \mathscr{E}\chi = B_{11} - B_{22}$ and $K_0(x)$ is the zero order Bessel function of the second kind.

Proof. Note that $\chi = (\xi_1 - \xi_2)(\xi_1 + \xi_2) = \varphi_1 \, . \, \varphi_2$ and (φ_1, φ_2) are also jointly normaly distributed with the density function

$$(16) \qquad f_\varphi(u, v) = \frac{1}{4\pi|B|} \exp\left[-\frac{1}{8|B|}(A_1u^2 + 2Duv + A_2v^2)\right],$$

where $A_1 = B_{11} + 2B_{12} + B_{22}$, $A_2 = B_{11} - 2B_{12} + B_{22}$ and D, $|B|$ as defined above.

Therefore,

$$(17) \qquad f_\chi(x) = \int_{-\infty}^{\infty} \frac{1}{|u|} f_\chi\left(u, \frac{x}{u}\right) du =$$

$$= \frac{\exp\left(-\frac{D}{4|B|}x\right)}{4\pi|B|^{1/2}} \int_0^{\infty} \frac{1}{|u|} \exp\left[-\frac{1}{8|B|}(A_1u^2 + A_2x^2u^{-2})\right] du =$$

$$= \frac{\exp\left(-\frac{D}{4|B|}x\right)}{4\pi|B|^{1/2}} \int_0^{\infty} y^{-1} \exp\left[-\frac{1}{8|B|}(A_1y + A_2x^2y^{-1})\right] dy \, .$$

From nonsingularity of the distribution of (ξ_1, ξ_2) we state $|B| > 0$, $A_1 > 0$, $A_2 > 0$ and the last integral in (17) may be represented [6] by means of the expression (15). The proof of lemma 3 is complete.

Corollary 3.1. $\mathscr{E}|\chi|^r < \infty$ for every $r = 1, 2, \ldots$

The proof follows as a result of a direct evaluation of $\mathscr{E}|\chi|^r$ that leads to a sum of two Gaussian hypergeometric series. One has only to prove its convergence.

Now we are able to infer a conditional density function of Q_u's given $u \leqq u_1$ or $u > u_1$. Let's define the linearly independent variables ξ_{1t} and ξ_{2t} by the equation

$$(18) \qquad \xi_{vt} = (X_t + a_1^{(v)}X_{t-1} + \ldots + a_{p_v}^{(v)}X_{t-p_v})/b_v \, , \quad v = 1, 2 \, .$$

For every fixed t, ξ_{1t} and ξ_{2t} are correlated because the model (2), (3) is considered. Evidently, $\mathscr{E}\xi_{vt} = 0$ and a joint distribution of every pair (ξ_{1t}, ξ_{2t}) is normal also. Let

$$(19) \qquad B_{v\mu}(t, \tau) = \mathscr{E}\{\xi_{vt}\xi_{\mu\tau}\} \, , \quad v, \mu = 1, 2 \, .$$

Now we can rewrite the formula (6) in the following way

$$(20) \qquad Q_u = q - \tfrac{1}{2}(\xi_{1u}^2 - \xi_{2u}^2) \, .$$

COROLLARY 3.2. The conditional probability densities $f_{Q_u}(x \mid u \leqq u_1)$ and $f_{Q_u}(x \mid u > u_1)$ of every random variable Q_u, $u = p_1 + 1, \ldots, u_1$ and Q_u, $u = u_1 + 1, u_1 + 2, \ldots, N - 1$ may be expressed by

$$f_{Q_u}(x \mid u \leqq u_1) = 2 f_\chi^{(1)}(2q - 2x)\,, \tag{21}$$

$$f_{Q_u}(x \mid u > u_1) = 2 f_{\chi_u}^{(2)}(2q - 2x)\,, \tag{22}$$

where $f_\chi^{(\nu)}(x)$, $\nu = 1, 2$, has the expression (15) with parameters $|B^{(1)}| = \sigma_{21} - e^{2q}$, $D^{(1)} = \sigma_{21} - 1$ and $|B_u^{(2)}|$, $D_u^{(2)}$ respectively.

If $N \to \infty$, then

$$\lim_{u \to \infty} f_{Q_u}(x \mid u > u_1) = 2 f_\chi^{(2)}(2q - 2x)\,, \tag{23}$$

the limiting density function $f_\chi^{(2)}(x)$ having the parameters $|B^{(2)}|$, $D^{(2)}$ independent of u:

$$|B^{(2)}| = \sigma_{12} - e^{2q}\,, \quad D^{(2)} = 1 - \sigma_{12}\,. \tag{24}$$

Proof. In order to apply the results of lemma 3 one has to calculate the covariance matrices of random variables (ξ_{1u}, ξ_{2u}) in the cases $u \leqq u_1$ and $u > u_1$ separately.

First let $u \leqq u_1$, then $\xi_{1u} = V_u$. From the definition (19) making use of the fact $\mathscr{E} X_s V_t = \begin{cases} b_1, & s = t; \\ 0, & s < t, \end{cases}$ we find

$$B_{11}(u, u) = \mathscr{E}\xi_{1u}^2 = \mathscr{E} V_u^2 = 1\,, \tag{25}$$

$$B_{12}(u, u) = \mathscr{E}\{\xi_{1u}(X_u + a_1^{(2)} X_{u-1} + \ldots + a_{p_2}^{(2)} X_{u-p_2})/b_2\} = \frac{b_1}{b_2}\,, \tag{26}$$

$$B_{22}(u, u) = b_2^{-2}\, \mathscr{E}\{(X_u + a_1^{(2)} X_{u-1} + \ldots + a_{p_2}^{(2)} X_{u-p_2})^2\} = \tag{27}$$

$$= b_2^{-2} \sum_{i,j=0}^{p_2} a_i^{(2)} a_j^{(2)} k_{i-j}^{(1)} = \sigma_{21}\,.$$

The determinant of the covariance matrix $B^{(1)}$ of random variables (ξ_{1u}, ξ_{2u}), $u \leqq u_1$ is

$$|B^{(1)}| = \sigma_{21} - \left(\frac{b_1}{b_2}\right)^2 > 0\,. \tag{28}$$

The inequality in (28) follows from (9) and the statement of lemma 3 is applicable to the random variable $\xi_{1u}^2 - \xi_{2u}^2$. The linear map (20) of this variable leads us to infer the density function (21) of Q_u's, $u \leqq u_1$.

Let $u > u_1$. In the same way that we got matrix $B^{(1)}$ we obtain the covariance matrix $B^{(2)}_{u_1+s}$ of random variables $(\xi_{1,u_1+s}, \xi_{2,u1+s})$ for every $s = 1, 2, \dots$ and its determinant

$$(29) \qquad \left|B^{(2)}_{u_1+s}\right| = b_1^{-2} \sum_{i,j=0}^{p_1} a_i^{(1)} a_j^{(1)} k_{u_1+s-i,u_1+s-j} - \left(\frac{b_2}{b_1}\right)^2 .$$

Using the property (12) we get $\lim_{s\to\infty} \left|B^{(2)}_{u_1+s}\right| = \left|B^{(2)}\right|$ and

$$(30) \qquad \left|B^{(2)}\right| = b_1^{-2} \sum_{i,j=0}^{p_1} a_i^{(1)} a_j^{(1)} k^{(2)}_{i-j} - \left(\frac{b_2}{b_1}\right)^2 = \sigma_{12} - \left(\frac{b_2}{b_1}\right)^2 > 0 .$$

Now it is easy to verify the statements (22)–(24) of corollary 3.2. The proof is complete.

THEOREM 1. Let $\hat{u}_1$ be the MLH estimate of the change point derived from the given set $x_1, \dots, x_N$ of the observation on the autoregressive sequence $\{X_t\}$ defined by equations (2), (3) with known parameters. Let a random variable $L_{\hat{u}_1}$ denote the maximum of the LLH function. Then for sufficiently large N and u_1 holds

$$(31) \qquad P\{L_{\hat{u}_1} < y\} = P\{\hat{u}_1 \leqq u_1\} \{\Phi[(y - n_1 m_1)(\mathcal{D}S_{n_1})^{-1/2}] + o(n_1^{-1/2})\} + + P\{\hat{u}_1 > u_1\} [1 - \delta e^{-\varkappa y}] ,$$

where $\Phi(x)$ is a normal distribution function, $n_1 = u_1 - p_1$, m_1 given by (10), $\mathcal{D}S_n$ is variance of the sum $\sum_{t=1}^{n} Q_t$ of the random variables Q_t (6) and $\delta, \varkappa > 0$ are some constants dependent on the parameters (24).

Proof. Consider the distribution function $P\{L_{\hat{u}_1} < x\}$ for large N and u_1 in the cases $\hat{u}_1 \leqq u_1$ and $\hat{u}_1 > u_1$:

$$(32) \qquad P\{L_{\hat{u}_1} < x\} = P\{\hat{u}_1 \leqq u_1\} P\{L_{\hat{u}_1} < x \mid \hat{u}_1 \leqq u_1\} + + P\{\hat{u}_1 > u_1\} P\{L_{\hat{u}_1} < x \mid \hat{u}_1 > u_1\} .$$

First let us study the case $\hat{u}_1 \leqq u_1$. Denote $n = u - p_1$, $n_1 = u_1 - p_1$ and $S_n = \sum_{t=p_1+1}^{u} Q_t$. Then

$$(33) \qquad L_{\hat{u}_1} = \max_{u=p_1+1,\dots,N-1} L_u = \max_{u=p_1+1,\dots,u_1} L_u = \max_{n=1,2,\dots,n_1} S_n ;$$

$$(34) \qquad P\{L_{\hat{u}_1} < x \mid \hat{u}_1 \leqq u_1\} = P\{\max_n S_n < x \mid n \leqq n_1\} .$$

It is necessary to obtain the asymptotic distribution function (34) where S_n is a sum of identicaly distributed but mutualy dependent random variables Q_t with $\mathscr{E}Q_t > 0$ and known density function (21). One approach to this problem would be to modify the known results [7, theorem 2] when applying them to the case of dependent Q_u's. On the basis of the results of lemma 3, corollaries 3.1, 3.2 it is easy to verify that conditions of the mentioned theorem are fulfiled except the fact that variance $\mathscr{D}S_n$ must be calculated otherwise. Now we are setting to this task.

$$(35) \qquad \mathscr{D}S_n = \mathscr{E}S_n^2 - n_1^2 m_1^2 \,, \quad \mathscr{E}S_n^2 = \sum_{s,t=1}^{n} \mathscr{E}\{Q_s Q_t\} \,.$$

As a result of the calculation $\mathscr{E}\{Q_s Q_t\}$ it is shown that the random variables Q_t, $t = p_1 + 1, \ldots, u_1$, are stationary correlated and $\mathscr{E}S_n^2$ has the expression

$$(36) \qquad \mathscr{E}S_n^2 = E_1 n + E_2 n(n-1) + \sum_{j=1}^{n-1} (n-j) \left[\sigma_{21}^2(j) + \tfrac{1}{2} d_j^2\right],$$

where E_1, E_2 are constants, the quantities $\sigma_{21}(s)$, d_s are defined by

$$(37) \qquad \sigma_{21}(s) = b_2^{-2} \sum_{i,j=0}^{p_2} a_i^{(2)} a_j^{(2)} k_{s+j-i}^{(1)} \,,$$

$$(38) \qquad d_s = b_2^{-1} \sum_{i=0}^{s} c_i^{(1)} a_{s-i}^{(2)} \,, \quad s = 1, 2, \ldots$$

There coefficients $c_i^{(1)}$ are from the Wold's decomposition of the sequence $\{X_t^{(1)}\}$. Using theorem 2, [7] we deduce the following result

$$(39) \qquad P\{L_{\hat{u}_1} < x \mid \hat{u}_1 \leqq u_1\} = \Phi[(x - n_1 m_1)(\mathscr{D}S_{n_1})^{-1/2}] + o(n_1^{-1/2}) \,,$$

where $\mathscr{D}S_n$ is given by (35)–(38).

Let's now consider the case $\hat{u}_1 > u_1$. Then

$$(40) \qquad L_{\hat{u}_1} = \max_{u = u_1 + 1, \ldots, N-1} L_u = \max\{S_0', S_n', n = 1, 2, \ldots, N - u_1 - 1\} \,,$$

where $S_0' = L_{u_1+1}$, $S_n' = S_{n-1}' + Q_n$, $n = 1, 2, \ldots, N - u_1 - 1$, $Q_s' = Q_{u_1+s+1}$ and S_n', $n \geqq 1$, represents a fluctuation tending to $-\infty$. Let $\gamma = P\{S_n' < 0\}$ then a probability distribution of Q_t's becomes a conditional one: $F(x) = P\{Q_t' < x \mid S_n' < 0\}$, and $(1-\gamma)F(x)$ is its non-conditional distribution with defect γ. In order to obtain and asymptotic estimate of $P\{L_{\hat{u}_1} < x \mid \hat{u}_1 > u_1\}$ we shall use a method of conjugate fluctuation [8]. For this purpose we have to prove the existence of $\varkappa > 0$ as a solution of the equation

$$(41) \qquad (1-\gamma) \int_{-\infty}^{\infty} e^{\varkappa y} F(dy) = 1 \,.$$

We integrate the left-hand side of (41) using expressions (23), (15) of the density function of Q_t's. That leads us to the equation

$$(1 - \gamma)^2 e^{2\varkappa q} = 1 - D^{(2)}\varkappa - |B^{(2)}| \varkappa^2 \tag{42}$$

which implies that the solution $\varkappa > 0$ exists and a conjugate fluctuation with a positive mean $\tilde{m}_2$,

$$\tilde{m}_2 = \frac{2|B^{(2)}| \varkappa + D^{(2)}}{(1 - \gamma)^3 e^{3\varkappa q}}, \tag{43}$$

may be used. Then according to ([8], chap. XI) an asymptotic estimate

$$P\{L_{\hat{u}_1} < y \mid \hat{u}_1 > u_1\} \sim 1 - \frac{\gamma(1 - \gamma)^3}{2|B^{(2)}| \varkappa^2 + D^{(2)}\varkappa} e^{\varkappa(3q - y)} \tag{44}$$

is valid. The expression (44) in addition to (39) leads us to infer the validity of (31). The proof of theorem 1 is complete.

Corollary 1.1. The null hypothesis $u_1 = \hat{u}_1$ at the significance level $\alpha + o[(\hat{u}_1 - p_1)^{-1/2}]$ must be rejected in favour of the alternatives $u_1 = \hat{u}_1 + r$, $r = 1, 2, \ldots$, if the observed value $l_{\hat{u}_1}$ of the maximum of LLH function satisfies the inequality

$$l_{\hat{u}_1} > z_{1-\alpha}(\mathscr{D}S_{\hat{u}_1 - p_1})^{1/2} + (\hat{u}_1 - p_1)\, m_1, \tag{45}$$

where $z_{1-\alpha}$ is the $(1 - \alpha)$-significance point of the standardized normal distribution and $\mathscr{D}S_n$, m_1 as defined above. The power $\beta(u_1)$ of that criterion approximately is

$$\beta(u_1) \approx \Phi[rm_1(\mathscr{D}S_{\hat{u}_1 + r})^{-1/2} - z_{1-\alpha}]. \tag{46}$$

Corollary 1.2. The null hypothesis $\hat{u}_1 > u_1$ at the significance level α must be rejected in favour of the alternative $\hat{u}_1 \leqq u_1$ if the observed value $l_{\hat{u}_1}$ of the maximum of LLH function satisfies the inequality

$$l_{\hat{u}_1} > \frac{1}{\varkappa} \ln \frac{\delta}{\alpha}, \quad \text{where} \quad \delta = \frac{\gamma(1 - \gamma)^3}{2|B^{(2)}| \varkappa^2 + D^{(2)}\varkappa} e^{3q\varkappa}. \tag{47}$$

The probability to accept the wrong hypothesis is

$$P\left\{L_{\hat{u}_1} \leqq \frac{1}{\varkappa} \ln \frac{\delta}{\alpha} \,\middle|\, \hat{u}_1 \leqq u_1\right\} = \Phi\left[\left(\frac{1}{\varkappa} \ln \frac{\delta}{\alpha} - \bar{n}_1 m_1\right) (\mathscr{D}S_{\bar{n}_1})^{-1/2}\right] + o(\bar{n}_1^{-1/2}), \tag{48}$$

where $\bar{n}_1 = \hat{u}_1 - p_1 + r$, $r = 1, 2, \ldots$

The proofs of both corollaries are evident.

4. CONCLUDING REMARKS

The change point of known parameters of an autoregressive sequence has been estimated but there is no explicit expression of its MLH estimate which complicated the study of the estimate's accuracy. The asymptotic distribution of the maximum of LLH function has been derived there and used to test hypotheses about the true value of the change point. It would be preferable to obtain a confidence interval for the change point but one can derive it from the available results only in the simplest case of the first order autoregression: $p_1 = p_2 = 1$.

The MLH estimate of the change point has been derived too [9] when changing parameters are not known or they are some known functions of time but then the problem of accuracy of the estimate is even more complicated.

REFERENCES

[1] D. V. Hinkley: Inference about the change-point in a sequence of random variables. Biometrika *57* (1970), 1, 1—17.
[2] K. P. S. Prabhu: On the detection of a sudden change in system parameters. IEEE Trans. on information theory *IT-16* (1970), 4, 497—500.
[3] L. Telksnys: Determination of changes in the properties and recognition of random process with a complicated structure. In: Proceedings of the 5th IFAC World Congress, Paris 1972.
[4] T. W. Anderson: The statistical analysis of time series. Wiley, New York 1971.
[5] N. Kligienė: Inference about the change point in an autoregressive sequence. In: Proceedings of the 2nd Seminar on Experimental Simulating and Solving of Probability Problems, Liblice—Prague 1973, 82—93.
[6] Г. Н. Ватсон: Теория бесселевых функций, ч. I. Изд. иностр. лит., Москва 1949.
[7] А. Алешкявичене: Асимптотические разложения для обобщенного процесса восстановления. Литовский матем. сборник *12* (1972), 1, 5—21.
[8] В. Феллер: Введение в теорию вероятностей и ее приложения. Т. 2, „Мир", Москва 1967.
[9] Н. Клигене: Оценка момента изменения параметров распределения случайных последовательностей, Теория вероятностей и ее применения *18* (1973), 3, 677—678.

Academy of Sciences of the Lithuanian SSR

Institute of Physics and Mathematics

STOCHASTIC PROCESSES AND PROPERTIES OF INVARIANCE FOR QUEUEING SYSTEMS WITH SPEEDS AND TEMPORARY INTERRUPTIONS

DIETER KÖNIG and UWE JANSEN

FREIBERG

1. INTRODUCTION

In queueing theory the question of invariance is well known: under what conditions the stationary probabilities of states of the queueing system are independent on the form of distribution functions of the elements (for example, of the service time) if there are given finite values of mathematical expectation? For the Erlang loss system with independent service times Sevastjanov [1] gave a complete answer; in the case of depending service times in the sense of stochastic point process see [2], [5], [6]. For other systems Kovalenko [3] proved a condition of invariance. A theory of invariance including algebraic criteria of invariance and connection with the so called set up of product of Sevastjanov [1] for a formalized class of queueing and reliability models is found in [4], [5], [6].

In this paper we generalize this class of systems taking into account possible speeds of realization of the random variables, where the speeds may depend on the index of the random variable and on the whole state of the system. In that way one can also seize certain dependences between elements of the system. Especially the value zero for the speed has the meaning of temporarily interrupting the realization of random variables (for example of service times in priority systems or of working-or lifetime in reliability systems).

Generalizing the definitions, ideas, construction of stochastic process and theorems of [4], [5], [6] we give necessary and sufficient algebraic conditions for property of invariance for a wide class of formalized queueing or reliability systems with speeds and temporary interruptions, which in the same way represent a practical method.

2. DEFINITION OF QUEUEING SYSTEM

2.1. Let J be a finite nonempty set of indices of objects of the system.

2.2. To each $j \in J$ two finite sets C_j, D_j are assigned, $C_j \cap D_j = \emptyset$, $C_j \cup D_j \neq \emptyset$; the elements $i_j \in C_j$ are called the active states of the object O_j, (in the following they have their own random lifetime) and $i_j \in D_j$ the passive (for example waiting) states of O_j.

2.3. A subset $G \subseteq \prod_{j \in J} (C_j \cup D_j)$, $G \neq \emptyset$, is fixed. Thus, each $g \in G$ is a family $g = (i_j)_{j \in J}$, $i_j \in C_j \cup D_j$, and means a macrostate of the system.

2.4. We denote $J_g := \{j \mid j \in J,\ i_j \in C_j\}$ for every $g \in G$.

2.5. To each $g \in G$ and $j \in J_g$ a non-negative finite real number c_{jg} is assigned, called the speed for the active state i_j by given macrostate g. The value $c_{jg} = 0$ corresponds to a temporary interruption of the realization of active state i_j.

2.6. We denote $J'_g := \{j \mid j \in J_g,\ c_{jg} > 0\}$.

2.7. A number $p(g, s, g')$ with $p(g, s, g') \geqq 0\ \forall g, g' \in G, \forall s \in J'_g$, $\sum_{q' \in G} p(g, s, g') = 1$ $\forall g \in G, \forall s \in J'_g$ is assigned to each $g \in G$, $g' \in G$ and $s \in J'_g$; $p(g, s, g')$ has the meaning of probability of jump $g \to g'$ from macrostate $g \in G$ into macrostate $g' \in G$ by finishing active state of object O_s, $s \in J'_g$.

We demand that a breaking off of an active state by reason of finishing another active state is not allowed.

DEFINITION 1. A *queueing system* is given by 2.1–2.7.

In [1], [2], [3] the case $c_{jg} = 1\ \forall j \in J_g, \forall g \in G$, – that means without speeds and temporary interruption – is investigated.

DEFINITION 2. A queueing system will be called *irreducible*, if a finite sequence $(g = g_0, s_0, g_1, s_1, \ldots, g_n, s_n, g_{n+1} = g')$ with $g_i \in G$, $s_i \in J'_{g_i}$, $i = 0, 1, \ldots, n$ for every pair $g, g' \in G$ exists with $p(g_i, s_i, g_{i+1}) > 0$, $i = 0, 1, \ldots, n$.

3. CONSTRUCTION OF A SUITABLE QUEUEING PROCESS

To every pair $[i_j, j]$ with $j \in J$, $i_j \in C_j$ we can associate distribution function (d.f.) $F_{i_j,j}$ with density $f_{i_j,j}$. $F_{i_j,j}$ denotes d.f. of a random quantity $\xi_{i_j,j}$ of work connected with the active state i_j of O_j. The random variables $\xi_{i_j,j}$ of different $i_j \in C_j$ and $j \in J$ are assumed to be independent.

For given macrostate $g = (i_j) \in G$ at time $t \geqq 0$ a microstate of the object O_j is denoted by

$$a_j := \begin{cases} [i_j, t_j] & \text{if} \quad i_j \in C_j, \\ i_j & \text{if} \quad i_j \in D_j \end{cases}$$

with $t_j \geqq 0$, $F_{i_j,j}(t_j) < 1$.

The supplementary variable t_j has the meaning of the accomplished quantity of work at time t on the active object O_j. A family $\boldsymbol{a} := (a_j)_{j \in J}$ denotes a microstate of the queueing system. With the notation $A_g = \{\boldsymbol{a} \mid \boldsymbol{a} = (a_j)_{j \in J} \text{ for given } g \in G\}$ we obtain $A := \bigcup_{g \in G} A_g$ as the set of states of our queueing system. The mapping $\boldsymbol{a} = (a_j)_{j \in J} \to (t_j)_{j \in J_g}$ for $\boldsymbol{a} \in A_g$ permits to transfer the topology of the open sets of the $|J_g|$ – dimensional Euclidean space to A_g. Thereby a σ-algebra $\mathfrak{A}$ of A is produced.

$[A, \mathfrak{A}]$ is the phase space of the process to be constructed. Let be given a macrostate $g \in G$ of the system and a state $\boldsymbol{a} \in A_g$ of the system. We interpret $\boldsymbol{a} \in A_g$ as the state of the system at time $t = 0$. Let $(\zeta_j)_{j \in J_g}$ be an independent family of positive random variables with d.f.

$$\frac{F_{i_j,j}(c_{jg}x + t_j) - F_{i_j,j}(t_j)}{1 - F_{i_j,j}(t_j)} \quad (x \geqq 0)$$

ζ_j means the remaining lifetime of the active state of object O_j, $j \in J_g$, under the condition of accomplished quantity t_j of work at time t and uninterrupted following sojourn of the system in macrostate g.

The next jump $g \to g'$ of the system occurs at random point $\tau = \min_{j \in J_g'} \zeta_j$. That point and the random index σ of the object originating the jump are determined with probability one (because of assumption of existence of densities).

If the random values τ and σ have the values t and s resp., so the system jumps with probability $p(g, s, g')$ into a new macrostate $g' = (i'_j)$. The new state $\boldsymbol{a}' = (a'_j) \in A_{g'}$ of the system at the point $t + 0$ is determined by

$$a'_j = \begin{cases} i'_j & \text{if} \quad j \notin J_{g'}, \\ [i_j, t_j + c_{jg}t] & \text{if} \quad j \in J_{g'} \quad \text{and} \quad j \in J_g \setminus \{s\}, \\ [i'_j, 0] & \text{if} \quad j \in J_{g'} \quad \text{and} \quad j \notin J_g \setminus \{s\}. \end{cases}$$

Further construction of stochastic process $(\psi(t))_{t \geqq 0}$ can be made analogously to [2], [3], regarding our speeds c_{jg}. One can prove that $(\psi(t))_{t \geqq 0}$ is a homogeneous Markov process with phase space $[A, \mathfrak{A}]$ and transition probability function $W(\boldsymbol{a}, t, X)$ for $\boldsymbol{a} \in A$, $t \geqq 0$, $X \in \mathfrak{A}$.

DEFINITION 3. The stochastic process $(\psi(t))_{t \geqq 0}$ is called the *queueing process* for a given queueing system.

DEFINITION 4. An initial distribution Q (that means an arbitrary distribution on $\mathfrak{A}$) is usually called a *stationary distribution*, if

$$Q_t(\,) = \int_A W(\boldsymbol{a}, t, (\,))\, Q(\mathrm{d}\boldsymbol{a}) = Q(\,)\,.$$

THEOREM 1. An initial distribution Q of a homogeneous Markov process with arbitrary phase space $[E, \mathfrak{M}]$ is a stationary distribution, if and only if, $\operatorname{Var}(Q - Q_t) = o(t)$ for $t \downarrow 0$ holds with $\operatorname{Var}(Q - Q_t) := \int_E |Q - Q_t|$.

This theorem (see [1], [2], [3]) suggests a method to prove stationarity of a given initial distribution.

DEFINITION 5. A queueing process with only one stationary initial distribution is called *primitive*.

4. REDUCED QUEUEING PROCESS AND THE SYSTEM Z OF EQUATIONS OF STATE

If d.f. $F_{i,j}$ for a pair $[i, j]$, $i \in C_j$, $j \in J$ is of the form

$$F_{i,j}(x) = 1 - \exp(-\varrho_{i,j} x)\,, \quad x \geqq 0\,, \tag{4.1}$$

then the values t_j are not necessary in definition of state a_j and in construction of the process. Therefore subsets C_j^* of C_j, $j \in J$, let be given with $F_{i,j}$ from type (4.1) for $i \in C_j^*$ and $F_{i,j}$ arbitrary (with our general assumptions) for $i \in C_j \setminus C_j^*$. Though $\boldsymbol{a}^* = (a_j^*)_{j \in J}$ with

$$a_j^* = \begin{cases} i_j & \text{if } \; i_j \in C_j^* \cup D_j\,, \\ [i_j, t_j] & \text{if } \; i_j \in C_j \setminus C_j^*\,, \quad t_j \geqq 0\,, \quad F_{i_j,j}(t_j) < 1\,, \end{cases}$$

is defined as state; and also we have $A_g^* = \{\boldsymbol{a}^* \mid \boldsymbol{a}^* = (a_j^*) \text{ for given } g \in G\}$. The quantities A^*, $\mathfrak{A}^*$ and the process $(\psi^*(t))_{t \geqq 0}$ now can be defined analogously to A and $\mathfrak{A}$ resp..

DEFINITION 6. The process $(\psi^*(t))_{t \geqq 0}$ is called the *queueing process reduced with regard to* $(C_j^*)_{j \in J}$.

By the mapping $\boldsymbol{a} \to \boldsymbol{a}^*$ a stationary initial distribution of the reduced process is received from a stationary initial distribution of the original process.

Let us consider the case $C_j^* = C_j$ for all $j \in J$, that means the completely reduced queueing process. That is a homogeneous continuous Markov chain with phase space $A = G$, $\mathfrak{A} = \mathfrak{P}(G)$ and all families $(p_g)_{g \in G}$ with $p_g \geqq 0$, $\sum_{g \in G} p_g = 1$ are initial distributions. Evaluation of transition probability functions $W(g, t, g')$ in sense of order of $o(t)$ for $t \downarrow 0$ and using Theorem 1 gives the

THEOREM 2. An initial distribution $(p_g)_{g \in G}$ of the completely reduced queueing process is a stationary distribution, if and only if, (p_g) is the solution of the following system $\mathbf{Z}$ of equations of states

$$\left(\sum_{j \in J_g} c_{jg} \varrho_{i_j, j}\right) p_g = \sum_{g' \in G} \sum_{s \in J'_{g'}} c_{sg'} \varrho_{i'_s, s} p(g', s, g) \, p_{g'} \quad \text{for each} \quad g \in G .$$

Apart from quantities included in definition of queueing system, only the quantities $\varrho_{i_j, j}$ are found in the system $\mathbf{Z}$ of equations.

So the following denotation is quite natural.

DEFINITION 7. A queueing system given by 2.1–2.7 is called *extended*, if a positive finite real number $\varrho_{i,j}$ is given for every pair $[i, j]$, $i \in C_j$, $j \in J$; $\varrho_{i,j}^{-1}$ signifies the mathematical expectation of $\xi_{i,j}$, that means for every $F_{i,j}$ holds

$$\int_0^\infty x f_{i,j}(x) \, \mathrm{d}x = \varrho_{i,j}^{-1} . \tag{4.2}$$

Based on well known ergodic theorems for homogeneous finite continuous Markov chains holds: System $\mathbf{Z}$ has an uniquely defined positive probability solution $(p_g)_{g \in G}$ if and only if the queueing system is irreducible.

5. PROPERTY OF INVARIANCE

Let be given an extended irreducible queueing system and a family $(I_j)_{j \in J}$ of subsets I_j from C_j, $j \in J$, with:

(5.1) $F_{i,j}$ are for $i \in I_j$, $j \in J$, arbitrary d.f. of positive random variables with (4.2) and $F_{i,j}$ are for $i \in C_j \setminus I_j$, $j \in J$, of the form (4.1).

DEFINITION 8. An extended irreducible queueing system possesses the *property* $\mathbf{E}((I_j)_{j \in J})$ for a family $(I_j)_{j \in J}$ of subsets $I_j \subseteq C_j$, $j \in J$, if for every family $F_{i,j}$ of d.f. with (5.1), for which the corresponding queueing process reduced with regard to $(C_j \setminus I_j)_{j \in J}$ is primitive with stationary initial distribution P, the stationary probabilities $p_g = P(A_g)$, $g \in G$, are a probability solution of the system $\mathbf{Z}$ of equations of states.

Property of complete invariance now means $\mathbf{E}((C_j)_{j \in J})$.

DEFINITION 9. A queueing system has *property* $\mathbf{e}_1((I_j)_{j\in J})$, if no active state $i_j \in I_j$ of any object O_j, $j \in J$, starts with speed zero. That means: $c_{jg} = 0$ for $i_j \in I_j$ at most if $p(g', j, g) = 0\ \forall g' \in G$ and $p(g', s, g) = 0\ \forall g' \in G,\ \forall s \in J'_{g'},\ s \neq j$, with $i'_j \neq i_j$. This property can always be obtained by defining suitable additional passive states of the objects when we construct the queueing system.

DEFINITION 10. A queueing system possesses *property* $\mathbf{e}((I_j)_{j\in J})$, if with probability one at a jump $g' \to g$ in the system two different active states i_{u_1}, i_{u_2} of objects O_{u_1}, O_{u_2} with $i_{u_1} \in I_{u_1}$, $i_{u_2} \in I_{u_2}$ can never start simultaneously.

6. CRITERIA OF INVARIANCE

Let be

$$U_g := \{j \mid j \in J_g,\ i_j \in I_j\} \quad \text{for given} \quad g = (i_j) \in G \quad \text{and} \quad (I_j)_{j\in J},$$

$$G_{X,g} := \{g' \mid g' \in G,\ i'_u = i_u,\ \forall u \in X\} \quad \text{for given} \quad g = (i_u) \in G \quad \text{and} \quad X \subseteq J.$$

THEOREM 3. Let an extended irreducible queueing system have the property $\mathbf{e}_1((I_j)_{j\in J})$ for a family $(I_j)_{j\in J}$ with $I_j \subseteq C_j$, $j \in J$. The queueing system obtains the property of invariance $\mathbf{E}((I_j)_{j\in J})$, if and only if, the queueing system has the property $\mathbf{e}((I_j)_{j\in J})$ and the following system of equations $\mathbf{Z}'((I_j)_{j\in J})$ has a positive probability solution (or, if and only if, the solution of $\mathbf{Z}'((I_j)_{j\in J})$ coincides with the unique solution of $\mathbf{Z}$). The system $\mathbf{Z}'((I_j)_{j\in J})$ has the form:

for all $g \in G$

$$\mathbf{Z}'.\ 1 \qquad \Big(\sum_{j\in J_g\setminus U_g} c_{jg}\varrho_{i_j,j}\Big)\, p_g = \sum_{g'\in G_{U_g,g}} \ \sum_{s\in J'_{g'}\setminus U_g} c_{sg'}\varrho_{i'_s,s}\, p(g', s, g)\, p_{g'}$$

and for all $u \in U_g$

$$\mathbf{Z}'.\ 2 \qquad c_{ug}\varrho_{i_u,u}p_g = \sum_{\substack{g'\in G_{U_g\setminus\{u\},g}\\ J'_{g'}\supseteq\{u\}}} c_{ug'}\varrho_{i'_u,u}\, p(g', u, g)\, p_{g'} + \\ + \sum_{g'\in G_{U_g\setminus\{u\},g}\setminus G_{U_g,g}} \ \sum_{s\in J'_{g'}\setminus U_g} c_{sg'}\varrho_{i'_s,s}\, p(g', s, g)\, p_{g'}\,.$$

Idea of proof. The extensive proof of the theorem is connected in the sufficient direction with the so called set up of product (this is a generalization of the set up of product of Sevastjanov [1])

$$P^*\Big(A_g^* \cap \bigcap_{u\in U_g} {}''0 \leq t_u < x''_u\Big) = p_g \prod_{u\in U_g} \varrho_{i_u,u} \int_0^{x_u} (1 - F_{i_u,u}(z))\, dz \quad (x_u \geq 0 \text{ for all } u \in U_g)\,.$$

We show that our set up of production gives a stationary initial distribution of the queueing process reduced with regard to $(C_j \setminus I_j)_{j \in J}$, if and only if, the queueing system has the properties $\mathbf{e}_1((I_j)_{j\in J})$ and $\mathbf{e}((I_j)_{j\in J})$ and the quantities p_g in the set up are a positive solution of $Z'((I_j)_{j\in J})$. Therefore we estimate the transition probabilities $W(a, t, (\,))$ and apply Theorem 1.

If we choose the set up

$$p_g = \prod_{j \in U_g} \varrho_{i_j,j}^{-1} q_g a \tag{6.1}$$

for all $g \in G$ with a positive constant a of normalization, then holds

THEOREM 4. An irreducible queueing system has the properties $\mathbf{e}_1((I_j)_{j\in J})$ and $\mathbf{E}((I_j)_{j\in J})$ for a family $(I_j)_{j\in J}$ with $I_j \subseteq C_j$ and for fixed chosen values $\varrho_{i,j}$ for $i \in$ $\in C_j \setminus I_j$, $j \in J$, if and only if, the probability solution of $\mathbf{Z}$ has for every choice of $\varrho_{i,j}$, $i \in I_j$, $j \in J$, the form (6.1), where the quantities q_g, $g \in G$, are independent of the values $\varrho_{i,j}$ with $i \in I_j$, $j \in J$, and a is a constant of normalization.

For the property of complete invariance $\mathbf{E}((C_j)_{j\in J})$ one can prove

THEOREM 5. An irreducible queueing system has the properties $\mathbf{e}_1((C_j)_{j\in J})$ and $\mathbf{E}((C_j)_{j\in J})$, if all J_g, $g \in G$, have the same numbers of elements and if the system of equations

$$c_{ug} q_g = \sum_{g' \in G_{U_g \setminus \{u\}, g}} \sum_{s \in J'_{g'} \setminus (J_g \setminus \{u\})} c_{sg'} p(g', s, g)\, q_{g'} \quad \text{for all} \quad g \in G \quad \text{and} \quad u \in U_g \tag{6.2}$$

has a positive solution. Such a solution is uniquely determined with the exception of a constant factor, and from that solution we receive the stationary probabilities p_g of macrostates in the form (6.1).

Thus Theorems 3, 4, 5 give a method to test whether a given queueing system with speeds and temporary interruptions modeled by 2.1–2.7 has for a family $(I_j)_{j\in J}$ with $I_j \subseteq C_j$, $j \in J$, with (5.1) the property of invariance or not; in the positive case the solution of $\mathbf{Z}'((I_j)_{j\in J})$ or (6.2) and (6.1) resp. are the wanted invariant stationary probabilities. Investigations of properties of invariance for different classes of concrete stochastic systems based on this method are found in [8]. The theorems of Kovalenko [3] and Guseinov [7] are special cases of our theory.

EXAMPLE. We regard a loss priority queueing system with one server and with preemptive resume discipline. There are two Poisson inputs with different kinds (types) of demands of service. Arriving demands of the first kind interrupt the service of demand of the second kind and this service will be continue at the point of service where it was interrupted, if the demand of higher priority is leaving the system after finishing service. An arriving demand is lost, if a demand of the same or higher priority is served at time of arrival.

The objects in sense of abstract queueing system are

O_1: input of demands of first kind,

O_2: input of demands of second kind,

O_3: service of demands of first kind,

O_4: service of demands of second kind,

with the sets of states

$C_1 = C_2 = \{1\}$ (there goes an interval between two arrivals of demands),

$D_1 = D_2 = \emptyset$,

$C_3 = C_4 = \{1\}$ (a demand of the corresponding kind is in the system),

$D_3 = D_4 = \{0\}$ (no demand of the corresponding kind is in the system).

Therefore we have $J = \{1, 2, 3, 4\}$. The macrostates g are given by $1 := (1, 1, 0, 0)$, $2 := (1, 1, 1, 0)$, $3 := (1, 1, 0, 1)$, $4 := (1, 1, 1, 1)$ and consequently $G = \{1, 2, 3, 4\}$.

For the speeds c_{jg} holds

$$c_{1g} = c_{2g} = 1 \ \forall g \in G, \quad c_{32} > 0, \quad c_{34} > 0, \quad c_{43} = 1 \quad \text{and} \quad c_{44} = 0 .$$

The speed $c_{44} = 0$ describes the temporary interruption of service demands of the second kind.

The transition probabilities $p(g, s, g')$ are defined by

$$\begin{aligned} p(1, 1, 2) &= p(1, 2, 3) = p(2, 1, 2) = p(2, 2, 2) = p(2, 3, 1) = \\ &= p(3, 1, 4) = p(3, 2, 3) = p(3, 4, 1) = p(4, 1, 4) = p(4, 2, 4) = \\ &= p(4, 3, 3) = 1 , \quad p(g, s, g') = 0 \quad \text{otherwise} . \end{aligned}$$

We denote the mathematical expections for service times and interarrival intervals for demands of the same kind with μ_i^{-1} or λ_i^{-1} resp., that means $\varrho_{11} = \lambda_1$, $\varrho_{12} = \lambda_2$, $\varrho_{13} = \mu_1$, $\varrho_{14} = \mu_2$. We accept $I_1 = \emptyset$, $I_2 = \emptyset$, $I_3 = C_3 = \{1\}$, $I_4 = C_4 = \{1\}$ and assert that the property $\mathbf{E}((I_j)_{j\in J}) = \mathbf{E}((\emptyset)_1, (\emptyset)_2, (\{1\})_3, (\{1\})_4)$ of invariance holds.

Then it is easy to see, that the properties $\mathbf{e}_1(\cdot)$ and $\mathbf{e}(\cdot)$ are satisfied. The system of equations $\mathbf{Z}'(\cdot)$ has the form:

$$\begin{aligned} (\lambda_1 + \lambda_2) \cdot p_1 &= c_{32} \cdot \mu_1 \cdot p_2 + \mu_2 \cdot p_3 , \\ (\lambda_1 + \lambda_2) \cdot p_2 &= (\lambda_1 + \lambda_2) \cdot p_2 , \\ (\lambda_1 + \lambda_2) \cdot p_3 &= \lambda_2 \cdot p_3 + c_{34} \cdot \mu_1 \cdot p_4 , \\ (\lambda_1 + \lambda_2) \cdot p_4 &= (\lambda_1 + \lambda_2) \cdot p_4 , \\ c_{32} \cdot \mu_1 \cdot p_2 &= \lambda_1 \cdot p_1 , \\ \mu_2 \cdot p_3 &= \lambda_2 \cdot p_1 , \\ c_{34} \cdot \mu_1 \cdot p_4 &= \lambda_1 \cdot p_3 , \\ 0 &= 0 . \end{aligned}$$

We receive the positive solution

$$p_2 = \frac{\lambda_1}{c_{32} \cdot \mu_1} p_1 , \quad p_3 = \frac{\lambda_2}{\mu_2} p_1 , \quad p_4 = \frac{\lambda_1 \cdot \lambda_2}{c_{34} \cdot \mu_1 \cdot \mu_2} p_1$$

with

$$p_1 = \frac{c_{32} \cdot c_{34} \cdot \mu_1 \cdot \mu_2}{c_{32} \cdot c_{34} \cdot \mu_1 \cdot \mu_2 + c_{34} \cdot \lambda_1 \cdot \mu_2 + c_{32} \cdot c_{34} \cdot \lambda_2 \cdot \mu_1 + c_{32} \cdot \lambda_1 \cdot \lambda_2} .$$

So it holds invariance of p_g with regard to d.f. of service times of the two kinds of demands.

REFERENCES

[1] B. A. Sevastjanov: A limit theorem for Markov processes with application on telephone loss systems. Theory of Probability and Applications *2* (1957), 106—116.

[2] D. König, K. Matthes: Verallgemeinerung der Erlangschen Formeln I. Mathematische Nachrichten *26* (1963), 45—56.

[3] I. N. Kovalenko: Sur la condition pour que, en régime stationnaire, la distribution soit indépendante des lois des durées de conservation. Ann. des télécommunications *17* (1962), 190—191.

[4] K. Matthes: Zur Theorie der Bedienungsprozesse. In: Trans. of the Third Prague Conference, Prague 1962. Academia, Prague 1964, 513—528.

[5] D. König, K. Matthes, K. Nawrotzki: Verallgemeinerungen der Erlangschen und Engsetschen Formeln (Eine Methode in der Bedienungstheorie). Akademie Verlag, Berlin 1967.

[6] D. König, K. Matthes, K. Nawrotzki: Unempfindlichkeitseigenschaften von Bedienungsprozessen. Appendix to the German edition of Gnedenko, Kovalenko: Introduction to queueing theory. Akademie Verlag, Berlin 1971, 356—445.

[7] B. T. Guseinov: Generalization of Kovalenko theorem on the invariance of probability of states of service system with respect to service time distribution. Sixth International Teletraffic Congress, München 1970.

[8] U. Jansen, D. König: Invariante stationäre Zustandwahrscheinlichkeiten für eine Klasse stochastischer Modelle mit funktionellen Abhängigkeiten. Mathem. Operationsforschung und Statistik 7 (1976), 497—522.

[9] D. König, U. Jansen, M. Kotzurek, H. Rabe: Ergebnisse von Invarianzuntersuchungen für ausgewählte Bedienungs- und Zuverlässigkeitssysteme. Mathem. Operationsforschung und Statistik 7 (1976), 523—526.

[10] D. König, U. Jansen: Eine Invarianzeigenschaft zufälliger Bedienungsprozesse mit positiven Geschwindigkeiten. Mathematische Nachrichten *70* (1976), 321—364.

Bergakademie Freiberg

Sektion Mathematik

GENTZEN-LIKE RANDOM AXIOMATIC SYSTEMS

IVAN KRAMOSIL

PRAGUE

An axiomatic system is investigated under the condition that some of its axioms are not valid with the probability one. The probability of the validity of axioms influences the probability of validity of conclusions derived from them. The notion of reliability of an axiomatic system is introduced and in the case of the so called Gentzen-like calculi an explicit estimation of this reliability is given.

1. INTRODUCTION

Until our days the investigation of axiomatic systems and formalized theories has been an exclusive matter of mathematical logic and some of its branches, e.g. proof theory or automatic theorem proving. The connections between a formalized theory and some other objects have been investigated only inside the framework of the so called "pure" mathematics, i.e. only connections between formalized theories and some other mathematical structures has been studied, model theory being the main field concentrating such investigations. The connections and links between a formalized theory and the surrouding us real world or environment has not been taken into consideration.

However, some recent achievements of applied mathematics and logic, namely robotics or, in other words, automaton-environment systems theory lead to new and quite non-traditional points of view toward these problems. An appropriate formal description of the state of the environment and the laws governing its development is a necessary condition under which any automated problem-solving procedure is able to work. We can even say that an appropriate formal representation of the environment is the most fundamental problem of artificial intelligence and the most necessary condition of its further development.

There are two quite natural conditions which ought to be satisfied by an ideal formal representation, namely:

(1) The representation should be a finite set of formulas and deduction rules from and by use of which every formula valid in the environment should be derivable.

(2) If a formula is not valid in the environment it should not be derivable in the formal representation.

Roughly speaking the representation should be an axiomatic and formalized theory the only categorical model of which is the environment itself.

However, it can be easily seen that such a representation is just the ideal one and it is not reasonable to expect that this case will be reachable practically. The number of axioms and the degree of complexity of such an ideal theory will be usually too great to be controlled practically, so some simplifications will be necessary. Moreover, the axioms forming the representation result from some empirical data and the process of observation and measurement of these data can be connected with a possibility of an incorrectness or error. It is generally known that these errors and incorrectness can be considered to be of statistical nature and this fact gives us the idea that all the procedure leading from the environment to its formal representation can be studied from the probabilistic and statistical point of view and that also the quality or adequacy of the obtained axiomatic system on which the formal representation is based can be judged from the same point of view. The rest of this paper is devoted to a more detailed investigation of this problem.

2. RANDOM AXIOMATIC SYSTEMS

Let us consider a formal language the set of all well-formed formulas of which will be denoted by $\mathscr{L}$. The language is chosen in such a way that it is adequate (or at least the most adequate among the languages being at our disposal) for describing the situations and the relations in the environment. Let us also consider a set of deduction rules adequate for this language. Usually the first order predicate language with appropriate functional and individual constants and the first order deduction rules are used to this goal. Under these conditions a formal representation is given by a set of axioms at least some of them being derived from empirical data and, hence, being representable by random variables. This explanation probably makes the following definition to be intuitively clear.

Definition 1. Let $(\Omega, \mathscr{S}, P)$ be a probability space, let $T = \langle 0, \infty)$ be a set of parameters, let N_n denote the set $\{1, 2, \ldots, n\}$ of integers. Then *random axiomatic system* of degree n, over the language $\mathscr{L}$ and with respect to the probability space $(\Omega, \mathscr{S}, P)$ is a mapping X of the Cartesian product $N_n \times T \times \Omega$ into $\mathscr{L}$ such that for every $i \leqq n$ and $t \geqq 0$ the mapping $X(i, t, \cdot)$ is a random variable defined on $(\Omega, \mathscr{S}, P)$ and taking its values in $\mathscr{L}$, i.e., because of the countability of $\mathscr{L}$, for every $p \in \mathscr{L}$

$$\{\omega : X(i, t, \omega) = p\} \in \mathscr{S} .$$

The parameter t, which can be interpreted as time, expresses the dynamics of a random axiomatic system, the possibility and necessity to modify the representation of the environment as the time passes with respect to its development and changes.

In order to be able to describe and judge somehow the quality of a random axiomatic system X we suppose that the state of the environment at the time instance t is represented by a subset $\mathscr{T}(t)$ of the set $\mathscr{L}$, namely by the subset of all formulas which are valid in the considered time instance. Supposing A is a subset of $\mathscr{L}$ we denote by $Cn(A)$ the set of all formulas derivable from the set A of formulas by the considered deduction rules, i.e. $A \subset Cn(A) \subset \mathscr{L}$. Using this notation we can easily see that a random axiomatic system X plays the role of an ideal representation of the environment if and only if for every $t \in T$

$$Cn(\{X(1, t, \omega), X(2, t, \omega), \ldots, X(n, t, \omega)\}) = \mathscr{T}(t) .$$

However, usually this is not the case, i.e. the symmetric difference of the two mentioned sets is not empty and we must find a way how to measure this difference.

DEFINITION 2. *Characteristic function* $p(i, t)$ *of the i-th axiom* $X(i, \cdot, \cdot)$ of a random axiomatic system X is defined as follows:

$$p(i, t) = P(\{\omega : X(i, t, \omega) \in \mathscr{L} - \mathscr{T}(t)\}) .$$

(As $\mathscr{L}$ is countable measurability of $X(i, t, \cdot)$ assures the correctness of this definition.)

For example, if $X(i, t, \omega)$ is a logical axiom or an axiom describing the fundamental time-and-space relations (and such axioms are necessary in any formalized representation) then $X(i, t, \omega)$ does not depend on ω and t and $p(i, t) \equiv 0$. A more detailed classification of random axioms can be found in [2]. Clearly the characteristic functions of random axioms can serve as a measure of the quality of axioms themselves. To be able to measure somehow the quality of a random axiomatic system in general we must suppose to have at our disposal a random variable measuring somehow the importance of particular formulas.

DEFINITION 3. Let X be a random axiomatic system with respect to the probability space $(\Omega, \mathscr{S}, P)$, let α be a random variable defined on $(\Omega, \mathscr{S}, P)$ and taking its values in the set $\mathscr{L}$ of formulas. *Reliability of X with respect to* α is defined as the conditional probability

$$R(t, \alpha)(X) = P(\{\omega : \alpha(\omega) \in \mathscr{T}(t)\} / \{\omega : \alpha(\omega) \in Cn(\{X(1, t, \omega) , X(2, t, \omega), \ldots, X(n, t, \omega)\}\}) .$$

This means that $R(t, \alpha)$ is the probability that a formula sampled at random with respect to α is valid in the environment under the condition that it is derivable from the random axioms. An explicit computation of this value can be, in general, very complicated and we shall do it only in the case of a Gentzen-like calculus in the next part of this paper. Now, we prove the following general theorem.

THEOREM 1. Let α be a random variable taking its values in the set $\mathscr{L}$. Let X be a random axiomatic system such that the random variables α, $X(i, t, \cdot)$, $i \leqq n$, $t \in T$ are mutually independent. Let $i_1 < i_2 < \ldots < i_k \leqq n$ be a sequence of integers, let $S(i_1, \ldots, i_k)$ be the set of all proper subsequences of $\langle i_1, i_2, \ldots, i_k \rangle$. Then

$$P(\{\omega : \alpha(\omega) \in \mathscr{T}(t)\} / \{\omega : \alpha(\omega) \in [Cn(\{X(i_1, t, \omega), \ldots, X(i_k, t, \omega)\}) - \\ - \bigcup_{\langle j_1, \ldots, j_l \rangle \in S(i_1, \ldots, i_k)} Cn(\{X(j_1, t, \omega), \ldots, X(j_l, t, \omega)\})]\}) = \\ = P(\{\omega : \alpha(\omega) \in \mathscr{T}(t)\} / \{\omega : \alpha(\omega) \in [Cn(\{X(i_1, t, \omega), \ldots, X(i_k, t, \omega)\}) - \\ - \bigcup_{j=1}^{k} Cn(\{X(i_1, t, \omega), \ldots, X(i_{j-1}, t, \omega), X(i_{j+1}, t, \omega), \ldots, X(i_k, t, \omega)\})]\}) \geqq \\ \geqq \prod_{j=1}^{k} (1 - p(i_j, t)) .$$

Proof. If $X(i_j, t, \omega) \in \mathscr{T}(t)$, $j = 1, 2, \ldots, k$ then $\alpha(\omega) \in \mathscr{T}(t)$, as $\mathscr{T}(t)$ is closed with respect to logical inferences. So the first conditional probability in the assertion is greater than or equal to

$$P(\{\omega : X(i_1, t, \omega) \in \mathscr{T}(t), \ldots, X(i_k, t, \omega) \in \mathscr{T}(t)\}) = \\ = \prod_{j=1}^{k} P(\{\omega : X(i_j, t, \omega) \in \mathscr{T}(t)\}) = \prod_{j=1}^{k} (1 - p(i_j, t)) .$$

The equality of the two conditional probabilities occuring in the assertion follows from the fact that for every proper subsequence of $\langle i_1, \ldots, i_k \rangle$, e.g. $\langle j_1, \ldots, j_l \rangle$, there exists at least one $j \leqq k$ such that $\langle j_1, \ldots, j_l \rangle$ is a subsequence of $\langle i_1, \ldots \ldots, i_{j-1}, i_{j+1}, \ldots, i_k \rangle$, hence

$$Cn(\{X(j_1, t, \omega), \ldots, X(j_l, t, \omega)\}) \subset \\ \subset Cn(\{X(i_1, t, \omega), \ldots, X(i_{j-1}, t, \omega), X(i_{j+1}, t, \omega), \ldots, X(i_k, t, \omega)\})$$

and the proof is closed.

3. GENTZEN-LIKE RANDOM AXIOMATIC SYSTEMS AND THEIR RELIABILITY

Let us consider a formalization of the first order predicate calculus having the following properties:

(1) No individual constants and no functional indeterminates occur, only a finite number of functional constants may occur.

(2) If $q_1, q_2, \ldots, q_n$ is a formalized proof and if q_j results from q_i, q_k, $i, k < j$ with respect to a deduction rule then to every occurence of a subformula of q_i or q_k an occurence of the same subformula (up to differences in indeterminates or terms)

in q_j can be shown (different occurence in q_j for different occurences in q_i or q_k). This is the so called subformula property or Gentzen property; for more details see, e.g., [4].

(3) The only operators are those of implication and general quantifier, $A \rightarrow B$ being written in the form $[A]\,[B]$, $\forall_x A$ in the form $x[A]$. There is one propositional constant F having the semantic interpretation "falsehood", hence non A is written in the form $[A]\,[F]$. For more details see [3].

Let us define a random variable G defined on $(\Omega, \mathscr{S}, P)$ and taking its values in the set of well-formed formulas of the theory just described. G is a slight generalization of the random variable F defined in [3].

Let K_0, M_0, N_0 be integers, let to the left bracket [integers $1, 2, \ldots, K_0$ be ascribed, to the right bracket] the integers $K_0 + 1, \ldots, 2K_0$, to F the integers $2K_0 + 1$, $2K_0 + 2, \ldots, 2K_0 + M_0$. To every individual indeterminate x_i, $i \leqq N_0$ the integer $2K_0 + M_0 + i$ is ascribed, to every elementary formula containing only indeterminates among $x_1, \ldots, x_{N_0}$ one integer starting from $2K_0 + M_0 + N_0 + 1$ is ascribed. Denote by N_2 the greatest integer used in this enumeration, let $N_3 > N_2$ be an integer, let θ be an auxiliary symbol not occuring in the formalized theory; the indices $N_2 + 1, \ldots, N_3$ are ascribed to θ.

Let $\beta_1, \beta_2, \ldots$ be a sequence of random variables defined on $(\Omega, \mathscr{S}, P)$, taking their values in the set $\{1, 2, \ldots, N_3\}$ of integers, mutually independent and equally distributed in such a way that

$$P(\{\omega : \beta_j(\omega) = i\}) = N_3^{-1}\,, \quad j = 1, 2, \ldots, \quad i = 1, 2, \ldots, N_3\,.$$

Now we define

$G(\omega) = \Lambda$ (the empty formula), if not occurence of an elementary formula or F preceeds the first occurence of θ in $\{\beta_1(\omega), \beta_2(\omega), \ldots\}$,

$G(\omega) = F(\beta_1(\omega), \ldots, \beta_k(\omega))$, if there is no occurence of θ and at least one occurence of F or of an elementary formula among $\beta_1(\omega), \ldots, \beta_k(\omega)$ and, at the same time, $\beta_{k+1}(\omega) = \theta$, i.e. $\beta_{k+1}(\omega) > N_2$, where F is the same mapping as in [3]. It means that G generalizes F in such a way that formulas of all lengths as well as the empty formula have a positive probability to be sampled. It follows from the properties of the mapping F that $G(\omega)$ is always the empty or a well-formed formula.

Denote

$$e = P(\{\omega : 2K_0 + M_0 + N_0 < \beta_1(\omega) \leqq N_2\}) = \frac{N_2 - 2K_0 - M_0 - N_0}{N_3}\,,$$

$$s = P(\{\omega : N_2 < \beta_1(\omega) \leqq N_3\}) = \frac{N_3 - N_2}{N_3}\,.$$

THEOREM 2. Consider a formalization of the first order predicate calculus satisfying (1)–(3). Let X be a random axiomatic system over this language such that the random variables $G, X(i, t, \cdot)$, $i \leqq n$, $t \in T$ are mutually independent. Let there exist a function

$\gamma(t)$ defined on T and taking its values in $\langle 0, 1\rangle$ such that for all $i \leqq n$ and $t \in T$ the inequality $p(i, t) \leqq 1 - \gamma(t)$ holds and, moreover, if $X(i, t, \omega)$ does not contain any elementary formula, then $p(i, t) \equiv 0$. Suppose, finally, that non-validity of the empty formula can be proved without any possibility of an error. Then

$$R(t, G) \geqq \frac{s}{e(1-\gamma)+s} + \left(\frac{e \cdot \gamma}{e+s}\right)^n \cdot \left(1 - \frac{s}{e(1-\gamma)+s}\right) \geqq \left(1 + \frac{e}{s}(1-\gamma)\right)^{-1}.$$

Proof. First, we have to prove the following: If $\mathscr{F}_n$, $n = 0, 1, 2, \ldots$ is the set of all well-formed formulas of our theory containing just n occurences of elementary formulas, then

$$P(\{\omega : G(\omega) \in \mathscr{F}_n\}) = \frac{s}{e+s} \cdot \left(\frac{e}{e+s}\right)^n.$$

Investigating in details the transformation F defined in [3] and admitting also the possibility that the empty formula Λ is formed we can see that during this transformation no elementary formula and no occurence of F is erased, added or changed. This gives that

$$P(\{\omega : G(\omega) \in \mathscr{F}_n\}) = \sum_{k=0}^{\infty} p_{nk}, \quad n = 0, 1, 2, \ldots$$

where p_{nk} denotes the probability that

$$k(\omega) = \min\{i : \beta_i(\omega) > N_2\} - 1 = k$$

and, at the same time, there are just n occurences of elementary formulas among $\beta_1(\omega), \beta_2(\omega), \ldots, \beta_k(\omega)$. This gives

$$p_{0n} = P\left(\bigcap_{i=1}^{n} \{\omega : \beta_i(\omega) \leqq 2K_0 + M_0 + N_0\} \cap \{\omega : \beta_{n+1}(\omega) > N_2\}\right) =$$

$$= \prod_{i=1}^{n} P(\{\omega : \beta_i(\omega) \leqq 2K_0 + M_0 + N_0\})\, P(\{\omega : \beta_{n+1}(\omega) > N_2\}) = (1 - e - s)^n \cdot s,$$

hence

$$\sum_{n=0}^{\infty} p_{0n} = \frac{s}{e+s}.$$

So $1 - s/(e+s) = e/(e+s)$ is the probability of the random event consisting in the fact that in the random sequence $\{\beta_1(\omega), \beta_2(\omega), \ldots\}$ first some elementary formula occurs, later θ occurs. However, this is a reccurent event, as after every occurence of an elementary formula or θ we can investigate, whether in the following part of the random sequence will occur or will not occur and its probability will be always the same, namely $e/(e+s)$. (For more details concerning reccurent events see, e.g. [5].) Now, it is clear that in $\{\beta_1(\omega), \ldots, \beta_k(\omega)\}$ an elementary formula occurs n-times if and only if the reccurent event defined above occurs just n-times. This gives

$$\sum_{k=0}^{\infty} p_{nk} = \frac{s}{e+s} \cdot \left(\frac{e}{e+s}\right)^n.$$

Now, let p be a formula derivable from $X(1, t, \omega), \ldots, X(n, t, \omega)$, let $p_1, p_2, \ldots$ $\ldots, p_l = p$ be a proof of p. The subformula property gives that if p_i is an axiom then there exists a one-to-one mapping of the set of all occurences of subformulas in p_i into the set of all occurences of subformulas in p_{i+1}. At the same time, this mapping, reduced to the elementary subformulas, again is one-to-one. Similarly, if p_{i+1} follows from $p_i, p_j, j < i$, using some deduction rule, again two one-to-one mappings exist, one of them mapping the occurences of elementary subformulas in p_i into the elementary subformulas of p_{i+1}, the other having the same property when p_j considered instead of p_i. Moreover, the two images generated in the set of all occurences of subformulas in p_{i+1} by these two mappings are disjoint. This gives that there exists a one-to-one mapping from the set of all occurences of elementary formulas occuring in the axioms which occur among $p_1, p_2, \ldots, p_l$ into the set of all occurences of elementary formulas in p. Hence, the number of axioms containing at least one elementary subformula and occuring among $p_1, p_2, \ldots, p_l$ is smaller than or equal to (in case $l = 1$) the number of occurences of elementary subformulas in p. In such a case Theorem 1 gives

$$P(\{\omega : G(\omega) \in \mathcal{T}\}/\{\omega : G(\omega) \in \mathcal{F}_n \cap Cn(\{X(1, t, \omega), \ldots, X(n, t, \omega)\})\}) \geqq \gamma^n$$

and so we obtain, computing the reliability

$$R(t, G) = \sum_{i=0}^{\infty} [P(\{\omega : G(\omega) \in \mathcal{T}\}/\{\omega : G(\omega) \in \mathcal{F}_i \cap Cn(\{X(1, t, \omega), \ldots, X(n, t, \omega)\})\}) .$$

$$. P(\{\omega : G(\omega) \in \mathcal{F}_i\})] =$$

$$= \sum_{i=0}^{n} \left[\left(\frac{s}{e+s} \cdot \left(\frac{e}{e+s} \cdot \gamma\right)^i\right]\right. + \sum_{i=n+1}^{\infty} \left[\left(\frac{s}{e+s}\right) \cdot \left(\frac{e}{e+s}\right)^i \gamma^n\right] =$$

$$= \frac{s}{e(1-\gamma)+s} + \left(\frac{e\gamma}{e+s}\right)^n \cdot \left(1 - \frac{s}{e(1-\gamma)+s}\right)$$

after an easy calculation, as in no proof in our theory more than n axioms occur. Moreover,

$$\sum_{i=0}^{n} \left[\left(\frac{s}{e+s}\right) \cdot \left(\frac{e}{e+s} \gamma\right)^i + \sum_{i=n+1}^{\infty} \left[\left(\frac{s}{e+s}\right) \cdot \left(\frac{e}{e+s}\right)^i \cdot \gamma^n\right]\right. \geqq$$

$$\geqq \sum_{i=0}^{\infty} \left[\left(\frac{s}{e+s}\right) \cdot \left(\frac{e}{e+s} \gamma\right)^i\right] = \left[1 + \frac{e}{s}(1-\gamma)\right]^{-1}$$

which proves also the other inequality of our theorem. The proof is completed.

COROLLARY. If $\gamma \to 1$ then $R(t, G) \to 1$.

4. CONCLUSIONS

The inequality in the theorem just proved is rather interesting because it does not depend on the number of axioms and tends to one if γ does. This fact offers a simple strategy when a formal representation is formed: it is better to have a great number of reliable axioms than a small number of less reliable ones. This agrees, in the case of Gentzen-like calculi, with the effort to atomize the information in the most possible way.

As far as the author knows the first attempt to introduce an aspect of uncertainty into the investigation of axiomatic systems was done in [1]. There the supposed uncertainty is formally expressed by the use of probabilistic or the so called fuzzy logics. We believe, however, that our way of expressing this aspect in the form of random variables is more close to the actual way of receiving and obtaining the uncertain form of information from the environment and it is why also the formal description of this information, i.e. axioms should be considered to be values of appropriate random variables. The both approaches lead to results which are from some points of view different. A more detailed discussion on this problem can be found in [2]. Some more investigation in this field seems to be hopeful and justified.

REFERENCES

[1] R. T. C. Lee: Fuzzy Logic and the Resolution Principle. Journal of the Association for Computing Machinery *19* (1972), 1, 109–119.

[2] I. Kramosil: Random Axiomatic Systems. Research Report no. 519, Institute of Information Theory and Automation, 1973.

[3] I. Kramosil: A Method for Random Sampling of Well-Formed Formulas. Kybernetika *8* (1972), 2, 135–148.

[4] H. B. Curry: Foundations of Mathematical Logic. Mc Graw-Hill Book Comp., New York—San Francisco—Toronto—London 1963.

[5] W. Feller: An Introduction to Probability Theory and its Applications I. John Wiley and Sons, New York—London—Sydney 1950.

Czechoslovak Academy of Sciences

Institute of Information Theory and Automation

PARAMETER ESTIMATION IN A LINEAR STOCHASTIC DIFFERENTIAL EQUATION

ALAIN LE BRETON

GRENOBLE

This paper is concerned with the problem of parameter estimation in a vector linear differential equation, with constant coefficients, of the form:

$$dX_t = \theta X_t\,dt + G\,d\xi_t\,; \quad t \geqq 0\,; \quad X_0 \equiv 0$$

where $(\xi_t : t \geqq 0)$ is a standard Wiener process and G is a non singular matrix, when one observes a trajectory of the solution process. We first show, using a strong limit theorem for Brownian motion, that we can assume the matrix GG' is completely known with the meaning that, with probability one, it may be computed on every finite time interval. Then we show that the statistical structure for the problem of estimation of the matrix θ, when GG' is known, is dominated and we compute the likelihood function with help of Girsanov's results [1]. We determine the maximum likelihood estimate of θ and study its asymptotic properties.

1. STATEMENT OF THE PROBLEM AND NOTATIONS

In [2] Shiryayev and [3] Novikov, the authors have studied the problem of estimation of the unknown parameters λ and (ω, μ) for equations of the form:

$$dX_t = \lambda A(t, X)\,dt + d\xi_t$$

and

$$dX_t = \begin{pmatrix} \omega & -\mu \\ \mu & -\omega \end{pmatrix} A(t, X)\,dt + d\xi_t$$

in $\boldsymbol{R}$ and $\boldsymbol{R}^2$ respectively. They propose a "sequential" estimation of the parameters, deduced from the maximum likelihood estimation, by introducing a stopping time.

We consider here the general case of a linear vector differential equation and use the following formalism:

Let $(\Omega, \mathscr{A})$ be a probabilisable space and $\{X_t : t \geqq 0\}$ a family of random vectors, defined on $(\Omega, \mathscr{A})$, with values in $\boldsymbol{R}^n$. We note, for every $t \geqq 0$, $\mathscr{A}_t$ the σ-algebra on Ω generated by $\{X_s : 0 \leqq s \leqq t\}$.

Θ (resp. $\mathscr{G}$) being a subset of the set $\mathscr{M}_n$ (resp. $\mathscr{I}_n$) of square matrices $n \times n$ (resp. non singular square matrices $n \times n$), we assume that there exists a matrix $\theta_0 \in \Theta$, a matrix $G_0 \in \mathscr{G}$ and a probability measure P_{θ_0,G_0} on $(\Omega, \mathscr{A})$ such that the process $(\Omega, (\mathscr{A}_t)_{t \geqq 0}, P_{\theta_0,G_0}, (\xi_t^{\theta_0,G_0})_{t \geqq 0})$ where

$$\xi_t^{\theta_0,G_0} = G_0^{-1}\left\{X_t - \int_0^t \theta_0 \,.\, X_s \,.\, \mathrm{d}s\right\}; \quad t \geqq 0,$$

is a standard Wiener process in $\boldsymbol{R}^n$. In other words the process $\{X_t : t \geqq 0\}$ is solution of the stochastic differential equation:

$$\mathrm{d}X_t = \theta_0 \,.\, X_t \,\mathrm{d}t + G_0 \,\mathrm{d}\xi_t^{\theta_0,G_0}; \quad t \geqq 0; \quad X_0 \equiv 0$$

or

$$X_t = \int_0^t \mathrm{e}^{(t-s)\theta_0} \,.\, G_0 \,\mathrm{d}\xi_s^{\theta_0,G_0}; \quad t \geqq 0$$

with respect to (w.r.t.) P_{θ_0,G_0}. This process is gaussian, zero-mean, with covariance function

$$r(s, t) = \mathrm{e}^{s\theta_0}\left\{\int_0^{s \wedge t} \mathrm{e}^{-u\theta_0} G_0 G_0' \,\mathrm{e}^{-u\theta_0'} \,\mathrm{d}u\right\} \mathrm{e}^{t\theta_0'}.$$

If one observes the process $\{X_t : t \geqq 0\}$ on the time interval $[0, T]$, we are naturally led to state the problem of estimation of matrices parameters θ and GG' in the statistical structure

$$(C(T; \boldsymbol{R}^n), \mathscr{S}_T, \{\mu_{\theta,G}^T; (\theta, G) \in \Theta \times \mathscr{G}\})$$

where

– $C(T; \boldsymbol{R}^n)$ is the space of continuous functions from $[0, T]$ to $\boldsymbol{R}^n$.

– $\mathscr{S}_T$ is the σ-algebra on this space generated by its coordinates Π_t for $0 \leqq t \leqq T$.

– for every $(\theta, G) \in \Theta \times \mathscr{G}$, $\mu_{\theta,G}^T$ is the measure on $\mathscr{S}_T$ induced of the Wiener measure μ_*^T on $\mathscr{S}_T$ by the process solution of the stochastic differential equation:

$$\mathrm{d}Z_t = \theta Z_t \,\mathrm{d}t + G \,\mathrm{d}\Pi_t; \quad t \geqq 0; \quad Z_0 \equiv 0$$

w.r.t. μ_*^T.

2. ESTIMATION OF THE MATRIX GG'

We first prove a proposition which says that for every $G \in \mathscr{G}$ the family $\{\mu_{\theta,G}^T; \theta \in \Theta\}$ of measures is dominated:

PROPOSITION 1. *For every $G \in \mathscr{G}$ and $\theta \in \Theta$, the measure $\mu_{\theta,G}^T$ is absolutely continuous w.r.t. the measure $\mu_{0,G}^T$ with Radon-Nikodym derivative*:

$$\frac{\mathrm{d}\mu_{\theta,G}^T}{\mathrm{d}\mu_{0,G}^T} = \exp\left\{\int_0^T \langle \theta\Pi_t, (GG')^{-1} \,\mathrm{d}\Pi_t\rangle - \frac{1}{2}\int_0^T \langle \theta\Pi_t, (GG')^{-1}\, \theta\Pi_t\rangle \,\mathrm{d}t\right\}$$

where the stochastic integral $\int_0^T \langle \theta\Pi_t, (GG')^{-1} \,\mathrm{d}\Pi_t\rangle$ is defined w.r.t. $\mu_{0,G}^T$.

Proof. Considering the Ito process $(Z_t^1;\ t \in [0, T])$ defined, w.r.t. μ_*^T by

$$dZ_t^1 = G\, d\Pi_t\,;\quad t \in [0, T]\,;\quad Z_0^1 \equiv 0$$

and the measure $\tilde{\mu}^T$ on $\mathscr{S}_T$, absolutely continuous w.r.t. μ_*^T, with Radon-Nikodym derivative

$$\frac{d\tilde{\mu}^T}{d\mu_*^T} = \exp\left\{\int_0^T \langle G^{-1}\theta Z_t^1, d\Pi_t\rangle - \frac{1}{2}\int_0^T \langle G^{-1}\theta Z_t^1, G^{-1}\theta Z_t^1\rangle\, dt\right\}$$

Girsanov's results [1] assure that, w.r.t. $\tilde{\mu}^T$, the process $\{\zeta_t;\ t \in [0, T]\}$ defined by

$$\zeta_t = \Pi_t - \int_0^t G^{-1}\theta Z_s^1\, ds$$

is a Wiener process and that

$$dZ_t^1 = \theta Z_t^1\, dt + G\, d\zeta_t\,;\quad t \in [0, T]\,;\quad Z_0^1 \equiv 0\,.$$

The measure on $\mathscr{S}_T$ induced by the process $(Z_t^1;\ t \in [0, T];\ \tilde{\mu}^T)$ is then absolutely continuous w.r.t. the measure induced by the process $(Z_t^1;\ t \in [0, T];\ \mu_*^T)$, with Radon-Nikodym derivative

$$\exp\left\{\int_0^T \langle G^{-1}\theta\Pi_t, G^{-1}\, d\Pi_t\rangle - \frac{1}{2}\int_0^T \langle G^{-1}\theta\Pi_t, G^{-1}\theta\Pi_t\rangle\, dt\right\}.$$

The result of the proposition is proved because of the unicity (in probability law) of the solution of equation $dZ_t = \theta Z_t\, dt + G\, dW_t$ where (W_t) is a Wiener process. □

We consider now the statistic $\hat{\sigma}_t$ defined on the relevant statistical structure by

$$\hat{\sigma}_t(x) = \lim_{N\to+\infty} \frac{1}{t} \sum_{i=1}^{2^N} (x_{it/2^N} - x_{(i-1)t/2^N})^{\otimes 2}$$

for $x = (x_t;\ t \in [0, T]) \in C(T; \boldsymbol{R}^n)$ and $t \in]0, T]$.

PROPOSITION 2. *For every $(\theta, G) \in \Theta \times \mathscr{G}$ one has, for every $t \in]0, T]$,*

$$\hat{\sigma}_t = GG' \quad \mu_{\theta,G}^T \text{ almost surely (a.s.)}\,.$$

Proof. One has

$$G^{-1}\hat{\sigma}_t G'^{-1} = \lim_{N\to+\infty} \frac{1}{t} \sum_{i=1}^{2^N} (G^{-1}\Pi_{it/2^N} - G^{-1}\Pi_{(i-1)t/2^N})^{\otimes 2}$$

where the process $(G^{-1}\Pi_t;\ t \in [0, T];\ \mu_{0,G}^T)$ is a standard Wiener process. Now, for a standard process $(W_t;\ t \geq 0)$ in $\boldsymbol{R}^n$, we study:

$$\lim_{N\to+\infty} \frac{1}{t} \sum_{i=1}^{2^N} (W_{it/2^N} - W_{(i-1)t/2^N})^{\otimes 2}\,.$$

We know that, in mean square, this limit is I_n (identity matrix). We also know (see for example Doob [4] p. 395]) that, a.s.,

$$\lim_{N\to+\infty} \frac{1}{t} \sum_{i=1}^{2^N} (W^k_{it/2^N} - W^k_{(i-1)t/2^N})^2 = 1 \quad \text{for} \quad k = 1, \ldots, n$$

where $W'_t = (W^1_t, \ldots, W^k_t, \ldots, W^n_t)$.

It easily follows that, a.s.,

$$\lim_{N\to+\infty} \frac{1}{t} \sum_{i=1}^{2^N} (W_{it/2^N} - W_{(i-1)t/2^N})^{\otimes 2} = I_n$$

and

$$\mu^T_{0,G} \quad \text{a.s.}, \quad G^{-1}\hat{\sigma}_t G'^{-1} = I_n$$

or

$$\hat{\sigma}_t = GG' .$$

Then, by proposition 1, we have

$$\hat{\sigma}_t = GG' \quad \mu^T_{\theta,G} \quad \text{a.s.} \qquad \square$$

So the matrix GG' can be computed, with probability one, with help of the observation of one trajectory of the process on every finite time interval. In the following, we assume that GG' is completely known and study the problem of estimation of θ alone. The results for the case $GG' = I_n$ have been announced in [5] Le Breton.

3. ESTIMATION OF THE MATRIX θ

The statistical structure for the problem is

$$(C(T; R^n), \mathcal{S}_T, \{\mu^T_{\theta,G}; \theta \in \Theta\}) .$$

From proposition 1, this structure is dominated by the measure $\mu^T_{0,G}$ with

$$\frac{d\mu^T_{\theta,G}}{d\mu^T_{0,G}} = \exp\left\{\int_0^T \langle\theta\Pi_t, (GG')^{-1}\, d\Pi_t\rangle - \frac{1}{2}\int_0^T \langle\theta\Pi_t, (GG')^{-1}\,\theta\Pi_t\rangle\, dt\right\}$$

the measure $\mu^T_{0,G}$ being completely defined by GG' (assumed known).

REMARK. It can be seen that the structure is of exponential type, admitting the sufficient statistic

$$\left(\int_0^T d\Pi_t \,.\, \Pi'_t, \quad \int_0^T \Pi_t \,.\, \Pi'_t\, dt\right)$$

where the stochastic integral is defined w.r.t. $\mu^T_{0,G}$. Further, in case $n = 1$, the statistic $(\Pi^2_T, \int_0^T \Pi^2_s\, ds)$ is sufficient.

The Log-likelihood function on the structure is

$$L_G(\theta; .) = \int_0^T \langle \theta \Pi_t, (GG')^{-1} \, d\Pi_t \rangle - \frac{1}{2} \int_0^T \langle \theta \Pi_t, (GG')^{-1} \, \theta \Pi_t \rangle \, dt$$

i.e.

$$L_G(\theta; x) = \left(\int_0^T \langle \theta \Pi_t, (GG')^{-1} \, d\Pi_t \rangle \right)(x) - \frac{1}{2} \int_0^T \langle \theta x_t, (GG')^{-1} \, \theta x_t \rangle \, dt$$

for $x \in C(T; \boldsymbol{R}^n)$, where the stochastic integral is defined w.r.t. $\mu_{0,G}^T$. One easily shows that the maximum likelihood equations system can be written:

$$\int_0^T d\Pi_t \, . \, \Pi_t' - \theta \int_0^T \Pi_t \, . \, \Pi_t' \, dt = 0 \, .$$

Then we have:

PROPOSITION 3. *The maximum likelihood estimate of θ is given by*

$$\hat{\theta}_G(T) = \left[\int_0^T d\Pi_t \, . \, \Pi_t' \right] . \left[\int_0^T \Pi_t \, . \, \Pi_t' \, dt \right]^{-1}$$

where the stochastic integral $\int_0^T d\Pi_t \, . \, \Pi_t'$ is defined w.r.t. $\mu_{0,G}^T$.

Proof. It consists in verifying that the matrix $\int_0^T \Pi_t \, . \, \Pi_t' \, dt$ is, $\mu_{0,G}^T$ a.s., non singular. Clearly it is sufficient to consider the case $GG' = I_n$ i.e. $\mu_{0,G}^T \equiv \mu_*^T$.

Let us define

$$A_0 = \left\{ x \in C(T; \boldsymbol{R}^n) : \exists h \in \boldsymbol{R}^n, \; h \neq 0 \Big/ \left\langle \left[\int_0^T x_s \, . \, x_s' \, ds \right] h, h \right\rangle = 0 \right\}$$

$$= \{ x \in C(T; \boldsymbol{R}^n) : \exists h \in \boldsymbol{R}^n, \; h \neq 0 / \forall s \in [0, T] \, \langle x_s, h \rangle = 0 \} \, .$$

For $\{s_1, \ldots, s_{n+1}\} \subset [0, T]$, $0 \leqq s_1 < \ldots < s_{n+1} \leqq T$, if $x \in A_0$ there exists $h \in \boldsymbol{R}^n$, $h \neq 0$ such that

$$\langle x_{s_{i+1}} - x_{s_i}, h \rangle = 0 \quad \text{for} \quad i = 1, \ldots, n \, .$$

Then

$$A_0 \subset A_0^* = \{ x \in C(T; \boldsymbol{R}^n) : \det [x_{s_2} - x_{s_1}, \ldots, x_{s_{n+1}} - x_{s_n}] = 0 \} \, .$$

One has $\mu_*^T(A_0^*) = P_\tau(\Delta)$ where P_τ is the probability law of the random element $\tau = (\Pi_{s_2} - \Pi_{s_1}, \ldots, \Pi_{s_{n+1}} - \Pi_{s_n})$ defined on $(C(T; \boldsymbol{R}^n), \mathscr{S}_T, \mu_*^T)$, and $\Delta = \{(V_1, \ldots$ $\ldots, V_n) \in (\boldsymbol{R}^n)^n : \det [V_1, \ldots, V_n] = 0\}$. P_τ being a gaussian non singular measure on $(\boldsymbol{R}^n)^n$ if we show that Δ has Lebesgue measure zero the proof shall be achieved. This fact results easily from the well-known theorem:

The range of $\boldsymbol{R}^p$ by a differentiable application from $\boldsymbol{R}^p$ in $\boldsymbol{R}^q$, with $p < q$, has Lebesgue measure zero in $\boldsymbol{R}^q$. □

REMARKS (a) Computing

$$\frac{\partial^2}{\partial\theta_{kl}\,\partial\theta_{ij}}\,L_G(\theta;\,.) = -((GG')^{-1})_{ki}\int_0^T \Pi_s^l \Pi_s^j\,\mathrm{d}s\,,$$

one can see that $\hat{\theta}_G(T)$ gives, $\mu_{0,G}^T$ a.s. a maximum for $L_G(\theta;\,.)$, $\theta \in \mathcal{M}_n$.

(b) Formally, one can write

$$L_G(\theta;x) = -\frac{1}{2}\left[\int_0^T \frac{|\mathrm{d}x_s - \theta x_s\,\mathrm{d}s|^2}{\mathrm{d}s} - \int_0^T \frac{|\mathrm{d}x_s|^2}{\mathrm{d}s}\right]$$

and $\hat{\theta}_G(T)$ appears also such as a "least-squares estimate".

(c) In case $n = 1$, we have, by use of Ito's formula:

$$\hat{\theta}_G(T)\,(x) = \frac{x_T^2 - G^2T}{2\displaystyle\int_0^T x_s^2\,\mathrm{d}s}\,.$$

4. ASYMPTOTIC PROPERTIES OF THE FAMILY $\{\hat{\theta}_G(T): T > 0\}$ OF ESTIMATES

We shall use the following considerations:

(i) $\hat{\theta}_G(T)$ is identified to the random element

$$\left(\int_0^T \mathrm{d}\hat{\Pi}_t\,.\,\hat{\Pi}_t'\right)\cdot\left(\int_0^T \hat{\Pi}_t\,.\,\hat{\Pi}_t'\,\mathrm{d}t\right)^{-1}$$

defined on $(C(\boldsymbol{R}^+;\boldsymbol{R}^n),\mathcal{S})$ where $C(\boldsymbol{R}^+;\boldsymbol{R}^n)$ is the space of continuous functions from $\boldsymbol{R}^+$ to $\boldsymbol{R}^n$ and $\mathcal{S}$ is the σ-algebra on this space generated by all its coordinates $\hat{\Pi}_t$, $t \geqq 0$. The stochastic integral $\int_0^T \mathrm{d}\hat{\Pi}_t\,.\,\hat{\Pi}_t'$ is here defined w.r.t. the Wiener measure $\mu_{0,G}$ with covariance GG' on $\mathcal{S}$.

(ii) If $\mu_{\theta,G}$ is the measure on $\mathcal{S}$ which, for every $T > 0$, coincides on $\mathcal{T}_T$ (σ-algebra on $C(\boldsymbol{R}^+;\boldsymbol{R}^n)$ generated by $\hat{\Pi}_t$, $t \in [0, T]$) with the measure

$$\exp\left\{\int_0^T \langle\theta\hat{\Pi}_s, (GG')^{-1}\,\mathrm{d}\hat{\Pi}_s\rangle - \frac{1}{2}\int_0^T \langle\theta\hat{\Pi}_s, (GG')^{-1}\,\theta\hat{\Pi}_s\rangle\,\mathrm{d}s\right\}.\,\mu_{0,G}$$

the process $(C(\boldsymbol{R}^+;\boldsymbol{R}^n),\mathcal{S},\mu_{\theta,G},(\eta_t^\theta)_{t\geqq 0})$ defined by

$$\eta_t^\theta = G^{-1}\hat{\Pi}_t - \int_0^t G^{-1}\theta\hat{\Pi}_s\,\mathrm{d}s\,;\quad t \geqq 0$$

is a Wiener process and

$$d\hat{\Pi}_t = \theta\hat{\Pi}_t + G\, d\eta_t^\theta\,; \quad t \geqq 0\,; \quad \hat{\Pi}_0 \equiv 0\,.$$

(iii) A Lyapunov's theorem (see for example Lee, Markus [6], p. 198) asserts that if θ is a real stability matrix (i.e. the eigenvalues of θ have negative real parts) then the equation

$$\theta Q + Q\theta' = -GG' \tag{1}$$

has for solution Q_θ the matrix of a definite positive quadratic form. One considers on $(\boldsymbol{R}^n \times C(\boldsymbol{R}^+; \boldsymbol{R}^n), \mathscr{B}_{\boldsymbol{R}^n} \otimes \mathscr{S})$, where $\mathscr{B}_{\boldsymbol{R}^n}$ is the Borel σ-algebra on $\boldsymbol{R}^n$, the probability $\tilde{\mu}_{\theta,G} = N(0, Q_\theta) \otimes \mu_{\theta,G}$ where $N(0, Q_\theta)$ is the normal distribution on $\boldsymbol{R}^n$ with mean zero and covariance Q_θ. One defines the random functions $(\tilde{\Pi}_t;\, t \geqq 0)$ and $\tilde{\eta}_t^\theta;\, t \geqq 0)$ on $(\boldsymbol{R}^n \times C(\boldsymbol{R}^+; \boldsymbol{R}^n), \mathscr{B}_{\boldsymbol{R}^n} \otimes \mathscr{S})$ by:

$$\tilde{\Pi}_t(a, u) = \hat{\Pi}_t(u) \quad \text{and} \quad \tilde{\eta}_t^\theta(a, u) = \eta_t^\theta(u) \quad \text{for} \quad (a, u) \in \boldsymbol{R}^n \times C(\boldsymbol{R}^+; \boldsymbol{R}^n)\,.$$

Clearly, the process $(\tilde{\Pi}_t;\, t \geqq 0;\, \tilde{\mu}_{\theta,G})$ verify

$$d\tilde{\Pi}_t = \theta\tilde{\Pi}_t\, dt + G\, d\tilde{\eta}_t^\theta\,; \quad t \geqq 0\,; \quad \tilde{\Pi}_0 \equiv 0$$

where the process $(\tilde{\eta}_t^\theta;\, t \geqq 0;\, \tilde{\mu}_{\theta,G})$ is a standard Wiener process.

We then consider, w.r.t. $(\boldsymbol{R}^n \times C(\boldsymbol{R}^+; \boldsymbol{R}^n), \mathscr{B}_{\boldsymbol{R}^n} \otimes \mathscr{S}, \tilde{\mu}_{\theta,G})$ the stochastic differential equation

$$dU_t = \theta U_t\, dt + G\, d\tilde{\eta}_t^\theta\,; \quad t \geqq 0\,; \quad U_0 = \Pi_0^*$$

where Π_0^* is the projection from $\boldsymbol{R}^n \times C(\boldsymbol{R}^+; \boldsymbol{R}^n)$ to $\boldsymbol{R}^n$. The solution of this equation is the gaussian zero-mean process defined by

$$\Pi_t^* = e^{t\theta}\Big(\Pi_0^* + \int_0^t e^{-s\theta} G\, d\tilde{\eta}_s^\theta\Big)\,; \quad t \geqq 0\,.$$

It is easy to show that this process is stationary with covariance function

$$R_\theta(h) = e^{h\theta} Q_\theta\,; \quad h \geqq 0\,. \tag{2}$$

4.1. Convergence of family $\{\hat{\theta}_G(T);\, T > 0\}$

We first prove some preliminary results:

LEMMA 1. *Let Θ be an $n \times n$ stability matrix and G an $n \times n$ non singular matrix. Let $(N, \mathscr{N}, \mu, (\gamma_t;\, t \geqq 0))$ be a standard Wiener process in $\boldsymbol{R}^n$. If $(Z_t;\, t \geqq 0)$ is solution of the stochastic differential equation*

$$dU_t = \theta U_t\, dt + G\, d\gamma_t\,; \quad t \geqq 0 \tag{E}$$

with initial condition $Z_0 \equiv 0$, *and* $(Z_t^*;\, t \geqq 0)$ *is a zero-mean second order stationary process verifying* (E), *then*

(a)
$$P \lim_{T \to +\infty} T^{-1/2} \left[\int_0^T Z_s^* \,.\, Z_s^{*\prime}\, \mathrm{d}s - \int_0^T Z_s \,.\, Z_s'\, \mathrm{d}s \right] = 0\,,$$

(b)
$$P \lim_{T \to +\infty} T^{-1/2} \left[\int_0^T \mathrm{d}Z_s^* \,.\, Z_s^{*\prime} - \int_0^T \mathrm{d}Z_s \,.\, Z_s' \right] = 0\,.$$

NOTES. (α) P lim means limit in the sense of the convergence in probability. $\mathcal{D}$ lim shall mean limit in the sense of the convergence in distribution.

(β) In the following, if B is a real matrix $n \times n$, $|B|$ denotes its Euclidian norm i.e.

$$|B|^2 = \operatorname{Tr}(BB') = \sum_{i,j} b_{ij}^2\,.$$

If B is a random matrix $n \times n$, with square integrable coefficients on $(N, \mathcal{N}, \mu)$ $\|B\|$ denotes its norm in $[L^2(N, \mathcal{N}, \mu)]^{n^2}$ defined by

$$\|B\|^2 = E|B|^2 = \operatorname{Tr} E[BB']\,.$$

(γ) For simplicity of proofs we shall assume that the eigenvalues of the concerned stability matrix θ are all different; if not, using the canonical Jordan decomposition of θ one could lead similar arguments.

Proof of Lemma 1. For $t \geqq 0$, one has:

$$Z_t = \int_0^t \mathrm{e}^{(t-u)\theta} G \,\mathrm{d}\gamma_u = Z_t^* - \mathrm{e}^{t\theta} Z_0^*\,.$$

Then

(3)
$$\int_0^T Z_s^* \,.\, Z_s^{*\prime}\, \mathrm{d}s - \int_0^T Z_s \,.\, Z_s'\, \mathrm{d}s =$$

$$= \int_0^T \mathrm{e}^{s\theta} Z_0^* \,.\, Z_s'\, \mathrm{d}s + \int_0^T Z_s \,.\, Z_0^{*\prime}\, \mathrm{e}^{s\theta'}\, \mathrm{d}s + \int_0^T \mathrm{e}^{s\theta} Z_0^* \,.\, Z_0^*\, \mathrm{e}^{s\theta'}\, \mathrm{d}s$$

where

$$\left\| \int_0^T \mathrm{e}^{s\theta} Z_0^* \,.\, Z_s'\, \mathrm{d}s \right\|^2 = \operatorname{Tr} \int_0^T \int_0^T \mathrm{e}^{u\theta} (Z_0^* \,.\, Z_u' \,.\, Z_s \,.\, Z_0^{*\prime})\, \mathrm{e}^{s\theta'}\, \mathrm{d}u\, \mathrm{d}s\,.$$

Every stationary process satisfying (E) has for covariance function R_θ defined by (2) (see for example Arato [7]. Lemma 3.1), so:

$$\left\| \int_0^T \mathrm{e}^{s\theta} Z_0^* \,.\, Z_s'\, \mathrm{d}s \right\|^2 = \operatorname{Tr} \int_0^T \int_0^T \mathrm{e}^{u\theta} Q_\theta\, \mathrm{e}^{s\theta'} E(Z_u' \,.\, Z_s)\, \mathrm{d}u\, \mathrm{d}s\,.$$

Using $E(Z_s . Z_s') = Q_\theta - e^{s\theta} Q_\theta e^{s\theta'}$, it comes:

$$|E(Z_u' . Z_s)| \leqq \operatorname{Tr} Q_\theta \quad \text{and}$$

$$\left\| \int_0^T e^{s\theta} Z_0^* . Z_s' \, ds \right\|^2 \leqq [\operatorname{Tr} Q_\theta] \int_0^T \int_0^T \left|\operatorname{Tr}\{e^{u\theta} Q_\theta e^{s\theta'}\}\right| du \, ds .$$

Writing $\theta = C\Lambda C^{-1}$ where Λ is the diagonal matrix with coefficients λ_i the eigenvalues of θ and C is a non singular matrix, one has:

$$e^{u\theta} = Ce^{u\Lambda} C^{-1}$$

and, if $\operatorname{Sup}\{\operatorname{Re}(\lambda_i) : i = 1, \ldots, n\} \leqq \lambda < 0$, there exists a constant K such that:

$$\operatorname*{Max}_{ij} |(e^{u\theta})_{ij}| \leqq K e^{\lambda u} ; \quad u \geqq 0 .$$

Then $|\operatorname{Tr}\{e^{u\theta} Q_\theta e^{s\theta'}\}| \leqq K' e^{\lambda(u+s)}$, which leads to a majoration of $\| \int_0^T e^{s\theta} Z_0^* . Z_s' \, ds \|^2$ uniformly in T.

The third part of the second member in (3) is semi-definite positive and

$$E \int_0^T e^{s\theta} Z_0^* . Z_0^{*\prime} e^{s\theta'} \, ds = \int_0^T e^{s\theta} Q_\theta e^{s\theta'} \, ds .$$

The trace is also less than a constant not depending on T.

We compute

$$\int_0^T dZ_s^* . Z_s^{*\prime} - \int_0^T dZ_s . Z_s' =$$

(4) $$= \theta \left\{ \int_0^T Z_s^* . Z_s^{*\prime} \, ds - \int_0^T Z_s . Z_s' \, ds \right\} + \int_0^T G \, d\gamma_s . (Z_s^* - Z_s)' .$$

Using the precedent, the norm of the first part in the second member of (4) is less than a constant independant of T. For the second part one has

$$\left\| \int_0^T d\gamma_s . Z_0^{*\prime} . e^{s\theta'} \right\|^2 = n \int_0^T E|e^{s\theta} Z_0^*|^2 \, ds$$

and the same is true. Using Tchebycheff inequality one easily derives assertions (a) and (b) in the lemma. □

LEMMA 2. *Under the assumptions of Lemma* 1, *one has*:

(a) $$P \lim_{T \to +\infty} T^{-1} \left[\int_0^T dZ_s^* . Z_s^{*\prime} - \theta \int_0^T Z_s^* . Z_s^{*\prime} \, ds \right] = 0 ,$$

(b) $$P \lim_{T \to +\infty} T^{-1} \left[\int_0^T dZ_s . Z_s' - \theta \int_0^T Z_s . Z_s' \, ds \right] = 0 .$$

Proof. The first assertion (a) can be obtained from the stationarity of $(Z_t^*;\ t \geqq 0)$. Assertion (b) results from the first and Lemma 1. □

We are now able to prove:

PROPOSITION 4. *If Θ is the subset of stability matrices in $\mathcal{M}_n$, the family $\{\hat{\theta}_G(T);\ T > 0\}$ of estimates is consistent in probability.*

Proof. We consider the processes $(\tilde{\Pi}_t;\ t \geqq 0;\ \tilde{\mu}_{\theta,G})$ and $(\Pi_t^*;\ t \geqq 0;\ \tilde{\mu}_{\theta,G})$ and set:

$$\tilde{\theta}_T(a, u) = \hat{\theta}_G(T)(u) \quad \text{for} \quad (a, u) \in R^n \times C(R^+; R^n)\,.$$

One has:

$$T^{-1}\left\{\int_0^T \mathrm{d}\tilde{\Pi}_s \,.\, \tilde{\Pi}_s' - \tilde{\theta}_T \int_0^T \tilde{\Pi}_s \,.\, \tilde{\Pi}_s' \,\mathrm{d}s\right\} = 0 \quad \tilde{\mu}_{\theta,G} \quad \text{a.s.}$$

By lemma 2, it comes,

$$P \lim_{T \to +\infty} T^{-1}\left\{(\tilde{\theta}_T - \theta)\int_0^T \tilde{\Pi}_s \,.\, \tilde{\Pi}_s' \,\mathrm{d}s\right\} = 0 \quad \text{w.r.t.} \quad \tilde{\mu}_{\theta,G}\,.$$

The process $(\Pi_t^*;\ t \geqq 0)$, being stationary gaussian zero-mean with covariance function R_θ, admitting a spectral density

$$f(\lambda) = \frac{1}{2\pi}(\theta - \mathrm{i}\lambda I_n)^{-1}\, GG'(\theta' + \mathrm{i}\lambda I_n)^{-1}\,,$$

an ergodic theorem (see Rozanov [8], p. 157 and 163) then asserts

$$P \lim_{T \to +\infty} T^{-1}\int_0^T \Pi_s^* \,.\, \Pi_s^{*\prime} \,\mathrm{d}s = R_\theta(0) = Q_\theta\,. \quad \text{w.r.t.} \quad \tilde{\mu}_{\theta,G}\,.$$

Lemma 1 gives:

$$P \lim_{T \to +\infty} T^{-1}\int_0^T \tilde{\Pi}_s \,.\, \tilde{\Pi}_s' \,\mathrm{d}s = Q_\theta\,.$$

Writing

$$(\tilde{\theta}_T - \theta) = (\tilde{\theta}_T - \theta)\left\{T^{-1}\int_0^T \tilde{\Pi}_s \,.\, \tilde{\Pi}_s' \,\mathrm{d}s\right\}\left\{T^{-1}\int_0^T \tilde{\Pi}_s \,.\, \tilde{\Pi}_s' \,\mathrm{d}s\right\}^{-1}$$

one can easily achieve the proof. □

4.2. Asymptotic law of the family $\{\hat{\theta}_G(T);\ T > 0\}$

LEMMA 3. *Under the assumptions of Lemma* 1, *one has:*

$$\mathscr{D} \lim_{N \to +\infty} N^{-1/2}\left\{\int_0^N \mathrm{d}Z_s^* \,.\, Z_s^{*\prime} - \theta\int_0^N Z_s^* \,.\, Z_s^{*\prime} \,\mathrm{d}s\right\} = \Gamma^\theta$$

where Γ^θ is a matrix with a gaussian zero-mean distribution with covariances defined by

$$E(\Gamma^\theta_{ij} \, . \, \Gamma^\theta_{kl}) = (GG')_{ik} \, . \, (Q_\theta)_{jl}$$

where Q_θ si given by (1).

Proof. For $p \in N^*$, set

$$Z_t^{(p)} = Z^*_{j/p} \quad \text{if} \quad \frac{j}{p} \leqq t < \frac{j+1}{p}, \quad j \in N$$

and

$$\varepsilon_j^{(p)} = \gamma_{j/p} - \gamma_{(j-1)/p}, \quad j \in N^* .$$

One computes, for $(N, p) \in N^{*2}$

$$Z(N, p) = \int_0^N G \, \mathrm{d}\gamma_s \, . \, Z_s^{(p)\prime} = G \sum_{j=0}^{pN-1} \varepsilon^{(p)}_{j+1} Z^{*\prime}_{j/p} .$$

Let

$$Z^*_{j/p} = \lambda_j^{(p)} + \mu_j^{(p)}$$

where

$$\lambda_0^{(p)} = Z_0^* \; ; \quad \lambda_j^{(p)} = \sum_{s=0}^{j-1} \exp\left(\frac{s}{p}\,\theta\right) G\varepsilon^{(p)}_{j-s} + \exp\left(\frac{j}{p}\,\theta\right) Z_0^* \; ; \quad j \in N^*$$

and

$$\mu_0^{(p)} = 0 \; ; \quad \mu_j^{(p)} = \sum_{k=0}^{j-1} \int_{k/p}^{(k+1)/p} \left[\exp\left(\left(\frac{j}{p} - t\right)\theta\right) - \right.$$

$$\left. - \exp\left(\left(\frac{j-k-1}{p}\right)\theta\right)\right] G \, \mathrm{d}\gamma_t \; ; \quad j \in N^* .$$

Then, for $(N, p) \in N^{*2}$

$$Z(N, p) = G \sum_{j=1}^{pN} \varepsilon_j^{(p)} \lambda^{(p)\prime}_{j-1} + G \sum_{i=1}^{pN} \varepsilon_j^{(p)} \mu^{(p)\prime}_{j-1} .$$

It is easy to show that, for every $p \in N^*$,

$$P \lim_{N \to +\infty} N^{-1/2} \sum_{j=1}^{N} \varepsilon_j^{(p)} \mu^{(p)\prime}_{j-1} = 0 .$$

We show also that, for every $p \in N^*$, the sequence $\{N^{-1/2} \sum_{j=1}^{N} G\varepsilon_j^{(p)} \lambda^{(p)\prime}_{j-1} ; \; N \in N^*\}$ converges in distribution: set $Q^\theta(p)$ the symetric definite positive matrix solution of

$$Q - \mathrm{e}^{\theta/p} Q \mathrm{e}^{\theta'/p} = \frac{1}{p} GG' . \tag{5}$$

Using a limit theorem (see Anderson [9], Ch. 5) one obtains:

$$\mathscr{D} \lim_{N \to +\infty} N^{-1/2} \sum_{j=1}^{N} G\varepsilon_j^{(p)} \lambda_{j-1}^{(p)\prime} = \Gamma_p^\theta$$

where Γ_p^θ is a gaussian matrix with mean zero and covariances

$$E(\Gamma_p^\theta(i, j) \,.\, \Gamma_p^\theta(k, l)) = \frac{1}{p} (GG')_{i,k} \cdot (Q_{(p)}^\theta)_{j,l} \,.$$

One easily deduces that

$$\mathscr{D} \lim_{N \to +\infty} N^{-1/2} \sum_{j=1}^{N} G\varepsilon_j^{(p)} Z_{(j-1)/p}^{*\prime} = \Gamma_p^\theta$$

and

$$\mathscr{D} \lim_{N \to +\infty} N^{-1/2} Z(N, p) = P^{1/2} \Gamma_p^\theta \,.$$

We then write

$$U_N = N^{-1/2} \int_0^N G \, \mathrm{d}\gamma_s \,.\, Z_s^{*\prime} = N^{-1/2} \int_0^N G \, \mathrm{d}\gamma_s \,.\, Z_s^{(p)\prime} + N^{-1/2} \int_0^N G \, \mathrm{d}\gamma_s \,.\, (Z_s^* - Z_s^{(p)})' =$$

$$= N^{-1/2} Z(N, p) + N^{-1/2} V(N, p) \,.$$

Using a Bernstein's Lemma (see for example Hannan [10], p. 242) the proof shall be achieved if we show that

(a) $\quad \left\{ \left\| \sqrt{\left(\frac{p}{N}\right)} V(N, p) \right\|^2 ; \quad p \in N^* \right\}$ is bounded uniformly in N,

(b) $$\lim_{p \to +\infty} Q^\theta(p) = Q_\theta \,.$$

Fact (a) can easily be proved. To prove (b) we write (5) in the form:

(5′) $$p(\mathrm{e}^{-\theta/p} - I_n) \, Q - Qp(\mathrm{e}^{\theta'/p} - I_n) = \mathrm{e}^{-\theta/p} GG' \,.$$

The coefficients $p(\mathrm{e}^{-\theta/p} - I_n)$, $p(\mathrm{e}^{\theta'/p} - I_n)$ and $\mathrm{e}^{-\theta/p}$ of (5′) converge respectively, when $p \to +\infty$, to $-\theta$, θ' and I_n. The equation

$$-\theta Q - Q\theta' = GG'$$

is nothing but (1). Because of unicity of the solutions in (5) and (1) one has

$$\lim_{p \to +\infty} Q^\theta(p) = Q_\theta \,. \qquad \square$$

Then we have

PROPOSITION 5. *Under the assumptions of Proposition* 4, *one has*

$$\mathscr{D} \lim_{T \to +\infty} T^{1/2}(\hat{\theta}_G(T) - \theta) = K^\theta$$

where K^θ *is a Gaussian matrix with mean zero and covariances defined by*

$$E(K^\theta_{ij} \, . \, K^\theta_{kl}) = (GG')_{ik} \, (Q_\theta^{-1})_{jl}$$

where Q_θ *is given by* (1).

Proof. We consider the processes $(\tilde{\Pi}_t;\ t \geqq 0)$, $(\Pi^*_t;\ t \geqq 0)$ and the family $\{\tilde{\theta}_T;\ T > 0\}$ introduced in the proof of Proposition 4. By Lemma 1 and 3 one can obtain that

$$\lim_{T \to +\infty} T^{1/2}(\tilde{\theta}_T - \theta) \left[T^{-1} \int_0^T \tilde{\Pi}_s \, . \, \tilde{\Pi}'_s \, ds \right] = \Gamma^\theta \quad \text{w.r.t.} \quad \tilde{\mu}_{\theta,G}$$

where $P \lim_{T \to +\infty} T^{-1} \int_0^T \tilde{\Pi}_s \, . \, \tilde{\Pi}'_s \, ds = Q_\theta$. Then, it follows that

$$\mathscr{D} \lim_{T \to +\infty} T^{1/2}(\tilde{\theta}_T - \theta) \, Q_\theta = \Gamma^\theta \quad \text{w.r.t.} \quad \tilde{\mu}_{\theta,G}$$

and

$$\mathscr{D} \lim_{T \to +\infty} T^{1/2}(\hat{\theta}_G(T) - \theta) \, Q_\theta = \Gamma^\theta \quad \text{w.r.t.} \quad \mu_{\theta,G} \, .$$

Q_θ being non singular the proof can easily be achieved. □

REFERENCES

[1] I. V. GIRSANOV: On transforming a certain class of stochastic processes by absolutely continuous substitution of measures. Theor. Prob. Appl. *5* (1960), 3, 285—301.

[2] A. N. SHIRYAYEV: Statistics of diffusion type processes. In: Proc. of the second Japan—USSR Symposium on Probability Theory, Kyoto 1972. Springer-Verlag, 1973, Lecture Notes in Mathematics 330, 397—412.

[3] A. A. NOVIKOV: Sequential estimation of parameters of diffusion processes. Theor. Prob. Appl. *16* (1971), 2, 391—393.

[4] J. L. DOOB: Stochastic Processes. Wiley, New York 1953.

[5] A. LE BRETON: Estimation des paramètres d'une équation différentielle stochastique vectoriel-le linéaire. C.R. Acad. Sci. Paris. Série A. *279* (1974), 289—292.

[6] E. B. LEE, L. MARKUS: Foundations of optimal control theory. Wiley, New York 1967.

[7] M. ARATO: The admissible estimation of the unknown mean of a stationary process with rational spectral density. Selected Transl. in Math. Statist. and Prob. *10* (1971), 211—223.

[8] YU. A. ROZANOV: Stationary Random Processes. Holden Day, San Francisco 1967.

[9] T. W. ANDERSON: The statistical analysis of time series. Wiley, New York 1971.

[10] E. J. HANNAN: Multivariate time series. Wiley, New York 1970.

UNIVERSITÉ SCIENTIFIQUE ET MÉDICALE DE GRENOBLE

INSTITUT DE RECHERCHE EN MATHÉMATIQUES AVANCÉES

РАСПРЕДЕЛЕНИЕ ВЕРОЯТНОСТЕЙ ТИПА CR*

Ю. П. ЛЕОНОВ

Москва

При выводе нормального закона распределения ошибок Гаусс сделал два предположения:

1. Средняя арифметическая является наиболее вероятным значением.
2. Отдельные ошибки наблюдения независимы между собой.

Эти предположения оказались в хорошем соответствии с наблюдениями. Однако, несмотря на большую гибкость и универсальность нормального закона, существует класс распределений, сохраняющий некоторые его свойства и обладающий еще одним важным качеством. Таким качеством является конечность функции распределения.

Определение. Распределение с плотностью вероятности $f(x)$, $x = (x_1, \ldots$ $\ldots, x_n)$, называется конечным, если существует ограниченное множество x вне которого плотность вероятности тождественно равна нулю.

Примеры. а) Равномерное распределение конечно в R_1, так как областью определения плотности вероятности $f(x)$ является отрезок. Равномерное распределение в R_n также является конечным.

б) Любое распределение с функцией плотности вероятности, удовлетворяющей условиям

$$(*) \qquad \begin{aligned} f(x) &\geqq 0\,, \quad x \subset (b, c)\,, \\ &\equiv 0\,, \quad x \not\subset (b, c) \end{aligned}$$

являются конечными в R_1.

* CR — начальные буквы „cramped".

в) Безгранично-делимые законы и, в частности, нормальное и пуасоновское распределения на являются конечными.

Всякое усеченное распределение с плотностью вероятности (*) является конечным (по определению).

Важно выяснить условия представления конечных распределений через „элементарные" законы. В частности, интересно знать при каких условиях конечные распределения также представляются композицией конечных распределений.

Оказывается, что конечное распределение можно получить как предельное для суммы независимых величин. Однако для этого недостаточно конечности распределений $f_k(x)$ этих величин. Необходимо, чтобы области определения $E_k \subseteq R_n$ плотностей вероятности $f_k(\cdot)$ стягивались в точку с определенной скоростью.

1. КОНЕЧНОЕ РАСПРЕДЕЛЕНИЕ ТИПА CR

Для того, чтобы получить плотность $f(x)$ типа CR рассмотрим сумму независимых случайных величин

$$X = \sum_0^\infty X_k\,, \tag{1}$$

распределенных с плотностями вероятности

$$f_k(x) = \frac{1}{\beta_k}\,\varphi\left(\frac{x}{\beta_k}\right), \quad |x| \leqq \beta_k\,, \tag{2}$$

где $\varphi(u)$ – заданная плотность вероятности $|u| \leqq 1$. Последовательность β_k такова, что ряд

$$L = \sum_0^\infty \beta_k\,. \tag{3}$$

сходится.

Пусть числа β_k образуют геометрическую прогрессию

$$\beta_k = aq^k \quad (0 \leqq q < 1,\ a > 0)\,. \tag{4}$$

Тогда

$$L = \frac{a}{1-q}\,.$$

При этих условиях будем говорить, что случайная величина X имеет конечное распределение типа CR.

Характеристическая функция (х.ф.) $g(\omega)$ распределения CR при $a = 1$ имеет вид

(5а) $$g(\omega) = \prod_0^\infty \mu(\omega q^n),$$

(5б) $$\mu(z) = \int_{-1}^{1} \exp(jzu)\, \varphi(u)\, du,$$

где $\varphi(u)$ – заданная плотность вероятности.

Бесконечное произведение в правой части (5а) при $0 \leqq q < 1$ сходится для $-\infty < \omega < \infty$.

а) Х.ф. $g(\omega)$ удовлетворяет функциональному уравнению

(6) $$g(\omega) = \mu(\omega)\, g(\omega q),$$

где $\mu(\cdot)$ х.ф. – соответствующая плотность вероятности $\varphi(\cdot)$.

б) Плотность вероятности случайной величины X является решением интегрального уравнения ($a = 1$)

(7) $$f(x) = \int_{(x-1)/q}^{(x+1)/q} \varphi(x - qz)\, f(z)\, dz, \quad |x| \leqq \frac{1}{1-q}.$$

Сходимость бесконечного произведения в (5а) является следствием сходимости ряда (3) для $0 \leqq q < 1$.

Уравнение (6) для х.ф. $g(\cdot)$ следует из (5а).

Интегральное уравнение (7) следует из (6) и равенств (3) и (4).

Теорема 1. Интегральное уравнение (7) для непрерывной $\varphi(u)$ имеет единственное решение для плотности вероятности $f(\cdot)$.

Это утверждение можно получить как следствие существования х.ф. $g(\cdot)$.

Основные свойства плотности вероятности $f(x)$ являются следствием уравнений (5) и (7).

На основании (5а) закон распределения типа CR не является безгранично-делимым. Плотность вероятности $f(x)$ зависит от параметра q, при изменении которого меняется область определения L(4). Можно „равномерно" деформировать $f(x)$ переходя к функции

(8) $$f^*(x) = \frac{1}{a} f\left(\frac{x}{a}\right), \quad |x| \leqq \frac{1}{1-q},$$

с сохранением площади под кривой $f^*(x)$, равной 1. Как следует из (6) при $q = 0$ случайная величина X имеет плотность $\varphi(x)$.

Сумма двух независимых случайных величин имеющих распределение каждое с параметром $L(4)$ имеет CR-распределение с параметром $2L$. Х.ф. суммы равна

$$g^2(\omega) = \prod_0^\infty \frac{\sin^2 q^n\omega}{(q^n\omega)^2} .$$

Нетрудно убедиться, что х.ф. $\sin^2 q^n\omega/(q^n\omega)^2$ соответствует плотность

$$f_n(x) = \frac{1}{4q^{2n}}(2q^n - |x|) , \quad |x| \leqq 2q^n .$$

Сумма m независимых CR-величин с параметрами L_i $(i = 1, \ldots, m)$ имеет CR-распределение с параметром

$$L = \sum_1^m L_i .$$

2.

Для того, чтобы получить какую-либо плотность вероятности $f(x)$ следует задать ядро $\varphi(\cdot)$ в уравнении (7).

Пусть, например:

$$\varphi(x) = \tfrac{1}{2} , \quad x \subset (-1, 1) , \qquad (9)$$

$$= 0 , \quad x \not\subset (-1, 1) .$$

Тогда х.ф. $g(\cdot)$ (5a) распределения CR будет иметь вид

$$g(\omega) = \prod_0^\infty \frac{\sin \omega q^n}{\omega q^n} . \qquad (10)$$

Интегральное уравнение (7) для плотности вероятности принимает наиболее простой вид

$$f(x) = \frac{1}{2}\int_{(x-1)/q}^{(x+1)/q} f(z)\,\mathrm{d}z . \qquad (11)$$

В данном случае плотность вероятности $f(\cdot)$ является четной функцией и имеет максимум при $x = 0$.

Непосредственное вычисление моментов α_{2n} случайной величины aX, равной согласно (1) сумме независимых равномерно-распределенных величин, дает

$$\alpha_{2n} = \int_{-L}^{L} x^{2n} f^*(x)\,\mathrm{d}x . \qquad (12)$$

В частности для моментов второго и четвертого порядков получаются следующие выражения:

$$\alpha_2 = \frac{a^2}{3(1-q^2)}, \quad \alpha_4 = \frac{3+10q^2(1+q^2)}{15(1-q^4)}\, a^4 .$$

Относительно случайной величины X с х.ф. (10) и плотностью $f(x)$ (11) можно доказать теорему о разложении CR-распределения. Эта теорема решает задачу поставленную вначале.

Теорема 2. Случайная величина с распределением (11) является суммой независимых дискретных величин X_{nk}

$$X = \sum_{n,k=1}^{\infty} X_{nk} \tag{13}$$

принимающих значения

$$X_{nk} = \begin{cases} q^n/2^k \\ -q^n/2^k \end{cases} \quad \text{и} \quad p\left(\frac{q^n}{2^k}\right) = p\left(-\frac{q^n}{2^k}\right) = \frac{1}{2} .$$

Доказательство. Согласно (1) случайная величина X является суммой равномерно-распределенных случайных величин X_n с плотностью вероятности

$$f_n(x) = \frac{1}{2q^n}, \quad x \in (-q^n, q^n) .$$

Но равномерно-распределенная случаймая величина X_n является суммой независимых дискретных величин X_{nk}. Действительно для х.ф. величины X_n имеем*

$$\frac{\sin q^n\omega}{q^n\omega} = \prod_{k=1}^{\infty} \cos\frac{q^n\omega}{2^k} . \tag{14}$$

Откуда следует утверждение теоремы.

Теорема 2 является основной в теории конечных распределений. Именно CR-распределение имеет сумма независимых дискретных случайных величин, принимающих различные равновероятные значения.

В отличие от центральной предельной теоремы здесь не нужно применять известную нормировку $1/\sqrt{n}$ использованную Moivre, Laplace et Gauss.

Теорема 2 указывает также весьма простой способ получения CR-распределения.

---

* Формула является следствием разложения

$$\frac{\sin\omega}{\omega} = \prod_{1}^{\infty} \cos\frac{\omega}{2^k} .$$

На основании численного решения интегрального уравнения (11) были получены плотность вероятности $f^*(x)$ и функция распределения $F^*(x)$ для $q = \frac{1}{2}$, $a = \frac{1}{2}$ (рис. 1).

Можно рассматривать решения управнения (7) также для других ядер $\varphi(\cdot)$. Удобство состоит в том, что уравнение допускает простое численное решение, позволяющее составить таблицы функций $f^*(x)$, $F^*(x)$. Приближенное решение (7) определяется рекуррентными соотношениями.*

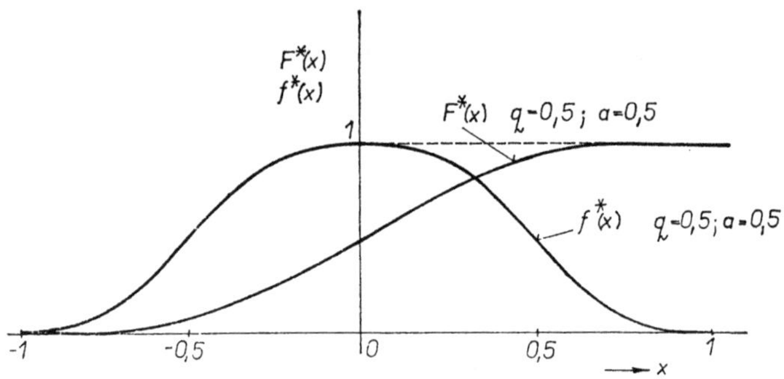

Рис. 1.

3.

Интересным свойством конечных распределений является восможность сколь-угодно точной аппроксимации конечными распределениями асимптотических законов. Такую аппроксимацию можно поллучить, выбирая различные функции $\varphi(\cdot)$ в уравнении (7). В частности CR-распределение с ядром (9) аппроксимирует нормальную плотность. Имеет место

Теорема 3. Случайная величина X/σ_q (1), (12) при $q \to 1$ асимптотически нормальна с параметрами N (0, 1)

(15) $$\sigma_q^2 = \frac{1}{3(1 - q^2)}\,; \quad \sigma_q^2 = \alpha_2 \quad \text{для} \quad a = 1\,.$$

Ясно, что величина X/σ_q имеет при любом $0 \leqq q < 1$ конечное распределение.

---

* Численные решения интегрального уравнения (11) были получены Ю. И. Марковым на машине М-220. Для построения графиков на рис. 1 использовалось 55-е приближение.

Теорема может выть доказана, если несколько видоизменить метод Лапласа [1].

На основании теоремы 3 и пользуясь (15) и таблицами нормального распределения легко вычислить приближенные значения плотности CR для значений q, близких к 1. В частности, хорошая для практических целей точность получается для значений $q \geqq 0{,}95$.

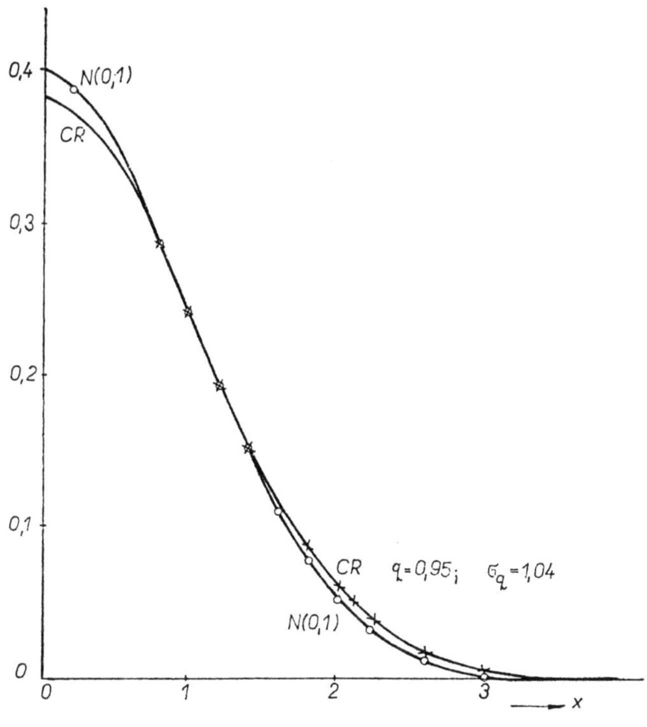

Рис. 2.

Приближениями будут усеченные нормальные кривые. Так для $q = 0{,}95$, $\sigma_q = 1{,}04$ (15) приближенной CR-плотностью является усеченная нормальная кривая $N(0; \sigma_q)$; $L \cong 20$. Значение L вычисляется из (4) при $a = 1$. На рис. 2 показана нормальная плотность N(0, 1) (o) и нормальная плотность $N(0, \sigma_q)$; $\sigma_q = 1{,}04$ ($\times$), являющаяся приближением CR-плотности для $q = 0{,}95$. Как следует из рис. 2 максимум CR-приближения несколько ниже максимума нормальной кривой, а для $x > 1{,}4$ значения CR-приближения лежат несколько выше кривой N(0, 1).

4.

С асимптотическими законами типа нормального (пуасоновского, экспоненциального и т.д.) связан парадокс бесконечнобольших значений. Например, для нормального закона отклонения, появляющиеся с вероятностью 10^{-100}, являются очень большими. Однако, так как такие отклонения не являются невозможным событием они могут появиться уже в первом опыте. При этом теория вероятностей не дает ответа в каком именно опыте должно произойти это событие. Поэтому выражение „событие, вероятность которого равна 10^{-100}, должно рассматриваться как невозможное в человеческом смысле“ (Борель) неверно по существу. Конечные распределения лишены этого недостатка. Поэтому можно надеяться, что они находятся в более точном соответствии с законами природы и аксиома конечности является существенной.

ЦИТИРОВАННАЯ ЛИТЕРАТУРА

[1] P. S. Laplace: Theorie analitique des Probabilités. Paris 1812.

Московский авто-дорожный институт

Кафедра высшей математики

A STABILITY THEOREM FOR A CHARACTERIZATION OF THE WIENER PROCESS

EUGENE LUKACS*

BOWLING GREEN and SHEFFIELD

1. INTRODUCTION

In an earlier paper, Laha and Lukacs [1] derived the following characterization for the Wiener process.

THEOREM A. *Let $X(t)$ be a homogeneous stochastic process with independent increments which is defined in a finite, closed interval $[A, B]$. Assume further that $X(t)$ is a second order process and that its mean value function $\mathscr{E}\, X(t) \equiv 0$ while its covariance function is of bounded variation. Also suppose that $a(t)$ and $b(t)$ are two continuous functions defined in $[A, B]$ such that $a(t)\, b(t) \neq 0$ for all $t \in [A_1, B_1]$ where $A \leqq A_1 < B_1 \leqq B$. Also assume that $a(t)$ is not proportional to $b(t)$. Let*

$$Y = \int_A^B a(t)\, \mathrm{d}X(t), \quad Z = \int_A^B b(t)\, \mathrm{d}X(t)$$

be two stochastic integrals defined as limits in the (quadratic) mean. The process $X(t)$ is a Wiener process if, and only if,

(i) *Y has linear regression on Z,*

(ii) *the conditional variance of Y, given Z, does not depend on Z.*

We then say that Y has constant scatter on Z.

As a consequence of Theorem A one obtains immediately the following statement.

THEOREM B. *Suppose that the conditions of Theorem A are satisfied, then the random variable $Z = \int_A^B b(t)\, \mathrm{d}X(t)$ is normally distributed with mean zero and variance $\mathscr{E}(Z^2) = \sigma^2 \int_A^B |b(t)|^2\, \mathrm{d}t$.*

* The preparation of this paper was partly supported by the National Science Foundation under Grant NSF-GP 35724X1.

In this paper we derive stability theorems for these results. Therefore we do not assume (i) and (ii) but suppose only that these conditions are 'almost' satisfied. Hence we must define what we mean, in this connection, by the word almost.

2. SOME DEFINITIONS, NOTATIONS AND STATEMENT OF RESULTS

We consider two random variables Y and Z with finite variance. We denote the expectation and the variance of the conditional distribution of Y, given Z, by $\mathscr{E}(Y|Z)$ and $\operatorname{Var}(Y|Z) = \mathscr{E}(Y^2|Z) - [\mathscr{E}(Y|Z)]^2$ respectively.

DEFINITION 1. We say that Y has ε-linear regression on Z if

(2.1) $$|\mathscr{E}(Y|Z) - \alpha - \beta Z| \leqq \varepsilon \quad \text{almost everywhere}.$$

DEFINITION 2. We say that Y has ε-scatter on Z if

(2.2) $$|\operatorname{Var}(Y|Z) - \sigma_0^2| \leqq \varepsilon \quad \text{almost everywhere}$$

Here α, β, σ_0^2 are constants $(\sigma_0^2 > 0)$ while $1 > \varepsilon > 0$.

Conditions (2.1) and (2.2) can be expressed in a slightly different form. Instead of (2.1) we can write

(2.1a) $$\mathscr{E}(Y|Z) = \alpha + \beta Z + R(Z)$$

where $|R(Z)| \leqq \varepsilon$.

Similarly one can write instead of (2.2)

(2.2a) $$\operatorname{Var}(Y|Z) = \sigma_0^2 + S(Z)$$

where $|S(Z)| \leqq \varepsilon$. Then

$$\mathscr{E}(Y^2|Z) = (\alpha^2 + \sigma_0^2) + \beta^2 Z^2 + 2\alpha\beta Z + 2\alpha\, R(Z) + 2\beta Z\, R(Z) + R^2(Z) + S(Z).$$

We put

(2.3) $$T(Z) = 2\alpha\, R(Z) + 2\beta Z\, R(Z) + R^2(Z) + S(Z)$$

and have

(2.4) $$\mathscr{E}(Y^2|Z) = \alpha^2 + \sigma_0^2 + \beta^2 Z^2 + 2\alpha\beta Z + T(Z).$$

It is also convenient to introduce the following notations. We write

(2.5) $$r(v) = \mathscr{E}(e^{ivZ} R(Z))$$

then

(2.6) $$|r(v)| \leqq \varepsilon.$$

Let

(2.7) $$z_1 = \mathscr{E}(|Z|), \quad z_2 = \mathscr{E}(Z^2)$$

be the first (respectively second) absolute moment of Z. Then

$$\mathscr{E}([Ze^{iv} ZR(Z)]) \leqq \varepsilon\mathscr{E}|Z| = \varepsilon z_1$$

and we see that $r(v)$ can be differentiated and that

(2.8) $$\left|\frac{dr(v)}{dv}\right| \leqq \varepsilon z_1 .$$

We also write

$$s(v) = \mathscr{E}[e^{ivZ} S(Z)]$$

then

(2.9) $$|s(v)| \leqq \varepsilon$$

and we see easily that it is possible to differentiate $s(v)$ two times, and that

(2.10) $$\begin{cases} \left|\dfrac{ds(v)}{dv}\right| \leqq \varepsilon z_1 , \\ \left|\dfrac{d^2 s(v)}{dv^2}\right| \leqq \varepsilon z_2 , \end{cases}$$

Finally, let

$$t(v) = \mathscr{E}(e^{ivZ} T(Z)) ,$$

then

(2.11) $$|t(v)| \leqq 2|\alpha|\,\varepsilon + 2|\beta|\, z_1\varepsilon + \varepsilon^2 + \varepsilon = O(\varepsilon) .$$

Let $g(u, v) = \mathscr{E}(e^{iuY + ivZ})$ be the characteristic function of the joint distribution of the random variables Y and Z and let $g(0, v)$ be the characteristic function of the (marginal) distribution of Z. Then

(2.12a) $$\frac{\partial g(u, v)}{\partial v} = i\mathscr{E}(Ze^{iuY + ivZ}) ,$$

(2.12b) $$\frac{\partial g(u, v)}{\partial u} = i\mathscr{E}(Ye^{iuY + ivZ}) ,$$

(2.12c) $$\frac{\partial^2 g(u, v)}{\partial u^2} = -\mathscr{E}(Y^2 e^{iuY + ivZ}) ,$$

(2.12d) $$\frac{dg(0, v)}{dv} = i\mathscr{E}(Ze^{ivZ}) ,$$

(2.12e) $$\frac{d^2 g(0, v)}{dv^2} = -\mathscr{E}(Z^2 e^{ivZ}) .$$

We formulate our results.

THEOREM 1. *Let $X(t)$ be a homogeneous stochastic process with independent increments which is defined on a finite closed interval $[A, B]$, Assume further that $X(t)$ is a second order process and that its mean value function is identically zero while its covariance function is of bounded variation.* Suppose also that $a(t)$ and $b(t)$ are two continuous functions defined on $[A, B]$ *and they are such that* $a(t)\,b(t) \neq 0$ *for all* $t \in [A_1, B_1]$ *where* $A \leqq A_1 < B_1 \leqq B$. *Suppose further that* $a(t)$ *is not proportional to* $b(t)$. *Let* $Y = \int_A^B a(t)\,\mathrm{d}X(t)$, $Z = \int_A^B b(t)\,\mathrm{d}X(t)$ *be two stochastic integrals defined as limits in the mean and assume that* Y *has* ε*-linear regression and* ε*-scatter on* Z *and denote the distribution function of* Z *by* F_1. *Then there exists a normal distribution* $\Phi_1 = \Phi[(x - a_1)/\sigma_1]$ *such that the Lévy distance* $L(F_1, \Phi_1) \leqq C\varepsilon_2$ *where* $\varepsilon_2 = O[(\ln\ln 1/\varepsilon)^{-1/11}]$, *provided* ε *is chosen sufficiently small. The mean* a_1 *and the variance* σ_1^2 *of* Φ_1 *are determined according to lemmas* 3.6 *and* 4.2, *while* C *is a constant.*

THEOREM 2. *Suppose that the conditions of theorem* 1 *are satisfied and let* $A = t_{n0} < t_{n1} < \ldots < t_{nn} = B$ $(n = 1, 2, \ldots)$ *be a sequence of sub-divisions of the interval* $[A, B]$ *such that* $\lim_{n\to\infty} \max_{1\leqq k\leqq n} (t_{nk} - t_{nk-1}) = 0$ *and denote the distribution function of the increments of the* $X(t)$ *process over the interval* $[t_{n,k-1}, t_{n,k}]$ by F_{kn}.

Then there exists a normal distribution $\Phi_3 = \Phi[(x - a_3)/\sigma_3]$ *such that, for* n *sufficiently large,* $L(F_{kn}\Phi_3) < C\varepsilon_3$ *where*

$$\varepsilon_3 = O\left[\left(\ln\ln\ln\frac{1}{\varepsilon}\right)^{-1/11}\right].$$

Here a_3 *and* σ_3 *are determined according to lemma* 3.6.

In the following we use the letter C for constants (not always the same) and we write often Φ_j for $\Phi[(x - a_j)/\sigma_j]$.

3. SOME LEMMAS

LEMMA 3.1. Let $X(t)$ be a homogeneous stochastic process with independent increments which is defined in a finite closed interval $[A, B]$. Assume further that $X(t)$ is a second order process whose mean value function and covariance function are of bounded variation in $[A, B]$. Denote the characteristic function of the increment $X(t + \tau) - X(t)$ by $\phi(u, \tau)$ and write $\psi(u) = \ln\phi(u, 1)$. Suppose that the functions $a(t)$ and $b(t)$ are continuous and let

$$Y = \int_A^B a(t)\,\mathrm{d}X(t), \quad Z = \int_A^B b(t)\,\mathrm{d}X(t).$$

Denote the characteristic functio᷉ of the random vector (Y, Z) by $g(u, v)$ and write $h(u, v) = \ln g(u, v)$. Then

$$h(u, v) = \int_A^B \psi[u\, a(t) + v\, b(t)]\, \mathrm{d}t\,. \tag{3.1}$$

For the proof we refer to Laha and Lukacs [1].

LEMMA 3.2. Let $g(u, v) = \mathscr{E}[\exp(\mathrm{i}uY + \mathrm{i}vZ)]$ be the joint characteristic function of the random variables Y and Z and suppose that Y has ε-linear regression on Z, then

$$\left.\frac{\partial g(u, v)}{\partial u}\right|_{u=0} = \mathrm{i}\alpha g(0, v) + \beta \frac{\mathrm{d}g(0, v)}{\mathrm{d}v} + \mathrm{i}r(v)\,.$$

Proof. It follows from (2.1a) that

$$\mathscr{E}(Y\mathrm{e}^{\mathrm{i}vZ}) = \alpha\, \mathscr{E}(\mathrm{e}^{\mathrm{i}vZ}) + \beta\, \mathscr{E}(Z\mathrm{e}^{\mathrm{i}vZ}) + r(v)\,.$$

We substitute formulas (2.12b) and (2.12d) into this equation and obtain the statement of the lemma.

LEMMA 3.3. Let again $g(u, v) = \mathscr{E}[\exp(\mathrm{i}uY + \mathrm{i}vZ)]$ and suppose that Y has ε-scatter on Z. Then

$$\left.\frac{\partial^2 g(u, v)}{\partial u^2}\right|_{u=0} = -(\alpha^2 + \sigma_0^2)\, g(0, v) + 2\mathrm{i}\alpha\beta \frac{\mathrm{d}g(0, v)}{\mathrm{d}v} + \beta^2 \frac{\mathrm{d}^2 g(0, v)}{\mathrm{d}v^2} - t(v)\,.$$

Proof. It follows from (2.4) that

$$\mathscr{E}(Y^2\mathrm{e}^{\mathrm{i}vZ}) = (\alpha^2 + \sigma_0^2)\, \mathscr{E}(\mathrm{e}^{\mathrm{i}vZ}) + \beta^2\, \mathscr{E}(Z^2\mathrm{e}^{\mathrm{i}vZ}) + 2\alpha\beta\, \mathscr{E}(Z\mathrm{e}^{\mathrm{i}vZ}) + t(v)$$

Using (2.12c), (2.12e) and (2.12d) we obtain the statement of the lemma.

We rewrite the equations of lemmas 3.2 and 3.3 in terms of the function $h(u, v) = \ln g(u, v)$. Therefore we must restrict $g(u, v)$ to a neighbourhood of the origin where $g(u, v) \neq 0$. After some elementary computations we obtain the following result.

LEMMA 3.4. Let $h(u, v) = \ln g(u, v) = \ln \mathscr{E}[\exp(\mathrm{i}uY + \mathrm{i}vZ)]$ and suppose that Y has ε-regression and ε-scatter on Z then

$$\left.\frac{\partial h(u, v)}{\partial u}\right|_{u=0} = \mathrm{i}\alpha + \beta \frac{\mathrm{d}h(0, v)}{\mathrm{d}v} + \frac{\mathrm{i}r(v)}{g(0, v)}\,, \tag{3.2}$$

$$\left.\frac{\partial^2 h(u, v)}{\partial u^2}\right|_{u=0} = -\sigma_0^2 + \beta^2 \frac{\mathrm{d}^2 h(0, v)}{\mathrm{d}v^2} + \left[\frac{r(v)}{g(0, v)}\right]^2 + \tag{3.3}$$

$$+ 2\alpha \frac{r(v)}{g(0, v)} - 2\mathrm{i}\beta \frac{\mathrm{d}h(0, v)}{\mathrm{d}v} \frac{r(v)}{g(0, v)} - \frac{t(v)}{g(0, v)}\,.$$

We simplify (3.3) by writing

$$s_1(v) = \left[\frac{r(v)}{g(0, v)}\right]^2 + 2\alpha \frac{r(v)}{g(0, v)} - 2i\beta \frac{dh(0, v)}{dv} \frac{r(v)}{g(0, v)} - \frac{t(v)}{g(0, v)} \tag{3.4}$$

and get

$$\left.\frac{\partial^2 h(u, v)}{\partial u^2}\right|_{u=0} = -\sigma_0^2 + \beta^2 \frac{d^2 h(0, v)}{dv^2} + s_1(v)\,. \tag{3.3a}$$

The most important tools in our work will be two results, due to Zolotarev. The first of these (lemma 3.5) estimates the Lévy distance* between two distribution functions by means of their characteristic functions (see Zolotarev [5] and [6]). The second result (lemma 3.6) is a stability theorem for Cramér's famous theorem. It states, again in terms of the Lévy distance, that a distribution function which is close to the normal distribution can only have factors which are also close to normal distributions. Estimates for the closeness of the factors are given (see Zolotarev [4]).

LEMMA 3.5. Let F and G be two distribution functions with characteristic functions f and g. Then the Lévy distance $L(F, G) \leqq B_r \Delta_r^{1/(1+r)}$ where

$$B_r < (2/r)(1 + r^2)\,.\,\pi^{-1/(1+r)} \text{ while } \Delta_r = \sup_{t>0} [|f(t) - g(t)|/t^r].$$

LEMMA 3.6. Let $\Phi(x) = (\sqrt{2\pi})^{-1} \int_{-\infty}^{x} e^{-y^2/2}\,dy$ be the standardized normal distribution, and let $F = F_1^* F_2$ be such that $L(F, \phi) \leqq \varepsilon < 1$. Then one can find a normal distribution ϕ_1 (with mean a_1 and variance σ_1^2) such that $L(F_1, \phi_1) \leqq C(\ln 1/\varepsilon)^{-1/11}$. Here $a_1 = \int_{|x| \leqq N} x\, dF_1(x)$, $\sigma_1^2 = \int_{|x| \leqq N} x^2\, dF_1(x) - a_1^2$ while $N = 1 + \sqrt{(2 \ln 1/\varepsilon)}$.

For the proof of this lemma we refer to the paper by Zolotarev [4].

4. PROOF OF THE THEOREMS

Let $g(u, v)$ be the joint characteristic function of the integrals $Y = \int_A^B a(t)\, dX(t)$ and $Z = \int_A^B b(t)\, dX(t)$. According to lemma 3.1 one has

$$h(u, v) = \ln g(u, v) = \int_A^B \psi[u\, a(t) + v\, b(t)]\, dt \tag{4.1}$$

where $\psi(u) = \ln \phi(u, 1)$ and $\phi(u, 1)$ is the characteristic function of the increment of the $X(t)$ process over an interval of length 1.

---

* The Lévy distance $L(F, G)$ between two distribution functions F and G is defined as $L(F, G) = \inf\{h : F(x - h) - h \leqq G(x) \leqq F(x + h) + h \text{ for all } x\}$.

Since $X(t)$ is a second order process we see that the functions $\psi(u)$, $\psi'(u)$ and $\psi''(u)$ are continuous. It follows from lemma 3.4 that $h(u, v)$ satisfies the relations (3.2) and (3.3a). We then obtain from (4.1) and (3.2) the equation

$$(4.2) \qquad \int_A^B a(t)\, \psi'[v\, b(t)]\, dt = i\alpha + \beta \int_A^B b(t)\, \psi'[v\, b(t)]\, dt + \frac{ir(v)}{g(0, v)}.$$

Similarly from (4.1) and (3.3a) we get

$$(4.3) \qquad \int_A^B a^2(t)\, \psi''[v\, b(t)]\, dt = -\sigma_0^2 + \beta^2 \int_A^B b^2(t)\, \psi''[v\, b(t)]\, dt + s_1(v).$$

We differentiate (4.2) once more and obtain

$$(4.3a) \qquad \int_A^B a(t)\, b(t)\, \psi''[v\, b(t)]\, dt = \beta \int_A^B b^2(t)\, \psi''[v\, b(t)]\, dt + i \frac{d}{dv}\left[\frac{r(v)}{g(0, v)}\right].$$

We use (4.3) and (4.3a) to form the expression

$$\int_A^B a^2(t)\, \psi''[v\, b(t)]\, dt - 2\beta \int_A^B a(t)\, b(t)\, \psi''[v\, b(t)]\, dt + \beta^2 \int_A^B b^2(t)\, \psi''[v\, b(t)]\, dt$$

and get

$$(4.4) \qquad \int_A^B [a(t) - \beta\, b(t)]^2\, \psi''[v\, b(t)]\, dt = -\sigma_0^2 + s_2(v)$$

where

$$(4.4a) \qquad s_2(v) = s_1(v) + 2i\beta \frac{d}{dv}\left[\frac{r(v)}{g(0, v)}\right].$$

We put here $v = 0$ and see that

$$(4.5) \qquad \int_A^B \{\psi''[v\, b(t)] - \psi''(0)\}\, [a(t) - \beta\, b(t)]^2\, dt = s_2(v) - s_2(0).$$

We introduce the function

$$\theta(u) = \psi(u) - \psi(-u) = \ln [\phi(u, 1)\, \phi(-u, 1)].$$

The function $\theta(u)$ is the logarithm of a real infinitely divisible characteristic function having finite second moments. It admits therefore the (Kolmogorov) canonical representation (see Lukacs [2])

$$\theta(v) = \int_{-\infty}^{\infty} (\cos vx - 1)\, \frac{d\tilde{K}(x)}{x^2}$$

where $\tilde{K}(x)$ is a bounded, non-decreasing function such that $\tilde{K}(-\infty) = 0$. It is then easily seen that

$$\left|\frac{\theta''(v)}{\theta''(0)}\right| \leqq 1 . \tag{4.6}$$

In (4.5) we replace the variable v by $(-v)$ and add the equation so obtained to (4.5) and get

$$\int_A^B [a(t) - \beta\, b(t)]^2 \{\theta''[v\, b(t)] - 2\psi''(0)\}\, dt = s_2(v) + s_2(-v) - 2s_2(0) .$$

We note that $\theta''(0) = 2\psi''(0) \neq 0$ (since the increments have a non-degenerate distribution) and obtain from the last equation the relation

$$\int_A^B [(a(t) - \beta\, b(t)]^2 \left\{1 - \frac{\theta''[v\, b(t)]}{\theta''(0)}\right\} dt = \tag{4.7}$$

$$= \frac{1}{\theta''(0)} [2\, s_2(0) - s_2(v) - s_2(-v)] .$$

It follows then from the generalized mean value theorem of integral calculus that there exists a value t_0, $A < t_0 < B$, such that

$$[a(t_0) - \beta\, b(t_0)]^2 \int_A^B \left\{1 - \frac{\theta''[v\, b(t)]}{\theta''(0)}\right\} dt = \tag{4.8}$$

$$= \frac{1}{\theta''(0)} [2\, s_2(0) - s_2(v) - s_2(-v)] .$$

Let $M_0 = [a(t_0) + \beta\, b(t_0)]^2 \geqq 0$. If $M_0 = 0$ then the right hand side of (4.8) and therefore also of (4.7) is zero and (4.7) reduces to

$$\int_A^B [a(t) - \beta\, b(t)]^2 \left\{1 - \frac{\theta''[v\, b(t)]}{\theta''(0)}\right\} dt = 0 .$$

This equation was treated in [1] and it was shown there that it led to the conclusion that $X(t)$ is a Wiener process. The statements of both theorems are then trivially true. We can therefore assume without loss of generality that $M_0 > 0$ and see that

$$\int_A^B \left[1 - \frac{\theta''[v\, b(t)]}{\theta''(0)}\right] dt = \frac{1}{M_0\, \theta''(0)} [2\, s_2(0) - s_2(v) - s_2(-v)] . \tag{4.9}$$

The presence of the terms on the right hand side of (4.9) is caused by the assumption that Y has ε-linear regression and ε-scatter on Z (and not, as in [1] linear regression and constant scatter). In order to assess the influence of our assumption, we have to estimate the right hand side of (4.9). Such an estimate will enable us to apply lemma 3.5.

We estimate first $s_1(v)$ and see from (3.4) that

$$\text{(4.9a)}\qquad |s_1(v)| \leqq \left|\left[\frac{r(v)}{g(0, v)}\right]\right|^2 + 2|\alpha| \left|\frac{r(v)}{g(0, v)}\right| + 2|\beta| \left|\frac{dg(0, v)}{dv}\right| \left|\frac{r(v)}{[g(0, v)]^2}\right| + \\ + \left|\frac{t(v)}{g(0, v)}\right| .$$

We select a real number δ so that $0 < \delta < \frac{1}{2}$ and define

$$\text{(4.10)}\qquad q = q(\varepsilon) = \sup \{\xi : |g(0, v)| > \varepsilon^\delta \text{ for } |v| < \xi\}$$

and we restrict v to the neighbourhood of the origin where $|v| < q$. Since $g(0, v) = \mathscr{E}(e^{ivZ})$ we see that

$$\text{(4.11a)}\qquad \left|\frac{dg(0, v)}{dv}\right| \leqq z_1 = \mathscr{E}(|Z|)$$

and conclude from (4.10), (2.6) and (2.8) that

$$\text{(4.11b)}\qquad \left|\frac{r(v)}{[g(0, v)]^2}\right| \leqq \varepsilon^{1-2\delta}, \quad \left|\frac{r(v)}{g(0, v)}\right| \leqq \varepsilon^{1-\delta}$$

and

$$\text{(4.11c)}\qquad \left|\frac{dr(v)}{dv}\right| \leqq \varepsilon z_1 .$$

It follows from (2.11) that

$$\text{(4.11d)}\qquad \left|\frac{t(v)}{g(0, v)}\right| \leqq \varepsilon^{1-2\delta}[1 + 2|\alpha| + 2|\beta| z_1 + \varepsilon]$$

or

$$|s_1(v)| \leqq 2|\beta| z_1 \varepsilon^{1-2\delta} \left\{1 + \frac{\varepsilon^\delta}{2|\beta| z_1} [2 + 4|\alpha| + 2|\beta| z_1 + 2\varepsilon]\right\} .$$

Since $(\varepsilon^\delta/(2|\beta| z_1))[2 + 4|\alpha| + 2|\beta| z_1 + 2\varepsilon] < 1$ provided ε is sufficiently small, we have

$$\text{(4.12)}\qquad |s_1(v)| \leqq 4|\beta| z_1 \varepsilon^{1-2\delta}$$

provided that $|v| < q$ and that ε is sufficiently small. We see from (4.4a) that

$$|s_2(v)| \leqq |s_1(v)| + 2|\beta| \left|\frac{1}{g(0, v)} \frac{dr(v)}{dv} - \frac{r(v)}{[g(0, v)]^2} \frac{dg(0, v)}{dv}\right|$$

and use (4.11a), (4.11b) and (4.12) to show that

$$|s_2(v)| \leqq 4|\beta| z_1 \varepsilon^{1-2\delta} + 2|\beta| [\varepsilon^{1-\delta} z_1 + \varepsilon^{1-2\delta} z_1] .$$

Hence

(4.13) $$|s_2(v)| \leqq 8|\beta|\, z_1 \varepsilon^{1-2\delta} \quad \text{for} \quad |v| < q \quad \text{and } \varepsilon \text{ sufficiently small}.$$

Therefore

$$|2s_2(0) - s(v) - s(-v)| \leqq 32|\beta|\, z_1 \varepsilon^{1-2\delta} \quad \text{for} \quad |v| < q \quad \text{and } \varepsilon \text{ sufficiently small}.$$

We put

$$\frac{32|\beta|\, z_1}{M_0|\theta''(0)|} = C$$

and see from (4.6) and (4.8) that

(4.14) $$0 \leqq \int_A^B \left\{1 - \frac{\theta''[v\, b(t)]}{\theta''(0)}\right\} dt \leqq C\varepsilon^{1-2\delta}.$$

We show, by means of an indirect proof, that (4.14) is only possible if

(4.15) $$0 \leqq 1 - \frac{\theta''[v\, b(t)]}{\theta''(0)} \leqq \frac{C}{B-A}\, \varepsilon^{1-2\delta}.$$

The left inequality of (4.14) is, according to (4.6), always true. To carry the indirect proof we assume therefore tentatively that

$$1 - \frac{\theta''[v\, b(t)]}{\theta''(0)} > \frac{C}{B-A}\, \varepsilon^{1-2\delta}.$$

Then

$$\int_A^B \left\{1 - \frac{\theta''[v\, b(t)]}{\theta''(0)}\right\} dt > C\varepsilon^{1-2\delta}.$$

This contradicts (4.14) so that (4.15) is valid. Since $\theta''(0) < 0$, it follows from (4.15) that

$$0 \geqq \theta''(0) - \theta''[v\, b(t)] \geqq \frac{C}{B-A}\, \theta''(0)\, \varepsilon^{1-2\delta}$$

or

(4.16) $$\theta''(0)\left[1 - \frac{C}{B-A}\, \varepsilon^{1-2\delta}\right] \geqq \theta''[v\, b(t)] \geqq \theta''(0)$$

for $|v| < q$ and ε sufficiently small.

We note that $\theta'(0) = \theta(0) = 0$ and integrate (4.16) with respect to v from 0 to u and get

$$\theta''(0)\left[1 - \frac{C}{B-A}\, \varepsilon^{1-2\delta}\right] u \geqq \int_0^u \theta''[v\, b(t)]\, dt \geqq \theta''(0)\, u$$

or

$$\theta''(0)\left[1 - \frac{C}{B-A}\, \varepsilon^{1-2\delta}\right] u \geqq \frac{1}{b(t)}\, \theta'[u\, b(t)] \geqq \theta''(0)\, u.$$

We integrate once more from 0 to w so that

$$\frac{w^2}{2}\,\theta''(0)\left[1-\frac{C}{B-A}\,\varepsilon^{1-2\delta}\right] \geqq \frac{1}{b^2(t)}\,\theta[w\,b(t)] \geqq \theta''(0)\,\frac{w^2}{2}$$

for $|w| < q$ and ε sufficiently small;
or if we write

(4.17)
$$C_1 = \frac{C}{B-A}, \quad C_2 = 1 - C_1\varepsilon^{1-2\delta},$$

$$\frac{w^2}{2}\,C_2\,\theta''(0)\,b^2(t) \geqq \theta[w\,b(t)] \geqq \theta''(0)\,b^2(t)\,\frac{w^2}{2}.$$

We integrate with respect to t and see that

$$-\gamma C_2\,\frac{w^2}{2} \geqq \int_A^B \theta[w\,b(t)]\,\mathrm{d}t \geqq -\gamma\,\frac{w^2}{2}$$

for $|w| < q$ and ε sufficiently small. Here

$$\gamma = -\theta''(0)\int_A^B b^2(t)\,\mathrm{d}t > 0\,.$$

Let

$$k(v) = \exp\left\{\int_A^B \theta[v\,b(t)]\,\mathrm{d}t\right\}$$

then

(4.18)
$$\mathrm{e}^{-\gamma C_2 v^2/2} \geqq k(v) \geqq \mathrm{e}^{-\gamma v^2/2}$$

for $|v| < q$ and ε sufficiently small.

Moreover, we see from (4.1) that

(4.19)
$$k(v) = g(0, v)\,g(0, -v) = |g(0, v)|^2$$

and it follows from (4.10) that

$$q = q(\varepsilon) = \sup\{\xi : |k(v)| > \varepsilon^{2\delta} \text{ for } |v| < \xi\}$$

We need now the following lemma.

LEMMA 4.1. Let $f(t)$ be a possibly complex valued continuous function of the real variable t and suppose that $f(0) \neq 0$ and that $|f(-t)| = |f(t)|$. Let $0 < \varepsilon < |f(0)|$ and suppose that

$$q(\varepsilon) = \sup\{\xi : |f(t)| > \varepsilon \text{ for } |t| < \xi\} < \infty\,.$$

Assume further that $|f(t)| > \eta$ for $|t| < q(\varepsilon)$ then $\varepsilon > \eta$.

For the proof we refer to E. Lukacs [3] (Lemma 2.2).

It follows from (4.18) and lemma 4.1 that for $|v| < q$ and ε sufficiently small

$$\varepsilon^{2\delta} > e^{-\gamma v^2/2},$$

so that

$$\varepsilon^{2\delta} \geqq e^{-\gamma q^2/2}$$

or

$$2\delta \ln \frac{1}{\varepsilon} \leqq \frac{\gamma}{2} q^2$$

hence

$$q^2 \geqq \frac{4\delta}{\gamma} \ln \frac{1}{\varepsilon}. \tag{4.20}$$

We rewrite (4.18) as

$$e^{-C_2 u^2/2} \geqq k\left(\frac{u}{\sqrt{\gamma}}\right) \geqq e^{-u^2/2} \quad \text{for} \quad |u| < q\sqrt{\gamma}. \tag{4.18a}$$

Let $K(x)$ be the distribution function whose characteristic function is $k(v)$. Then $k(u/\sqrt{\gamma})$ is the characteristic function of $K_\gamma = K(x\sqrt{\gamma})$.

Our next aim is to estimate the Lévy distance $L(K_\gamma, \Phi)$ between K_γ and the standardised normal distribution

$$\Phi = \Phi(x) = \frac{1}{\sqrt{2\pi}} \int_{-\infty}^{x} e^{-y^2/2}\, dy.$$

In order to do this we have to obtain an upper bound for $|k(u/\sqrt{\gamma}) - e^{-u^2/2}|$. We conclude from (4.18a) after an elementary computation

$$0 \leqq k\left(\frac{u}{\sqrt{\gamma}}\right) - e^{-u^2/2} \leqq e^{-u^2/2} \left\{\exp\left(C_1 \frac{u^2}{2} \varepsilon^{1-2\delta}\right) - 1\right\}.$$

Using the inequality

$$|e^w - 1| \leqq |w|\, e^{|w|}$$

we get

$$\left|k\left(\frac{u}{\sqrt{\gamma}}\right) - e^{-u^2/2}\right| \leqq C_1 \frac{u^2}{2} \varepsilon^{1-2\delta} \exp\left[-\frac{u^2}{2}\left(1 - C_1\varepsilon^{1-2\delta}\right)\right]$$

so that

$$\left|k\left(\frac{u}{\sqrt{\gamma}}\right) - e^{-u^2/2}\right| \leqq C_1 \varepsilon^{1-2\delta} u^2/2 \quad \text{for} \quad |u| < q\sqrt{\gamma}. \tag{4.21}$$

In order to apply lemma 3.5 we have to determine

$$\Delta_r = \sup_{u>0} \left| \frac{k\left(\dfrac{u}{\sqrt{\gamma}}\right) - e^{-u^2/2}}{u^r} \right| .$$

We write

$$\Delta_r^{(1)} = \sup_{0<u<q\sqrt{\gamma}} \left| \frac{k\left(\dfrac{u}{\sqrt{\gamma}}\right) - e^{-u^2/2}}{u^r} \right| ,$$

and

$$\Delta_r^{(2)} = \sup_{u \geqq q\sqrt{\gamma}} \left| \frac{k\left(\dfrac{u}{\sqrt{\gamma}}\right) - e^{-u^2/2}}{u^r} \right|$$

so that $\Delta_r = \max(\Delta_r^{(1)}, \Delta_r^{(2)})$.

It follows from (4.21) that, for ε sufficiently small,

$$\Delta_r^{(1)} \leqq C_1 \varepsilon^{1-2\delta} u^{2-r}/2 .$$

For $r > 2$ the right hand side of this inequality tends to infinity as u goes to zero; we select therefore $r = 2$ and have

$$\Delta_2^{(1)} \leqq C_1 \varepsilon^{1-2\delta}/2 .$$

We also see – using (4.20) – that

$$\Delta_2^{(2)} \leqq \frac{2}{\gamma q^2} \leqq \frac{1}{2\delta \ln \dfrac{1}{\varepsilon}}$$

so that

$$\Delta_2 \leqq \max \left[C_1 \varepsilon^{1-2\delta}/2 , \quad \frac{1}{2\delta \ln \dfrac{1}{\varepsilon}} \right] .$$

Using de l'Hospital's rule one see that

$$\lim_{\varepsilon \to 0} \left[\frac{C_1}{2} \varepsilon^{1-2\delta} \Big/ \frac{1}{2\delta \ln \dfrac{1}{\varepsilon}} \right] = 0$$

so that

$$\frac{C_1}{2} \varepsilon^{1-2\delta} < \frac{1}{2\delta \ln \dfrac{1}{\varepsilon}}$$

for ε sufficiently small, or

$$\Delta_2 \leqq \frac{1}{2\delta \ln \frac{1}{\varepsilon}} .$$

We also have

$$B_2 = 9\pi^{-1/3}$$

and conclude from lemma 3.5 that

$$L(K_\gamma, \Phi) \leqq 9 \left(\frac{1}{2\pi\delta \ln \frac{1}{\varepsilon}} \right)^{1/3}$$

or

$$L(K_\gamma, \Phi) \leqq \varepsilon_1 \tag{4.22}$$

where

$$\varepsilon_1 = 9 \left(\frac{1}{2\pi\delta \ln \frac{1}{\varepsilon}} \right)^{1/3} = \mathrm{O} \left[\frac{1}{\ln \frac{1}{\varepsilon}} \right]^{1/3} . \tag{4.22a}$$

The distribution function $K_\gamma = K(x \sqrt{\gamma})$ has the characteristic function $k(u/\sqrt{\gamma}) = g(0, u/\sqrt{\gamma})\, g(0, -u/\sqrt{\gamma})$. The function $g(0, u/\sqrt{\gamma})$ is the characteristic function of the random variable $(1/\sqrt{\gamma})\, Z = (1/\sqrt{\gamma}) \int_A^B b(t)\, \mathrm{d}X(t)$; we denote the distribution function of this random variable by G_γ. According to lemma 3.6 it is possible to find a normal distribution

$$\Phi_2 = \Phi \left(\frac{x - a_2}{\sigma_2} \right)$$

such that

$$L(G_\gamma, \Phi_2) \leqq \varepsilon_2 \tag{4.23}$$

where

$$\varepsilon_2 = C \left(\ln \frac{1}{\varepsilon_1} \right)^{-1/11} = \mathrm{O} \left[\left(\ln \ln \frac{1}{\varepsilon} \right)^{-1/11} \right] . \tag{4.23a}$$

To complete the proof of theorem 1 we need the following lemma which is due to Thompson [7].

LEMMA 4.2. Let $F(x)$ and $G(x)$ be two distribution functions and write $F_a = F(x/a)$, $G_a = G(x/a)$. Then $\min(a, 1)\, L(F_1, G_1) \leqq L(F_a, G_a) \leqq \max(a, 1)\, L(F_1, G_1)$. It follows from (4.23), (4.23a) and lemma 4.2 that there exists a normal distribution $\Phi_1 = \Phi[(x - a_1)/\sigma_1]$ such that

$$L(F_1, \Phi_1) \leqq \varepsilon_2 , \tag{4.24}$$

q.e.d.

To prove theorem 2 we note that the random variable Z can be represented as the limit in the quadratic mean (written l.i.m.) of a sequence of Riemann sums. Let

$$A = t_{n0} < t_{n1} < \dots < t_{nn} = B \quad (n = 1, 2, \dots)$$

be a sequence of subdivisions of the interval $[A, B]$ such that $\lim_{n\to\infty} \max_{1 \leq k \leq n} (t_{nk} - t_{nk-1}) = 0$ and let $S_n = \sum_{k=1}^{n} g(t_{nk}) [X(t_{nk}) - X(t_{nk-1})]$. Then $\operatorname{l.i.m.}_{n\to\infty} S_n = Z$.

It is known that convergence in the quadratic mean implies the weak convergence of the corresponding distribution functions. We write Lim for the weak convergence of distribution functions and denote the distribution function of the random variable S_n by H_n. Then

$$\operatorname*{Lim}_{n\to\infty} H_n = F_1 .$$

Since weak convergence is equivalent to convergence in the Lévy metric, it follows that

$$\lim_{n\to\infty} L(H_n, F_1) = 0 .$$

We see therefore that for n sufficiently large, $n > n_0 = n_0(\varepsilon_2)$,

$$L(H_n, F_1) < \varepsilon_2 . \tag{4.25}$$

Using the triangle inequality we see from (4.25) and (4.24) that

$$L(H_n, \Phi_1) \leqq O(\varepsilon_2) \quad (n > n_0)$$

or, written in greater detail

$$L\left((H_n(x), \Phi\left(\frac{x - a_1}{\sigma_1}\right)\right) \leqq O(\varepsilon_2) .$$

Using again lemma 4.2 we see that

$$L(H_n(a_1 + \sigma_1 y), \Phi(y)) \leqq O(\varepsilon_2) . \tag{4.26}$$

Let H_{jn} be the distribution function of the random variable $g(t_{nj}) [X(t_{nj}) - X(t_{nj-1})]$. Then $H_n = H_{kn} * \tilde{H}_{kn}$ where $\tilde{H}_{kn}$ is the convolution of the $(n - 1)$ distribution functions H_{jn} with $j \neq k$. We write $H_n^1(y)$ and $H_{kn}^1(y)$ for $H_n(a_1 + \sigma_1 y)$ and $H_{kn}(a_1 + \sigma_1 y)$ respectively and see from the preceding equation that

$$H_n^1 = H_{kn}^1 * \tilde{H}_{kn}^1 . \tag{4.27}$$

We conclude then from lemma 3.6 and equations (4.26) and (4.27) that there exists a normal distribution Φ_4 such that

$$L(H_{kn}^1, \Phi_4) \leqq O(\varepsilon_3)$$

where

$$\varepsilon_3 = O\left[\left(\ln \ln \ln \frac{1}{\varepsilon}\right)^{-1/11}\right] .$$

Using again lemma 4.2 we see that there exists a normal distribution Φ_5 such that

$$L(H_{kn}, \Phi_5) \leqq O(\varepsilon_3).$$

We have denoted the distribution function of the increment $[X(t_{nk}) - X(t_{nk-1})]$ by F_{kn}, then

$$H_{kn}(x) = F_{kn}\left(\frac{x}{g(t_{nk})}\right)$$

so that

$$L\left(F_{kn}\left(\frac{x}{g(t_{nk})}\right), \quad \Phi_5\right) \leqq O(\varepsilon_3)$$

therefore there exists a normal distribution Φ_3 such that

$$L(F_{kn}, \Phi_3) \leqq O(\varepsilon_3).$$

REFERENCES

[1] R. G. Laha, E. Lukacs: On a characterization of the Wiener process. Trans. Second Prague Conference on information theory, statistical decision functions, random processes, held in 1959, pp. 307—312. Publishing House of the Czechoslovak Academy of Sciences, Prague 1960.

[2] E. Lukacs: Characteristic Functions (2nd edition). Charles Griffin & Co., London 1970.

[3] E. Lukacs: Stability theorems for characterisations by constant regression. Periodica Math. Hung. *2* (1972), 111—128.

[4] V. M. Zolotarev: On the stability of the decomposition of the normal law into components. Teoriya Veroyatnosti i ee primenenia *13* (1968), 738—742 (English translation: Theory of probability and applications *13* (1968), 697—700).

[5] V. M. Zolotarev: Some inequalities in probability connected with the Lévy metric. Dokl. A.N. SSSR *190* (1970), 1019—1021. (English translation: Soviet Mathematics Dokladi *11* (1970), 231—234.)

[6] V. M. Zolotarev: Estimates of the difference between distributions in the Lévy metric. Trudi Steklov *112* (1971), 224—231 (English translation: Proc. Steklov Inst. Math. *112* (1971), 232—240; American Math. Soc. Providence R.I. (1973).

[7] J. W. Thompson: A note on the Lévy distance. Journal of Applied Probability *12* (1975), 412—414.

Bowling Green State University

and

University of Sheffield

THE FERMION PROCESS — A MODEL OF STOCHASTIC POINT PROCESS WITH REPULSIVE POINTS

ODILE MACCHI

ORSAY

1. INTRODUCTION

This paper presents a model of stochastic point process (P.P.) whose occurrences are repulsive to one another. The origin of the model arises in the theory of quantum particles and describes the statistical distribution of a fermion system in thermal equilibrium. For such a system, two P.P. can be defined which describe the places of the particles at a given time or the instants when any particle passes in a given point of space. Then physical arguments lead to the value of coincidence densities (C.D.) of these P.P.

In section 2, we briefly recall how the system of C.D. can define the statistics of a P.P. In section 3, we give the mathematical model of a fermion process (F.P.), starting from these C.D. and we point out repulsivity. Then we prove the existence of the F.P., give some details about the physical process, and illustrate the F.P. with an example.

2. PRELIMINARIES: REGULAR POINT PROCESSES

For the sake of clarity, we must recall some useful material about P.P. These ideas can be found with more details in [1], particularly a definition of P.P. suitable for our purposes. We consider P.P. in $\mathbf{R}^M$ that is, the state space for the occurrences is a Borel-set X in $\mathbf{R}^M$. The reference measure is the usual Lebesgue-measure.

2.1. Exlusion Probability Densities

Let $p_n(t_1, \ldots, t_n)$, $n = 0, 1, \ldots$, be a sequence of non-negative, symmetrical and measurable functions respectively defined on X^n.

We assume that the convergence radius r_p of the power series

(1) $$\Phi_p(z) = \sum_{n=0}^{\infty} \frac{z^n}{n!} \int_{X^n} p_n(t_1, \ldots, t_n) \, dt_1 \ldots dt_n ,$$

is greater than one

(2) $$r_p > 1 ,$$

(r_p is called the convergence radius of the sequence p_n). Moreover suppose that this sequence is a unit, meaning that

(3) $$\Phi_p(1) = 1 .$$

Then, the sequence p_n defines an unique P.P. in X. The statistical features of that process are described by the following: for distinct points $t_1, \ldots, t_n$, $p_n(t_1, \ldots, t_n) \cdot dt_1 \ldots dt_n$ is the probability that the P.P. has exclusively n occurrences within X, respectively located in the infinitesimal volumes $\{t_1, dt_1\}, \ldots, \{t_n, dt_n\}$. Therefore the functions $p_n(t_1, \ldots, t_n)$ are called the exclusion probability densities (E.P.D.) of the P.P. They entirely define the statistics of that P.P., which is called regular because all exclusion probabilities admit densities with respect to the Lebesgue-measure on X^n.

2.2. Coincidence Densities

For regular P.P., one can define another significant sequence of functions, $k = 0, 1, \ldots$

(4) $$h_k(t_1, \ldots, t_k) = \sum_{j=0}^{\infty} \frac{1}{j!} \int_{X^j} p_{k+j}(t_1, \ldots, t_k, u_1, \ldots, u_j) \, du_1 \ldots du_j .$$

Their statistical meaning is the following: for distinct points $t_1, \ldots, t_k$, $h_k(t_1, \ldots, t_k) \cdot dt_1 \ldots dt_k$ is the probability that the P.P. has one occurrence within each infinitesimal volume $\{t_1, dt_1\}, \ldots, \{t_k, dt_k\}$, no matter how many occurrences there are within the whole volume X. Therefore the functions $h_k(t_1, \ldots, t_k)$ are called coincidence densities (C.D.). They are also called product density functions by Russian litterature. It can also be shown that they are the densities of the factorial moments of the number of occurrences of the P.P. [2]. They are of great practical importance because coincidence events can be simply detected by counting experiments. On the other hand, measuring an exclusion probability involves not only ensuring that there are actually occurrences within each concerned volume, but also that there is no other occurrence within the whole state space X, which is much more difficult. Hence C.D. are practically more valuable than E.P.D.

2.3. Completely Regular P.P.

The system of C.D. of a regular point process thoroughly characterises this process, as well as the system of its E.P.D. Moreover, one can build a P.P. through its C.D., thanks to the following construction theorem proved in [1].

THEOREM 1. *The sequence of non-negative, symmetrical and measurable functions* $\{h_k\}$ *is the set of* C.D. *of an unique regular* P.P. *if its convergence radius* r_h *is greater than one and if all the functions*

$$p_n(t_1, \ldots, t_n) = \sum_{j=0}^{\infty} \frac{(-1)^j}{j!} \int_{X^j} h_{n+j}(t_1, \ldots, t_n, u_1, \ldots, u_j)\, du_1 \ldots du_j \tag{5}$$

are almost everywhere non-negative. Then $\{p_n\}$ *is the set of* E.P.D. *of the process.*

Notice that (4) and (5) are formally equivalent. Because this inversion formalism leading from the set of C.D. to that of E.P.D. is valid, the process is called completely regular. Now, we shall use theorem 1 to build the F.P. on the basis of its C.D.

3. DEFINITION AND MAJOR PROPERTY OF THE FERMION PROCESS

3.1. Definition

Let $\mathbf{P}_k$ be the set of all permutations $\alpha_1, \ldots, \alpha_k$ of the integers $1, \ldots, k$, and for any function $G(\alpha_1, \ldots, \alpha_k)$ of such a permutation, let us denote

$$\textstyle\sum_k^* G(\alpha_1, \ldots, \alpha_k) = \sum_{\alpha_1, \ldots, \alpha_k \in \mathbf{P}_k} \operatorname{sign}(\alpha_1, \ldots, \alpha_k)\, G(\alpha_1, \ldots, \alpha_k)\,.$$

DEFINITION: *A* F.P. *is a completely regular* P.P. *whose* C.D. *have the form*

$$h_k(t_1, \ldots, t_k) = \textstyle\sum_k^* \prod_{j=1}^{k} C(t_j, t_{\alpha_j})\,, \tag{6}$$

where the function $C(t, u)$ *is continuous over* X^2.

We call fermions the occurrences of that P.P.; the physical context is given in Section 5.

That function h_k is symmetrical and measurable. As for non-negativity, notice that (6) is the Fredholm determinant built upon the function $C(t, u)$ at points $t_1, \ldots, t_k$.

It is known ([3], p. 59) that this determinant is non-negative for all k, and all $t_1, \ldots, t_k$ iff $C(t, u)$ is non-negative definite. Suppose that $C(t, u)$ is so. Then notice that

$$\sum_k^* \prod_{j=1}^{k} C(t_j, t_{\alpha_j}) \leqq \prod_{j=1}^{k} C(t_j, t_j) . \tag{7}$$

Now consider the convergence radius r_h. It follows from (7) that

$$|\Phi_h(z)| \leqq \sum_{k=0}^{\infty} \frac{|z|^k}{k!} \left[\int_X C(t, t)\,\mathrm{d}t\right]^k . \tag{8}$$

Since $C(t, t)$ is continuous, for a bounded state-space X, the set $\{h_k\}$ has infinite convergence radius r_h.

In order to prove the existence of the F.P. by application of theorem 1, it remains to prove the non-negativity of (5). This is the difficult part of the paper. It will be outlined in section 4–3. At the moment, we take it for granted.

3.2. Repulsivity

The basic feature of the F.P. is that its occurrences tend to exclude one another. For $k = 2$, $h_2(t_1, t_2)$ is the conditional probability density of the presence of a fermion in t_2, given that one fermion has occurred in t_1. Now (6) shows that for a given t_1, this density vanishes as t_2 approaches t_1. Hence the fermion in t_1 is repulsive to other fermions. Moreover it derives from (6) that all C.D. vanish when their arguments are not distinct, which reinforces the repulsive character of the fermions. This property implies a greater degree of regularity in the process than in a Poisson process, at least over short intervals.

4. EXISTENCE OF THE FERMION PROCESS

4.1. The characteristic function of the F.P.

It can be shown [4] that the characteristic function (C.F.) $\varphi(u)$ of the number N of occurrences of a P.P. is related to the C.D. through

$$\varphi(u) = \sum_{k=0}^{\infty} \frac{(\mathrm{e}^{\mathrm{i}u} - 1)^k}{k!} \int_{X^k} h_k(t_1, \ldots, t_k)\,\mathrm{d}t_1 \ldots \mathrm{d}t_k . \tag{9}$$

Therefore, for the F.P., $\varphi(u)$ is the value at point

$$z = 1 - \mathrm{e}^{\mathrm{i}u} , \tag{10}$$

of the Fredholm determinant of the second kind integral equation

(11) $$f(t) - z \int_X C(t, u) f(u) \, du = g(t), \; t \in X .$$

This determinant is identical to the function $\Phi_h(-z)$ and its zeros are the eigenvalues z_j of the first kind integral equation which corresponds to (11).

Owing to the non-negative definite character of $C(t, u)$, the z_j are positive. By a simple study of the genus of sequence z_j, the order of the entire function $\Phi_h(z)$ and by application of Hadamard's theorem, it follows that [5]

(13) $$\Phi_h(-z) = \prod_{j=1}^{\infty} \left(1 - \frac{z}{z_j}\right),$$

or

(14) $$\varphi(u) = \prod_{j=1}^{\infty} \left[\left(1 - \frac{1}{z_j}\right) + \frac{e^{iu}}{z_j}\right].$$

If one can find the eigenvalues z_j of the integral kernel associated to $C(t, u)$ over the bounded Borel-set X, then the probability law of the fermion number is a straightforward consequence of (14).

4.2. The Necessary Condition

In order that (6) actually describes a P.P., it is necessary that (14) be the C.F. of an integer non-negative random variable. Such is the case if

(15) $$z_j \geqq 1 , \quad j = 1, 2, \dots ,$$

since each factor of (14) is the C.F. of a random variable which is 0 with probability $(1 - 1/z_j)$ and 1 with probability $1/z_j$. Conversely, if (15) does not hold, the Fourier transform of $\varphi(u)$ takes on negative values. Thus (15) is a necessary condition.

4.3. The Sufficient Condition

Let us assume that the integral equation

(16) $$g(t, u) - \int_X g(t, v) \, C(v, u) \, dv = C(t, u) , \quad \forall (t, u) \in X^2 ,$$

has an unique root $g(t, u)$. In particular it implies that (15) is replaced by the stronger condition (which is shown in [1] to be necessary)

(15)bis $$z_j > 1 , \quad j = 1, 2, \dots$$

Now the integral kernel associated to $g(t, u)$ over X, has eigenvalues $(z_j - 1)$, hence all positive. Thus $g(t, u)$ is non-negative definite. In that case, it is shown in [1] that the functions of (5), that is, the E.P.D. of the presumed P.P. (if it is such) are related to function $g(t, u)$ through formula

$$(17) \qquad p_n(t_1, \ldots, t_n) = \prod_{j=1}^{\infty} \left(1 - \frac{1}{z_j}\right) \sum_n^* \prod_{l=1}^{n} g(t_l, t_{\alpha_l}) .$$

Therefore the p_n are all non-negative. This finishes the proof that under condition (15)bis, model (6) is valid as the C.D. of a regular P.P., thus settling the existence of the F.P. and the following theorem

THEOREM 2. *The sequence of functions*

$$h_k(t_1, \ldots, t_k) = \sum_k^* \prod_{j=1}^{k} C(t_j, t_{\alpha_j})$$

is the set of C.D. *of a* F.P. *defined on the bounded space X, if the eigenvalues of the integral kernel associated to* $C(t, u)$ *over X are all greater than one.*

5. PHYSICAL CONTEXT

The origin of the F.P. is in Statistical Optics. One considers a particle system within a given volume X, using the wave-packet formalism that is, associating a random function to each particle. Each random function depends on the particle position and the time, and all have the same distribution. Their number is stochastic as is the number of particles within X. The statistics of the system are described by the superposition of these wave-packets.

Recently such statistics have been investigated theoretically [6] for a fermion system in thermal equilibrium. For the space F.P., that is, for the P.P. which consists of the locations of all fermions within X at a given time instant, it has been shown that the C.D. are given by (6), where $C(t, u)$ is related to the – space – covariance function $g(t, u)$ of the wave-packets through Fredholm equation (16). The time F.P. (that is, the random instants at which any particle passes in a given point of space) is completely similar under conditions of stationary speed for the particles.

Since $g(t, u)$ is a covariance function, it must be non-negative definite. Hence the eigenvalues of the associated kernel are all non-negative. It then derives from (16) that those associated to $C(t, u)$ are at least 1. Hence condition (15) is a physical necessity. Moreover the physical meaning of the z_j is that z_j^{-1} is the average number of fermions in the j-th mode of the system (a mode can be thought of as a frequency characteristic of the system). Thus condition (15) is simply the expression of the Pauli exclusion principle "at most one fermion per quantum state" which is set up as foundations in the physical model, and which accounts for the intrinsic repulsivity of fermions.

6. EXAMPLE

In this example the state space X is a bounded part of $\mathbf{R}$; for instance, the considered P.P. could be the ideal detection of physical fermions during time. We assume that the function $C(t, u)$ of model (6) satisfies the following

(18) $$C(t_1, t_2)\, C(t_2, t_3) = C(t_1, t_3)\, C(t_2, t_2)\,, \quad t_1 \leqq t_2 \leqq t_3\,.$$

It follows from (18) and (6) that C.D. can be written for increasing $t_1, \ldots, t_k$

(19) $$h_k(t_1, \ldots, t_k) = \prod_{i=1}^{k-1} \frac{h_2(t_i, t_{i+1})}{h_1(t_i)\, h_1(t_{i+1})} \prod_{i=1}^{k} h_1(t_i)\,.$$

This recursive form of C.D. is typical of a kind of renewal. In fact, it can be shown [4] on the basis of (19), that the random variables which represent the distances between successive fermions are independent, although not necessarily identically distributed. Let us assume moreover, that the F.P. is stationary. That means that $C(t, u)$ is a function of $(t - u)$ only. Then (18) is equivalent to

(20) $$C(t, u) = a \exp \frac{-2|t - u|}{L}\,.$$

Moreover, distances between fermions are now identically distributed. Thus the P.P. is an ordinary stationary renewal process. It can then be characterised by the probability density $v(t)$ of lifetime (distance from one fermion to the following one). For a stationary regular P.P., it is easily seen that $v(t)$ is related to the first order C.D. $h_1(0)$ and to the second order E.P.D. $p_2(0, t)$ over the closed interval $[0, t]$ by

(21) $$v(t) = \frac{p_2(0, t)}{h_1(0)}\,.$$

Therefore, by introducing (20) into (6) and (16), then evaluating (17) for $n = 2$ and $X = [0, t]$, the lifetime probability density follows. After some calculations, again involving the use of Hadamard's theorem, one finds

(22) $$v(t) = 2a(1 - aL)^{-1/2}\, \mathrm{e}^{-2t/L} \sinh\left[\frac{2t}{L}(1 - aL)^{-1/2}\right].$$

Obviously, the model cannot be valid unless

(23) $$aL \leqq 1\,.$$

It is easily checked that condition (23) is satisfied iff (15)bis is satisfied for all bounded intervals X of $\mathbf{R}$. Now L is the coherence time of each wave-packet, and aL is the average number of fermions during that time. Hence (23) is again the expression of the repulsivity of fermions.

7. CONCLUSION

We have presented a new P.P. model, originated in physics. Its basic property is repulsivity between occurrences. In the physical context it derives from Pauli exclusion principle. However, repulsivity may hold in many other statistical problems. Then the F.P. may turn out to be very usefull.

REFERENCES

[1] O. Macchi: The coincidence approach to stochastic point processes. Adv. Appl. Proba. *7* (1975), 83–122.

[2] J. E. Moyal: The general theory of stochastic population processes. Acta Math. *108* (1962), 1–31.

[3] E. F. Beckenbach, R. Bellman: Inequalities. Second Edition. Springer Verlag, New York 1965.

[4] O. Macchi: Stochastic point processes and multicoincidences. IEEE Trans. I. T. *17* (1971), 1, 2–7.

[5] R. P. Boas: Entire functions. Academic Press, New York 1954.

[6] C. Benard, O. Macchi: Detection and emission processes of quantum particles in a chaotic state. J. Math. Phys. *14* (1973), 2, 155–167.

Laboratoire des Signaux et Systemes

Centre National de la Recherche Scientifique

SOME RESULTS IN THE ADAPTIVE CONTROL OF LINEAR SYSTEMS

PETR MANDL

PRAGUE

A paper read by the present author at the Sixth Prague Conference dealt with the following adaptive policy to control a Markov chain with unknown parameters: "The trajectory $\{X_n, n = 0, 1, \ldots\}$ is observed, and the unknown parameters are estimated at times $n = 0, 1, 2, \ldots$ Thereafter, the optimal stationary policy is computed as if the estimates were the exact values of the unknown parameters. The value of the estimated optimal stationary policy at the state X_n gives the control parameter value for the transition to X_{n+1}". In the past three years this adaptive policy has been further investigated, since it represents a counterpart to the Bayesian approach to the adaptive control. Results of varying degree of completeness and generality were obtained for controlled Markov chains and processes with a finite state space, for processes of the diffusion type, and for discrete time linear systems ([4], [3], [2], [5]). In Sections 1 and 2 of the present paper we exemplify briefly the essential lines of the research on the case of discrete time linear systems. This makes possible a comparison with other approaches to adaptive control, since linear systems are the mostly explored subject of stochastic control theory. In Sections 3, 4, 5 we demonstrate the law of the iterated logarithm for quadratic functionals under adaptive control policies based on least squares estimation.

Besides usual denotations we use the following ones. The transpose of a vector or matrix M is denoted by M'. $\|M\|$ is the square root of the sum of squares of the elements of M. tr( ) is trace ( ). I( ) is the indicator of the random event in the brackets. const is a generic symbol for finite positive constants. If $\{a_n, n = 0, 1, \ldots\}$ and $\{b_n, n = 0, 1, \ldots\}$ are two sequences of random vectors, then $a_n = o(b_n)$, $n \to \infty$, means $\lim_{b \to \infty} \|a_n\|/\|b_n\| = 0$ a.s. (almost surely). Equalities involving conditional expectations are supposed to hold almost surely.

1.

Consider a controlled system, the output $\{X_n, n = 0, 1, \ldots\}$ of which fulfils

(1) $$X_{n+1} = \Phi_0 X_n + \Psi U_n + e_{n+1}\,, \quad n = 0, 1, \ldots$$

X_n is an r-dimensional column vector. U_n is the (s-dimensional) control signal at time n. The disturbances $\{e_n, n = 1, 2, \ldots\}$ are a sequence of mutually independent, identically distributed random r-vectors with zero mean, with covariance matrix Q, and with finite fourth moments. Φ_0 and Ψ are constant matrices of type $r \times r$ and $r \times s$, respectively. X_0 is assumed to be non-random for the sake of simplicity. Moreover,

(2) $$U_n = k_n X_n\,, \quad n = 0, 1, \ldots,$$

where the control matrices k_n of type $s \times r$ are for $n = 0, 1, \ldots$ functions of $\{X_0, \ldots, X_n\}$,

$$k_n = k_n(X_0, \ldots, X_n)\,, \quad n = 0, 1, \ldots$$

The sequence of matrix valued functions

(3) $$\{k_n, n = 0, 1, \ldots\}$$

is the *control policy*. If

$$k_n \equiv K\,, \quad n = 0, 1, \ldots,$$

where K is a constant matrix, the control policy is called a *stationary control* K. It is called *stable* if all eigenvalues of $\Phi_0 + \Psi K$ have modulus less than 1.

To define a criterion for comparing the control policies introduce the sum

(4) $$C_N = \sum_{n=0}^{N-1} (X_n' R X_n + U_n' S U_n)\,,$$

where R is a non-negative definite $r \times r$-matrix, and S is a positive definite $s \times s$-matrix. The most common way is to assume N fixed and to minimize the risk EC_N. This criterion is used in connection with the Bayesian method in the adaptive control (see e.g. [1]). An alternative approach, which will be followed here, is to assume N large, and to minimize $\lim_{N\to\infty} C_N$.

A stationary control K_0 is said to be *optimal*, if it is stable, and if, under K_0,

(5) $$\lim_{N\to\infty} N^{-1} C_N = \Theta \quad \text{a.s.}\,,$$

where Θ is a constant such that under any stable stationary control

$$\lim_{N\to\infty} N^{-1} C_N \geqq \Theta \quad \text{a.s.}$$

The optimality of K_0 is preserved, if we take non-stationary policies into consideration.

THEOREM 1. *Let Θ be defined as above and let the policy* (3) *be such that*

$$\sum_{n=1}^{\infty} E\|X_n\|^2/n^2 < \infty .$$

Then

$$\limsup_{N\to\infty} N^{-1}C_N \geqq \Theta \quad a.s.$$

Let us make two hypotheses.

ASSUMPTION 1. For a positive integer l the matrix

$$\sum_{k=0}^{l-1} (\Phi_0^k)' R\Phi_0^k$$

is positive definite.

ASSUMPTION 2. There exists an $s \times r$ matrix K for which $\Phi_0 + \Psi K$ has all eigenvalues inside the unit circle.

Under Assumptions 1 and 2 the optimal stationary control K_0 can be computed from the equations

$$W = R + K_0'SK_0 + (\Phi_0 + \Psi K_0)' W(\Phi_0 + \Psi K_0), \tag{6}$$

$$K_0 = -(S + \Psi' W\Psi)^{-1} \Psi' W\Phi_0 . \tag{7}$$

W is an $r \times r$ symmetric matrix. (6) and (7) can be solved using an iterative method. Furthermore,

$$\Theta = \mathrm{tr}(WQ) . \tag{8}$$

Note that (6), (7), and consequently K_0, do not depend on Q.

The class of non-stationary policies, under which (5) holds, is rather broad.

THEOREM 2. *Let, under the policy* (3),

$$\lim_{n\to\infty} k_n = K_0 \quad a.s. ,$$

then (5) *is valid.*

For better grading of the control policies we have to introduce additional criteria. The validity of the central limit theorem or of the law of the iterated logarithm with the same parameters as under K_0 are the most obvious choices.

THEOREM 3. *Under the optimal stationary control* K_0, $(C_N - N\Theta)/\sqrt{N}$ *is asymptotically normal* $N(0, \zeta)$, *as* $N \to \infty$. *Further*

(9) $$\limsup_{n\to\infty} (C_N - N\Theta)/\sqrt{(2N \log\log N)} = \sqrt{\zeta} \quad a.s.,$$

where

(10) $$\zeta = E(e_1' W e_1 - \Theta)^2 + 4\operatorname{tr}(HQ),$$

and H is the $r \times r$*-matrix, satisfying*

(11) $$H = (\Phi_0 + \Psi K_0)' (WQW + H)(\Phi_0 + \Psi K_0).$$

Section 2 contains a result about the validity of the central limit theorem. Section 3–5 are devoted to the law of the iterated logarithm.

2.

Let us now pass to the proper subject of the adaptive control theory which are the *systems with unknown parameters.* We assume that

$$\Phi_0 = \Phi(\alpha_0),$$

where $\Phi(\alpha)$ is a linear combination of known matrices $F_0, \ldots, F_q$,

(12) $$\Phi(\alpha) = F_0 + \alpha^1 F_1 + \ldots + \alpha^q F_q.$$

The true value α_0 of the q-dimensional parameter $\alpha = (\alpha^1, \ldots, \alpha^q)'$ is unknown. Ψ is assumed to be known, the knowledge of the autocovariance matrix Q of the disturbances is not required, since optimal stationary controls do not depend on Q.

The hypotheses made imply that to each α from a neighbourhood of α_0 the controller is able to compute a stationary control $\hat{K}(\alpha)$ which is optimal for the input-output relation

$$X_{n+1} = \Phi(\alpha) X_n + \Psi U_n + e_{n+1}, \quad n = 0, 1, \ldots$$

Hence, we may suppose that the controller knows a continuously differentiable $r \times s$-matrix function $K(\alpha)$ on the parameter space R^q, for which

$$K(\alpha_0) = K_0.$$

When controling the system by the method described in the introduction, the controller makes a series of estimates of α_0

(13) $$\{\alpha_N^*,\ N = 0, 1, \ldots\}\,.$$

In (13) α_N^* is a function of $X_0, \ldots, X_N$ taking values in R^q. Further, he sets

(14) $$k_n = K(\alpha_n^*)\,,\quad U_n = k_n X_n\,,\quad n = 0, 1, \ldots$$

We consider here *least squares estimates* obtained by minimizing

(15) $$L_N(\alpha) = \sum_{n=0}^{N-1} (X_{n+1} - \Phi(\alpha)\,X_n - \Psi U_n)'\;T(X_{n+1} - \Phi(\alpha)\,X_n - \Psi U_n)\,.$$

In (15) T is a positive semi-definite matrix such that the following is true.

ASSUMPTION 3. Let

(16) $$d_{ij} = \operatorname{tr}(W^{ij}Q)\,,\quad i, j = 1, \ldots, q\,,$$

where W^{ij} is the matrix satisfying

(17) $$W^{ij} = F_i' T F_j + (\Phi_0 + \Psi K_0)'\; W^{ij}(\Phi_0 + \Psi K_0)\,,$$

then the matrix $\|d_{ij}\|_{i,j=1}^q$ is non-singular.

REMARK 1. From (32) and (33) it follows that (18) implies

$$\lim_{N\to\infty} \frac{1}{2N} \frac{\partial^2}{\partial\alpha^i\,\partial\alpha^j} L_N(\alpha) = d_{ij} \quad \text{a.s.}\,,\quad i, j = 1, \ldots, q\,.$$

THEOREM 4. *Let* (14) *hold, and let*

(18) $$\lim_{N\to\infty} \alpha_N^* = \alpha_0 \quad a.s.\,,$$

(19) $$\frac{\partial}{\partial\alpha^i} L_N(\alpha_N^*) = 0\,,\quad i = 1, \ldots, q\,,$$

with probability 1 *for* N *sufficiently large. Then* $(C_N - N\Theta)/\sqrt{N}$ *is asymptotically normal* $N(0, \zeta)$, *as* $N \to \infty$.

Theorem 4 states that the control policy, which uses optimal stationary controls computed from strongly consistent least squares estimates, fulfils the postulate of asymptotic normality of C_N. Sufficient conditions for the validity of (18) and (19) are given in [5].

3.

Further we investigate the validity of (9) under the adaptive control (14). We retain the assumptions and denotations made previously with the following exceptions. The disturbances are subject to a somewhat stronger condition on moments, namely

$$E\|e_n\|^{4+\varepsilon}, \quad n = 1, 2, \ldots, \tag{20}$$

for an $\varepsilon > 0$.

Theorem 5 stated next and demonstrated in Section 5 shows that under (14), (18), (19) quadratic functionals of the type as in (4) fulfil the *law of the iterated logarithm* with the same parameters as under K_0.

THEOREM 5. *Let* (14) *hold, and let*

$$\lim_{N\to\infty} \alpha_N^* = \alpha_0 \quad a.s., \tag{21}$$

$$\frac{\partial}{\partial \alpha^i} L_N(\alpha_N^*) = 0, \quad i = 1, \ldots, q, \tag{22}$$

with probability one for N sufficiently large. Then

$$\limsup_{N\to\infty} (C_N - N\Theta)/\sqrt{(2N \log\log N)} = \sqrt{\zeta} \quad a.s. \tag{23}$$

with Θ *defined by* (8), (6), ζ *defined by* (10), (11).

4.

This section contains auxiliary assertions. In the proof of Theorem 5 occur the martingales

$$M_N = \sum_{n=0}^{N-1} Y_n, \quad N = 1, 2, \ldots, \quad M_0 = 0, \tag{24}$$

with respect to the sequence $\{\mathscr{F}_N, N = 0, 1, \ldots\}$ of Borel fields. $\mathscr{F}_0$ is the trivial field, and, for $N = 1, 2, \ldots, \mathscr{F}_N$ is the Borel field generated by $\{e_1, e_2, \ldots, e_N\}$. The following version of the law of the iterated logarithm is a corollary to a theorem of Stout ([6], Theorem 3).

LEMMA 1. *Let (24) be a martingale with respect to* $\{\mathscr{F}_N, N = 0, 1, \ldots\}$, $EY_n^2 < \infty$, $n = 0, 1, \ldots$ Further let

$$\lim_{N\to\infty} N^{-1} \sum_{n=0}^{N-1} E\{Y_n^2 \mid \mathscr{F}_n\} = \eta \quad a.s.,$$

where η *is a constant, and let*

(25) $$\sum_{n=3}^{\infty} \frac{\log\log n}{nc_n} E\left\{Y_n^2 I\left(Y_n^2 > \frac{nc_n}{\log\log n}\right) \mid \mathscr{F}_n\right\} < \infty \quad a.s.$$

for a sequence of positive numbers $\{c_n, n = 3, 4, \ldots\}$ *converging to zero. Then*

$$\limsup_{N\to\infty} M_N/\sqrt{(2N \log\log N)} = \sqrt{\eta} \quad a.s.$$

Lemmas 2 and 3 are also given without proofs. The former is demonstrated in a similar way as Lemma 1 in [5] using (20). The latter is a slight modification of Theorem 2 and of (70) in [5].

LEMMA 2. *Let (1) and (2) hold, and let*

(26) $$\lim_{n\to\infty} k_n = K_0 \quad a.s.$$

Then

$$\limsup_{N\to\infty} N^{-1} \sum_{n=0}^{N-1} \|X_n\|^{4+\varepsilon} < \infty \quad a.s.$$

LEMMA 3. *Let A be a symmetric* $r \times r$*-matrix. Under the hypotheses of Lemma 2*

$$\lim_{N\to\infty} N^{-1} \sum_{n\to\infty}^{N-1} X_n' A X_n = \vartheta \quad a.s.,$$

ϑ *is a constant,*

$$\vartheta = \operatorname{tr}(VQ),$$

where V is the (unique and symmetric) matrix, satisfying

$$V = A + (\Phi_0 + \Psi K_0)' V (\Phi_0 + \Psi K_0).$$

Furthermore,

$$\lim_{N\to\infty} N^{-1} \sum_{n=0}^{N-1} X_n = 0 \quad a.s.$$

The following formula for $C_N - N\Theta$ will be used.

LEMMA 4. *Let (1) and (2) hold, and let W and Θ be defined by (6) and by (8), respectively. Then, for $N = 1, 2, \ldots$,*

$$(27) \qquad C_N - N\Theta = X_0' W X_0 - X_N' W X_N + \\ + \sum_{n=1}^{N-1} \{e_{n+1}' W e_{n+1} - \operatorname{tr}(WQ) + 2e_{n+1}' W(\Phi_0 X_n + \Psi U_n) + \\ + X_n'(k_n - K_0)'(S + \Psi' W \Psi)(k_n - K_0) X_n\} .$$

Proof. Consider the expression in braces. The sum of the first and of the third term equals in virtue of (1), (2)

$$X_{n+1}' W X_{n+1} - X_n'(\Phi_0 + \Psi k_n)' W(\Phi_0 + \Psi k_n) X_n .$$

The last term equals

$$X_n' k_n' S k_n X_n + X_n(\Phi_0 + \Psi k_n)' W(\Phi_0 + \Psi k_n) X_n - \\ - 2X_n'(k_n - K_0)' [SK_0 + \Psi' W(\Phi_0 + \Psi K_0)] X_n - \\ - X_n' K_0' S K_0 X_n - X_n'(\Phi_0 + \Psi K_0)' W(\Phi_0 + \Psi K_0) X_n .$$

The third term in this expression vanishes because of (7), and the last two terms give because of (6)

$$X_n' R X_n - X_n' W X_n .$$

Hence, with regard to (8) we conclude that the expression in braces equals

$$U_n' S U_n + X_n' R X_n - \Theta + X_{n+1}' W X_{n+1} - X_n' W X_n .$$

From here, (27) immediately follows. □

We conclude this section by a lemma on real sequences.

LEMMA 5. *Let $\{a_n, n = 1, 2, \ldots\}$, $\{b_n, n = 1, 2, \ldots\}$ be sequences of non-negative numbers. Then*

$$(28) \qquad (a_1 + \ldots + a_N)^{2+\gamma} \leqq N^{1+\gamma}(a_1^{2+\gamma} + \ldots + a_N^{2+\gamma}), \quad N = 1, 2, \ldots, \quad \gamma > 0 .$$

If

$$\limsup_{N\to\infty} N^{-1} \sum_{n=1}^{N} a_n < \infty , \quad \lim_{n\to\infty} \sqrt{\left(\frac{n}{\log\log n}\right)} b_n = 0 ,$$

then

$$\sum_{n=1}^{\infty} \frac{a_n}{n^{1+\gamma}} < \infty , \quad \gamma > 0 ; \quad \lim_{N\to\infty} \frac{1}{\sqrt{(N \log\log N)}} \sum_{n=1}^{N} a_n b_n = 0 .$$

5. PROOF OF THEOREM 5

For the proof we may without loosing generality assume

$$\|\alpha_n^*\| \leqq \text{const} < \infty\,, \quad n = 1, 2, \ldots\,, \quad \text{a.s.}\,, \tag{29}$$

since from (21) follows that the probability of (29) is arbitrarily close to one if the constant is large enough. Note also that

$$K(\alpha) - K_0 = \sum_{j=1}^{q} (\alpha^j - \alpha_0^j) \frac{\partial}{\partial \alpha^j} K(\alpha_0) + o(\alpha - \alpha_0)\,, \quad \alpha \to \alpha_0\,. \tag{30}$$

Hence, (21) implies

$$\lim_{n\to\infty} k_n = K_0 \quad \text{a.s.} \tag{31}$$

Consider $L_N(\alpha)$. It holds

$$L_N(\alpha) = L_N(\alpha_0) + \sum_{n=0}^{N-1} X_n'(\Phi(\alpha) - \Phi_0)'\, T(\Phi(\alpha) - \Phi_0)\, X_n - $$
$$-2 \sum_{n=0}^{N-1} e_{n+1}'\, T(\Phi(\alpha) - \Phi_0)\, X_n\,.$$

From (12) follows

$$\frac{\partial}{\partial \alpha^i} L_N(\alpha) = 2 \sum_{j=1}^{q} (\alpha^j - \alpha_0^j) \sum_{n=0}^{N-1} X_n' F_i' T F_j X_n - 2 \sum_{n=0}^{N-1} e_{n+1}' T F_i X_n\,, \quad i = 1, 2, \ldots, q\,. \tag{32}$$

From Lemma 3 and from (16), (17) one gets

$$\lim_{N\to\infty} N^{-1} \sum_{n=0}^{N-1} X_n' F_i' T F_j X_n = d_{ij} \text{ a.s.}\,, \quad i, j = 1, \ldots, q\,. \tag{33}$$

Let $\|d_{ij}^{(-1)}\|_{i,j=1}^{q}$ be the inverse matrix to $\|d_{ij}\|_{i,j=1}^{q}$ (cf. Assumption 3). Define

$$y_n^i = \sum_{j=1}^{q} d_{ij}^{(-1)} e_{n+1}' T F_j X_n\,, \quad l_N^i = \sum_{n=0}^{N-1} y_n^i\,, \quad i = 1, \ldots, q\,, \tag{34}$$
$$l_0 = (0, \ldots, 0)\,, \quad l_N = (l_N^1, \ldots, l_N^q)\,, \quad N = 1, 2, \ldots$$

From (19), (32), and (33) it is not difficult to obtain

$$N(\alpha_N^* - \alpha_0) = l_N + o(l_N)\,, \quad N \to \infty\,. \tag{35}$$

Obviously, for $i = 1, \ldots, q$, $\{l_N^i, N = 0, 1, \ldots\}$ are martingales with respect to $\{\mathscr{F}_N, N = 0, 1, \ldots\}$. The martingales fulfil the hypotheses of Lemma 1. We shall not verify it, since an analogous verification will be made for the martingale $\{\sum_{n=0}^{N-1} Z_n, N = 1, 2, \ldots\}$. Lemma 1 implies

$$\limsup_{N\to\infty} \|l_N\|/\sqrt{(N \log\log N)} \leqq \text{const} \quad \text{a.s.}, \tag{36}$$

hence, from (35),

$$N(\alpha_N^* - \alpha_0) = l_N + o(\sqrt{(N \log\log N)}), \quad N \to \infty,$$

and, from (30),

$$k_n - K_0 = \sum_{i=1}^{q} N^{-1} l_N^i \frac{\partial}{\partial \alpha^i} K(\alpha_0) + o\left(\sqrt{\left(\frac{\log\log N}{N}\right)}\right), \quad N \to \infty. \tag{37}$$

Let us turn now to (27). The first two terms on the right-hand side are $o(\sqrt{(N \log\log N)})$ as $N \to \infty$. In fact, from Lemmas 2 and 5 follows

$$\sum_{n=1}^{\infty} n^{-(1+\varepsilon/4)} \|X_n\|^{4+\varepsilon} < \infty \quad \text{a.s.},$$

and hence,

$$\limsup_{N\to\infty} \frac{|X_N' W X_N|}{\sqrt{(N \log\log N)}} \leqq \|W\| \lim_{N\to\infty} \frac{\|X_N\|^2}{\sqrt{N}} = 0 \quad \text{a.s.} \tag{38}$$

The sum of the last term in (27) is negligible as well. Namely

$$\begin{aligned}
&\Big|\sum_{n=0}^{N-1} X_n'(k_n - K_0)' (S + \Psi' W \Psi)(k_n - K_0) X_n\Big| \leqq \\
&\leqq \|S + \Psi' W \Psi\| \sum_{n=0}^{N-1} \|X_n\|^2 \|k_n - K_0\|^2 .
\end{aligned}$$

(36) and (37) imply

$$\|k_n - K_0\|^2 = o\left(\sqrt{\left(\frac{\log\log n}{n}\right)}\right), \quad n \to \infty .$$

From here and from Lemmas 2 and 5 follows

$$\sum_{n=0}^{N-1} X_n'(k_n - K_0)' (S + \Psi' W \Psi)(k_n - K_0) X_n = o(\sqrt{(N \log\log N)}), \quad N \to \infty .$$

Suming the results obtained, we conclude from (27) that

$$C_N - N\Theta = \sum_{n=0}^{N-1} Z_n + o(\sqrt{(N \log\log N)}), \quad N \to \infty, \tag{39}$$

where

$$Z_n = e_{n+1}' W e_{n+1} - \operatorname{tr}(WQ) + 2e_{n+1}' W(\Phi_0 X_n + \Psi U_n) .$$

Let us verify that the martingale $\{\sum_{n=0}^{N-1} Z_n, N = 1, 2, \ldots\}$ fulfils the hypotheses of Lemma 1.

It holds

$$E\{Z_n^2 \mid \mathscr{F}_n\} = E(e'_{n+1}We_{n+1} - \operatorname{tr}(WQ))^2 + 4(\Phi_0 X_n + \Psi U_n)' WQW(\Phi_0 X_n + \Psi U_n) + \\ + 4E(e'_{n+1}We_{n+1}e'_{n+1})\, W(\Phi_0 X_n + \Psi U_n).$$

With regard to (31) and to (14) we get from Lemma 3

$$\lim_{N\to\infty} N^{-1} \sum_{n=0}^{N-1} E\{Z_n^2 \mid \mathscr{F}_n\} = \zeta \quad \text{a.s.},$$

where ζ is defined by (10).

Proceed now to the verification of (25). Set $\delta = \frac{1}{2}\varepsilon$. Using (28) one obtains

$$E\{|Z_n|^{2+\delta} \mid \mathscr{F}_n\} \leqq \text{const}\,[E\|e_{n+1}\|^{4+2\delta} + |\operatorname{tr} WQ|^{2+\delta} + \\ + (E\|e_{n+1}\|^{2+\delta})(\|X_n\|^{2+\delta} + \|U_n\|^{2+\delta})], \quad n = 0, 1, \ldots$$

Hence, from Lemmas 2 and 5,

$$\sum_{n=1}^{\infty} n^{-(1+\gamma)} E\{|Z_n|^{2+\delta} \mid \mathscr{F}_n\} < \infty \quad \text{a.s.} \tag{40}$$

for arbitrary $\gamma > 0$. Further,

$$\sum_{n=3}^{\infty} \frac{\log\log n}{nc_n} E\left\{Z_n^2 I\left(Z_n^2 > \frac{nc_n}{\log\log n}\right) \mid \mathscr{F}_n\right\} \leqq \\ \leqq \sum_{n=3}^{\infty} \left(\frac{\log\log n}{nc_n}\right)^{1+\delta/2} E\{|Z_n|^{2+\delta} \mid \mathscr{F}_n\},$$

which for

$$c_n = n^{-\varrho} \log\log n, \quad \varrho > 0,$$

equals

$$\sum_{n=3}^{\infty} n^{-(1-\varrho)(1+\delta/2)} E\{|Z_n|^{2+\delta} \mid \mathscr{F}_n\}. \tag{41}$$

In virtue of (40), the series (41) converges almost surely, if ϱ is chosen so that $(1 - \varrho) \, . \, (1 + \frac{1}{2}\delta) > 1$.

The hypotheses of Lemma 1 are thus verified, and it follows that

$$\limsup_{N\to\infty} \sum_{n=0}^{N-1} Z_n / \sqrt{(2N \log\log N)} = \sqrt{\zeta} \quad \text{a.s.}$$

This together with (39) implies (23), which was to be established. □

REFERENCES

[1] K. J. ÅSTRÖM, B. WITTENMARK: Problems of identification and control. J. of Math. Anal. Appl. *34* (1971), 2, 90—113.
[2] P. MANDL: An application of Itô's formula to stochastic control systems. In: Stability of Stoch. Dynamical Systems, Proc. Int. Symposium, Coventry 1972. Springer-Verlag, Heidelberg 1972, 8—13.
[3] P. MANDL: On the adaptive control of finite state Markov processes. Z. Wahrscheinlichkeitstheorie *27* (1973), 4, 263—276.
[4] P. MANDL: Estimation and control in Markov chains. Adv. in Appl. Prob. *6* (1974), 1, 40—60.
[5] P. MANDL: The use of optimal stationary policies in the adaptive control of linear systems. In: Proc. Symposium to Honour J. Neyman, Warsaw 1974.
[6] W. F. STOUT: A martingale analogue of Kolmogorov's law of the iterated logarithm. Z. Wahrscheinlichkeitstheorie *15* (1970), 4, 279—290.

CZECHOSLOVAK ACADEMY OF SCIENCES

INSTITUTE OF INFORMATION THEORY AND AUTOMATION

COMPETITIVE EQUILIBRIUM IN A MARKET WITH LIMITED COOPERATION

MILAN MAREŠ

PRAGUE

The well-known notion of competitive equilibrium (cf. [1], [2], [3]) can be defined in rather modificated way, which enables us to study more analogies between the competitive equilibrium and some game-theoretical bargaining solutions. These analogies were investigated in author's contribution at "Computing Equilibria: How and Why?" Conference, July 1974, Toruń, Poland. The presented paper is devoted to a particular problem connected with that field, how the modified concept of equilibrium depends on the form of the coalitions class which is respected in its definition. It means, how the set of equilibrial market states changes if we consider different classes of coalitions being relevant from the point of view of the equilibrium determination.

1. MARKET AND EQUILIBRIUM

In this part we prepare the concepts and notations, used in the whole paper, and we introduce the main formal results.

Let m be a natural number. Then R, R^+, R_m, R_m^+ are the sets of all real numbers, non-negative real numbers, m-dimensional real-valued vectors, and m-dimensional real-valued vectors with non-negative coordinates, respectively. If $\mathcal{M}$ is a class of sets, then we denote

$$(1.1) \qquad \bigcup\mathcal{M} = \{i : \exists(K \in \mathcal{M})\, i \in K\}\,, \quad \bigcap\mathcal{M} = \{i : \forall(K \in \mathcal{M})\, i \in K\}\,;$$

we say that $\mathcal{M}$ is a *partition* of a set M if $M = \bigcup\mathcal{M}$ and $K \cap L = \emptyset$ for any $K, L \in \mathcal{M}$, $K \neq L$; and we denote

$$(1.2) \quad \langle\mathcal{M}\rangle = \{K \in \mathcal{M}\text{: there is no } \mathcal{L} \subset \mathcal{M}\text{, where } \mathcal{L} \text{ is a partition of } K,\ \mathcal{L} \neq \{K\}\}\,.$$

A *market* $\mathfrak{m}$ is a quintuple

$$(1.3) \qquad \mathfrak{m} = (I, \mathcal{K}, R \times R_m^+, (U_i)_{i \in I}, (a^i)_{i \in I})\,,$$

where I is a non-empty finite set of *agents*; $\mathcal{K}$ is a union of positive finite number of partitions of I, elements of $\mathcal{K}$ are called *coalitions*; $R \times R_m^+$ is a set of real valued vectors $\boldsymbol{x} = (x_0, x_1, \ldots, x_m)$ where x_0 represents a quantity of *money* and x_j, $j = 1, \ldots, m$ represent quantities of *regular goods* existing in the market; vectors $\boldsymbol{a}^i \in R \times R_m^+$, $i \in I$, represent *initial quantities of money and goods*, owned by agents at the beginning of the bargaining process; finally, continuous mappings $U_i : R_m^+ \to R$, $i \in I$, represent the *utilities* of goods quantities from R_m^+ for agents $i \in I$.

We introduce the mappings $u_i : R \times R_m^+ \to R$, $i \in I$, such that $u_i(x_0, x_1, \ldots, x_m) = x_0 + U_i(x_1, \ldots, x_m)$, for $i \in I$, $(x_0, x_1, \ldots, x_m) \in R \times R_m^+$, which represent the *global utilities* of money and goods.

In order to simplify further explanation we introduce the following notations.

$$\mathbf{X} = \{X = (\boldsymbol{x}^i)_{i \in I} : \boldsymbol{x}^i \in R \times R_m^+,\ i \in I\}, \tag{1.4}$$

$$\mathbf{X}_K = \{X = (\boldsymbol{x}^i)_{i \in I} \in \mathbf{X} : \sum_{i \in K} \boldsymbol{x}^i \leqq \sum_{i \in K} \boldsymbol{a}^i\}, \quad K \subset I, K \neq \varnothing. \tag{1.5}$$

All goods in the considered market, even money, have their *prices*. We denote them by $\boldsymbol{p} = (p_0, p_1, \ldots, p_m) \in R_m$, $p_j > 0$, $j = 1, \ldots, m$; by $\boldsymbol{P}$ we denote the set of all pricevectors which are admissible in the considered market. We assume that $\boldsymbol{p} \in \boldsymbol{P}$ are row vectors and $\boldsymbol{x} \in R \times R_m^+$ are column vectors, so that the scalar product $\boldsymbol{px}$ can be considered. Then we denote

$$\mathbf{A}_K(\boldsymbol{p}) = \{X \in \mathbf{X} : \sum_{i \in K} \boldsymbol{px}^i \leqq \sum_{i \in K} \boldsymbol{pa}^i\} \quad \text{for} \quad \boldsymbol{p} \in \boldsymbol{P}, \quad K \in \mathcal{K}. \tag{1.6}$$

REMARK 1. For any $K \in \mathcal{K}$, $\boldsymbol{p} \in \boldsymbol{P}$, the sets $\mathbf{A}_K(\boldsymbol{p})$ and $\mathbf{X}_K$ are non-empty, and $\mathbf{A}_K(\boldsymbol{p}) \supset \mathbf{X}_K$.

The state of market in any moment can be described by the amounts of money and goods owned by players, and by prices which are prescribed to them. So we call the pair $(X, \boldsymbol{p})$, $X \in \mathbf{X}$, $\boldsymbol{p} \in \boldsymbol{P}$ a *market state*.

DEFINITION 1. Let $\mathcal{M} \subset \mathcal{K}$, $X = (\boldsymbol{x}^i)_{i \in I} \in \mathbf{X}$, $\boldsymbol{p} \in \boldsymbol{P}$. We call the market state $(X, \boldsymbol{p})$ an $\mathcal{M}$-equilibrium iff $X \in \mathbf{X}_I$ and for any $K \in \mathcal{M}$

$$X \in \mathbf{A}_K(\boldsymbol{p}),$$

$$\sum_{i \in K} u_i(\boldsymbol{x}^i) \geqq \sum_{i \in K} u_i(\boldsymbol{y}^i) \quad \text{for any} \quad Y = (\boldsymbol{y}^i)_{i \in I} \in \mathbf{A}_K(\boldsymbol{p}).$$

The previous Definition differs from the classical one, well-known from the literature. The difference is caused by two facts mainly. First: we suppose a more general model of market where the one-element coalitions need not be admissible. Second: the concept of equilibria formulated in such a way points more distinctly the existing analogies between the market and a coalition-game, and enables us to transfer many game-theoretical solutions into the language of a market theory without troubles.

The analogy between a market and a coalition-game was investigated by author in [4] and [5]. This work deals with a particular problem – how the existence of an $\mathcal{M}$-equilibrium in a given market depends on the form of the lcass $\mathcal{M}$, and on the form of $\mathcal{K}$.

Lemma 1. If $\boldsymbol{p} \in \boldsymbol{P}$, $\mathcal{M} \subset \mathcal{K}$, $\langle\mathcal{M}\rangle$ is a partition of I, and if $\boldsymbol{X} \in \boldsymbol{A}_K(\boldsymbol{p})$ for some $K \in \mathcal{M}$ then there exists at least one $J \in \langle\mathcal{M}\rangle$, $J \subset K$, such that $\boldsymbol{X} \in \boldsymbol{A}_J(\boldsymbol{p})$.

Proof. The statement follows from

$$\sum_{J\in\langle\mathcal{M}\rangle, J\subset K} \sum_{i\in J} \boldsymbol{p}\boldsymbol{x}^i = \sum_{i\in K} \boldsymbol{p}\boldsymbol{x}^i \leqq \sum_{i\in K} \boldsymbol{p}\boldsymbol{a}^i = \sum_{J\in\langle\mathcal{M}\rangle, J\subset K} \sum_{i\in J} \boldsymbol{p}\boldsymbol{a}^i . \qquad \square$$

Lemma 2. Let $\boldsymbol{p} \in \boldsymbol{P}$, $\mathcal{M} \subset \mathcal{K}$, $\langle\mathcal{M}\rangle$ be a partition of I. Let $\boldsymbol{X} \in \boldsymbol{A}_J(\boldsymbol{p})$ for all $J \in \langle\mathcal{M}\rangle$. Then $\boldsymbol{X} \in \boldsymbol{A}_K(\boldsymbol{p})$ for all $K \in \mathcal{M}$.

Proof. The statement is an immediate consequence of

$$\sum_{i\in K} \boldsymbol{p}\boldsymbol{x}^i = \sum_{J\in\langle\mathcal{M}\rangle, J\subset K} \sum_{i\in I} \boldsymbol{p}\boldsymbol{x}^i \leqq \sum_{J\in\langle\mathcal{M}\rangle, J\subset K} \sum_{i\in J} \boldsymbol{p}\boldsymbol{a}^i = \sum_{i\in K} \boldsymbol{p}\boldsymbol{a}^i. \qquad \square$$

Lemma 3. Let $\boldsymbol{p} \in \boldsymbol{P}$, $\mathcal{M} \subset \mathcal{K}$, $\langle\mathcal{M}\rangle$ be a partition of I. Let for all $J \in \langle\mathcal{M}\rangle$ and all $\boldsymbol{Y} = (y^i)_{i\in I} \in \boldsymbol{A}_J(\boldsymbol{p})$ the inequality

$$\sum_{i\in J} u_i(x^i) \geqq \sum_{i\in J} u_i(y^i)$$

holds. Then

$$\sum_{i\in K} u_i(x^i) \geqq \sum_{i\in K} u_i(y^i)$$

for all $K \in \mathcal{M}$ and all $\boldsymbol{Y} = (y^i)_{i\in I} \in \boldsymbol{A}_K(\boldsymbol{p})$.

Proof. Let $K \in \mathcal{M}$, $\boldsymbol{Y} = (y^i)_{i\in I} \in \boldsymbol{A}_K(\boldsymbol{p})$, and let

$$\sum_{i\in K} u_i(x^i) < \sum_{i\in K} u_i(y^i) . \tag{1.7}$$

We construct $\boldsymbol{Z} = (z^i)_{i\in I} \in \boldsymbol{X}$ such that

$$\begin{aligned} z^i &= y^i && \text{for} \quad i \in I - K , \\ z^i_j &= y^i_j && \text{for} \quad i \in K , \quad j = 1, \dots, m , \\ z^i_0 &= y^i_0 + d_i && \text{for} \quad i \in K , \end{aligned}$$

where d_i, $i \in K$, are such that

$$\sum_{i\in K} d_i = 0 \quad \text{and} \quad u_i(x^i) < u_i(z^i) \quad \text{for all} \quad i \in K .$$

It means also that

(1.8) $$\sum_{i\in J} u_i(x^i) < \sum_{i\in J} u_i(z^i) \quad \text{for all} \quad J \in \langle \mathcal{M} \rangle, \quad J \subset K.$$

Inequality (1.7) implies that such Z exists. Moreover, $Z \in \mathbf{A}_K(p)$, as

$$\sum_{i\in K} pz^i = \sum_{i\in K} py^i \leqq \sum_{i\in K} pa^i .$$

According to Lemma 1 there exists $J \in \langle \mathcal{M} \rangle$, $J \subset K$, such that $Z \in \mathbf{A}_J(p)$. This fact, together with (1.8), contradicts to the assumptions of the lemma. Consequently, (1.7) can not hold for any $K \in \mathcal{M}$ and $Y \in \mathbf{A}_K(p)$. □

REMARK 2. If $\mathcal{L} \subset \mathcal{M} \subset \mathcal{K}$ and if (X, p), $X \in \mathbf{X}$, $p \in \boldsymbol{P}$, is an $\mathcal{M}$-equilibrium then Definition 1 implies that it is also an $\mathcal{L}$-equilibrium.

THEOREM 1. Let $\langle \mathcal{K} \rangle$ be a partition of I, let $\mathcal{M} \subset \mathcal{K}$, $X \in \mathbf{X}$, $p \in \boldsymbol{P}$. If (X, p) is an $\langle \mathcal{K} \rangle$-equilibrium then it is also an $\mathcal{M}$-equilibrium.

Proof. If (X, p) is an $\langle \mathcal{K} \rangle$-equilibrium then it is also an $\mathcal{K}$-equilibrium, as follows from Lemma 2 and Lemma 3. By Remark 2 it is also an $\mathcal{M}$-equilibrium for any $\mathcal{M} \subset \mathcal{K}$. □

COROLLARY 1. If $\langle \mathcal{K} \rangle$ is a partition of I, $\langle \mathcal{K} \rangle \subset \mathcal{M} \subset \mathcal{K}$, $\langle \mathcal{K} \rangle \subset \mathcal{N} \subset \mathcal{K}$, $X \in \mathbf{X}$, $p \in \boldsymbol{P}$, then (X, p) is an $\mathcal{M}$-equilibrium if and only if it is an $\langle \mathcal{K} \rangle$-equilibrium, and if and only if it is an $\mathcal{N}$-equilibrium, as follows from Theorem 1 and Remark 2.

REMARK 3. If $X \in \mathbf{X}$, $p \in \boldsymbol{P}$, $\mathcal{M} \subset \mathcal{K}$, and if $\langle \mathcal{M} \rangle$ is a partition of I then (X, p) is an $\mathcal{M}$-equilibrium if and only if it is an $\langle \mathcal{M} \rangle$-equilibrium, and if (X, p) is an $\langle \mathcal{M} \rangle$-equilibrium then it is also an $\mathcal{L}$-equilibrium for any $\mathcal{L} \subset \mathcal{M}$ as follows from Lemma 2, Lemma 3 and Remark 2.

If all one-element coalitions are admissible, i.e. if $\{i\} \in \mathcal{K}$ for all $i \in I$, then we denote by

(1.9) $$\mathcal{I} = \{\{i\}\}_{i\in I}$$

the class of exactly all one-element coalitions.

REMARK 4. The competitive equilibrium defined in literature (e.g. in [1]) is identical with $\mathcal{I}$-equilibrium in the sense of Definition 1. It follows from Theorem 1 and Remark 2 that if $\mathcal{I} \subset \mathcal{M} \subset \mathcal{K}$ then any market state (X, p) is an $\mathcal{M}$-equilibrium (or, especially, a $\mathcal{K}$-equilibrium) if and only if it is an $\mathcal{I}$-equilibrium.

REMARK 5. If $K \in \mathcal{K}$, $p \in \boldsymbol{P}$, $X = (x^i)_{i\in I} \in \mathbf{X}$ then (X, p) is an $\{K\}$-equilibrium if and only if $X \in \mathbf{A}_K(p) \cap \mathbf{X}_I$, and

$$\sum_{i\in K} u_i(x^i) = \max \{ \sum_{i\in K} u_i(y^i) : Y = (y^i)_{i\in I} \in \mathbf{A}_K(p)\} ,$$

and it is not difficult to prove that for any $\boldsymbol{p} \in \boldsymbol{P}$ there exists $X \in \mathbf{X}$ such that $(X, \boldsymbol{p})$ is a $\{K\}$-equilibrium (c. f. [4]).

REMARK 6. It follows from Definition 1 immediately that if $(X, \boldsymbol{p})$, $X \in \mathbf{X}$, $\boldsymbol{p} \in \boldsymbol{P}$, is an $\mathcal{M}$-equilibrium and an $\mathcal{N}$-equilibrium then it is also an $(\mathcal{M} \cup \mathcal{N})$-equilibrium.

LEMMA 4. Let $X \in \mathbf{X}$, $\boldsymbol{p} \in \boldsymbol{P}$, $\mathcal{M} \subset \mathcal{K}$, $\mathcal{N} \subset \mathcal{K}$, and let $(X, \boldsymbol{p})$ be an $\mathcal{M}$-equilibrium. If $(X, \boldsymbol{p})$ is an $(\mathcal{N} - \mathcal{M})$-equilibrium then it is also an $\mathcal{N}$-equilibrium.

Proof. Let $(X, \boldsymbol{p})$ be an $\mathcal{M}$-equilibrium and an $(\mathcal{N} - \mathcal{M})$-equilibrium. Then $(X, \boldsymbol{p})$ is an $(\mathcal{N} \cup \mathcal{M})$-equilibrium, by Remark 6, and Remark 2 implies that it is also an $\mathcal{N}$-equilibrium. □

2. MAIN RELATIONS

In this part, all main relations between the form of the class $\mathcal{M}$ and the form of the set of $\mathcal{M}$-equilibria, following from the previous statements, are gathered. Let us suppose that $X \in \mathbf{X}$, $\boldsymbol{p} \in \boldsymbol{P}$, and let us denote

$$\mathsf{K} = \{\mathcal{M} : \mathcal{M} \subset \mathcal{K}\},$$

$$\mathsf{K}(X, \boldsymbol{p}) = \{\mathcal{M} \in \mathsf{K} : (X, \boldsymbol{p}) \textit{ is an } \mathcal{M}\textit{-equilibrium}\},$$

$$\mathcal{O} \text{ is an empty subset of } \mathcal{K}\,.$$

Then

$$\mathcal{O} \in \mathsf{K}(X, \boldsymbol{p}),$$

$$\forall(K \in \mathcal{K})\, \forall(\boldsymbol{p} \in \boldsymbol{P})\, \exists(X \in \mathbf{X}) \quad \text{such that} \quad \{K\} \in \mathsf{K}(X, \boldsymbol{p}),$$

$$(\mathcal{J} \subset \mathcal{L} \subset \mathcal{K},\ \mathcal{L} \in \mathsf{K}(X, \boldsymbol{p})) \Rightarrow \mathcal{J} \in \mathsf{K}(X, \boldsymbol{p}),$$

$$(\mathcal{M} \in \mathsf{K}(X, \boldsymbol{p}) \quad \text{and} \quad \mathcal{N} \in \mathsf{K}(X, \boldsymbol{p})) \Rightarrow$$

$$\Rightarrow (\mathcal{M} \cup \mathcal{N} \in \mathsf{K}(X, \boldsymbol{p}),\ \mathcal{M} \cap \mathcal{N} \in \mathsf{K}(X, \boldsymbol{p}),$$

$$\mathcal{M} - \mathcal{N} \in \mathsf{K}(X, \boldsymbol{p}),\ \mathcal{M} \div \mathcal{N} \in \mathsf{K}(X, \boldsymbol{p})),$$

$$(\mathcal{N} \in \mathsf{K} - \mathsf{K}(X, \boldsymbol{p}),\ \mathcal{M} \in \mathsf{K}) \Rightarrow \mathcal{N} \cup \mathcal{M} \in \mathsf{K} - \mathsf{K}(X, \boldsymbol{p}),$$

$$(\mathcal{M} \in \mathsf{K}(X, \boldsymbol{p}) \quad \text{and} \quad \mathcal{N} \in \mathsf{K} - \mathsf{K}(X, \boldsymbol{p})) \Rightarrow$$

$$\Rightarrow (\mathcal{M} \cup \mathcal{N} \in \mathsf{K} - \mathsf{K}(X, \boldsymbol{p}),\ \mathcal{M} \cap \mathcal{N} \in \mathsf{K}(X, \boldsymbol{p}),$$

$$\mathcal{M} - \mathcal{N} \in \mathsf{K}(X, \boldsymbol{p}),\ \mathcal{N} - \mathcal{M} \in \mathsf{K} - \mathsf{K}(X, \boldsymbol{p}),$$

$$\mathcal{M} \div \mathcal{N} \in \mathsf{K} - \mathsf{K}(X, \boldsymbol{p}))\,.$$

If, moreover, $\langle \mathscr{K} \rangle$ is a partition of I then

$$\langle \mathscr{K} \rangle \in \mathbf{K}(X, p) \Rightarrow \mathbf{K}(X, p) = \mathbf{K} ,$$

$$\langle \mathscr{K} \rangle \subset \mathscr{M} \subset \mathscr{K} \Rightarrow (\langle \mathscr{K} \rangle \in \mathbf{K}(X, p) \Leftrightarrow \mathscr{M} \in \mathbf{K}(X, p)) .$$

3. AGGREGATED MARKET

The investigated model of a market enables us to consider even such a case where the one-element coalitions are not generally admissible, i.e. where $\mathscr{I} \not\subset \mathscr{K}$, cf. (1.9), and where also the single agents do not influence the bargaining. It is possible, in such a case, to find the smallest groups of agents, which are the intersections of admissible coalitions, and to consider these groups to be "collective" or aggregated agents. It could be intuitively expected that such aggregation of agents does not influence the equilibria. In this brief conclusive part of the presented work we formulate that intuitive idea in more exact way.

Let us consider a market $\mathfrak{m} = (I, \mathscr{K}, R \times R_m^+, (U_i)_{i\in I}, (\boldsymbol{a}^i)_{i\in I})$ with prices from set $\boldsymbol{P}$, and with the sets $\mathbf{X}, \mathbf{X}_K, \mathbf{A}_K(\boldsymbol{p})$, $K \in \mathscr{K}$, $\boldsymbol{p} \in \boldsymbol{P}$, defined by (1.4), (1.5) and (1.6). Let us introduce the notation

$$\mathscr{J} = \{J \subset I : \exists(\mathscr{L} \subset \mathscr{K})\,(J = \bigcap \mathscr{L}) \text{ and } \forall(K \in \mathscr{K})\ (K \cap J \neq \emptyset \Rightarrow J \subset K)\}$$

for the class of atomic subsets of I according to $\mathscr{K}$,

$$\tilde{\mathscr{K}} = \{\mathscr{L} : \mathscr{L} \subset \mathscr{J},\ \bigcup \mathscr{L} \in \mathscr{K}\} ,$$

and for any $X = (\boldsymbol{x}^i)_{i\in I} \in \mathbf{X}$ we denote

$$\tilde{\boldsymbol{x}}^J = \sum_{i\in J} \boldsymbol{x}^i , \quad \tilde{\boldsymbol{a}}^J = \sum_{i\in J} \boldsymbol{a}^i , \quad J \in \mathscr{J} , \quad \tilde{X} = (\tilde{\boldsymbol{x}}^J)_{J\in\mathscr{J}} . \tag{3.1}$$

Furthermore, we introduce the symbols

$$\tilde{\mathbf{X}} = \{\tilde{X} = (\tilde{\boldsymbol{x}}^J)_{J\in\mathscr{J}} : \exists(X = (\boldsymbol{x}^i)_{i\in I} \in \mathbf{X})\ \forall(J \in \mathscr{J})\,(\tilde{\boldsymbol{x}}^J = \sum_{i\in J} \boldsymbol{x}^i)\} ,$$

$$\tilde{\mathbf{A}}_{\mathscr{L}}(\boldsymbol{p}) = \{\tilde{X} \in \tilde{\mathbf{X}} : \sum_{J\in\mathscr{L}} \boldsymbol{p}\tilde{\boldsymbol{x}}^J \leqq \sum_{J\in\mathscr{L}} \boldsymbol{p}\tilde{\boldsymbol{a}}^J\} , \quad \mathscr{L} \in \tilde{\mathscr{K}} ,$$

$$\tilde{U}_J(\tilde{x}_1^J, \ldots, \tilde{x}_m^J) = \sum_{i\in J} U_i(x_1^i, \ldots, x_m^i) , \quad \tilde{u}_J(\tilde{\boldsymbol{x}}^J) = \sum_{i\in J} u_i(\boldsymbol{x}^i)$$

$$\tilde{x}_j^J = \sum_{i\in J} x_j^i , \quad j = 1, \ldots, m , \quad \tilde{\boldsymbol{x}}^J = \sum_{i\in J} \boldsymbol{x}^i , \quad J \in \mathscr{J} .$$

Then we introduce a new market

$$\tilde{\mathfrak{m}} = (\mathscr{I}, \tilde{\mathscr{K}}, R \times R_m^+, (\tilde{U}_J)_{J\in\mathscr{I}}, (\tilde{a}^J)_{J\in\mathscr{I}}),$$

which we call an *aggregated market* associated with $\mathfrak{m}$. The mutual relations between equilibria in the markets $\mathfrak{m}$ and $\tilde{\mathfrak{m}}$ are described by the following theorem.

THEOREM 2. Let $\mathfrak{m}$ be a market and let $\tilde{\mathfrak{m}}$ be an aggregated market associated with $\mathfrak{m}$. Let $p \in \boldsymbol{P}$, $X = (x^i)_{i\in I} \in \boldsymbol{X}$, $\tilde{X} = (\tilde{x}^J)_{J\in\mathscr{I}} \in \tilde{\boldsymbol{X}}$, where (3.1) is fulfilled, let $\mathscr{M} \subset \mathscr{K}$, $\tilde{\mathscr{M}} \subset \tilde{\mathscr{K}}$, $\tilde{\mathscr{M}} = \{\mathscr{L} : \mathscr{L} \subset \mathscr{I}, \bigcup \mathscr{L} \in \mathscr{M}\}$. Then (X, p) is an $\mathscr{M}$-equilibrium in $\mathfrak{m}$ if and only if $(\tilde{X}, p)$ is an $\tilde{\mathscr{M}}$-equilibrium in $\tilde{\mathfrak{m}}$.

Proof. Let (X, p) be an $\mathscr{M}$-equilibrium. It means that

$$X \in \boldsymbol{X}_I, \ \sum_{i\in L} px^i \leqq \sum_{i\in L} pa^i \ \text{ and } \ \sum_{i\in L} u_i(x^i) \geqq \sum_{i\in L} u_i(y^i) \tag{3.2}$$

$$\text{for all } \ L \in \mathscr{M}, \quad Y = (y^i)_{i\in I} \in \boldsymbol{A}_L(p).$$

Let us consider an arbitrary $\mathscr{L} \in \tilde{\mathscr{M}}$, and denote $L = \bigcup \mathscr{L} \in \mathscr{M}$. Then

$$\sum_{J\in\mathscr{I}} p\tilde{x}^J = \sum_{i\in L} px^i \leqq \sum_{i\in L} pa^i = \sum_{J\in\mathscr{L}} p\tilde{a}^J, \tag{3.3}$$

and, consequently, $\tilde{X} \in \tilde{\boldsymbol{A}}_{\mathscr{L}}(p)$. Let us suppose that there exists $\tilde{Y} = (\tilde{y}^J)_{J\in\mathscr{I}} \in \tilde{\boldsymbol{A}}_{\mathscr{L}}(p)$ such that

$$\sum_{J\in\mathscr{L}} \tilde{u}_J(\tilde{x}^J) < \sum_{J\in\mathscr{L}} \tilde{u}_J(\tilde{y}^J). \tag{3.4}$$

Then there exists $Y = (y^i)_{i\in I} \in \boldsymbol{X}$ such that

$$\tilde{y}^J = \sum_{i\in J} y^i \quad \text{for all} \quad J \in \mathscr{L}, \tag{3.5}$$

and

$$\sum_{i\in L} py^i = \sum_{J\in\mathscr{L}} p\tilde{y}^J \leqq \sum_{J\in\mathscr{L}} p\tilde{a}^J = \sum_{i\in L} pa^i, \tag{3.6}$$

so that $Y \in \boldsymbol{A}_L(p)$, and, moreover,

$$\sum_{i\in L} u_i(y^i) = \sum_{J\in\mathscr{L}} \tilde{u}_J(\tilde{y}^J) > \sum_{J\in\mathscr{L}} \tilde{u}_J(\tilde{x}^J) = \sum_{i\in L} u_i(x^i),$$

which is a contradiction with (3.2). Consequently, (3.4) can not be true for any $\tilde{Y} \in \tilde{\boldsymbol{A}}_{\mathscr{L}}(p)$, and $(\tilde{X}, p)$ is an $\tilde{\mathscr{M}}$-equilibrium.

Let us suppose, now, that $(\tilde{X}, p)$ is an $\mathscr{M}$-equilibrium. Then

$$\sum_{J\in\mathscr{I}} \tilde{x}^J \leqq \sum_{J\in\mathscr{I}} \tilde{a}^J, \ \sum_{J\in\mathscr{L}} p\tilde{x}^J \leqq \sum_{J\in\mathscr{L}} p\tilde{a}^J \ \text{ and } \ \sum_{J\in\mathscr{L}} \tilde{u}_J(\tilde{x}^J) \geqq \sum_{J\in\mathscr{L}} \tilde{u}_J(\tilde{y}^J) \tag{3.7}$$

$$\text{for all } \ \mathscr{L} \in \tilde{\mathscr{M}}, \quad \tilde{Y} = (\tilde{y}^J)_{J\in\mathscr{I}} \in \tilde{\boldsymbol{A}}_{\mathscr{L}}(p).$$

It is possible to prove, analogously to (3.3) or (3.6), that for arbitrary $L \in \mathcal{M}$ is $X \in \in \boldsymbol{A}_L(p)$, and if we suppose that there exists $\boldsymbol{Y} = (y^i)_{i \in I} \in \boldsymbol{A}_L(p)$ such that

$$\sum_{i \in L} u_i(x^i) < \sum_{i \in L} u_i(y^i) \tag{3.8}$$

then there exists $\tilde{\boldsymbol{Y}} = (\tilde{y}^J)_{J \in \mathcal{J}} \in \tilde{\boldsymbol{X}}$, such that (3.5) is true for $\mathcal{L} \in \tilde{\mathcal{M}}$, $L = \bigcup \mathcal{L}$. Then also $\tilde{\boldsymbol{Y}} \in \tilde{\boldsymbol{A}}_{\mathcal{L}}(p)$ and

$$\sum_{J \in \mathcal{L}} \tilde{u}_J(\tilde{x}^J) = \sum_{i \in L} u_i(x^i) < \sum_{i \in L} u_i(y^i) = \sum_{J \in \mathcal{L}} \tilde{u}_J(\tilde{y}^J),$$

which is a contradiction with (3.7), so that (3.8) can not be true, and $(\boldsymbol{X}, p)$ is an $\mathcal{M}$-equilibrium in $\mathfrak{m}$. □

4. ON COOPERATION

The last, brief part of the presented work introduces a few comments concerning the investigated model. Their purpose is to interpret the concept of cooperation in coalitions with side-payments. This concept is familiar in the coalition-games theory, but it seems that it is not so commonly known in the market models. It could cause certain difficulties in understanding the sense of Definition 1, and, generally, in understanding the author's motivation for the presented attitude.

We suppose that the utility of agents is transferable, it means that each agent can trasmit an arbitrary amount of his utility to any other agent. The mean which enables this activity is money with common utility value for all players. In this work we assume, for simplification, that the utility of money is linear, and the same for all agents. This assumption enables to shorten some steps of our considerations, and to make the main idea more visible. But, such a strong form of the assumption is generally not necessary, and similar results could be obtained for more general monotonously increasing continuous utilities of money or for games without any money (c. f. [6]).

When assuming the transferability of utility among agents, we can easily seen that the members of each realized coalition, if they are rational enough, try to maximize the common utility of the whole coalition, and then to distribute this utility among them in certain, just, way. How they are to distribute the utility, it is a problem of arbitrage theory which was developed inside the game theory. The problem, which is rather more interesting for us is, how they manage to increase their common utility as much as possible. The simpliest, and most effective, way is to gather their goods together, without respect to prices, then to realize the process of exchange with other coalitions in such a way that the common utulity will be maximal, and, finally, to re-distribute the obtained goods in such a way that the summ of utilities is still maximal (i.e. the goods will be distributed to the agents for which they cause maximal increase of utility). After that the players with increased utility pay finantial compensations (called also side-payments) to other members of coalition. Those side-payments are such that the final utility of each agent is greater than befor the beginning of the whole process.

In the last step the close relation between this market model and the game theoretical solutions is visible. The exchange process between coalitions represents a problem of market theory, where prices are respected. The re-distribution of utility inside the coalition, and especially the payment of compensations, represents a typical game theoretical problem, namely the construction of an arbitrage schema, or, in a more general case, the construction of a core of some new coalition game played inside the investigated coalition.

REFERENCES

[1] J. Rosenmüller: Kooperative Spiele und Märkte. Springer, Berlin 1971.

[2] L. S. Shapley, M. Shubik: On Market Games. Journal of Economic Theory *1* (1969), 1.

[3] K. Vind: Equlibirium with Respect to a Simple Market. Research report of University of Copenhagen, Institute of Economics.

[4] M. Mareš: Suggestion of a Cooperative Market Model. Kybernetika *12* (1976), 6, 439–450.

[5] M. Mareš: On Cooperative Games Connected with Markets. Kybernetika *12* (1976), 6, 451–461.

[6] T. Q. Chien: Generalized Cooperative Games and Markets. Kybernetika *12* (1976), 5, 328–354.

Czechoslovak Academy of Sciences

Institute of Information Theory and Automation

ADVANCES IN THE THEORY OF STOCHASTIC INTEGRATION AND APPLICATIONS

MICHEL MÉTIVIER

RENNES

1. PURPOSE OF THE PAPER

We intend to mention and comment on a few recent results in the field of stochastic integration, related with the two following directions:

– Deepennings and simplifications of basic results, which make quite general situations accessible without using a big machinery of preliminary sophisticated notions on martingales.

– Integration with respect to Hilbert-valued martingales and applications to stochastic evolution equations.

In the first place we describe the general scheme involved in any construction of stochastic integrals.

Next we sketch an L^2-Theory of integration with respect to square integrable Hilbert-valued martingales, telling in this case how Ito's Formulas read.

In the third part we describe in which context stochastic evolution equations arise and state a few results.

2. A GENERAL SCHEME FOR BUILDING STOCHASTIC INTEGRALS

2.1. Rough description of the problem

The problem of stochastic integration arises from the fact that many processes $(Z_t)_{t\in\mathbf{R}^+}$ have paths $t \leftrightarrow Z_t(\omega)$ which are not with bounded variation, and that, consequently, except for the case of very particular processes $(\Phi_t)_{t\in\mathbf{R}^+}$, the expression

(2.1) $$\int_0^t \Phi_s(\omega)\,\mathrm{d}Z_s(\omega)$$

has no meaning for a particular path ω, according to the classical integration theory.

The expression (2.1) has a natural meaning nevertheless when, for every ω, $s \mapsto \Phi_s(\omega)$ is a step function on intervals of $\boldsymbol{R}^+$. Suppose for example that

$$\Phi_s(\omega) = \sum_{k=1}^{n} \varphi_k(\omega) \cdot 1_{]s_k, t_k]}(s) \tag{2.2}$$

then the natural value for (2.1) is

$$\int_0^t \Phi_s(\omega)\, \mathrm{d}Z_s(\omega) = \sum_{k=1}^{n} \varphi_k(\omega) \cdot (Z_{t_k \wedge t} - Z_{s_k \wedge t}) . \tag{2.3}$$

The theory built by Ito, when Z is a finite dimensional brownian motion consisted precisely of taking (2.3) as the defining expression when Φ is a process of the form (2.2) (with some more measurability condition to be made precise later), to put a topology on the vector spaces of these processes, for which the mapping $\Phi \mapsto \int_0^t \Phi_s(.)\, \mathrm{d}Z_s(.)$, taking its values in a suitable space of random variables, is continuous, and to extend it by continuity to a space of stochastic processes, which happens to be easily characterized.

When you want to consider more general processes Z taking their values in a vector space $\boldsymbol{H}$, you follow the same procedure: start with a class of step processes taking their values in a vector space $\boldsymbol{F}$, and defined through a formula of type (2.2). Then if $(x, y) \mapsto x \circ y$ is a bilinear mapping of $(F \times \boldsymbol{H})$ into $\boldsymbol{G}$, the natural value of $\int_0^t \Phi_s \, \mathrm{d}Z_s$ is the random vector

$$\sum_{k=1}^{n} \varphi_k(.) \circ (Z_{t_k \wedge t} - Z_{s_k \wedge t}) .$$

The problem amounts then to defining topologies on spaces of stochastic processes and random variables in such a way that $\Phi \mapsto \int_0^t \Phi \, \mathrm{d}Z$ be continuous and to characterize the closure for this topology of the initial set of considered processes.

We will make those considerations more precise now.

2.2. General definitions and basic results

Let $(\Omega\, (\mathscr{F}_t), P)$ be the classical setting, $(\mathscr{F}_t)_{t \in \boldsymbol{R}^+}$ beeing an increasing family of σ-algebras of subsets of Ω, and P a probability on the σ-algebra $\mathscr{F}_\infty$ generated by the $\mathscr{F}_t$'s.

We will assume as usually done that $\mathscr{F}_\infty$ is P-complete, and that any P-null set in $\mathscr{F}_\infty$ belongs to $\mathscr{F}_0$.

When considering vector valued processes (Z_t) we will assume, when not otherwise specified, that Z_t is strongly measurable with respect to $\mathscr{F}_t$.

As we want to consider only non anticipative processes, the departing class $\mathscr{E}$ of step processes which we will consider consists of those processes of the form (2.2) with the extra property: $\forall k \varphi_k$ is $\mathscr{F}_{s_k}$-measurable.

This leads us to introduce the following class of subsets of $\boldsymbol{R}^+ \times \Omega$.

DEFINITION 1. A subset $]s, t] \times F$, $s < t$, of $\boldsymbol{R}^+ \times \Omega$ is called a predictable rectangle if $F \in \mathscr{F}_s$. We denote by $\mathscr{R}$ the class of predictable rectangles, $\mathscr{A}$ the generated boolean ring and $\mathscr{P}$ the generated σ-algebra, which is called the σ-algebra of predictable sets.

It is known (and easy to check) that the σ-algebra of predictable sets is generated by processes with left-continuous paths, and by stochastic intervals $]\sigma, \tau] = \{(t, \omega) : : \sigma(\omega) < t \leqq \tau(\omega)\}$ where σ and τ are two bounded stopping times of the family $(\mathscr{F}_t)$ as well.

We denote by $\mathscr{E}_{\boldsymbol{F}}$ the vector space of processes

$$\Phi = \sum_{k=1}^{n} 1_{]s_k, t_k] \times F_k} \cdot a_k \quad \text{where} \quad a_k \in \boldsymbol{F} \quad \text{and} \quad]s_k, t_k] \times F_k \in \mathscr{R} .$$

One way generally used to build the integral consists of finding a real measure λ on $\mathscr{P}$, depending on Z, in such a way that for some $p \in [0, \infty[$, the mapping $\Phi \leftrightarrow \int_0^t \Phi \, dZ$ from $\mathscr{E}_{\boldsymbol{F}}$ into $L^p_{\boldsymbol{G}}(\Omega, \mathscr{F}_t, P)$ defined by

$$\text{(2.2.1)} \qquad \int_0^t \Phi \, dZ = \sum_{k=1}^{n} 1_{F_k} \cdot a_k \circ (Z_{t_k \wedge t} - Z_{s_k \wedge t})$$

be continuous for the $L^p_{\boldsymbol{F}}(R^+ \times \Omega, \mathscr{P}, \lambda)$ metric structure on $\mathscr{E}_{\boldsymbol{F}}$.

With respect to this general question the following result is of interest:

THEOREM 1. If, for some real measure λ on $\mathscr{P}$, the mapping $\mathscr{E}_{\boldsymbol{R}} \ni \Phi \leftrightarrow \int_0^t \Phi \, dZ$ is continuous for the $L^p_{\boldsymbol{R}}(\boldsymbol{R}^+ \times \Omega, \mathscr{P}, \lambda)$ and $L^p_{\boldsymbol{H}}(\Omega, \mathscr{F}_t, P)$ norms $p \geqq 1$, then $Z = M + V$ where M is a martingale and V a process, the paths of which are with bounded variation.

Proof. It is clear that the continuity condition implies that for any $(\Phi_n) \uparrow \Phi \in \mathscr{E}$, $\int_0^t \Phi_n \, dZ \xrightarrow[n \to \infty]{L^p_{\boldsymbol{H}}(\Omega, \mathscr{F}_t, P)} \int_0^t \Phi \, dZ$. The proposition is then deduced from [14] (theorem 3).

This proposition tells us about the widest class of processes for which a stochastic integral can be defined along the classical lines.

In case Z is an m-dimensional brownian motion B, we are provided with a λ by taking $d\lambda = dt \otimes dP$ on $\mathscr{P}$. Taking for $\boldsymbol{F}$ the space of bounded linear operators $\mathscr{L}(\boldsymbol{R}^m; \boldsymbol{R}^n)$, with the bilinear mapping $u \circ x = u(x)$ from $(\boldsymbol{F} \times \boldsymbol{R}^m)$ into $\boldsymbol{R}^n$, the mapping $\Phi \leftrightarrow \int_0^t \Phi \circ dB$ is a contraction for the norms $L^2_{\boldsymbol{F}}(\boldsymbol{R}^+ \times \Omega, \mathscr{P}, \lambda)$ and $L^2_{\boldsymbol{R}^n}(\Omega, \mathscr{F}_t, P)$. It is even an isometry for $m = n = 1$. Moreover every process X which is non anticipating and such that $\int_{]0,t] \times \Omega} |X(u, \omega)|^2 \, du \, dP(u) < \infty$, is λ-equivalent to a process in $L^2_{\boldsymbol{F}}(R^+ \times \Omega, \mathscr{P}, \lambda)$, and the class of integrable processes is thus immediately extended.

When Z is a square integrable margingale M, with right continuous paths, the measure λ is usually built with the help of the "natural increasing process" of the submartingale $(\|M_t\|^2)$ as occuring in the Doob-Meyer decomposition of $\|M\|^2$ (see [8]). A simple measure theoretic argument shows that the same measure λ can be defined directly from M. Actually we have:

THEOREM 2. 1°) If $(S_t)_{t\in \boldsymbol{R}^+}$ is an integrable real sub-martingale with right continuous paths, the set function defined on $\boldsymbol{R}$ by

$$]s, t] \times F \leftrightarrow E[1_F . (S_t - S_s)]$$

can be extended into a σ-additive positive measure on $\mathscr{P}$, if and only if for all $\alpha > 0$, the family of random variables $\{S_\tau : \tau \text{ stopping time } \leqq \alpha\}$ is uniformly integrable.

2°) This is the case in particular if $S_t = \|M_t\|^2$ where (M_t) is a square integrable right continuous martingale taking its values in a Hilbert space $\boldsymbol{H}$. If we take $\boldsymbol{F} = \mathscr{L}(\boldsymbol{H}; \boldsymbol{G})$ where $\boldsymbol{G}$ is another Hilbert space and $\boldsymbol{F}$ is endowed with the operator norm, the mapping $\Phi \leftrightarrow \int_0^t \Phi \, dM$ can be extended into a contraction of $L^2_{\boldsymbol{F}}(\boldsymbol{R}^+ \times \Omega, \mathscr{P}, \lambda)$ into $L^2_{\boldsymbol{G}}(\Omega, \mathscr{F}_t, P)$.

3°) In the situation of 2°), the process $(\int_0^t \Phi \circ dM)_{t\in \boldsymbol{R}^+}$ is a martingale with a right continuous version.

Proof. C. Doleans proved the direct part of 1°) (see [7]). For the reciprocal statement see [17]. The fact that it applies to $\|M_t\|^2$ comes from uniform integrability properties of integrable sub-martingales. The remaining of 2°) is deduced from trivial verifications.

The property in 3°) is easily checked for $\Phi \in \mathscr{E}_{\boldsymbol{F}}$ and the general case can be reduced to this one by a limit procedure.

The stochastic integral which is thus built is the stochastic integral III of Kunita (See [12]), which includes all the L^2-stochastic integrals of the literature.

Before improving on this construction in § 3, we will close this § 2 with a few remarks.

2.3. Remarks

REMARK 1. For an L^p, $p \neq 2$ theory, see [21] and [18].

REMARK 2. Theorem 1 shows that if we want to define stochastic integral with respect to processes which are not quasimartingales (see [17] for example for the definition) we have to look for an entirely different extension procedure. This has been done for example in [22] where, to every square integrable process Z, is associated the simple additive extension of $]\sigma, \tau] \leftrightarrow Z_\tau - Z_\sigma$ to the boolean ring $\bar{\mathscr{A}}$ generated by the bounded stochastic intervalls $]\sigma, \tau]$. An integral (in the measure theoretic sense) is immediately available for those processes X which are uniform limits of step processes on $\bar{\mathscr{A}}$: this class includes all the left continuous processes.

REMARK 3. If for an increasing sequence of stopping times (σ_n) such that $\sigma_n^{\uparrow\infty}$ a.s. the processes $1_{]0,\sigma_n]} \cdot \Phi$ are in $L_F^2(R^+ \times \Omega, \mathscr{P}, \lambda)$ it is possible to prove, as in the classical finite dimensional case, that there exists one right continuous process (Y_t), still denoted by $(\int_0^t \Phi \, dM)$, unique up to a zero process (i.e. with path a.s. identically zero), such that

$$\forall_{t,n} \quad Y_{t \wedge \sigma_n} = \int_0^t 1_{]0,\sigma_n]} \Phi \, dM .$$

The process Y is a local martingale. For convenience, in the sequel the process $(\int_0^t \Phi \, dM)$ will be called the local integral of Φ with respect to M.

3. MORE ON L^2-THEORY OF STOCHASTIC INTEGRATION WITH RESPECT TO HILBERT VALUED MARTINGALES

3.3. Extension of the class of Integrable processes

In Kunita's Integral as exposed above the processes which can be integrated are strongly measurable with respect to the σ-algebra of predictable sets (we will abbreviate: strongly predictable). In some applications this is too much to ask for. (For example if e^{-tA} is a semi-group of operators on a Hilbert space $\boldsymbol{H}$, with infinitesimal generator A, we would like to write $\int_0^t e^{-sA} \, dM_s$, without knowning whether $s \mapsto e^{-sA}$ is measurable for the uniform operator norm in $\mathscr{L}(\boldsymbol{H}; \boldsymbol{H})$).

We want to mention here an interesting extension which is developed in [24], and which at the same time exhibits a norm on processes, such that the mapping $\Phi \mapsto \int_0^t \Phi \, dM$ becomes an isometry into $L_G^2(\Omega, \mathscr{F}_t, P)$ as in the real case.

THEOREM 3. Let $\boldsymbol{H} \otimes_1 \boldsymbol{H}$ be the projective tensor product.

If M is a square integrable right continuous martingale, with values in the Hilbert space $\boldsymbol{H}$, λ the real measure associated with $\|\boldsymbol{M}\|^2$ by Theorem 2, then, there exists a strongly predictable process Q, unique up to the λ-equivalence, such that

$$\forall]s, t] \times F \in \mathscr{R} \quad E[1_F \cdot (M_t - M_s)^{\otimes 2}] = \int_{]s,t] \times F} Q \, d\lambda .$$

Proof. It is easily checked that

$$\|E(1_F \cdot (M_t - M_s)^{\otimes 2})\|_{H \otimes_1 H} \leqq \lambda]s, t] \times F .$$

As $\boldsymbol{H} \otimes_1 \boldsymbol{H}$ is a separable dual of a Banach space (Shatten's theorem) one may apply the classical Radon-Nikodym theorem for vector valued measures. This gives Q.

An elementary but longer argument may be given to prove this theorem (see [24]),

We recall that $\boldsymbol{H} \otimes_1 \boldsymbol{H}$ is isomorphic to the vector space of trace-class operators $\mathscr{L}_1(\boldsymbol{H})$ with the trace norm, if $\tilde{u}$ is the trace operator associated with u, and $\tilde{b}$ the bounded operator associated with a bilinear form b by $b(x, y) = (\tilde{b}(x), y)$, we have (denoting still by b the linear extension of b to $\boldsymbol{H} \otimes_1 \boldsymbol{H}$):

$$b(u) = \mathrm{Tr}(\tilde{u} \circ \tilde{b}) .$$

We then have the following:

THEOREM 4. Let $\Lambda_t(M)$ be the class of processes taking their values in $\mathscr{L}(\boldsymbol{H}; \boldsymbol{G})$, such that for every $h \in \boldsymbol{H}$ and $g \in \boldsymbol{G}$, the real process $(Xh, g)_G$ is predictable and

$$\int_{]s,t] \times \Omega} \mathrm{Tr}(XQX^*)\, \mathrm{d}\lambda < \infty .$$

Then $\Lambda_t(M)$ is complete for the hilbertian semi-norm $X \leftrightarrow (\int_{]0,t] \times \Omega} \mathrm{Tr}(XQX^*)\, \mathrm{d}\lambda)^{1/2}$ and the mapping $\Phi \leftrightarrow \int_0^t \Phi\, \mathrm{d}M$ on $\mathscr{E}$ extends uniquely into an isometry of $\Lambda_t(M)$ into $L^2_G(\Omega, \mathscr{F}_t, P)$.

The process $(\int_0^t \Phi\, \mathrm{d}M)$ is a square integrable martingale which has a right continuous version.

Proof. We refer to [24] for the proof.

3.2. Quadratic variation

Even though the notion of natural process associated with the Doob-Meyer Decomposition theorem of $(\|M_t\|^2)$ is superfluous to build the stochastic integral with respect to a general square integrable martingale, this notion or some equivalent substitute is necessary to build the stochastic differential calculus, in other words to get Ito's Formulas.

When (M_t) is continuous, the natural process is the quadratic variation which is easily directly defined, as we will see. When (M_t) is not continuous the process which enters the Ito's formula is in some sense the quadratic variation of the "continuous projection" (M_t^c) (see [12], theorem 3) of M.

When M is real an easy way of getting the notion of quadratic variation is the one used by Millar (cf. [18]). Writte, for any sequence $\{0 < t_1^n < \ldots < t_k^n < \ldots\}$,

$$|M_t|^2 - |M_0|^2 = \sum_k |M_{t_{k+1}^n \wedge t} - M_{t_k^n \wedge t}|^2 - 2 \sum_k M_{t_k^n} \cdot (M_{t_{k+1}^n \wedge t} - M_{t_k^n \wedge t}) \tag{3.2.1}$$

and remark that, if we denote by Φ^n the left continuous process:

$$\Phi^n = \sum_k M_{t_k^n} 1_{]t_k^n, t_{k+1}^n]}$$

the formula (3.2.1) can be written

$$(3.2.2) \qquad |M_t|^2 - |M_0|^2 = \sum_k |M_{t_{k+1}^n \wedge t} - M_{t_k^n \wedge t}|^2 - 2\int_0^t \Phi^n \, \mathrm{d}M .$$

Using available continuity properties of $\int_0^t \Phi^n \, \mathrm{d}M$, when $\lim_{n\to\infty} \sup_k |t_{k+1}^n - t_k^n| = 0$, we get the existence of a process denoted $[M_t]$, and called the quadratic variation, with right continuous increasing paths such that

$$(3.2.3) \qquad |M_t^2| - |M_0|^2 = [M_t] - 2\int_0^t M_{u-} \, \mathrm{d}M_u .$$

When M is vector valued, the quadratic variation is a process with values in $\boldsymbol{H} \otimes_1 \boldsymbol{H}$ (if a basis in $\boldsymbol{H}$ is taken, this is a matrix valued process) defined as the limit (see which limit below) of the sums $\sum_k (M_{t_{k+1}^n \wedge t} - M_{t_k^n \wedge t})^{\otimes 2}$. We have immediately a formula analogous to (3.2.2):

$$(3.2.4) \quad M_t^{\otimes 2} - M_0^{\otimes 2} = \sum_k (M_{t_{k+1}^n} - M_{t_k^n})^{\otimes 2} - \int_0^t \Phi^n \otimes \mathrm{d}M - \int_0^t \mathrm{d}M \otimes \Phi^n .$$

More precisely, denoting by $\boldsymbol{H} \otimes_2 \boldsymbol{H}$ the Hilbert-Schmidt tensor product, we have the following:

THEOREM 5. *M being a square integrable martingale with values in the Hilbert space $\boldsymbol{H}$, and (t_n^k) a sequence of divisions with the already mentioned property, then*

$$\sum_k \|M_{t_{k+1}^n \wedge t} - M_{t_k^n \wedge t}\|^2 \xrightarrow[n \to \infty]{L^1_{\boldsymbol{R}}(\Omega, \mathscr{F}_t, P)} [\![M_t]\!] ,$$

$$\sum_k (M_{t_{k+1}^n \wedge t} - M_{t_k^n \wedge t})^{\otimes 2} \xrightarrow[n \to \infty]{L^1_{\boldsymbol{H} \otimes_2 \boldsymbol{H}}(\Omega, \mathscr{F}_t, P)} [M_t] ,$$

where $[\![M_t]\!]$ and $[M_t]$ are right continuous processes (continuous if M is), $[\![M_t]\!]$ is real and increasing, $[M_t]$ is with values in $\boldsymbol{H} \otimes_1 \boldsymbol{H}$ and such that for every ω, the interval function $]s, t] \mapsto [M_t] - [M_s]$ extends into a σ-additive $\boldsymbol{H} \otimes_1 \boldsymbol{H}$-valued measure with bounded variation.

Moreover $\|M_t\|^2 - [\![M_t]\!]$ and $M_t^{\otimes 2} - [M_t]$ are martingales and

$$(3.2.5) \qquad \|M_t\|^2 - \|M_0\|^2 = [\![M_t]\!] - 2\int_0^t (M_{u-} \mid \mathrm{d}M_u) ,$$

$$(3.2.6) \qquad M_t^{\otimes 2} - M_0^{\otimes 2} = [M_t] - \int_0^t M_{u-} \otimes \mathrm{d}M_u - \int_0^t \mathrm{d}M_u \otimes M_{u-} .$$

The above convergences are in $L^2_{\boldsymbol{R}}$ and $L^2_{\boldsymbol{H} \otimes_2 H}$ respectively if $E\|M_t\|^4 < \infty$.

Proof. We refer to [15] and [24].

3.3. Ito's formula

THEOREM 6. Let $(x, y) \mapsto \varphi(x, y)$ be a mapping from $(\boldsymbol{H} \times K)$ into $\boldsymbol{G}$ (Hilbert spaces), with bounded weakly continuous directional partial derivatives $D_x\varphi$ and $D_x^2\varphi$, $D_y\varphi$.

Let M be a right continuous square integrable martingale with values in $\boldsymbol{H}$, and V a right continuous square integrable process, with values in K, and with bounded variations.

1°) The process

$$S(t) = [M_t] - \sum_{s \leq t} (M_s - M_{s^-})^{\otimes 2}$$

is a $\boldsymbol{H} \otimes_1 \boldsymbol{H}$-valued continuous process with paths of bounded variation.

2°) For every t

$$\begin{aligned} \text{(3.3.1)} \qquad \varphi(M_t, V_t) - \varphi(M_0, V_0) &= \sum_{s \leq t} [\varphi(X_s) - \varphi(X_{s^-}) - D_x\varphi(M_{s^-}, V_{s^-}) \,. \\ &\quad . (M_s - M_{s^-}) - D_y\varphi(M_{s^-}, V_{s^-})(V_s - V_{s^-})] + \\ &\quad + \int_0^t D_x\varphi(M_{u^-}, V_{u^-}) \, dM_u + \int_0^t D_y\varphi(M_{u^-}, V_{u^-}) \, dV_u + \\ &\quad + \frac{1}{2} \int_0^t D_x^2\varphi(M_{u^-}, V_{u^-}) \, dS_u \quad \text{a.s.} \end{aligned}$$

where the last integral is to be understood as the classical integral on each path ω, of the operator valued function $u \mapsto D_x^2\varphi(M_{u^-}(\omega), V_{u^-}(\omega)) \in \mathscr{L}(\boldsymbol{H} \otimes_1 \boldsymbol{H}; \boldsymbol{G})$, with respect to the vector valued measure induced on $\boldsymbol{R}^+$ by $u \mapsto S_u(\omega)$.

Proof. We refer to [10] and [24]. In the continuous case the first term in the right hand side of (3.3.1) vanishes and $S = [M]$. In this case a proof has been published in [14] and [21].

4. STOCHASTIC EVOLUTION EQUATIONS

4.1. Problem

Many practical problems lead to models of systems which are governed by equations of the following type:

$$\text{(4.1.1)} \qquad \frac{\partial}{\partial t} f(t, x) = L_x^t f(t, x) + g(t, x)$$

where for each t L_x^t denotes an operator on functions of x, and g is a perturbation function. Such an equation can be read

$$\frac{\mathrm{d}}{\mathrm{d}t} X_t = A_t X_t + g(t, .) \tag{4.1.2}$$

where $t \leftrightarrow X_t = f(t, .)$ can be interpreted as a mapping in some Hilbert space $\boldsymbol{H}$ (for example some Sobolev space) of functions and A_t is an operator with domain in $\boldsymbol{H}$.

If the perturbation is random $t \leftrightarrow g(t, .)$ becomes a process M with values in $\boldsymbol{H}$, and equation (4.1.2) leads to a stochastic differential equation for processes X with values in $\boldsymbol{H}$, which can be written

$$X_t = X_0 + \int_0^t A_u X_u \,\mathrm{d}u + M_t \tag{4.1.3}$$

where M_t is a random process with values in $\boldsymbol{H}$, or, more abruptly in the "differential" form

$$\mathrm{d}X_t = A_t . X_t \,\mathrm{d}t + \mathrm{d}M_t . \tag{4.1.4}$$

Such problems have been studied in the linear case (i.e. A_t is a linear operator) by Curtain and Falb [4], Dawson [3] and the non-linear case has been considered in particular by Daleckii, Bensoussan and Temam [1], Pardoux [20] and Viot [26]. We do not pretend to give an account of the already numerous results in the subject as appearing in [1], [3], [4], [5], [19], [20], [26], [27] but to illustrate this introduction by slightly modified versions of some of them. The proof of the here stated theorems will be found in a forthcomming short thesis work by Pistone [24].

4.2. Linear Case

We consider the case of a linear operator A (we take the homogeneous case to simplify the statement, that is A is independant of t), with domain $\boldsymbol{D} \subset \boldsymbol{H}$. We assume that A is the infinitesimal generator of a strongly continuous semi group R_t on $\boldsymbol{H}$.

THEOREM 7.

1°) Let M_t be a right continuous, square integrable martingale with values in $\boldsymbol{D}$.* Then for any X_0 in $\boldsymbol{D}$, there exists a unique (up to a zero process) right continuous X_t with values in $\boldsymbol{D}$, solution of (4.1.3) given by

$$X_t = R_t X_0 + \int_0^t R_{t-u} \,\mathrm{d}M_u . \tag{4.2.1}$$

* Null in 0 and such that, a.s., $u \rightarrow AM_u$ $L^\infty([0, 1]; \boldsymbol{D})$.

2°) If A is such that $R_t(\boldsymbol{H}) \subset \boldsymbol{D}$ for all t, and $(A_t z \mid z) \leqq 0$ $\forall x \in \boldsymbol{D}$, then, for any martingale M, square integrable and right continuous, for any X_0 in $\boldsymbol{D}$, there exists a unique right continuous solution of (4.1.3) given still by the formula (4.2.1).

Proof. The proof of 1°) consists in checking that (4.2.1) gives the solution by using an integration by part formula, derived from the Ito's formula.

Part 2°) is reduced to part 1°) by approximating a martingale with values in $\boldsymbol{H}$ by a sequence of martingales with values in $\boldsymbol{D}$.

4.3. Non linear case

We will state a variation of one Pardoux's result. The situation is the usual one when dealing with variational methods in partial differential equations: there are three Hilbert spaces V, $\boldsymbol{H}$, V' with continuous injections $V \to \boldsymbol{H} \to V'$, V' being the dual space of V. (We note $(\, , \,)$ the scalar product in $\boldsymbol{H}$, $|\cdot|$ the corresponding norm, and $((\, , \,))$ the scalar product in V, $\|\cdot\|$ the corresponding norm.)

A_t is a family of continuous mapping of V into V' with the following properties: coercivity: $\exists \alpha > 0$ and $\lambda \in \boldsymbol{R}$ such that $\forall v \in V$ and t

$$2\langle A(t)\, v, v\rangle \leqq \lambda |v|^2 - \alpha \|v\|^2 \; ;$$

strict monotony: $\forall u, v \in V$, $\forall t$

$$2\langle A(t)\, v - A(t)\, u, v - u\rangle \leqq \lambda |v - u|^2 \, ,$$

$$\forall y \in V \; \|A_t(y)\|_{V'} \leqq \beta \|y\|_V \, .$$

Theorem 8. If A verifies the preceeding conditions and M_t is a square integrable right continuous martingale, with values in V, then, for every initial condition X_0 in H, there exists a unique process X_t with values in V', right continuous, such that $\forall \omega \; X_t(\omega) \in V$ a.e. in t, $X \in L^2_V([0, 1] \times \Omega, \mathrm{d}t \times \mathrm{d}P)$ and X is solution of (4.1.3).*

The method of proof is based on a finite dimensional approximation of the problem, and on the fact that the solutions of equation (4.1.3) depend "continuously" on the martingales M as a consequence of the following fundamental lemma (consequence of Ito's formula):

Lemma (Inequality of energy). Let M be a martingale in V and X a solution of (4.2.3), then:

1°) $$|X_t|^2 + \alpha \int_0^t \|X_u\|^2 \, du \leqq |X_0|^2 + 2 \int_0^t (X_u, \mathrm{d}M_u) + \mathrm{Tr}[M_t] + \int_0^t \lambda |X_u|^2 \, du \, ,$$

2°) $$E|X_t|^2 + E \int_0^t \alpha \|X_u\|^2 \, du \leqq E|X_0|^2 + E|M_t|^2 + \lambda \int_0^t E|X_u|^2 \, du \, .$$

* This means that both members of (4.1.3) are well defined processes, equal up to a zero process.

5. EXTENSIONS

An interesting extension of the problem of the preceeding § has been worked out by Viot [26] who replaces monotony assumptions by the hypothesis of compacity of $V \to \boldsymbol{H}$. He gets a solution in law for equation (4.1.2), when $M_t = hB_t$, where $h \in \boldsymbol{H}$ and B is a one-dimensional Brownian motion. The unicity is deduced from monotony assumption again.

Pardoux [20] studied the infinite dimensional extension of the Ito's equations under lipschitzian conditions on the coefficients with the same coercivity and monotony assumption as above.

In all these situations, M_t is a process with values in $\boldsymbol{H}$. For example M_t is a Brownian motion (B_t) with values in $\boldsymbol{H}$ as in [4]. This means that B has a nuclear covariance. In some cases it is natural [3] to work with a perturbation which is a "white noise in space" that is some generalized process with covariance the identity operator in $\boldsymbol{H}$. This gives rise to the notion of cylindrical martingales which can be defined as linear functionals $M_t : h \leftrightarrow M_t(h) \in L^2_{\boldsymbol{R}}(\Omega, \mathscr{F}_t, P)$ such that for every $h \in \boldsymbol{H} (M_t(h))_{t \in \boldsymbol{R}^+}$ is a martingale. This situation has been considered in particular cases (see [9] and [27]).

REFERENCES

[1] A. Bensoussan, R. Temam: Equations aux dérivées partielles stochastiques non linéaires. Isr. J. of Math. *11* (1972), 95–129.

[2] E. Cabana: Stochastic integration in separable Hilbert spaces. Publ. Inst. de Matematica y Estadistica, Uruguay *IV* (1966), 49–79.

[3] D. A. Dawson: Stochastic evolution equations. Mathematical biosciences *15* (1972), 287–316.

[4] R. F. Curtain, P. L. Falb: Stochastic differential equations in Hilbert spaces. Journal of Differential equations *10* (1971), 412–430.

[5] Ya. L. Daleckii: Differential equations with functional derivatives and stochastic equations for generalized random processes (English translation). Soviet Math. Dokl. *7* (1966), 220–223.

[6] Ya. L. Daleckii: Infinite dimensional elliptic operators and parabolic equations connected with them (English translation). Usp. Math. Nauk *22* (1967).

[7] C. Doleans: Existence du processus croissant naturel associé à un potentiel de la classe (D). Zeit. W. Theorie *9* (1968), 309–314.

[8] C. Doleans, P. A. Meyer: Intégrales stochastiques par rapport aux martingales locales. Séminaire de probabilité IV — Lecture Notes in Math. Vol. 24, Springer Verlag, Berlin 1970.

[9] B. Gaveau: Intégrale stochastique radonifiante. C.R. Acad. Sc. Paris *276*, (19 février 1973).

[10] B. Gravereau, J. Pellaumail: Formule de Ito pour des processus non continus à valeurs dans un espace de Banach. Ann. Inst. Poincaré *10* (1974), 399–422.

[11] L. GROSS: Potential Theory on Hilbert spaces. J. Funct. Analys. (1968), 1—2, 123—181.
[12] H. KUNITA: Stochastic integrals based on martingales taking values in Hilbert spaces. Nayoya Math. J. *38* (1970), 41—52.
[13] A. H. KUO: Stochastic integrals in abstract Wiener spaces. Pacific J. of Math. *41*, (1972), no 2.
[14] M. MÉTIVIER: Intégrale stochastique par rapport à des processus à valeurs dans un espace de Banach réflexif. Teoria Veroyatnostei XIX (1974), 787—816.
[15] M. MÉTIVIER: Approximation des solutions des équations différentielles stochastiques. Séminaire Rennes (1974).
[16] M. MÉTIVIER: Intégrale stochastique par rapport à des martingales hilbertiennes. C.R. Acad. Sc. Paris *276*, 1009—1012.
[17] M. MÉTIVIER, J. PELLAUMAIL: On Doleans-Föllmer's measure for quasi-martingales. Séminaires Rennes (1974).
[18] P. W. MILLAR: Martingale Integrals. Trans. Am. Math. Soc. *133* (1968), 145—166.
[19] SH. OGAWA: A partial differential equation with the white Noise as coefficient. Z. W. Theorie *28* (1973), 53—71.
[20] E. PARDOUX: Sur des équations aux dérivéespartielles stochastiques monotones. C.R. Acad. Sci. Paris *275* (1972), Serie A, 101—103.
[21] J. PELLAUMAIL: Sur l'intégrale stochastique et la décomposition de Doob-Meyer. Astérisque n° 9, Soc. Math. de France, 1973.
[22] J. PELLAUMAIL: Formule de Ito dans le cas réel non continu. Séminaire Rennes (1973).
[23] J. PELLAUMAIL: Un lemme élémentaire de théorie de la mesure. Séminaire Rennes (1973).
[24] G. PISTONE: Thèse 3ème cycle. Université de Rennes.
[25] STROOCK, VARADHAN: Diffusion processes with continuous coefficients. Comm. in Pure and appl. Math. *22* (1969), 345—400.
[26] M. VIOT: Solution en loi d'une équation aux dérivées partielles stochastique non linéaire: méthode de compacité. C.R. Acad. Sc. Paris *278*, 1185—1188.
[27] M. YOR: Le problème des martingales dans un espace de Hilbert. Thèse 3 me cycle. Paris 1973.

Basic references on classical stochastic integrations:

GIHMAN, SKOROKHOD: Stochastic differential equations. Springer, 1972.
K. ITO: On stochastic differential equations. Mem. A. M. S. *4* (1951).

UNIVERSITÉ DE RENNES

and

CENTRE NATIONAL DE LA RECHERCHE SCIENTIFIQUE

LABORATOIRE DE PROBABILITÉS

SECOND ORDER STOCHASTIC PROCESSES AND SELF-ADJOINT OPERATORS

JIŘÍ MICHÁLEK

PRAGUE

The author studies in this paper a spectral resolution of second order stochastic processes continuous in the quadratic mean derived from a resolution of the identity in Hilbert space.

Let $x(t)$, $t \in \mathscr{R}$, be a stochastic process with finite second moments i.e. $\mathsf{E}|x(t)|^2 < \infty$ for every real t and we can put for the sake of simplicity $\mathsf{E}\, x(t) = 0$. Let λ be a real number and let us denote $\mathscr{L}_\lambda$ the closed linear hull by the all random variables $x(t)$, where $t \leqq \lambda$. The closure is considered in the topology defined by the norm $\|x(t)\|^2 = \mathsf{E}|x(t)|^2$. Clearly, $\mathscr{L}_\lambda \subset \mathscr{L}(x(\cdot))$, where $\mathscr{L}(x(\cdot))$ is the Hilbert space by the all random variables $x(t)$, where $t \in \mathscr{R}$. Further, let P_λ denote the orthogonal projector on the subspace $\mathscr{L}_\lambda$ in the space $\mathscr{L}(x(\cdot))$. For our purposes the family $\{P_\lambda\}_{\lambda \in \mathscr{R}}$ of orthogonal projectors in the space $\mathscr{L}(x(\cdot))$ must be a resolution of the identity in the space $\mathscr{L}(x(\cdot))$. Hence, in the first place we must consider such a stochastic process $x(\cdot)$, which defines the family $\{P_\lambda\}_{\lambda \in \mathscr{R}}$ as a resolution of the identity in the space $\mathscr{L}(x\cdot))$.

THEOREM 1. Let $x(t)$, $t \in \mathscr{R}$, be a second order stochastic process which is left-continuous in the quadratic mean and $\lim_{\lambda \to -\infty} \mathsf{E}\|P_\lambda\, x(t)\|^2 = 0$ for every $t \in \mathscr{R}$. Then the projector family $\{P_\lambda\}_{\lambda \in \mathscr{R}}$ is a resolution of the identity in the space $\mathscr{L}(x(\cdot))$.

Proof. We must prove that $P_\lambda P_\mu = P_{\min(\lambda,\mu)}$, $P_{-\infty} = 0$, where $P_{-\infty}\xi = \lim_{\lambda \to -\infty} P_\lambda \xi$, $P_{+\infty} = I$, where $P_{+\infty}\xi = \lim_{\lambda \to +\infty} P_\lambda \xi$ and I is the identity operator in the space $\mathscr{L}(x(\cdot))$ and $P_{\lambda - 0} = P_\lambda$ for every $\lambda \in \mathscr{R}$ and every $\xi \in \mathscr{L}(x(\cdot))$. If $\lambda \leqq \mu$, then $\mathscr{L}_\lambda \subset \mathscr{L}_\mu$ and hence $P_\lambda P_\mu = P_\lambda$ in the space $\mathscr{L}(x(\cdot))$. Let ξ be an arbitrary element in the space $\mathscr{L}(x(\cdot))$. Then a sequence $\{\xi_n\} \subset \mathscr{L}(x(\cdot))$ must exist, where $\xi_n = \sum c_i^n\, x(t_i^n)$, c_i^n are complex numbers, such that $\|\xi - \xi_n\| \underset{n}{\to} 0$. It is obvious that $\xi_n \in \mathscr{L}_{\max(t_i^n)}$ for every $n \in \mathscr{N}$. The operators P_λ, $\lambda \in \mathscr{R}$, are linear and continuous, hence $\|P_\lambda \xi - P_\lambda \xi_n\| \underset{n}{\to} 0$

for every $\lambda \in \mathcal{R}$. The last fact holds in the following sence: $\forall \varepsilon > 0\ \exists n_\varepsilon\ \forall n \geqq n_\varepsilon\ \forall \lambda \in \mathcal{R} \Rightarrow \|P_\lambda \xi - P_\lambda \xi_n\| \leqq \varepsilon$ because $\|P_\lambda\| = 1$ for every $\lambda \in \mathcal{R}$. From the inequality $\|P_\lambda \xi\| \leqq \|P_\lambda \xi - P_\lambda \xi_{n_\varepsilon}\| + \|P_\lambda \xi_{n_\varepsilon}\|$ it follows that it is sufficient to prove the convergence $\|P_\lambda \xi\| \xrightarrow[\lambda \to -\infty]{} 0$ for every linear combination $\xi = \sum_1^n c_i\, x(t_i)$, where $t_i \in \mathcal{R}$. So, we assume that $\lim\limits_{\lambda \to -\infty} \|P_\lambda\, x(t)\| = 0$, i.e. $\forall \varepsilon > 0\ \exists \lambda_\varepsilon\ \forall \lambda \leqq \lambda_\varepsilon \Rightarrow \|P_\lambda\, x(t)\| \leqq \varepsilon$. Let us set $\lambda_\varepsilon \geqq \lambda$ and consider the norm $\|P_\lambda \xi\|$ where $\xi = \sum_1^n c_i\, x(t_i)$. Because the operator P_λ is a orthogonal projector on the subspace $\mathcal{L}_\lambda$ then $P_\lambda \xi \in \mathcal{L}_\lambda$. So, it is $P_\lambda \xi = \sum_1^n c_i P_\lambda\, x(t_i)$, hence $\|P_\lambda \xi\| \leqq \sum_1^n |c_i|\, \|P_\lambda\, x(t_i)\| \leqq \varepsilon \sum_1^n |c_i|$, because $\|P_\lambda\, x(t_i)\| \leqq \varepsilon$ for every $\lambda \leqq \lambda_\varepsilon$ and every $i = 1, 2, \ldots, n$. Further, we must prove the left-continuity of the family $\{P_\lambda\}_{\lambda \in \mathcal{R}}$ at every $\lambda \in \mathcal{R}$. It means that the equality $P_{\lambda - 0} = P_\lambda$ must hold for every $\lambda \in \mathcal{R}$. This equality corresponds to the relation $\bigcup\limits_{\mu < \lambda} \mathcal{L}_\mu = \mathcal{L}_\lambda$, where $\bigcup\limits_{\mu < \lambda} \mathcal{L}_\mu$ is the smallest subspace by the all subspaces $\mathcal{L}_\mu$, $\mu < \lambda$. It is clear that the inclusion $\bigcup\limits_{\mu < \lambda} \mathcal{L}_\mu \subset \mathcal{L}_\lambda$ holds everytime. Under the condition of the left-continuity in the quadratic mean of the process $x(\cdot)$ we can prove the opposite inclusion. Let $\xi \in \mathcal{L}_\lambda$. Then a sequence $\{\xi_n\}$ must exist, $\xi_n \in \mathcal{L}_\lambda$ for every $n \in \mathcal{N}$, such that $\|\xi - \xi_n\| \xrightarrow[n]{} 0$. The random variables ξ_n are in the form $\xi_n = \sum_i c_i^n\, x(t_i^n)$, where $t_i^n \leqq \lambda$ for every $n \in \mathcal{N}$ and every $i = 1, 2, \ldots, k_n$. Because the process $x(\cdot)$ is left-continuous in the quadratic mean the $x(\lambda) \in \bigcup\limits_{\mu < \lambda} \mathcal{L}_\mu$. This fact completes the prof of the property $P_{\lambda - 0} = P_\lambda$. In the end, it must hold $\lim\limits_{\lambda \to \infty} P_\lambda \xi = \xi$ for every $\xi \in \mathcal{L}(x(\cdot))$. Let ξ be an arbitrary random variable in the space $\mathcal{L}(x(\cdot))$ and $\{\xi_n\}$ be a sequence of linear combinations for which $\|\xi - \xi_n\| \xrightarrow[n]{} 0$ holds. Because $\xi_n = \sum_i c_i^n\, x(t_i^n)$ and $\forall \varepsilon > 0\ \exists n_\varepsilon\ \forall n \geqq n_\varepsilon\ \forall \lambda \in \mathcal{R} \Rightarrow \|P_\lambda \xi - P_\lambda \xi_n\| \leqq \varepsilon$, then for $n = n_\varepsilon$ and $\lambda_\varepsilon = \min\limits_i \{t_i^{n_\varepsilon}\}$. $\cdot\ \|P_\lambda \xi - \xi\| \leqq \|P_\lambda \xi - P_\lambda \xi_{n_\varepsilon}\| + \|P_\lambda \xi_{n_\varepsilon} - \xi\| \leqq \varepsilon + \|\sum_i c_i^{n_\varepsilon}\, x(t_i^{n_\varepsilon}) - \xi\| \leqq 2\varepsilon$. It completes the proof.

REMARK 1. The condition $\|P_\lambda\, x(t)\| \xrightarrow[\lambda \to -\infty]{} 0$ for every $t \in \mathcal{R}$ seems to be too general. But the statistical sense of this condition is very clear: this condition says that the random variables $x(t)$ and $x(\lambda)$ are "almost" orthogonal if λ is sufficiently small. There exists a wide class of stochastic processes for which this condition is fulfilled. If $x(\cdot)$ is a martingal in the wide sense and $E|x(t)|^2 \xrightarrow[t \to -\infty]{} 0$, then $P_\lambda\, x(t) = x(\lambda)$ a.s. and hence $\|P_\lambda\, x(t)\| \xrightarrow[\lambda \to -\infty]{} 0$ for every $t \in \mathcal{R}$. Another class of stochastic processes for which this condition is valid is the class of second-order processes defined at a closed interval. If a process $x(\cdot)$ is defined at the interval $\langle a, b \rangle$, then we can consider without loss of generality that $x(a) = 0$ a.s. and define $x(t) = 0$ for $t < a$ and $x(t) = x(b)$ for $t > b$. Now, if we define subspaces $\mathscr{L}_\lambda$, $\lambda \in \mathscr{R}$, as above, we obtain: $\mathscr{L}_\lambda = \{0\}$ for $\lambda \leqq a$, $\mathscr{L}_\lambda = \mathscr{L}_b = \mathscr{L}(x(\cdot))$ for $\lambda \geqq b$. Under the condition of the left-continuity in the quadratic mean it can be easily proved that the orthogonal projectors on the subspaces $\mathscr{L}_\lambda$ in this Hilbert space $\mathscr{L}_b$ form a resolution of the identity.

Further, in this paper we shall study such stochastic processes for which following both conditions are valid:

A) the process $x(\cdot)$ is left-continuous in the quadratic mean,

B) for every $x(t)$ $\lim_{\lambda \to -\infty} \|P_\lambda x(t)\| = 0$.

Due to Theorem 1 we can construct a resolution of the identity defined in the pertaining Hilbert space $\mathscr{L}(x(\cdot))$ to every stochastic process $x(\cdot)$ satisfying the conditions A, B.

Every resolution of the identity $\{P_\lambda\}_{\lambda \in \mathscr{R}}$ corresponding to a process $x(t)$, $t \in \mathscr{R}$ satisfying the conditions A, B defines in the due space $\mathscr{L}(x(\cdot))$ a self-adjoint operator $H = \int \lambda \, \mathrm{d}P_\lambda$ which is defined at everywhere dense subset $\mathscr{D}(H) = \{\xi \in \mathscr{L}(x(\cdot)) : : \int \lambda^2 \mathrm{d}\langle P_\lambda \xi, \xi \rangle < +\infty\}$. If such a process $x(t)$ is defined at the interval $\langle a, b \rangle$ then $H = \int_a^b \lambda \, \mathrm{d}P_\lambda$ and the operator H corresponding to this stochastic process is bounded and defined everywhere in the space $\mathscr{L}(x(\cdot))$.

The existence of the following integrals follows from the spectral decomposition of the operator H:

$$\int \lambda \mathrm{d}\langle P_\lambda \xi, \xi \rangle \quad \text{and} \quad \int \lambda \, \mathrm{d}P_\lambda \xi \quad \text{exist for every} \quad \xi \in \mathscr{D}(H).$$

Definition 1. The measure $\mu(\cdot, \xi)$ defined by the function $\|(P_\mu - P_\lambda)\xi\|^2$ with the finite variation $\|\xi\|^2$, where $\lambda \leqq \mu$, $\lambda, \mu \in \mathscr{R}$, we shall call the spectral measure with respect to the random variable $\xi \in \mathscr{L}(x(\cdot))$.

Let us denote $\mathscr{L}_2(\mu(\cdot, \xi))$ the Hilbert space of all complex functions defined at the real line for which the integral $\int |f(\lambda)|^2 \, \mathrm{d}\mu(\lambda, \xi)$ exists.

Theorem 2 (Properties of stochastic integral). For every $f, g \in \mathscr{L}_2(\mu(\cdot, \xi))$ the following properties hold:

1. $E \int f(\lambda) \, \mathrm{d}P_\lambda \xi = 0$,
2. $E|\int f(\lambda) \, \mathrm{d}P_\lambda \xi|^2 = \int |f(\lambda)|^2 \, \mathrm{d}\mu(\lambda, \xi)$,
   $E(\int f(\lambda) \, \mathrm{d}P_\lambda \xi)(\int g(\lambda) \, \mathrm{d}P_\lambda \xi)^* = \int f(\lambda))^* \, \mathrm{d}\mu(\lambda, \xi)$ where $*$ denotes the complex adjoint number,
3. $\int f(\lambda) \langle \mathrm{d}P_\lambda \xi, \xi \rangle = \langle \int f(\lambda) \, \mathrm{d}P_\lambda \xi, \xi \rangle$ where $\langle \cdot, \cdot \rangle$ denotes the inner product in the space $\mathscr{L}(x(\cdot))$,

4. $\int \Psi_{\langle a,b \rangle}(\lambda)\, \mathrm{d}P_\lambda \xi = P_b \xi - P_a \xi$ where $\Psi_{\langle a,b \rangle}(\cdot)$ is the indicator of the interval $\langle a, b \rangle$,
5. $\int_a^b f(\lambda)\, \mathrm{d}P_\lambda \xi$ is a bilinear operator which is continuous at the Cartesian product $\mathscr{L}_2(\mu(\cdot, \xi)) \times \mathscr{L}(x(\cdot))$ and

$$\left\| \int\!\!\int_a^b f(\lambda)\, \mathrm{d}P_\lambda \xi \right\| \leqq \|\xi\| \, \|f\|_{\mathscr{L}_2(\mu(\cdot,\xi))} \, .$$

Proof. It immediately follows from the spectral decomposition of the self-adjoint operator H in the space $\mathscr{L}(x(\cdot))$ and from the properties of the process $x(\cdot)$.

THEOREM 3. A real number λ_0 is an eigenvalue of the operator H due to a process $x(\cdot)$ if and only if there exists such a random variable $\xi_0 \in \mathscr{L}(x(\cdot))$ its spectral measure is discontinuous at λ_0.

Proof. The eigenvalues of a self-adjoint operator can be easily recognized from its spectral decomposition. A real number λ_0 is an eigenvalue of the operator H if and only if $P_{\lambda_0+0} \neq P_{\lambda_0}$, e.i. $\mathscr{L}_{\lambda_0} \neq \bigcap_{\lambda > \lambda_0} \mathscr{L}_\lambda$. In this case a random variable $\xi_0 \in \mathscr{L}_{\lambda_0+0} - \mathscr{L}_{\lambda_0}$, $\xi_0 \neq 0$, must exist such that $(P_{\lambda_0+0} - P_{\lambda_0})\, \xi_0 = \xi_0$. It implies that $\lim_{\lambda \downarrow \lambda_0} \langle (P_\lambda - P_{\lambda_0})\, \xi_0, \xi_0 \rangle = \langle \xi_0, \xi_0 \rangle = \|\xi_0\|^2 > 0$ i.e., the spectral measure $\mu(\cdot, \xi_0)$ is discontinuous at the point λ_0. Conversely, if the spectral measure $\mu(\cdot, \xi_0)$ is discontinuous at a point λ_0 i.e. $\lim_{\lambda \downarrow \lambda_0} \langle (P_\lambda - P_{\lambda_0})\, \xi_0, \xi_0 \rangle > 0$, it means that also $\lim_{\lambda \downarrow \lambda_0} \mathsf{E} |P_\lambda \xi_0 - P_{\lambda_0} \xi_0|^2 > 0$ hence the resolution of the identity of the operator H is discontinuous at λ_0.

Since this moment we shall assume that a process $x(t)$, $t \in \mathscr{R}$, satisfying the condition B is continuous at every point. In this case it can be easily proved the separability of the due Hilbert space $\mathscr{L}(x(\cdot))$. The continuity in the quadratic mean implies the continuity of the covariance function $R(s, t) = \mathsf{E}\, x(s)\, x(t)^*$. Let ξ be any element in the space $\mathscr{L}(x(\cdot))$ and consider the continuous linear functional $F_t(\xi) = \mathsf{E}\xi\, x(t)^*$ defined on the space $\mathscr{L}(x(\cdot))$. The equality $\|F_t\| = \|x(t)\|$ holds obviously. For every fixed $\xi \in \mathscr{L}(x(\cdot))$ we can regard $F_t(\xi)$ as a function of the argument $t \in \mathscr{R}$, then for every $\xi \in \mathscr{L}(x(\cdot))$ $F_t(\xi)$ is a continuous function in $(-\infty, +\infty)$ because

$$|F_t(\xi) - F_s(\xi)|^2 = |\mathsf{E}\xi(x(t) - x(s))^*|^2 \leqq \|\xi\|^2\, \mathsf{E}|x(t) - x(s)|^2 \to 0$$

if $s \to t$. So we have constructed a mapping I defined by the relation $I\xi = F_t(\xi)$. The mapping I is one-to-one mapping because the random variables $x(t)$, $t \in \mathscr{R}$, form the space $\mathscr{L}(x(\cdot))$. Denote RI the range of the mapping I. The set RI is linear and we introduce a inner product (F_t, G_t) in this set by the relation $(F_t, G_t) = \mathsf{E}\xi\eta^*$ where $F_t(\xi) = \mathsf{E}\xi\, x(t)^*$, $G_t(\eta) = \mathsf{E}\eta\, x(t)^*$. In this moment the mapping I is an isometric mapping and the space RI is complete in the topology induced by inner product $(\cdot, \cdot)$. Because the space $\mathscr{L}(x(\cdot))$ is a separable the space $(RI, (\cdot, \cdot))$ is also separable. The space $(RI, (\cdot, \cdot))$ is called the reproducing kernel Hilbert space due to the stochastic process $x(\cdot)$. The most important property of this space is the property of the reproducing kernel by the covariance function $R(s, t) : (F_s, R(s, t)) = F_t$ because $(F_s, R(s, t)) = E\xi\, x(t)^* = F_t(\xi)$ (see [1]).

Now let us consider the self-adjoint operator H corresponding to the process $x(t)$, $t \in \mathcal{R}$. Using the mapping $I : \mathcal{L}(x(\cdot)) \to RI$ we can transform this operator H in the space $(RI, (\cdot, \cdot))$ by the relation

$$H_R\, F_t(\xi) = EH\xi\, x(t)^*\,, \quad t \in \mathcal{R}\,.$$

The operator H_R is defined on everywhere dense subset $\mathcal{D}(H_R) = \{F_t : F_t = I\xi, \xi \in \mathcal{D}(H)\}$. The operator H_R is obviously linear, continuous and symmetric at the set $\mathcal{D}(H_R)$. On the other hand, we can also transform the projectors P_λ, $\lambda \in \mathcal{R}$, in the space $(RI, (\cdot, \cdot))$

$$P_\lambda^R\, F_t(\xi) = EP_\lambda\xi\, x(t)^*\,, \quad t \in \mathcal{R}\,.$$

Each operator P_λ^R is defined everywhere in the space $(RI, (\cdot, \cdot))$ and is linear and continuous. Operators P_λ^R, $\lambda \in \mathcal{R}$, are projectors in the space $(RI, (\cdot, \cdot))$ because

$$P_\lambda^R(P_\lambda^R\, F_t(\xi)) = EP_\lambda^2\xi\, x(t)^* = EP_\lambda\xi\, x(t)^* = P_\lambda^R F_t\,.$$

Similarly, it can be proved that the operator family $\{P_\lambda^R\}$, $\lambda \in \mathcal{R}$, forms a resolution of the identity in the space $(RI, (\cdot, \cdot))$. This resolution of the identity defines a self-adjoint operator H_R' in the space $(RI, (\cdot, \cdot))$. It is obvious that $H_R = H_R'$.

Now we use Hellinger-Hahn theorem (see [2]) which ensures for every self-adjoint operator defined on the Hilbert space $\mathcal{L}(x(\cdot))$ (separable) the existence of a finite or an infinite sequence of mutually orthogonal random variables $\{\eta_n\}$ such that

$$\mathcal{L}(x(\cdot)) = \mathcal{M}(\eta_1) \oplus \mathcal{M}(\eta_2) \oplus \ldots \oplus \mathcal{M}(\eta_n) \oplus \ldots$$

where $\mathcal{M}(\eta_n) = \mathrm{cl}\{\sum c_i P_{\lambda_i}\eta_n\}$. Let us denote $\Delta\eta_n = (P_\mu - P_\lambda)\,\eta_n$, where $\Delta = \langle\lambda, \mu)$. This system of random variables has obviously the following properties:

1. The system $\{\Delta\eta_n\}$ forms a complete system of random variables in the space $\mathcal{L}(x(\cdot))$.
2. For every couple $m, n \in \mathcal{N}$, $m \neq n$ and any intervals $\Delta_1, \Delta_2 \langle\Delta_1\eta_m, \Delta_2\eta_n\rangle = 0$ holds.
3. For every $n \in \mathcal{N}$ and any disjoint intervals $\Delta_1, \Delta_2 \langle\Delta_1\eta_n, \Delta_2\eta_n\rangle = 0$ holds.

Now we shall define a set function $\mu_n(\Delta) = E|\Delta\eta_n|^2$ for every random variable η_n.

It is easily seen that the set function μ_n is a measure on Borel sets in the real line for every $n \in \mathcal{N}$, hence we can consider integrals $\int\Phi(\lambda)\,\mathrm{d}\mu_n(\lambda)$, where $\Phi \in \mathcal{L}_1(\mu_n)$, $n \in \mathcal{N}$. For every $\xi \in \mathcal{L}(x(\cdot))$ and every $n \in \mathcal{N}$ the inner product $\langle\xi, \Delta\eta_n\rangle$ defines a finite signed measure on Borel sets in the real line and the inequality $|\langle\xi, \Delta\eta_n\rangle|^2 \leqq \leqq \|\xi\|^2\, \|\Delta\eta\|^2 = \|\xi\|^2\, \mu_n(\Delta)$ implies that this signed measure $\langle\xi, \Delta\mu_n\rangle$ is absolutely continuous with respect to the measure μ_n. There exists Radon-Nikodym derivative $\mathrm{d}\langle\xi, \eta_n\rangle(\cdot) = \Phi_n(\cdot)\,\mathrm{d}\mu_n(\cdot)\,[\mu_n]$. In this way we obtain an expression of the inner product in the space $\mathscr{L}(x(\cdot))$:

$$\langle\xi_1, \xi_2\rangle = \sum_{n=1}^{\infty} \int \Phi_n^1(\lambda)\, \Phi_n^2(\lambda)^*\, \mathrm{d}\mu_n(\lambda)$$

where

$$\Phi_n^i(\lambda) = \frac{\mathrm{d}\langle\xi_i, \eta_n(\lambda)\rangle}{\mathrm{d}\mu_n(\lambda)}, \quad i = 1, 2\,.$$

Because for every fixed $t \in \mathscr{R}$ the covariance function $R(s, t)$ of the process $x(t)$, $t \in \mathscr{R}$ belongs to the space $(RI, (\cdot, \cdot))$ the limit

$$\Psi_n(t, \lambda) = \lim_{\Delta\to\lambda} \frac{(R(s, t), \Delta f_n(s))}{F_n(\Delta)} = \lim_{\Delta\to\lambda} \frac{\Delta f_n(t)^*}{F_n(\Delta)}\,[F_n]$$

must exist, where $\Delta f_n = I(\Delta\eta_n)$, $n = 1, 2, \ldots$ and $F_n(\Delta) = \|\Delta f_n\|_{RI}^2$. It is easily to see that the functions $\{f_n\}$ form Hellinger-Hahn system in the space $(RI, (\cdot, \cdot))$. Due to the reproducing kernel property $(R(s, z), R(t, z)) = R(s, t)$ the expression of the covariance function $R(\cdot, \cdot)$

$$R(s, t) = \sum_{n=1}^{\infty} \int \Psi_n(s, \lambda)\, \Psi_n(t, \lambda)^*\, \mathrm{d}F_n(\lambda)$$

must hold.

So we have proved the following theorem.

THEOREM 4. Let $x(t)$, $t \in \mathscr{R}$, be a second order stochastic process continuous in the quadratic mean, $\mathsf{E}\, x(t) = 0$ and satisfying the condition B. Then its covariance function $R(\cdot, \cdot)$ can be expressed in the form

$$R(s, t) = \sum_{n=1}^{\infty} \int \Psi_n(s, \lambda)\, \Psi_n(t, \lambda)^*\, \mathrm{d}F_n(\lambda)$$

where $\Psi_n(s, \cdot) \in \mathscr{L}_2(F_n)$ for every $s \in \mathscr{R}$ and every $n \in \mathscr{N}$.

REMARK 2. If the system of the orthogonal measures $\{\Delta f_n\}$ in the space $(RI, (\cdot, \cdot))$ is represented by the only measure $\{\Delta f_{n_0}\}$ then the spectral representation of the covariance function $R(s, t)$ is well-known Karhunen representation

$$R(s, t) = \int \Psi(s, \lambda)\, \Psi(t, \lambda)^*\, \mathrm{d}F(\lambda)$$

It implies that such a process can be expressed in the form of stochastic integral

$$x(t) = \int \Psi(t, \lambda)\, \mathrm{d}\mu(\lambda)\,.$$

THEOREM 5. A second order stochastic process $x(t)$, $t \in \mathscr{R}$ continuous in the quadratic mean, $E\,x(t) = 0$ and satisfying condition B can be expressed in Karhunen representation if and only if a random variable $\xi_0 \in \mathscr{L}(x(\cdot))$ exists such that the set of all finite linear combinations $\sum_i c_i P_{\lambda_i} \xi_0$ is everywhere dense in the space $\mathscr{L}(x(\cdot))$.

Proof. It immediately follows from the properties of a self-adjoint operator with a simple spectrum.

THEOREM 6. A second order stochastic process $x(t)$, $t \in \mathscr{R}$ continuous in the quadratic mean, $E\,x(t) = 0$ and satisfying the condition B can be expressed in the form of a sum of stochastic integrals

$$x(t) = \sum_{n=1}^{\infty} \int \Psi_n(t, \lambda)\, d\mu_n(\lambda).$$

Proof. It follows from Theorem 4 and Karhunen representation.

THEOREM 7. Let $x(t)$, $t \in \mathscr{R}$ be a second order Markov process continuous in the quadratic mean, $E\,x(t) = 0$ and satisfying the following conditions:

1. there exists $\lim_{t \to \infty} E|x(t)|^2 = K < +\infty$,

2. $\lim_{s \to -\infty} r(t, s) = 0$ for every $t \in \mathscr{R}$, $\lim_{t \to \infty} r(t, t + h) = 1$ uniformly in $h > 0$ where

$$r(s, t) = \frac{E\,x(t)\,x(s)^*}{E|x(s)|^2}, \quad r(s, t) = 0 \quad \text{for} \quad E|x(s)|^2 = 0.$$

Then the process $x(\cdot)$ can be written in the form of a stochastic integral

$$x(t) = \int \Psi(t, \lambda)\, d\mu(\lambda).$$

Proof. Because the process $x(\cdot)$ is continuous in the quadratic mean it must be $P_\lambda\, x(t) = r(\lambda, t)\, x(\lambda)$ a.s. From the condition 2 follows that $\lim_{\lambda \to -\infty} \|P_\lambda\, x(t)\|^2 = \lim_{\lambda \to -\infty} \|r(\lambda, t)\, x(\lambda)\|^2 = \lim_{\lambda \to -\infty} |r(\lambda, t)|^2\, \|x(\lambda)\|^2 = \lim_{\lambda \to -\infty} r(\lambda, t)\, R(\lambda, t) \leqq \lim_{\lambda \to -\infty} \|x(t)\|^2 \,.\, r(\lambda, t) = 0$ for every $t \in \mathscr{R}$. But it is clear, that $x(t + h) - r(t, t + h)\, x(t) \perp x(t)$ for every $t \in \mathscr{R}$ then $\|x(t + h)\|^2 = \|x(t + h) - r(t, t + h)\, x(t)\|^2 + |r(t, t + h)|^2 \,.\, \|x(t)\|^2$. Conditions 1 and 2 imply that $\lim_{t \to \infty} \|x(t + h) - r(t, t + h)\, x(t)\|^2 = 0$. Then the sequence $\{x(t)\}$ is fundamental in the quadratic mean because $\|x(t + h) - x(t)\|^2 = \|x(t + h) - r(t, t + h)\, x(t)\|^2 + |r(t, t + h) - 1|^2\, \|x(t)\|^2$. The space

$\mathscr{L}(x(\cdot))$ is complete hence a random variable $x(+\infty)$ must exist for which $\lim_{t\to\infty} x(t) =$ $= x(+\infty)$. It is clear that $P_\lambda x(+\infty) = \lim P_\lambda x(t) = \lim r(\lambda, t)\, x(\lambda) = x(\lambda) \lim r(\lambda, t)$ $= x(+\infty)$. It is clear that $P_\lambda\, x(+\infty) = \lim_{t\to\infty} P_\lambda\, x(t) = \lim_{t\to\infty} r(\lambda, t)\, x(\lambda) = x(\lambda)\, .$ $.\lim_{t\to\infty} r(\lambda, t) = r(\lambda, +\infty)\, x(\lambda)$. By using Theorem 5 we can immediately complete the proof.

REFERENCES

[1] N. ARONSZJAN: Theory of reproducing kernels. Trans. Amer. Math. Soc. *68* (1950), 337—404.

[2] M. H. STONE: Linear transformations in Hilbert space. Amer. Math. Soc. Colloquium Publications, Vol. XV.

CZECHOSLOVAK ACADEMY OF SCIENCES

INSTITUTE OF INFORMATION THEOR AND AUTOMATION

THE VELOCITY OF STOCHASTIC PROCESSES

DENIS MOLLISON

EDINBURGH

The purpose of this paper is to present a theorem (2.1) which says, roughly speaking, that the velocity of any Markovian stochastic process which spreads by contact (on an n-dimensional regular lattice, or in continuous n-dimensional Euclidean space) is finite if the contact distribution is exponentially bounded; and (2.3) that this is also a necessary condition for populations whose total numbers grow exponentially. The first section (para. 1) describes the class of processes (including epidemic, and birth death and migration models) to which these theorems apply ("contact processes", based on the idea of "pseudoepidemics" in Mollison [2] and affording a generalisation of the "percolation processes" of Hammersley [1]). The main purpose of this theoretical section is to provide a basis for comparison between models (1.3), so that the main theorem (2.1) need only be proved for a simple "contact birth process" or "simple epidemic with an unlimited pool of susceptibles at each point" model.

Although this result agrees with that for the corresponding deterministic model, I at least find it rather surprising, since the most widely believed conjecture was that finite variance would at least turn out to be sufficient. Finite variance of the contact distribution does imply finite variance, proportional to t, for the population as a whole; this shows the danger of extrapolating from properties of the distribution of the whole population to the furthest-scattered individuals (another example is furnished by the simple one-way nearest-neighbour birth and migration model, where the expected population distribution is Poisson, $\rightarrow$ normal as $t \rightarrow \infty$, but the velocity of the population front is not that given by the normal approximation (this is easily proved using Stirling's formula to approximate the Poisson distribution).

1. CONTACT PROCESSES

1.0

What do simple epidemics, percolation processes, 'birth death and migration' etc. models have in common? Although at first sight it looks as though what happens to one individual is inextricably dependent on what happens to others, this dependence can in fact be split off into a set of rules, leaving the actual probability space (Ω, F, p) equal to a product $\prod_i(\Omega_i, F_i, p_i)$, where the product is taken over all individuals i. Not only does this give a lucid description which may be of help in considering individual models, but it also provides a powerful rigorous way of comparing models: this is because we can often choose the *same* probability space for two models, all the differences being expressed in the rules. If the rules of one model which say whether an individual i is 'susceptible' at time t are stricter than those of the second model, it will follow that $S_1(\omega) \subseteq S_2(\omega)$ for the sets of susceptibles *for all* $\omega \in \Omega$; whence, e.g., for any parameter such as a velocity c which is monotone decreasing with S, it will follow that $E(c_1) \geqq E(c_2)$.

1.1

With the above motivation, define a *contact process* on J to consist of

(1) $(\Omega, F, p) = \prod_{i \in J} (\Omega_i, F_i, p_i)$, where the points of an individual sample space Ω_i are sequences of times: $\omega_i = \{t_i^k(j): k = 1, 2, \ldots\}$: a particular t_i^k may be a 'contact time' for another individual or location j, or relate only to i ($j = \emptyset$ conventionally). [For example, in the general epidemic $G(\alpha, \mu, v)$ t_i^k is either $t_i^k(j)$, a time at which 'i infects j': these occur as a Poisson process of density $\alpha v(|j - i|)$; or a 'death time' $t_i^k(\emptyset)$: these occur as an independent Poisson process of density μ. (The *rules* will tell us to ignore all t_i^k after the first death-time).]

(2) rules which, at least a.s., specify, given the sets of various types of individual (e.g. infected, dead) at time 0, the sets of each type at times $t > 0$ as functions of ω.

[This definition could be extended to allow for the possibility of a time $t^*(\omega)$ beyond which these sets are undefined, as e.g. if the number of one type may $\to\infty$ in finite time and this causes the rules to break down.]

1.2. Further examples

(a) The simple and general epidemics in (say) 2 dimensions, with population density ϱ at each point of a square lattice: these may be called $S_2^\varrho(\alpha, v)$ and $G_2^\varrho(\alpha, \mu, v)$, or S and G for short. The intuitive description here is that each infected individual produces germs at rate α from the time of his infection onwards, until (in the case of G) his death after a life distributed as $1 - e^{-\mu t}$; these germs are distributed independently according to the contact distribution $v(i, j)$, usually taken to depend only on the distance $|j - i|$ from i to j, and infect j if he is not already infected or dead. In order to allow a comparison of the sort described above, formal definitions as contact processes take the same probability space $\prod_i(\Omega_i, F_i, p_i)$, where each individual space is the product of one Poisson process of density μ, and one of density $\alpha v(i, j)$ for each j. The rules for G are then as follows:

(1) The time at which j is infected,

$$t_0(j) = \inf\{t : \exists \text{a "}\textit{chain of infection}\text{ from } i \text{ to } j \text{ with } \textit{transition time } t\text{" for some } i \in I(0)\}$$

where a "chain of infection from i to j with transition time t" means a sequence of times t_n such that each

$$t_n = t_{i_n}^{k_n}(i_{n+1}) < \inf\{t_{i_n}^k(\emptyset) \quad (0 \leqq n \leqq N - 1)\}$$

and

$$i_0 = i\,, \quad i_N = j\,.$$

(2) The time at which j dies $= t_0(j) + \inf_k \{t_j^k(\emptyset)\}$.

The rules for S differ simply in that the death-process is made 'inoperative' by removing the condition that $t_n < \inf\{t_{i_n}^k(\emptyset)\}$. Then it is immediate that the set of infected plus dead for G, $I_G(\omega, t) + D_G(\omega, t)$, is contained in $I_s(\omega, t)$, a.s. on Ω ("a.s." is necessary, since there may be ω for which I and D are undefined; but the set of such ω will be of measure zero).

(For other examples involving $S_1^e(\alpha, v)$, see [2]).

(b) $S_n^*(\alpha, v)$ "simple epidemic with unlimited pool of susceptibles at each point" or "contact birth process" model. This can be formulated in terms of contacts between individuals, like $S_n^e(\alpha, v)$, but only at the cost of a complicated population set, each 'k^{th} order' individual i having infinitely many $k + 1^{\text{st}}$ order individuals available to infect at each location. It is much simpler to define $\omega_i = t_i^\omega(s)$, where $t_i^\omega(s)$ is an infection time for location s. Then the rule for $S_n^*(\alpha, v)$ defines $t_0(j)$ for $j = n^{\text{th}}$ individual at s as the nth smallest transition time for a chain of infection from $I(0)$ to s.

$G_n^*(\alpha, \mu, v)$ is then defined similarly with the addition of a Poisson 'death-process' for each i.

If we then change the rules to make i inoperative after his first infection time involving a different location, we obtain the birth death and migration model $M_n(\lambda, \mu, \nu, v')$, where $\lambda = \alpha\, v(O)$, $\nu = \alpha - \alpha\, v(O)$, $v'(O) = (\alpha/\nu)\, v(s)$ for $s \neq O$; and conversely, given $M_n(\lambda, \mu, \nu, v)$, we can set $\alpha = \lambda + \nu$, $v(O) = \lambda/\alpha$, $v^*(s) = (\nu/\alpha)\, v(s)$ for $s \neq O$, to obtain $G_n^*(\alpha, \mu, v^*)$ which differs from M only in having more generous rules of infection. Hence we have established the top line of:

1.3

$$\begin{array}{ccccc} M_n(\lambda, \mu, \nu, v) & \leqq & G_n^*(\alpha, \mu, v^*) & \leqq & S_n^*(\alpha, v^*) \\ & & \text{VII} & & \text{VII} \\ & & G_n^{\varrho}(\alpha, \mu, v^*) & \leqq & S_n^{\varrho}(\alpha, v^*) . \end{array}$$

To complete the proof of this diagram, only the vertical inequalities are needed. These are supplied by reformulating G^{ϱ} and S^{ϱ} in terms of locations – the same (Ω, F, p) can be taken as for G^* and S^*, but with all individuals after the ϱth at each location declared inoperative.

(Note that several different probability spaces may represent the same contact process. $A \leqq B$ means that the set of susceptibles $S_A(\omega, t) \supseteq S_B(\omega, t)$ for all t a.s. on Ω, *for at least one* probability space (Ω, F, p) which can be chosen for A & B, not necessarily for all such formulations.)

(Note that my examples have only been of Markovian processes; for a non-Markovian example see [1].)

2. VELOCITY OF STOCHASTIC PROCESSES

2.0

From 1.3 it follows that $S_n^*(\alpha, v)$, the simple epidemic with an infinite pool of susceptibles at each point on an n-dimensional (regular) lattice, is the fastest of what may be called Markovian contact processes, where the individual processes are products of Poisson processes describing birth, migration, infection, death, etc. Thus a bound on v which implies that the expected velocity of S_n^* is finite will apply to all these processes, and in particular to the birth, death and migration model and the 'general epidemic'; and to the simple and general epidemics with bounded population density.

2.1

Main Theorem. *If v is exponentially bounded, the expected velocity of $S(\alpha, v)$ is finite.*

More precisely, let $v(n)$, the probability that a contact is made at distance n (measured along the edges of the lattice) be $\leqq ae^{-b(n+1)}$ for some $a, b > 0$ [it is easily seen that this is equivalent to "$\sum v(n)\, e^{b'n} < \infty$, some $b' > 0$" or to "$v(s) \leqq$ $\leqq a'e^{-b''|s|}$, $a', b'' > 0$" for any finite-dimensional lattice]. Let N_t denote the distance of the furthest infected individual from 0, when $I(0)$ consists of a single individual at 0; and let $M_t = E(N_t)$. Then:

$$\limsup_t M_t/t \leqq C_{\alpha,v} = \alpha C_v < \infty .$$

Proof. (a) it is trivial that any bound $C_{\alpha,v}$ will be proportional to α; so we may w.l.o.g.* take $\alpha = 1$. The first part of the proof then follows the corresponding part of Hammersley's proof of the corresponding result for percolation processes.

$$M_t = \mathrm{E}(N_t) = \sum_{n=1}^{\infty} \Pr(N_t \geqq n) = \sum_{1}^{ct} \Pr(N_t \geqq n) + \sum_{ct}^{\infty} \Pr(N_t \geqq n)$$

where c is an arbitrary positive constant.

Now $\Pr(N_t \geqq n) \leqq 1$ for any n; and $\Pr(N_t \geqq n) \leqq \Pr(N_{n/c} \geqq n)$ if $n/c \geqq t$ (since, for $S(\alpha, v)$, the set of infected is monotone non-decreasing with t).

Hence

$$M_t \leqq ct + \sum_{ct}^{\infty} P_n$$

where

$$P_n \equiv \Pr(N_{n/c} \geqq n) .$$

Let us try to show that $P_n \leqq Ae^{-\varepsilon n}$ for some A, $\varepsilon > 0$. It will then follow that $M_t/t \leqq c + (1 - \mathrm{e}^{-\varepsilon n})^{-1} Ae^{-ct})/t = c + o(t)$, whence the result desired.

(b) If $N_{n/c} \geqq n$, there must be a sequence of distances $\{n_j : 1 \leqq j \leqq k\}$ $(1 \leqq \leqq k < \infty)$ such that $\sum n_j \geqq n$ and a chain of infection with 'links' of length n_j and total 'transition time' $\leqq n/c$ exists. To take account of the possibility that $n_j = 0$, it is convenient to treat n_j as though it were $n_j + 1$ (this is why the bound for $v(n)$ has been taken of the form $v(m) \leqq ae^{-b(m+1)}$). Thus, a fortiori, there must be a sequence of distances $\{n_j : 1 \leqq j \leqq k\}$ $(1 \leqq k \leqq n)$ such that $\sum_1^k n_j = n - k$, and a chain of infection with links n_j $(1 \leqq j \leqq k - 1)$ and n_k^* $(\geqq n_k)$ and transition time $\leqq n/c$ exists. The probability of this, for a particular sequence $\{n_i\}$, is

$$P_{\{n_i\}} = \iint \cdots \int \prod_1^k (r_i \, \mathrm{d}t_i) ,$$

$$0 \leqq t_1 \leqq t_2 \ldots \leqq t_k \leqq n/c ,$$

where $r_i = v(n_i)$ for $1 \leqq i \leqq k - 1$

$$r_k = \sum_{n_k}^{\infty} v(m) .$$

* Without loss of generality.

Now by hypothesis $v(m) \leqq ae^{-b(m+1)}$, whence

$$r_i \leqq ae^{-b(n_i+1)} \quad (1 \leqq i \leqq k-1),$$

$$r_k \leqq (1 - e^{-b})^{-1} ae^{-b(n_k+1)},$$

$$\therefore P_{\{n_i\}} = \left(\prod_1^k r_i\right) \underset{0 \leqq t_1 \leqq t_2 \ldots \leqq t_k \leqq n/c}{\iint \cdots \int} \prod_1^k \mathrm{d}t_i \leqq ((1-e^{-b})^{-1} a^k e^{-bn}) \left(\frac{(n/c)^k}{k!}\right) = q_k, \quad \text{say,}$$

independent of the particular sequence $\{n_i\}$.

(c) Now the number of such sequences $\{n_i\}$ for fixed k is given by the binomial coefficient $\binom{n-1}{k-1}$.

Hence

$$P_n \leqq \sum_{k=1}^{n} \binom{n-1}{k-1} q_k \leqq \max_{1 \leqq k \leqq n} f(k)$$

where

$$f(k) = n \binom{n-1}{k-1} q_k = \frac{k \times n!}{(n-k)!\, k!} \times (1 - e^{-b})^{-1} e^{-bn} (an/c)^k / k!\,.$$

Now $m! = \left(\frac{m}{e}\right)^m \sqrt{(2\pi m)} \times h(m)$, where $h(m)$ is bounded $(0 < h_1 \leqq h(m) \leqq$ $\leqq h_2 < \infty)$ (and $\to 1$ as $m \to \infty$) [Stirling's formula]

$$\therefore f(k) = \left[\frac{d}{2\pi}(1 - e^{-b})^{-1} \sqrt{n/(n-k)}\right] \times e^{-bn} n^n g(k),$$

where $d(k, n)$ is a bounded term (which $\to 1$ as $k; n-k \to \infty$), so that the term in square brackets is $\leqq D_\gamma e^{\gamma n}$ where γ may be chosen arbitrarily small; and

$$g(x) = (n-x)^{-(n-x)} x^{-2x} \left(\frac{ane}{c}\right)^x.$$

(d) Now $g(x) > 0$ on $(0, n)$, $g'(0+) = +\infty$, $g'(n-) = -\infty$, where

$$g'(x) = g(x)\left((1 + \log(n-x) - 2(1 + eg(x)) + \log \frac{ane}{c}\right).$$

Hence $g(x)$ attains its supremum on $[0, n]$ at the positive root of

$$x^2 = \frac{an}{c}(n-x).$$

This root $= \delta n$, where $\delta \to 0$ as $c \to \infty$ (note that $a/c = \delta^2/(1-\delta)$).

Thus

$$\max_{1 \leqq k \leqq n} g(k) \leqq \sup_{0 \leqq x \leqq n} g(x) = (n(1-\delta))^{-n(1-\delta)}(\delta n)^{-2\delta n}\left(\frac{ane}{c}\right)^{\delta n} = n^{-n}e^{\eta n}$$

where

$$\eta = (1-2\delta)\log(1-\delta) + \delta \to 0 \quad \text{as} \quad \delta \to 0\,.$$

Then, for any $\beta < b$, we may choose c sufficiently large that $\eta < \beta$; and $\gamma < b - \beta$, $A = D_\gamma$, so that

$$P_n \leqq \max_{1 \leqq k \leqq n} f(k) \leqq (Ae^{\gamma n})\, e^{-bn} n^n (n^{-n} e^{\eta n}) \leqq Ae^{-\varepsilon n}$$

where

$$\varepsilon = b - \beta - \gamma > 0\,,$$

from which, as remarked at the end of (a), the desired result follows.

2.2

Corollary. The proof can easily be adapted to the case where the population space is k-dimensional euclidean space instead of a discrete lattice. The only important change is the replacement of the multiplicity $\binom{n-1}{k-1}$ (see (c) above) by an integral

$$\iint_{\sum_1^k x_i \leqq n} \cdots \int \mathrm{d}x_i = \frac{n^k}{k!}\,.$$

This leads to

$$g(x) = n^{-n} x^{-2x}\left(\frac{an^2e^2}{c}\right)^x\,,$$

whence a bound $Ae^{-\varepsilon n}$ can again be found for c sufficiently large.

2.3 Converse (of (2.1))

If $v(n) > ae^{-bn}$ for some n, all $a, b > 0$, then for any Markovian contact process in which the total population, conditional on non-extinction, grows exponentially,

$$\limsup_t N_t/t = \infty \quad \text{a.s. (conditional on non-extinction)}\,.$$

NOTES. (1) This therefore applies to the birth, death and migration model $M(\lambda, \mu, \nu, v)$ and the epidemics with unlimited susceptibles G^* and S^*, but not to G^ϱ and S^ϱ (see (2.4) below).

(2) I shall assume for simplicity that v is symmetric; some condition on $v(n)$ may be necessary, but I hope it will be evident that something much weaker would do.

(3) Again this result applies equally well to a continuous population space.

Proof. (a) for all $c < \infty$, there exist infinitely many n such that

$$\beta e^{(\gamma/c)n}\, v(n) \geqq \log 4$$

for all $\beta, \gamma > 0$, and in particular when β is the contact rate for our process, and γ is such that the total population at time t is (conditional on non-extinction) $\geqq e^{\gamma(+1)}$ for all $t \geqq t_*(\omega)$, a.s. on Ω. Choose a subsequence of such n, n_m such that the sequence $t_m = n_m/c$ satisfies:

$$t_{m+1} - t_m \geqq 1\,.$$

Let E_m denote "a contact distant $>n_m$ from the origin is made in $(t_m - 1, t_m)$". Thus if E_m occurs for infinitely many m, $N_t/t > c$ for some t in $(t_m - 1, t_m)$ for each such m, whence $\limsup_t N_t/t > c$. But c was arbitrary, so our result will follow.

(b) Now let F_m denote "total population during $(t_m - 1, t_m)$ is $\geqq e^{\gamma(t+1)}$". The probability of a contact to distance n_m in any subinterval of $(t_m - 1, t_m)$ of length dt is, given F_m,

$$\geqq e^{\gamma(t_m)} \times \beta\, v(n_m) \times dt\,,$$

and the probability that this gives $N_t \geqq n_m$ is at least $\frac{1}{2}$ because of the assumed symmetry of v. Hence

$$\Pr(E_m \mid F_m) \geqq 1 - \exp\left[-(\tfrac{1}{2}e^{\gamma t_m}\beta\, v(n_m)) \times 1\right] \geqq$$
$$\geqq 1 - \exp(-\tfrac{1}{2}\log 4) = \tfrac{1}{2} \text{ by choice of } n_m, t_m\,.$$

Now if G_{m-1} denotes any condition on the E_i's $(1 \leqq i \leqq m-1)$ $\Pr(E_m \mid F_m)$ will depend on G_{m-1}, but we shall still have

$$\Pr(E_m \mid F_m \cap G_{m-1}) \geqq \tfrac{1}{2}\,.$$

Hence by the Borel-Cantelli lemma, infinitely many E_m's occur if infinitely many F_m's do. But, conditional on non-extinction, a.s. on Ω $\exists t^*(\omega)$ s.t. total population is $\geqq e^{\gamma(t+1)}$ for $t \geqq t_*$, so that F_m happens for all m with $t_m \geqq t_*$.

Therefore infinitely many E_m's occur (a.s. on $\Omega|$ conditional on non-extinction), whence the result follows.

2.4

What happens in the case of, eg. S_n^q? Here the total population at time t is at least $O(t^n)$, and cannot be more if the velocity is to be finite. From the results of Mollison [2] for one dimension we might conjecture that a necc. and suff. condition for finite velocity is that $\sum_0^\infty m^{(n+k_n)} v(m) < \infty$ where k_n perhaps $=1$ for all n. Extending (2.3) to show that $E(M_{t/t}) \to \infty$ if $\sum_0^\infty m^{n+k} v(m)$ diverges for some k is easy; but it is difficult to obtain a 'best possible' k; and the converse (an equivalent of Th. 2.1) seems likely to need some new approach.

REFERENCES

[1] J. M. Hammersley: First-passage percolation. J. Roy. Statist. Soc. B *28* (1966), 491—496.

[2] Denis Mollison: The rate of spatial propagation of simple epidemics. Proc. 6th Berkeley Symp. on Math. Statist. and Prob. *3*, 579—614.

Heriot-Watt University

Department of Actuarial Mathematics and Statistics

A MATHEMATICAL MODEL OF HELMINTHIASIS WITH EXTERNAL INFECTION*

INGEMAR NÅSELL

STOCKHOLM

A hybrid model is established for the spread of helminthic infections in a community. The qualitative behavior of a resulting system of differential equations is given epidemiological interpretations. A threshold effect is shown to exist for isolated communities. The threshold effect disappears if the community is exposed to even a small amount of continuous external infection.

1. INTRODUCTION

Helminthic infections are caused in human beings and in other vertebrates by a variety of parasites, called flatworms, that belong to the phylum Platyhelminthes. We consider here flatworms of the class Trematoda and the order Digenea. The members of this class are commonly referred to as flukes. The digenetic trematodes have some of the most complex life cycles of any animal. Each of them utilizes a vertebrate as a definitive host, in which sexual reproduction takes place, and at least one intermediate host, usually a mollusk, for asexual reproduction. The life cycle of the parasite is of importance for the establishment of a mathematical model. We give therefore below a description of the life cycle of a specific parasite species, *Fasciolopsis buski*. This parasite causes the disease fasciolopsiasis, which is common in eastern Asia and the southwestern Pacific.

Sexually mature forms of *F. buski* are hermaphroditic. They are large and fleshy and reside in man's small intestine. Eggs are fertilized and laid. Some of them leave the body of the host with the feces. If they reach fresh water, small free-swimming larvae called miracidia develop and escape. If during its short lifetime a miracidium

* Research supported by the Swedish Natural Sciences Research Council.

reaches a freshwater snail of suitable species, it penetrates the snail and initiates an asexual phase of reproduction. Cercariae (a second small free-swimming larval form) develop in the infected snail and are liberated into the surrounding water. When a cercaria reaches equatic vegetation it changes form into an immobile cyst called metacercaria that attaches itself to the plant. Man becomes infected by eating these plants (water caltrop, water chestnut).

The sexually mature form of the parasite is attached to man's intenstine by two powerful suckers that cause inflammation and ulceration. Other symptoms of the disease are toxic diarrhea and edema. The disease is diagnosed by the recovery of eggs in the feces.

The transmission of helminthic infections in a community involves an interaction between two host populations and several life cycle stages of a parasite population. The transmission pattern for helminthic infections is therefore more complicated than for infections that are transmitted directly from infectious to susceptible individuals. A hybrid model (i.e. a mathematical model with both stochastic and deterministic ingredients) was shown in [1] to be a powerful tool in studying helminthic infections. The model was used to establish some important threshold results for the transmission of the infection.

The model in [1] deals with helminthiasis in an isolated community. We proceed here to study the influence on the transmission pattern of a continuous exposure of the community to external infection. We shall find that even a small amount of external infection causes the threshold effect to disappear. Thus, the external infection leads to a qualitative change in the model's predictions. We note, however, that the quantitative change is small if the amount of external infection is small.

2. MODEL FORMULATION

We establish a model for the transmission of helminthiasis in a community. The model ignores birth, aging, and death in the host populations. These populations are therefore constant in time; we denote the number beings of human in the community by N_1 and the number of snails by N_2. The model consists of a collection of $N_1 + 1$ Markov chains

$$\{F_1(t), F_2(t), \ldots, F_{N_1}(t), S(t)\}\,, \quad t \geqq 0\,.$$

Here, $F_k(t)$ corresponds to the number of flukes in definitive host k ($k = 1, 2, \ldots, N_1$) at time t, and $S(t)$ corresponds to the number of infected snails at time t. The random variables $F_k(t)$ take nonnegative integral values and $S(t)$ takes integral values from 0 to N_2.

Each of the processes F_k is an immigration-death process. The immigration of one individual in the F_k process has its counterpart in the infection of host k by one metacercaria. Such infection can be due either to infected snails in the community or to the infection imposed on the community from outside its boundaries. The immigration rate at time t for each of the processes F_k is therefore expressed as the sum of the two terms

$$p_1 r_1 E[S(t)] + e_1 .$$

Here, r_1 is the rate of cercarial shedding per infected snail and p_1 is the probability that a cercaria will infect a given human host. (We note that the success probability for a cercaria equals $p_1 N_1$.) The factor $E[S(t)]$ denotes the expected number of infected snails at time t. It is by taking the expectation here (and at a corresponding point in the infection rate below) that the model becomes hybrid rather than fully stochastic. The external infection is assumed to give a constant contribution e_1 to the immigration rate. The death rate per individual in each of the processes F_k is assumed constant and denoted by d_1.

The process S is, like the processes F_k, a birth-death process. We shall refer to it in a descriptive way as an infection-recovery process. Infection is due either to eggs from flukes in the community or it is imposed from external sources. The infection rate when i snails are infected at time t is therefore expressed as the sum of the two terms

$$p_2 r_2 (N_2 - i)\, E[\sum_{k=1}^{N_1} F_k(t)] + e_2(N_2 - i) .$$

Here, r_2 is the rate of egg-laying per fluke and p_2 is the probability that an egg will lead to infection of a given snail. We note that the success probability of an egg at time t is $p_2(N_2 - S(t))$. Hence it depends on the number of infected snails $S(t)$, and can therefore not be used as a parameter in the establishment of the mathematical model. The external infection rate per uninfected snail is assumed constant and denoted by e_2. The recovery rate per infected snail is constant and denoted d_2.

With the above hypotheses we have introduced a total of ten parameters (N_1, N_2, r_1, r_2, p_1, p_2, d_1, d_2, e_1, e_2) to quantify biological and environmental influences on the parasite population. For our further discussion it is convenient to introduce two so-called transmission factors as follows:

$$T_1 = \frac{p_1 r_1 N_2}{d_1}, \quad T_2 = \frac{p_2 r_2 N_1}{d_2} .$$

It is readily seen that large values of either of the transmission factors T_1 or T_2 favor a large parasite population.

3. MATHEMATICAL DEVELOPMENT

The Markov chains F_k, $k = 1, 2, \ldots, N_1$, all have the same transition probability functions $P_{mn}(s, t)$, $m, n = 0, 1, \ldots, 0 \leqq s < t$, while the Markov chain S has the transition probability function $\Pi_{ij}(s, t)$, $i, j = 0, 1, \ldots, N_2$, $0 \leqq s < t$. The initial distributions of F_k and S are assumed given. A standard argument using the Chapman-Kolmogorov equations and the above hypotheses leads to an infinite system of differential equations (the backward equations) for the functions P_{mn} and Π_{ij}. This system is nonlinear because of our hybrid assumptions. The initial values are given by $P_{mn}(t, t) = \delta_{mn}$ and $\Pi_{ij}(t, t) = \delta_{ij}$. The proof of existence and uniqueness of the solution of this initial value problem is similar to the corresponding proof in [1]. The mathematical development proceeds by establishing the asymptotic properties as $t \to \infty$ of the solution of the initial value problem. Again, the development is similar to that in [1]. We quote here one of the essential results. We denote the expected total number of flukes at time t by $X(t)$ and the expected number of infected snails at time t by $Y(t)$, i.e. $X(t) = E[\sum_{k=1}^{N_1} F_k(t)]$ and $Y(t) = E[S(t)]$. It can then be shown that the functions X, Y satisfy the following autonomous system of differential equations:

$$
\begin{aligned}
X' &= p_1 r_1 N_1 Y + e_1 N_1 - d_1 X\,, \\
Y' &= (p_2 r_2 X + e_2)(N_2 - Y) - d_2 Y\,.
\end{aligned}
\tag{1}
$$

The initial values are given by

$$
\begin{aligned}
X(0) &= X_0\,, \\
Y(0) &= Y_0\,.
\end{aligned}
\tag{2}
$$

Here, $X_0 = E[\sum_{k=1}^{N_1} F_k(0)]$ and $Y_0 = E[S(0)]$.

We note that a natural deterministic counterpart to our hybrid model would lead to the same initial value problem (1), (2). (This assumes proper modifications of the interpretations of parameters and functions.) In spite of this, the hybrid model is preferred over its deterministic counterpart. Among other things, the hybrid model is more realistic since it allows for variation between the members of the population of definitive hosts. Such variation must be accounted for if one is to be able to deal with the prevalence, i.e. the proportion of definitive hosts that are infected at a given instant.

It is convenient to normalize the functions X and Y by putting $x(t) = X(t)/N_1$, $y(t) = Y(t)/N_2$, $x_0 = X_0/N_1$, $y_0 = Y_0/N_2$. The function x then represents the average fluke load per definitive host, and the function y represents the expected proportion of infected snails. By introducing the transmission factors T_1 and T_2

we find from (1), (2) that the functions x, y satisfy the following initial value problem:

$$(3)\qquad \begin{aligned} x' &= T_1 d_1 y + e_1 - d_1 x\,, \\ y' &= (T_2 d_2 x + e_2)(1 - y) - d_2 y\,, \end{aligned}$$

$$(4)\qquad \begin{aligned} x(0) &= x_0\,, \\ y(0) &= y_0\,. \end{aligned}$$

Here, the initial values obey the restrictions $0 \leqq x_0$, $0 \leqq y_0 \leqq 1$.

It can be shown that the initial value problem (3), (4) has a unique solution that exists for all $t \geqq 0$, that $0 \leqq x(t) \leqq \max(T_1 + T_1 d_1, x_0)$, $t \geqq 0$, $0 \leqq y(t) \leqq 1$, $t \geqq 0$, and that the solution approaches a critical point of the system (3) as $t \to \infty$. The critical points of (3) are found by setting the left hand sides of the two equations equal to 0. Thus the coordinates $(\mathring{x}, \mathring{y})$ of any critical point satisfy the two equations

$$(5)\qquad \mathring{y} = \frac{d_1\mathring{x} - e_1}{T_1 d_1}\,,$$

$$(6)\qquad \mathring{y} = \frac{T_2 d_2 \mathring{x} + e_2}{T_2 d_2 \mathring{x} + e_2 + d_2}$$

and the requirement

$$(7)\qquad \mathring{x} \geqq 0\,.$$

We consider first the case $e_1 = e_2 = 0$. We find then from (5), (6), (7) that (0, 0) is a critical point for all values of T_1 and T_2, and that there exists an additional critical point, $(\bar{x}, \bar{x}/T_1)$ with $\bar{x} = T_1 - 1/T_2$, for $T_1 T_2 > 1$. Furthermore, it is readily shown by linearization of (3) that the critical point (0, 0) is asymptotically stable for $T_1, T_2 < 1$ and unstable for $T_1 T_2 > 1$, while $(\bar{x}, \bar{x}/T_1)$ is asymptotically stable for $T_1 T_2 > 1$. The phenomenon that the number of critical points depends on the values of the parameters T_1 and T_2 is denoted bifurcation. It accounts for the threshold result discussed in [1] and mentioned in Section 4 below.

We proceed to consider the case when $\max(e_1, e_2) > 0$. By eliminating $\mathring{y}$ from (5), (6) and applying the requirement (7) we find that

$$(8)\qquad \mathring{x} = \tfrac{1}{2}(\bar{x} + e_3) + \frac{1}{2}\left((\bar{x} + e_3)^2 + \frac{4e_3}{T_2}\right)^{1/2} - \frac{e_2}{T_2 d_2}\,,$$

where

$$\bar{x} = T_1 - \frac{1}{T_2}$$

and

$$e_3 = \frac{e_1}{d_1} + \frac{e_2}{T_2 d_2}\,.$$

In this case, then, there exists only one critical point of the system (3). It is readily shown that $\mathring{x} > \max(0, \bar{x})$ and that the difference $\mathring{x} - \max(0, \bar{x})$ approaches 0 as e_1 and e_2 approach 0.

4. EPIDEMIOLOGICAL INTERPRETATIONS

The diseases that we study here result from a long exposure of parasites and hosts to rather constant biological and environmental influences. The steady state response is therefore of interest. In particular, we find that the asymptotic one dimensional distributions of the stochastic processes F_k and S can be given specific epidemiological interpretations.

It can be shown that, as $t \to \infty$, each random variable $F_k(t)$ converges weakly to a Poisson variable with parameter $\mathring{x}$, and that $S(t)$ converges weakly to a binomially distributed random variable with parameters N_2 and $(d_1\mathring{x} - e_1)/T_1 d_1$. In particular, it follows that, in the steady state, the average number of flukes per definitive host is $\mathring{x}$. The functional relationship between the steady state fluke load $\mathring{x}$ and the biological-environmental parameters that have been introduced through our hypotheses is given in relation (8). Figure 1 shows the dependence of $\mathring{x}$ on T_2 for constant $T_1 = 2$ both in the case $e_1 = e_2 = 0$ and in the case $e_1 = 0{,}05 T_1 d_1$, $e_2 = 0$. The factor 0,05 in this expression for e_1 can be interpreted to equal the increase in the proportion of infected snails that has the same effect on the immigration rate as the external infection. (This interpretation holds for sufficiently small T_2-values so that the resulting proportion does not exceed 1.)

There is a qualitative difference between a community that is continuously exposed to external infection ($\max(e_1, e_2) > 0$) and one where such exposure is absent ($e_1 = e_2 = 0$). The latter type of community was dealt with in [1]. It exhibits a threshold phenomenon that is absent in a community of the former type.

To describe the threshold phenomenon, we consider a community with $e_1 = e_2 = 0$. If $T_1 T_2 \leqq 1$, then the steady state fluke load is 0 regardless of the conditions that were prevailing at time $t = 0$. Thus, any amount of infection that may have been present at $t = 0$ will ultimately die out of its own. If $T_1 T_2 > 1$, then the steady state fluke load is either 0 or it equals $T_1 - (1/T_2) > 0$. It is 0 only if no infection was present in the community at $t = 0$; if any infection at all was present at $t = 0$, then it results in a build-up of fluke load toward the value $T_1 - 1/T_2$. Ultimate eradication of an existing infection in a community can be achieved by reducing $T_1 T_2$ to a value not larger than 1 and maintaining the transmission factors at their new values.

No threshold effect exists in a community where at least one of the external infection rates e_1 and e_2 is strictly positive (see Figure 1). In particular, this means that complete ultimate eradication of an infection in such a community can not be achieved unless the external infection is stopped. On the other hand, the quantitative difference between the steady state fluke load in a community with external infection and the steady state fluke load in a community without external infection is small if the rates of external infection e_1 and e_2 are small.

The mathematical model allows a study of other manifestations of a helminthic infection in a community than the fluke load. We give three examples below.

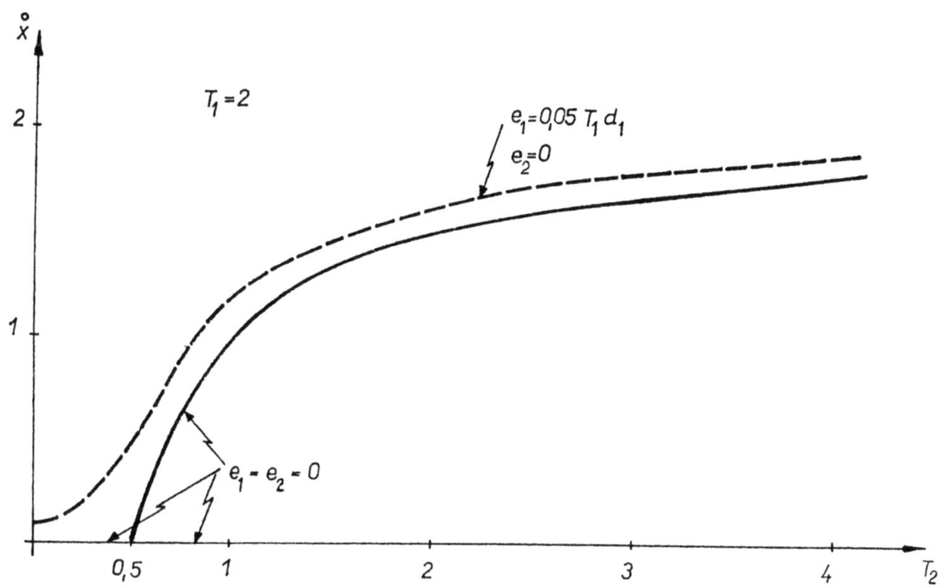

Fig. 1.

The steady state prevalence is measured by the probability that a definitive host contains at least one fluke. Since in the limit as $t \to \infty$ the number of flukes per host has a Poisson distribution with parameter $\mathring{x}$, we find that the steady state prevalence equals

$$1 - e^{-\mathring{x}} .$$

The steady state proportion of intermediate hosts that are infected is approximated by its expectation

$$\frac{d_1\mathring{x} - e_1}{T_1 d_1} .$$

A measure of steady state public health conditions in the community is given by the so-called public health factor, which equals

$$p_1 r_1 N_2 \frac{d_1\mathring{x} - e_1}{T_1 d_1} + e_1 = d_1\mathring{x} .$$

This expression measures the risk of infection in the sense that the probability for any definitive host to become infected by one parasite during a short time interval is asymptotically equal to the public health factor multiplied by the length of the time interval.

The results presented here can be used to evaluate the efficiencies of various control procedures. For this purpose we note from (5), (6) that $\mathring{x}$ is the unique positive solution of the equation $f(x) = 0$ (we assume that $\max(e_1, e_2) > 0$), where the function f is defined through the relation

$$f(x) = \frac{x - e_1/d_1}{T_1} - \frac{T_2 x + e_2/d_2}{T_2 x + e_2/d_2 + 1}.$$

We note that f depends on our ten biological-environmental parameters through T_1, T_2, e_1/d_1, e_2/d_2, although the dependence is not explicit in the notation. We note also that $f'(\mathring{x}) > 0$. Denoting any parameter or function of parameters by p, we find that

$$\frac{\partial \mathring{x}}{\partial p} = -\frac{\partial f(\mathring{x})}{\partial p} \frac{1}{f'(\mathring{x})}.$$

Thus we get

$$\frac{\partial \mathring{x}}{\partial T_1} = \frac{\mathring{x}}{T_1^2} \frac{1}{f'(\mathring{x})}$$

and

$$\frac{\partial \mathring{x}}{\partial T_2} = \frac{\mathring{x}}{(T_2 \mathring{x} + e_2/d_2 + 1)^2} \frac{1}{f'(\mathring{x})}.$$

We proceed now to compare the efficiencies of two control programs. One of the programs serves to reduce T_1 while keeping T_2, e_1/d_1, e_2/d_2 constant. The second program serves to reduce T_2 while keeping T_1, e_1/d_1, e_2/d_2 constant. The more efficient of the two programs is that one which for a given relative decrease in the control parameter gives the larger reduction in the fluke load $\mathring{x}$. Thus, the efficiencies are mesured by

$$T_1 \frac{\partial \mathring{x}}{\partial T_1} = \frac{\mathring{x}}{T_1} \frac{1}{f'(\mathring{x})}$$

and

$$T_2 \frac{\partial \mathring{x}}{\partial T_2} = \frac{T_2 \mathring{x}}{(T_2 \mathring{x} + e_2/d_2 + 1)^2} \frac{1}{f'(\mathring{x})},$$

respectively.

We note now that

$$(T_2 \mathring{x} + 1)^2 > T_1 T_2.$$

This inequality follows for $T_1 T_2 \leqq 1$ from $\mathring{x} > 0$ and for $T_1 T_2 > 1$ from $\mathring{x} > T_1 - 1/T_2$.

Thus we conclude that

$$T_1 \frac{\partial \mathring{x}}{\partial T_1} > T_2 \frac{\partial \mathring{x}}{\partial T_2} .$$

This inequality can be interpreted as saying that control of the fluke load through reduction of T_1 is more efficient than through reduction of T_2.

Additional studies of efficiencies of control programs are readily carried out on the basis of the above results. These studies can turn to a comparison of efficiencies of other pairs of control parameters and they can use other criteria than reduction of the fluke load, e.g. reduction of prevalence, of the proportion of infected intermediate hosts, or of the public health factor.

5. CONCLUDING REMARKS

The model for helminthiasis that we have discussed here represents a simplification of the conditions in an actual focus of infection. Added realism in the model will require additional epidemiological features to be modeled and studied mathematically. Mathematical epidemiology has the potential of contributing to the solution of some important public health problems for the many parasitic diseases that plague mankind mainly in the tropical areas of the world. Large and exciting areas of work are awaiting those applied mathematicians who are willing to devote themselves to this brand of mathematical epidemiology.

REFERENCE

[1] I. NÅSELL, W. M. HIRSCH: A mathematical model of some helminthic infections. Comm. Pure Appl. Math. *25* (1972), 459–477.

ROYAL INSTITUTE OF TECHNOLOGY

DEPARTMENT OF MATHEMATICS

ZUR OPERATORENTHEORETISCHEN CHARAKTERISIERUNG SYMMETRISCH STABILER PROZESSE

Volker Nollau

Dresden

In der vorliegenden Mitteilung beschäftigen wir uns mit den infinitesimalen Generatoren von Halbgruppen linearer Kontraktionen, die von den Übergangswahrscheinlichkeiten symmetrisch stabiler Prozesse erzeugt werden. Dabei interessieren wir uns insbesondere für die von Bochner [1] sowie von Butzer und Trebels [2] in diesem Zusammenhang angegebenen Interpretationen des Operators

$$-\left(-\frac{d^2}{dx^2}\right)^{\alpha/2}, \quad 0 < \alpha < 2 \quad (\text{vgl. Satz 2}).$$

Ausgehend von einem allgemeinen Kalkül für gebrochene Potenzen und den Logarithmus linearer abgeschlossener Operatoren in Banachschen Räumen (vgl. [4], [5]) werden hier die gebrochenen Potenzen und der Logarithmus des (negativen) Laplace-Operators definiert und zur Charakteriseirung von Eigenschaften stabiler Prozesse – speziell auch bei zufälligen Zeitsubstitutionen (vgl. z. B. [3]) – herangezogen.

Es bezeichne $C_0(-\infty, \infty)$ den Banachraum der auf $(-\infty, \infty)$ stetigen Funktionen f, die der Bedingung $\lim_{x\to-\infty} f(x) = \lim_{x\to\infty} f(x) = 0$ genügen, mit $\|f\| = \sup_{-\infty<x<\infty} |f(x)|$ und für $\varrho > 0$

$$[L_x f](\varrho) = \int_0^\infty e^{-\xi\varrho}\, \tfrac{1}{2}(f(x-\xi) + f(x+\xi))\, d\xi \quad (-\infty < x < \infty).$$

Dann gilt der

Satz 1. Der durch

$$\frac{2 \sin \pi\alpha}{\pi} \int_0^\infty \varrho^{2\alpha-1}\{f(x) - \varrho[L_x f](\varrho)\}\, d\varrho$$

auf $\{f \in C_0(-\infty, \infty) : f'' \in C_0(-\infty, \infty)\}$ definierte lineare Operator ist für jedes $0 < \alpha < 1$ auf $C_0(-\infty, \infty)$ stark abschließbar. Diese Abschließung, die mit $(-d^2/dx^2)^\alpha$ bezeichnet wird, besitzt dabei die folgende Eigenschaften:

(1.1) $$\left(\left(-\frac{d^2}{dx^2}\right)^\alpha\right)^\beta = \left(-\frac{d^2}{dx^2}\right)^{\alpha\beta} \quad (0 \leqq \alpha,\ \beta \leqq 1).$$

(Dabei bezeichnet $(-d^2/dx^2)^1 = -d^2/dx^2$ und $(-d^2/dx^2)^0$ ist durch $((-d^2/dx^2)^0 f)(x) = f(x)$ definiert.);

(1.2) $$\left(-\frac{d^2}{dx^2}\right)^\alpha \left(-\frac{d^2}{dx^2}\right)^\beta = \left(-\frac{d^2}{dx^2}\right)^{\alpha+\beta} \quad (0 \leqq \alpha,\ \beta \leqq 1 \text{ und } 0 \leqq \alpha + \beta \leqq 1);$$

(1.3) $$\boldsymbol{D}\left(\left(-\frac{d^2}{dx^2}\right)^\alpha\right) \supseteq \boldsymbol{D}\left(\left(-\frac{d^2}{dx^2}\right)^{\alpha'}\right) \quad \text{und} \quad \boldsymbol{R}\left(\left(-\frac{d^2}{dx^2}\right)^\alpha\right) \supseteq \boldsymbol{R}\left(\left(-\frac{d^2}{dx^2}\right)^{\alpha'}\right)$$

für $0 \leqq \alpha < \alpha' \leqq 1$. (Dabei bezeichnen $\boldsymbol{D}(.)$ bzw. $\boldsymbol{R}(.)$ den Definitionsbereich bzw. den Rang eines Operators.)

und für $f \in D\left(\left(-\frac{d^2}{dx^2}\right)^{\alpha'}\right)$

$$\left(\left(-\frac{d^2}{dx^2}\right)^\alpha f\right)(x) = \frac{2\sin \pi\alpha}{\pi}\int_0^\infty \varrho^{2\alpha-1}\{f(x) - \varrho[L_x f](\varrho)\}\, d\varrho,$$

(1.4) $$\sigma\left(\left(-\frac{d^2}{dx^2}\right)^\alpha\right) \subseteq \{z : |\arg z| \leqq \pi\alpha\},$$

(1.5) die Gleichung $\mathscr{D}^n = -d^2/dx^2$ besitzt für $n \geqq 2$ genau eine Lösung, die der Bedingung $\sigma(\mathscr{D}) \subseteq \{z : |\arg z| \leqq \pi/n\}$ genügt, diese Lösung ist $\mathscr{D} = (-d^2/dx^2)^{1/n}$;

(1.6) die Gleichung

$$\left(\left(-\frac{d^2}{dx^2}\right)^\alpha\right) f(x) + \eta f(x) = g(x) \quad (g \in C_0(-\infty, \infty))$$

besitzt für jedes $\eta > 0$ eine eindeutig bestimmte Lösung

$$f(x) = \frac{2\sin \pi\alpha}{\pi}\int_0^\infty \frac{\varrho^{2\alpha}[L_x g](\varrho)}{\varrho^{4\alpha} + 2\eta\varrho^{2\alpha}\cos\pi\alpha + \eta^2}\, d\varrho.$$

Weiter seien $P(t, x, \Gamma) = \int_\Gamma p(t, x - y)\, dy$ $(t > 0,\ x \in R^1,\ \Gamma \in \mathfrak{B}^1)$ mit $p(t, \eta) = [2\sqrt{(\pi t)}]^{-1} e^{-\eta^2/4t}$ $(\eta > 0)$ die Übergangswahrscheinlichkeiten eines Weiner-

Prozesses* $\{X(t) : t \geqq 0\}$ und $\{T_\alpha(t) : t \geqq 0\}$ ein stabiler stochastischer Prozeß mit nicht-negativen Zuwächsen, der von $\{X(t) : t \geqq 0\}$ unabhängig ist. Die Übergangswahrscheinlichkeiten von $\{T_\alpha(t) : t \geqq 0\}$ besitzen also (vgl. z. B. [7]) die Form

$$Q_t^{(\alpha)}(x) = \int_0^x f_{t,\alpha}(\lambda)\, d\lambda$$

mit

$$f_{t,\alpha}(\lambda) = \begin{cases} \dfrac{1}{2\pi i} \displaystyle\int_{\sigma - i\infty}^{\sigma + i\infty} e^{z\lambda - tz^\alpha}\, dz & (\lambda > 0), \\ 0 & (\lambda \leqq 0) \end{cases}$$

($\sigma > 0$ bel.). Dann ist $\{X(T_\alpha(t)) : t \geqq 0\}$ ein Markoffscher Prozeß (vgl. [3], [6]) der als gerichteter oder dem Wiener-Prozeß $\{X(t) : t \geqq 0\}$ untergeordneter (engl.: subordinated [3]) Prozeß bezeichnet wird. Für diesen durch eine „zufällige Zeitsubstitution" entstandenen Prozeß erhält man die folgenden Aussagen in

SATZ 2. Es bezeichne $p_\alpha(t, \eta)$, $0 < \alpha < 1$, die Dichten der Übergangswahrscheinlichkeiten $P_\alpha(t, x, \Gamma)$ von $\{X(T_\alpha(t)) : t \geqq 0\}$, d. h.

$$P_\alpha(t, x, \Gamma) = \int_\Gamma p_\alpha(t, x - y)\, dy \quad (t > 0,\ x \in R^1,\ \Gamma \in \mathfrak{B}^1)$$

mit

$$p_\alpha(t, \eta) = \int_0^\infty f_{t,\alpha}(\lambda)\, e^{-\eta^2/4\lambda} \frac{d\lambda}{2\sqrt{(\lambda\pi)}}$$

dann gilt:

(2.1) der Operator $-(-d^2/dx^2)^\alpha$ ist der infinitesimale Generator der durch

$$(T_{t,\alpha} f)(x) = \int_{-\infty}^\infty p_\alpha(t, x - y) f(y)\, dy$$

auf $C_0(-\infty, \infty)$ definierten Halbgruppe $\{T_{t,\alpha} : t \geqq 0\}$ von Kontraktionen

(2.2) (vgl. [3])

$$\int_{-\infty}^\infty e^{i\eta x} p_\alpha(t, \eta)\, d\eta = e^{-t|x|^{2\alpha}} \quad (t > 0,\ x \in R^1)$$

d. h. $\{X(T_\alpha(t)) : t \geqq 0\}$ ist symetrisch stabil mit dem Stabilitätskoeffizienten 2α.

Als Folgerung der Sätze 1 und 2 erhält man dann u. a. die nachstehenden Ausagen.

---

* Im Unterschied z. B. zu [3] definieren wir $p(t, \eta)$ nicht in der Form $\bar{p}(t, \eta) = (2\pi t)^{-1/2} \cdot e^{-\eta^2(2t)^{-1}}$, allerdings nur aus dem Grunde einer einfacheren Beschreibung der folgenden Ergebnisse, die sich leicht auf $\bar{p}(t, \eta)$ übertragen lassen.

FOLGERUNG 1. Es sei $\{S_\alpha(t) : t \geqq 0\}$, $0 < \alpha < 1$ ein symmetrisch stabiler Prozeß mit dem Stabilitätskoeffizienten 2α und $\{T_\beta(t) : t \geqq 0\}$, $0 < \beta < 1$ ein stabiler Prozeß mit nicht-negativen Zuwächsen, der von $\{S_\alpha(t) : t \geqq 0\}$ unabhängig ist. Dann ist $\{S_\alpha(T_\beta(t)) : t \geqq 0\}$ symetrisch stabil mit dem Stabilitätskoeffizienten $2\alpha\beta$.

FOLGERUNG 2. Genügt der infinitesimale Generator $\mathscr{G}$ der durch die Übergangswahrscheinlichkeiten

$$P(t, x, \Gamma) = \int_\Gamma p(t, x - y)\, \mathrm{d}y \quad (t > 0,\ x \in R^1,\ \Gamma \in \mathfrak{B}^1)$$

eines Markoffschen Prozesses $\{X(t) : t \geqq 0\}$ erzeugten Halbgruppe

$$(T_t f)(x) = \int_{-\infty}^{\infty} p(t, x - y) f(y)\, \mathrm{d}y \quad (f \in C_0(-\infty, \infty))$$

den Bedingungen

$$(-\mathscr{G})^n = -\frac{\mathrm{d}^2}{\mathrm{d}x^2} \quad \text{und} \quad \sigma(-G) \subseteq \left\{z : |\arg z| \leqq \frac{\pi}{n}\right\}$$

$$\text{für ein} \quad n \geqq 2,$$

so ist $p(t, \eta) = p_{1/n}(t, \eta)$, d. h. $\{X(t) : t \geqq 0\}$ ist symmetrisch stabil mit dem Stabilitätskoeffizienten $2/n$, und es gilt $\mathscr{G} = -(-\mathrm{d}^2/\mathrm{d}x^2)^{1/n}$.

Schließlich geben wir eine einheitliche Darstellung der Logarithmen der infinitesimalen Generatoren $A(\gamma)$ von Halbgruppen linearer Kontraktionen auf $C_0(-\infty, \infty)$, die durch die Übergangswahrscheinlichkeiten

$$P_{(\gamma)}(t, x, \Gamma) = \int_\Gamma p_{(\gamma)}(t, x - y)\, \mathrm{d}y \quad (t > 0,\ x \in R^1,\ \Gamma \in \mathfrak{B}^1)$$

von symmetrisch stabilen Prozessen $\{S_{(\gamma)}(t) : t \geqq 0\}$ mit dem Stabilitätskoeffizienten γ $(0 < \gamma \leqq 2)$ erzeugt werden. Dabei bezeichne

$$\mathbf{M}\left(-\frac{\mathrm{d}^2}{\mathrm{d}x^2}\right) = \bigcup_{\alpha > 0} \boldsymbol{D}\left(\left(-\frac{\mathrm{d}^2}{\mathrm{d}x^2}\right)^\alpha\right) \cap \bigcup_{\beta > 0} \boldsymbol{R}\left(\left(-\frac{\mathrm{d}^2}{\mathrm{d}x^2}\right)^\beta\right).$$

Wie in [5] angegeben wurde, ist durch

$$\lim_{h \downarrow 0} \frac{\left(\left(-\frac{\mathrm{d}^2}{\mathrm{d}x^2}\right)^h f\right)(x) - f(x)}{h}$$

auf $M(-\mathrm{d}^2/\mathrm{d}x^2)$ ein abschießbarer Operator definiert, der als Logarithmus

$\log(-\,\mathrm{d}^2/\mathrm{d}x^2)$ des (negativen) Laplace-Operators bezeichnet wird. Mit Hilfe der in [5] angegebenen Eigenschaften von $\log(-\,\mathrm{d}^2/\mathrm{d}x^2)$ – wie z. B.

$$\log\left(-\frac{\mathrm{d}^2}{\mathrm{d}x^2}\right)^{\alpha} = \alpha\log\left(-\frac{\mathrm{d}^2}{\mathrm{d}x^2}\right) \quad (0 \leqq \alpha \leqq 1)$$

und Integraldarstellungen für diesen Operator und dessen Resolvente – erhält man auf Grund von $A_{(\gamma)} = -(-\,\mathrm{d}^2/\mathrm{d}x^2)^{\gamma/2}$ (vgl. (2.1)) den

SATZ 3. Es sei $\{S_{(\gamma)}(t) : t \geqq 0\}$ ein symmetrisch stabiler Prozeß mit dem Stabilitätskoeffizienten γ $(0 < \gamma \leqq 2)$.
Dann gilt:

(3.1) $$\log(-A_{(\gamma)}) = \frac{\gamma}{2}\log\left(-\frac{\mathrm{d}^2}{\mathrm{d}x^2}\right);$$

(3.2) $$(\log(-A_{(\gamma)})f)(x) = \gamma\int_0^\infty \left\{\frac{\varrho}{1+\varrho^2}f(x) - [L_x f](\varrho)\right\}\mathrm{d}\varrho$$

für $f \in M(-\,\mathrm{d}^2/\mathrm{d}x^2)$;

(3.3) die Gleichung

$$(\log(-A_{(\gamma)})f)(x) - z f(x) = g(x), \quad g \in C_0(-\infty, \infty),$$

besitzt für jedes z mit $|\mathrm{Im}\, z| > \pi\gamma/2$ eine eindeutig bestimmte Lösung

$$f(x) = \frac{1}{\gamma}\int_0^\infty \frac{[L_x g](\varrho)}{\left(\log\varrho - \frac{z}{\gamma}\right)^2 + \frac{\pi^2}{4}}\,\mathrm{d}\varrho\,.$$

6

LITERATURVERZEICHNIS

[1] S. Bochner: Diffusion equation and stochastic processes. Proc. Nat. Acad. Sci. *35* (1949), 368—370.

[2] P. L. Butzer, W. Trebels: Hilberttransformation, gebrochene Integration und Differentiation. Westdeutscher Verlag, Köln und Opladen 1968.

[3] W. Feller: An Introduction to Probability Theory and its Applications. Vol. II. Wiley, New York 1966.

[4] V. Nollau: Über Potenzen von linearen Operatoren in Banachschen Räumen. Acta Sci. Math. *28* (1967), 107—121.

[5] V. Nollau: Über den Logarithmus abgeschlossener Operatoren in Banachschen Räumen. Acta Sci. Math. *30* (1969), 161—174.

[6] V. Nollau: Über gebrochene Potenzen infinitesimaler Generatoren Markoffscher Übergangswahrscheinlichkeiten I und II. Math. Nachr. *65* (1975), 235—246 und *72* (1976), 99—107.

[7] K. Yosida: Functional Analysis. Springer-Verlag, Berlin—Göttingen—Heidelberg 1965.

Technische Universität Dresden

Sektion Mathematik

A CONTRIBUTION TO THE TIME SERIES DESIGN PROBLEMS

ANDREJ PÁZMAN

BRATISLAVA

The main aim of this paper is to help to clarify the situation which occurs in time series design problems when an infinite-dimensional regression experiment is to be performed. The notion of an approximate estimability of a functional defined on the space of regression functions, which arises in the case of an infinite-dimensional regression experiment, is discussed here (Theorem 2). Theorem 3 gives a condition under which designs optimal for finite-dimensional regression problems may be considered as approximately optimal also in the infinite-dimensional case. In section 3 some remarks about the possibility of comparing two designs with uncorrelated observations are given.

1. INTRODUCTION

A standard model of a random process usually called "time series" is considered. A set T and a probability space $(\Omega, \mathscr{S}, \mu)$ are given. The studied random process is a set $\{X(t); t \in T\}$ of real random variables $X(t) \in L_2(\Omega, \mathscr{S}, \mu)$. We denote by

$$(1) \qquad \mathscr{K}(t, s) = E(X(t) - E\,X(t))\,(X(s) - E\,X(s))\;; \quad t, s \in T$$

the covariance function of the process. We shall suppose that

$$EX(.) \in M\,,$$

where M is a given set of real functions defined in T. The function $\mathscr{K}$ and the set M constitute the a priori information that we have before observing the process.

We denote by $H(T)$ the reproducing kernel Hilbert space with the kernel $\mathscr{K}$ (for the definition see section 2). It is known [9] that if $\{X(t); t \in T\}$, $\{X'(t); t \in T\}$ are Gaussian random processes with the same covariance function $\mathscr{K}$ and such that $EX(.) - EX'(.) \notin H(T)$, then the processes induce orthogonal probability measures in the sample space. The discrimination of the two alternatives $EX(.)$ and $EX'(.)$ in this case is no more a statistical problem. Therefore we shall make the assumption: $M \in H(T)$, throughout the paper.

The aim of the experimenter observing the process is the estimation of one or several functionals on M in an optimal way. For that purpose he has to consider different designs of observations of the process $\{X(t);\ t \in T\}$. If $A \subset T$ and if he observes the set of random variables $\{X(t);\ t \in A\}$ simultaneously (i.e. from one realization of the process), we say that the design is *simple* (and we identify it with the set A). A *multiple design* is a set of simple designs such that observations corresponding to different simple designs are uncorrelated. One extreme case of a multiple design is the simple design. The other extreme case arises when the comparing simple designs are one-point subsets of T; such multiple designs are *designs with uncorrelated observations.*

For the case of a finite-dimensional set M (the finite-dimensional regression) simple designs have been studied in [10], [11], but especially designs with uncorrelated observations received the attention of many authors [2], [3], [5], [6] etc).

2. SIMPLE DESIGNS

In this section A is a fixed but arbitrary subset of T (the simple design). We denote by $H(A)$ the reproducing kernel Hilbert space [1] with the kernel $\mathscr{K}^A$:

$$\mathscr{K}^A(t, s) = \mathscr{K}(t, s)\,; \quad t, s \in A$$

(the restriction of $\mathscr{K}$ to $A \times A$). $H(A)$ is the Hilbert space defined by the following properties:

a) the elements of $H(A)$ are real functions defined in A,

(2) b) $$\mathscr{K}^A(., t) \in H(A)\,; \quad t \in A\,,$$

(3) c) $$(h, \mathscr{K}^A(., t))_A = h(t)\,; \quad t \in A\,, \quad h \in H(A)\,,$$

where $(.,.)_A$ is the inner product in $H(A)$. It is known [1] that the set $\{\mathscr{K}^A(., t);\ t \in A\}$ spans $H(A)$.

We shall use the following notation. If Q is a subset of a linear space we denote by $\mathscr{L}Q$ the set of all finite linear combinations of elements from Q. If $Q \in L_2(\Omega, \mathscr{S}, \mu)$, then L_2Q will be the (closed) subspace of $L_2(\Omega, \mathscr{S}, \mu)$ spanned by Q. If $Q \subset H(T)$, then we denote by Q^- the closure of Q in the strong topology of $H(T)$ and by $Q^\perp$ the orthogonal complement of Q in $H(T)$.

For any $h \in H(T)$, h^A will denote the restriction of h to the set A. By $H_M(A)$ we denote the (closed) subspace of $H(A)$ spanned by the set $\{m^A : m \in M\}$. $N(A)$ will be the set of all elements of $H(T)$ which are zero in A. It may be shown that $N^{\perp}(A) = [\mathscr{L}\{\mathscr{K}(\cdot, t);\ t \in A\}]^{-}$. Both $N(A)$ and $N^{\perp}(A)$ are closed subspaces of $H(T)$. We denote by P_A the projection of $H(T)$ onto $N^{\perp}(A)$.

LEMMA 1. *The following is valid:*

1. $H(A) = \{h^A : h \in H(T)\}$.

2. *The operator* $W_A : H(A) \to N^{\perp}(A)$ *defined by:* $W_A h^A = P_A h;\ h \in H(T)$, *is an unitary operator. Therefore*

$$(h^A, l^A)_A = (h, P_A l)_T\ ; \quad h, l \in H(T)\,.$$

3. *The set* $\{h^A : h \in H_M(T)\}$ *is a dense subset of* $H_M(A)$.

4. *The set* $P_A\, H_M(T)$ *is a dense subset of* $W_A\, H_M(A)$.

Proof. 1) and 2): Evidently $h^A = l^A \Leftrightarrow h - l \in N(A) \Leftrightarrow P_A h = P_A l$. This implies that $(P_A h)^A = h^A$ and that $W_A : \{h^A : h \in H(T)\} \to N^{\perp}(A)$ is one-to-one and onto. We define

$$(h^A, l^A)' = (h, P_A l)_T\ ; \quad h, l \in H(T)\,.$$

We have: $\mathscr{K}^A(\cdot, t) \in \{h^A : h \in H(T)\}$ and $(h^A, \mathscr{K}^A(\cdot, t))' = (P_A h, \mathscr{K}(\cdot, t))_T = (P_A h)(t) = h^A(t)$. Hence, according to (3), $\{h^A : h \in H(T)\}$ with the inner product $(\cdot, \cdot)'$ is a dense subset of $H(A)$ and $(\cdot, \cdot)' = (\cdot, \cdot)_A$. Moreover W_A is then an isometric operator and $\{h^A : h \in H(T)\}$ being the domain of W_A is closed, therefore equal to $H(A)$.

3) The set $\{h^A : h \in \mathscr{L}M\}$ is a dense subset of $H_M(A)$ because $H_M(A)$ is spanned by $\{m^A : m \in M\}$. If $h \in H_M(T)$, then there is a sequence $h_1, h_2, \ldots \in \mathscr{L}M$ such that $\lim_{n \to \infty} \|h - h_n\|_T = 0$. Therefore also $\|h^A - h_n^A\|_A = \|P_A(h - h_n)\|_T \to 0$, i.e. $h \in H_M(A)$. This proves that also $\{h^A : h \in H_M(T)\}$ is a dense subset of $H_M(A)$.

4) From the unitarity of W_A and from 3) we obtain that $P_A\, H_M(T) = W_A\{h^A : h \in H_M(T)\}$ is a dense subset of $W_A\, H_M(A)$.

We shall denote by $\{Y(t);\ t \in T\}$ the centred process with the covariance function $\mathscr{K}(Y(t) = X(t) - E\,X(t);\ t \in T)$.

LEMMA 2 [7]. *There is a unitary operator* $V_A : L_2\{Y(t);\ t \in A\} \to H(A)$ *such that*

$$V_A[X(t) - E\,X(t)] = \mathscr{K}^A(\cdot, t)\ ; \quad t \in A \tag{4}$$

and

$$EX = (V_A(X - EX), E^A X(\cdot))_A\ ; \quad X \in L_2\{X(t);\ t \in A\}\,. \tag{5}$$

For any $h \in H(T)$ we denote by the symbolic notation $(h^A, X^A(\cdot))_A$ the random variable $V_A^{-1} h \in L_2\{X(t);\ t \in A\}$. According to (5)

$$E(h^A, X^A(\cdot))_A = (h^A, E^A X(\cdot))_A\,. \tag{6}$$

Let g be a real functional defined on the set M, which is to be estimated linearly from the process $\{X(t); t \in A\}$. A *linear unbiassed estimate* (*LUE*) for g on A is a random variable $(h^A, X^A(.))_A \in L_2\{X(t); t \in A\}$ which satisfies the equations:

$$(h^A, E^A X(.))_A = E(h^A, X^A(.))_A = g[EX(.)] ; \quad EX(.) \in M . \tag{7}$$

If there is a *LUE* for g on A, we say that g is estimable on A. The variance of the $LUE(h^A, E^A X(.))_A$ is $\|h^A\|_A^2$ (Lemma 1 and 2). The *LUE* with the minimal variance is the best linear estimate (*BLUE*) for g on A.

We shall denote by IA the *information kernel* corresponding to the simple design A:

$$IA(m, l) = (m^A, l^A)_A ; \quad m, l \in M . \tag{8}$$

IA is a positive definite function in $M \times M$, Therefore [1] there is a unique reproducing Hilbert space with the kernel IA, which we denote by $H(IA)$. The ideas of [7] and Lemma 1 lead to the following theorem.

THEOREM 1. *The following five statements are equivalent*:

1. *There is a LUE for g on A.*
2. *There is a BLUE for g on A.*
3. $g \in H(IA)$.
4. *There is an $h \in H(T)$ such that*

$$g(m) = (h, P_A m)_T ; \quad m \in M .$$

5. *There is an $l \in [P_A H_M(T)]^-$ such that*

$$g(m) = (l, m)_T ; \quad m \in M .$$

The element l is unique. The variance of the BLUE for g on A is equal to

$$var_A\, g = \|g\|_{IA}^2 = \|l\|_T^2 . \tag{9}$$

Proof. It was shown already that 1 and 4 are equivalent. The variance of the *BLUE* for g on A has to attain the infimum $\inf\{\|h^A\|_A : h \in H(T), (h^A, m^A)_A = g(m)$ for $m \in M\}$. Therefore if $(h^A, X^A(.))_A$ is a *LUE* for g in A, then the projection l^A of h^A ibto $H_M(A)$ defines the *BLUE*: $(l^A, X^A(.))_A$. According to Lemma 1 we may take $l \in [P_A H_M(T)]^-$. Thus 2, 4, 5 are equivalent, l is unique and $var_A\, g = \|l\|_T^2$. Finally we may show that $H(IA)$ is the set of all bounded linear functionals on $H_M(A)$ restricted to the set M. Thus, according to the Riesz representation theorem [4], $g \in H(IA)$ iff there is an $h^A \in H_M(A)$ such that $g(m) = (h^A, m^A)_A = (h, P_A m)_T$ for every $m \in M$. That means 1 and 4 are equivalent.

COROLLARY. *$g(.) = (l, .)_T$ is estimable on A iff the expression*

$$\sup_{\substack{k\in H_M(T)\\ P_A k \neq 0}} \frac{(l, k)_T^2}{\|P_A k\|_T^2} \tag{10}$$

is finite. In such a case (10) is equal to $var_A\, g$.

Proof. According to Theorem 1, g is estimable on A iff for some $h^A \in H_M(A)$

$$g(m) = (h^A, m^A)_A\,; \quad m \in M\,,$$

i.e. iff g can be extended to a bounded linear functional on $H_M(A)$. Hence g is estimable on A iff the expression

$$var_A\, g = \sup_{\substack{k^A\in H_M(A)\\ k^A \neq 0}} \frac{(h^A, k^A)_A^2}{\|k^A\|_A^2} \tag{11}$$

is finite. Since the set $\{k^A : k \in H_M(T)\}$ is a dense subset of $H_M(A)$ (Lemma 1), we may write (10) instead of (11).

REMARK. To find the optimum simple design for estimating g means to solve the game-theoretic problem which leads to

$$\inf_{A\in\mathscr{A}} \sup_{\substack{k\in H_M(T)\\ P_A k \neq 0}} \frac{(l, k)_T^2}{\|P_A k\|_T^2}\,,$$

where $\mathscr{A}$ is the set of allowed simple designs. (Compare with the proof of Elfving's theorem in [5]).

In the case of an infinite-dimensional set M a concept of approximate estimability may be introduced in the following way. A functional g on M is *approximately estimable* on A if there is a sequence $g_1, g_2, \ldots$ of functionals on M which are estimable on A and such that

(12) a) $$g(m) = \lim_{n\to\infty} g_n(m)\,; \quad m \in M\,,$$

b) the limit

$$\lim_{n\to\infty} \frac{g_n^{(m)}}{\|m\|_T} \tag{13}$$

is uniform in the set $\{m : m \in M,\, m \neq 0\}$.

THEOREM 2. 1. *g is approximately estimable on A iff g is estimable on T and g is zero in $N(A) \cap M$.*

2. *If g is approximately estimable on A and $\sup_n var_A\, g_n < \infty$, then g is estimable on A.*

Proof. 1. According to Theorem 1 we may write

$$g_n(\cdot) = (h_n, \cdot)_T\,; \quad n = 1, 2, \ldots, \tag{14}$$

for some $h_1, h_2, \ldots \in [P_A\, H_M(T)]^-$. Hence, if g is approximately estimable on A, then for every $m \in M$, $m \neq 0$ we have

$$\frac{|g(m)|}{\|m\|_T} \leqq \left|\frac{g_n(m)}{\|m\|_T} - \frac{g(m)}{\|m\|_T}\right| + \|h_n\|_T\,.$$

Thus g is bounded on M, i.e. $g(\cdot) = (l, \cdot)_T$ for some $l \in H_M(T)$. From (12) and (14) we obtain that g is zero on $N(A) \cap M$.

On the other hand if g is estimable on T, then $g(\cdot) = (l, \cdot)_T$ for some $l \in H_M(T)$. If moreover $g(\cdot)$ is zero in $N(A) \cap M$, then $l \in [N(A) \cap M]^\perp$. Hence there are sequences $l_1, l_2, \ldots \in H_M^\perp(T)$ and $h_1, h_2, \ldots \in N^\perp(A)$ such that $\lim\limits_{n\to\infty} \|l - h_n - l_n\|_T = 0$. Therefore also $\lim\limits_{n\to\infty} \|l - Ph_n\|_T = \lim\limits_{n\to\infty} \|P(l - l_n - h_n)\|_T = 0$, where P is the projection of $H(T)$ onto $H_M(T)$. We may write $g(m) = \lim\limits_{n\to\infty} (Ph_n, m)_T = \lim\limits_{n\to\infty} (h_n^A, m^A)_A$. We define $g_n(m) = (h_n^A, m^A)_A$ and we obtain (12). Moreover $|g_n(m) - g(m)| \leqq \leqq \|Ph_n - l\|_T\, \|m\|_T$, hence also (13) is valid.

2. According to (12) the sequence $\{(h_n^A, k^A)_A\}_{n=1}^\infty$ is convergent for every $k \in M$ and evidently also for every $k \in \mathscr{L}\{m^A : m \in M\}$. The same is true also for every $k \in H_M(A) \equiv [\mathscr{L}\{m^A : m \in M\}]^-$, since

$$2 \sup_n \|h_n^A\|_A\, \|k^A - l^A\|_A + |(h_n^A, l^A)_A - (h_r^A, l^A)_A| \geqq |(h_n^A, k^A)_A - (h_r^A, k^A)_A|\,;$$
$$n, r = 1, 2, \ldots, \quad l^A \in \mathscr{L}\{m^A : m \in M\}\,.$$

Hence $\{h_n^A\}_{n=1}^\infty$ is a weakly convergent sequence; we denote by h^A its limit. Evidently $h^A \in H_M(A)$ and $g(\cdot) = (h^A, \cdot)_A$.

COROLLARY. *If g is approximately estimable but not estimable on A, then $var_A\, g$ does not exist (is infinite). We may then use the expression*

$$\inf_{q \in H(IA)} \{(var_A\, q)^{1/2} + \sup_{m \in M, \|m\| = 1} |q(m) - g(m)|\}^2$$

instead of $var_A\, g$.

We shall denote by $\mathscr{A}$ the class of all allowed simple designs for observing the process $\{X(t);\, t \in T\}$. We shall choose n-dimensional subsets M_n of M; $n = 1, 2, \ldots$ so that $M_1 \subset M_2 \subset \ldots$ An important question is if for some n we can approximate the optimum simple design taken from $\mathscr{A}$ by the simple design which is optimal under the a priori information: $EX(\cdot) \in M_n$.

We denote by $N(M_n)$ the set of all $q \in H(IT)$ which are zero on M_n and by $N^A(M_n)$ the subspace of $H(IA)$ spanned by the set $\{(m^A, \cdot)_A : m \in M_n\}$. $N^A(M_n)$ is the orthogonal complement of $N(M_n) \cap H(IA)$ in $H(IA)$. We denote by R_{nA} the projection of $H(IA)$ onto $N^A(M_n)$. Finally we use the notation $var_A(g \mid M_n)$ for the variance of the $BLUE$ for g on A under the a priori information: $EX(.) \in M_n$. Evidently

$$var_A(g \mid M_1) \leqq var_A(g \mid M_2) \leqq \dots \leqq var_A g .$$

THEOREM 3. *If g is estimable for every $A \in \mathscr{A}$ $\bigcup_i M_i = M$ and the mappings*

$$\varphi_n : A \in \mathscr{A} \to var_A(g \mid M_n) ; \quad n = 1, 2, \dots ,$$

$$\varphi \;: A \in \mathscr{A} \to var_A g$$

are continuous in a topology on $\mathscr{A}$ in which $\mathscr{A}$ is compact, then

$$\lim_{n \to \infty} \inf_{A \in \mathscr{A}} var_A(g \mid M_n) = \inf_{A \in \mathscr{A}} var_A g .$$

Moreover, if the simple designs $An \in \mathscr{A}$; $n = 1, 2, \dots$ are such that

$$var_{An}(g \mid M_n) \inf_{A \in \mathscr{A}} var_A(g \mid M_n) + \frac{1}{2^n} ,$$

then

$$\lim_{n \to \infty} var_{An} g = \inf_{A \in \mathscr{A}} var_A g .$$

Proof. We have $var_A(g \mid M_n) = \|R_{nA} g\|_{IA}$. Evidently $\|R_{nA} g - g\|_{IA} \to 0$ with $n \to \infty$. Therefore

(17) $$\lim_{n \to \infty} \varphi_n(A) = \varphi(A) ; \quad A \in \mathscr{A} .$$

The sequence $\{\varphi_n(A)\}_{n=1}^{\infty}$ is nondecreasing (see (16)). Hence from the assumptions of the Theorem it follows that the convergence (17) is uniform in $\mathscr{A}$. This implies that to every $\varepsilon > 0$ there is an n_0 such, that for $n > n_0$

(18) $$|var_{An} g - var_{An}(g \mid M_n)| < \varepsilon .$$

On the other hand it follows from (16) that

(19) $$\lim_{n \to \infty} \inf_{A \in \mathscr{A}} var_A(g \mid M_n) \leqq \inf_{A \in \mathscr{A}} var_A g \leqq var_{Ak} g ; \quad k = 1, 2, \dots .$$

Comparing (18) and (19) we obtain the needed results.

COROLLARY. *The conditions of Theorem 3 are satisfied if $\mathscr{A}$ is a finite set. Then every optimal simple design of an infinite-dimensional regression experiment may be approximated by optimal designs of finite-dimensional regression experiments.*

3. REMARKS ON MULTIPLE DESIGNS AND DESIGNS WITH UNCORRELATED OBSERVATIONS

A multiple design may be described by a discrete probability measure ξ on $\mathscr{A}$ ($\mathscr{A}$ is the set of simple designs which may be the components of the multiple design). $\xi(A)$ is then the relative number of independent repetitions of the simple design A in the multiple design. The information kernel I_ξ corresponding to the multiple design ξ is given by

$$I_\xi(m, m') = \sum_{A \in \mathscr{A}} \xi(A)\, IA(m, m')\,; \quad m, m' \in M\,.$$

We define a bounded operator P_ξ in $H(T)$ by

$$(h, P_\xi l)_T = \sum_{A \in \mathscr{A}} \xi(A)\,(h, P_A l)_T\,; \quad l, h \in H(T)\,.$$

Then we may write

$$I_\xi(m, m') = (m, P_\xi m')_T\,; \quad m, m' \in M\,.$$

Theorem 1 is valid also for multiple designs if we write P_ξ instead of P_A and I_ξ instead of IA.

If $A = \{\{t\} : t \in T\}$, then ξ is a design with uncorrelated observations. It is used to define a topology on T and to consider as a design any probability measure ξ on the Borel σ-algebra $\mathscr{B}$ on T [5], [6]. Suppose that T is compact and $\mathscr{K}$ is continuous in $T \times T$. Without restriction on generality we may suppose that $\mathscr{K}(t, t) = 1$; $t \in T$. Then, as an analogy to Theorem 1 we have that the following five statements are equivalent:

1. There is a *LUE* for g.
2. There is a *BLUE* for g.
3. $g \in H(I_\xi)$, where

$$I_\xi(m, m') = \int_T mm'\, d\xi\,; \quad m.\, m' \in M\,.$$

4. There is an $h \in L_2(T, \mathscr{B}, \xi)$ such that

$$g(m) = \int_T hm\, d\xi\,; \quad m \in M\,.$$

5. There is an $l \in L_2(M; \xi)$ (the subspace of $L_2(T, \mathscr{B}, \xi)$ spanned by M) such that

$$g(m) = \int_T lm\, d\xi\,; \quad m \in M\,.$$

The variance of the *BLUE* for g under ξ is equal to $\int_T l^2\, d\xi$.

(For a detailed proof see [8].)

If G is the set of functionals on M which are to be estimated, then the covariance kernel D_ξ for the $BLUE$-s for g $(g \in G)$ is equal to

$$D_\xi(g, g') = (g, g')_{I_\xi}; \quad g, g' \in G.$$

To compare the two designs ξ and η we have to consider the operator $\mathbf{D}$ in the Hilbert space $H(D_\eta)$ (with the reproducing kernel D_η), which is defined by

$$(\mathbf{D}v)(g) = (v, D_\xi(., g))_{D_\eta}; \quad g \in G.$$

The operator $\mathbf{D}$ may not be defined. It is defined in the important case when

$$\xi = \alpha\eta + (1 - \alpha)\eta'$$

for a number $\alpha \in (0, 1)$ and a probability measure η', i.e. when ξ is obtained by adding new points (simple designs) to the design η.

What is intesting is that in the case when $\xi = \alpha\eta + (1 - \alpha)\eta'$, $\alpha \in (0, 1))$ $M = H(IT)$ (i.e. when we have no a priori information about the mean $EX(.)$ of the process) and when $G = \{g : g(.) = \int_B . \, d\eta; \; B \in \mathscr{B}\}$ (i.e. when the aim of the experimenter is to know the mean $EX(.)$ up to the precision given by η), then the operator $\mathbf{D}$ is unitary equivalent to the operator of multiplication by the function $d\eta/d\xi$ in $L_2(T, \mathscr{B}, \eta)$ [8]. As a consequence of this [8] the comparison of such designs with respect to the entropy

$$\int_T \left(\ln \frac{d\eta}{d\xi}\right) d\eta$$

is a generalization of the comparison of experiments with respect to the D-optimality criterion [6]. (The D-optimality criterion is based on the generalized variance of the estimated functionals in the finite – dimensional case.)

REFERENCES

[1] N. Aronszajn: Theory of reproducing kernels. Trans. Amer. Math. Soc. *68* (1950), 337—404.

[2] C. L. Atwood: Sequences converging to D-optimal designs of experiments. Ann. of Statist. *1* (1973), 2, 342—352.

[3] V. V. Fedorov, A. Pázman: Design of Physical Experiments (Statistical methods). Fortschritte d. Physik *16* (1968), 325—355.

[4] P. R. Halmos: Introduction to Hilbert space. Chelsea, New York 1957.

[5] S. Karlin, W. J. Studden: Optimal experimental designs. Ann. Math. Statist. *39* (1966), 4, 783—815.

[6] J. Kiefer, J. Wolfowitz: Optimum designs in regression problems. Ann. Math. Statist. *30* (1959), 271—294.

[7] E. Parsen: Regression analysis of continuous parameter time series. In: Proc. Fourth Berkeley Symp. Math. Statist. Prob., 1961. Univ. of California, Berkeley, 1963, 469—490.

[8] A. Pázman: The ordering of experimental designs. A Hilbert space approach. Kybernetika (1974) (to appear).

[9] J. A. Rozanov: Gaussovskije beskonečnomernyje raspredelenija. Nauka, Moskva 1968.

[10] J. Sacks, D. Ylvisaker: Designs for regression problems with correlated errors. Ann. Math. Statist. *37* (1966), 1, 66—89.

[11] J. Sacks, D. Ylvisaker: Designs for regression problems with correlated errors; many parameters. Ann. Math. Statist. *39* (1968), 1, 49—69.

Slovak Academy of Sciences

Institute for Measurement Theory

A CHARACTERIZATION OF BOUNDED COMPLETENESS IN THE UNDOMINATED CASE

DETLEF PLACHKY

MÜNSTER

It is proved that a (not necessarily dominated) family of probability measures is boundedly complete iff the closed linear hull of this family in the weak star topology coincides with the set of all finite and signed measures, which are continuous with respect to this family. In the dominated case this characterization reduces to a well known result. As an application it is shown that bounded completeness is preserved with respect to families of product probability measures without any additional assumptions.

1. NOTATIONS AND TERMINOLOGY

Let $\mathfrak{P}$ be an arbitrary family of probability measures on a σ-algebra $\mathfrak{A}$ of subsets of a non void set Ω. Then $B(\Omega, \mathfrak{A})$ resp. ba$(\Omega, \mathfrak{A})$ denotes in the notation of [1] the set of all (realvalued) bounded and $\mathfrak{A}$-measurable functions on Ω resp. the family of all bounded and finitely additive (signed) set functions on $\mathfrak{A}$. The corresponding set of all finite, signed measures on $\mathfrak{A}$ is denoted by ca$(\Omega, \mathfrak{A})$. Furthermore $L(\mathfrak{P})$ resp. ba$(\mathfrak{P})$ resp. ca$(\mathfrak{P})$ is defined to be the linear subspace of ca$(\Omega, \mathfrak{A})$ spanned by $\mathfrak{P}$ resp. the set of all $\mu \in$ ba$(\Omega, \mathfrak{A})$ resp. $\mu \in$ ca$(\Omega, \mathfrak{A})$ with $\mu \ll \mathfrak{P}$, i.e. $P(A) = 0$ for all $P \in \mathfrak{P}$ implies $\mu(A) = 0$, $A \in \mathfrak{A}$. In the following this property of μ is called $\mathfrak{P}$-continuity. Finally $\mathfrak{P}$ is said to be boundedly complete if $\int f \,dP = 0$ for all $P \in \mathfrak{P}$ implies $f = 0$ $\mathfrak{P}$-a.e., i.e. $f = 0$ up to a measurable set of P-measure zero for all $P \in \mathfrak{P}$ (or equivalently $f = 0$ P-a.e. for all $P \in \mathfrak{P}$), where $f \in B(\Omega, \mathfrak{A})$.

2. MAIN RESULTS

First it will be proved, that bounded completeness of $\mathfrak{P}$ is equivalent with the following approximation property of $\mathfrak{P}$: For every $\mu \in$ ca$(\mathfrak{P})$ there exists a generalized sequence $\mu_\alpha \in L(\mathfrak{P})$, which converges to μ in the weak topology of ba$(\Omega, \mathfrak{A})$. For this purpose the following variant of the theorem of Hahn-Banach is needed:

LEMMA 1. For every $\mu_0 \in \text{ba}(\Omega, \mathfrak{A})$ not in the closure (with respect to the weak topology of $\text{ba}(\Omega, \mathfrak{A})$) of a linear subspace L there exists a $f_0 \in B(\Omega, \mathfrak{A})$, such that $\int f_0 \, d\mu = 0$ for all $\mu \in L$ and $\int f_0 \, d\mu_0 \neq 0$ holds.

Proof. Because of the representation theorem IV.5.1 in [1] of the dual space of $B(\Omega. \mathfrak{A})$ and the separation theorem V.2.7 in [1] together with the fact, that linear functionals, which are continuous in the weak topology, are evaluation maps (V.3.9 in [1]), there is a $f_0 \in B(\Omega, \mathfrak{A})$, such that $\int f_0 \, d\mu \leqq \int f_0 \, d\mu_0 \neq 0$ holds for all $\mu \in L$. This implies the statement of Lemma 1, since L is a linear subspace.

If $\overline{L(\mathfrak{P})}$ denotes the closure of $L(\mathfrak{P})$ in the weak topology of $\text{ba}(\Omega, \mathfrak{A})$, Lemma 1 yields:

THEOREM 1. $\mathfrak{P}$ is boundedly complete iff $\overline{L(\mathfrak{P})} \cap \text{ca}(\Omega, \mathfrak{A}) = \text{ca}(\mathfrak{P})$.

Proof. $\int f \, dP = 0$ for all $P \in \mathfrak{P}$ together with $\overline{L(\mathfrak{P})} \cap \text{ca}(\Omega, \mathfrak{A}) = \text{ca}(\mathfrak{P})$ implies $\int_A f \, dP = 0$ for all $A \in \mathfrak{A}$ and all $P \in \mathfrak{P}$, from which $f = 0$ P-a.e. for all $P \in \mathfrak{P}$ or equivalently $f = 0$ $\mathfrak{P}$-a.e. follows. If on the other hand $\overline{L(\mathfrak{P})} \neq \text{ba}(\mathfrak{P})$ holds, then there exists because of Lemma 1 for $\mu_0 \in \text{ba}(\mathfrak{P}) \setminus \overline{L(\mathfrak{P})}$ a function $f_0 \in B(\Omega, \mathfrak{A})$, such that $\int f_0 \, d\mu = 0$ holds for all $\mu \in L(\mathfrak{P})$ and $\int f_0 \, d\mu_0 \neq 0$. Hence bounded completeness of $\mathfrak{P}$ would imply $f_0 = 0$ $\mathfrak{P}$-a.e., from which the contradiction $f_0 = 0$ μ_0-a.e. would follow, since μ_0 is $\mathfrak{P}$-continuous.

REMARK 1. If a family $\mathfrak{P}$ of probability contents on an algebra $\mathfrak{A}$ of subsets of a non void set Ω is said to be boundedly complete if $\int f \, dP = 0$ for all $P \in \mathfrak{P}$ implies f is a P-null function in the sense of ([1], III.2.) for all $P \in \mathfrak{P}$, where $f \in B(\Omega, \mathfrak{A})$ and $B(\Omega, \mathfrak{A})$ is the space of all uniform limits of simple functions in the sense of ([1], IV.2.10), then the proof of Theorem 1 shows: $\mathfrak{P}$ is boundedly complete iff $\overline{L(\mathfrak{P})}$ is equal to the set of all $\mu \in \text{ba}(\Omega, \mathfrak{A})$ with the property, that $P(N) = 0$ for all $P \in \mathfrak{P}$ implies $\mu(N) = 0$, where $N \in \mathfrak{A}$.

REMARK 2. If the boundedly complete family $\mathfrak{P}$ is dominated by a σ-finite measure μ, which can be chosen $\mathfrak{P}$-continuous, then $\nu \to d\nu/d\mu$ is a homeomorphism between $\text{ca}(\mathfrak{P})$ and $L_1(\Omega, \mathfrak{A}, \mu)$, if $\text{ca}(\mathfrak{P})$ resp. $L_1(\Omega, \mathfrak{A}, \mu)$ is equipped with the relative weak topology of $\text{ba}(\Omega, \mathfrak{A})$ resp. weak topology of $L_1(\Omega, \mathfrak{A}, \mu)$. This implies, that $\mathfrak{P}$ is boundedly complete in the dominated case iff the closure in the norm topology of $L_1(\Omega, \mathfrak{A}, \mu)$ of the subspace spanned by $\{dP/d\mu \mid P \in \mathfrak{P}\}$ coincides with $L_1(\Omega, \mathfrak{A}, \mu)$, since the closures in the weak resp. strong topology of a convex set of a Banach space are equal (V.3.13 in [1]). This necessary and sufficient condition for bounded completeness in the dominated case is well known [3].

In the following an application of Theorem 1 will be given, which shows, that bounded completeness is a hereditary property with respect to products. For this purpose, the following auxiliary result is proved:

LEMMA 2. Let $\mathfrak{P}_i$ be a family of probability measures on the σ-algebra $\mathfrak{A}_i$ of subsets of a set Ω_i, $i = 1, 2$, and $\mathfrak{Q}$ the set of all product probability measures $Q_1 \otimes Q_2$ on the product σ-algebra $\mathfrak{A}_1 \otimes \mathfrak{A}_2$, which are $\mathfrak{P}_1 \otimes \mathfrak{P}_2$-continuous, where $\mathfrak{P}_1 \otimes \mathfrak{P}_2 = \{P_1 \otimes P_2 \mid P_i \in \mathfrak{P}_i,\ i = 1, 2\}$. Then $\overline{L(\mathfrak{Q})} \cap \mathrm{ca}(\Omega_1 \times \Omega_2,\ \mathfrak{A}_1 \otimes \mathfrak{A}_2) = \mathrm{ca}(\mathfrak{P}_1 \otimes \mathfrak{P}_2)$.

Proof. If the statement of Lemma 2 is false, there would exist $Q_0 \in \mathrm{ca}(\mathfrak{P}_1 \otimes \mathfrak{P}_2)$ and because of Lemma 1 a function $g_0 \in B(\Omega_1 \times \Omega_2, \mathfrak{A}_1 \otimes \mathfrak{A}_2)$, such that $\int g_0 \, \mathrm{d}Q_0 \neq 0$ and $\int g_0 \, \mathrm{d}(Q_1 \otimes Q_2) = 0$ for all $Q_1 \otimes Q_2 \in \mathfrak{Q}$ holds. For fixed $P_i \in \mathfrak{P}_i$, $i = 1, 2$, this implies $\int_{A_1 \times A_2} g_0 \, \mathrm{d}(P_1 \otimes P_2) = 0$ for all $A_i \in \mathfrak{A}_i$, $i = 1, 2$, and therefore $\int_A g_0 \, \mathrm{d}(P_1 \otimes P_2) = 0$ for all $A \in \mathfrak{A}_1 \otimes \mathfrak{A}_2$. Hence $g_0 = 0$ $P_1 \otimes P_2$-a.e. for all $P_1 \otimes P_2 \in \mathfrak{P}_1 \otimes \mathfrak{P}_2$. The $\mathfrak{P}_1 \otimes \mathfrak{P}_2$-continuity of Q_0 now implies $g_0 = 0$ Q_0-a.e., which is a contradiction to $\int g_0 \, \mathrm{d}Q_0 \neq 0$.

With the help of Lemma 2 it is now easy to prove, that bounded completeness is preserved with respect to products.

THEOREM 2. Let $\mathfrak{P}_i$ be boundedly complete families of probability measures, $i = 1, 2$. Then $\mathfrak{P}_1 \otimes \mathfrak{P}_2$ is boundedly complete.

Proof. Because of Lemma 2 and Theorem 1 it is enough to prove $\overline{L(\mathfrak{P}_1 \otimes \mathfrak{P}_2)} = \overline{L(\mathfrak{Q})}$. If this equality is false, there would exist $Q_1 \otimes Q_2 \in \mathfrak{Q}$ and because of Lemma 1 a function $g_0 \in B(\Omega_1 \times \Omega_2. \mathfrak{A}_1 \otimes \mathfrak{A}_2)$, such that $\int g_0 \, \mathrm{d}(Q_1 \otimes Q_2) \neq 0$ and $\int g_0 \, \mathrm{d}(P_1 \otimes P_2) = 0$ for all $P_1 \otimes P_2 \in \mathfrak{P}_1 \otimes \mathfrak{P}_2$ holds. This implies $\int g_0(\omega_1, \omega_2) \, . \, P_1(\mathrm{d}\omega_1) = 0$ for P_2-almost all $\omega_2 \in \Omega_2$ and all $P_2 \in \mathfrak{P}_2$ and hence for Q_2-almost all $\omega_2 \in \Omega_2$, since Q_2 is $\mathfrak{P}_2$-continuous. Therefore $\int g_0 \, \mathrm{d}(P_1 \otimes Q_2) = 0$ for all $P_1 \in \mathfrak{P}_1$, from which $\int g_0(\omega_1, \omega_2) \, Q_2(\mathrm{d}\omega_2) = 0$ for P_1-almost all $\omega_1 \in \Omega_1$ follows, since Q_1 is $\mathfrak{P}_1$-continuous. This yields the contradiction $\int g_0 \, \mathrm{d}(Q_1 \otimes Q_2) = 0$.

REMARK. In the literature the corresponding result for complete families of probability measures is proved under the additional assumption, that one of the families is strongly complete [2].

3. GENERALIZATIONS

Theorem 2 remains true, if $\mathfrak{P}_1$ is a family of transition probability measures on $\Omega_2 \times \mathfrak{A}_1$ and $\mathfrak{P}_2$ a family of probability measures on $\mathfrak{A}_2$, where $P_1 \otimes P_2$ is defined by $P_1 \otimes P_2(A) = \int P_1(\omega_2, A_{\omega_2}) \, P_2(\mathrm{d}\omega_2)$, $\mathfrak{P}_1 \otimes \mathfrak{P}_2$ by $\{P_1 \otimes P_2 \mid P_i \in \mathfrak{P}_i,\ i = 1, 2\}$, and where $\mathfrak{P}_1$ is called boundedly complete for $\mathfrak{P}_2$-almost all $\omega_2 \in \Omega_2$ if $\int f(\omega_1) \, . \, P_1(\omega_1, \mathrm{d}\omega_2) = 0$ for all $P_1 \in \mathfrak{P}_1$ and $\mathfrak{P}_2$-almost all $\omega_2 \in \Omega_2$ implies $f(\omega_1) = 0$ for $P_1(\omega_2, \cdot)$-almost all $\omega_1 \in \Omega_1$ and $\mathfrak{P}_2$-almost all $\omega_2 \in \Omega_2$ with $f \in B(\Omega_1, \mathfrak{A}_1)$.

This version of Theorem 2 is proved with the help of a variant of Lemma 2, where $\mathfrak{Q}$ is the set of all probability measures $Q_1 \otimes Q_2$ which are $\mathfrak{P}_1 \otimes \mathfrak{P}_2$-continuous with Q_1 being a transition probability measure on $\Omega_2 \times \mathfrak{A}_1$ and Q_2 being a probability measure. This yields the generalization of Theorem 2, because $\iint g_0(\omega_1, \omega_2) \cdot$ $\cdot P_1(\omega_2, d\omega_1) P_2(d\omega_2) = 0$ for all $P_1 \in \mathfrak{P}_1$ and $P_2 \in \mathfrak{P}_2$ implies $g_0(\omega_1, \omega_2) = 0$ for $P_1(\omega_2, \cdot)$-almost all $\omega_1 \in \Omega_1$ and P_2-almost all $\omega_2 \in \Omega_2$ and for all $P_1 \otimes P_2 \in$ $\in \mathfrak{P}_1 \otimes \mathfrak{P}_2$. From this $\iint g_0(\omega_1, \omega_2) Q_1(\omega_2, d\omega_1) Q_2(d\omega_2) = 0$ follows for $Q_1 \otimes Q_2 \in$ $\in \mathfrak{Q}$, since the $\mathfrak{P}_1 \otimes \mathfrak{P}_2$-continuity of $Q_1 \otimes Q_2$ implies $\int g_0(\omega_1, \omega_2) Q_1(\omega_2, d\omega_1) = 0$ for P_2-almost all $\omega_2 \in \Omega_2$ for all $P_2 \in \mathfrak{P}_2$, which implies $\iint g_0(\omega_1, \omega_2) Q_1(\omega_2, d\omega_1) \cdot$ $\cdot P_2(d\omega_2) = 0$ for all $P_2 \in \mathfrak{P}_2$ and therefore $\int g_0 \, d(Q_1 \otimes Q_2) = 0$.

APPLICATION. Let $\mathfrak{P}_1$ resp. $\mathfrak{P}_2$ be families of probability measures on the set of onedimensional Borel sets, such that $\{P_1 * \delta_x \mid x \in \boldsymbol{R}\}$ resp. $\mathfrak{P}_2$ is boundedly complete for all $P_1 \in \mathfrak{P}_1$, then $\mathfrak{P}_1 * \mathfrak{P}_2 = \{P_1 * P_2 \mid P_i \in \mathfrak{P}_i, i = 1, 2\}$ is boundedly complete, if the support of P_2 is $\boldsymbol{R}$ for every $P_2 \in \mathfrak{P}_2$ and if $\mathfrak{P}_1$ is continuous with respect to the Lebesgue measure. The bounded completeness of $\{P_1 * \delta_x \mid x \in \boldsymbol{R}\}$ is equivalent with $\int e^{itx} P_1(dx) \neq 0$ for all $t \in \boldsymbol{R}$ in the dominated case [3]. If $\mathfrak{P}_i$ is boundedly complete for $i = 1, 2$, the bounded completeness of $\mathfrak{P}_1 * \mathfrak{P}_2$ follows from Theorem 2, because $P_1 \otimes P_2$ together with the mapping $(x, y) \to x + y$, $x, y \in \boldsymbol{R}$, induces the convolution $P_1 * P_2$.

I wish to thank H. Luschgy for valuable discussions.

REFERENCES

[1] N. DUNFORD, T. J. SCHWARTZ: Linear Operators. Part I. Interscience Publishers, New York 1964.

[2] D. A. S. FRASER: Nonparametric Methods in Statistics. Wiley, New York 1957.

[3] J. K. GHOSH, R. SINGH: Unbiased estimation of location and scale parameters. Ann. Math. Stat. *37* (1966), 1671—1675.

UNIVERSITÄT MÜNSTER

INSTITUT FÜR MATHEMATISCHE STATISTIK

CONTINUOUS FILTERING PROBLEM FOR ITÔ-MCSHANE STOCHASTIC DIFFERENTIAL EQUATION

ZORAN R. POP-STOJANOVIC

GAINESVILLE

INTRODUCTION

In this paper we present a continuous filtering problem for Itô-McShane stochastic differential equation. The concept of Itô-McShane integral has been introduced by the second mentioned author in [3]. The basic motivation for this concept is to present a sort of an unified theory of Stochastic integration [2], [3].

THE SETTING. We are going to deal with real-valued functions defined over intervals of real numbers.

DEFINITION (McShane [3]). A gauge δ in a set $E \subseteq R^1$ is a function δ on E such that for every $t \in E$, $\delta(t)$ is a neighborhood of t. We say that δ is a constant size gauge if there is a positive number γ such that $\delta(t) = (t - \gamma, t + \gamma)$ for all t in E.

Let $[a, b)$ be a right-open interval in R^1. A partition Π *in* $[a, b)$ is a finite set $\Pi = \{(\tau_1, A_1), \ldots, (\tau_k, A_k)\}$ of ordered pairs in which every $\tau_i \in [a, b]$, $i = 1, 2, \ldots, k$, and A_i's are pairwise disjoint subintervals of $[a, b)$. If $\bigcup_{i=1}^{k} A_i = [a, b)$, we say that Π is a partition *of* $[a, b)$.

DEFINITION (McShane [3]). Let δ be a gauge on an interval $[a, b)$, and let δ^* be a positive number. A partition $\Pi = \{(\tau_1, A_1), \ldots, (\tau_k, A_k)\}$ *in* $[a, b)$ is a (δ, δ^*) belated partition in $[a, b)$ if:

(i) $$A_i \subseteq \delta(\tau_i) \cap [\tau_i, b], \quad i = 1, 2, \ldots, k,$$

(ii) $$b - a - \sum_{i=1}^{k} m(A_i) < \delta^*$$

(m denotes the Lebesgue measure).

Let $(\Omega, \mathscr{F}, \mathscr{P})$ be a probability space and let $\{f_t;\ t \in [a. b]\}$, $\{z_t;\ t \in [a, b]\}$ be stochastic processes on $[a, b]$, i.e., for every $t \in [a, b]$, f_t, z_t are random variables on Ω. If $\Pi = \{(\tau_1, A_1), \ldots, (\tau_k, A_k)\}$ is a partition in $[a, b)$ with $A_j = [\alpha_j, \beta_j)$, $j = 1, 2, \ldots, k$, we introduce the "Riemann Sum" corresponding to Π:

$$S(\Pi; f; z) = \sum_{j=1}^{k} f_{\tau_j}(z_{\beta_j} - z_{\alpha_j}).$$

DEFINITION [3]. Itô-McShane Stochastic integral $I_{a,b} = \int_a^b f_t \, dz_t$ is defined in the following way: For every $\varepsilon > 0$ there exists a gauge δ in $[a, b)$ and a positive number δ^* such that whenever Π is a (δ, δ^*) belated partition in $[a, b)$ one has that:

$$I_{a,b} = \operatorname*{l.i.m.}_{\varrho \to 0} S(\Pi; f, z),$$

where ϱ = mesh Π.

REMARK. Note that in this definition $[\alpha_j, \beta_j)$'s come "later" than their corresponding τ_j's – hence came the name: belated partition. This innovation which has been introduced for the first time by E. J. McShane in [4], makes possible an unified theory of Stochastic integration. The concept of Itô-McShane integral generalizes Itô's Stochastic integral [2], McShane's belated Stochastic integral [4], [6] and in deterministic case, i.e., when Ω consists of a single point it reduces to Lebesgue integral [3].

Let $\{x_t; t \in [a, b]\}$ be a Stochastic process satisfying the following Itô-McShane Stochastic differential equation:

$$\mathrm{d}x_t = f(t, x_t)\, \mathrm{d}t + g(t, x_t)\, \mathrm{d}z_t, \tag{1}$$

which can be written in the equivalent integral form:

$$x_t = x_a + \int_a^t f(s, x_s)\, \mathrm{d}s + \int_a^t g(s, x_s)\, \mathrm{d}z_s, \quad t \in [a, b], \tag{2}$$

where the first integral in (2) is a mean square Riemann integral and the second integral is Itô-McShane Stochastic integral.

Note that in (1) process $\{z_t; t \in [a, b]\}$ is no longer the Brownian motion process as in Itô's Stochastic integral. Since we are going to investigate a continuous filtering problem for (1), we need to find out under which conditions (1) has a solution and what kind of properties does this solution possess.

In order to do so, let $\mathscr{F}_t = \mathscr{B}(\{x_s; s \leqq t\})$, $t \in [a, b]$, where $\mathscr{B}(\{\cdot\})$ denotes a Borel field generated by $\{\cdot\}$. It follows that if $s \leqq t$, then $\mathscr{F}_s \subseteq \mathscr{F}_t$, for every $s, t \in [a, b]$.

Now we can state the following existence theorem for Itô-McShane Stochastic differential equation (1).

THEOREM. Let $\{z_t; t \in [a, b]\}$ be a second order process adapted to $\mathscr{F}_t$ for every $t \in [a, b]$ and such that almost all sample functions z_t are continuous on $[a, b]$.

Furthermore, let us assume that there exists a constant $K > 0$ such that for every $s, t \in [a, b]$,

$$a \leqq s \leqq t \leqq b, \quad E(|z_t - z_s| \mid \mathscr{F}_s) \leqq K(t - s)$$

$$\text{and} \quad E((z_t - z_s)^2 \mid \mathscr{F}_s) \leqq K(t - s).$$

Suppose the functions f and g satisfy the hypotheses:

A_1: There is a constant $K_1 > 0$ such that

$$|f(t, x)| \leqq K_1(1 + |x|^2)^{1/2}$$

$$|g(t, x)| \leqq K_1(1 + |x|^2)^{1/2}.$$

A_2: For every t,

$$|f(t, x_2) - f(t, x_1)| \leqq K_1|x_2 - x_1|,$$

$$|g(t, x_2) - g(t, x_1)| \leqq K_1|x_2 - x_1|.$$

A_3: For every x,

$$|f(t_2, x) - f(t_1, x)| \leqq K_1|t_2 - t_1|,$$

$$|g(t_2, x) - g(t_1, x)| \leqq K_1|t_2 - t_1|.$$

A_4: x_a is a random variable such that $E(|x_a|^2) < +\infty$, and being independent of $\{dz_t; t \in [a, b]\}$.

Then (1) has a solution $\{x_t; t \in [a, b]\}$ in the mean square sense which has the following properties:

(I) x_t is mean square continuous in $[a, b]$.

(II) $E(|x_t|^2) < +\infty$, for every $t \in [a, b]$.

(II_1) $E(|x_t|^2) \leqq M < +\infty$ on $[a, b]$.

(III) $\int_a^b E(|x_t|^2)\, dt < +\infty$.

(IV) $x_t - x_a$ is independent of $\{dz_\tau; \tau \geqq t\}$ for every $t \in [a, b]$.

Proof. Let us define a new process $\{Y_t; t \in [a, b]\}$ by the following relation:

$$Y_t = \int_a^t f(s, x_s)\, ds + \int_a^t g(s, x_s)\, dz_s, \quad t \in [a, b]. \tag{3}$$

Let us show now that if an $\{x_t\}$ process has properties (I), (II), (III), (IV), then $\{Y_t\}$ process has properties (I), (II_1), (III), (IV). Indeed, by A_1 and (II),

$$E(|f(t, x_t)|^2 \leqq K_1^2(1 + E(|x_t|^2)) < +\infty,$$

$$E(|g(t, x_t)|^2) \leqq K_1^2(1 + E(|x_t|^2)) < +\infty \quad \text{for all} \quad t \in [a, b].$$

By A_2, A_3 and (I),

$$E(|f(t+h, x_{t+h}) - f(t, x_t)|^2) \leqq 2K_1 E(|x_{t+h} - x_t|^2) + 2K_1^2 h^2 \to 0 \quad \text{as} \quad h \to 0,$$

and similarly for g. Thus f, g are mean square continuous. This implies that f is mean square Riemann integrable which further implies that the first integral in (2) is a mean square continuous function of its upper limit. Now, let us show that Itô-McShane integral $\int_a^t g(s, x_s)\, \mathrm{d}z_s$ in (2) is a mean square continuous function of its upper limit. Since $g(t, x_t)$ is dt. $P(\mathrm{d}\omega)$ measurable, and $\int_a^b E(|g(t, x_t)|^2)\, \mathrm{d}t < +\infty$ from A_1 and (II), it follows by the theorem in ([3], p. 259) that Itô-McShane integral $\int_a^t g(s, x_s)\, \mathrm{d}z_s$ exists for every $t \in [a, b]$. Moreover, using ([3], p. 259), one can obtain the following inequality:

$$(*) \qquad E\left(\left|\int_t^{t+h} g(s, x_s)\, \mathrm{d}z_s\right|^2\right) \leqq C \int_t^{t+h} E(|g(s, x_s)|^2)\, \mathrm{d}s,$$

where C is a constant.

It follows from (*) that $E(|\int_t^{t+h} g(s, x_s)\, \mathrm{d}z_s|^2) \to 0$ as $h \to 0$ since the right-hand side of (*) is a continuous function of its upper limit. Thus, both integrals in (3) are mean square continuous with respect to t which implies that $\{Y_t; t \in [a, b]\}$ has property (I). It is obvious that $\{Y_t\}$ has property (IV). Moreover, (I) $\Rightarrow$ (III). Finally, let us show that $\{Y_t\}$ has property (II_1).

$$E(|Y_t|^2) \leqq 2E\left(\left|\int_a^t f(s, x_s)\, \mathrm{d}s\right|^2\right) + 2E\left(\left|\int_a^t g(s, x_s)\, \mathrm{d}z_s\right|^2\right) \leqq$$

$$\leqq 2\int_a^t \int_a^t E(f_\tau f_s)\, \mathrm{d}\tau\, \mathrm{d}s + 2C\int_a^t E(|g_s|^2)\, \mathrm{d}s \leqq$$

$$\leqq 2(b-a)\int_a^b E(|f_s|^2)\, \mathrm{d}s + 2C\int_a^t E(|g_s|^2)\, \mathrm{d}s \leqq$$

$$\leqq 2K_1^2(b-a)\int_a^b (1 + E(|x_s|^2))\, \mathrm{d}s + 2CK_1^2 \int_a^b (1 + E(|x_s|^2))\, \mathrm{d}s < M < +\infty,$$

where we used (III).

Now let $\{x_t^{(0)}\}$ be any process satisfying (I), (II), (III), (IV). Let us construct the successive approximations:

$$(4) \qquad \text{For} \quad n \geqq 1: x_t^{(n)} = x_a + \int_a^t f(s, x_s^{(n-1)})\, \mathrm{d}s + \int_a^t g(s, x_s^{(n-1)})\, \mathrm{d}z_s.$$

Then, according to the preceding, the process $\{x_t^{(n)}\}$ has properties (I), (II_1), (III), (IV), for $n \geqq 1$.

By using inequality $(*)$ one can prove that

(i) $$\underset{n\to+\infty}{\text{l.i.m.}}\, x_t^{(n)} = x_t\,, \quad \text{for every} \quad t \in [a, b]\,, \quad \text{uniformly in } t$$

and that the process $\{x_t;\, t \in [a, b]\}$ defined in (i) has properties (I), (II), (III), (IV). Finally, we prove that

(ii) $$\underset{n\to\infty}{\text{l.i.m.}} \int_a^t f(s, x_s^{(n)})\, ds = \int_a^t f(s, x_s)\, ds\,, \quad t \in [a, b]\,,$$

and

(iii) $$\underset{n\to+\infty}{\text{l.i.m.}} \int_a^t g(s, x_s^{(n)})\, dz_s = \int_a^t g(s, x_s)\, dz_s\,, \quad t \in [a, b]\,,$$

which terminates the proof of the theorem.

REMARK. Using the inequality $(*)$ one can show the uniqueness (in mean square sense) of the solution of (1).

CONTINUOUS FILTERING PROBLEM

Since Itô-McShane Stochastic integral in a natural way generalizes Itô's integral, it is relatively easy to obtain equations for the conditional density connected with a non-linear continuous filtering problem generated by Stochastic differential equation (1). The method is essentially the same as obtained earlier for Itô's equation in [1] and [5].

Let us assume that the state of a system evolves according to Itô-McShane Stochastic differential equation

(5) $$dx_t = f(t, x_t)\, dt + g(t, x_t)\, dz_t\,, \quad t \geqq a\,,$$

and the observed process is the solution of Itô-McShane equation

(6) $$du_t = h(t, x_t)\, dt + d\eta_t\,, \quad t \geqq a\,,$$

where $\{\eta_t;\, t \geqq a\}$ is a Brownian motion process with unit parameter and independent of the process $\{z_t;\, t \geqq a\}$. In this case, the equation (6) is Itô's equation since Itô's integral and Itô-McShane integral are overlapping in this case [4]. Let $U_\tau = \mathscr{B}(\{u_s;\, a \leqq s \leqq \tau\})$. In order to solve a filtering problem we have to find the conditional density $f(x_t \mid U_t)$ of x_t. (Here we assume that all conditional densities mentioned in the sequel do exist!) Let us assume that (5) satisfies the hypotheses of the existence theorem given before. Furthermore, let us assume that the unconditional density $f(x_t)$

of x_t is differentiable and that h is a continuous function in both arguments. If, moreover, $E((\int_a^t h^2(s, x_s)\,ds)^2) < +\infty$, for every $t \geqq a$, then

$$(7)\qquad f(x_t \mid U_t) = \frac{f(x_t)\, E^t\left\{\exp\left[-\frac{1}{2}\int_a^t h^2(s, x_s)\,ds + \int_a^t h(s. x_s)\,du_s\right] \middle| x_t\right\}}{E^t\left\{\exp\left[-\frac{1}{2}\int_a^t h^2(s, x_s)\,ds + \int_a^t h(s, x_s)\,du_s\right]\right\}},$$

for every $t \geqq a$, where $E^t(\cdot) = E(\cdot \mid U_t)$. Derivation of (7) is basically the same as that for Itô's equation obtained in [1] and [5].

Outline of the proof. First, the assumption $E((\int_a^t h^2(s, x_s)\,ds)^2) < +\infty$ for every $t \geqq a$ implies the existence of all integrals in (7). Furthermore, the second integral in (7) can be written as

$$\int_a^t h(s, x_s)\,du_s = \int_a^t h(s, x_s)\,ds + \int_a^t h(s, x_s)\,d\eta_s$$

where the last integral is an Itô integral. By Bayes' theorem one obtains that

$$(8)\qquad f(x_t \mid u_{x_0}, \ldots, u_{s_n}) = \frac{f(u_{x_0}, \ldots, u_{s_n} \mid x_t)\, f(x_t)}{f(u_{x_0}, \ldots, u_{s_n})}.$$

Also, one has that

$$(9)\qquad f(u_{x_0}, \ldots, u_{s_n} \mid x_t) = E^t\{f(u_{x_0}, \ldots, u_{s_n} \mid u_{x_0}, \ldots, x_{s_{n-1}}, x_t) \mid x_t\}$$

and

$$(10)\qquad f(u_{x_0}, \ldots, u_{s_n}) = E^t\{f(u_{x_0}, \ldots, u_{s_n} \mid x_t)\}.$$

Our goal is to compute $f(u_{x_0}, \ldots, u_{s_n} \mid u_{x_0}, \ldots, x_{s_n})$. Integrating (6) and assuming a fine partition, we have

$$(11)\qquad u_{s_{k+1}} - u_{s_k} = h(s_k, x_{s_k})(s_{k+1} - s_k) + (\eta_{s_{k+1}} - \eta_{s_k}) + 0(\varrho).$$

Therefore, using (11) one gets that

$$f(u_{s0}, \ldots, u_{s_n} \mid x_{s0}, \ldots, x_{s_n}) =$$
$$= f(u_{s_n} \mid u_{s_{n-1}}, x_{s0}, \ldots, x_{s_n})\, f(u_{s_{n-1}} \mid u_{s_{n-2}}, x_{s0}, \ldots, x_{s_n}) \ldots f(u_{s0}),$$

since $\{\eta_t\}$ has independent increments. Now, using (11) one gets

$$(13)\qquad f(u_{s0}, \ldots, u_{s_n} \mid x_{s0}, \ldots, x_{s_n}) =$$
$$f(u_{s0})\, K(\varrho) \exp\left[-\tfrac{1}{2}\sum_{k=0}^{n-1}(s_{k+1} - s_k)^{-1}\{(u_{s_{k+1}} - u_{s_k}) - h(s_k, x_{s_k})(s_{k+1} - s_k)\}^2\right].$$

Finally, after one substitutes (11) into (9) and (10), and subsequently into (8), and after cancellation of terms such as $f(u_{s0})$, constant $K(\varrho)$ and $(u_{s_{k+1}} - u_{s_k})^2$, one gets that

$$(14) \qquad f(x_t \mid u_{s_0}, \ldots, u_{s_n}) = ME^t\{\exp\left[-\tfrac{1}{2}\sum_{k=0}^{n-1} h^2(s_k, x_{s_k})(s_{k+1} - s_k) + \right.$$
$$\left. + \sum_{k=0}^{n-1} h(s_k, x_{s_k})(u_{s_{k+1}} - u_{s_k})\right] \mid x_t\} f(x_t),$$

where M is a normalizing constant to be determined.

On the other hand

$$\underset{\substack{\varrho\to 0\\ n\to+\infty}}{\text{l.i.m.}} \sum_{k=0}^{n-1} h^2(s_k, x_{s_k})(s_{k+1} - s_k) = \int_a^t h^2(s, x_s)\, ds,$$

$$\underset{\substack{\varrho\to 0\\ n\to+\infty}}{\text{l.i.m.}} \sum_{k=0}^{n-1} h(s_k, x_{s_k})(u_{s_{k+1}} - u_{s_k}) = \int_a^t h(s, x_s)\, du_s .$$

Therefore,

$$\underset{\substack{\varrho\to 0\\ n\to+\infty}}{\text{l.i.m.}} f(x_t \mid u_{s_0}, \ldots, u_{s_n}) =$$

$$= M f(x_t)\, E_t \left\{\exp\left[-\frac{1}{2}\int_a^t h^2(s, x_s)\, ds + \int_a^t h(s, x_s)\, du_s\right] \mid x_t\right\}.$$

It is obvious that $1/M$ is the denominator in (7) which completes the outline of the proof.

REFERENCES

[1] R. S. Bucy: Nonlinear Filtering Theory. IEEE Trans. Automatic Control *10* (1965), 198—201.

[2] K. Itô: Stochastic Integration. Nagoya Math. Journal *3* (1951), 212—231.

[3] E. J. McShane: Stochastic Integration. Vector and Operator valued measures and Application. Academic Press, New York 1973, 247—281.

[4] E. J. McShane: Stochastic integrals and Stochastic functional equations. SIAM Journal Appl. Math. *17* (1969), 287—306.

[5] R. E. Mortensen: Optimal Control of Continuous-Time Stochastic System. Ph. D. thesis, Univ. of California, Berkeley, California, 1966.

[6] Z. Z. Pop-Stojanovic: On McShane's belated Stochastic integral. SIAM J. Appl. Math. *22* (1972), 1, 87—92.

University of Florida

Department of Mathematics

HOST-VECTOR EPIDEMICS

JOHN RADCLIFFE

LONDON

1. INTRODUCTION

The recurrent and periodic behaviour of measles has been studied in some detail, [4], and theoretical expressions obtained for the recurrence time between outbreaks in small communities and the period of oscillation in large communities. Deterministic and stochastic models have also been used to describe the initial spatial spread of a direct infection, [1], [2], [4], [5], [7], [12], [13], [19].

This paper shows how an analogous theory can be developed for asexual host-vector diseases. Host-vector diseases can be broadly divided into virus and parasitic diseases. An example of the former is yellow fever. Parasitic diseases can be further divided, [16], into asexual diseases such as malaria and sexual diseases such as schistosomiasis, the behaviour of which has been outlined in [15], [17]. We focus our attention on the two diseases, malaria and yellow fever. The essential difference between the two diseases is that a person who recovers from a virus infection has an immunity and it is unlikely that he will be reinfected, but malaria is a parasitic disease and reinfection and superinfection are common occurrences, [9], [14]. So that as far as the model for a virus infection is concerned it is unnecessary to distinguish between recovery and death of an infective. The usual model for measles, [5], divides the population into susceptibles, infected and removed. However in the model for malaria the population is divided into susceptibles and infectives only, and it is necessary to distinguish between deaths form the disease and recoveries. Host-vector epidemics will tend to have different characteristics dependent on the infectious periods for the diseases. A person infected with malaria will tend to remain infectious for some time or become reinfected, which will result in the disease having an endemic nature. However, the infectious period of a person suffering from yellow fever is only of the order of a couple of days. Thus yellow fever should behave in some respects like measles in small communities, where there will be a tendency for the infection to die out altogether. Then following isolated instances of infection brought in from outside, there will, after sufficient time in which the number of susceptibles has had time to build up, be another major outbreak of the disease.

In section 2 an approximate expression is derived for the probability density of the recurrence time between major outbreaks of a yellow fever epidemics in urban populations. In many areas the level of infection of malaria exhibits a regular cycle. In section 3 a model is formulated to describe endemic malaria, and an expression is obtained for the period of oscillation of the infection. In section 4 a model is formulated as a Markov population process, to take into accout the topographical factor, and the forward equations for the model for a virus disease are written down, and the modification for an asexual parasitic disease noted. If infection is introduced into a previously uninfected area, the approximation that the numbers of susceptibles remain constant can be used, to specify a model for the initial spread of the infection as a position dependent Markov branching process which gives a more mathematically tractable model which can be used to obtain a theoretical expression for the velocity of propagation of the wave of infection.

2. THE RECURRENCE TIME BETWEEN MAJOR OUTBREAKS OF YELLOW FEVER

Yellow fever is basically a host-vector disease involving monkeys and mosquitos in a forest cycle. In Central and South America, and Africa the mosquitos concerned are Haemagogus sp. and Aedes africans respectively. The tangential spread to man occurs when the forest cycle spills over to infect man. The urban cycle then acts as a host-vector epidemic involving man and the mosquito, Aedes aegypti.

The initial development of the epidemic arising from a particular host who enters the population at time s may be approximated by a bivariate birth-death process, [10], [11], [22]. Let $X(t)$ and $X'(t)$ be the number of infectious hosts and vectors respectively at time t. In the time interval $(t, t + \delta t)$ the possible transitions are:

$$(x, x') \to (x + 1, x') \quad \text{with probability} \quad n\lambda x' \,\delta t + o(\delta t)\,,$$

$$(x, x') \to (x, x' + 1) \quad \text{with probability} \quad n'\lambda' x \,\delta t + o(\delta t)\,,$$

$$(x, x') \to (x - 1, x') \quad \text{with probability} \quad \mu x \,\delta t + o(\delta t)\,,$$

$$(x, x') \to (x, x' - 1) \quad \text{with probability} \quad \mu' x' \,\delta t + o(\delta t)\,,$$

where n and n' are the numbers of hosts and vectors respectively.

Then if the increase in the number of susceptibles due to births after time s is neglected, it is well known, [1], [7], that the probability that this individual does not initiate a major epidemic is

$$1 \quad \text{if} \quad s \leqq \tau$$

and

$$(1)\qquad \left(1 + \frac{\mu'}{n\lambda}\right)\bigg/\left(1 + \frac{n'\lambda'}{\mu}\right) \quad \text{if} \quad s > \tau\,,$$

where $\tau = \mu\mu'/(vn'\lambda\lambda')$ and v is the birth rate for new individuals.

Time is measured from the point at which the last major epidemic died out. The times at which infected hosts enter the population are taken to be the epochs of a Poisson process. It is assumed that the rate is dependent on the number of susceptibles in the population capable of acting as a link host, and is taken to be ωt, since the number of susceptibles is supposed to increase linearly due to the births of new individuals. Any individual immigrating in $[0, \tau]$ will with probability one not initiate a major epidemic. Let $t_1, t_2, \ldots, t_k$ be the epochs in the interval $(\tau, t]$ at which infectious hosts enter the process. If $E(t)$ is the event that a major epidemic has not occurred by time t, then

$$(2)\qquad P[E(t) \mid t_1, t_2, \ldots, t_k] = \prod_{i=1}^{k} \left(1 + \frac{\mu'}{\lambda v t_i}\right)\bigg/\left(1 + \frac{\lambda' n'}{\mu}\right).$$

Since conditional on k events in $(\tau, t]$, the unordered points of arrival of link hosts can be regarded as independent random variables with probability density

$$(3)\qquad f(s) = 2s/(t^2 - \tau^2)\,, \quad \tau < s \leqq t\,,$$

and the number of epochs in $(\tau, t]$ is a Poisson variable with mean $\frac{1}{2}\omega(t^2 - \tau^2)$

$$(4)\qquad P(E(t)) = \exp\left[-\frac{1}{2}\alpha\tau\left(\frac{t}{\tau} - 1\right)^2\right],$$

where

$$\alpha = \omega\mu'\bigg/\left(v\lambda\left(1 + \frac{n'\lambda}{\mu}\right)\right).$$

The probability density of the time until a major outbreak is given by

$$(5)\qquad p(t) = \alpha\left(\frac{t}{\tau} - 1\right)\exp\left[-\frac{1}{2}\alpha\tau\left(\frac{t}{\tau} - 1\right)^2\right], \quad \tau \leqq t < \infty\,.$$

3. THE PERIOD OF ENDEMIC MALARIA

A deterministic model is adequate to obtain an expression for the period of oscillation of infection, since the possibility of extinction is not considered, [6]. Let $S_1(t), I_1(t), S_2(t)$ and $I_2(t)$ be the number of susceptible hosts, infectious hosts, susceptible vectors and infectious vectors respectively. The death rates for susceptible and infectious hosts are taken to be the same, [21]. The same is supposed true for the vector population.

The deterministic equations are

$$\frac{dS_1}{dt} = v_1 - \mu_1 S_1 - \lambda_{12} S_1 I_2 + \beta I_1 , \tag{6}$$

$$\frac{dI_1}{dt} = \lambda_{12} S_1 I_2 - \mu_1 I_1 - \beta I_1 , \tag{7}$$

$$\frac{dS_2}{dt} = v_2 - \mu_2 S_2 - \lambda_{21} S_2 I_1 , \tag{8}$$

$$\frac{dI_2}{dt} = \lambda_{21} S_2 I_1 - \mu_2 I_2 . \tag{9}$$

In a stable population where $S_1 + I_1$ is constant we have the relationship

$$v_1 - \mu_1(S_1 + I_1) = 0 . \tag{10}$$

It is also assumed that the region in question is not subject to a very great seasonal variation, so that the vector population can also be assumed constant. Then

$$v_2 - \mu_2(S_2 + I_2) = 0 . \tag{11}$$

From (7) and (10) the equation

$$\frac{dI_1}{dt} = \frac{\lambda_{12}}{\mu_1}(v_1 - \mu_1 I_1) I_2 - (\mu_1 + \beta) I_1 \tag{12}$$

is obtained.

From (9) and (11) we obtain the equation

$$\frac{dI_2}{dt} = \frac{\lambda_{21}}{\mu_2}(v_2 - \mu_2 I_2) I_1 - \mu_2 I_2 . \tag{13}$$

The equilibrium or endemic levels m and n of I_1 an I_2 are given by

$$m = \frac{\lambda_{12}\lambda_{21} v_1 v_2 - \mu_1(\mu_1 + \beta)\mu_2^2}{\mu_1 \lambda_{21}[\lambda_{12} v_2 + (\mu_1 + \beta)\mu_2]} \tag{14}$$

and

$$n = \frac{\lambda_{12}\lambda_{21} v_1 v_2 - \mu_1(\mu_1 + \beta)\mu_2^2}{\mu_2 \lambda_{12}[\lambda_{21} v_1 + \mu_1 \mu_2]} . \tag{15}$$

The method of small oscillations, [1], [8], about m and n can be used to show, [21] that the period of oscillation of the system is $2\pi/\zeta$ where $\zeta = (c - b^2)^{1/2}$,

$$b = \tfrac{1}{2}(\mu_1 + \mu_2 + \beta + \lambda_{12}n + \lambda_{21}m)$$

and

$$c = (\mu_1 + \beta + \lambda_{12}n)(\mu_2 + \lambda_{21}m) - nm\lambda_{12}\lambda_{21}\left(\frac{v_1}{\mu_1 m} - 1\right)\left(\frac{v_2}{\mu_2 n} - 1\right).$$

4. THE TOPOGRAPHICAL FACTOR

Using Moyal's formulation of a population process, [18], the forward equation, expressed in generating functional form, is derived for a Markov population process with several types of individual where the individuals can branch as in a multiplicative process and also interact pairwise between types. This is then used to write down the forward equation for host-vector epidemics such as yellow fever and malaria. These equations are the analogous of the equations for measles in [3], [4].

Consider a population process as described in [18]. Each individual is in a state x belonging to a fixed set X. Let $I = \{1, 2, \ldots, n\}$ and the individual state space is $W = I \times X$, so that each individual is described by $w = (i, x)$, where i denotes its type. The population state space is then $\sum_{j=0}^{\infty} W^j$.

Let $P(A, s)$ be the probability of an event A at time s, $P^{(n)}$ be the restriction of P on W^n and $P(A, s \mid w^k, u)$ be the conditional probability of an event A at time s given $(w_1, w_2, \ldots, w_k)$ at time u. Let ζ be a bounded measurable function on W. The Chapman-Kolmogorov equation implies that the probability generating functional satisfies the equation

$$(16) \qquad G(\zeta, t + \delta t \mid w_0^k, t_0] = \\ = \sum_{j=0}^{\infty} \int_{W^j} G[\zeta, t + \delta t \mid w^j, t]\, P^{(j)}(dw^j, t \mid w_0^k, t_0)\,.$$

So that

$$(17) \qquad \frac{\partial}{\partial t} G[\zeta, t \mid w_0^k, t_0] = \sum_{j=0}^{\infty} \int_{W^j} K[\zeta \mid w^j, t]\, P^{(j)}(dw^j, t \mid w_0^k, t_0)\,,$$

where

$$K[\zeta \mid w^j, t] = \lim_{\delta t \downarrow 0} \frac{1}{\delta t} \{G[\zeta, t + \delta t \mid w^j, t] - \prod_{i=1}^{j} \zeta(w_i)\}\,.$$

Suppose that in an interval $(t, t + \delta t)$ there are three types events that can occur:

(i) $E(w)$ is the event that the individual at w dies and is replaced by a random number of offspring with states in W.

(ii) $E(w, z)$ is the event that the individuals at w and z interact resulting in the disappearance of these two individuals and their replacement by a random number of individuals with states is W.

(iii) F_i is the event that a new individual of type i immigrates into the system $(i = 1, 2, \ldots, n)$.

The events $E(w)$, $E(w, z)$ and F_i have rates $b(w)$, $c(w, z)$ and v_i respectively with associated conditional probability generating functionals $f(\zeta \mid w)$, $f(\zeta \mid w, z)$ and $h_i(\zeta)$ $(i = 1, 2, \ldots, n)$.

Let

$$v = (v_1 + v_2 + \ldots + v_n)$$

and

$$h(\zeta) = (v_1 h_1(\zeta) + v_2 h_2(\zeta) + \ldots + v_n h_n(\zeta))/v\,.$$

Then is follows that

$$K[\zeta \mid w^j, t] \Big/ \prod_{i=1}^{j} \zeta(w_i) = \sum_{i=1}^{j} b(w_i)\left[\frac{f(\zeta \mid w_i)}{\zeta(w_i)} - 1\right] + \sum_{i<l} c(w_i, w_l)\left[\frac{f(\zeta \mid w_i, w_l)}{\zeta(w_i)\,\zeta(w_l)} - 1\right] + v[h(\zeta) - 1]\,. \tag{18}$$

The notation is now changed. If $w = (i, x)$ and $z = (l, y)$, we write $b(w) = b_i(x)$, $c(w, z) = c_{il}(x, y)$ $f(\zeta \mid w) = f_i(\zeta \mid x)$ and $f(\zeta \mid w, z) = f_{il}(\zeta \mid x, y)$. The forward equation is then

$$\frac{\partial G(\zeta, t)}{\partial t} = \sum_{i=1}^{n} \int_X b_i(r)\,[f_i(\zeta \mid r) - \zeta_i(r)]\,\frac{\partial G(\zeta, t)}{\partial \zeta_i(r)\,\mathrm{d}r}\,\mathrm{d}r + \sum_{i<l} \int_{X^2} c_{il}(r, s)\,[f_{il}(\zeta \mid r, s) - \zeta_i(r)\,\zeta_l(s)]\,\frac{\partial^2 G(\zeta, t)}{\partial \zeta_i(r)\,\mathrm{d}r\,\partial \zeta_l(s)\,\mathrm{d}s}\,\mathrm{d}r\,\mathrm{d}s + v[h(\zeta) - 1]\,. \tag{19}$$

Type one and two individuals represent susceptible and infectious hosts respectively, and types three and four represent susceptible and infectious vectors respectively. The set X is taken to be R^2. For a virus disease, the assumptions regarding infection are taken to be

(i) The probability that a specific susceptible host in δs is infected by a specific infective vector in δr in $(t, t + \delta t)$ is

$$\lambda_{14}(s - r)\,\delta t + o(\delta t)$$

and

(ii) The probability that a specific susceptible vector in δs is infected by a specific infective host in δr in $(t, t + \delta t)$ is

$$\lambda_{32}(s - r)\,\delta t + o(\delta t)\,.$$

It is also assumed that both vectors and hosts can move and die subject to the following axioms, which refer to a single individual:

(iii) The probability that a type i individual moves from δr to δs in $(t, t + \delta t)$ is

$$\alpha_i(s - r)\,\delta s\,\delta t + o(\delta t) \quad (i = 1, 2, 3, 4)$$

and

(iv) The probability that a type i individual dies in $(t, t + \delta t)$ is

$$\mu_i\,\delta t + o(\delta t) \quad (i = 1, 2, 3, 4)\,.$$

Immigration occurs subject to the following axiom:

(v) The probability that a type i individual immigrates into the process in $(t, t + \delta t)$ is

$$v_i\,\delta t + o(\delta t)$$

and the probability density of its position given that a type i arrives in $(t, t + \delta t)$ is

$$p_i(r) \quad (i = 1, 2, 3, 4)\,.$$

From the above axioms it is observed that

(20) $$b_i(r) = \mu_i + \int_{R^2} \alpha_i(s - r)\,ds\,,$$

(21) $$f_i(\zeta \mid r) = \frac{1}{b_i(r)}\left(\mu_i + \int_{R^2} \alpha_i(s - r)\,\zeta_i(s)\,ds\right) \quad (i = 1, 2, 3, 4)\,,$$

(22) $$c_{14}(r, s) = \lambda_{14}(r - s)\,,$$

(23) $$f_{14}(\zeta \mid r, s) = \zeta_2(r)\,\zeta_4(s)\,;$$

(24) $$c_{23}(r, s) = \lambda_{32}(s - r)\,,$$

(25) $$f_{23}(\zeta \mid r, s) = \zeta_2(r)\,\zeta_4(s)\,,$$

and

(26) $$h_i(\zeta) = \int_{R^2} p_i(r)\,\zeta_i(r)\,dr \quad (i = 1, 2, 3, 4)\,.$$

The forward equation for this model is

$$\frac{\partial G(\zeta, t)}{\partial t} = \tag{27}$$

$$= \sum_{i=1}^{4} \int_{R^2} \left[\mu_i + \int_{R^2} \alpha_i(s - r)\, \zeta_i(s)\, ds - \zeta_i(r) \left(\mu_i + \int_{R^2} \alpha_i(s - r)\right) ds\right] \frac{\partial G(\zeta, t)}{\partial \zeta_i(r)\, dr}\, dr +$$

$$+ \int_{R_4} \lambda_{14}(r - s)\, [\zeta_2(r)\, \zeta_4(s) - \zeta_1(r)\, \zeta_4(s)]\, \frac{\partial^2 G(\zeta, t)}{\partial \zeta_1(r)\, dr\, \partial \zeta_4(s)\, ds}\, dr\, ds +$$

$$+ \int_{R^4} \lambda_{32}(s - r)\, [\zeta_2(r)\, \zeta_4(s) - \zeta_2(r)\, \zeta_3(s)]\, \frac{\partial^2 G(\zeta, t)}{\partial \zeta_2(r)\, dr\, \partial \zeta_3(s)\, ds}\, dr\, ds +$$

$$+ \sum_{i=1}^{4} v_i \left[\int_{R^2} p_i(r)\, \zeta_i(r)\, dr - 1\right] G(\zeta, t)\,.$$

For a parasitic disease such as malaria it is necessary to distinguish between deaths and recoveries of infective hosts, [21], and to include the additional axiom

(vi) The probability that a specific infective host recovers and becomes a susceptible host again in $(t, t + \delta t)$ is

$$\beta\, \delta t + o(\delta t)\,.$$

This will change $b_2(r)$ and $f_2(\zeta \mid r)$ and the appropriate expressions are

$$b_2(r) = \mu_2 + \beta + \int_{R^2} \alpha_2(s - r)\, ds \tag{28}$$

and

$$f_2(\zeta \mid r) = \frac{1}{b_2(r)} \left[\mu_2 + \beta \zeta_1(r) + \int_{R^2} \alpha_2(s - r)\, \zeta_2(s)\, ds\right].$$

Thus the only change in equation (27) is that the term

$$\int_{R^2} \left[\mu_2 + \int_{R^2} \alpha_2(s - r)\, \zeta_2(s)\, ds - \zeta_2(r) \left(\mu_2 + \int_{R^2} \alpha_2(s - r)\, ds\right)\right] \frac{\partial G(\zeta, t)}{\partial \zeta_2(r)\, dr}\, dr$$

is replaced by

$$\int_{R^2} \left[\mu_2 + \beta \zeta_1(r) + \int_{R^2} \alpha_2(s - r)\, \zeta_2(s)\, ds -\right.$$

$$\left. - \zeta_2(r) \left(\mu_2 + \beta + \int_{R^2} \alpha_2(s - r)\, ds\right)\right] \frac{\partial G(\zeta, t)}{\partial \zeta_2(r)\, dr}\, dr\,.$$

If infection is introduced into a previously uninfected area, the approximation that the numbers of susceptible hosts and vector remain constant can be used to simplify equation (19) to study the initial spread of the epidemic. Equation (27) reduces to equation (2.8) of [20], from which equations for the mean distributions can be obtained and used to obtain an expression for the velocity of propagation of the infection.

REFERENCES

[1] N. T. J. BAILEY: The mathematical theory of epidemics. Griffin, London 1957.

[2] N. T. J. BAILEY: The simulation of stochastic epidemics in two dimensions. In: Proc. 5th Berkeley Symp. Math. Statist. and Prob. *4*, Berkeley 1965. University of California Press, Berkeley and Los Angeles 1967, 237—257.

[3] M. S. BARTLETT: Processus stochastiques ponctuels. Ann. Inst. Poincare *14* (1954), 35—60.

[4] M. S. BARTLETT: Deterministic and stochastic models for recurrent epidemics. In: Proc. 3rd Berkeley Symp. Math. Statist. and Prob. *4*, Berkeley 1955. University of California Press, Berkeley and Los Angeles 1956, 81—109.

[5] M. S. BARTLETT: Stochastic population models in ecology and epidemiology. Methuen, London 1960.

[6] M. S. BARTLETT: The relevance of stochastic models for large-scale epidemiological phenomena. Applied Statistics *13* (1964), 2—8.

[7] M. S. BARTLETT: Some notes on epidemiological theory. In: Research Papers in Statistics, Festschrift for J. Neyman. Wiley London 1966, 25—35.

[8] M. S. BARTLETT: An introduction to stochastic processes. Cambridge University Press, London 1966.

[9] L. J. BRUCE-CHWATT: Quantitative epidemiology of tropical diseases. Transactions of the Royal Society of Tropical Medicine and Hygiene *63* (1969), 131—143.

[10] D. A. GRIFFITHS: A bivariate birth-death process which approximates to the spread of a disease involving a vector. J. Appl. Prob. *9* (1972), 65—75.

[11] D. A. GRIFFITHS: Multivariate birth-and-death processes as approximations to epidemic processes. J. Appl. Prob *10* (1973), 15—26.

[12] D. G. KENDALL: Contribution to the discussion in Barlett, measles periodicity and community size J. R. Statist. Soc. A. *120* (1957), 48—70.

[13] D. G. KENDALL: Mathematical models of the spread of infection. In: Mathematics and Computer Science in Biology and Medicine. H.M.S.O., London 1965, 213—225.

[14] G. MACDONALD: The mechanism of infection with malaria in children living under endemic and hyperendemic conditions. Indian Journal of Medical Research *18* (1931), 1347—1372).

[15] G. MACDONALD: Epidemiologic models in studies of vector-borne diseases. Public Health Reports, Washington *76* (1961), 753—764.

[16] G. MACDONALD: In the scientific basis of tropical hygiene. Transactions of the Royal Society of Tropical Medicine and Hygiene *59* (1965), 611—620.

[17] G. MACDONALD: The dynamics of helminth infections, with special reference to schistosomes. Transactions of the Royal Society of Tropical Medicine and Hygiene *59* (1965), 489—506).

[18] J. E. MOYAL: The general theory of stochastic population processes. Acta Math *108* (1962), 1—31.

[19] J. NEYMAN, E. SCOTT: A stochastic model of epidemics. In: Stochastic Models in Medicine and Biology. University of Wisconsin Press, 1964, 45—85.
[20] J. RADCLIFFE: The initial geographical spread of host-vector and carrier-borne epidemics. J. Appl. Prob. *10* (1973), 703—717.
[21] J. RADCLIFFE: The periodicity of endemic malaria. J. Appl. Prob (1974, to appear).
[22] J. RADCLIFFE: A note on the recurrence of yellow-fever epidemics in urban populations. J. Appl. Prob. *11* (1974), 170—173.
Papers [9; 14—17] are reprinted in Dynamics of Tropical Disease, ed. L. J. Bruce-Chwatt and V. J. Glanville. Oxford University Press, London 1973.

UNIVERSITY OF LONDON
QUEEN MARY COLLEGE

REPLICATION AND STOCHASTIC PROCESSES IN SPACE

DANIEL RICHARDSON

BRISTOL

A mathematical model is defined in which some of the processes of molecular biology can be simulated. Then a definition of replication is given and some techniques are described which have been developed to prove that certain things do in fact replicate.

Let G be a graph in three dimensional space which is obtained by connecting some of the points in the cubic lattice by edges. A screw symmetry of G is a translation followed by a sequence of rotations which takes points into points and edges into edges. The points of G will be considered as possible positions for objects and the edges will be considered as possible positions for bonds between the objects. Let Ob and B be finite sets. If R is a region of space, an R configuration will be an assignment of an element of Ob or 0 to each point of R and an element of B or 0 to each edge of R. If we think of Ob and B as colors, an R configuration may be visualized as a coloring (with blanks) of the points and edges of R. A state of a region is defined to be a probability distribution over its configurations. A patch is an R configuration where R is finite. If P and Q are patches with the same domain, a pair $P \to Q$ will be called an operation. A patch process is given by a finite list of operations with positive real numbers attached to them:

$$\begin{array}{c} P_1 \xrightarrow{\lambda_1} Q_1 \\ \vdots \\ P_n \xrightarrow{\lambda_n} Q_n \end{array}$$

For each region R, a patch process is intended to be a Markov process in continuous time transforming R configurations into R configurations. For each i, the number λ_i is intended to be the rate at which the operation $P_i \to Q_i$ is applied. The operation $P_i \to Q_i$ may be understood as removing P_i (or any screw symmetry of P_i) and replacing it by Q_i (or a corresponding screw symmetry of Q_i). If R is finite, let $X_1 \ldots, X_m$ be a list of all R configurations. The patch process applied to R gives a finite Markov chain in continuous time. X_i is transformed into X_j at rate $\sum_{k=1}^{n} N_k \lambda_k$ where N_k is the number of screw symmetries s of G so that $s(P_k)$ agrees with X_i on the domain of $s(P_k)$ and if we remove $s(P_k)$ and replace it by $s(Q_k)$, we change X_i to X_j. (For any patch P and screw symmetry s, $s(P)$ is the patch so that for each point or edge, z, the coloring of P at z is the same as the coloring of $s(P)$ at $s(z)$.)

For infinite regions, R, a patch process also gives a Markov process. This can be defined as the limit of the processes applied to an increasing sequence of finite subregions of R which tends to R. Another and more direct definition, which turns out to be equivalent, has been given by Peter Clifford.

The chrystal soups, a certain subclass of patch processes applied to regions, are considered as possible models of molecular interaction. We define a chrystal soup to be a region R together with a patch process which satisfies the following two conditions:

1. Every operation in the process is one of the three following types:

A. Elementary motion;

B. Bond creation or destruction;

C. Thing creation or destruction.

2. If the operations of type C) are dropped, all the equilibrium states of the new process are time reversible.

The three types of operation are described below.

A. An operation $P \to Q$ is an elementary motion if there is a one to one correspondence f between the non 0 objects and bonds of P and Q so that f preserves the type of each object and bond and if b connects x to y in P, then $f(b)$ connects $f(x)$ to $f(y)$ in Q. The operation must leave fixed any objects or bonds on the boundary of P which are connected to something outside of P.

B. An operation $P \to Q$ is a bond creation or destruction if the only difference between P and Q is one bond.

C. We define a *thing* to be a collection of objects connected by bonds. An operation $P \to Q$ is a thing creation or destruction if one of the patches is identically 0.

It is supposed that "energy" is added to the soup only by the creation and destruction of things. For this reason, in the absence of these operations, the soup is assumed to tend to a time reversible equilibrium state.

There are several natural but non-equivalent intuitive concepts of replication. One of these is as follows. A thing H is replicating in a chrystal soup with environment S if when we start two copies of the soup in state S and add H to the second soup, the waiting time for the first appearance of H in the first soup is relatively long and the waiting time for two simultaneous appearances of H in the second soup is relatively short.

The above can be made into a precise definition by giving any precise notion of short and long. This can be done arbitrarily in various ways by assigning numbers, or short and long can be treated as limiting concepts. The latter approach is followed below, after the example.

The simplest type of replication is replication by template in which a chain of objects acts as a catalyst for a similar or complementary chain.

EXAMPLE. Let the graph G be obtained by connecting nearest neighbours in the cubic lattice. There are four types of objects, w, x, y, and z, and two types of bonds, strong bonds, – – –, which may exist between objects of different types and weak bonds, ∿∿, which may exist between objects of the same type. No object may be involved in more than two strong bonds, and all bond angles must be ninety degrees. Strong bonds form at rate F between neighbouring objects of different type and break at rate B. Weak bonds form at rate f between neighbouring objects of the same type and break at rate b. No object may be involved in more than one weak bond. Elementary motions move one object or two bonded objects simultaneously, keeping all bond angles at ninety degrees, and not moving any object more than distance two. The rate at which an elementary motion occurs is $m \exp(-d)$ where d is the sum of the distances moved by objects in the motion. There are no creation and destruction of thing operations.

The possibly replicating thing H is a loop of objects of size four connected by strong bonds:

```
x----y
:    :
w----z
```

The region R is supposed to be a large cube and the initial state S is that there are N unbonded objects of each type in R. We suppose that m is extremely large, f is large, b and F are small, and B is very small. The intuitive idea is that H will collect loose objects by weak bonds. This increases the local density of appropriate objects and thus encourages the formation of strong bonds.

When an attempt is made to analyse intuitions about events in soups, the concepts of fast and slow, large and small seem to occur naturally. Instead of thinking, for example, of B, F b, f, m, R, N as quantities which have not yet been specified, we can think of them as variables which are tending to satisfy some limiting condition of the form $C \to 0$. (C is a function of the variables whose range includes a sequence tending to 0.) Instead of studying a single soup, it is often easier and more interesting to study a sequence of soups which tend to satisfy some such limiting condition. It is sometimes useful to imagine an "ideal soup" at, or near, the end of the sequence. This also gives a way of avoiding the artificiality of the cubic lattice. It is possible to imagine continuous space as an ideal soup at the end of a certain sequence. (Ideal soups seem to have an interesting mathematical structure in relation to their sequences. For example, the study of how ideal soups vary in response to variation in their sequences includes the subject of phase transition which has attracted a great deal of attention in mathematical physics.)

If A and B are functions of the variables, we will write

$$A \sim_C B \quad \text{if} \quad (\forall\, \varepsilon > 0\, \exists \delta > 0)\left(|C| < \delta \rightarrow |A - B| < \varepsilon\right),$$

$$A \simeq_C B \quad \text{if} \quad A \sim_C B \quad \text{and} \quad (A/B) \sim_C 1\,.$$

DEFINITION. Let $*$ be a collection of chrystal soups with the same operations. Let $C \rightarrow 0$ be a limiting condition on the parameters of $*$, i.e. the rates and the region. Let H be a thing and S a state of $*$. Suppose we have two copies of $*$ in state S and add H to the second copy. Let T_1 be the waiting time until the appearance of the first H in the first soup. Let T_2 be the waiting time until the simultaneous appearance of two copies of H in the second soup. Then H is replicating in $*$ in state S relative to $C \rightarrow 0$ if for any $k > 0$, Prob $(kT_2 < T_1) \sim_C 1$.

If this definition is satisfied, it implies that given a particular k and a particular number $p < 1$, there exists a particular soup so that Prob $(kT_2 < T_1) > p$. In the ideal soup, replication occurs infinitely faster than spontaneous generation.

The idea of a strongly meta stable set is useful in proving that something is replicating in the sense of the definition above. Using the same notation, let $\Gamma = \{X_1, \ldots, X_n\}$ be a set of configurations of $*$ which is intercommunicating. If we eliminate all exists from Γ and treat it by itself as a finite Markov chain, it will have a unique equilibrium state $(\hat{X}_1, \ldots, \hat{X}_n)$. Now, thinking of Γ again as a part of $*$, we will say that Γ is strongly meta stable relative to the conditions $C \rightarrow 0$ if there is a function t of the parameters of $*$ so that:

1. If we start anywhere in Γ at time 0, the probability of an exit from Γ before t is $\sim_C 0$.

2. If we start anywhere in Γ at time 0, then, for all i, $X_i(t) \simeq_C \hat{X}_i$, where $X_i(t)$ means the probability of being in X_i at time t.

The idea is that in the ideal soup Γ almost alwys goes into the equilibrium state before an exit from Γ occurs; so we may replace Γ by a random configuration corresponding to the equilibrium state.

Now returning to the example, we will find conditions on the parameters which imply that the loop is replicating in the sense of the definition.

We don't want H to fall apart before replication occurs. The rate of replication is bounded by F so we must have to begin with $B/F \rightarrow 0$.

Let R be a cube with V points in it. Pick N and V so that $6(4N)^2/V$ is small. Assume m is sufficiently large relative to f, b, F, B, N, V so that if x is any configuration in our two soups, then Γ_x, is strongly meta stable, where Γ_x is the set of configurations obtained from x by elementary motions. All configurations in Γ_x are equally likely since the elementary motions go forwards and backwards at the same rate.

If x is any configuration of the first soup with no strong bonds, it can be shown that in Γ_x, the probability of an unlike pair being neighbours is $<6(4N)^2/V$. Strong bonds only form between unlike pairs so the first strong bond appears in the first soup at rate $<F\,6(4N)^2/V$.

Now for the second soup. Until all possible weak bonds have been formed, new ones appear at rate at least f/V. Assume that f/V is large relative to b and F. Then before any strong bonds appear in the first soup, almost all objects are weakly bonded in the second soup. Because of the presence of H, the second soup has a neigh bouring unlike pair with probability $>\frac{1}{2}$. The first strong bond forms at rate $>F/2$. The new pair is attached to H by two weak bonds α_1 and α_2:

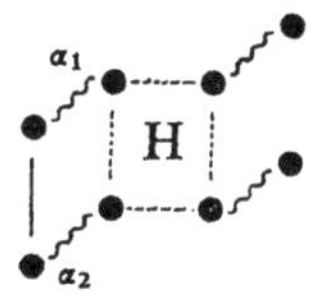

The pair is extended to a triple at rate $>F/2$. It will be convenient to impose some condition so that the extension almost always occurs before the pair becomes detached from H. Let us look at the bond α_1 in isolation. α_1 breaks at rate b. Having broken, it can either reform (at rate $\frac{1}{4}f$), or the other bond α_2 can break, or α_1 can be preemted by the formation of a new weak bond with one of its end points. The probability that α_2 will break before α_1 reforms is $b/(b+\frac{1}{4}f)$. The probability that α_1 will be preempted is $<N/V$. Assume that $4b/f$ is small relative to N/V. Then the rate of detachment is $<bN/V$. So if F is large relative to bN/V, we can suppose that extension occurs before detachment with high probability. Continuing, we have that replication occurs in the second soup after four steps, each happening at rate $>F/2$.

Since $6(4N)^2/V$ was small, replication occurs in the second soup before any strong bond is formed in the first soup.

Listing the conditions used, we have:

1. $B/F \to 0$,
2. all Γ_x strongly meta stable,
3. $6(4N)^2/V \to 0$,
4. $Vb/f \to 0$, $VF/f \to 0$,
5. $bV/fN \to 0$, $bN/VF \to 0$,
6. $F \to \infty$.

To check that these conditions are consistent, it is enough to see that they would be satisfied if $m = \infty$, $N = 10$, $B = 0$, $b = 0$, $V = F$, $f = F^3$, $F \to \infty$.

Using these ideas, and some work which has been done on Gibbs states, and the work of von Neumann and others in the field of finite automata theory, it is possible to design self replicating systems in the context of a chrystal soup which approach the sophistication of biological entities. It should also be possible to study in a mathematical way some of the simpler ideas which have been expressed about the early stages in the development of life. As much as possible of this subject will be discussed in a book which is now being completed.

REFERENCES

[1] R. M. Baer, H. M. Martinez: Automata and biology. Annual Review of Biophysics, to appear.
[2] F. J. Dyson: A model for pre-biotic evolution. To appear.
[3] J. Monod: Chance and Necessity. W. Collins, 1974.
[4] C. J. Preston: Gibbs States on Countable Sets. Cambridge University Press, 1974.
[5] D. Richardson: Spatial Processes and Replication. To appear.
[6] H. Quastler: The Emergence of Biological Organization. Yale University Press, New Haven 1964.
[7] J. von Neumann: Theory of Self Reproducing Automata (A. Burks, editor). University of Illinois Press, 1970.

University of Bristol

School of Mathematics

RÉNYI-MIXING AND FUNCTIONAL LIMIT THEOREMS FOR MARTINGALES

HOLGER ROOTZÉN

LUND

Let $\{S_n\}_{n=0}^{\infty}$ be a martingale on a probability space $(\Omega, \mathscr{B}, P)$ such that $E(S_n^2) = s_n^2 < \infty$ with $s_n^2 \to \infty$ when $n \to \infty$. Put $Y_n(t, \omega) = s_n^{-1} S_k(\omega)$ for $t \in [0, 1]$ and $s_k^2 \leqq s_n^2 t < s_{k+1}^2$, $k = 0, \ldots, n$. It is known that, under a Lindeberg type condition, $\{Y_n\}_1^{\infty}$ converges weakly in $D(0, 1)$ to a Wiener process. It is proved that this convergence also holds on the probability space $(\Omega, \mathscr{B}, P(\cdot \mid B))$ for any event B that has positive probability, i.e. the sequence of random elements $\{Y_n\}$ is mixing in the sense of Rényi. Two consequences of this are given. Firstly let N' be an infinite set of positive integers. Then, for ω outside a null set, the set of functions $\{Y_n(\cdot, \omega); n \in N'\}$ is dense in the set of continuous functions on [0, 1] which vanish at zero. Secondly, the mixing property is used to prove a random-indices central limit theorem for the sequence $\{Y_n\}$.

1. INTRODUCTION

Let $\{S_n\}_{n=1}^{\infty}$ be a martingale defined on a probability space $(\Omega, \mathscr{B}, P)$ such that $E(S_n^2) = s_n^2 < \infty$. Define for $n \geqq 1$ and $\omega \in \Omega$ random functions $Y_n(\cdot, \omega)$ on $[0, 1]$ by

$$(1) \qquad Y_n(t, \omega) = s_n^{-1} S_k(\omega) \quad \text{for} \quad t \in [0, 1] \quad \text{such that} \quad s_k^2 \leqq s_n^2 t < s_{k+1}^2 ,$$
$$k = 0, \ldots, n \quad (S_0 = s_0 = 0).$$

Then $Y_n(\cdot, \omega)$ belongs to $D = D(0, 1)$, the space of real functions on $[0, 1]$ which are right continuous and have left-hand limits. It is known that if $\{S_n\}$ satisfies a Lindeberg type condition then Y_n converges weakly in (D, d_0) to Wiener measure, where d_0 is Billingsley's modification of the Skorokhod metric ([1], p. 112).

In the present note we show that also a little bit more is true, in that we prove that the sequence of random elements $\{Y_n\}_1^{\infty}$ is mixing in the sense of Rényi (Theorem 2). This is then used to prove two more results. Firstly, let C_0 be the space of continuous functions on $[0, 1]$ that vanish at zero and let N' be an infinite set of integers. Then, for ω outside a null set, the set of functions $\{Y_n(\cdot, \omega); n \in N'\}$ is dense in C_0 (Theorem

3). Secondly, under a condition on the variation of the s_n's, mixing still holds if the sequence $\{Y_n\}$ is randomly indexed (Theorem 5). One-dimensional versions of Theorems 2 and 3 are proved in [10] by the present author. For references on random limit theorems, see e.g. [6] and [12]. In [7] a random limit theorem is proved for a stationary, ergodic martingale by using a strong-mixing assumption.

2. RESULTS

Rényi's concept of mixing sequences of random variables ([8], [9]) immediately carries over to sequences of random elements in metric spaces. Let $\{Z_n\}_{n=1}^{\infty}$ be a sequence of measurable mappings from $(\Omega, \mathscr{B}, P)$ to a metric space S with its Borel σ-algebra $\mathscr{S}$, i.e. the Z_n's are random elements in $(S, \mathscr{S})$. The sequence $\{Z_n\}_1^{\infty}$ converges in distribution to a probability distribution G on $(S, \mathscr{S})$ (notation: $Z_n \xrightarrow{d} G$) if $P(Z_n \in A) \to G(A)$ for all G-continuity sets $A \in \mathscr{S}$.

DEFINITION 1. The sequence $\{Z_n\}_1^{\infty}$ is mixing with limiting distribution G if

$$P(\{Z_n \in A\} \mid B) \to G(A)$$

for every G-continuity set $A \in \mathscr{S}$ and every $B \in \mathscr{B}$ with $P(B) > 0$.

Thus the sequence $\{Z_n\}$ is mixing with limiting distribution G if it converges in distribution to G on every fixed subset of Ω that has positive probability.

Now, as in the introduction, $\{S_n\}_1^{\infty}$ is a sequence of random variables on $(\Omega, \mathscr{B}, P)$ with $E(S_{n+1} \parallel S_n, \ldots, S_1) = S_n$ and $E(S_n^2) = s_n^2 < \infty$, and the random functions $\{Y_n\}$ are defined by (1). Let

$$X_i = S_i - S_{i-1}, \quad i \geqq 1,$$

be the martingale differences, so that $S_n = \sum_{i=1}^{n} X_i$. We assume that $E(S_1) = 0$ and thus that $E(S_n) = 0$ for all $n \geqq 1$, and that $s_n^2 \to \infty$ as $n \to \infty$. Further W denotes Wiener measure on $D(0, 1)$. In the sequel we will use the metric d_0 on $D(0, 1)$ defined on p. 112 of [1]. Under d_0 the space $D(0, 1)$ is separable and complete.

THEOREM 2. If $\{S_n\}_1^{\infty}$ is a martingale as above, the random elements $\{Y_n\}_1^{\infty}$ in $(D(0, 1), d_0)$ are defined by (1) and furthermore

$$s_n^{-2} \sum_{i=1}^{n} X_i^2 \xrightarrow{P} 1 \quad \text{as} \quad n \to \infty \tag{2}$$

and

$$s_n^{-2} \sum_{i=1}^{n} X_i^2 I(|X_i| > \varepsilon s_n) \xrightarrow{P} 0 \quad \text{as} \quad n \to \infty, \quad \forall \varepsilon > 0,$$

then $\{Y_n\}_1^{\infty}$ is mixing with limiting distribution W.

Proof. Fix an event $B \in \mathscr{B}$ which has positive probability. To prove mixing we have to show that $\{Y_n\}_1^\infty$ converges in distribution to W also when we use the probability measure $P(\cdot \mid B)$, which will follow if we prove tightness and that the finite-dimensional distributions converge. However, by Theorem 1 of [11], $\{Y_n\}$ converges in distribution to W when using the uniform metric, and thus also when using d_0. Hence $\{Y_n\}$ is relatively compact, and since (D, d_0) is separable and complete it is also tight. It then follows from Theorem 15.2 of [1] that $\{Y_n\}$ is tight also when we use the probability $P(\cdot \mid B)$.

It remains to prove that the finite-dimensional distributions converge, which we do by means of the Cramér-Wold Theorem. Let the real numbers $\{a_i\}_1^l$ and $\{t_i\}_0^l$ satisfy $a_i \neq 0$, $i = 1, \ldots, l$ and $0 = t_0 < t_1 < \ldots < t_l \leqq 1$ but otherwise be arbitrary and set

$$Z_n = \sum_{i=1}^{l} a_i \left(Y_n(t_i) - Y_n(t_{i-1})\right).$$

If Y_n were a Wiener process then Z_n would have the variance

$$v = \sum_{i=1}^{l} a_i^2 (t_i - t_{i-1}).$$

By using the Skorokhod representation theorem precisely as in the proof of Theorem 2.4 of [10] it follows that the conditional characteristic functions converge, i.e. for k fixed

$$E(e^{iuZ_n} \parallel S_1, \ldots, S_k) \xrightarrow{P} \exp\left(-\tfrac{1}{2} v u^2\right),$$

as $n \to \infty$, for every real u. By Lemma 4.1 of the cited paper it then follows that the sequence $\{Z_n\}_1^\infty$ is mixing with limiting distribution (function) $\Phi(\cdot / v^{1/2})$, where Φ is the standard normal distribution function. Recalling the definition of mixing and of v and Z_n this means that

$$P(\{\sum_{i=1}^{l} a_i (Y_n(t_i) - Y_n(t_{i-1})) \leqq x\} \mid B) \to \Phi(x / (\sum_{i=1}^{l} a_i^2 (t_i - t_{i-1}))^{1/2})$$

as $n \to \infty$, for every real x. Since $\{a_i\}_1^l$ are arbitrary this implies by the Cramér-Wold Theorem that the distribution of $(Y_n(t_1), \ldots, Y_n(t_l))$ considered as a random vector on $(\Omega, \mathscr{B}, P(\cdot \mid B))$ converges to the distribution of a Wiener process taken at the same time points. As also $\{t_i\}_1^l$ are arbitrary it follows that the finite-dimensional distributions converge, which concludes the proof of the theorem. □

Mixing in the sense of Rényi entails that the realizations must fluctuate strongly; in fact by Lemma 4 below they are dense in the support of the limiting distribution. (If A and B are subsets of a topological space we say that A is dense in B if $B \subset \subset Cl(A)$.) We apply this to the martingale considered above.

THEOREM 3. If $\{Y_n\}_1^\infty$ satisfies the conditions of Theorem 2 and N' is an infinite set of integers, then for ω outside a null set the set of functions $\{Y_n(\cdot, \omega);\ n \in N'\}$ is dense in C_0, the set of continuous functions on $[0, 1]$ that vanish at zero.

Proof. Since C_0 is the support of Wiener measure (see e.g. [5]) the theorem follows immediately from Theorem 2 and Lemma 4 below. (Note that a subsequence of a mixing sequence also is mixing with the same limiting distribution.) □

LEMMA 4. Let the random elements $\{Z_n\}_1^\infty$ on $(\Omega, \mathscr{B}, P)$ take values in the metric space S with its Borel σ-algebra $\mathscr{S}$. If $\{Z_n\}_1^\infty$ is mixing with limiting distribution G and the support of G is separable, then $\{Z_n(\omega);\ n \geqq 1\}$ is dense in the support of G, for ω outside a null set.

Proof. Use the notation $B(x, \varepsilon)$ for the open sphere in S with center x and radius ε. The support of G (notation: supp G) is the set $\{s \in S;\ G(B(s, \varepsilon)) > 0,\ \forall \varepsilon > 0\}$. Let $\{r_i\}_1^\infty$ be dense in supp G and such that $r_i \in$ supp G for $i \geqq 1$. Fix i and take a sequence of positive numbers $\varepsilon_k \downarrow 0$ when $k \to \infty$ such that $B(r_i, \varepsilon_k)$ is a continuity set of G for every k. Put for k fixed

$$I_k = \{\omega \in \Omega;\ Z_n(\omega) \in B^*(r_i, \varepsilon_k) \text{ for all } n \geqq 1\}\,.$$

Then $P(I_k) = 0$, since if not, we have by the definition of mixing that

$$P(\{Z_n \in B(r_i, \varepsilon_k)\} \mid I_k) \to G(B(r_i, \varepsilon_k))$$

as $n \to \infty$, and further $G(B(r_i, \varepsilon_k)) > 0$ since $r_i \in$ supp G. Now this is a contradiction as $P(\{Z_n \in B(r_i, \varepsilon_k)\} \cap I_k) = 0$ for $n \geqq 1$ by definition. Thus for ω outside the null set $\bigcup_{k=1}^\infty I_k$ we have that r_i belongs to the closure of $\{Z_n(\omega);\ n \geqq 1\}$, and for ω outside the null set N consisting of the countable union of the null sets thus obtained for each i it holds that each of the r_i's belongs to the closure of $\{Z_n(\omega);\ n \geqq 1\}$. Since $\{r_i\}_1^\infty$ is dense in supp G this proves that for every $\omega \notin N$ each point of supp G belongs to the closure of $\{Z_n(\omega);\ n \geqq 1\}$. □

REMARK. Let x be a fixed real number and put $f(t) = x\sqrt{t}$ for $t \in [0, 1]$. By applying Theorem 3 to the subsequences $\{Y_k, Y_{k+1}, \ldots\}$ for $k \geqq 1$ it follows that for any $\varepsilon > 0$ we have $d_0(Y_n, f) < \varepsilon$ infinitely often a.s. Assume for simplicity that $s_n = \sqrt{n}$. We then get that for arbitrary $\varepsilon, \varepsilon' > 0$

$$\frac{1}{n}\sum_{i=1}^{n} I_{(x-\varepsilon,\, x+\varepsilon)}(S_i/\sqrt{i}) > 1 - \varepsilon'$$

infinitely often a.s., and it follows that also the events

$$\frac{1}{n}\sum_{i=1}^{n} I_{(x-\varepsilon,\, x+\varepsilon)}(S_i/\sqrt{i}) < \varepsilon'$$

occur infinitely often a.s. Thus if one computes the empirical distribution function of $S_1/\sqrt{1}, \dots S_n/\sqrt{n}$, then the probability it will assign to any interval will fluctuate between zero and one almost surely. (This remark is due to I. Berkes.)

We will now use Rényi-mixing for a different purpose, namely to prove a random-indices limit theorem for martingales. Assume that s_n^2 does not grow too irregularly, so that

$$(3) \qquad S(1+\delta) = \limsup_{n\to\infty} \left(s^2_{[n(1+\delta)]}/s_n^2\right) < \infty \quad \text{for every} \quad \delta \geqq 0$$

and

$$\lim_{\delta \downarrow 0} S(1+\delta) = 1\,,$$

where $[x]$ denotes the integer part of x. Let $\{\nu_n\}_{n=1}^\infty$ be a sequence of random indices, i.e. positive integer-valued random variables, which may very well depend on $\{Y_n\}_1^\infty$. We will use a similar approach as in [3] to prove the theorem below. (See also [12].)

THEOREM 5. If $\{Y_n\}_1^\infty$ satisfies the hypothesis of Theorem 2 and the condition (3) is fulfilled and furthermore the random indices $\{\nu_n\}_{n=1}^\infty$ satisfy the condition

$$(4) \qquad \frac{\nu_n}{n} \xrightarrow{P} \nu \quad \text{where the random variable } \nu \text{ is strictly positive}\,,$$

then $\{Y_{\nu_n}\}_{n=1}^\infty$ is mixing with limiting distribution W. In particular $Y_{\nu_n} \xrightarrow{d} W$.

In the proof we are going to use the following two lemmas. Lemma 6 takes the place of the Lemma in [3], while Lemma 7 is a slight modification of a well-known result of Guiasu.

LEMMA 6. If $\{Y_n\}_1^\infty$ satisfies the hypothesis of Theorem 2 and the condition (3) is fulfilled and furthermore $\nu_n/n \xrightarrow{P} \nu > 0$ as $n \to \infty$, then

(5) for every $\varepsilon > 0$ and $\eta > 0$ there is a δ, $0 < \delta < 1$ and an integer n_0 such that for $n > n_0$ it holds that

$$P(\sup\{\min(|Y_{\nu_n}(t_1) - Y_{\nu_n}(t)|, |Y_{\nu_n}(t_2) - Y_{\nu_n}(t)|)\,; \quad 0 \leqq t_1 < t < t_2 \leqq 1\,,$$
$$t_2 - t_1 < \delta\} > \varepsilon) < \eta\,.$$

Proof. Set $Y(t) = S_k$ for $s_k^2 \leqq t < s_{k+1}^2$ and furthermore let $w_n(\delta) = \sup\{\min(|Y(t_1) - Y(t)|, |Y(t_2) - Y(t)|);\ 0 \leqq t_1 < t < t_2 \leqq s_n^2,\ t_2 - t_1 < s_n^2\delta\}$. Then $Y_n(t) = s_n^{-1}\, Y(s_n^2 t)$ and further the probability in (5) equals $P(w_{\nu_n}(\delta) > s_{\nu_n}\varepsilon)$. Since $\nu_n/n \xrightarrow{P} \nu > 0$ we can choose $1 < a < \infty$ so that $P(n/a < \nu_n < na - 1) > 1 - \eta/2$ for all n greater than some integer n_1. Moreover s_n^2 and thus also $w_n(\delta)$ is increasing in n. Putting this together we see that the probability in (5) is bounded from above by

$$(6) \qquad P(w_{[na]}(\delta) > s_{[n/a]}\varepsilon) + \eta/2\,.$$

By (3) we can find an integer n_2 and a constant $c > 0$ such that $s_{[n/a]} > cs_{[na]}$ for $n > n_2$ and hence (6) is estimated by

$$P(w_{[na]}(\delta) > s_{[na]}c\varepsilon) + \eta/2\,, \tag{7}$$

for $n > \max(n_1, n_2)$. Now $Y_n \xrightarrow{d} W$ (Theorem 2) and hence $\{Y_{[na]}\}_{n=1}^{\infty}$ is tight. By Theorem 15.3 of [1] this implies that there exist an n_0 (we may choose $n_0 > \max(n_1, n_2)$) and a $\delta > 0$ such that

$$P(w_{[na]}(\delta) > s_{[na]}c\varepsilon) < \eta/2\,,$$

for $n > n_0$, and inserting this into (7) completes the proof. □

LEMMA 7. If $\{v_n\}_{n=1}^{\infty}$ satisfies (4) and the sequence of random variables $\{Z_n\}_1^{\infty}$ is mixing with limiting distribution G and furthermore

(8) for every $\varepsilon > 0$, $\eta > 0$ there is a positive number $\delta = \delta(\varepsilon, \eta)$ so that for every event A with $P(A) > 0$ which belongs to the σ-algebra generated by $Z_1, Z_2, \ldots$ there is an integer $n_0 = n_0(\varepsilon, \eta, A)$ such that for $n > n_0$ it holds that

$$P(\{\max_{n<k<n(1+\delta)} |Z_k - Z_n| > \varepsilon\} \mid A) \leqq \eta\,,$$

then $\{Z_{v_n}\}_{n=1}^{\infty}$ is mixing with limiting distribution G. (Note that δ is not allowed to depend on A.)

Proof. Let $\mathscr{C}$ be the σ-algebra generated by $Z_1, Z_2, \ldots$ We first prove that if for $A \in \mathscr{C}$ with $P(A) > 0$ we have

$$\limsup_{n\to\infty} P(\{\max_{n<k<n(1+\delta)} |Z_k - Z_n| > \varepsilon\} \mid A) \leqq \eta\,, \tag{9}$$

then the same inequality holds for arbitrary events A. Let H_n be the indicator function of $\{\max_{n<k<n(1+\delta)} |Z_k - Z_n| > \varepsilon\}$. Then (9) can be put in the following form: if ξ is the indicator function of an event in $\mathscr{C}$ then

$$\limsup_{n\to\infty} E(H_n\xi) \leqq \eta\, E(\xi)\,. \tag{10}$$

By approximation with step functions in the usual way it follows that (10) holds also when ξ is an arbitrary bounded and positive function which is measurable with respect to $\mathscr{C}$. Now let ξ_A be the indicator function of an arbitrary event A. Then

$$\limsup_{n\to\infty} E(H_n\xi_A) = \limsup_{n\to\infty} E(H_n\, E(\xi_A \parallel \mathscr{C})) \leqq \eta\, E(E(\xi_A \parallel \mathscr{C})) = \eta\, P(A)$$

since $E(\xi_A \parallel \mathscr{C})$ is a $\mathscr{C}$-measurable, bounded, and positive. Dividing through with $P(A)$ shows that (9) holds for arbitrary events which have positive probability.

It then follows from Theorem 3 of [6] that $Z_{\nu_n} \xrightarrow{d} G$. Moreover we have that $Z_{\nu_n} \xrightarrow{d} G$ on $(\Omega, \mathscr{B}, P(\cdot \mid B))$ too if $P(B) > 0$, since the conditions of the cited theorem are satisfied also when we use $P(\cdot \mid B)$. Thus we have proved that $\{Z_{\nu_n}\}_{n=1}^{\infty}$ is mixing with limiting distribution G. □

Proof of Theorem 5. By Lemma 6 above (5) holds, and it is easy to see that this implies that (5) holds also when we use the probability $P(\cdot \mid B)$ if $P(B) > 0$. Thus, to prove mixing, it is by Theorem 15.4 of [1] enough to prove that the finite-dimensional distributions converge when using $P(\cdot \mid B)$. Let $\{a_i\}_1^l, \{t_i\}_1^l, \{Z_n\}_1^{\infty}$, and v be as in the proof of Theorem 2. If we prove that

(11) $\{Z_{\nu_n}\}_{n=1}^{\infty}$ is mixing with limiting distribution $\Phi(\cdot/v^{1/2})$

then, as in the proof of Theorem 2, it will follow by the Cramér-Wold Theorem that the finite-dimensional distributions converge when using $P(\cdot \mid B)$, for every B with $P(B) > 0$.

We use Lemma 7 to prove that (11) holds. The Z_n's and the ν_n's satisfy the first two conditions of the Lemma by hypothesis and by Theorem 2 and hence it only remains to prove (8).

Assume that the event A belongs to the σ-algebra generated by $S_1, \ldots, S_{n_0}$ for some integer n_0, and write P_A for $P(\cdot \mid A)$ and E_A for expectations taken with respect to P_A. Now, putting $a_{l+1} = 0$,

$$P_A\Big(\max_{n<k<n(1+\delta)} |Z_k - Z_n| > \varepsilon\Big) = \tag{12}$$

$$= P_A\Big(\max_{n<k<n(1+\delta)} \Big|\sum_{i=1}^{l} a_i(Y_k(t_i) - Y_k(t_{i-1})) - \sum_{i=1}^{l} a_i(Y_n(t_i) - Y_n(t_{i-1}))\Big| > \varepsilon\Big) \leqq$$

$$\leqq \sum_{i=1}^{l} P_A\Big(\max_{n<k<n(1+\delta)} |Y_k(t_i) - Y_n(t_i)| > \varepsilon(l|a_i - a_{i+1}|)^{-1}\Big),$$

so it suffices to prove that each of the terms in the sum can be made smaller than η/l by choosing n large enough. This however follows if we show that (8) holds for $Z_n = Y_n(t)$, for fixed t $(0 < t \leqq 1)$. As in the proof of Lemma 6 set $Y(t) = S_k$ for $s_k^2 \leqq t < s_{k+1}^2$. Then

$$|Y_k(t) - Y_n(t)| = |s_k^{-1} Y(s_k^2 t) - s_n^{-1} Y(s_n^2 t)| \leqq$$
$$\leqq s_k^{-1}|Y(s_k^2 t) - Y(s_n^2 t)| + (s_n^{-1} - s_k^{-1})\,|Y(s_n^2 t)|$$

and further if $n < k < n(1 + \delta)$ then $s_n \leqq s_k \leqq s_{[n(1+\delta)]}$ so that

$$P_A\Big(\max_{n<k<n(1+\delta)} |Y_k(t) - Y_n(t)| > \varepsilon\Big) \leqq \tag{13}$$

$$\leqq P_A\big(s_n^{-1}(s_{[n(1+\delta)]} - s_n)\,|Y_n(t)| > \varepsilon/2\big) +$$

$$+ P_A\Big(\max_{n<k<n(1+\delta)} |Y(s_k^2 t) - Y(s_n^2 t)| > s_n\varepsilon/2\Big) = P_1 + P_2\,, \quad \text{say}.$$

Here P_1 is bounded by

$$P_A(|Y_n(t)| > \varepsilon/(2 \limsup_{k\to\infty} ((s_{[k(1+\delta)]} - s_k)/s_k))) + o(1) \to \tag{14}$$

$$\to 2(1 - \Phi(\varepsilon/(2\sqrt{t} \limsup_{k\to\infty} ((s_{[k(1+\delta)]} - s_k)/s_k))))$$

when $n \to \infty$ since $\{Y_n(t)\}$ is mixing with limiting distribution $\Phi(\cdot/\sqrt{t})$. Furthermore, for n such that $s_n^2 t > s_{n_0}^2$ the quantity $Y(s_k^2 t) - Y(s_n^2 t)$ appearing in P_2 is the sum of martingale differences X_i with index $i > n_0$, and for all such i it holds that $E_A(X_i \parallel X_{i-1}, \ldots, X_1) = 0$. Hence we can apply Kolmogorov's inequality for martingales ([4], p. 105) on the probability space $(\Omega, \mathscr{B}, P_A)$ to get that if $s_n^2 t > s_{n_0}^2$ then

$$P_2 \leqq (s_n \varepsilon/2)^{-2} E_A(\sum_k \{X_k^2;\, s_n^2 t < s_k^2 \leqq s_{[n(1+\delta)]}^2 t\}) \,. \tag{15}$$

Furthermore, in the same way as in the proof of Lemma 1 of [2], it follows from (2) that $E|s_n^{-2} \sum_{k=1}^{n} X_k^2 - 1| \to 0$ when $n \to \infty$ and thus also that $E_A|s_n^{-2} \sum_{k=1}^{n} X_k^2 - 1| \to 0$. An application of this to (15) shows that

$$P_2 \leqq (2/\varepsilon)^2 \limsup_{n\to\infty} (s_{[n(1+\delta)]}^2 - s_n^2)/s_n^2) + o(1). \tag{16}$$

It is now by (3) possible to choose $\delta > 0$ small enough to make

$$(2/\varepsilon)^2 \limsup_{n\to\infty} ((s_{[n(1+\delta)]}^2 - s_n^2)/s_n^2) \leqq \eta/2$$

and

$$2(1 - \Phi(\varepsilon/(2\sqrt{t} \limsup_{n\to\infty} ((s_{[n(1+\delta)]} - s_n)/s_n)))) \leqq \eta/2$$

and by (13), (14), and (16) this proves that (8) holds if the event A belongs to the σ-algebra generated by $S_1, \ldots, S_{n_0}$ for n_0 finite. Now any event A belonging to the σ-algebra generated by $S_1, S_2, \ldots$ can be arbitrarily well approximated by an event belonging to the σ-algebra generated by $S_1, \ldots, S_{n_0}$ if n_0 is taken large enough, and it follows that (8) holds also for arbitrary events A belonging to the σ-algebra generated by $S_1, S_2, \ldots$ This proves that the conditions of Lemma 7 are satisfied, and thus that (11) holds, which concludes the proof of the theorem.

REFERENCES

[1] P. Billingsley: Convergence of probability measures. Wiley, New York 1968.

[2] B. M. Brown: Martingale central limit theorems. Ann. Math. Statist. *42* (1971), 1, 59—66.

[3] M. Csörgö & S. Csörgö: On weak convergence of randomly selected sums. Acta Sci. Math. Hungar. *34* (1973), 53—60.

[4] J. L. Doob: Stochastic processes. Wiley, New York 1953.

[5] D. Freedman: Brownian motion and diffusion. Holden-Day, San Fransisco 1970.

[6] S. Guiasu: On the asymptotic distribution of sequences of random variables with random indices. Ann. Math. Statist. *42* (1971), 6, 2018—2028.

[7] B. L. S. Prakasa Rao: Random central limit theorems for martingales. Acta Math. Acad. Sci. Hungar. *20* (1969), 1—2, 217—222.

[8] A. Rényi: On mixing sequences of sets. Acta Math. Acad. Sci. Hungar. *9* (1958), 1—2, 215—228.

[9] A. Rényi & P. Révész: On mixing sequences of random variables. Acta Math. Acad. Sci. Hungar. *9* (1958), 3—4, 389—393.

[10] H. Rootzén: Fluctuations of sequences which converge in distribution. Techn. Report 1974: 1 Dept. of Math. Statist. Univ. of Lund, Sweden.

[11] D. J. Scott: Central limit theorems for martingales and proceses with stationary increments using a Skorokhod representation approach. Adv. Appl. Probability *5* (1973), 1, 119—137.

[12] D. Szász: Limit theorems of randomly indiced sequences of random variables. Preprint 58/1973 of the Mathematical institute of the Hungarian Academy of Sciences.

University of Lund

Department of Mathematical Statistics

ON THE MOMENTS OF A CERTAIN CLASS OF RANDOM PERMANENTS

WOLFGANG SENDLER

DORTMUND

1. INTRODUCTION

Let $A = ((a_{ij}))$, $1 \leqq i \leqq n$, $1 \leqq j \leqq n$ be an $n \times n$ random matrix. In connection with certain statistical questions one is interested to have an idea of the distribution of the Permanent Per A, i.e. the quantity defined by

$$\text{Per } A := \sum_{p \in \mathscr{S}_n} a_{1p(1)} \cdot a_{2p(2)} \cdots a_{np(n)},$$

the summation to be taken over the set $\mathscr{S}_n$ of all permutations of the numbers 1, 2, ..., n. A possible approach to this problem is to investigate all moments of the distribution of Per A. While the exact determination of these moments is only possible in special cases (cf. [1]), it seems reasonable to try to get results on the asymptotic behaviour of these moments for certain general classes of random matrices.

In this paper we consider a special class of matrices whose entries are independently identically distributed.

2. ASYMPTOTIC EXPRESSIONS FOR THE MOMENTS OF THE PERMANENT

For each integer $n \geqq 1$, let $A^{(n)} := ((a_{ij}^{(n)}))$, $1 \leqq i \leqq n$, $1 \leqq j \leqq n$, be an $n \times n$ matrix where the $a_{ij}^{(n)}$ are independently identically distributed according to the distribution of a certain real valued random variable a which does not depend on n. Moreover, we suppose

(1) $$\mathrm{P}(a \geqq 0) = 1 \quad \text{and} \quad \mathrm{E}(a) > 0\,.$$

We introduce the following notation:

$$\mu_j := \mathrm{E}(a^j) = \mathrm{E}(|a|^j)$$

(the last equality due to (1)),

$$\alpha_j := \frac{\mu_j}{\mu_1^j}, \quad j \geqq 1 .$$

The k-th moment of Per $A^{(n)}$ is the expectation of

$$(*) \qquad \Big(\sum_{p \in \mathscr{S}_n} a_{1p(1)}^{(n)} \cdot a_{2p(2)}^{(n)} \cdots a_{np(n)}^{(n)}\Big)^k = \sum \big(a_{1p_1(1)}^{(n)} \cdots a_{np_1(n)}^{(n)}\big) \cdots \big(a_{1p_k(1)}^{(n)} \cdots a_{np_k(n)}^{(n)}\big)$$

(the summation on the right side is to be taken over all k-tuples of permutations of the numbers $1, \ldots, n$).

We intend to compute $M_k^{(n)} := \mathrm{E}[(\text{Per } A^{(n)})^k]$ as a function of $\mu_1, \ldots, \mu_k$; thus from our independence assumption it is clear that we have to know how many equal factors each summand on the right side of $(*)$ contains.

Let h be an integer with $1 \leqq h \leqq k$. For one arbitrary summand of $(*)$, let us consider the h factors $a_{i_1 p_{j_1}(i_1)}, \ldots, a_{i_h p_{j_h}(i_h)}$ (we drop the upper index n for sake of simplicity when no confusion is to be expected). These h factors are identical iff

$$(2) \qquad i_1 = \ldots = i_h, \quad p_{j_1}(i_1) = \ldots = p_{j_h}(i_h) .$$

Therefore, writing the permutations $p_1, \ldots, p_k$ under each other

$$(3) \qquad \begin{matrix} p_1(1), p_1(2), \ldots, p_1(n) \\ p_2(1), p_2(2), \ldots, p_2(n) \\ \vdots \\ p_k(1), p_k(2), \ldots, p_k(n) \end{matrix}$$

(2) is fulfilled iff the i_1-th column of the matrix (3) contains at least h equal numbers. The problem to solve is thus of primarily combinatorial nature. To give a precise description of this situation, we introduce some definitions (cf. [2]).

Definition 1. Let j be fixed, $1 \leqq j \leqq n$; a number ϱ, $1 \leqq \varrho \leqq n$, is called fixpoint of order σ in the j-th column if ϱ occurs exactly σ times in the j-th column of (3). If no confusion is to be expected, we simply speak of a σ-fixpoint.

Let $h_\sigma^{(j)}$ be the number of (different) σ-fixpoints in the j-th column (we again drop the n-dependence). Clearly, we have $\sum_{\sigma=1}^{k} \sigma \,.\, h_\sigma^{(j)} = k$ for each j, whence, denoting the number of all σ-fixpoints occurring in (3) by h_σ, we have

$$\sum_{\sigma=1}^{k} \sigma h_\sigma = \sum_{j=1}^{n} \sum_{\sigma=1}^{k} \sigma h_\sigma^{(j)} = n \,.\, k \,.$$

Thus, to each batch of the form (3), and therefore to each summand of the right side of (2), there corresponds a certain fixpointvector $(h_1, \ldots, h_k)$. Denoting by $g(n, k; h_1, \ldots, h_k)$ the number of all batches of the form (3) with fixpointvector $(h_1, \ldots, h_k)$, one concludes

$$M_k^{(n)} = \sum_{(h_1, \ldots, h_k)} g(h, k; h_1, \ldots, h_k) \,.\, \mu_1^{h_1} \ldots \mu_k^{h_k}, \tag{4}$$

the sum to be taken over all possible fixpointvectors, i.e. over all possible k-tuples $(h_1, \ldots, h_k)$ of nonnegative integers with $\sum_{\sigma=1}^{k} \sigma h_\sigma = nk$. We call $\mu_1^{h_1} \ldots \mu_k^{h_k}$ the weight of $g(n, k; h_1, \ldots, h_k)$ (cf. [1]).

Definition 2. Let r be fixed, $0 \leqq r \leqq n$; a batch of form (3) is called r-essential, if $h_\sigma = 0$ for $\sigma > 2$ and if it has exactly r columns, each of which contains exactly one fixpoint of order 2. We call a batch essential, if it is r-essential for some r.

Roughly speaking, an r-essential batch is "nearly" a Latin rectangle, except of r columns, each of which contains a certain number twice, all other numbers being different from each other. Denoting by $f(n, k; r)$ the number of r-essential batches, we have

Lemma 1.

$$\frac{f(n, k; r)}{n!^k} \to \frac{1}{r!} \mathrm{e}^{-\binom{k}{2}} \,.\, \binom{k}{2}^r \quad \text{for} \quad n \to \infty$$

(r, k fixed).

For the Proof see [2], Theorem 3 and the remarks following Theorem 4.

The basic idea is now to show, that only summands of (3) corresponding to an essential batch give an asymptotically relevant contribution to $M_k^{(n)}$. For this purpose we turn from our original matrix $A^{(n)}$ to a normalized matrix $\bar{A}^{(n)} = ((\bar{a}_{ij}^{(n)}))$ whose entries $\bar{a}_{ij}^{(n)}$ are independently identically distributed according to the random variable $\bar{a}^{(n)} := (a/\mu_1)\, n!^{-1/n}$. We again define $\mathrm{E}(|\bar{a}^{(n)}|^j) = \mathrm{E}((\bar{a}^{(n)})^j) = \bar{\mu}_j$, whence $\bar{\mu}_j/\bar{\mu}^j = \mu_j/\mu^j = \alpha_j$. Note, that α_j is independent of n, while μ_j and μ^j depend on n, $1 \leqq j \leqq k$, this last dependence being dropped for sake of simplicity.

Next we compute the asymptotic value of the contribution of the essential batches (essential contribution).

Lemma 2. Let $k \geqq 2$; for the essential contributions $C_{\text{ess}}(k, n)$ to $\bar{M}_k^{(n)} := \mathrm{E}[(\mathrm{Per}\, \bar{A}^{(n)})^k]$ we have

$$C_{\text{ess}}(k, n) = \sum_{r=0}^{n} f(n, k; r)\, \bar{\mu}_2^r \bar{\mu}^{nk-2r} \to \mathrm{e}^{\binom{k}{2}(\alpha_2 - 1)} \quad \text{for} \quad n \to \infty\,. \tag{5}$$

Proof. Clearly, the weight of $f(n, k; r)$ is

$$\bar{\mu}_2^r \cdot \bar{\mu}_1^{r(k-2)+(n-r)k} = \bar{\mu}_2^r \bar{\mu}_1^{nk-2r} = \alpha_2^r \cdot \bar{\mu}_1^{nk} .$$

Since $\bar{\mu}_1 = n!^{-1/n}$, we have

$$C_{\text{ess}}(k, n) = \sum_{r=0}^{n} \frac{f(n, k; r)}{n!^k} \alpha_2^r . \tag{6}$$

From Lemma 1, for fixed r the summand corresponding to this r will tend to $\frac{1}{r!} e^{-\binom{k}{2}} \binom{k}{2}^r \alpha_2^r$ for $n \to \infty$. Thus (5) will be proved if we construct a sequence γ_r, $\gamma_r = 0, 1, \ldots$, independent of n with $\gamma_r > f(n, k; r)/n!^k$ and $\sum_{r=0}^{\infty} \gamma_r \alpha_2^r < \infty$. For this, we claim

$$f(n, k; r) \leqq n!^k \binom{k}{2}^r \cdot \frac{1}{r!} . \tag{7}$$

For the proof, consider for fixed $r \geqq 0$ an r-essential $k \times n$-batch of form (3) which has exactly one 2-fixpoint in each of the first r columns. We intend to give an upper bound for the number of possibilities to construct such a batch (cf. [2], Lemma 2).

Each position of the first column, occupied by a fixpoint of order 1, can be filled in at most n different ways. One of the two positions, occupied by the 2-fixpoint, can also be filled in at most n different ways; however, once this position is filled, the second position can be filled in exactly one way. Thus, the first column can be generated in at most n^{k-1} ways. Hereby, we obviously count also columns which never can be part of an essential batch, but in view of giving an upper bound this will not bother us.

Turning to the second column, we can fill $k - 1$ positions in at most $n - 1$ ways by the same argument as before and the additional condition, that each number, once used in the first column, can not occur in the same row in the second column. The remaining position, corresponding to the 2-fixpoint, can be filled in exactly one way again.

By induction and taking into account that the positions of the 2-fixpoint can be chosen in each column in at most $\binom{k}{2}$ ways, an upper bound for the number of possibilities for generating the first r columns is given by

$$n^{k-1}(n-1)^{k-1} \ldots (n-r+1)^{k-1} \binom{k}{2}^r = (n)_r^{k-1} \binom{k}{2}^r .$$

We again count too much when arguing that the remaining $n - r$ columns can be generated by permuting the $n - r$ symbols left in each row, i.e. in $(n-r)!^k$ ways.

Since the described procedure can be applied for each of the $\binom{n}{r}$ possible choices of r columns, the upper bound has the form

$$(n)_r^{k-1}\binom{k}{2}^r (n-r)!^k\binom{n}{r} = \frac{n!^{k-1}(n-r)!^k \,.\, n!}{(n-r)!^{k-1}\,(n-r)!\,r!}\binom{k}{2}^r = n!^k\binom{k}{2}^r\cdot\frac{1}{r!}$$

as to be proved.

Thus, taking $\gamma_r = \binom{k}{2}^r \frac{1}{r!}$, we have $\frac{f(n, k; r)}{n!^k} < \gamma_r$ and $\sum_{r=0}^{\infty} \gamma_r \alpha_2^r = e^{\alpha_2 \binom{k}{2}} < \infty$,

hence the lemma is proved.

The following considerations will serve to estimate the contribution of the non-essential batches. We have to give an upper bound for the number of non-essential batches. A batch is obviously non-essential iff at least one of the following statements is true:

a) there is at least one column containing at least one fixpoint of order >2;

b) there is at least one column containing at least 2 different fixpoints of order 2.

We call a batch r-non-essential, if it has exactly r columns with fixpoints of order >1; by $\varphi_a(n, k; r)$ we denote the number of r-non-essential batches fulfilling a), with $\varphi_b(n, k; r)$ those fulfilling b); moreover, we set $\varphi_a(n, k) = \sum_{r=1}^{n} \varphi_a(n, k; r)$, $\varphi_b(n, k) = \sum_{r=1}^{n} \varphi_b(n, k; r)$. Clearly, the sets of batches described by a) and b) are not disjoint.

We first consider a). To estimate $\varphi_a(n, k; r)$, we assume without loss of generality the fixpoints of order >1 to be located in the first r columns; further, we assume the first column to contain a fixpoint of order >2. By an argument completely analogous to the proof of (7) we can fill each of at most $k - 2$ positions of the first column in at most n ways; the remaining two positions corresponding to the fixpoint of order >2 can be filled in exactly one way. We can restrict ourselves to the "leastfavourable" case where the first column contains a fixpoint of order 3 (each other "fixpoint-situation" of the first column, listed under a), is obviously included in our counting procedure due to the fact, that we fill each of the $k - 1$ positions independently of each other) and the columns $2, 3, \ldots, r$ contain each a fixpoint of order 2. (For related and more detailed investigations in this direction see [2]). Thus, the first column can be generated in at most $\binom{k}{3} n^{k-2}$ ways, column 2 in at most $\binom{k}{2}$. $.(n-1)^{k-1}$ ways, and so on, column r in at most $\binom{k}{2}(n-r+1)^{k-1}$ ways. As before, the remaining $n - r$ columns can be generated in at most $(n-r)!^k$ ways.

Since we can choose a column with a fixpoint of order >2 among the first r columns in exactly r ways, and we can choose r columns in $\binom{n}{r}$ ways, we get

$$\varphi_a(n, k; r) \leqq n^{k-2}(n-1)^{k-1} \dots (n-r+1)^{k-1} (n-r)!^k \binom{n}{r} r \binom{k}{3} \binom{k}{2}^{r-1} =$$

$$= \frac{1}{n} \cdot n!^k \binom{k}{3} \binom{k}{2}^{r-1} \frac{1}{(r-1)!}$$

whence

(8) $$\frac{\varphi_a(n, k)}{n!^k} \leqq \frac{1}{n} \binom{k}{3} \sum_{r=1}^{n} \binom{k}{2}^{r-1} \cdot \frac{1}{(r-1)!} \leqq \frac{1}{n} \binom{k}{3} e^{\binom{k}{2}}.$$

Considering b), we again assume without loss of generality the first r columns to contain fixpoints of order >1; we further assume the first column to contain two different fixpoints of order $\geqq 2$. As under a) we need only considering the "least favourable" case, where the first column contains two different fixpoints of order 2, each of the remaining $r-1$ columns containing at least one fixpoint of order 2.

Arguing as under a), in column 1 we can fill up each of $k-2$ positions in at most n ways, having $\binom{k}{2}\binom{k-2}{2}$ possibilities to choose two pairs of positions for the two 2-fixpoints; the remainder of the counting procedure is the same as under a), whence

$$\varphi_b(n, k; r) \leqq n^{k-2}(n-1)^{k-1} \dots (n-r+1)^{k-1} (n-r)!^k \binom{n}{r} r \cdot \binom{k-2}{2} \binom{k}{2}^{r} =$$

$$= \frac{1}{n} n!^k \binom{k-2}{2} \binom{k}{2} \binom{k}{2}^{r-1} \frac{1}{(r-1)!}$$

and

(9) $$\frac{\varphi_b(n, k)}{n!^k} \leqq \frac{1}{n} \binom{k-2}{2} \binom{k}{2} \sum_{r=1}^{n} \binom{k}{2}^{r-1} \frac{1}{(r-1)!} \leqq \frac{1}{n} \binom{k-2}{2} \binom{k}{2} e^{\binom{k}{2}}.$$

Next we turn to the weights of the non-essential batches.

Obviously, the weight of an arbitrary batch can be written as a product of the weights corresponding to the single columns of the batch (cf. [1]). Considering an r-non-essential batch, there are exactly r columns whose weights contain other factors than μ_1, whence $h_1 \geqq (n-r)k$. From $\bar{\mu}_1 \leqq \bar{\mu}_2^{1/2} \leqq \dots \leqq \bar{\mu}_k^{1/k}$ we have $\bar{\mu}_j^{h_j} \leqq \bar{\mu}_k^{jh_j/k}$; setting $\sum_{j=2}^{k} jh_j = nk - h_1 = \varrho$, this yields

(10) $$\bar{\mu}_1^{h_1} \dots \bar{\mu}_k^{h_k} \leqq \bar{\mu}_1^{h_1} \cdot \bar{\mu}_k^{2h_2/k} \dots \bar{\mu}_k^{kh_k/k} = \bar{\mu}_1^{h_1} \bar{\mu}_k^{\sum_{j=2}^{k} jh_j/k} =$$

$$= \bar{\mu}_1^{nk-\varrho} \bar{\mu}_k^{\varrho/k} = (\bar{\mu}_1^k)^{(n-\varrho)/k} \bar{\mu}_k^{\varrho/k} = \bar{\mu}_1^{kn} \cdot \alpha_k^{\varrho/k}.$$

From $\varrho/k = n - h_1/k \leqq n - [(n-r)\,k]/k = r$ and $\alpha_k \geqq 1$ we find the right side of (10) to be $\leqq \bar{\mu}_1^{kn} \,.\, \alpha_k^r$, whence finally

$$\bar{\mu}_1^{h_1} \dots \bar{\mu}_k^{h_k} \leqq \bar{\mu}_1^{kn} \,.\, \alpha_k^r \,. \tag{11}$$

Denoting the non-essential contribution to $M_k^{(n)}$ by $R(n, k)$, we have by using eqs. (8) to (11)

$$R(n, k) \leqq n!^{-k} \sum_{r=1}^{n} [\varphi_a(n, k; r) + \varphi_b(n, k; r)] \,.\, \alpha_k^r \leqq$$

$$\leqq \frac{1}{n} \alpha_k \left[\binom{k}{3} \sum_{r=1}^{n} \binom{k}{2}^{r-1} \alpha_k^{r-1} \,.\, \frac{1}{(r-1)!} + \right.$$

$$\left. + \binom{k-2}{2} \binom{k}{2} \sum_{r=1}^{n} \binom{k}{2}^{r-1} \alpha_k^{r-1} \frac{1}{(r-1)!} \right] \leqq$$

$$\leqq \frac{1}{n} \alpha_k \left[\binom{k}{3} + \binom{k-2}{2} \binom{k}{2} \right] e^{\alpha_k \binom{k}{2}} \to 0 \quad \text{for} \quad n \to \infty \,.$$

We thus have proved

LEMMA 3.

$$\bar{M}_k^{(n)} \to e^{\binom{k}{2}(\alpha_2 - 1)} \quad \text{for} \quad n \to \infty \,.$$

Finally we will drop the initial condition $P(a \geqq 0) = 1$ and replace $E(a) > 0$ by $E(a) \neq 0$.

THEOREM 1. Let a be a real random variable; let the absolute moments μ_j of order j exist for each j and denote by ν_j the ordinary moment of the order j. Let $E(a) = \nu_1 \neq 0$, let $A^{(n)}$, $\bar{A}^{(n)}$, $\bar{\mu}_j$, α_j, $j = 1, 2, \dots$, be defined as before, denote by ν_j the ordinary moment of $\bar{a}^{(n)}$ and write $\beta_j := \bar{\nu}_j/\bar{\nu}_1^j$. Then

1. $E(a) > 0$ implies $\bar{M}_k^{(n)} \to e^{\binom{k}{2}(\beta_2 - 1)}$;
2. $E(a) < 0$ implies

a)
$$\bar{M}_k^{(n)} \to e^{\binom{k}{2}(\beta_2 - 1)} \quad \text{for even } k \,,$$

b)
$$\left.\begin{aligned} \bar{M}_k^{(2p)} &\to e^{\binom{k}{2}(\beta_2 - 1)}, \quad p \to \infty \\ \bar{M}_k^{(2p+1)} &\to -e^{\binom{k}{2}(\beta_2 - 1)}, \quad p \to \infty \end{aligned}\right\} \text{for odd } k$$

Proof. Since $|\bar{\nu}_j| \leqq \bar{\mu}_j$ and $\beta_2 = \bar{\nu}_2/\bar{\nu}_1^2 \geqq \bar{\mu}_2/\bar{\mu}_1^2 = \alpha_2 \geqq 1$ the absolute value of the non-essential contributions to $\overline{M}_k^{(n)}$ is dominated by $R(n, k)$ with α_2 replaced by β_2. The essential contribution is clearly $C_{ess}(n, k)$ with α_2 replaced by β_2, whence in case 1 the proof of Lemma 2 is applicable. In case 2 note that $2kp$ is always even, $k \,.\, (2p + 1)$ always odd, which concludes the proof.

REMARK. We emphasize that the case $E(a) = 0$ cannot be covered by the method described here. Each summand of (2) corresponding to a batch having a fixpointvector $(h_1, \ldots, h_k)$ with $h_1 \neq 0$ contributes nothing to $M_k^{(n)}$, i.e. only batches, subsumed in our previous context as "non-essential", contribute a relevant amount. Thus, one needs probably much finer estimations for the different fixpointsituations giving a relevant contribution. Moreover, another normalisation procedure is need to get an appropriate matrix $\overline{\overline{M}}_k^{(n)}$ playing the role of $\overline{M}_k^{(n)}$.

A similar argument shows that this method cannot be used to give asymptotic results for the moments of the random determinant of $A^{(n)}$.

REFERENCES

[1] A. PRÉKOPA: On Random Determinants I. Studia Sci. Math. Hungarica *2* (1967), 125—132.

[2] R. HAFNER, W. SENDLER: Some Asymptotic Results on Multiple Matching. J. Appl. Prob. *11* (1974), 479—492.

UNIVERSITÄT DORTMUND

ABTEILUNG STATISTIK

ESTIMATION OF THE PARAMETERS AND RECOGNITION OF STOCHASTIC PROCESSES

EDWARD SHPILEWSKI

VILNIUS

1. INTRODUCTION

While processing the information there arises the necessity to solve such important problems as – recognition, detection, estimation of the parameters, determination of the momenta of time changes in the properties of stochastic signals. When solving these problems the correlation methods for Gaussian models have been considered. A new approach based on Markov models of stochastic processes has been developed [1]. On the basis of the above model the solution of such problems as filtering, extrapolation, interpolation has been obtained [1], [2]. In the present communication the problem of estimation of the parameters and that of recognition of multi-dimensional stochastic processes under noise conditions is being considered on the basis of the theory of conditional Markov processes [3].

2. STATEMENT OF PROBLEM

Let $H = \{h_i : i = 1, 2, \ldots, N\}$ be a set N out of the possible random processes $X(t)$. The k-dimensional process $X(t)$ is observed under noise conditions. According to the observable l-dimensional signal

$$y_0^t = \{y_\tau, 0 \leqq \tau \leqq t\} \tag{1}$$

the process $X(t)$ is to be recognized under the condition of the minimum of the average risk with a given loss function. For the minimization of the average risk the minimization of the a-posteriori risk is sufficient. Optimum decision procedure with the loss function l_{ij} consists of decision making that we get the process h_k if

$$\sum_{i=1}^{N} l_{ik} \mathbf{P}(h_i | y_0^t) \leqq \sum_{i=1}^{N} l_{ij} \mathbf{P}(h_i | y_0^t), \quad j = 1, 2, \ldots, N. \tag{2}$$

Under this criterion the Bayes procedure is reduced to the determination of the a-posteriori probability $\mathbf{P}(h_i|y_0^t)$, or the ratio of the a-posteriori probability $\mu_{ij}(t)$, or the likelihood ratio $\lambda_{ij}(t)$. In particular using the criteria of the minimum of the probability of misclassification the solution of the problem follows from the expression

$$\arg\max_i \mathbf{P}(h_i|y_0^t) = \arg\max_i F_i[y_0^t]\,. \tag{3}$$

If the a-priori probability p_i of the process h_i is not known, the problem of recognition is solved by using either the likelihood ratio or the criterion of maximum likelihood. In case the process is determined by the parameters, the problem is reduced to that of estimation of the parameters according to the signal observed.

3. RECOGNITION OF CONTINUOUS STOCHASTIC PROCESSES

Let the process h_i be given to be defined by the system of stochastic differential equations

$$\mathrm{d}x_t = a_0^i(y_t, t)\,\mathrm{d}t + a_1^i(y_t, t)\,x_t\,\mathrm{d}t + b^i(y_t, t)\,\mathrm{d}w_1(t) \tag{4}$$

an the signal observed be linked with the process x_t by means of stochastic equations

$$\mathrm{d}y_t = A_0(y_t, t)\,\mathrm{d}t + A_1(y_t, t)\,x_t\,\mathrm{d}t + B(y_t, t)\,\mathrm{d}w_2(t)\,, \tag{5}$$

where $w_1(t)$, $w_2(t)$ are independent Wiener processes, a_0, a_1, A_0, A_1, b, B are matrices of the $[k \times 1]$, $[k \times k]$, $[l \times 1]$, $[l \times k]$, $[k \times k]$, $[l \times l]$ order. The solution of the recognition problem considered may be obtained on the basis of the following theorem.

THEOREM. *The a-posteriori probability $\mathbf{P}(h_i \mid y_0^t)$ under the observation of a multi-dimensional process* (5) *satisfies such a stochastic differential equation*

$$\mathrm{d}\mathbf{P}(h_i \mid y_0^t) = \tag{6}$$
$$= \mathbf{P}(h_i \mid y_0^t)\,[(BB')^{-1}\,A_1(m_i(t) - m(t))]\,[\mathrm{d}y_t - A_0\,\mathrm{d}t - A_1\,m(t)\,\mathrm{d}t]\,,$$

with initial conditions $\mathbf{P}(h_i) = p_i$, where

$$m(t) = \sum_{i=1}^{N} m_i(t)\,\mathbf{P}(h_i \mid y_0^t)$$

and $m_i(t)$ satisfies the filtering equation

(7) $$dm_i(t) = [a_0^i(y_t, t) + a_1^i(y_t, t)\, m_i(t)]\, dt + \\ + \Gamma_i(t)\, A_1'(BB')^{-1}\, [dy_t - A_0\, dt - A_1 m_i(t)\, dt]\,,$$

$$d\Gamma_i(t) = [a_1^i \Gamma_i(t) + \Gamma_i(t)\,(a_1^i)' - \Gamma_i(t)\, A_1'(BB')^{-1}\, A_1\, \Gamma_i(t) + b_i b_i']\, dt\,.$$

The proof of the theorem may be obtained by means of the theory of conditional Markov processes [2].

COROLLARY 1. *The ratio of a-posteriori probabilities is equal to*

(8) $$\mu_{ij}(t) = \frac{p_i}{p_j} \exp\left\{\int_0^t [(BB')^{-1}\, A_1\, m_i(t)]'\, [dy_t - A_0\, dt - \tfrac{1}{2}A_1\, m_i(t)\, dt] - \\ - \int_0^t [(BB')^{-1}\, A_1\, m_j(t)]'\, [dy_t - A_0\, dt - \tfrac{1}{2}A_1\, m_j(t)\, dt]\right\}.$$

Proof. Using K. Ito's formula for complicated functions we obtain

(9) $$d \ln \mathbf{P}(h_i \mid y_0^t) = [(BB')^{-1}\, A_1(m_i(t) - m(t))]' \times \\ \times [dy_t - A_0\, dt - \tfrac{1}{2}A_1(m_i(t) - m(t))\, dt]\,.$$

Subtracting the equation for $\ln \mathbf{P}(h_j \mid y_0^t)$ from equation for $\ln \mathbf{P}(h_i \mid y_0^t)$ we get

(10) $$d \ln \frac{\mathbf{P}(h_i \mid y_0^t)}{\mathbf{P}(h_j \mid y_0^t)} = [(BB')^{-1}\, A_1\, m_i(t)]'\, [dy_t - A_0\, dt - \tfrac{1}{2}A_1\, m_i(t)\, dt] - \\ - [(BB')^{-1}\, A_1\, m_j(t)]'\, [dy_t - A_0\, dt - \tfrac{1}{2}A_1\, m_j(t)\, dt]\,.$$

By integrating (10) with initial conditions $\mu_{ij}(0) = p_i/p_j$ we obtain that the ratio of a-posteriori probabilities is equal to (8).

REMARK. The likelihood ratio $\lambda_{ij}(t)$ will be obtained from (8) with the help of the following formulae

(11) $$\lambda_{ij}(t) = \frac{p_j}{p_i}\, \mu_{ij}(t)\,.$$

COROLLARY 2. *The decisive rule (3) minimizing the probability of misclassification lies in the decision making that the process is h_k for which the logarithm of the likelihood function*

(12) $$\Phi_i[y_0^t] = \int_0^t [(BB')^{-1}\, A_1\, m_i(t)]'\, [dy_t - A_0\, dt - \tfrac{1}{2}A_1\, m_i(t)\, dt] + \ln p_i$$

is the largest.

The proof of formulae (12) is easily obtained from (3) and (8), having in mind that the maximum of the a-posteriori probability is identical with that of the logarithm of the likelihood function.

The problem of recognition gets considerably complicated if the processes are determined by (4) and (5) with the accuracy of unknown parameters. In this case there arises the problem of estimation of the parameters u according to the realization observed. If it is known a-priori that the parameter may assume the finite set of values $u_1, u_2, \ldots, u_N$ then the problem of estimation of the parameters may be regarded as that of recognition of N hypothesis according to the signal observed. The Bayes solution may be obtained with the help of the above techniques. If the parameter is continuous in the system (5) then equation (7) can be used for its estimation.

4. CONCLUSIONS

The obtained non-linear methods of recognition are optimal in the sense of the minimization of the average risk or the criterion of maximum likelihood.

The equations and formulae have been obtained for computing an a-posteriori probability as well as for likelihood ratios and likelihood functions, convenient for realization by a computer. They may also be used for the solution of the problem of estimation of parameters, determination of momenta of time changes in the properties of a stochastic signal.

REFERENCES

[1] R. E. Kalman, R. C. Bucy: New results in linear filtering and prediction theory. Trans. ASME, J. Basic Engr. *83D* (1961), 95–107.

[2] Р. Ш. Липцер, А. Н. Ширяев: Нелинейная фильтрация диффузионных марковских процессов. Труды Матем. инст. им. В. А. Стеклова 104, Москва 1968, 135–180.

[3] R. L. Stratonovich: Conditional Markov Processes and Their Application to the Theory of Optimal Control. Modern Analytic and Computational Methods in Science and Mathematics, No 7. New York 1968.

Academy of Sciences of the Lithuanian SSR

Institute of Physics and Mathematics

DISTANCE DISTRIBUTIONS

Robin Sibson

Cambridge*

The squared Euclidean distance between two multivariate observations is a statistic of them, and its distribution is thus determined by that of the observations. If many observations are taken independently, a large number of distinct pairs are available on which interpoint distances can be calculated. It is shown that despite the non-independence of these distances, their empirical distribution function almost surely converges to the distribution of a single squared distance. This result is of interest in the theory of multidimensional scaling.

Let vectors x and y be independent observations from a fixed K-variate distribution. We can form the squared Euclidean distance

$$d^2(x, y) = \sum(x_k - y_k)^2$$

as a statistic of the pair (x, y), and this will have a distribution as an ordinary real random variable. In some cases this distribution is easily calculated; for example, if the parent distribution is unit spherical normal, then $d^2(x, y)$ is distributed as $2\chi_K^2$.

Let Δ be the distribution function of a unit point mass at the origin on the line. If $\xi_1, \ldots, \xi_N$ are real numbers, define their *empirical distribution function* (EDF) to be

$$F_N(\xi) = \frac{1}{N} \sum \Delta(\xi - \xi_n) .$$

It is a well-known corollary of the strong law of large numbers that if $\xi_1. \ldots, \xi_N$ are independent observations from a fixed distribution, F, say, then $F_N \to F$ in distribution, the convergence taking place almost surely as $N \to \infty$. Now suppose that $x_1, \ldots, x_n$ are independent observations from the fixed K-variate distribution. It is possible to construct $\frac{1}{2}n(n - 1)$ distinct unordered pairs from $x_1, \ldots, x_n$, and hence $N = \frac{1}{2}n(n - 1)$ squared distances. These squared distances can all be regarded as

* Present address: University of Bath.

observations from a single distribution, that of any one squared distance, but they are not all independent. Nevertheless, we can construct an EDF from them and ask whether it converges, as $n \to \infty$, to the DF F of a single squared distance between two independent points. The familiar result does not guarantee this, because of the lack of independence of the N squared distances.

Nevertheless the result is still true, and is proved by McGinley and Sibson [1] in a general context. This brief communication reports these results and their application to the case of distances. The application is of particular interest in the context of data analysis, particularly in the area of multidimensional scaling, where it is often of value to know that the EDF of all the interpoint distances from a configuration is likely to follow approximately a distribution which may be obtainable in advance of the extraction from the data of the detailed shape of the configuration.

The technique of proof essentially follows that of the familiar case of the EDF of independent observations. The strong law has to be established in a context of non-independence of a specific type, and consequently requires the imposition of some compensatory extra conditions for its validity; in the version given the condition is the familiar fourth-moment one encountered in the proof of the easiest version of the strong law, and since the application is to uniformly bounded random variables this is quite acceptable, although it would be of interest to investigate to what extent it can be weakened.

We begin with a counting result. An $l \times j$ block of cells we call a *mosaic*; a *colouring* of the mosaic with n colours is an assignment of one colour to each cell – not all the colours need be used, and different cells may be of the same colour, so the total number of colourings is n^{lj}. A colouring is *exochrome* if every row has the property that at least one of the colours used to colour its cells is re-used in some other row of the mosaic. We shall be concerned only with the case $l = 4$.

LEMMA. The number of exochrome colourings of a $4 \times j$ mosaic is

$$3j^4 \,.\, n^{4j-2} + O(n^{4j-3})\,.$$

Proof. The leading term is contributed by the most polychrome colourings which are exochrome; these are of the following form:

$$\begin{matrix} a & * & * & * & \dots & * \\ a & * & * & * & \dots & * \\ b & * & * & * & \dots & * \\ b & * & * & * & \dots & * \end{matrix}$$

and similar patterns obtained by permuting cells within each row and changing the pairing of the rows. Cells which are lettered are to be the same colour as one another; apart from this all cells are coloured differently. The usual combinatorial arguments give the term stated.

We now give a precise definition of the kind of weakly dependent system to which our version of the strong law applies. A family of random variables is called *off-diagonal* if it is subscripted by integer vectors of fixed length such that no two entries in any one subscript vector are equal. j will denote the length of the subscripting vector. An off-diagonal family is *dissociated* if, loosely speaking, members of it can only be dependent if their subscripts contain a common entry. A precise definition is that an off-diagonal family is dissociated if a sufficient condition for the joint distributions of two subfamilies to be independent of one another is that the sets of values occurring in subscripts of the two families be disjoint. McGinley and Sibson give the following simple example of a dissociated off-diagonal family.

Let $Y_1, \ldots, Y_n$ be independent random variables; let $X_r = \max\{Y_i : i = r_1, r_2, \ldots \ldots, r_j\}$, where $r = (r_1, r_2, \ldots, r_j)$. Then $\{X_r\}$ is dissociated if r runs through vectors whose entries are distinct within any one vector and taken from the range $1, \ldots, n$. Note that in the case of a singly-subcripted family independence and dissociatedness are the same thing, but that, as the example shows, dissociatedness is a weaker condition in general. The following version of the strong law holds for dissociated families.

THEOREM [Off-diagonal SLLN]. Let $\{X_r\}$ be an off-diagonal dissociated family of random variables such that $\mathbf{E}(X_r) = 0$ for all r and there exists $K < \infty$ such that $\mathbf{E}(X_r^4) \leqq K$ for all r. Define

$$S_n = \sum\{X_r : r_i \in 1, \ldots, n \text{ for } i = 1, \ldots, j\}\,.$$

Then $S_n/N \to 0$ almost surely as $n \to \infty$, where $N = n(n-1)\ldots(n-j+1)$.

Proof. $\mathbf{E}(S_n^4)$ consists of N^4 terms, and by dissociatedness all of these terms vanish except for those where the integers $1, \ldots, n$ form an exochrome colouring of the $4 \times j$ mosaic of subscript positions in the term (4 factors each j-fold subscripted); the nonvanishing terms are all bounded by K, and there are certainly not more of them than the total number of exochrome colourings. Thus $\mathbf{E}(S_n^4) = O(n^{4j-2})$, from the lemma. The proof is completed by way of the usual arguments based on Chebyshev's inequality and the first Borel-Cantelli lemma.

Now suppose that f is a real statistic of j observations, and suppose $\omega_1, \ldots, \omega_n$ are n independent observations from a probability space $(\Omega, \mathcal{J}, P)$. Then $N = n(n-1)\ldots(n-j+1)$ ordered j-tuples can be selected from $\omega_1, \ldots, \omega_n$ without repetition within a j-tuple. Let F_N be the EDF of the values of f on these, and let F be the DF of f.

THEOREM. $F_N \to F$ in distribution, almost surely as $n \to \infty$.

Proof. Define

$$X_r(\alpha) = 0 \quad \text{if} \quad f(\omega_{r_1}, \omega_{r_2}, \ldots, \omega_{r_j}) < \alpha\,,$$
$$= 1 \quad \text{otherwise}\,.$$

Then $E(X_r) = F(\alpha)$ and $\{X_r - F(\alpha)\}$ is an off-diagonal dissociated family of uniformly bounded random variables with zero expectation. It follows from the off-diagonal strong law of large numbers that $F_N(\alpha) \to F(\alpha)$ almost surely as $n \to \infty$. By applying this result at a countable dense set of continuity points of F, we obtain the conclusion.

We now return to the specific consideration of distances. The pairwise squared distances $d^2(x_i, x_j)$ form a dissociated family. It is in fact a symmetrical family, with $d^2(x_i, x_j) = d^2(x_j, x_i)$, but this makes no difference to the argument. Applying the theorem on convergence in distribution with $j = 2$ gives the desired result, namely:

COROLLARY. The EDF F_N of all $N = \frac{1}{2}n(n - 1)$ pairwise squared distances in a configuration of n independently chosen points from a fixed multivariate distribution almost surely converges in distribution as $n \to \infty$ to the DF F of the pairwise squared distance of two points chosen independently.

REFERENCE

[1] W. G. McGinley, R. Sibson: Dissociated random variables. Proc. Camb. Phil. Soc. (in press).

UNIVERSITY OF CAMBRIDGE

KING'S COLLEGE

MISCELLANEOUS TOPICS IN MARKOV CHAINS WITH A GENERAL STATE SPACE

ZBYNĚK ŠIDÁK

PRAGUE

0. INTRODUCTION AND SUMMARY

This paper contains some improvements and complements of the author's older results, and some corrections. The point of view is analytic (or functional-analytic) rather than probabilistic.

We have a general abstract space X of points x, and a σ-algebra $\mathscr{X}$ of sets in X. Throughout the paper we suppose that all subsets of X introduced here belong to $\mathscr{X}$, that all functions on X introduced here are $\mathscr{X}$-measurable, and similarly. Our basic notion is a *transition function* p, that is a function $p(\cdot, \cdot)$ of two variables $x \in X$, $A \in \mathscr{X}$, which is not identically zero, and which enjoys the following properties:

(i) $p(x, \cdot)$ is a σ-additive non-negative measure on $\mathscr{X}$ for each $x \in X$, and $p(x, X) \leqq 1$,

(ii) $p(\cdot, A)$ is an $\mathscr{X}$-measurable function on X for each $A \in \mathscr{X}$.

Further, as usual, define the iterates $p^{(n)}$ by

$$p^{(n)}(x, A) = \int_X p^{(n-1)}(y, A)\, p(x, \mathrm{d}y), \quad \text{with} \quad p^{(1)} = p;$$

and also $p^{(0)}(x, A) = 1$ for $x \in A$, $p^{(0)}(x, A) = 0$ for $x \notin A$.

The first part of the paper deals with irreducible chains. Some errors in the literature are corrected, and it is further shown that for every irreducible Markov chain there always exists a sub-invariant measure. It is recalled that irreducible Markov chains can be classified into three types: positive-recurrent, null-recurrent, transient. Some complements for transient chains are presented, namely, that $\sum_{n=1}^{\infty} p^{(n)}(x, A) < \infty$ need not be true for all x, and that they have always "many" sub-invariants measures.

The second part of the paper is devoted to certain functional-analytic representations of systems of arbitrary transition functions. Some auxiliary results are given on operators in the space of continuous functions, and then it is mentioned that every system of transition functions can be represented by such another system in a compact Hausdorff space, where the basic σ-algebra consists either of Baire sets or of Borel sets. Some applications are presented, namely to ergodic and spectral properties of transition operators in the space of bounded functions.

1. IRREDUCIBLE TRANSITION FUNCTIONS

Throughout this whole first part we shall assume that the transition function p is *irreducible* (cf. [8], [16]), i.e. that all of the measures $\nu_x = \sum_{n=1}^{\infty} 2^{-n} p^{(n)}(x, .)$ for all $x \in X$ have the same null sets. For simplicity, we shall denote by ν any fixed measure ν_x.

1.1. Stochasticity of transition functions

First, we present a simple auxiliary result.

THEOREM 1.1. *If*

$$\sum_{n=1}^{\infty} p^{(n)}(x, A) = \infty \tag{1.1}$$

for all x and all A such that $\nu(A) > 0$, then $p(x, X) = 1$ for ν-almost all x.

Proof is similar to that of Lemma 10 in [17]. Setting $g(x) = 1 - p(x, X)$, we have $g(x) \geqq 0$ for all x. Integrating this equality with respect to $p^{(m)}(y, .)$, $m = 1, 2, \ldots, n$, we get

$$\int_X g(x)\, p^{(m)}(y, \mathrm{d}x) = p^{(m)}(y, X) - p^{(m+1)}(y, X)\,, \quad m = 1, 2, \ldots, n\,.$$

On summing all equalities we obtain

$$\sum_{m=0}^{n} \int_X g(x)\, p^{(m)}(y, \mathrm{d}x) = 1 - p^{(n+1)}(y, X)\,.$$

If we put $N_k = \{x; g(x) \geqq 1/k\}$, where k is an arbitrary natural number, we see that

$$\sum_{m=0}^{n} \int_X g(x)\, p^{(m)}(y, \mathrm{d}x) \geqq \sum_{m=0}^{n} \int_{N_k} g(x)\, p^{(m)}(y, \mathrm{d}x) \geqq \sum_{m=0}^{n} p^{(m)}(y, N_k)/k\,,$$

and therefore

$$\sum_{m=0}^{n} p^{(m)}(y, N_k) \leqq k[1 - p^{(n+1)}(y, X)] \leqq k < \infty\,.$$

Letting $n \to \infty$, and making use of (1.1), we get obviously $\nu(N_k) = 0$ for each k, which implies $g(x) = 0$ for ν-almost all x. □

It is very easy to see that in Theorem 1.1 the assertion cannot be strengthened to $p(x, X) = 1$ for all x. E.g., we can mention the following

EXAMPLE 1.1. Consider an arbitrary irreducible (in the classical sense) recurrent Markov chain in the space $X_0 = \{0, 1, 2, \ldots\}$ with transition probabilities $p(i, j)$. Adjoin a further point Δ to X_0, and define $p(\Delta, 0) = \delta$ with $0 < \delta < 1$, $p(\Delta, j) = 0$ for $j \neq 0$, $j \in X_0$, $p(\Delta, \Delta) = 0$, $p(i, \Delta) = 0$ for each $i \in X_0$. It is obviously seen that such an extended p is irreducible in the sense of our definition (and $\nu(j) > 0$ for $j \in X_0$, $\nu(\Delta) = 0$), and it satisfies (1.1) but $p(\Delta, X_0 \cup \Delta) < 1$.

1.2. Correction

For a while, we need the Markov chain generated in the usual way by the transition function p, and we denote the random variables (in the abstract space X) of this chain by $\xi_0, \xi_1, \xi_2, \ldots$ If the initial probability distribution for ξ_0 concentrates in one point $x \in X$ then the probability for the chain $\xi_0, \xi_1, \xi_2, \ldots$ is denoted by P_x.

THEOREM 1.2. *Suppose that the transition function p is stochastic, i.e.* $p(x, X) = 1$ *for all x. Then* (1.1) *holds if and only if*

$$P_x\{\xi_n \in A \text{ for at least one } n = 1, 2, \ldots\} = 1 \tag{1.2}$$

for all x and all A such that $\nu(A) > 0$.

Proof of this theorem was given by Nelson [8], Theorem 4.1. However, Nelson's Theorem 4.1 contains an error, since it is formulated for any sub-stochastic transition function p. The falseness of such a more general theorem is seen from our Example 1.1, where (1.1) is satisfied but $P_\Delta\{\xi_n \in X_0$ for at least one $n = 1, 2, \ldots\} = \delta < 1$. The error in Nelson's proof is on p. 684, line 8 from the bottom, where $f_E(y) = 1$ a.e. is said to imply $P f_E(y) = 1$ for all y; but this is not true generally for sub-stochastic p. Thus, our Theorem 1.2 presents a corrected version of Nelson's Theorem 4.1.

Unfortunately, Nelson's erroneous Theorem 4.1 was quoted in my paper [16, p. 551], and it was used in proofs of Theorems 1, 2 and Corollary 1 in [16]. However, the damage brought into my papers by this error is quite unimportant: Theorem 1 in [16] (and Lemma 3 in [17]) is false, as we see by our present Example 1.1, but this Theorem had only an auxiliary character. On the other hand, Theorem 2 and Corollary 1 in [16] (and Theorem 1 in [15]) are correct, but their proofs require a minor complement, cf. Theorem 1.3 and Corollary of the present paper. Further, in Corollary 2 in [16] (and in Corollary 1 in [15]) the phrase "all measures μ_x, $x \in X$" in the assertion must be replaced only by "ν-almost all measures μ_x, $x \in X$".

As for Theorem 16 in [16], quoted from [8] without a proof, I am not able to see at present whether it is correct. All remaining theorems in my papers [15], [16], [17] are correct; in fact, at all further "critical" points where Nelson's erroneous theorem was used it is sufficient to use only the equality $p(x, X) = 1$ for ν-almost all x, so that it is sufficient to employ Theorem 1.1 of the present paper in place of Theorem 1 from [16].

1.3. Existence of a sub-invariant measure

A σ-additive non-negative measure μ on $\mathscr{X}$ is called *sub-invariant* for p if it is σ-finite, not identically zero, and if

$$\int_X p(x, A)\, \mu(dx) \leqq \mu(A) \quad \text{for all} \quad A \in \mathscr{X}\,. \tag{1.3}$$

If in (1.3) the equality occurs for all $A \in \mathscr{X}$ then μ is called *invariant* for p.

We shall show that for every irreducible p there exists a sub-invariant measure μ. The following results stregthen Theorem 1 in [15], Theorem 2 and Corollary 1 in [16], where we had to suppose that $\mathscr{X}$ is generated by a denumerable class of sets; now this restriction may be removed due to the results in [6]. At the same time, the present proof is a corrected version of that in [16] (cf. Section 1.2 above). The following result is essentially due to Harris [4].

THEOREM 1.3. *If the transition function p satisfies* (1.1) *there exists an invariant measure μ for p.*

Proof. If p is stochastic, by Theorem 1.2 we have (1.2). By Corollary 4.2 in [8] this entails $P_x\{\xi_n \in A$ for infinitely many $n\} = 1$ for all x and A with $\nu(A) > 0$, which is a special case of Condition C in [4]. Therefore, by [4], there is an invariant μ for p provided that $\mathscr{X}$ is generated by a denumerable class of sets. However, this last condition may be omitted as shown in [6].

If p is sub-stochastic put $N = \{x;\, p(x, X) < 1\}$. By Theorem 1.1 we have $\nu(N) = 0$ so that especially $p(x, M) = 0$ for all x and all $M \subset N$. Define a transition function p_0 in $X_0 = X - N$ by $p_0(x, A) = p(x, A)$ for $x \in X_0$, $A \subset X_0$. Since p_0 in X_0 is stochastic, by the first part of the proof there is an invariant μ_0 for p_0 in X_0. If we define now μ by $\mu(A) = \mu_0(A \cap X_0)$ for $A \subset X$, it is easily seen that μ is invariant for p. □

Further, if p does not satisfy (1.1), Theorem 3 in [16] gives the existence of a sub-invariant measure for p, so that we have the following

COROLLARY. *For any irreducible transition function p there exists a sub-invariant measure.*

1.4. Classification of transition functions

We suppose now that μ is some sub-invariant measure for p, and that μ is equivalent to all v_x, $x \in X$. It can be seen that this condition does not impose any restriction. Let us recall that the following classification of transition functions has been given in [15], [16].

THEOREM 1.4. *An irreducible transition function p belongs precisely to one of the three following types:* (i) *either*

$$\sum_{n=1}^{\infty} p^{(n)}(x, A) = \infty \quad \textit{and} \quad \lim_{n\to\infty} n^{-1} \sum_{m=1}^{n} p^{(m)}(x, A) > 0 \tag{1.4}$$

for all x and all A such that $0 < \mu(A) < \infty$, *in which case p is called positive-recurrent;* (ii) *or*

$$\sum_{n=1}^{\infty} p^{(n)}(x, A) = \infty \quad \textit{and} \quad \lim_{n\to\infty} n^{-1} \sum_{m=1}^{n} p^{(m)}(x, A) = 0 \tag{1.5}$$

for all x and all A such that $0 < \mu(A) < \infty$, *in which case p is called null-recurrent;* (iii) *or, for each set A such that* $0 < \mu(A) < \infty$,

$$\sum_{n=1}^{\infty} p^{(n)}(x, A) < \infty \tag{1.6}$$

holds for μ*-almost all x, in which case p is called transient.*

In [16, p. 561] the question has been raised whether for transient chains (1.6) holds for all x. Now we present here an example answering this question in negative.

EXAMPLE 1.2. Let X consist of all integers and of a further point Δ, and let p be defined as follows:

$$p(j, j+1) = \tfrac{1}{3}, \quad p(j, j-1) = \tfrac{2}{3} \quad \text{for} \quad j = \ldots, -1, 0, 1, \ldots,$$

$$p(\Delta, j) = 6/(\pi^2 j^2) \quad \text{for} \quad j = 1, 2, \ldots,$$

$p(i, j) = 0$ for all remaining i, j. It is then easy to see that μ given by $\mu(j) = 2^{-j}$ for $j = \ldots, -1, 0, 1, \ldots$, $\mu(\Delta) = 0$, defines a sub-invariant measure, that p is irreducible transient, and that the set $A = \{1, 2, 3, \ldots\}$ has the measure $\mu(A) = \sum_{j=1}^{\infty} 2^{-j} = 1$. However, we shall prove that

$$\sum_{n=1}^{\infty} p^{(n)}(\Delta, A) = \infty, \tag{1.7}$$

so that (1.6) does not hold for all points. Actually,

$$\sum_{n=1}^{\infty} p^{(n)}(\Delta, A) = \sum_{n=1}^{\infty} \sum_{j=1}^{\infty} p(\Delta, j) p^{(n-1)}(j, A) =$$

$$= 6\pi^{-2} \sum_{m=0}^{\infty} \sum_{j=1}^{\infty} j^{-2} \sum_{k=1}^{\infty} p^{(m)}(j, k) \geqq 6\pi^{-2} \sum_{j=1}^{\infty} j^{-2} \sum_{k=1}^{j} \sum_{m=0}^{\infty} p^{(m)}(j, k) .$$

Denoting $H_{jk} = P_j\{\xi_n = k \text{ for any } n = 0, 1, 2, \ldots\}$, Propositions 4–12 and 4–15 in § 4.6 in [7] show that

$$\sum_{m=0}^{\infty} p^{(m)}(j, k) = H_{jk} \sum_{m=0}^{\infty} p^{(m)}(k, k) .$$

Obviously, $\sum_{m=0}^{\infty} p^{(m)}(k, k) = \sum_{m=0}^{\infty} p^{(m)}(0, 0)$ for all $k = 1, 2, \ldots$; further, for $k \leqq j$ we have clearly $H_{jk} = 1$. On summarizing, we get

$$\sum_{n=1}^{\infty} p^{(n)}(\Delta, A) \geqq 6\pi^{-2} \sum_{j=1}^{\infty} j^{-2} \sum_{k=1}^{j} H_{jk} \sum_{m=0}^{\infty} p^{(m)}(k, k) =$$

$$= 6\pi^{-2} \sum_{j=1}^{\infty} j^{-2} j \sum_{m=0}^{\infty} p^{(m)}(0, 0) = 6\pi^{-2} \sum_{m=0}^{\infty} p^{(m)}(0, 0) \sum_{j=1}^{\infty} j^{-1} = \infty . \qquad \square$$

1.5. The number of sub-invariant measures

It is well known that a sub-invariant measure for a recurrent p is actually invariant and essentially unique (up to a multiplying constant). We shall now present an "opposite" result for a transient p, saying roughly that for such a p there are at least "almost as many" different sub-invariant measures as is the number of "distinguishable" points in the whole state space.

First, we define: A point $z \in X$ is said to be *distinguishable* from the point $x \in X$ if there is a set $A \in \mathscr{X}$ such that $x \in A$, $z \notin A$.

THEOREM 1.5. *Let p be a transient transition function and μ a sub-invariant measure for it equivalent to the measure ν. Then there exists a set $X_1 \in \mathscr{X}$, $\mu(X - X_1) = 0$, such that for each $x \in X_1$ the measure*

$$\mu_x = \sum_{n=0}^{\infty} p^{(n)}(x, .) \tag{1.8}$$

is sub-invariant for p, and these measures μ_x enjoy the following properties (analogous to linear independence):

(a) *If $x \in X_1$, then μ_x cannot be expressed in the form*

$$\mu_x = \sum_{k=1}^{\infty} c_k \mu_{z_k} \tag{1.9}$$

where $c_1, c_2, \ldots$ *is any sequence of positive constants, and* $z_1, z_2, \ldots$ *is any sequence of points from* X_1 *distinguishable from* x.

(b) *If* $x \in X_1$, *then* μ_x *cannot be expressed in the form*

$$\mu_x = \int_{X_1 - E} \mu_z \, \lambda(\mathrm{d}z) \tag{1.10}$$

where λ *is any measure, and* E *is any set such that* $x \in E$, $E \in \mathscr{X}$. (*The integral in* (1.10) *is understood in the sense of the integration of the function* $\mu_z(A)$ *of the argument* z, *i.e. defined by* $\mu_x(A) = \int_{X_1 - E} \mu_z(A) \, \lambda(\mathrm{d}z)$ *for* $A \in \mathscr{X}$.

Proof. By σ-finiteness of μ, there is a set $A \in \mathscr{X}$ such that $0 < \mu(A) < \infty$. By Theorem 1.4, we have (1.6) for all $x \in X_1$ with $\mu(X - X_1) = 0$. Now, for this x and A we can proceed as in the proof of Theorem 3 in [16], showing in this way the σ-finiteness of μ_x given by (1.8). Sub-invariance of μ_x is then obvious since

$$\begin{aligned} \int_X p(y, A) \, \mu_x(\mathrm{d}y) &= \sum_{n=0}^{\infty} \int_X p(y, A) \, p^{(n)}(x, \mathrm{d}y) = \\ &= \sum_{n=0}^{\infty} p^{(n+1)}(x, A) = \mu_x(A) - p^{(0)}(x, A) \leqq \mu_x(A) . \end{aligned} \tag{1.11}$$

To prove (a), denote the right-hand side of (1.9) by μ^*. Since $z_1, z_2, \ldots$ are distinguishable from x, and μ_x is σ-finite, there exists a set $A \in \mathscr{X}$ such that $x \in A$, $z_1, z_2, \ldots \notin A$, $\mu_x(A) < \infty$. Then (1.11) implies

$$\int_X p(y, A) \, \mu_x(\mathrm{d}y) < \mu_x(A) ,$$

$$\int_X p(y, A) \, \mu^*(\mathrm{d}y) = \sum_{k=1}^{\infty} c_k \int_X p(y, A) \, \mu_{z_k}(\mathrm{d}y) = \sum_{k=1}^{\infty} c_k \, \mu_{z_k}(A) = \mu^*(A)$$

so that $\mu_x = \mu^*$ cannot be true.

The proof of (b) is similar, considering now A such that $x \in A$, $A \subset E$, $\mu_x(A) < \infty$. □

2. REPRESENTATIONS IN COMPACT HAUSDORFF SPACE

If S is a compact Hausdorff space, we shall denote by $\mathscr{S}(\mathscr{S}_0)$ the σ-algebra of all its Borel sets (Baire sets, respectively), i.e. the smallest σ-algebra containing all closed subsets in S (closed G_δ-subsets in S, respectively). Further, let **rca**$(S, \mathscr{S})$ (or **rca**$(S, \mathscr{S}_0)$, respectively) be the Banach space of all regular σ-additive set functions on $\mathscr{S}(\mathscr{S}_0$, respectively) with a finite total variation which is taken to be the norm in these spaces. (Of course, in the second case the word "regular" in the definition need not be explicitely mentioned, since all Baire measures are regular.) Finally, let $\mathbf{C}(S)$ denote the Banach space of all bounded continuous real functions on S with the norm $\|f\| = \sup_{s\in S} |f(s)|$.

If X is an arbitrary abstract space with a σ-algebra $\mathscr{X}$, let $\mathbf{B}(X, \mathscr{X})$ denote the Banach space of all bounded $\mathscr{X}$-measurable real functions on X with the norm $\|f\| = \sup_{x\in X} |f(x)|$.

2.1. Some functional-analytic results

The following Theorem is well known; assertions (a), (b), (c) follow e.g. from Theorem VI.7.1 in [3], assertion (d) was proved in [14].

THEOREM 2.1. *Let S be a compact Hausdorff space, and V a linear continuous operator in $\mathbf{C}(S)$. Then there exists a function $v(.,.)$ of two variables $s \in S$, $E \in \mathscr{S}$ having the following properties:*

(a) $v(s, .) \in \mathbf{rca}(S, \mathscr{S})$ *for each* $s \in S$,

(b) $(Vf)(s) = \int_S f(s_1)\, v(s, \mathrm{d}s_1)$ *for each* $f \in \mathbf{C}(S)$, $s \in S$,

(c) $\|V\| = \sup_{s\in S} \|v(s, .)\|$,

(d) $v(., E) \in \mathbf{B}(S, \mathscr{S})$ *for each* $E \in \mathscr{S}$.

We shall need two more auxiliary results. The idea of the proof of the following Theorem 2.2 was borrowed from Rosenblatt [9].

THEOREM 2.2. *Let $v(.,.)$ be the function from Theorem 2.1. If $\mu \in \mathbf{rca}(S, \mathscr{S})$ then also*

$$\int_S v(s, .)\, \mu(\mathrm{d}s) \in \mathbf{rca}(S, \mathscr{S}) . \tag{2.1}$$

Proof. It suffices to show that the left-hand side of (2.1) is a regular set function, the remaining assertions being obvious. First, suppose that v and μ are non-negative.

Denote by λ the set function from $\mathbf{rca}(S, \mathscr{S})$ such that

$$\lambda(E) = \int_S v(s, E)\, \mu(\mathrm{d}s) \quad \text{for each} \quad E \in \mathscr{S}_0 . \tag{2.2}$$

Our aim is to prove (2.2) for each $E \in \mathscr{S}$.

Let us start with an arbitrary open $G \in \mathscr{S}$, and denote by $C[G]$ the system of all functions $f \in \mathbf{C}(S)$ for which $0 \leqq f \leqq 1$, $f(s) = 0$ for $s \in S - G$. Since λ and $v(s, .)$

are regular we have

$$(2.3)\qquad \lambda(G) = \sup_{f\in C[G]} \int_S f(s_1)\left\{\int_S v(s, ds_1)\,\mu(ds)\right\} \leqq$$

$$\leqq \int_S \left\{\sup_{f\in C[G]} \int_S f(s_1)\, v(s, ds_1)\right\} \mu(ds) = \int_S v(s, G)\,\mu(ds)\,.$$

Choose an arbitrary $\varepsilon > 0$ and some integer n such that $2^{-n+1}\,\mu(S) < \varepsilon/2$. Define the sets $A_k = \{s;\, 2^{-n}k \leqq v(s, G) < 2^{-n}(k+1)\}$, $k = 0, 1, \ldots, 2^n K(=L)$, where K is an integer, $\|V\| \leqq K$. Since $0 \leqq v(s, G) \leqq \|V\| \leqq K$, we have $\bigcup_{k=0}^{L} A_k = S$, and also

$$(2.4)\qquad \left|\int_S v(s, G)\,\mu(ds) - \sum_{k=0}^{L} 2^{-n}k\,\mu(A_k)\right| \leqq 2^{-n}\,\mu(S)\,.$$

Since μ is regular there is a closed $F_k \subset A_k$ such that

$$(2.5)\qquad \mu(A_k) - 2^{-n-1}\varepsilon K^{-2} < \mu(F_k)\,,$$

and since $v(s, .)$ is regular there is, for each $s \in F_k$, an $f_s \in C[G]$ such that $v(s, G) - 2^{-n} < \int_S f_s(s_1)\, v(s, ds_1)$. However, $\int_S f_s(s_1)\, v(., ds_1) = Vf_s \in \mathbf{C}(S)$, and therefore the set

$$B_s = \left\{s_0;\, \int_S f_s(s_1)\, v(s_0, ds_1) > 2^{-n}(k-1)\right\}$$

is an open set containing s. Thus $\bigcup_{s\in F_k} B_s \supset F_k$, but the compactness of F_k implies the existence of a finite number of points $s_1, \ldots, s_q$ such that $\bigcup_{i=1}^{q} B_{s_i} \supset F_k$. Denoting $\bar{f}_k(s) = \max_{1\leqq i\leqq q} f_{s_i}(s)$, we have $\bar{f}_k \in C[G]$, and

$$(2.6)\qquad \int_S \bar{f}_k(s_0)\, v(s, ds_0) > 2^{-n}(k-1) \quad \text{for each} \quad s \in F_k\,.$$

Put finally $\bar{f}(s) = \max_{0\leqq k\leqq L} \bar{f}_k(s)$. Again $\bar{f} \in C[G]$, and by (2.6), (2.5), (2.4) we get successively

$$\int_S \left\{\int_S \bar{f}(s_0)\, v(s, ds_0)\right\} \mu(ds) \geqq \sum_{k=0}^{L} \int_{F_k} \left\{\int_S \bar{f}(s_0)\, v(s, ds_0)\right\} \mu(ds) \geqq$$

$$\geqq \sum_{k=0}^{L} 2^{-n}(k-1)\,\mu(F_k) \geqq \sum_{k=0}^{L} 2^{-n}k\,\mu(F_k) - 2^{-n}\,\mu(S) \geqq \sum_{k=0}^{L} 2^{-n}k\,\mu(A_k) -$$

$$- \sum_{k=0}^{L} 2^{-n}k \,.\, 2^{-n-1}\varepsilon K^{-2} - 2^{-n}\,\mu(S) \geqq \int_S v(s, G)\,\mu(ds) - 2^{-n}\,\mu(S) -$$

$$- 2^{-2n-1}\varepsilon K^{-2}L^2 - 2^{-n}\,\mu(S) \geqq \int_S v(s, G)\,\mu(ds) - \varepsilon\,.$$

However, making use of (2.3), this implies

$$\int_S v(s, G)\,\mu(\mathrm{d}s) - \varepsilon \leqq \int_S \left\{\int_S \bar{f}(s_0)\, v(s, \mathrm{d}s_0)\right\} \mu(\mathrm{d}s) \leqq$$

$$\leqq \sup_{f \in C[G]} \int_S \left\{\int_S f(s_0)\, v(s, \mathrm{d}s_0)\right\} \mu(\mathrm{d}s) = \lambda(G) \leqq \int_S v(s, G)\,\mu(\mathrm{d}s)\,.$$

Since $\varepsilon > 0$ was arbitrary, we get $\lambda(G) = \int_S v(s, G)\,\mu(\mathrm{d}s)$. However, if two Borel measures coincide for all open sets G, they coincide for all sets $E \in \mathscr{S}$, which proves (2.2) for each $E \in \mathscr{S}$.

In the case when $v(s, .)$ and μ are general signed measures we decompose them into their positive and negative variations and use the preceding part of the proof. □

Now, denote by V^* the operator adjoint to V; it is defined on $\mathbf{C}^*(S)$, the space of all linear continuous functionals on $\mathbf{C}(S)$. It is well known that there is an isometric isomorphism U of $\mathbf{C}^*(S)$ onto $\mathbf{rca}(S, \mathscr{S})$, and we define $V_1^*\mu = UV^*U^{-1}\mu$ for each $\mu \in \mathbf{rca}(S, \mathscr{S})$; essentially, V_1^* may be considered as the adjoint of V.

THEOREM 2.3. *For each* $\mu \in \mathbf{rca}(S, \mathscr{S})$ *we have*

$$(V_1^*\mu)(E) = \int_S v(s, E)\,\mu(\mathrm{d}s) \quad \textit{for each} \quad E \in \mathscr{S}\,.$$

Proof follows by the present Theorem 2.2 and by Theorem 2 and its Corollary in [14].

2.2. Representations of systems of transition functions in a compact Hausdorff space

Let R be some set (of parameters) on the real line satisfying $r_1 + r_2 \in R$ whenever $r_1, r_2 \in R$. By a *Markov system* in a space X with a σ-algebra $\mathscr{X}$ we understand a system of functions $p^{(r)}(.,.)$ of variables $x \in X$, $E \in \mathscr{X}$, for $r \in R$, such that $p^{(r)}(.,.)$ is a transition function for each $r \in R$, and so called transition relations

$$p^{(r_1+r_2)}(x, E) = \int_X p^{(r_2)}(x_1, E)\, p^{(r_1)}(x, \mathrm{d}x_1)$$

hold for each $x \in X$, $E \in \mathscr{X}$, $r_1, r_2 \in R$. Now, the main result reads as follows.

THEOREM 2.4. *Let* $p^{(r)}(.,.)$, $r \in R$, *be an arbitrary Markov system in a space* X *with a* σ*-algebra* $\mathscr{X}$. *Then there exists a topological space* S *and a Markov system* $v^{(r)}(.,.)$, $r \in R$, *in the space* S *with the* σ*-algebra of Baire sets* $\mathscr{S}_0$ *(or with the* σ*-algebra of Borel sets* $\mathscr{S}$*) enjoying the following properties:*

(a) *S is a totally disconnected compact Hausdorff space,*

(b) $v^{(r)}(s, .) \in \mathbf{rca}(S, \mathscr{S}_0)$ *(or* $\in \mathbf{rca}(S, \mathscr{S})$*) for each* $s \in S$, $r \in R$,

(c) *the corresponding operators* $V^{(r)}$, $r \in R$, *defined by*

$$(V^{(r)}f)(s) = \int_S f(s_1)\, v^{(r)}(s, \mathrm{d}s_1)$$

map $f \in \mathbf{C}(S)$ *onto* $V^{(r)}f \in \mathbf{C}(S)$,

(d) *the functions* $v^{(r)}(., E)$ *are lower (upper) semi-continuous for each open (closed, respectively)* $E \in \mathscr{S}$, *and continuous for each closed-open* $E \in \mathscr{S}$,

(e) *there exist a mapping* ψ *of* X *into* S *and an isomorphic mapping* Ψ *of* $\mathscr{X}$ *onto the σ-algebra* $\mathscr{S}_2$ *of all closed-open sets in* S *(i.e.* $\Psi(E \cup F) = \Psi(E) \cup \Psi F)$, $\Psi(E \cap F) = \Psi(E) \cap \Psi(F)$, $\Psi(X - E) = S - \Psi(E)$*) such that for each* $x \in X$, $E \in \mathscr{X}$, $r \in R$ *we have*

$$p^{(r)}(x, E) = v^{(r)}(\psi(x), \Psi(E)),$$

(f) *if* $p^{(r)}$ *is stochastic then also* $v^{(r)}$ *is stochastic.*

The proof for the case of $\mathscr{S}_0$ was given in [14]. The proof for the case of $\mathscr{S}$ is almost the same except that we use the present Theorem 2.1 and in proving the transition relations the present Theorem 2.3 in place of Theorem 5 in [14]. □

The case of $\mathscr{S}$ forms an addition to the result of [14].

2.3. Applications

Now, we shall restrict ourselves to the case where $p^{(r)}$ are stochastic and $R = \{0, 1, 2, \ldots\}$, and we shall work with complex spaces $\mathbf{B}(X, \mathscr{X})$ and $\mathbf{C}(S)$ (which can be accomplished by the common simple complexification). The letter T will denote the operator in $\mathbf{B}(X, \mathscr{X})$ defined by

$$Tf = \int_X f(x)\, p(., \mathrm{d}x) \quad \text{for} \quad f \in \mathbf{B}(X, \mathscr{X}),$$

and the letter V the corresponding operator in $\mathbf{C}(S)$ defined by

$$Vf_1 = \Phi T \Phi^{-1} f_1 \quad \text{for} \quad f_1 \in \mathbf{C}(S),$$

where Φ is the isometric and isomorphic mapping of $\mathbf{B}(X, \mathscr{X})$ onto $\mathbf{C}(S)$ used in the proof of Theorem 2.4 (cf. Theorem IV.6.18 or IV.6.20 in [3] which was used in this proof).

We shall briefly, for illustration, present some applications. The theorems to follow could very probably be proved also directly but our considerations show a simple and immediate possibility of "translating" some results on Markov operators in $\mathbf{C}(S)$ into the results on general Markov transition functions.

THEOREM 2.5. *Let, for each $f \in \mathbf{B}(X, \mathscr{X})$, the set $\{T^n f;\ n = 1, 2, \ldots\}$ have a strongly compact closure in $\mathbf{B}(X, \mathscr{X})$. Then the sequence T^n, $n = 1, 2, \ldots$, converges strongly in $\mathbf{B}(X, \mathscr{X})$ if and only if 1 is the only eigenvalue of T on the unit circle.*

Proof. The assumption implies that, for each $f_1 \in \mathbf{C}(S)$, the set $\{V^n f_1;\ n = 1, 2, \ldots\} = \{\Phi T^n \Phi^{-1} f_1;\ n = 1, 2, \ldots\}$ has a strongly compact closure in $\mathbf{C}(S)$ (i.e. the semigroup $\{V^n;\ n = 1, 2, \ldots\}$ is almost periodic, cf. [10, p. 838]). By Theorem IV.6.7 in [3] the set $\{V^n f_1;\ n = 1, 2, \ldots\}$ is bounded and equicontinuous, so that V satisfies the assumption of Theorem 6 in [9]. Further, it is easily seen that T^n, $n = 1, 2, \ldots$, converge strongly in $\mathbf{B}(X, \mathscr{X})$ if and only if V^n, $n = 1, 2, \ldots$, converge strongly in $\mathbf{C}(S)$, and moreover, that 1 is the only eigenvalue of T on the unit circle if and only if the same is true for V. Theorem 2.5 now follows from Theorem 6 in [9]. □

THEOREM 2.6. *Let the resolvent $R(\lambda; T)$ of the operator T satisfies, for $\lambda > 1$, the growth condition of the first order (i.e. $(\lambda - 1) R(\lambda; T)$ is uniformly bounded for all $\lambda > 1$). If the number α, $|\alpha| = 1$, is an eigenvalue of T, then each number α^n, n integer, belongs to the spectrum of T.*

Proof. Since p is stochastic, the spectral radius of both T and V is 1. Further, obviously, the resolvent $R(\lambda; V)$ of V is given by $R(\lambda; V) = \Phi R(\lambda; T) \Phi^{-1}$, and the spectra of T and V coincide. Theorem 2.6 follows at once from Theorem 2 in [11]. □

Another possibility of application is "translating" some results of Schaefer [12], [13] concerning invariant ideals of positive operators in $\mathbf{C}(S)$. Again only some illustrations, but we must start with some definitions.

DEFINITION. By an *ideal* in $\mathbf{B}(X, \mathscr{X})$ we call a closed vector subspace J of $\mathbf{B}(X, \mathscr{X})$ such that $f \in J$, $g \in \mathbf{B}(X, \mathscr{X})$, $|g| \leq |f|$ implies $g \in J$. Further, a *T-ideal* is a proper ideal in $\mathbf{B}(X, \mathscr{X})$ which is invariant under T. A T-ideal is *maximal* if it is not properly contained in any other T-ideal. Finally, the *radical* of T is the intersection of all maximal T-ideals. An analogous terminology is introduced for operators V in $\mathbf{C}(S)$ (cf. [12], [13]).

THEOREM 2.7. *If the operator T is ergodic (i.e. $n^{-1} \sum_{k=0}^{n-1} T^k f$ converges strongly for $n \to \infty$ for each $f \in \mathbf{B}(X, \mathscr{X})$ to Pf where P is a positive projection on a T-invariant subspace), then f belongs to the radical of T if and only if $\lim_{n \to \infty} n^{-1} \sum_{k=0}^{n-1} T^k |f| = 0$. If T is uniformly ergodic (i.e. $n^{-1} \sum_{k=0}^{n-1} T^k$ converges for $n \to \infty$ in the uniform operator topology to P), then f belongs to the radical of T if and only if $\lim_{n \to \infty} T^n |f| = 0$.*

Proof. Since the mapping Φ is isometric, preserves the order and has the property $\Phi|f| = |\Phi f|$, it is easy to see that the maximal V-ideals have precisely the form ΦJ where J are maximal T-ideals, and therefore the radical of V is equal to ΦR where R is the radical of T. Theorem 2.7 is then an obvious consequence of Theorem 4 in [13]. □

THEOREM 2.8. *Let the radical of T contain only the null function, and let α, $|\alpha| = 1$, be an eigenvalue of T with an eigenfunction $f = |f|\, g$. Then α^n, for each integer n, is an eigenvalue of T with an eigenfunction $|f|\, g^n$.*

Proof. It rests on similar considerations as before, and on the fact that Φ is an algebraic isomorphism. The Theorem follows then from Theorem 5 in [13]. □

In connection with Theorem 2.7 and the preceding Definition we can quickly and roughly mention the following. One of the basic tasks in the theory of Markov chains with a general state space X is to decompose X in such a way that the behavior of the Markov chain is different in different parts of X. The first method of doing this may be called probabilistic and it rests on a direct analysis of transition probabilities. A pioneering work in this direction is due to Doeblin [2] (cf. also [1] for a revised and complemented version). Here the basic notions are stochastically closed sets, essential and inessential sets, etc. The second method is based on the study of the Markov operator T in $L_\infty(\mu)$ and of the operator in $L_1(\mu)$ to which T is adjoint. A fundamental result in this direction is the decomposition of X into conservative and dissipative parts, as it was done by Hopf [5]. Finally, we could propose a third method (purely analytic, not probabilistic, because it disregards measure, but we think it to be a mathematically interesting problem) which would be based on an analysis of the operator T in $\mathbf{B}(X, \mathscr{X})$. The author is not aware of any essential work in this direction. A suitable tool could be here the study of the maximal T-ideals in $\mathbf{B}(X, \mathscr{X})$, similarly as it was started by Schaefer [12], [13] in $\mathbf{C}(S)$. Specifically, we could expect e.g. that the union of supports of all functions from the radical of T could play in some sense a similar role as the Hopf dissipative part, the complement of this union a similar role as the conservative part.

REFERENCES

[1] K. L. CHUNG: The general theory of Markov processes according to Doeblin. Z. Wahrscheinlichkeitstheorie *2* (1964), 230—254.

[2] W. DOEBLIN: Éléments d'une théorie générale des chaînes simples constantes de Markoff. Ann. Sci. École Norm. Sup. (3) *57* (1940), 2, 61—111.

[3] N. DUNFORD, J. T. SCHWARTZ: Linear Operators I. Interscience Publishers, New York—London 1958.

[4] T. E. HARRIS: The existence of stationary measures for certain Markov processes. In: Proc. Third Berkeley Symp. 1956, vol. II, 113—124.

[5] E. HOPF: The general temporally discrete Markoff process. J. Rat. Mech. and Anal. *3* (1954), 13—45.
[6] B. JAMISON, S. OREY: Markov chains recurrent in the sense of Harris. Z. Wahrscheinlichkeitstheorie *8* (1967), 41—48.
[7] J. G.KEMENY, J. L. SNELL, A. W. KNAPP: Denumerable Markov Chains. Van Nostrand, Princeton 1966.
[8] E. NELSON: The adjoint Markoff process. Duke Math. J. *25* (1958), 671—690.
[9] M. ROSENBLATT: Equicontinuous Markov operators. Teorija verojatnostej *9* (1964), 205 to 222.
[10] M. ROSENBLATT: Almost periodic transition operators acting on the continuous functions on a compact space. J. Math. Mech. *13* (1964), 837—847.
[11] H. H. SCHAEFER: Spektraleigenschaften positiver linearer Operatoren. Math. Zeitschrift *82* (1963), 303—313.
[12] H. H. SCHAEFER: Invariant ideals of positive operators in $C(X)$. I. Illinois J. Math. *11* (1967), 703—715.
[13] H. H. SCHAEFER: Invariant ideals of positive operators in $C(X)$. II. Illinois J. Math. *12* (1968), 525—538.
[14] Z. ŠIDÁK: Operators in the space of continuous functions and the representation of Markov processes in a compact Hausdorff space. (In Russian.) Czech. Math. J. *13* (*88*) (1963), 37—50.
[15] Z. ŠIDÁK: Classification of Markov chains with a general state space. Bull. Amer. Math. Soc. *72* (1966), 149—152.
[16] Z. ŠIDÁK: Classification of Markov chains with a general state space. In: Trans. Fourth Prague Conference, Prague 1965. Academia, Prague 1967, 547—571.
[17] Z. ŠIDÁK: Eigenvalues of operators in L_p-spaces in Markov chains with a general state space. Czech. Math. J. *17* (*92*) (1967), 148—157.

CZECHOSLOVAK ACADEMY OF SCIENCES

MATHEMATICAL INSTITUTE

NOT NECESSARILY ADDITIVE REALIZATIONS OF COMPARATIVE PROBABILITY RELATIONS – A BASIS OF DECISION THEORY

HEINZ J. SKALA

HEIDELBERG

The realization of comparative probability relations is, besides of utility theory, one of the cornerstones of decision theory. Indicating that the class of reasonable comparative probability relations is not restricted to the class of relations which can be realized by finitely additive probability measures we are led to investigate not necessarily additive realizations. Moreover our measures are not restricted to real valued ones but may take values in ordered extension fields of the reals.

0. INTRODUCTION

Let $\mathfrak{F}$ be a field of sets on a non-empty set X and let $\lesssim$ be a binary (comparative probability) relation on $\mathfrak{F}$. If $x \lesssim y$ and $y \lesssim x$ we shall write $x \sim y$. The system $\langle X, \mathfrak{F}, \lesssim \rangle$ is called a comparative (or qualitative) probability system if it satisfies the following axioms:

(1) $$\text{not } X \lesssim \emptyset \,,$$

(2) $$\emptyset \lesssim x \,,$$

(3) $$x \lesssim y \quad \text{or} \quad y \lesssim x \,,$$

(4) $$x \lesssim y \quad \text{and} \quad y \lesssim z \quad \text{implies} \quad x \lesssim z \,,$$

(5) $$x \cap y = x \cap z = \emptyset \quad \text{implies} \quad y \lesssim z \quad \text{if and only if} \quad x \cup y \lesssim x \cup z$$

for all x, y, z in $\mathfrak{F}$.

The above five axioms have been stated for the first time by deFinetti [2] who proposed also the question whether all comparative probability relations satisfying (1)–(5) can be realized by finitely additive probability measures, i.e. whether there exists a finitely additive real valued probability measure p on $\mathfrak{F}$ such that

$$x \lesssim y \quad \text{if and only if} \quad p(x) \leqq p(y) \quad \text{for all} \quad x, y \in \mathfrak{F} \,.$$

It has been shown by Kraft/Pratt/Seidenberg [12] that even for the finite case this question cannot always be answered in the positive. We restate their famous counter-example: Let $X = \{p, q, r, s, t\}$ and let $\mathfrak{F}$ denote the set of all subsets of X. Consider the following ordering

$$\begin{gathered}\emptyset \prec \{p\} \prec \{q\} \prec \{r\} \prec \{p, q\} \prec \{p, r\} \prec \{s\} \prec \{p, s\} \prec \\ \prec \{q, r\} \prec \{t\} \prec \{p, q, r\} \prec \{q, s\} \prec \{r, s\} \prec \{p, t\} \prec \{p, q, s\} \prec \\ \prec \{q, t\} \prec \{p, r, s\} \prec \{r, t\} \prec \{q, r, s\} \prec \{p, q, t\} \prec \{p, r, t\} \prec \{s, t\} \prec \\ \prec \{p, q, r, s\} \prec \{p, s, t\} \prec \{q, r, t\} \prec \{p, q, r, t\} \prec \{q, s, t\} \prec \\ \prec \{r, s, t\} \prec \{p, q, s, t\} \prec \{p, r, s, t\} \prec \{q, r, s, t\} \prec \{p, q, r, s, t\}\,.\end{gathered}$$

It is easily verified that this ordering satisfies all the axioms (1)–(5); however there exists no finitely additive probability measure p realizing this ordering. To see this assume on the contrary that p realizes the above ordering. Hence we get:

$$\begin{aligned} p(p) + p(r) &< p(s)\,, \\ p(p) + p(s) &< p(q) + p(r)\,, \\ p(r) + p(s) &< p(p) + p(t)\,. \end{aligned}$$

Adding these three inequalities and cancelling out we get

$$p(p) + p(r) + p(s) < p(q) + p(t)\,.$$

On the other hand we have $\{q, t\} \prec \{p, r, s\}$; hence there cannot exist a finitely additive probability measure realizing $\prec$.

There arise the following important questions:

1. Which comparative probability relations satisfying the axioms (1)–(5) can be realized by finitely additive probability measures?

2. Are there empirically relevant comparative probability relations which cannot be realized by finitely additive probability measures? If the answer is yes, what can be said about weaker forms of realization.

3. Which binary relations on fields of sets not necessarily satisfying (1)–(5) can be realized in some sense.

Using in essence the Theorem of the Alternative the first question has been answered by Kraft/Pratt/Seidenberg [12] and Scott [20] for the finite case.

From an empirical point of view the insistence on additive realizations is reasonable only if every empirically relevant comparative probability relation must be compatible with some finitely additive measure. That this is not the case and that the counter-example of Kraft/Pratt/Seidenberg can be obtained by estimating the relation from relative-frequency data has been shown by Fine [5]. This motivates the investigation of not necessarily additive realizations.

As to the third question we should like to quote a paper of Fishburn [6] where in particular semiorders on $\mathfrak{F}$ are studied. We shall not go into this further for the interested reader may obtain corresponding generalizations for himself.

Let us close this section by some remarks concerning the numerical structure used for realizing the comparative probability relations. That we shall not confine ourselves to the reals, as is usually done, but allow the (not necessarily additive) realizations to take values in ordered extension fields of the reals has several reasons.

1. If $\mathfrak{F}$ is infinite we must introduce an Archimedean axiom in order to do with the reals. It is however known (see for example [22]) that Archimedean axioms cannot be empirically tested in any conceivable way.

2. Measures with values in an ordered extension field of the reals occur quite naturally if one tries to avoid some paradoxical results in game theory (see [7], [22], [23]).

3. It is well known that not every field of sets admits a strictly positive real valued measure. Luxemburg [13] and Nikodym [16] however proved that strictly positive measures with values in an ordered extension field of the reals always exist. See in this connection also deFinetti [3] where the comparability of zero probabilities is discussed. Many difficulties occuring there may be avoided by not restricting oneself to real valued probabilities.

4. Already in Kraft/Pratt/Seidenberg [12] we find the following remark: "In a general axiomatic theory of probability it would undoubtedly be of significance to let the values or measures be elements of an arbitrary simply ordered group, because such groups are capable of accomodating events p, q with p more probable than q but only by an infinitely small amount."

1. PRELIMINARIES

DEFINITION 1. Let I be a non-empty set. A non-empty set $\mathfrak{F}$ of subsets of I is called a filter if:

(1) $$A \in \mathfrak{F} \quad \text{and} \quad A \subseteq B, \quad \text{then} \quad B \in \mathfrak{F},$$

(2) $$\text{if} \quad A, B \in \mathfrak{F}, \quad \text{then} \quad A \cap B \in \mathfrak{F},$$

(3) $$\emptyset \notin \mathfrak{F}.$$

DEFINITION 2. A filter $\mathfrak{F}$ is called free if $\bigcap\{A : A \in \mathfrak{F}\} = \emptyset$.

DEFINITION 3. A filter $\mathfrak{F}$ on I is called an ultrafilter if there exists no filter $\mathfrak{F}'$ on I which is strictly finer than $\mathfrak{F}$.

Definition 4. A non-empty set $\{A_i : i \in I\}$ of subsets of the set A is said to have the finite intersection property if for every finite subset I' of I it holds that

$$\bigcap\{A_i : i \in I'\} \neq \emptyset .$$

Lemma 1. Every filter $\mathfrak{F}$ on I can be extended to an ultrafilter $\mathfrak{F}$ on I.

Lemma 2. A non-empty set $\{A_i : i \in I\}$ of subsets of the set A which has the finite intersection property can be extended to a filter on A.

Lemma 3. $\mathfrak{U}$ is an ultrafilter on I if and only if for every $A \subseteq I$ either $A \in \mathfrak{U}$ or $I - A \in \mathfrak{U}$.

Definition 5. A non-empty set A together with a specified set of relations R_k, $k = 1, \ldots, n$, on A will be called a relational system. We write $\mathfrak{A} = \langle A, R_1, \ldots, R_n \rangle$.

Obviously an m-ary operation O defined on A can be treated as $(m + 1)$-ary relation R' on A by defining $\langle a_1, \ldots, a_{m+1} \rangle \in R'$ if and only if $O(a_1, \ldots, a_m) = a_{m+1}$.

For the following we shall restrict ourselves to binary relational systems. It will be clear that everything we do can be extended to arbitrary relational system.

Given an index set I, for every $i \in I$ let $\mathfrak{A}_i = \langle A_i, R_i \rangle$ be a binary relational system. Let $A = \prod_{i \in I} A_i$ be the Cartesian product of the sets A_i and denote the elements of A by $f, g, h, \ldots$ The i-th component of $f \in A$ is written as $f(i)$.

Lemma 4. Let $\mathfrak{F}$ be a filter on I. The relation $\sim_{\mathfrak{F}}$ on A defined by

$$f \sim_{\mathfrak{F}} g \quad \text{if and only if} \quad \{i \in I : f(i) = g(i)\} \in \mathfrak{F}$$

is an equivalence relation on A.

We define a relation S on A by

$$f \, S \, g \quad \text{if and only if} \quad \{i \in I : f(i) \, R_i \, g(i)\} \in \mathfrak{F} .$$

Lemma 5. $\sim_{\mathfrak{F}}$ is a congruence relation with respect to S, i.e. if $f \sim_{\mathfrak{F}} f'$ and $g \sim_{\mathfrak{F}} g'$, then $f \, S \, g$ implies $f' \, S \, g'$.

For each $f \in A$ we denote by $f/\mathfrak{F}$ the equivalence class to which f belongs under $\sim_{\mathfrak{F}}$ and we let

$$\prod_{i \in I} A_i/\mathfrak{F} = \{f/\mathfrak{F} : f \in A\} .$$

Defining $f/\mathfrak{F} \, R_{\mathfrak{F}} \, g/\mathfrak{F}$ if and only if $f \, S \, g$, a relation $R_{\mathfrak{F}}$ on $\prod A_i/\mathfrak{F}$ is induced by S.

Definition 6. We write $\prod_{i \in I} \mathfrak{A}_i/\mathfrak{F}$ for the relational system $\langle \prod_{i \in I} A_i/\mathfrak{F}, R_{\mathfrak{F}} \rangle$ and call $\prod \mathfrak{A}_i/\mathfrak{F}$ the reduced product of $\{\mathfrak{A}_i : i \in I\}$ over the filter $\mathfrak{F}$. If $\mathfrak{F}$ is an ultrafilter then the reduced product is called an ultraproduct. If $\mathfrak{A}_i = \mathfrak{A}$ for every $i \in I$ we write $\mathfrak{A}^I/\mathfrak{F}$ and call this system reduced power of $\mathfrak{A}$ over $\mathfrak{F}$ or, if $\mathfrak{F}$ is an ultrafilter, ultrapower of $\mathfrak{A}$.

Let $\mathfrak{U}$ be an ultrafilter on an infinite set I and let $\mathfrak{R}$ denote the reals.

LEMMA 6. If $\mathfrak{U}$ is a free ultrafilter on I then $\mathfrak{R}^* = \mathfrak{R}^I/\mathfrak{U}$ is a nonstandard model of $\mathfrak{R}$. $\mathfrak{R}^*$ is an ordered extension field of $\mathfrak{R}$ which is non-Archimedean and in which equality $(=^*)$, ordering $(\geqq^*)$ and addition $(+^*)$ (relative to $\mathfrak{U}$) is defined as follows: Let $r, s, t \in \mathfrak{R}^*$

$$r =^* s \text{ if and only if there exist } r', s' \in \mathfrak{R}^I, \quad r' \in r, \quad s' \in s$$
$$\text{such that } \{i \in I : r'(i) = s'(i)\} \in \mathfrak{U};$$
$$r \geqq^* s \text{ if and only if there exist } r', s' \in \mathfrak{R}^I, \quad r' \in r, \quad s' \in s$$
$$\text{such that } \{i \in I : r'(i) \geqq s'(i)\} \in \mathfrak{U};$$
$$r +^* s =^* t \text{ if and only if there exist } r', s', t' \in \mathfrak{R}^I, \quad r' \in r,$$
$$s' \in s, \quad t' \in t \text{ such that } \{i \in I : r'(i) + s'(i) = t'(i)\} \in \mathfrak{U}.$$

For more details and proofs the reader is referred to Luxemburg [14], Robinson [19] and Bell/Slomson [1].

A very useful tool for showing that a certain realization exists in the finite case is the following theorem which has essentially been used by Kraft/Pratt/Seidenberg [12], Fishburn [6] and others.

THEOREM 1 (Theorem of the Alternative). Let A be a real (m, n)-matrix, $m \geqq 1$, and let B be a real (p, n)-matrix. Denote by A^T and B^T the transposed matrix of A and B respectively. There exists an $x \in \mathfrak{R}^n$ such that

(i) $$Ax > 0 \quad \text{and} \quad Bx \geqq 0$$

if and only if there do not exist $y \in \mathfrak{R}^m$, $z \in \mathfrak{R}^p$ such that

(ii) $$A^T y + B^T z = 0, \quad z \geqq 0, \quad y \geqq 0 \quad \text{and} \quad y_i > 0 \quad \text{for at least one } i = 1, \ldots, m.$$

Proof. See for example Motzkin [15] or Pfanzagl [17].

If one is able to prove, for example by means of the Theorem of the Alternative, that for every finite subsystem of a given relational system there exists an embedding into another relational system the following theorem may be used in order to show the embeddability of the whole relational system into an ultrapower of the other relational system.

THEOREM 2. Let A be a non-empty set and let $R \subseteq A^n$ be an n-ary relation. Let I be the index set of all finite subsystems of $\mathfrak{A} = \langle A, R \rangle$. There exists an ultrafilter $\mathfrak{U}$ on I such that $\mathfrak{A}$ is embeddable into $\prod_{i \in I} \mathfrak{A}_i/\mathfrak{U}$.

Proof. See Frayne/Morel/Scott [8].

2. NOT NECESSARILY ADDITIVE REALIZATIONS

We have already remarked that Fine [5] showed that the class of reasonable comparative probability relations is significantly larger than the class of relations which can be realized by a finitely additive probability measure. This motivated the following

THEOREM 3 (Fine [4], [5]). Let $\langle X, \mathfrak{F}, \lesssim \rangle$ be a comparative probability system, i.e. $\lesssim$ satisfies (1)–(5). For uncountable $\mathfrak{F}/\sim$ there exists a (not necessarily additive) real valued p on $\mathfrak{F}$ realizing $\lesssim$ (i.e. $a \lesssim b$ if and only if $p(a) \leqq p(b)$) if and only if $\mathfrak{F}$ contains a countable order-dense subset.

It is obvious that for at most countable $\mathfrak{F}/\sim$ such a p always exists. Thus, by an easy application of Theorem 2 we get

THEOREM 4. Let I be the index set of all finite subsets of $\mathfrak{F}/\sim$. If $\lesssim$ satisfies (1)–(5) then there exists an ultrafilter $\mathfrak{U}$ on I and a (not necessarily additive) $\mathfrak{R}^I/\mathfrak{U} = \mathfrak{R}^*$--valued p on $\mathfrak{F}$ realizing $\lesssim$.

Theorem 4 is a trivial consequence of a result in utility theory which has been stated independently by Richter [18] and Skala [21]. It should be noted that a similar statement holds if, for example, $\lesssim$ is a semiorder.

DEFINITION 7 (Hausdorff [9]). Let α be any ordinal. A strictly ordered set C is called an η_α-set provided that whenever A and B are substs of C of cardinality $< \aleph_\alpha$ with $a < b$ for all $a \in A$ and $b \in B$, there exists a $c \in C$ such that $a < c < b$ for all $a \in A$ and $b \in B$. (A well known η_0-set of cardinality $\aleph_0$ form the naturally ordered rationals.)

The interest in η_α-sets is due to the following facts:

1. Uniqueness: Any two η_α-sets of cardinality $\aleph_\alpha$ are isomorphic.

2. Universality: Any strictly ordered set of cardinality $\leq \aleph_\alpha$ can be embedded into an η_α-set.

3. η_α-sets may be "constructed" as ultrapowers of the naturally ordered rationals (reals).

The following statement serves as demonstration of the last fact:

THEOREM 5 (Kochen [11]). Denote by $\mathfrak{Q}$ the naturally ordered rationals (an η_0-set). If $\mathfrak{U}$ is a free ultrafilter on the denumerable set I, then $\mathfrak{Q}^* = \mathfrak{Q}^I/\mathfrak{U}$ is an η_1-set.

An immediate consequence of the above theorem is

COROLLARY 1. Let $\mathfrak{U}$ be a free ultrafilter on the denumerable set I. If $\lesssim$ satisfies (1)–(5) and $\mathfrak{F}/\sim$ is of cardinality $\leqq \aleph_1$, then there exists a (not necessarily additive) $\mathfrak{Q}^I/\mathfrak{U} = \mathfrak{Q}^*$-valued p on $\mathfrak{F}$ realizing $\lesssim$.

In [22] related problems in measurement and utility theory have been studied. It should be noted that one may obtain results quite similar to Corollary 1 if the cardinality of $\mathfrak{F}/\sim$ is greater than $\aleph_1$.

Up to now our results are not very intersting as essentially p on $\mathfrak{F}$ has only been assumed to be orderpreserving. The question therefore arises what can be said if p is assumed to be superadditive, i.e.

$$x \cap y = \emptyset \quad \text{implies} \quad p(x) + p(y) \leqq p(x \cup y)\,.$$

The first thing to observe is that axiom (5) of a comparative probability system which is a necessary condition for a finitely additive realization p to exist no longer remains necessary if p is assumed to be only subadditive. We therefore replace (5) by

$$(5') \qquad \text{if} \quad y \subseteq x \quad \text{then} \quad \emptyset \prec x - y \quad \text{implies} \quad y \prec x \quad \text{and}$$
$$\emptyset \sim x - y \quad \text{implies} \quad y \lesssim x\,.$$

That (1)–(5′) is necessary for a superadditive realization p of $\lesssim$ to exist is easily verified. For let $y \subseteq x$, $\emptyset \prec x - y$ but $y \sim x$, it is obvious that there cannot exist a p such that $p(x - y) + p(y) \leqq p(x)$, $p(x) = p(y)$ and $p(x - y) > 0$. A similar argument holds for the second part of (5′).

On the other hand an application of the Theorem of the Alternative (Theorem 1) shows that for finite fieds of sets (1)–(5′) is also sufficient for a superadditive realization p of $\lesssim$ to exist. This generalizes a result of Kirsch [10].

In order to state the following theorem we remember the fact that a field of sets generated by finitely many elements is necessarily finite. Moreover we denote by $\mathfrak{F}_F$ and $\lesssim_F$ the subfield of $\mathfrak{F}$ generated by the set F of elements of $\mathfrak{F}$ and the restriction of $\lesssim$ to $\mathfrak{F}_F$ respectively.

THEOREM 6. Let $\mathfrak{F}$ be a field of sets on X and let $\lesssim$ be a binary (comparative probability) relation on $\mathfrak{F}$. There exists an ordered extension field $\mathfrak{R}^*$ of the reals $\mathfrak{R}$ and a superadditive $\mathfrak{R}^*$-valued p^* on $\mathfrak{F}$ realizing $\lesssim$ if and only if for every finite set F of elements of $\mathfrak{F}$ the relation $\lesssim_F$ on $\mathfrak{F}_F$ satisfies (1)–(5′).

Proof. Assume that the $\mathfrak{R}^*$-valued superadditive p^* realizes $\lesssim$. Obviously the restriction of p^* to $\mathfrak{F}_F$ realizes $\lesssim_F$. For finite F there must also exist a real valued superadditive p' on $\mathfrak{F}_F$ realizing $\lesssim_F$. This is a consequence of Corollary 2 stated in Kraft/ /Pratt/Seidenberg [12].

On the other hand let us assume that for every finite set F of elements of $\mathfrak{F}$ the relation $\lesssim_F$ on $\mathfrak{F}_F$ satisfies (1)–(5′). Hence for every finite F there exists at least one real valued superadditive p^F realizing $\lesssim_F$. Let $\{p_i : i \in I\}$ be the set of all real valued superadditive measures on $\mathfrak{F}$ with the property that for every $i \in I$ there exists at least one finite F of elements of $\mathfrak{F}$ such that the restriction of p_i to $\mathfrak{F}_F$ realizes $\lesssim_F$. Denote by I_F the set of all $i \in I$ such that the restriction of p_i to $\mathfrak{F}_F$ realizes $\lesssim_F$. If the $\lesssim_F$ satisfies (1)–(5′) then for every finite F the set $I_F \subseteq I$ is nonempty and $\{I_F : F$ is a finite set of elements of $\mathfrak{F}\}$ has the finite intersection property. Hence, as is well known, $\{I_F\}$ can be extended to an ultrafilter $\mathfrak{U}$ on I. Remembering that the ultrapower $\mathfrak{R}^* = \mathfrak{R}^I/\mathfrak{U}$ of the reals $\mathfrak{R}$ has all the first-order properties of $\mathfrak{R}$ we have "constructed" an ordered extension field of the reals. It is now possible to define the required $\mathfrak{R}^*$--valued superadditive p^* on $\mathfrak{F}$.

For every $a \in \mathfrak{F}$ let $p^*(a)$ be that element of $\mathfrak{R}^* = \mathfrak{R}^I/\mathfrak{U}$ which is represented by $\bar{p}(a) \in \mathfrak{R}^I$ defined as follows:

$$\bar{p}_i(a) = \begin{cases} p_i(a) & \text{if} \quad i \in I_{\{a\}}, \\ c & \text{else}, \end{cases}$$

where c is an arbitrary fixed constant.

Let $a \lesssim b$, $a, b \in \mathfrak{F}$, then for every $i \in I_{\{a,b\}}$ we have:

$$p_i(a) \leqq p_i(b),$$

thus

$$I_{\{a,b\}} \subseteq \{i \in I : \bar{p}_i(a) \leqq \bar{p}_i(b)\} \in \mathfrak{U}$$

and hence $p^*(a) \leqq^* p^*(b)$. On the other hand $p^*(a) \leqq^* p^*(b)$ implies the existence of some $I' \subseteq I$ such that

$$I' = \{i \in I : \bar{p}_i(a) \leqq \bar{p}_i(b)\} \in \mathfrak{U}.$$

Now one can choose an $I_F \in \mathfrak{U}$ such that $a, b \in F$. As $\emptyset \neq I_F \cap I' \in \mathfrak{U}$ there exists an $k \in I_F \cap I'$ such that $p_k(a) \leqq p_k(b)$ and hence $a \lesssim b$.

In order to prove superadditivity assume $a \cap b = \emptyset$ and $c = a \cup b$. For every $i \in I_{\{a,b,c\}}$ we have $p_i(a) + p_i(b) \leqq p_i(c)$, thus

$$I_{\{a,b,c\}} \subseteq \{i \in I : \bar{p}_i(a) + \bar{p}_i(b) \leqq \bar{p}_i(c)\} \in \mathfrak{U}$$

and hence $p^*(a) +^* p^*(b) \leqq^* p^*(c)$.

For the following three theorems assume that $\langle X, \mathfrak{F}, \lesssim \rangle$ satisfies (1)–(5). If X is finite Kraft/Pratt/Seidenberg [12] and Scott [20] showed that a finitely additive real valued realization p of $\lesssim$ exists if and only if for all sequences $a_0, \ldots, a_{n-1}$, $b_0, \ldots, b_{n-1} \in \mathfrak{F}$ if $a_i \lesssim b_i$ for all i, $0 < i < n$, and if for all $x \in X$

$$\sum_{i=0}^{n-1} \hat{a}_i(x) = \sum_{i=0}^{n-1} \hat{b}_i(x), \quad \text{then} \quad b_0 \lesssim a_0 .$$

(The symbol $\hat{a}$ is used here to denote the characteristic function of a, i.e. $\hat{a}(x) = 1$ means $x \in a$ and $\hat{a}(x) = 0$ means $x \notin a$.) It should be remarked that for finite X a subadditive realization obviously exists if (1)–(5) is satisfied. By a proof similar to the one of Theorem 6 we get

THEOREM 7. Let $\langle X, \mathfrak{F}, \lesssim \rangle$ satisfy (1)–(5). There exists an ordered extension field $\mathfrak{R}^*$ of the reals $\mathfrak{R}$ and a finitely additive $\mathfrak{R}^*$-valued measure p^* on $\mathfrak{F}$ realizing $\lesssim$ if and only if for every finite set F of elements of $\mathfrak{F}$ the relation $\lesssim_F$ on $\mathfrak{F}_F$ satisfies Scott's condition stated above.

Theorem 7 was also given in [18].

A connection between finitely additive $\mathfrak{R}^*$-valued realizations (also called agreeing $\mathfrak{R}^*$-valued measures) and almost agreeing $\mathfrak{R}$-valued measures is given by

THEOREM 8. Let $\langle X, \mathfrak{F}, \lesssim \rangle$ satisfy (1)–(5) and let p^* be an $\mathfrak{R}^*$-valued finitely additive measure realizing $\lesssim$. There exists a finitely additive $\mathfrak{R}$-valued measure p such that

$$a \lesssim b \quad \text{implies} \quad p(a) \leqq p(b) \quad \text{(almost agreement)}.$$

Proof. The only thing to show is the finite additivity of p. Let $\mathfrak{R}^*$ be an ordered extension field of $\mathfrak{R}$. We call a number $r \in \mathfrak{R}^*$ an infinitesimal if and only if $|r| < 1/n$ for all natural numbers n. A number $r \in \mathfrak{R}^*$ is called finite if and only if there exists a natural number n such that $|r| < n$. Denoting the set of all finite numbers by $\mathfrak{M}_0$ and the set of all infinitesimals by $\mathfrak{M}_1$ we observe $\mathfrak{R} \subset \mathfrak{M}_0 \subset \mathfrak{R}^*$, $\mathfrak{M}_0$ is an ordered integral domain, $\mathfrak{M}_1$ is a maximal ideal in $\mathfrak{M}_0$, $\mathfrak{M}_0/\mathfrak{M}_1$ is isomorphic to $\mathfrak{R}$ and every $r \in \mathfrak{M}_0$ determines a cut in $\mathfrak{R}$. (Proofs may be found in [19] and [14].)

Let $r \in \mathfrak{M}_0$ and denote the real cut number which is determined by r with the symbol $\mathrm{re}(r)$. It is an easy consequence of the fact that $\mathfrak{M}_1$ is a maximal ideal in $\mathfrak{M}_0$ that for every $r_1, r_2 \in \mathfrak{M}_0$ the relation $\mathrm{re}(r_1 + r_2) = \mathrm{re}(r_1) + \mathrm{re}(r_2)$ holds. Define $p(a) = \mathrm{re}(p^*(a))$. If a and b are disjoint we have $p(a \cup b) = \mathrm{re}(p^*(a \cup b)) = \mathrm{re}(p^*(a) +^* p^*(b)) = \mathrm{re}(p^*(a)) + \mathrm{re}(p^*(b)) = p(a) + p(b)$ as required.

THEOREM 9. Let $\langle X, \mathfrak{F}, \lesssim \rangle$ satisfy (1)–(5) and let $\mathfrak{Q}^*$ be the η_1-set of Theorem 5. If $\mathfrak{F}/\sim$ is of cardinality $\leqq \aleph_1$, then there exists a (not necessarily additive) $\mathfrak{Q}^*$-valued p realizing $\lesssim$ and a $\mathfrak{Q}^*$-valued q such that

$$a \cap b = \emptyset \quad \text{implies} \quad p(a \cup b) = q(p(a), p(b)).$$

Proof. By Corollary 1 the existence of the desired p is obvious. In order to verify that an appropriate q exists we have only to show that $a \cap b = c \cap d = \emptyset$, $p(a) = p(c)$ and $p(b) = p(d)$ implies $p(a \cup b) = p(c \cup d)$. This however is a routine matter.

REMARKS. 1. For real valued p and q the analogous theorem has been stated by Fine [5] where also further properties of q may be found.

2. For cardinalities greater than $\aleph_1$ similar theorems hold. See in this connection also [22].

REFERENCES

[1] J. L. Bell, A. B. Slomson: Models and Ultraproducts. North-Holland, Amsterdam (1969).

[2] B. deFinetti: La prévision: ses lois logiques, ses sources subjectives. Ann. Inst. H. Poincaré *7* (1937), 1—68.

[3] B. deFinetti: Theory of Probability. Wiley, London—New York 1974.

[4] T. L. Fine: A note on the existence of quantitative probability. Ann. Math. Stat. *42* (1971), 1182—1186.

[5] T. L. Fine: Theories of Probability. Academic Press, New York 1973.

[6] P. C. Fishburn: Weak qualitative probability on finite sets. Ann. Math. Stat. *40* (1969), 2118—2126.

[7] P. C. Fishburn: On the foundation of game theory: The case of non-Archimedean utilities. Int. J. Game Theory *2* (1972), 65—71.

[8] T. Frayne, A. C. Morel, D. S. Scott: Reduced direct products. Fund. Math. *51* (1962), 195—228.

[9] F. Hausdorff: Grundzüge der Mengenlehre. Veit, Leipzig 1914.

[10] A. Kirsch: Über lineare Ordnungen endlicher Boolescher Verbände. Arch. Math. *17* (1966), 489—491.

[11] S. Kochen: Ultraproducts in the theory of models. Ann. Math. *74* (1962), 221—261.

[12] C. H. Kraft, J. W. Pratt, A. Seidenberg: Intuitive probability on finite sets. Ann. Math. Stat. *30* (1959), 408—419.

[13] W. A. J. Luxemburg: Two applications of the method of construction by ultrapowers to analysis. Bull. Am. Math. Soc. *68* (1962), 416—419.

[14] W. A. J. Luxemburg: Non-Standard Analysis. C.I.T., Pasadena 1962.

[15] T. Motzkin: Beiträge zur Theorie der linearen Ungleichungen. Diss., Basel 1963.

[16] O. Nikodym: On extension on a given finitely additive field-valued non-negative measure on a finitely additive Boolean tribe to another tribe more ample. Rend. Sem. Math. Univ. de Padova *26* (1956), 232—327.

[17] J. Pfanzagl: Theory of Measurement. Physica, Würzburg—Wien 1971.

[18] M. Richter: Rational choice. In: J. S. Chipman et al (Eds.): Preferences, Utility, and Demand. Harcourt Brace Jovanovich, New York 1971.

[19] A. Robinson: Non-Standard Analysis. North-Holland, Amsterdam 1966.

[20] D. Scott: Measurement structures and linear inequalities. J. Math. Psychology *1* (1964), 233—247.

[21] H. J. Skala: An application of model theory to the theory of measurement. 2nd Int. Game Theory Workshop, Berkeley 1970 (mimeographed).

[22] H. J. Skala: Über einige Grundprobleme der Nutzentheorie — Nichtarchimedische Nutzentheorie. Univ. Heidelberg 1973. A revised English version is to appear in Reidel.

[23] H. J. Skala: Nonstandard utilities and the foundations of game theory. To appear in Int. J. Game Theory.

Universität Heidelberg

Institut für Statistik

ON THE OPTIMALITY CONDITIONS FOR SEMI-MARKOV DECISION PROCESSES

KAREL SLADKÝ

PRAGUE

The paper presents a recurrence formula for the difference between expected rewards and sojourn times generated by N transitions of a semi-Markov decision process with finite state space. Using the recurrence formula convergence of policy iteration method can be easily verified and also necessary and sufficient optimality conditions for average optimal and more selective average overtaking optimal policies are established.

1. INTRODUCTION AND NOTATIONS

The paper extends the method and some of the results established in [4], [6] to semi-Markov decision processes (SMDP). Notations used in [4], [6] are followed as close as possible.

We consider a SMDP $\{X(t), t \geqq 0\}$ with state space $I = \{1, \ldots, r\}$ and the action set $J = \{1, \ldots, s\}$ in any state. Choosing action z in state j state k will be reached in the next transition with given probability $p(j, k; z)$ and the time until this transition occurs is a random variable with probability distribution $F(.; j, k; z)$. The transition from state j into state k is associated with given value $r(j, k)$ and, if action z is selected, reward rate $\varrho(j; z)$ is being obtained in state j per time unit. Denoting $c(j; z)$ (resp. $d(j; z)$) expected reward (resp. sojourn time) in state j if action z is selected then, obviously, $c(j; z) = \sum_{k \in I} p(j, k; z) [r(j, k) + \varrho(j; z) \int_0^\infty \tau \, dF(\tau; j, k; z)]$ (resp. $d(j, z) = \sum_{k \in I} p(j, k; z) \int_0^\infty \tau \, dF(\tau; j, k; z)$). We suppose that $0 < d' = \min_{j \in I, z \in J} d(j; z) \leqq \leqq \bar{d} = \max_{j \in I, z \in J} d(j; z) < \infty$.

A policy ω controlling the considered SMDP is given by a sequence of actions $z \in J$ selected with respect to the complete history of the SMDP. So we shall write

$\omega = \{z_n(j_0, z_0, t_0; \ldots; j_{n-1}, z_{n-1}, t_{n-1}; j_n), n = 0, 1, \ldots\}$ where $z_n(\ldots)$ is the action chosen at the n-th transition of the SMDP following the occurrence of states $j_0, \ldots$ $\ldots, j_{n-1}, j_n$, selected actions $z_0, \ldots, z_{n-1}$ and the associated sojourn times $t_0, \ldots, t_{n-1}$. (For shortening we shall often delete the arguments in $z_n(\ldots)$, so we shall write only $\omega \sim z_n$ to denote the policy ω selecting action z_n at the n-th transition of the SMDP). ω will be called stationary policy if for any $n = 0, 1, \ldots z_n(\ldots) = z(j_n)$. For stationary policy we shall write $\omega \sim z(j)$ (where $z(j)$ is the action used in state j) and by $\pi(j, k; \omega)$ we shall denote the limit transition probability considered in Cesaro sense associated with $\|p(j, k; z(j))\|$.

Denoting X_n (for $n = 0, 1, \ldots$) the random state of the SMDP at the n-th transition, C_N (resp. D_N) the random reward (resp. elapsed time) considered up to the N transitions, $C(t)$ the random reward incurred up to time t and reserving symbol E_j^ω for mathematical expectation if $X_0 = j$ and policy ω is selected, the quality of control ω of the SMDP is usually considered according to

$$\limsup_{t \to \infty} \frac{1}{t} E_j^\omega C(t) \quad \text{or} \quad \limsup_{N \to \infty} \frac{E_j^\omega C_N}{E_j^\omega D_N} .$$

We shall consider the latter criterion as it is easier to work with it. (Moreover, it can be shown (see [5]) if for a stationary policy ω $E_j^\omega[\min t > 0 : X(t) = j, X(t-) \neq j] < \infty$ both criteria are equal.)

A policy $\bar{\omega}$ will be called average optimal if for an arbitrary policy ω and all $j \in I$

$$g_j^* = \lim_{N \to \infty} \frac{E_j^{\bar{\omega}} C_N}{E_j^{\bar{\omega}} D_N} \geqq \limsup_{N \to \infty} \frac{E_j^\omega C_N}{E_j^\omega D_N} .$$

As it will be shown the class of average optimal policies contains also the policies using "sufficiently rarely" unoptimal decisions. To exclude such a class of "optimal" policies a more selective average overtaking optimality criterion is suggested and studied (this optimality criterion is an extension of the optimality criterion suggested by Veinott for Markov chains [7]). A policy $\bar{\omega}^*$ will be called average overtaking optimal if for an arbitrary policy ω and all $j \in I$

$$\liminf_{N \to \infty} \frac{1}{N} \sum_{n=0}^{N-1} [E_j^{\bar{\omega}^*} C_n - E_j^\omega C_n - g_j^*(E_j^{\bar{\omega}^*} D_n - E_j^\omega D_n)] \geqq 0 .$$

Throughout the paper we shall make

Assumption A. For any stationary policy $\tilde{\omega} \sim \tilde{z}(j)$ the resulting transition probability matrix $\|p(j, k; \tilde{z}(j))\|$ contains only one class of recurrent states.

2. RECURRENCE RELATION FOR THE DIFFERENCE OF EXPECTED REWARDS AND SOJOURN TIMES

As under Assumption A $\pi(j_1, k; \tilde{\omega}) = \pi(j_2, k; \tilde{\omega})$ for any $j_1, j_2 \in I$ we shall only write $\pi(k; \tilde{\omega})$ instead of $\pi(j, k; \tilde{\omega})$. Similarly we abbreviate g_j^* by g^*. We shall need the following lemma.

2.1. Lemma. For any stationary policy $\tilde{\omega} \sim \tilde{z}(j)$ there exists unique solution denoted $g^{\tilde{\omega}}$, $w_j^{\tilde{\omega}}$ (resp. $u_j^{\tilde{\omega}}$) to the set of equations

$$w_j = c(j; \tilde{z}(j)) + \sum_{k \in I} p(j, k; \tilde{z}(j))\, w_k - d(j; \tilde{z}(j))\, g\,, \tag{2.1}$$

$$\sum_{k \in I} \pi(k; \tilde{\omega})\, w_k = 0 \quad (j = 1, \ldots, r) \tag{2.2}$$

resp.

$$u_j = -w_j^{\tilde{\omega}} + \sum_{k \in I} p(j, k; \tilde{z}(j))\, u_k \tag{2.3}$$

$$\sum_{k \in I} \pi(k; \tilde{\omega})\, u_k = 0)\,. \tag{2.4}$$

Proof. Multiplying (2.1) by $\pi(j; \tilde{\omega})$ and summing for $j = 1, \ldots, r$ after simple manipulations we obtain (as $\sum_{j \in I} \pi(j; \tilde{\omega})\, p(j, k; \tilde{z}(j)) = \pi(k; \tilde{\omega})$)

$$g^{\tilde{\omega}} = \frac{\sum_{k \in I} \pi(k; \tilde{\omega})\, c(k; \tilde{z}(k))}{\sum_{k \in I} \pi(k; \tilde{\omega})\, d(k; \tilde{z}(k))}\,. \tag{2.5}$$

Setting (2.5) into (2.1) and adding (2.2) as rank $\|\delta_{jk} - p(j, k; \tilde{z}(j))\| +$ $+$ rank $\|\pi(k; \tilde{\omega})\| = r$ (where δ_{jk} is Kronecker's symbol), the existence of unique solution $w_j^{\tilde{\omega}}$ is obvious. The same can be applied also to (2.3), (2.4). □

If $\tilde{\omega} \sim \tilde{z}(j)$ is a stationary policy, let us set for $j \in I$, $z \in J$

$$\varphi^{\tilde{\omega}}(j; z) = c(j; z) + \sum_{k \in I} p(j, k; z)\, w_k^{\tilde{\omega}} - w_j^{\tilde{\omega}} - d(j; z)\, g^{\tilde{\omega}}\,, \tag{2.6}$$

$$\gamma^{\tilde{\omega}}(j; z) = -w_j^{\tilde{\omega}} + \sum_{k \in I} p(j, k; z)\, u_k^{\tilde{\omega}} - u_j^{\tilde{\omega}}\,. \tag{2.7}$$

By $J^{\tilde{\omega}}(j) \subset J$ we shall denote the set of all actions in state j for which $\varphi^{\tilde{\omega}}(j; z) = 0$.

In the further text the following recurrence formulas will play an important role.

2.2. THEOREM. Let $\tilde{\omega} \sim \tilde{z}(j)$ be a stationary policy. Then for an arbitrary policy $\omega \sim z_n$ and all $j \in I$, $N = 0, 1, \ldots$

$$E_j^{\omega} C_N - g^{\tilde{\omega}} E_j^{\omega} D_N = \sum_{n=0}^{N-1} E_j^{\omega} \varphi^{\tilde{\omega}}(X_n; z_n) + w_j^{\tilde{\omega}} - E_j^{\omega} w_{X_N}^{\tilde{\omega}}, \tag{2.8}$$

$$\sum_{n=0}^{N-1} [E_j^{\omega} C_n - g^{\tilde{\omega}} E_j^{\omega} D_n - w_j^{\tilde{\omega}} + E_j^{\tilde{\omega}} w_{X_n}^{\tilde{\omega}}] = \tag{2.9}$$

$$= \sum_{n=0}^{N-1} [(N-1-n) E_j^{\omega} \varphi^{\tilde{\omega}}(X_n; z_n) + E_j^{\omega} \gamma^{\tilde{\omega}}(X_n; z_n)] - (E_j^{\omega} u_{X_N}^{\tilde{\omega}} - E_j^{\tilde{\omega}} u_{X_N}^{\tilde{\omega}}).$$

Proof. As

$$E_j^{\omega} \varphi^{\tilde{\omega}}(X_n; z_n) = E_j^{\omega}[c(X_n; z_n) + w_{X_{n+1}}^{\tilde{\omega}} - w_{X_n}^{\tilde{\omega}} - d(X_n; z_n) g^{\tilde{\omega}}] \tag{2.10}$$

summing (2.10) for $n = 0, 1, \ldots, N-1$ we obtain (2.8).

From (2.8) we have

$$\sum_{n=0}^{N-1} [E_j^{\omega} C_n - g^{\tilde{\omega}} E_j^{\omega} D_n - w_j^{\tilde{\omega}} + E_j^{\tilde{\omega}} w_{X_n}^{\tilde{\omega}}] = \tag{2.11}$$

$$= \sum_{n=0}^{N-1} [(N-1-n) E_j^{\omega} \varphi^{\tilde{\omega}}(X_n; z_n) - (E_j^{\omega} w_{X_n}^{\tilde{\omega}} - E_j^{\tilde{\omega}} w_{X_n}^{\tilde{\omega}})].$$

Employing (2.7) we obtain

$$E_j^{\omega} \gamma^{\tilde{\omega}}(X_n; z_n) = -E_j^{\omega} w_{X_n}^{\tilde{\omega}} + E_j^{\omega} u_{X_{n+1}}^{\tilde{\omega}} - E_j^{\omega} u_{X_n}^{\tilde{\omega}} \tag{2.12}$$

and summation of (2.12) for $n = 0, 1, \ldots, N-1$ yields (as $\gamma^{\tilde{\omega}}(j; \tilde{z}(j)) = 0$)

$$-\sum_{n=0}^{N-1} (E_j^{\omega} w_{X_n}^{\tilde{\omega}} - E_j^{\tilde{\omega}} w_{X_n}^{\tilde{\omega}}) = \sum_{n=0}^{N-1} E_j^{\omega} \gamma^{\tilde{\omega}}(X_n; z_n) - E_j^{\omega} u_{X_N}^{\tilde{\omega}} + E_j^{\tilde{\omega}} u_{X_N}^{\tilde{\omega}}. \tag{2.13}$$

Inserting (2.13) into (2.11) we obtain (2.9). □

2.3. COROLLARY. As

$$\lim_{N \to \infty} \frac{1}{N} \sum_{n=0}^{N-1} E_j^{\tilde{\omega}} w_{X_n}^{\tilde{\omega}} = 0 \tag{2.14}$$

from (2.9) for $\omega \equiv \tilde{\omega}$ we obtain

$$\lim_{N \to \infty} \frac{1}{N} \sum_{n=0}^{N-1} (E_j^{\tilde{\omega}} C_n - g^{\tilde{\omega}} E_j^{\tilde{\omega}} D_n) = w_j^{\tilde{\omega}}. \tag{2.15}$$

3. NECESSARY AND SUFFICIENT OPTIMALITY CONDITIONS

3.1. Lemma. There exists stationary policy $\hat{\omega} \sim \hat{z}(j)$ (resp. $\omega^* \sim z^*(j)$) so that for all $j \in I$

$$\varphi^{\hat{\omega}}(j; z) \leqq 0 \quad \text{for any} \quad z \in J \tag{3.1}$$

(resp.

$$\varphi^{\omega^*}(j; z) \leqq 0 \quad \text{for any} \quad z \in J \tag{3.2}$$

and

$$\gamma^{\omega^*}(j; z) \leqq 0 \quad \text{for any} \quad z \in J^{\omega^*}(j)) . \tag{3.3}$$

Proof. The proof can be performed by policy iteration method. Let ω_1 be an arbitrary stationary policy. If (3.1) does not hold for $\hat{\omega} \equiv \omega_1$, let us construct stationary policy $\omega_2 \sim z_2(j)$ so that $\varphi^{\omega_1}(j; z_2(j)) = \max_{z \in J} \varphi^{\omega_1}(j; z)$ (note that a strict inequality will hold at least for one $j \in I$). Letting $N \to \infty$ from (2.8) we have

$$g^{\omega_2} = g^{\omega_1} + \lim_{N \to \infty} \frac{1}{E_j^{\omega_2} D_N} \sum_{n=0}^{N-1} E_j^{\omega_2} \varphi^{\omega_1}(X_n; z_2(X_n)) \geqq g^{\omega_1} \tag{3.4}$$

and a strict inequality holds if $\varphi^{\omega_1}(j; z_2(j)) > 0$ at least for one state j recurrent with respect to $\|p(j, k; z_2(j))\|$. If $g^{\omega_2} = g^{\omega_1}$ then it can be shown from (2.6) that $w_j^{\omega_2} \geqq w_j^{\omega_1}$ for all $j \in I$ and a strict inequality holds at least for one j. Repeating the above procedure we can construct a sequence of stationary policies $\{\omega_1, \omega_2, \ldots\}$ the terms of which cannot recur. As there exists a finite number of stationary policies $\hat{\omega}$ must be found.

In case that the obtained policy $\hat{\omega}$ does not satisfy conditions (3.2), (3.3), setting $\hat{\omega}_1 \equiv \hat{\omega}$ let us construct stationary policy $\hat{\omega}_2 \sim \hat{z}_2(j)$ so that $\gamma^{\hat{\omega}_1}(j; \hat{z}_2(j)) = \max_z \gamma^{\omega_1}(j; z)$ for $z \in J^{\hat{\omega}_1}(j)$ (so at least one $\gamma^{\omega_1}(j; \hat{z}_2(j)) > 0$). As $g^{\hat{\omega}_2} = g^{\hat{\omega}_1}$ from (2.9) (where $\tilde{\omega} \equiv \hat{\omega}_1$, $\omega \equiv \hat{\omega}_2$) and (2.15) (where $\tilde{\omega} \equiv \hat{\omega}_2$) we obtain

$$w_j^{\hat{\omega}_2} = w_j^{\hat{\omega}_1} + \lim_{N \to \infty} \frac{1}{N} \sum_{n=0}^{N-1} E_j^{\hat{\omega}_2} \gamma^{\hat{\omega}_1}(X_n; z_2(X_n)) \geqq w_j^{\hat{\omega}_1} . \tag{3.5}$$

Using the similar arguments as in the first part of the proof (compare (3.4), (3.5) and (2.6), (2.7)) we can construct a sequence of stationary average optimal policies $\{\hat{\omega}_1, \hat{\omega}_2, \ldots\}$ converging to ω^*. □

3.2. THEOREM. A policy $\bar{\omega} \sim \bar{z}_n$ is average optimal if and only if for any $j \in I$

$$(3.6) \qquad \lim_{N \to \infty} \frac{1}{N} \sum_{n=0}^{N-1} E_j^{\bar{\omega}} \varphi^{\hat{\omega}}(X_n; z_n) = 0 .$$

Then

$$(3.7) \qquad \lim_{N \to \infty} \frac{E_j^{\bar{\omega}} C_N}{E_j^{\bar{\omega}} D_N} = g^* .$$

Proof. The proof immediately follows from (2.8) (where $\tilde{\omega} \equiv \hat{\omega}$) as

$$d' \leqq \frac{1}{N} E_j^{\bar{\omega}} D_N \leqq \bar{d} \quad \text{for any} \quad N . \qquad \square$$

The following lemma will be very useful in the proof of Theorem 3.4.

3.3. LEMMA. Let for $n = 0, 1, \ldots$ $\{a_n\}, \{b_n\}$ (where $a_n \leqq 0$, $|a_n|, |b_n| < K < \infty$) be given sequences fulfilling condition $a_n H \geqq b_n$ for given $H < 0$. Let for certain $n = n_0$ $a_{n_0} < 0$ and $a_n = 0$ for any $n < n_0$. Then

$$(3.8) \qquad \limsup_{N \to \infty} \frac{1}{N} \sum_{n=0}^{N-1} [(N - 1 - n) a_n + b_n] \leqq a_{n_0} .$$

Proof. Notice that $a_n = 0 \Rightarrow b_n \leqq 0$. Obviously, it holds for any integer m_0 (where $0 < m_0 < N$)

$$(3.9) \qquad \frac{1}{N} \sum_{n=0}^{N-1} [(N - 1 - n) a_n + b_n] \leqq \frac{1}{N} [(N - 1 - n_0) a_{n_0} + b_{n_0}] +$$

$$+ \frac{1}{N} \sum_{n=n_0+1}^{N-m_0} (N - 1 - n + H) a_n + \frac{1}{N} \sum_{n=N-m_0+1}^{N-1} (n - 1 - n + H) a_n .$$

Choosing $m_0 \geqq 1 - H$ then for any $n \leqq N - m_0$ $N - 1 - n + H \geqq 0$ and so $\sum_{n=n_0+1}^{N-m_0} (N - 1 - n + H) a_n \leqq 0$. As the last sum on the righthand side of (3.9) contains only a finite number of terms, from (3.9) for $N \to \infty$ immediately follows (3.8). $\square$

3.4. THEOREM. A policy $\bar{\omega}^* \sim \bar{z}_n^*$ is average overtaking optimal if and only if the following conditions are fulfilled for all $n = 0, 1, \ldots$ and any $j \in I$

$$(3.10) \qquad E_j^{\bar{\omega}^*} \varphi^{\omega^*}(X_n; z_n) = 0 ,$$

$$(3.11) \qquad \lim_{N \to \infty} \frac{1}{N} \sum_{n=0}^{N-1} E_j^{\bar{\omega}^*} \gamma^{\omega^*}(X_n; z_n) = 0 .$$

Proof. As $E_j^{\omega}\varphi^{\omega^*}(X_n; z_n) = 0 \Rightarrow E_j^{\omega}\gamma^{\omega^*}(X_n; z_n) \leqq 0$ and the action set is finite setting $a_n = E_j^{\omega}\varphi^{\omega^*}(X_n; z_n)$, $b_n = E_j^{\omega}\gamma^{\omega^*}(X_n; z_n)$ the values a_n, b_n satisfy the assumptions of Lemma 3.3. So if for certain $n = n_0$ $E_j^{\omega}\varphi^{\omega^*}(X_{n_0}; z_{n_0}) < 0$ using the results of Lemma 3.3 we obtain

$$(3.12) \qquad \limsup_{N\to\infty} \frac{1}{N} \sum_{n=0}^{N-1} [(N - 1 - n) E_j^{\omega}\varphi^{\omega^*}(X_n; z_n) + E_j^{\omega}\gamma^{\omega^*}(X_n; z_n)] < 0 .$$

In virtue of (3.7), (3.12) from (2.9), (2.15) (where $\tilde{\omega} \equiv \omega^*$) we have

$$(3.13) \qquad \limsup_{N\to\infty} \frac{1}{N} \sum_{n=0}^{N-1} [E_j^{\omega} C_n - E_j^{\omega^*} C_n - g^*(E_j^{\omega} D_n - E_j^{\omega^*} D_n)] < 0$$

and ω cannot be average overtaking optimal. As under condition (3.10) $E_j^{\omega}\gamma^{\omega^*}(X_n; z_n) \leqq 0$ from (2.9), (2.15) we can infer that

$$(3.14) \qquad \liminf_{N\to\infty} \frac{1}{N} \sum_{n=0}^{N-1} [E_j^{\omega} C_n - E_j^{\omega^*} C_n - g^*(E_j^{\omega} D_n - E_j^{\omega^*} D_n)] =$$

$$= \liminf_{N\to\infty} \frac{1}{N} \sum_{n=0}^{N-1} E_j^{\omega}\gamma^{\omega^*}(X_n; z_n) = 0$$

if and only if (3.11) is satisfied for $\omega \equiv \bar{\omega}^*$. Combining (3.13), (3.14) (written for lim sup) we can show that for an arbitrary policy $\omega \sim z_n$

$$(3.15) \qquad \limsup_{N\to\infty} \frac{1}{N} \sum_{n=0}^{N-1} [E_j^{\omega} C_n - E_j^{\omega^*} C_n - g^*(E_j^{\omega} D_n - E_j^{\omega^*} D_n)] \leqq 0$$

and

$$(3.16) \qquad \lim_{N\to\infty} \frac{1}{N} \sum_{n=0}^{N-1} [E_j^{\bar{\omega}^*} C_n - E_j^{\omega^*} C_n - g^*(E_j^{\bar{\omega}^*} D_n - E_j^{\omega^*} D_n)] = 0$$

holds if and only if conditions (3.10), (3.11) are satisfied. From (3.15), (3.16) immediately follows that

$$(3.17) \qquad \liminf_{N\to\infty} \frac{1}{N} \sum_{n=0}^{N-1} [E_j^{\bar{\omega}^*} C_n - E_j^{\omega} C_n - g^*(E_j^{\bar{\omega}^*} D_n - E_j^{\omega} D_n)] \geqq 0 .$$ □

3.5. Corollary. Stationary policy $\omega^* \sim z^*(j)$ is not only average optimal but even average overtaking optimal.

3.6. Remark. If the considered SMDP reduces to a Markov chain the obtained optimality conditions are equivalent with the optimality conditions established by Denardo and Miller in [1].

4. CONCLUSION

We shall indicate the difficulties arising if Assumption A is not fulfilled. It can be shown that for any stationary policy $\tilde{\omega} \sim \tilde{z}(j)$ there exists unique solution $g_j^{\tilde{\omega}}, w_j^{\tilde{\omega}}$ $(j = 1, \ldots, r)$ to the set of equations (2.1), (2.2) if we write g_j instead of g in the j-th equation (2.1) and set $g_j = \sum_{k \in I} p(j, k; \tilde{z}(j))\, g_k$ for all $j \in I$. Then for an arbitrary policy $\omega \sim z_n$ and all $N = 0, 1, \ldots$ recurrence formula

$$E_j^{\omega} C_N - g_j^{\tilde{\omega}} E_j^{\omega} D_N = \sum_{n=0}^{N-1} E_j^{\omega}(g_{X_{n+1}}^{\tilde{\omega}} - g_{X_n}^{\tilde{\omega}}) \sum_{m=n+1}^{N-1} d(X_m; z_m) +$$

$$+ \sum_{n=0}^{N-1} E_j^{\omega} \varphi^{\tilde{\omega}}(X_n; z_n) + w_j^{\tilde{\omega}} - E_j^{\omega} w_{X_N}^{\tilde{\omega}}$$

can be established. But only in case of a Markov chain (where $D_N = N$, $\sum_{m=n+1}^{N-1} d(X_m; z_m) = N - 1 - n$) we can successfully employ the above recurrence relation for establishing convergence of policy iteration method and for investigating the optimality conditions. Such an analysis was performed in the author's paper [6].

REFERENCES

[1] E. V. Denardo, B. L. Miller: An Optimality Condition for Discrete Dynamic Programming with No Discounting. Annals Mathem. Statistics *39* (1968), 4, 1220–1227.

[2] R. A. Howard: Dynamic Programming and Markov Processes. M.I.T. and Wiley Press, New York 1960.

[3] W. S. Jewell: Markov Renewal Programming I, II. Operations Research *11* (1963), 6, 938–971.

[4] P. Mandl: On the Variance in Controlled Markov Chains. Kybernetika *7* (1971), 1, 1–12.

[5] S. M. Ross: Applied Probability Models with Optimization Applications. Holden-Day, San Francisco 1970.

[6] K. Sladký: Necessary and Sufficient Optimality Conditions for Average Reward of Controlled Markov Chains. Kybernetika *9* (1973), 2, 124–137.

[7] A. F. Veinott: On Finding Optimal Policies in Discrete Dynamic Programming with No Discounting. Annals Mathem. Statistics *37* (1966), 5, 1284–1294.

Czechoslovak Academy of Sciences

Institute of Information Theory and Automation

THE BROWNIAN TABOO BRIDGE WITH AN APPLICATION IN THE THEORY OF NONPARAMETRIC STATISTICS

GÁBOR J. SZÉKELY

BUDAPEST

The notion of Markov chain with taboo states turned out to be a very interesting one [1]. In this paper we discuss some properties of the most fundamental Markov processes (the Wiener process and the Brownian bridge) with taboo states.

EXAMPLE 1. If $w(t)$ $t \geqq 0$ denotes the Wiener process $(w(0) = 0)$, $\varepsilon > 0$ and $\mathrm{d}x$ is an intervall about $x > 0$, then by the reflection principle [2] the conditional density function

$$P(w(1) \in \mathrm{d}x \mid w(t) > -\varepsilon,\ 0 \leqq t \leqq 1) = \frac{\varphi(x) - \varphi(x + 2\varepsilon)}{P(w(t) > -\varepsilon,\ 0 \leqq t \leqq 1)}$$

(where $\varphi(x)$ is the density function of the standard normal distribution). Here the right side tends to

$$\frac{\varphi'(x)}{\int_0^\infty \varphi'(x)\,\mathrm{d}x} = x\mathrm{e}^{-x^2/2}$$

as ε tends to zero. This limit relation can be denoted briefly as follows:

$$P(w(1) \in \mathrm{d}x \mid w(t) \geqq 0,\quad 0 \leqq t \leqq 1) = x\mathrm{e}^{-x^2/2}.$$

In this conditional probability the probability of the condition is zero.

EXAMPLE 2. As in the above example easy to show that

$$\lim_{\substack{\varepsilon \to 0 \\ T \to +\infty}} P(w(1) \in \mathrm{d}x \mid w(t) > -\varepsilon,\quad 0 \leqq t \leqq T) = \sqrt{\frac{2}{\pi}}\, x^2 \mathrm{e}^{-x^2/2} \quad (x > 0).$$

REMARK 1. Let us denote by $w_d(t)$ the d-dimensional Wiener process starting from the zero vector. The absolute value of $w_d(t)$ is (by definition) the Bessel process:

$$r_d(t) = |w_d(t)| \quad t \geqq 0.$$

The transition function of this process is the following [3]

$$g(t, x, y) = t^{-1} e^{-(x^2+y^2)/2t}(xy)^{1-(d/2)} I_{(d/2)-1}\left(\frac{xy}{t}\right) y^{d-1},$$

$t > 0$, $x, y > 0$, $I_{(d/2)-1}$ is the Bessel function. It is interesting to mention that the density function of $r_2(1)$ and $r_3(1)$ is $xe^{-x^2/2}$ and $\sqrt{(2/\pi)}\, x^2 e^{-x^2/2}$ $(x > 0)$ as in Examples 1 and 2.

A deeper result is also true (e.g. in the second case): if

$$0 < t_1 < t_2 < \ldots < t_n$$

then

$$\lim_{\substack{\varepsilon \to 0 \\ T \to +\infty}} P(w(t_1) < x_1,\ w(t_2) < x_2, \ldots, w(t_n) < x_n \mid w(t) > -\varepsilon,\ 0 \leqq t \leqq T) =$$

$$= P(r_3(t_1) < x_1,\ r_3(t_2) < x_2, \ldots, r_3(t_n) < x_n)$$

i.e. the Wiener process whose taboo states are the negative numbers at every $t > 0$ is the 3-dimensional Bessel process. The proof of this fact is somewhat longer than that of Example 1, but it doesn't require any further idea, therefore we don't go into details.

A Gauss process $b(t)$ $0 \leqq t \leqq 1$ with zero expectation and covariances $\operatorname{cov}(b(t), b(s)) = t(1 - s)$, $0 \leqq t \leqq s \leqq 1$ will be called Brownian bridge. (Roughly speaking $b(t)$ is a Wiener process whose taboo states are the nonzero numbers at $t = 1$.) Many interesting results on $b(t)$ can be found e.g. in [4]. Here we prove a theorem on the Brownian bridge whose taboo states are the negative numbers at every $t \in [0, 1]$.

THEOREM.

$$\lim_{\varepsilon \to 0} P(\sup_{0 \leqq t \leqq 1} b(t) < x \mid b(t) > -\varepsilon \textit{ for every } t \in [0, 1]) = \sum_{k=-\infty}^{+\infty} (1 - k^2x^2)\, e^{-k^2x^2/2}.$$

Proof. Let $S_0 = 0$, $S_n = \sum_{i=1}^{n} \xi_i$ $(n = 1, 2, \ldots)$ where $\{\xi_i\}_{i=1}^{\infty}$ are mutually independent random variables,

$$P(\xi_i = \pm 1) = \tfrac{1}{2} \quad (i = 1, 2, \ldots).$$

Because of the invariance principle it is enough to prove that

$$\lim_{n \to \infty} P(\max_{0 \leqq i \leqq 2n} S_i < x\sqrt{(2n)} \mid S_1 > 0, S_2 > 0, \ldots, S_{2n-1} > 0, S_{2n} = 0) =$$

$$= \sum_{k=-\infty}^{+\infty} (1 - k^2x^2)\, e^{-(k^2x^2)/2}.$$

If $2c$ is the least even (integer) number which is not less than $x\sqrt{2n}$ then by the discrete reflection principle [5]:

$$P(\max_{0 \leqq i \leqq 2n} S_i < x\sqrt{(2n)} \mid S_1 > 0, S_2 > 0, \ldots, S_{n-1} > 0, S_{2n} = 0) =$$

$$= \frac{\sum_{k=-\infty}^{+\infty} \left[\binom{2n-2}{n-1-kc} - \binom{2n-2}{n-1-kc+1} \right]}{n^{-1} \binom{2n-2}{n-1}} =$$

$$= \binom{2n = 1}{n}^{-1} \sum_{k=-\infty}^{+\infty} (1 + 2kc) \binom{2n-1}{n+kc} =$$

$$= \binom{2n-1}{n}^{-1} \sum_{k=-\infty}^{+\infty} \binom{2n-1}{n+kc} + \binom{2n-1}{n}^{-1} \sum_{k=1}^{+\infty} 2kc \left[\binom{2n-1}{n+kc} - \binom{2n-1}{n-kc} \right] =$$

$$= \binom{2n-1}{n}^{-1} \sum_{k=-\infty}^{+\infty} \binom{2n-1}{n+kc} + \binom{2n-1}{n}^{-1} \sum_{k=1}^{+\infty} \binom{2n-1}{n+kc} \frac{-4k^2c^2}{n-kc}$$

and this tends to $\sum_{k=-\infty}^{+\infty} (1 - k^2x^2)\, e^{-(k^2x^2)/2}$ as n tends to $+\infty$, q.e.d.

REMARK 2. In the theorem the probability of the condition tends to 0 as ε tends to 0. This type of conditional distributions can be well interpreted in the non-standard analysis [6], [7].

REMARK 3. The distribution function which is on the right side of the theorem, appeared also in [8] in another problem.

REMARK 4. The well-known theorem of Smirnov states that $P(\sup_{0 \leqq t \leqq 1} b(t) < x) = 1 - e^{-2x^2}$ $(x > 0)$ and Example 1 shows that this probability is the same as $P(w(1) < 2x \mid w(t) \geqq 0, 0 \leqq t \leqq 1)$.

COROLLARY OF THEOREM. *Let $F_n(x)$ and $G_n(x)$ be two empirical distribution functions belonging to the continuous distribution functions $F(x)$ and $G(x)$. If $F(x) \equiv G(x)$ then Theorem implies that*

$$\lim_{n \to \infty} P\left(\sqrt{\frac{n}{2}} \sup_{-\infty < x + \infty} (F_n(x) - G_n(x) < z \mid F_n(x) > G_n(x) \text{ for every } x \in (-\infty, +\infty) \right) = \sum_{k=-\infty}^{+\infty} (1 - k^2z^2)\, e^{-(k^2z^2)/2}.$$

Other modifications and generalisations of Smirnov type theorems can be found e.g. in [9].

REFERENCES

[1] K. L. CHUNG: Markov Chains with Stationary Transition Probabilities. Springer-Verlag, Berlin—Göttingen—Heidelberg 1960.
[2] J. L. DOOB: Stochastic Processes. John Wiley and Sons, Inc., New York 1953.
[3] K. ITO, H. P. MCKEAN, JR.: Diffusion Processes and Their Sample Path. Academic Press, New York 1964.
[4] J. HÁJEK, Z. ŠIDÁK: Theory of Rank Statistics. Academia Praha 1967.
[5] W. FELLER: Introduction to Probability Theory and Its Applications, Vol. I. (Third Edition). John Wiley and Sons, Inc., New York 1968.
[6] W. A. J. LUXEMBURG: Applications of Model Theory to Algebra, Analysis and Probability. Holt, Rinehart and Winston, New York 1969.
[7] A. ROBINSON: Non-standard Analysis. North-Holland Publ. Co., Amsterdam 1966.
[8] H. N. KUIPER: Tests Concerning Random Points on a Circle Proc. Koninkl. Nederl. Akad. van Wetenschappen Ser. A. *63* (1960), 38—47.
[9] I. VINCZE: On Some Questions Connected with Two-sample Tests of Smirnov type. Fifth Berkeley Symp. on Math. Stat. and Prob. Vol. I., 1967, p. 657.

EÖTVÖS L. UNIVERSITY

DEPARTMENT OF PROBABILITY THEORY

STOCHASTIC PROCESSES OF SECOND ORDER WITH VALUES IN BANACH SPACES

ALEKSANDER WERON

WROCŁAW

1. INTRODUCTION

In [6] Kolmogorov studied second order stationary one-dimensional processes as curves in Hilbert space. The idea again occured in the paper [4] of Cramér for non-stationary processes. By generalizing the concept of Hilbert space in such a way that the inner product takes values which are no longer scalars but elements of more general topological space, it is possible to extend the above model to infinite-dimensional stochastic processes of second order with values in a suitable abstract space. This idea was studied in [1] for stationary processes, where a pseudo-hilbertian space (LVH-space which first appeared in the paper [7] of Loynes) had been used as the generalized Hilbert space. Independently Saworotnow in [10] proposed a Hilbert module as such a space. We remark that the situation in [1] is in contrast with another generalization of second order stationarity given in [8], where LVH-space valued stationary processes are studied.

We shall present some results on Banach space valued stochastic processes of second order (non-stationary in general) and their linear prediction as an application of the general model given in [1]. The model presented in [10] may not be used for this aim. The stationary case is considered in detail in [2].

2. SECOND ORDER PROCESSES

Fix a probability system $(\Omega, \mathscr{B}, \mathscr{P})$. Let E be a complex Banach space with the topological dual space E^*. A function $x : \Omega \to E$ is said to be a *random element* (*random E-variable*) if for every linear functional $f \in E^*$ the scalar function $f(x(\omega))$ is measurable. The mathematical expectation of a random element is defined by the Pettis integral. If E is a separable Banach space and a random E-variable $x(\omega)$ has a weak p-order $p > 1$ (i.e. $f(x(\omega)) \in L_p(\Omega, \mathscr{B}, \mathscr{P})$ for each $f \in E^*$) then the mathematical expectation of $x(\omega)$ exists. With each random E-variable $x(\omega)$ we may relate in a natural way a linear mapping X on the dual space E^* into the space of all measurable functions:

$$(Xf)(\omega) = f(x(\omega)).$$

The operator X generated by a random element $x(\omega)$ of a weak p-order $(1 \leqq p < \infty)$ is a linear continuous on the dual space E^* to $L_p(\Omega, \mathscr{B}, \mathscr{P})$ (cf. [2, Proposition 1.2]). But we remark that there exist continuous linear operators which are not generated by any random elements. Sufficient conditions for an operator to be generated by a random element are given in [2, Section 1].

In this paper we consider only random elements of second order. We use the covariance theory of Banach space valued random elements which first appeared by Vakhania (cf. [11, Section 4]).

A linear continuous operator $X : E^* \to L_2(\Omega, \mathscr{B}, \mathscr{P})$ is said to be a *generalized random element* on E. The mathematical expectation of it is defined as an element $m \in E$ (if it there exists) such that

$$\int_\Omega (Xf)(\omega)\, \mathscr{P}(\mathrm{d}\omega) = f(m)$$

for each $f \in E^*$. Following Vakhania (cf. [11, p. 137]) we obtain that there exists an antilinear continuous operator R on E^* into E^{**} defined by the relation:

$$(Rf)(g) = \int_\Omega \overline{(Xf)(\omega)}\,(Xg)(\omega)\, \mathscr{P}(\mathrm{d}\omega), \quad f, g \in E^*. \tag{1}$$

The operator R defined in (1) is called the *covariance operator* of X. It is easy to see that this operator may be given by the equivalent form $R = X^*X$, where X^* is the adjoint operator of X. We may define the *correlation operator* of two generalized random elements by the following relation which is analogous to (1)

$$(R_{XY}f)(g) = \int_\Omega \overline{(Xf)(\omega)}\,(Yg)(\omega)\, \mathscr{P}(\mathrm{d}\omega), \quad f, g \in E^*, \tag{2}$$

or simply by $R_{XY} = Y^*X$. We remark that the operator R defined in (1) and the operator R_{XY} defined in (2) belong to the space $\bar{L}(E^*, E^{**})$ – of all antilinear continuous operators on E^* into E^{**}.

Let us denote

$$E^* = B \quad \text{and} \quad L_2(\Omega, \mathscr{B}, \mathscr{P}) = H.$$

We observe that a set of all generalized random elements may be considered as the linear space $L(B, H)$. If we define an inner product in $L(B, H)$ by the following relation

$$[X, Y] = R_{XY} = Y^*X \tag{3}$$

then $L(B, H)$ with the strong operator topology is an LVH-space (cf. [1, Theorem 2]). We remark that the inner product defined in (3) takes its values in the so called admissible space $\bar{L}(B, B^*)$.

DEFINITION. By a stochastic process of second order with values in a Banach space (or simply by a *stochastic process*) we mean a family $(X_t)_{t \in T}$ of elements from $L(B, H)$ indexed by a set T. Its *correlation function* is defined by the relation

$$K(t, s) = [X_t, X_s],$$

where $[\cdot, \cdot]$ denotes the inner product in the LVH-space $L(B, H)$ defined by the correlation operator in (3).

If T is a locally compact abelian group and the stochastic process $(X_t)_{t \in T}$ is *stationary*, i.e. $K(t, s) = K(t - s)$, then spectral representations for the process and for its correlation function may be obtained (cf. [2]). We remark only that analogous spectral representations are known also for special classes of non-stationary processes. For example, for oscilatory processes (see Section 4).

An $\bar{L}(B, B^*)$-valued function $K(t, s)$ defined on $T \times T$ is the correlation function of a stochastic process if and only if it is positive definite i.e. if

$$\sum_{i,j=1}^{N} (K(t_i, t_j)\, b_i)\,(b_j) \geqq 0$$

for each N, $t_1, t_2, \ldots, t_N \in T$ and $b_1, b_2, \ldots, b_N \in B$. The proof of this fact is analogous to the one presented in [2, Theorem 5.2] for stationary processes and hence omitted.

EXAMPLE. Let B be a Banach space and $H = L_2(\Omega, \mathscr{B}, \mathscr{P})$. A mapping $\xi : B \to H$ which satisfies the condition $\int_\Omega \xi(x)\, \mathscr{P}(d\omega) = 0$ for every $x \in B$ is called (one-dimensional) random field of second order over the Banach space B. If a field ξ is differentiable in the Fréchet sense at every point $x \in B$, i.e. if there exists a linear operator $\delta\, \xi(x) \in L(B, H)$ such that

$$\xi(x + h) - \xi(x) = \delta\, \xi(x)\, h + o(\|h\|) \quad \text{as} \quad h \to 0,$$

then the derivative $\delta\xi$ is a stochastic process in our sense defined on $T = B$. More exactly $\delta\xi : B \to L(B, H)$. The correlation function of the process $\delta\xi$ is the $\bar{L}(B, B^*)$-valued function defined on $B \times B$ as follows

$$K(x, y) = \delta\, \xi(y)^*\, \delta\, \xi(x) \quad \text{for} \quad x, y \in B.$$

Random fields over Banach spaces and their derivatives were studied by Jajte in [5].

3. LINEAR PREDICTION PROBLEM

Let $\mathcal{M}_X(S)$ denote a subspace of $L(B, H)$ spanned by the generalized random elements X_t, $t \in S$ and $S \subseteq T$, with operator coefficients from $L(B, B)$. It can be said that $\mathcal{M}_X(S)$ contains all the information available when we know the development of all the components X_t from the set of observations S.

Let $H_X(S)$ denote the subspace of the Hilbert space H spanned by the vectors $X_t(b)$, where $t \in S$ and $b \in B$. In the one-dimensional theory the best possible prediction can be expressed with the aid of the orthogonal projection. For an LVH-space there are no theorems guaranteeing the existence of the orthogonal projection. This is the main difference from Hilbert space situation (cf. [7]). But using idea announced in [1] we obtain

THEOREM 1. *There exists the best linear least squares prediction $\hat{X}_{t,S}$ of the stochastic process $(X_s)_{s \in T}$ at point $t \in T$, based on a subset $S \subseteq T$. It has the following form*

$$\hat{X}_{t,S} = P_S X_t ,$$

where P_S denotes the orthogonal projection on the closed subspace $H_X(S)$ in the Hilbert space H.

Proof. The subspace $\mathcal{M}_X(S)$ is a right closed ideal in $L(B, H)$ i.e., if $Z \in \mathcal{M}_X(S)$ then also $ZA \in \mathcal{M}_X(S)$ for $A \in L(B, B)$. Therefore from the Yood's theorem ([12], Theorem 5.2) follows that the right ideal $\mathcal{M}_X(S)$ contains only operators which the range is included in $H_X(S)$ i.e.,

$$\mathcal{M}_X(S) = \{Z \in L(B, H);\ Z(b) \in H_X(S),\ b \in B\} .$$

Let P_S denote the orthogonal projection into the space $H_X(S)$ in the Hilbert space H. If we put $\mathcal{P}(Z) = P_S Z$ for all $Z \in L(B, H)$ then the operator

$$\mathcal{P} : L(B, H) \to \mathcal{M}_X(S) \subset L(B, H)$$

is linear, continuous, idempotent ($\mathcal{P}^2 = \mathcal{P}$) and hermitian ($[\mathcal{P}(Z), V] = [Z, \mathcal{P}(V)]$, $Z, V \in L(B, H)$). This means that $\mathcal{P}$ is the orthogonal projection onto the closed right ideal $\mathcal{M}_X(S)$ in $L(B, H)$. Hence

$$\hat{X}_{t,S} = \mathcal{P}(X_t) = P_S X_t$$

is the best possible prediction of X_t in terms of elements X_s with $s \in S$.

As an immediate consequence of Theorem 1 we obtain

COROLLARY 1. *If $\hat{X}_{t,S}$ is the best possible prediction then*

$$\hat{X}_{t,S} = \lim_{N \to \infty} \sum_{k=1}^{N} X_{t_k} A_k ,$$

where $t_k \in S$, $A_k \in L(B, B)$ *and the limit is meant in the strong operator topology in* $L(B, H)$. *The operator* $Q_{t,S} \in \bar{L}(B, B^*)$ *defined by the following relation*

$$Q_{t,S} = [X_t - \hat{X}_{t,S}, X_t - \hat{X}_{t,S}]$$

is the coresponding "error" of the prediction.

Let J be a family of subsets of T. We shall say that the stochastic process $(X_t)_{t\in T}$ is

$$J\text{-}singular \quad \text{if} \quad \bigcap_{S\in J} \mathcal{M}_X(S) = \mathcal{M}_X(T)$$

and

$$J\text{-}regular \quad \text{if} \quad \bigcap_{S\in J} \mathcal{M}_X(S) = (0)\,.$$

Let $L_X = \bigcap_{S\in J} H_X(S)$. It is easy to verify that a stochastic process is J-singular if and only if $L_X = H_X(T)$ and is J-regular if and only if $L_X = (0)$.

Let us consider, for example, a process with discrete time, i.e. T is the set of all integers. Let J_1 be the family of sets: $(-\infty, n]$, J_2 the family of complements of all finite subsets and J_3 the family of complements of all singletons of T. We observe that the family J_1 arises in the linear extrapolation problem and the families J_2 and J_3 in the linear interpolation problem. With the usual terminology borrowed from the classical theory we may say that process is J_1-regular or purely nondeterministic, J_1-singular or deterministic, J_2-singular or interpolable and not J_3-singular or minimal, respectively.

Let $(X_t)_{t\in T}$ and $(Y_t)_{t\in T}$ be stochastic processes. $(Y_t)_{t\in T}$ is said to be *J-subordinate* to $(X_t)_{t\in T}$ if

(i) $$\mathcal{M}_Y(S) \subseteq \mathcal{M}_X(S) \quad \text{for all} \quad S \in J\,,$$

(ii) $$\mathcal{M}_Y(T) \subseteq \mathcal{M}_X(T)\,.$$

We can now prove the analogue of the *Wold decomposition* for stochastic processes of second order with values in Banach spaces. Cramér in [3] first proved the Wold decomposition theorem for non-stationary processes.

THEOREM 2. *Let J be a family of subsets of T such that for each $t \in T$ there exists $S \in J$ that $t \in S$. For any given stochastic process $(X_t)_{t\in T}$ there is a uniquely determined decomposition*

$$X_t = V_t + W_t$$

having following properties:

(i) *$(V_t)_{t\in T}$ and $(W_t)_{t\in T}$ are orthogonal stochastic processes i.e. $[V_t, W_s] = (0)$ for each $t, s \in T$,*

(ii) *$(V_t)_{t\in T}$ and $(W_t)_{t\in T}$ are J-subordinate to $(X_t)_{t\in T}$,*

(iii) *$(V_t)_{t\in T}$ is J-singular and $(W_t)_{t\in T}$ is J-regular.*

Proof. Let for each $t \in T$

$$X_t = PX_t + (I - P) X_t = V_t + W_t , \tag{4}$$

where P denotes the orthogonal projection onto L_X in the space H and I – the identical operator in H. By (4) stochastic processes $(V_t)_{t\in T}$ and $(W_t)_{t\in T}$ are well defined and satisfying (i). It is easy to verify that $H_V(S) \subseteq L_X \subseteq H_X(S)$ for $S \in J$ and for $S = T$. Consequently the process $(V_t)_{t\in T}$ is J-subordinate to $(X_t)_{t\in T}$. It implies that $(W_t)_{t\in T}$ as a sum of two J-subordinate processes is J-subordinate to $(X_t)_{t\in T}$ also. So property (ii) is also satisfied.

By (4) for $S \in J$ we have

$$H_X(S) = L_X \oplus H_W(S) \tag{5}$$

and

$$H_X(S) = H_V(S) \oplus H_W(S) . \tag{6}$$

From (5) and (6) we obtain

$$L_V = H_V(S) = H_V(T) = L_X , \quad S \in J . \tag{7}$$

Relation (7) shows that $(V_t)_{t\in T}$ is J-singular. But $L_W \subseteq L_X = L_V$ and $L_W \perp L_V$. We conclude that $(W_t)_{t\in T}$ is J-regular.

We remark that for any decomposition of $(X_t)_{t\in T}$ into $(V_t)_{t\in T}$ and $(W_t)_{t\in T}$ satisfying conditions (i)–(iii), we have that $L_V = L_X$. This relation makes the decomposition unique.

4. ASSOCIATED PROCESSES

From a given stochastic process $(X_t)_{t\in T}$ we may obtain a new one in several ways. Let $P_t \in L(B, B)$ and $Q_t \in L(H, H)$ for $t \in T$. Let

$$Y_t = X_t P_t \quad \text{and} \quad Z_t = Q_t X_t .$$

We shall say that the process $(Y_t)_{t\in T}$ is *r-associated* and $(Z_t)_{t\in T}$ is *l-associated* with $(X_t)_{t\in T}$. If for example, $P_t \equiv P$, where P is a bounded projection onto a finite-dimensional subspace in B then the r-associated process is a simple finite-dimensional stochastic process. Let S be arbitrary subset in T. Then the process $(\hat{X}_{t,S})_{t\in T}$ in view of Theorem 1 is l-associated with $(X_t)_{t\in T}$.

Now we consider another l-associated processes. Let T be the set of real numbers. $(Z_t)_{t\in T}$ is said to be an *oscillatory stochastic process* if

$$Z_t = \int_{-\infty}^{+\infty} e^{itu}\, a_t(u)\, \Phi(du) ,$$

where

(i) Φ is an orthogonal $L(B, H)$-valued strongly countable additive set function,

(ii) the integral in the last formula is the stochastic integral (cf. [2, Theorem 4.7]),

(iii) for each $t \in T \int_{-\infty}^{+\infty} |a_t(u)|^2 F(du)$ exists, where $F(A) = [\Phi(A), \Phi(A)]$.

In the following theorem a characterization of oscillatory stochastic processes is given. This is a generalization of the main result of [9] to the infinite-dimensional case. The proof of [9] can be adapted to establish this result and hence omitted.

THEOREM 3. *$(Z_t)_{t\in T}$ is an oscillatory stochastic process if and only if it is l-assotiated with a stationary stochastic process such that for each $t \in T$ $Q_t U_s(h) = U_s Q_t(h)$, for $h \in \mathscr{D}(Q_t)$ and for all $s \in T$, where U_s is the shift group defined by the stationary process.*

In the following theorem we study a connection between J-regularity and J-singularity of a process $(X_t)_{t\in T}$ and of processes associated with $(X_t)_{t\in T}$.

THEOREM 4. *Let $(X_t)_{t\in T}$ be a stochastic process. Then*

(i) if $(X_t)_{t\in T}$ is J-regular then all r-associated processes are also J-regular if P_t are bounded projections in the Banach space B,

(ii) if for each $b \in B$ r-associated (one-dimensional) processes $(X_t(b))_{t\in T}$ are J-singular then $(X_t)_{t\in T}$ is J-singular,

(iii) if $(X_t)_{t\in T}$ is J-regular and for each $t \in S$ and $S \in J$ $Q_t(H_X(S)) \subseteq H_X(S)$ then the l-associated process is J-regular.

Proof. (i) Let $Y_t = X_t P_t$, where P_t is a bounded projection in B. For each set $S \in J$ we have $H_Y(S) \subseteq H_X(S)$. Since $(X_t)_{t\in T}$ is J-regular hence $L_Y \subseteq L_X = (0)$. This means that the r-associated process $(Y_t)_{t\in T}$ is J-regular.

(ii) Let each one-dimensional process $(X_t(b))_{t\in T}$ be J-singular and suppose that process $(X_t)_{t\in T}$ is not J-singular. Then there exists a set $S_0 \in J$ such that $H_X(S_0) \neq H_X(T)$, i.e. we may find $t_0 \in T$ and $b_0 \in B$ such that $X_{t_0}(b_0) \notin H_X(S_0)$. But it contradicts with assumption that all one-dimensional processes are J-singular.

(iii) If $Z_t = Q_t X_t$, where $Q_t(H_X(S)) \subseteq H_X(S)$ for $t \in S$, then $H_Z(S) \subseteq H_X(S)$ for $S \in J$. Therefore $L_Z \subseteq L_X = (0)$ and consequently $(Z_t)_{t\in T}$ is J-regular.

As a simple application of the part (ii) of this result we obtain the following

COROLLARY 2. *Let $(X_t)_{t\in T}$ be a stationary stochastic process with continuous time. If for each $b \in B$*

$$\int_{-\infty}^{+\infty} \frac{\ln((dF_a b)(b)/du)}{1 + u^2} du = -\infty ,$$

where F_a is absolutely continuous part of the spectral measure, then the process $(X_t)_{t\in T}$ is J_1-singular.

REFERENCES

[1] S. A. Chobanyan, A. Weron: Stationary processes in pseudo-hilbertian space (in Russian). Bull. Acad. Polon. Sci., Sér. Sci. Math. Astronom. Phys. *21* (1973), 847—854.

[2] S. A. Chobanyan, A. Weron: Banach space valued stationary processes and their linear prediction. Diss. Math. *125* (1975), 1—50.

[3] H. Cramér: On some classes of non-stationary stochastic processes. In: Proc. of the Fourth Berkeley Sympos. Math. Statist. and Prob. II, Berkeley 1960. University of California Press, Berkeley and Los Angeles 1961, 57—77.

[4] H. Cramér: Stochastic processes as curves in Hilbert space. Teor. Verojatnost. i Primenen. *9* (1964), 193—204.

[5] R. Jajte: On derivatives of random fields over a Banach space. Bull. Acad. Polon. Sci., Sér. Sci. Math. Astronom. Phys. *22* (1974), 305—311.

[6] A. N. Kolmogorov: Stationary sequences in Hilbert's space (in Russian). Bull. Moskov. Gos. Univ. Mat. *2* (1941), 1—40.

[7] R. M. Loynes: Linear operators in VH-spaces. Trans. Amer. Math. Soc. *116* (1965), 167 to 180.

[8] R. M. Loynes: On a generalization of second-order stationarity. Proc. Lond. Math. Soc. *15* (1965), 385—398.

[9] V. Mandrekar: A characterization of oscillatory processes and their prediction. Proc. Amer. Math. Soc. *32* (1972), 280—284.

[10] P. P. Saworotnow: Abstract stationary processes. Proc. Amer. Math. Soc. *40* (1973), 585—589.

[11] N. N. Vakhania: Probability distributions on linear spaces (in Russian). Mecniereba, Tbilisi 1971.

[12] B. Yood: Additive groups and linear manifolds of transformations between Banach spaces. Amer. J. Math. *71* (1949), 663—677.

Wrocław Technical University

Institute of Mathematics and Theoretical Physics

STOPPING TIMES AND MARKOV PROGRAMMING

J. WESSELS

EINDHOVEN

0. INTRODUCTION

In this paper we consider finite state Markov decision processes with finite decision spaces for each state. The optimality criterion will be total expected discounted reward over an infinite time horizon. For these problems a great variety of optimization procedures has been developed. We divide them in two classes:

policy improvement procedures and

policy improvement-value determination procedures.

For procedures of the first class the main part of each iteration step is a policy improvement procedure ([1], [2], [3], [9]). For procedures of the second class each iteration step contains a policy improvement part and a part in which the values for the new strategy are estimated or computed ([4], [5], [6], [7], [8], [3]). As a matter of fact it is possible to expand any procedure of the first class to a procedure of the second class. For different procedures this has been proved in [3]. A general approach will be presented in a forthcoming paper.

In this paper a unifying approach will be given for all known policy improvement procedures. At the same time a number of new policy improvement procedures is generated. It is proved that, in a way, our generating principle is exhaustive.

We use stopping times for the generation of policy improvement procedures. Actually the choice of a stopping time (or equivalently a go ahead set) will determine a procedure. We will present sufficient and necessary conditions for the stopping time which guarantee the convergence of the procedure (non-zero stopping times) and which guarantee that the procedure only requires the use of stationary Markov or memoryless decision rules (stationary second order Markov or transition memoryless stopping times).

The main tool in this paper consists of the theory of monotone contraction mappings, hich has been used intensively in the past in this type of problems ([10], [11]).

1. STOPPING TIMES AND CONTRACTION IN STOCHASTIC PROCESSES

We consider a stochastic process in a finite state space $S = \{1, \ldots, N\}$ and in discrete time $(n = 0, 1, 2, \ldots)$. S^∞ is the set of allowed paths.

DEFINITION 1.1. a) The function τ on S^∞ with integer values between 0 and ∞ (bounds included) is called a *stopping time* if and only if

$$\tau^{\leftarrow}(n) = B \times S^\infty , \quad \text{with} \quad B \subset S^{n+1} .$$

b) A nonempty subset A of $\bigcup_{k=0}^{\infty} S^k$ is called a *go ahead set*, if and only if

$$(\alpha, \beta) \in A \Rightarrow \alpha \in A \quad \text{for all} \quad \alpha, \beta \in \bigcup_{k=0}^{\infty} S^k .$$

(S^0 only contains a null-tuple which concatenates to α with any α; our definitions imply that any go ahead set contains this null-tuple).

NOTATIONS.

$$A_n = \bigcup_{k=0}^{n} S^k \ (0 \leqq n \leqq \infty) ;$$

the i-th component of $\alpha \in S^n$ $(n \geqq 1)$ is denoted by $[\alpha]_{i-1}$;

if $\alpha \in S^n$, k_α is defined to be n;

hence $\alpha \in S^n$ $(n \geqq 1)$ may be written as $([\alpha]_0, [\alpha]_1, \ldots, [\alpha]_{k_\alpha - 1})$;

hence $k_\gamma = k_\alpha + k_\beta$ if $\gamma = (\alpha, \beta)$;

$A(i) = \{\alpha \in A \mid [\alpha]_0 = i \text{ if } k_\alpha \geqq 1\}$.

There is a one-to-one correspondence between stopping times and go ahead sets:

$$A = \bigcup_{n=0}^{\infty} \{\alpha \in S^n \mid \forall_{\beta \in S^\infty} \ \tau(\alpha, \beta) \geqq n\} .$$

The correspondence between stopping times and go ahead sets may be represented by:

$$\alpha \in A , \quad (\alpha, l) \notin A , \quad l \in S \Leftrightarrow \tau(\alpha, l, s_1, s_2, \ldots) = k_\alpha ,$$

$$(l) \notin A , \quad l \in S \Leftrightarrow \tau(l, s_1, s_2, \ldots) = 0 .$$

We will apply the concepts of stopping time and go ahead set at will, always with this correspondence in mind.

DEFINITION 1.2. A stopping time τ (or its go ahead set A) is said to be *nonzero* if and only if $\tau(\alpha) \geqq 1$ for all $\alpha \in S^{\infty}$, or (equivalently) $S \subset A$.

The only nonzero stopping time which is an entrytime (memoryless) is $\tau \equiv \equiv \infty$ $(A = A_{\infty})$.

Examples of nonzero stopping times.

1.1. A_n $(1 \leqq n \leqq \infty)$;

1.2. A_H defined by: $A_H(i) = S^0 \cup \{(i)\} \cup \{(i, \alpha) \mid \alpha \in \bigcup_{j=1}^{i-1} A_H(j)\}$;

1.3. A_R defined by: $A_R(i) = \bigcup_{n=0}^{\infty} \{\alpha \in S^n \mid [\alpha]_j = i, j = 0, \ldots, n-1, \text{ if } n \geqq 1\}$;

1.4. A_E with E a subset of S, defined by:

$$A_E = \bigcup_{n=2}^{\infty} \{\alpha \in S^n \mid [\alpha]_j \in E, j = 1, \ldots, n-1\} \cup S \cup S^0$$

$(E = S \Rightarrow A_E = A_{\infty};\ E = \emptyset \Rightarrow A_E = A_1)$.

We now suppose that a reward structure has been given: at each time instant n a reward is earned. This reward $q(\alpha)$ depends on the history until that time: $\alpha \in S^{n+1}$. So the reward structure is a function q on A_{∞}. q is supposed to be bounded and (without loss of generality) to be zero on S^0. Rewards are discounted with discount-factor β $(0 < \beta < 1)$. We further denote the state of the stochastic process at time n by the random variable x_n and reward at time n by the random variable q_n. The probability of path $\alpha \in S^n$ is denoted by $P(\alpha)$. $P(\alpha \mid i)$ denotes the probability of α given $x_0 = i$. All such conditional probabilities are supposed to be defined properly. Defined on the process, the stopping time is a random variable.

DEFINITION 1.3. A is a go ahead set, τ its corresponding stopping time. The operator L_A (or L_τ) on R^N is defined by:

$$(L_\tau v)(i) = E(\sum_{k=0}^{\tau-1} \beta^k q_k + \beta^\tau v(x_\tau) \mid x_0 = i)$$

(where E denotes expectation), or equivalently:

$$(L_A v)(i) = \sum_{\alpha \in A(i)} P(\alpha \mid i)\, \beta^{k_\alpha - 1} q(\alpha) + \sum_{\substack{\alpha \in A(i) \\ l \in S \\ (\alpha, l) \notin A(i)}} P(\alpha, l \mid i)\, \beta^{k_\alpha} v(l)\,.$$

THEOREM 1.1 (τ is a stopping time).

a) L_τ is a monotone mapping: $v \geqq w \Rightarrow L_\tau v \geqq L_\tau w$;

b) L_τ is strictly contracting with respect to supnorm in R^N if and only if τ is nonzero;

c) the contraction radius ϱ_τ (or ϱ_A) lies for nonzero τ between 0 and β (bounds included):

$$\varrho_\tau = \max_{i \in S} E(\beta^\tau \mid x_0 = i);$$

d) $\varrho_A \leqq \varrho_B$ if A and B go ahead sets with $A \supset B$, or $\varrho_\tau \leqq \varrho_\sigma$ if $\tau \geqq \sigma$.

Proof. The proof follows straightforward.

A natural question arises after the observation that strictly contracting mappings on R^N possess a unique fixed point: which point is mapped on itself by L_τ if τ is nonzero?

Lemma 1.1. If the stochastic process $\{x_n \mid n = 0, 1, \ldots\}$, the nonzero stopping time τ, and the reward function q satisfy

$$E(\beta^\tau q_{\tau+l} \mid x_0 = i,\ \tau < \infty,\ x_\tau = j) = E(\beta^\tau \mid x_0 = i,\ \tau < \infty,\ x_\tau = j)\, E(q_l \mid x_0 = j)$$

$$\text{for} \quad l = 1, 2, \ldots, i, j \in S \quad \text{with} \quad P(x_0 = i,\ \tau < \infty,\ x_\tau = j) > 0,$$

then L_τ possesses the unique fixed point $L_{A_\infty} 0$ (where 0 denotes the nullvector in R^N).

Proof. Since L_τ possesses a unique fixed point if τ is nonzero, we only have to verify whether:

$$L_\tau L_{A_\infty} 0 = L_{A_\infty} 0, \quad \text{where} \quad L_{A_\infty} 0(i) = E(\sum_{l=0}^{\infty} \beta^l q_l \mid x_0 = i).$$

Using

$$(L_\tau v)(i) = E(\sum_{k=0}^{\tau-1} \beta^k q_k \mid x_0 = i) +$$

$$+ \sum_{j \in S} P(\tau < \infty,\ x_\tau = j \mid x_0 = i)\, E(\beta^\tau \mid x_0 = i,\ \tau < \infty,\ x_\tau = j)\, v(j)$$

the proof follows straightforward.

Theorem 1.2. Suppose:

1. the stopping time τ is nonzero,
2. the random variables τ and $q_{\tau+l}$ are conditionally independent (condition:

$$x_0 = i,\ \tau < \infty,\ x_\tau = j) \quad \text{for all} \quad l \in N,\ i, j \in S,$$

3.

$$E(q_{\tau+l} \mid x_0 = i,\ \tau < \infty,\ x_\tau = j) = E(q_l \mid x_0 = j) \quad \text{for all} \quad l \in N,\ i, j \in S,$$

then $L_{A_\infty} 0$ is the unique fixed point of L_τ.

Proof.

$$E(\beta^\tau q_{\tau+l} \mid x_0 = i,\ \tau < \infty,\ x_\tau = j) =$$
$$= E(\beta^\tau \mid x_0 = i,\ \tau < \infty,\ x_\tau = j)\, E(q_{\tau+l} \mid x_0 = i,\ \tau < \infty,\ x_\tau = j) =$$
$$= E(\beta^\tau \mid x_0 = i,\ \tau < \infty,\ x_\tau = j)\, E(q_l \mid x_0 = j)\,.$$

The statement is now implied by the foregoing lemma.

COROLLARY 1.2. If $\{x_n \mid n = 0, 1, 2, \ldots\}$ is a homogeneous Markov chain and $q(\alpha) = r([\alpha]_{k_\alpha - 1})$ for $k_\alpha \geqq 1$, then $L_{A_\infty}0$ is the unique fixed point for L_τ with τ a nonzero stopping time.

2. STOPPING TIMES AND CONTRACTION IN MARKOV DECISION PROCESSES

From this section on we will treat Markov decision processes with state space S as described below.

DEFINITION 2.1. A *decision rule D* is a function on $\bigcup_{k=1}^{\infty} S^k$ with values in a given set K (here supposed to be finite and nonempty);

the decision rule D is said to be *memoryless* (stationary Markov) if

$$D(\alpha) = D([\alpha]_{k_\alpha - 1}) \quad \text{for each} \quad \alpha \in \bigcup_{k=1}^{\infty} S^k\,;$$

the set of decision rules is denoted by $\mathscr{D}$, the set of memoryless decision rules by $\mathscr{M}$;

the sequence $\{D_n\}_{n=1}^{\infty}$ of decision rules is said to converge to $D \in \mathscr{D}$, if and only if for each $\alpha \in \bigcup_{k=1}^{\infty} S^k$ holds:

$$\lim_{n\to\infty} D_n(\alpha) = D(\alpha)\,.$$

LEMMA 2.1. $\mathscr{D}$ is compact in the topology induced by the limit concept in $\mathscr{D}$.

Proof. The proof proceeds exactly as in [12]: the topology induced by the limit concepts of componentwise convergence is the same as the product topology in the countably infinite topological product of the sets of all maps of S^n into K ($n = 1, 2, \ldots$) (see e.g. Kelley [13]). Hence the compactness of $\mathscr{D}$ follows by Tychonov's theorem.

We suppose that each decision rule D determines a probability law for the stochastic process $\{x_n \mid n = 0, 1, \ldots\}$ in the following way:

$$P_D(\alpha, j, l \mid i) = P_D(\alpha, j \mid i)\, p_{jl}^{D(\alpha, l)} \quad \text{for} \quad \alpha \in A_\infty\,, \quad i, j, l \in S\,,$$

here p_{jl}^k $(j, l \in S,\ k \in K)$ are supposed to be given numbers satisfying: $p_{jl}^k \geqq 0$, $\sum_{l \in S} p_{jl}^k = 1$.

A visit to state i with a decision k gives the reward $r^k(i)$.

Under these conditions a decision rule determines a stochastic process on S with the reward function q defined by:

$$q(\alpha, l) = r_{(l)}^{D(\alpha, l)} \quad (\alpha \in A_\infty,\ l \in S)\,.$$

Since we wish to consider such processes for different decisions rules D, we use D as a subscript or a superscript: $P_D(\alpha)$, L_τ^D, $E_D(\beta^\tau \mid x_0 = i)$, etc.

LEMMA 2.2. If D_1 and D_2 coincide on A: $L_A^{D_1} = L_A^{D_2}$.

LEMMA 2.3. $L_\tau^D v$ is a continuous function of D (τ, v fixed).

From the foregoing section (Theorem 1.1), it follows that L_τ^D is monotone and (if τ is nonzero) strictly contracting with contraction radius:

$$\varrho_\tau^D = \max_{i \in S} E_D(\beta^\tau \mid x_0 = i)\,.$$

The following theorem gives an assertion about the fixed point.

THEOREM 2.1. If D is memoryless, then L_τ^D has the fixed point $L_{A_\infty}^D 0$ for all nonzero τ; if D is not memoryless, then there exists a situation $(\{p_{ij}^k, r_{(i)}^k\})$ such that $L_{\tau_1}^D$ and $L_{\tau_2}^D$ possess different fixed points for certain τ_1 and τ_2.

Proof. If D is memoryless, then $D(\alpha, l) = D(\alpha)$. Hence the process $\{x_n \mid n \geqq 0\}$ then becomes a Markov chain with rewards only depending on the current state. Corollary 1.2 now produces the result.

D not memoryless implies: $\# K > 1$. τ_1 and τ_2 may be chosen identical to 1 and ∞ respectively.

Now: $L_{\tau_1}^D$ possesses the fixed point $L_{A_\infty}^{D_0} 0$ with $D_0(\alpha, l) := D(l)$ for any $\alpha \in A_\infty$, $l \in S$;

$$L_{\tau_2}^D \text{ possesses the fixed point } L_{A_\infty}^D 0\,.$$

It is not difficult to find p_{ij}^k's and $r_{(i)}^k$'s, such that the two fixed points are different: $r_{(i)}^k = \delta_{kD(i)}$ or all $i \in S$, $k \in K$, then $(L_{A_\infty}^D 0)(i) = (1 - \beta)^{-1}$, while $(L_{A_\infty}^D 0)(i) < (1 - \beta)^{-1}$ for at least one $i \in S$, if $p_{jl}^{D(j)} > 0$ for all $j, l \in S$.

Example 2.1. For A_R we get:

$$\varrho^D_{A_R} = \max_i \left[\beta(1 - p^{D(i)}_{ii}) + \beta^2 p^{D(i)}_{ii}(1 - p^{D(i,i)}_{ii}) + \ldots\right] \leqq \beta \frac{1 - p}{1 - \beta p},$$

$$\text{with} \quad p := \min_{i,k} p^k_{ii}.$$

If D memoryless: $\varrho^D_{A_R} = \beta[(1 - q)/(1 - \beta q)]$, with $q := \min_i p^{D(i)}_{ii}$.

Lemma 2.4. (τ is an arbitrary stopping time). For any $v \in R^N$, there exists a decision rule D_0, such that $L^{D_0}_\tau v \geqq L^D_\tau v$ (componentwise) for all D.

Notation. The vector $L^{D_0}_\tau v$ of the foregoing lemma, will be denoted by

$$\max_D L^D_\tau v,\ U_\tau v,\ \max_D L^D_A v,\ U_A v.$$

3. A SET OF OPTIMIZATION PROCEDURES

The operator U_τ serves for some specific choices of τ to construct optimization procedures, which aim actually at finding $U_{A_\infty}0$. In the set-up of this paper the question now arises how generally it is true that U_τ induces a procedure.

Theorem 3.1. a) The operator U_τ on R^N is strictly contracting, if and only if τ is nonzero;

b) the contraction radius ν_τ of U_τ satisfies: $\nu_\tau = \max_D \varrho^D_\tau$;

c) for all nonzero τ the operators U_τ posses the fixed point $U_{A_\infty}0$;

d) $\nu_{\tau_1} \leqq \nu_{\tau_2}$ if $\tau_1 \geqq \tau_2$.

Proof. a) Suppose

$$L^{D_1}_\tau v = U_\tau v, \quad L^{D_2}_\tau w = U_\tau w,$$

$$L^{D_2}_\tau v - L^{D_2}_\tau w \leqq U_\tau v - U_\tau w \leqq L^{D_1}_\tau v - L^{D_1}_\tau w.$$

This implies (as in Theorem 1.1):

$$\|U_\tau v - U_\tau w\| \leqq \|v - w\| \max\{\varrho^{D_1}_\tau, \varrho^{D_2}_\tau\} \leqq \|v - w\| \max_D \varrho^D_\tau.$$

The existence of the last maximum is a consequence of lemma 2.4. Hence $\nu_\tau \leqq \leqq \max_D \varrho^D_\tau$, which already implies the strict contractness property of U_τ for nonzero τ.

That U_τ, if τ is not nonzero, is not strictly contracting follows easily from the fact that any such τ possesses an $i \in S$ with $(U_\tau v)(i) = v(i)$.

b) Take $v(i) = V > 0$, $w(i) = 0$ (all $i \in S$), then:

$$(U_\tau v)(i) - (U_\tau w)(i) = \max_D [(L_\tau^D 0)(i) + VE_D(\beta^\tau \mid x_0 = i)] - \max_D (L_\tau^D 0)(i)$$

$$= V\left\{\max_D \left[\frac{1}{V}(L_\tau^D 0)(i) + E_D(\beta^\tau \mid x_0 = i)\right] - \max_D \frac{1}{V}(L_\tau^D 0)(i)\right..$$

Choose $V > 0$, such that $(1/V)\,\|L_A^D 0\| < \varepsilon$ for all D $(\varepsilon > 0)$.

Then $\|U_\tau v - U_\tau w\| \geqq V\{-\varepsilon + \max_D \max_i E_D(\beta^\tau \mid x_0 = i) - \varepsilon\}$.

Hence $v_\tau \geqq \max_D \varrho_\tau^D$.

c) Suppose A is nonzero, hence U_A posseses a unique fixed point. Consider $U_A U_{A_\infty} 0$.

Its i-th component is equal to:

$$\max_D \Big[\sum_{\alpha \in A(i)} \beta^{k_\alpha - 1} P_D(\alpha \mid i)\, r^{D(\alpha)}([\alpha]_{k_\alpha - 1}) + \sum_{(\alpha, l) \in B} P_D(\alpha, l \mid i)\, \beta^{k_\alpha} \,.$$

$$.\,\{\max_{D_1} \sum_{\gamma \in A_\infty(l)} \beta^{k_\gamma - 1} P_{D_1}(\gamma \mid l)\, r^{D_1(\gamma)}([\gamma]_{k_\gamma - 1})\}\Big] =$$

$$= \max_{D, D_1} \Big[\sum_{\alpha \in A(i)} \beta^{k_\alpha - 1} P_D(\alpha \mid i)\, r^{D(\alpha)}([\alpha]_{k_\alpha - 1}) + \sum_{(\alpha, l) \in B} \beta^{k_\alpha + k_\gamma - 1} \,.$$

$$P_D(\alpha, [\gamma]_0 \mid i)\, P_{D_1}(\gamma \mid [\gamma]_0)\, r^{D_1(\gamma)}([\gamma]_{k_\gamma - 1})\Big] =$$

$$= \max_{D, D_1} (L_{A_\infty}^{(D, D_1)} 0)(i) = \max_D (L_{A_\infty}^D 0)(i)\,,$$

where (D, D_1) denotes the decision rule defined by: $(D, D_1)(\alpha) = D(\alpha)$, if $\alpha \in A$, $(D, D_1)(\alpha, \gamma) = D_1(\gamma)$, if $\alpha \in A$, $(\alpha, [\gamma]_0) \notin A$ and B contains the elements (α, γ) with $\alpha \in A(i)$, $\gamma \in A_\infty$, $(\alpha, [\gamma]_0) \notin A$. $l \in S \subset A_\infty$. The last equality holds since the class of decision rules $\{(D, D_1)\}$ contains $\mathcal{M}$ and

$$\max_D L_{A_\infty}^D 0 = \max_{D \in \mathcal{M}} L_{A_\infty}^D 0$$

(e.g. [6], [10], [12]).

EXAMPLES

3.1. $$v_{A_k} = \beta^k \quad (1 \leqq k \leqq \infty)\,;$$

3.2. $$v_{A_H} = \beta\,;$$

3.3. $$v_{A_R} = \beta \frac{1 - p}{1 - \beta p}\,.$$

In principle this theorem makes it possible to construct an infinite number of procedures for finding $U_{A_\infty}0$, namely choose a nonzero stopping time τ, choose a starting guess $v_0 \in R^N$, and define:

$$v_n = U_\tau v_{n-1} \quad (n = 1, 2, \ldots).$$

Then v_n converges to $U_{A_\infty}0$.

Regrettably the computation of $U_\tau v$ may be equally cumbersome as the computation of $U_{A_\infty}0$. So the following problem for investigation is the characterization of the nonzero stopping times which allow easy computation of $U_\tau v$. The two theorems in the sequel of this section provide the main step for such a characterization.

DEFINITION 3.1. A stopping time τ (and its corresponding go ahead set A) is said to be *transition memoryless*, if and only if there is a subset T of S^2 and a subset S_0 of S, such that:

$$\tau(\alpha) = 0 \Leftrightarrow [\alpha]_0 \in S_0,$$

$$\tau(\alpha) = n > 0 \Leftrightarrow ([\alpha]_k, [\alpha]_{k+1}) \notin T \quad \text{for} \quad k = 0, \ldots, n-2, \quad ([\alpha]_{n-1}, [\alpha]_n) \in T.$$

LEMMA 3.1. Memoryless stopping times are transition memoryless.

THEOREM 3.2. If τ is a transition memoryless stopping time:

$$U_\tau = \max_{D \in \mathcal{M}} L_\tau^D .$$

Proof.

$$(L_\tau^D v)(i) = \begin{cases} v(i) \quad \text{if} \quad i \in S_0, \\ r_{(i)}^{D(i)} + \beta \sum\limits_{j \in T(i)} p_{ij}^{D(i)}\, v(j) + \beta \sum\limits_{\substack{j \notin T(i) \\ j \notin S_0}} p_{ij}^{D(i)} (L_\tau^{D_{ij}} v)(j) + \\ \quad + \beta \sum\limits_{\substack{j \notin T(i) \\ j \in S_0}} p_{ij}^{D(i)} (L_{\tau_1}^{D_{ij}} v)(j) \quad \text{if} \quad i \notin S_0 . \end{cases}$$

Here: $T(i) = \{j \in S \mid (i, j) \in T\}$,

D_{ij} is a decision rule with $D_{ij}(\alpha) = D(i, \alpha)$ if $[\alpha]_0 = j$,

τ_1 is the transition memoryless stopping time with the same T as τ, but with an empty S_0.

It is possible to define a new Markov decision process with essentially the same decision rules in such a way that $L_\tau^D v$ is exactly the vector of total expected discounted rewards. Hence for this new Markov decision process attention may be restricted to memoryless strategies (e.g. [6], [10], [12]), which implies the same for the original problem. This new Markov decision process is defined in the following way: $\bar{S}$, the new set of states, consists of S_0 and two representations of $S : S^* = \{s^* \mid s \in S\}$ and $S_* = \{s_* \mid s \in S\}$. So some states of S are three times represented in $\bar{S}$ and others two times.

For the states $\bar{s} \in S_* \cup S_0$ we define:

$$p^k_{\bar{s}\bar{s}} = 1\,, \quad r^k_{(\bar{s})} = (1 - \beta)\, v(s) \quad (k \in K)\,.$$

For the states $s^* \in S^*$ we define:

$$p^k_{s^* s_1{}^*} = p^k_{ss_1} \quad \text{if} \quad (s, s_1) \notin T \qquad (k \in K)\,,$$

$$p^k_{s^* s_1{}_*} = p^k_{ss_1} \quad \text{if} \quad (s, s_1) \in T \qquad (k \in K)\,,$$

$$r^k_{(s^*)} = r^k_{(s)} \quad (k \in K)\,.$$

Transition memoryless stopping times are the only stopping times, for which restriction to memoryless decision rules is always allowed:

THEOREM 3.3. Suppose the stopping time τ for the state set S is not transition memoryless, then there exists a Markov decision process with state set S (i.e. there exists a set K, and numbers $\{p^k_{ij}\}$, $\{r^k_{(i)}\}$) such that for this Markov decision process

$$U_\tau \neq \max_{D \in \mathcal{M}} L^D_\tau\,.$$

REMARK. In fact, $\max_{D \in \mathcal{M}} L^D_\tau$ may not be defined.

Proof. Representing τ by its go ahead set A, its not being transition memoryless implies the existence of a state i and two paths α, γ such that $B = \{j \mid (\alpha, i, j) \in A\} \neq \{j \mid (\gamma, i, j) \in A\} = C$, while (α, i), $(\gamma, i) \in A$. This implies that in determining U_A the following forms have to be maximized with respect to k and l respectively:

$$(*) \qquad r^k_{(i)} + \beta \sum_{j \notin B} p^k_{ij}\, v(j) + \beta \sum_{j \in B} p^k_{ij} (U_{\tau_1} v)\,(j)\,,$$

$$(**) \qquad r^l_{(i)} + \beta \sum_{j \notin C} p^l_{ij}\, v(j) + \beta \sum_{j \in C} p^l_{ij} (U_{\tau_2} v)\,(j)\,,$$

where $\tau_1(\delta) = \tau(\alpha, i, \delta)$ and $\tau_2(\delta) = \tau(\gamma, i, \delta)$.

By investigating different possibilities for the relation between B and C examples can be constructed for which the maximizing k and l in (*) and (**) are different.

CONCLUSION. It will be clear that the existing policy improvement procedures [1], [8] follows directly from our unifying approach by choosing the corresponding stopping time. While the policy improvement procedure introduced by Reets [2] can be achieved by a slight extension of the set of allowed stopping times.

REFERENCES

[1] J. MacQueen: A modified dynamic programming method for Markovian decision problems. J. Math. Anal. Appl. *14* (1966), 38—43.

[2] D. Reetz: Solution of a Markovian decision problem by successive overrelaxation. Z.f. Oper. Res. *17* (1973), 29—32.

[3] J. A. E. E. v. Nunen: Improved successive approximation methods for discounted Markov decision processes, p. 667—682 in A. Prékopa (ed), Progress in Operations Research, Colloquia Mathematica Societatis János Bolyai, vol. *12* (1975), North Holland Publ. Co, Amsterdam.

[4] R. A. Howard: Dynamic programming and Markov processes. MIT-Press, Cambridge 1960.

[5] G. T. de Ghellinck, G. D. Eppen: Linear programming solutions for separable Markovian decision problems. Man. Sc. *13* (1967), 371—394.

[6] J. Wessels, J. A. E. E. v. Nunen: Discounted semi-Markov decision processes: linear programming and policy iteration. Statistica Neerlandica *29* (1975), 1—7.

[7] J. A. E. E. v. Nunen: A set of successive approximation methods for discounted Markovian decision problems. Zeitschrift für Operations Research *20* (1976), 203—208.

[8] N. Hastings: Some notes on dynamic programming and replacement. Oper. Res. Q. *19* (1963), 453—464.

[9] H. Schellhaas: Zur Extrapolation in Markoffschen Entscheidungsmodellen mit Diskontierung. Zeitschrift für Operations Research *18* (1974), 91—104.

[10] D. Blackwell: Discounted dynamic programming. Ann. Math. Statist. *36* (1965), 226 to 234.

[11] E. V. Denardo: Contraction mappings in the theory underlying dynamic programming. SIAM-Review *9* (1967), 165—177.

[12] J. Wessels: Decision rules in Markovian decision processes with incompletely known transition probabilities. Technological University Eindhoven, 1968.

[13] J. L. Kelley: General topology. D. van Nostrand Comp., New York 1955.

[14] H. Mine, S. Osaki: Markovian decision processes. Elsevier Int. Publ., New York 1970.

Technological University Eindhoven

Department of Mathematics

DAMS WITH VARIABLE RELEASE RATES

GEOFFREY YEO

LINKÖPING

A dam of capacity $K \leqq \infty$ is considered with independents and indentically distributed inputs occuring in a renewal process, and in particular a Poisson process, with a general instantaneous release rate $r(\cdot)$ depending on the content. This is related to a GI/G/1 queue with service times dependent on the waiting time. Some results may be obtained for the limiting content distribution when it exists; the results are more complete for some special release rates such as $r(x) = \mu x^{\alpha}$ $(0 < \mu < \infty,\ 0 \leqq \alpha < \infty)$ and $r(x) = a + \mu x$, and input size distributions such as deterministic and exponential. For $r(x) = \mu x$ the moments of the limiting content distribution may be determined by recurrence for inputs occurring in a renewal process.

Consider a stochastic process $X(t)$, $0 \leqq t < \infty$, called the content of a dam of capacity $K \leqq \infty$, defined on $[0, K]$. Inputs, at $(0 = t_0 < t_1 < t_2 < \ldots)$ occur in a renewal process with $\tau_i = t_{i+1} - t_i$, $i = 1, 2, \ldots$, being independently and identically distributed (i.i.d.) random variables with $P(\tau_i \leqq x) = A(x)$, $0 \leqq x < \infty$, $E(\tau_i) = 1/\lambda$, $0 < \lambda < \infty$, and with Laplace transform (LT) $a^*(\theta)$. The inputs S_n, $n = 1, 2, \ldots$, are i.i.d. random variables with $P(S_n \leqq x) = G(x)$, $0 \leqq x < \infty$, $G(0) = 0$, $E(S_n) = \beta < \infty$, and LT $g^*(\theta)$. When $A(x) = 1 - \exp(-\lambda x)$ there is a compound Poisson input process.

There is an instantaneous release rate $r(X(t))$, which may be a function of the content $X(t)$ at time t, such that $r(0-) = 0$ and $r(x)$ is positive and continuous on $(0, \infty)$. Various aspects of a dam with this general release rule have been considered by a number of authors [1], [2], [4], [5], [7], [8]; in the special case $r(x) = 1$, $0 < x < \infty$, $X(t)$ is the virtual waiting time in a GI/G/1 queue [6] (possibly with a maximum waiting time).

By making the transformation $Z(t) = D(X(t))$, where [5]

$$D(x) - D(z) = \int_z^x \frac{1}{r(y)}\,\mathrm{d}y\,, \quad 0 < z < x < \infty\,, \tag{1}$$

and $X(t_n-) = X_n$, $Z(t_n-) = Z_n$, then the sequence Z_n, $n = 1, 2, \ldots$, may be regarded as the virtual waiting time of successive customs in a generalized GI/G/1 queueing system, where the service time of a customer depends on its waiting time. The service time of the nth customer is $S_n^*(Z_n) = D(S_n + D^{-1}(Z_n)) - Z_n$, where $D^{-1}(x)$ is the unique inverse function of $D(x)$ and its conditional distribution function (d.f.) is

$$P(S_n^*(Z_n) \leqq x \mid Z_n = z) = G(D^{-1}(x + z) - D^{-1}(z)),$$
$$0 < x < \infty, \quad 0 < z < \infty.$$

Furthermore the content at successive input times is related by

$$X_{n+1} = [D^{-1}\{D(X_n + S_n) - \tau_n\}]^+ . \tag{2}$$

For $D(x) = x$, which is GI/G/1, this is Lindley's form [6], and the limiting d.f. $H(x) = \lim_{n \to \infty} P(X_n \leqq x \mid X(0) = x_0)$ may be determined by Wiener-Hopf methods, whenever it exists, i.e. for $E(S_n) < E(\tau_n)$. The present problem is much more complex.

For a compound Poisson input process some results may be obtained for the limiting d.f. $F(x) = \lim_{t \to \infty} P(X(t) \leqq x \mid X(0) = x_0) = P(X \leqq x)$, whenever the limiting random variable X exists with a proper d.f., i.e. always for $K < \infty$ and when $\lim_{x \to \infty} r(x) > \lambda\beta$ for $K = \infty$. Formally [2]

$$r(x)\, F'(x) = \lambda F(x) - \lambda \int_{y=0}^{x} F(x - y)\, \mathrm{d}\, G(y), \quad 0 < x < K, \tag{3}$$

and for the transformed process $Z = D(X)$ with d.f. $L(z) = P(Z \leqq z)$

$$L'(z) = \lambda L(z) - \lambda \int_{y=0}^{z} L(y)\, \mathrm{d}\, G(D^{-1}(z) - D^{-1}(y)), \quad 0 < z < D(K). \tag{4}$$

This has been solved for some special cases of $r(x)$, e.g. $r(x) = 1$ [6], $r(x) = \mu x$ [3] and $r(x) = a + \mu x$ [8], and may be solved for some special input size distributions. If $G(x) = 1 - \exp(-\nu x)$, $0 \leqq x < \infty$, $0 < \nu < \infty$, then

$$F(x) = \frac{\mathrm{e}^{Q(x)} + \nu \int_{y=0}^{x} \mathrm{e}^{Q(y)}\, \mathrm{d}y}{\mathrm{e}^{Q(K)} + \nu \int_{y=0}^{K} \mathrm{e}^{Q(y)}\, \mathrm{d}y}, \quad 0 \leqq x \leqq K,$$

where $Q(x) = -\nu x + \lambda\, D(x)$, which has been obtained in [4] by another method. Some further results for this case and for inputs bounded away from zero are given in [8].

When $r(x) = \mu x$ there is an exponential decay except at times of inputs, so $X_{n+1} = \{\min(K, X_n + S_n)\} \exp(-\mu\tau_n)$, and it is possible to use the properties of independence between the variables to obtain further results, even for inputs occurring in a renewal process. For example with $K = \infty$ all moments of X can be obtained iteratively from

$$E(X^m) = \frac{\lambda}{\mu m} \sum_{j=0}^{m-1} \binom{m}{j} E(S^{m-j})\, E(X^j)\, a^*(j\mu)\,, \quad m = 1, 2, 3, \ldots$$

REFERENCES

[1] E. CINLAR, M. PINSKY: On dams with additive inputs and a general release rule. J. Appl. Prob. *9* (1972), 422–429.

[2] D. P. GAVER, R. G. MILLER: Limiting distributions for some storage problems. In: K. J. Arrow, S. Karlin and H. Scarf, ed. Studies in Applied Probability and Management Science. Stanford U.P., 1962.

[3] J. KEILSON, N. D. MERMIN: The second-order distribution of integrated shot noise. IRE Trans. on Information Theory *IT-5* (1959), 75–77.

[4] D. R. MCNEIL: A simple model for a dam in continuous time with Markovian inputs. Z. Wahrscheinlichkeitstheorie verw. Geb. *21* (1972), 241–254.

[5] P. A. P. MORAN: A theory of dams with continuous input and a general release rule. J. Appl. Prob. *6* (1969), 88–93.

[6] N. U. PRABHU: Queues and Inventories. Wiley, New York 1965.

[7] G. F. YEO: A finite dam with variable release rate. J. Appl. Prob. *12* (1975), 205–211.

[8] G. F. YEO: A dam with general release rule. Linköping University, Department of Mathematics, Research Report 1974.

ODENSE UNIVERSITY

DEPARTMENT OF MATHEMATICS

ON EXPECTED FIGURES AND A STRONG LAW OF LARGE NUMBERS FOR RANDOM ELEMENTS IN QUASI-METRIC SPACES

HERBERT ZIEZOLD

BIELEFELD

One can find in nature various sorts of figures which challenge one to study them by probabilistic methods. In this paper we are concerned with the notion of expected figures based on ideas of Fréchet [5] on expected elements ('positions moyennes') in metric spaces.

After having introduced some notions and definitions a strong law of large numbers for random elements in separable, finite quasi-metric spaces is proved. It follows the study of random figures being random n-tuples of points in $\mathbf{R}^N$, $N \in \mathbf{N}$, or respectively random equivalence classes of such n-tupels with respect to a suitably chosen quasi-metric in $(\mathbf{R}^N)^n$. For $N = 1$ the connection to order statistics is pointed out, and for $N = 2$ some relations are shown which serve to demonstrate two counter-examples against uniqueness of expected figures in $(\mathbf{R}^N)^n$ and against a relation like $\mathrm{E}(\mathrm{E}(X \mid \mathscr{G})) = \mathrm{E}(X)$.

1. NOTIONS AND DEFINITIONS

Let $(\Omega, \mathscr{F}, \mathrm{P})$ be a probability space (for denotations see e.g. [6]) and $(\mathscr{X}, d)$ a quasi-metric space, i.e. $d(x, y) = 0$, $x, y \in \mathscr{X}$, need not imply $x = y$ and $d(x, y)$ may be infinite (see e.g. [8]). Let $X : \Omega \to \mathscr{X}$ be a random element in $(\mathscr{X}, d)$, i.e. the map $\omega \to X(\omega)$, must be Borel-measurable. If then for $a \in \mathscr{X}$

$$(1) \qquad \mathrm{E}(d(X, a)^2) = \inf_{\alpha \in \mathscr{X}} \mathrm{E}(d(X, \alpha)^2) < \infty ,$$

a is called an *expected element of* X (Fréchet [5]). The set of all expected elements of X is denoted by $\mathbf{E}(X)$.

Analogically, an element $\boldsymbol{a} \in \mathscr{X}$ is called a *mean of* $\boldsymbol{x}_1, \ldots, \boldsymbol{x}_n \in \mathscr{X}$, if

$$\sum_{i=1}^{n} d(\boldsymbol{x}_i, \boldsymbol{a})^2 = \inf_{\alpha \in \mathscr{X}} \sum_{i=1}^{n} d(\boldsymbol{x}_i, \boldsymbol{\alpha})^2 < \infty . \tag{2}$$

The set of all means of $\boldsymbol{x}_1, \ldots, \boldsymbol{x}_n$ is denoted by $\mathbf{M}(\boldsymbol{x}_1, \ldots, \boldsymbol{x}_n)$.

Obviously these definitions are natural generalizations of expectations of real-valued random variables with existing second moments.

Another generalization of expectation for random elements X in metric spaces $(\mathscr{X}, d)$ was given by Doss [2]: $\boldsymbol{a} \in \mathscr{X}$ is an expected element of X if $d(\boldsymbol{a}, \boldsymbol{\alpha}) \leqq \mathrm{E}(d(X, \boldsymbol{\alpha}))$ for all $\boldsymbol{\alpha} \in \mathscr{X}$.

These two definitions of expected elements are in general not equivalent. In this paper we are only concerned with the expected elements in the sense of M. Fréchet.

In what follows we shall only consider finite quasi-metrics. By considering equivalence-classes one could of course reduce these to metrics, but to simplify denotations later on we shall formulate our results for finite quasi-metric spaces.

2. A STRONG LAW OF LARGE NUMBERS

The following theorem is a deeper justification of the notion of expected elements in quasi-metric spaces than a formal generalization can be. It can be regarded as a strong law of large numbers. Similar results seem to be known only for expected elements in the sense of Doss [2] and for random elements in Banach spaces (see e.g. [3], [4], [7]).

THEOREM. *Let* $(\Omega, \mathscr{F}, \mathrm{P})$ *be a probability space and* $(\mathscr{X}, d)$ *a separable quasi-metric space with finite quasi-metric d. Let* $X_1, X_2, \ldots$ *be independent, identically distributed random elements* $X_i : \Omega \to \mathscr{X}$, $i = 1, 2, \ldots$, *such that* $\mathrm{E}(d(X_1, \boldsymbol{\alpha})^2) < \infty$ *for at least one* $\boldsymbol{\alpha} \in \mathscr{X}$. *Then for almost all* $\omega \in \Omega$

$$M(\omega) = \bigcap_{k=1}^{\infty} \overline{\bigcup_{n=k}^{\infty} \mathbf{M}(X_1(\omega), \ldots, X_n(\omega))} \subseteq \mathbf{E}(X_1) , \tag{3}$$

where $\bar{D}$ *denotes the closure of the set D.*

REMARK 1. Identical distributions of $X_1, X_2, \ldots$ in $(\mathscr{X}, d)$ entail identical distributions of $d(X_1, \boldsymbol{\alpha})^2, d(X_2, \boldsymbol{\alpha})^2, \ldots$ in $\mathbf{R}$, $\boldsymbol{\alpha} \in \mathscr{X}$. Therefore $\mathbf{E}(X_1) = \mathbf{E}(X_2) = \ldots$ and

$\mathrm{E}(d(X_1, \alpha)^2) < \infty$ implies $\mathrm{E}(d(X_i, \alpha)^2) < \infty$, $i = 2, 3, \ldots$. As d is supposed to be a finite quasi-metric, $\mathrm{E}(d(X_1, \alpha)^2) < \infty$ for one $\alpha \in \mathscr{X}$ implies the same relation for all $\alpha \in \mathscr{X}$.

REMARK 2. If $(\mathscr{X}, d)$ is the real line $\mathbf{R}$ with the usual metric, then the well-known strong law of large numbers for real-valued random variables states $M(\omega) = \{\mathrm{E}(X_1)\} = \mathrm{E}(X_1)$ for almost all $\omega \in \Omega$, but this is not true for all metric spaces as we shall see later on.

Proof of Theorem. Because of the supposed separability of $(\mathscr{X}, d)$, there is a dense sequence $\alpha_1, \alpha_2, \ldots$ in $\mathscr{X}$. For such a sequence we define a set A by

$$A = \left\{\omega \in \Omega \mid \lim_{n\to\infty} \frac{1}{n} \sum_{i=1}^{n} d(X_i(\omega), \alpha_k)^m = \mathrm{E}(d(X_1, \alpha_k)^m) \quad \text{for} \quad m = 1, 2, \text{ and } k = 1, 2, \ldots\right\}.$$

From the well-known strong law of large numbers for real-valued random variables it follows by simple measure-theoretic conclusions that A is measurable and $\mathrm{P}(A) = 1$. We shall show that for $\omega \in A$

$$\lim_{n\to\infty} \frac{1}{n} \sum_{i=1}^{n} d(X_i(\omega), \alpha)^2 = \mathrm{E}(d(X_1, \alpha)^2) \tag{4}$$

is even valid for all $\alpha \in \mathscr{X}$:

Given $\omega \in A$ and $\alpha \in \mathscr{X}$ let $\alpha_{k_1}, \alpha_{k_2}, \ldots$ be a subsequence of $\alpha_1, \alpha_2, \ldots$ converging to α. Then for $j = 1, 2, \ldots$

$$\left|\frac{1}{n} \sum_{i=1}^{n} \left(d(X_i(\omega), \alpha_{k_j})^2 - d(X_i(\omega), \alpha)^2\right)\right| \leqq$$

$$\leqq \frac{1}{n} \sum_{i=1}^{n} \left(d(X_i(\omega), \alpha_{k_j}) + d(X_i(\omega), \alpha)\right) \left|d(X_i(\omega), \alpha_{k_j}) - d(X_i(\omega), \alpha)\right| \leqq$$

$$\leqq \frac{1}{n} \sum_{i=1}^{n} \left(d(X_i(\omega), \alpha_{k_j}) + d(X_i(\omega), \alpha_{k_j}) + d(\alpha_{k_j}, \alpha)\right) d(\alpha_{k_j}, \alpha) =$$

$$= d(\alpha_{k_j}, \alpha)^2 + 2d(\alpha_{k_j}, \alpha) \frac{1}{n} \sum_{i=1}^{n} d(X_i(\omega), \alpha_{k_j}),$$

thus

$$\frac{1}{n}\sum_{i=1}^{n} d(X_i(\omega), \alpha_{k_j})^2 - d(\alpha_{k_j}, \alpha)^2 - 2d(\alpha_{k_j}, \alpha)\frac{1}{n}\sum_{i=1}^{n} d(X_i(\omega), \alpha_{k_j}) \leqq$$

$$\leqq \frac{1}{n}\sum_{i=1}^{n} d(X_i(\omega), \alpha)^2 \leqq$$

$$\leqq \frac{1}{n}\sum_{i=1}^{n} d(X_i(\omega), \alpha_{k_j})^2 + d(\alpha_{k_j}, \alpha)^2 + 2d(\alpha_{k_j}, \alpha)\frac{1}{n}\sum_{i=1}^{n} d(X_i(\omega), \alpha_{k_j}) .$$

Because of $\omega \in A$, we get for $n \to \infty$

$$\mathrm{E}(d(X_1, \alpha_{k_j})^2) - d(\alpha_{k_j}, \alpha)^2 - 2d(\alpha_{k_j}, \alpha)\,\mathrm{E}(\mathrm{d}(X_1, \alpha_{k_j})) \leqq$$

$$\leqq \varliminf_{n\to\infty} \frac{1}{n}\sum_{i=1}^{n} d(X_i(\omega), \alpha)^2 \leqq \varlimsup_{n\to\infty} \frac{1}{n}\sum_{i=1}^{n} d(X_i(\omega), \alpha)^2 \leqq$$

$$\leqq \mathrm{E}(d(X_1, \alpha_{k_j})^2) + d(\alpha_{k_j}, \alpha)^2 + 2d(\alpha_{k_j}, \alpha)\,\mathrm{E}(d(X_1, \alpha_{k_j})) ,$$

$j = 1, 2, \ldots$, and because of the continuity of $\mathrm{E}(d(X_1, \cdot))$ and $\mathrm{E}(d(X_1, \cdot)^2)$, we get (4) for $j \to \infty$.

To end the proof of the theorem we consider for $\omega \in A$ an element $z \in M(\omega)$, the case $M(\omega) = \emptyset$ is trivial, and show that $z \in \mathbf{E}(X_1)$. For this purpose we choose for every $k = 1, 2, \ldots$ an element $z_k \in \bigcup_{n=k}^{\infty} \mathbf{M}(X_1(\omega), \ldots, X_n(\omega))$ with $d(z_k, z) \leqq 1/k$ and define $n_k = \min\{n \in \mathbf{N} \mid n \geqq k,\ z_k \in \mathbf{M}(X_1(\omega), \ldots, X_n(\omega))\}$. From

$$\left(\frac{1}{n_k}\sum_{i=1}^{n_k} d(X_i(\omega), z)^2\right)^{1/2} \leqq \left(\frac{1}{n_k}\sum_{i=1}^{n_k} d(X_i(\omega), z_k)^2\right)^{1/2} + \left(\frac{1}{n_k}\sum_{i=1}^{n_k} d(z_k, z)^2\right)^{1/2} \leqq$$

$$\leqq \left(\frac{1}{n_k}\sum_{i=1}^{n_k} d(X_i(\omega), z_k)^2\right)^{1/2} + \frac{1}{k}$$

it follows for $k \to \infty$

$$(\mathrm{E}(d(X_1, z)^2))^{1/2} \leqq \varliminf_{k\to\infty} \left(\frac{1}{n_k}\sum_{i=1}^{n_k} d(X_i(\omega), z_k)^2\right)^{1/2}, \tag{5}$$

while because of $z_k \in \mathbf{M}(X_1(\omega), \ldots, X_{n_k}(\omega))$ the inequality

$$\frac{1}{n_k}\sum_{i=1}^{n_k} d(X_i(\omega), z_k)^2 \leqq \frac{1}{n_k}\sum_{i=1}^{n_k} d(X_i(\omega), \alpha)^2 \tag{6}$$

holds for any $\alpha \in \mathscr{X}$, $k \in \mathbf{N}$. (5) and (6) yield

$$\mathrm{E}(d(X_1, z)^2) \leqq \mathrm{E}(d(X_1, \alpha)^2)$$

for every $\alpha \in \mathscr{X}$, which implies $z \in \mathrm{E}(X_1)$.

The relation $\mathbf{E}(X_1) = M(\omega)$ for almost every $\omega \in \Omega$ is not true in general, as the following example shows.

EXAMPLE. Let $0'$ be an element not contained in $\mathbf{R}$ and $\mathscr{X} = \mathbf{R} \cup \{0'\}$. We extend the Euclidean metric on $\mathbf{R}$ to $\mathscr{X}$ by defining $d(x, 0') = |x|$ for $|x| \geqq 1$, and $d(x, 0') = 2 - |x|$ for $|x| \leqq 1$. Let $X_1, X_2, \ldots$ be a sequence of independent random elements in $\mathscr{X}$, being uniformly distributed in $[-2, -1] \cup [1, 2]$ with respect to Lebesgue measure in $\mathbf{R}$. Obviously $\mathbf{E}(X_1) = \{0, 0'\}$, but for almost all $\omega \in \Omega$ $0' \notin \overline{\bigcup_{n=1}^{\infty} \mathbf{M}(X_1(\omega), \ldots, X_n(\omega))}$, thus $0' \notin M(\omega)$ and therefore $\mathbf{E}(X_1) \neq M(\omega)$ for almost all $\omega \in \Omega$.

3. EXPECTED FIGURES IN $(\mathbf{R}^N)^n$

In this section a way of studying figures by probabilistic methods will be pointed out. For this purpose we consider as figures n-tupels of points in the N-dimensional Euclidean space $\mathbf{R}^N$, $n, N \in \mathbf{N} = \{1, 2, \ldots\}$, or respectively the equivalence classes of these n-tupels with respect to some closed subgroup $\mathscr{B}$ of the group of motions in $\mathbf{R}^N$ and some subgroup Ξ of the symmetric group S_n of the permutations of the set $\{1, 2, \ldots, n\}$. Therefore we use the quasi-metric space $(\mathscr{X}, d)$ with $\mathscr{X} = (\mathbf{R}^N)^n$ and

$$(7) \qquad d((x_1', \ldots, x_n'), (x_1'', \ldots, x_n'')) = \inf_{\pi \in \Xi, B \in \mathscr{B}} \left(\sum_{j=1}^{n} (Bx'_{\pi(j)} - x''_j)^2\right)^{1/2}.$$

This quasi-metric d will be denoted by $d_{\mathscr{B},\Xi}$. Unless indicated otherwise, we shall always suppose that $\mathscr{B}$ contains the group $\mathscr{T}_N$ of all translations in $\mathbf{R}^N$.

Given d by (7), we say that $(x_1^{(1)}, \ldots, x_n^{(1)}), \ldots, (x_1^{(r)}, \ldots, x_n^{(r)}) \in \mathscr{X}$ are *in optimal position to each other* (*with respect to* $d_{\mathscr{B},\Xi}$), if

$$(8) \qquad \sum_{1 \leq i < k \leq r} \sum_{j=1}^{n} (x_j^{(i)} - x_j^{(k)})^2 = \inf_{\substack{B_1, \ldots, B_r \in \mathscr{B} \\ \pi_1, \ldots, \pi_r \in \Xi}} \sum_{1 \leq i < k \leq r} \sum_{j=1}^{n} (B_i x^{(i)}_{\pi_i(j)} - B_k x^{(k)}_{\pi_k(j)})^2 .$$

PROPOSITION 1. *If* $x_i = (x_1^{(i)}, \ldots, x_n^{(i)}) \in \mathscr{X} = (\mathbf{R}^N)^n$, $i = 1, \ldots, r$, $\mathscr{X}$ *being endowed with a quasi-metric* $d = d_{\mathscr{B},\Xi}$, *are in optimal position to each other, then the centres of gravity* $1/n \sum_{j=1}^{n} x_j^{(i)}$, $i = 1, \ldots, r$, *are all equal and a mean is given by* $(1/r \sum_{i=1}^{r} x_1^{(i)}, \ldots, 1/r \sum_{i=1}^{r} x_n^{(i)})$.

Conversely, if $(x_1^{(0)}, \ldots, x_n^{(0)})$ *is a mean of* $(x_1^{(1)}, \ldots, x_n^{(1)}), \ldots, (x_1^{(r)}, \ldots, x_n^{(r)}) \in \mathscr{X}$, *and if for* $B_1^0, \ldots, B_r^0 \in \mathscr{B}$, $\pi_1^0, \ldots, \pi_r^0 \in \Xi$

$$(9) \qquad \sum_{i=1}^{r} \sum_{j=1}^{n} (B_i^0 x_{\pi_i^0(j)}^{(i)} - x_j^{(0)})^2 = \inf_{\substack{B_1, \ldots, B_r \in \mathscr{B} \\ \pi_1, \ldots, \pi_r \in \Xi}} \sum_{i=1}^{r} \sum_{j=1}^{n} (B_i x_{\pi_i(j)}^{(i)} - x_j^{(0)})^2 ,$$

then $(B_i^0 x_{\pi_i^0(1)}^{(i)}, \ldots, B_i^0 x_{\pi_i^0(n)}^{(i)})$, $i = 1, \ldots, r$, *are in optimal position to each other and* $x_j^{(0)} = 1/r \sum_{i=1}^{r} B_i^0 x_{\pi_i^0(j)}^{(i)}$, $j = 1, \ldots, r$.

Furthermore

$$(10) \qquad \inf_{\alpha \in X} \sum_{i=1}^{r} d(x_i, \alpha)^2 = \inf_{\substack{B_1, \ldots, B_r \in \mathscr{B} \\ \pi_1, \ldots, \pi_r \in \Xi}} \frac{1}{r} \sum_{1 \leq i < k \leq r} \sum_{j=1}^{n} (B_i x_{\pi_i(j)}^{(i)} - B_k x_{\pi_k(j)}^{(k)})^2 .$$

For proving this proposition we need the following lemma.

LEMMA 1. *Let X and Y be identically distributed, independent random variables in* $\mathbf{R}^N$, $N \in \mathbf{N}$, *with existing second moments. Then*

$$(11) \qquad \mathrm{E}((X - Y)^2) = 2\mathrm{E}((X - \mathrm{E}(X))^2) ,$$

especially

$$(12) \qquad \frac{1}{r} \sum_{1 \leq i < k \leq r} (x_i - x_k)^2 = \sum_{i=1}^{r} \left(x_i - \frac{1}{r} \sum_{l=1}^{r} x_l\right)^2$$

for any $x_1, \ldots, x_r \in \mathbf{R}^N$.

Proof. Because of the independence and the identical distribution of X and Y we get

$$\mathrm{E}((X - Y)^2) = \mathrm{E}(X^2 + Y^2 - 2XY) = \mathrm{E}(X^2) + \mathrm{E}(Y^2) - 2\mathrm{E}(XY) =$$
$$= 2\mathrm{E}(X^2) - 2\mathrm{E}(X)\,\mathrm{E}(X) = 2\mathrm{E}((X - \mathrm{E}(X))^2) .$$

(12) is a trivial application of (11).

Proof of Proposition 1. To show the first part, let $\boldsymbol{x}_i = (x_1^{(i)}, \ldots, x_n^{(i)}) \in (\mathbf{R}^N)^n$, $i = 1, \ldots, r$, be in optimal position to each other. As $\mathscr{B}$ contains $\mathscr{T}_N$ and therefore

$$\sum_{1 \leq i < k \leq r} \sum_{j=1}^{n} (x_j^{(i)} - x_j^{(k)})^2 = \inf_{t_1, \ldots, t_r \in \mathbf{R}^N} \sum_{1 \leq i < k \leq r} \sum_{j=1}^{n} ((x_j^{(i)} - t_i) - (x_j^{(k)} - t_k))^2 ,$$

and as for every pair (i, k), $1 \leq i < k \leq r$, the infimum of $\sum_{j=1}^{n} (x_j^{(i)} - x_j^{(k)} - t)^2$,

$t \in \mathbf{R}^N$, is only attained for $t_{ik} = 1/n \sum_{j=1}^{n} (x_j^{(i)} - x_j^{(k)})$, the centres of gravity of the figures $\boldsymbol{x}_i$, $i = 1, \ldots, r$, are all equal.

Furthermore we have, by using Lemma 1,

$$\sum_{j=1}^{n} \sum_{i=1}^{r} \left(x_j^{(i)} - \frac{1}{r} \sum_{l=1}^{r} x_j^{(l)} \right)^2 = \frac{1}{r} \sum_{j=1}^{n} \sum_{1 \leq i < k \leq r} (x_j^{(i)} - x_j^{(k)})^2 =$$

$$= \inf_{\substack{B_1, \ldots, B_r \in \mathscr{B} \\ \pi_1, \ldots, \pi_r \in \Xi}} \frac{1}{r} \sum_{1 \leq i < k \leq r} \sum_{j=1}^{n} (B_i x_{\pi_i(j)}^{(i)} - B_k x_{\pi_k(j)}^{(k)})^2 =$$

$$= \inf_{\substack{B_1, \ldots, B_r \in \mathscr{B} \\ \pi_1, \ldots, \pi_r \in \Xi}} \sum_{j=1}^{n} \sum_{i=1}^{r} \left(B_i x_{\pi_i(j)}^{(i)} - \frac{1}{r} \sum_{l=1}^{r} B_l x_{\pi_l(j)}^{(l)} \right)^2 =$$

$$= \inf_{\substack{B_1, \ldots, B_r \in \mathscr{B} \\ \pi_1, \ldots, \pi_r \in \Xi}} \sum_{j=1}^{n} \left(\inf_{x_j^{(0)} \in \mathbf{R}^N} \sum_{i=1}^{r} (B_i x_{\pi_i(j)}^{(i)} - x_j^{(0)})^2 \right) =$$

$$= \inf_{(x_1^{(0)}, \ldots, x_n^{(0)}) \in \mathscr{X}} \inf_{\substack{B_1, \ldots, B_r \in \mathscr{B} \\ \pi_1, \ldots, \pi_r \in \Xi}} \sum_{i=1}^{r} \sum_{j=1}^{n} (B_i x_{\pi_i(j)}^{(i)} - x_j^{(0)})^2 =$$

$$= \inf_{(x_1^{(0)}, \ldots, x_n^{(0)}) \in \mathscr{X}} \sum_{i=1}^{r} \left(\inf_{B_i \in \mathscr{B}, \pi_i \in \Xi} \sum_{j=1}^{n} (B_i x_{\pi_i(j)}^{(i)} - x_j^{(0)})^2 \right) =$$

$$= \inf_{(x_1^{(0)}, \ldots, x_n^{(0)}) \in \mathscr{X}} \sum_{i=1}^{r} d((x_1^{(i)}, \ldots, x_n^{(i)}), (x_1^{(0)}, \ldots, x_n^{(0)}))^2 \leq$$

$$\leq \sum_{i=1}^{r} d\left((x_1^{(i)}, \ldots, x_n^{(i)}), \left(\frac{1}{r} \sum_{l=1}^{r} x_1^{(l)}, \ldots, \frac{1}{r} \sum_{l=1}^{r} x_n^{(l)} \right) \right)^2 \leq$$

$$\leq \sum_{i=1}^{r} \sum_{j=1}^{n} \left(x_j^{(i)} - \frac{1}{r} \sum_{l=1}^{r} x_j^{(l)} \right)^2.$$

This shows that $\left(1/r \sum_{l=1}^{r} x_1^{(l)}, \ldots, 1/r \sum_{l=1}^{r} x_n^{(l)}\right)$ is a mean to $(x_1^{(1)}, \ldots, x_n^{(1)}), \ldots, (x_1^{(r)}, \ldots, \ldots, x_n^{(r)}) \in \mathscr{X}$.

The other assertions are proved in a similar way.

Corollary. *Whatever* $\boldsymbol{x}_1, \ldots, \boldsymbol{x}_r \in \mathscr{X} = (\mathbf{R}^N)^n$, $\mathbf{M}(\boldsymbol{x}_1, \ldots, \boldsymbol{x}_r) \neq \emptyset$.

Proof. Without loss of generality we can suppose that the centres of gravity of $\boldsymbol{x}_i = (x_1^{(i)}, \ldots, x_n^{(i)}) \in \mathscr{X}$ lie in the origin, $i = 1, \ldots, r$. Then we get an optimal position of these $\boldsymbol{x}_1, \ldots, \boldsymbol{x}_r$ to each other by minimizing the continuous map

$$(B_1, \ldots, B_r, \pi_1, \ldots, \pi_r) \to \sum_{1 \leq i < k \leq r} \sum_{j=1}^{n} (B_i x_{\pi_i(j)}^{(i)} - B_k x_{\pi_k(j)}^{(k)})^2$$

on the product of the r-fold product of the compact set of all rotations round the origin, which are also elements of the closed set $\mathscr{B}$, with the r-fold product of Ξ equipped with the discrete topology.

REMARK. Proposition 1 can be generalized for expected figures of discrete distributed random figures in $\mathscr{X} = (\mathbf{R}^N)^n$ instead of means, but we shall not go into the details here.

Now we are able to prove an analogue to the well-known fact that for every bounded real-valued random variable the expectation exists.

PROPOSITION 2. *Let X be a random element in $\mathscr{X} = (\mathbf{R}^N)^n$, $\mathscr{X}$ being endowed with a quasi-metric $d = d_{\mathscr{B},\Xi}$, and suppose that a compact set $K \subset \mathbf{R}^N$, $\mathbf{R}^N$ being endowed with the Euclidean metric, exists, such that the values of X almost surely have representatives in K^n. Then there exists at least one expected element of X, i.e.* $\mathrm{E}(X) \neq \emptyset$.

Proof. Let $X_1, X_2, \ldots$ be a sequence of independent random elements in $\mathscr{X}$, distributed like X and defined on a probability space $(\Omega, \mathscr{F}, \mathrm{P})$. Without loss of generality, we can assume that for all $\omega \in \Omega$ and all $i \in \mathbf{N}$ $X_i(\omega)$ has a representative in K^n. (We say that $\boldsymbol{x} \in \mathscr{X}$ is a representative to $\boldsymbol{y} \in \mathscr{X}$ if $d(\boldsymbol{x}, \boldsymbol{y}) = 0$.) Let K be contained in the closed ball $Q_R(0)$ with radius R and with the origin 0 as midpoint.

Now let $\omega \in \Omega$ be fixed. From the Corollary to Proposition 1 we know that for all $r = 1, 2, \ldots$ $M(X_1(\omega), \ldots, X_r(\omega))$ is not empty. It is readily seen that for every $r \in \mathbf{N}$ we can choose $z_r = (z_1^{(r)}, \ldots, z_n^{(r)}) \in \mathbf{M}(X_1(\omega), \ldots, X_r(\omega))$ with $z_i^{(r)} \in Q_{2R}(0)$, $i = 1, \ldots, n$. As $Q_{2R}(0)$ is compact, there is a subsequence $z_{k_1}, z_{k_2}, \ldots$ of $z_1, z_2, \ldots$ such that the corresponding sequences of the components $z_i^{(k_1)}, z_i^{(k_2)}, \ldots$ converge, $i = 1, \ldots, n$. This implies the convergence of $z_{k_1}, z_{k_2}, \ldots$ in $(\mathscr{X}, d)$, and thus $\mathbf{M}(\omega) \neq \emptyset$. As furthermore $\mathrm{E}(d(X, \boldsymbol{\alpha})^2) < \infty$ for any $\boldsymbol{\alpha} \in \mathscr{X}$ and $(\mathscr{X}, d)$ is separable, we can conclude from the Theorem in Section 2 that $\mathbf{M}(\omega) \subseteq \mathbf{E}(X)$ for almost all $\omega \in \Omega$, and thus $\mathbf{E}(X) \neq \emptyset$, too.

The computation of expected figures in $(\mathbf{R}^N)^n$ is in general far from trivial. An exception is

PROPOSITION 3. *Let $X = (X_1, \ldots, X_n)$ be a random element in $\mathscr{X} = (\mathbf{R}^N)^n$, $\mathscr{X}$ being endowed with the quasi-metric $d = d_{\mathscr{B},\Xi}$, where $\mathscr{B} = \mathscr{T}_N$ is the group of all translations in $\mathbf{R}^N$ and Ξ contains only the identity. If $\mathrm{E}(X_i^2) < \infty$, $i = 1, \ldots, n$, then $\mathbf{E}(X)$ equals the equivalence class of $(\mathrm{E}(X_1), \ldots, \mathrm{E}(X_n))$ modulo d.*

Proof. We have to find $\boldsymbol{a} = (a_1, \ldots, a_n) \in \mathscr{X}$ with

$$\mathrm{E}(d(X, \boldsymbol{a})^2) = \inf_{\boldsymbol{\alpha} \in \mathscr{X}} \mathrm{E}(d(X, \boldsymbol{\alpha})^2) .$$

By the first part of Proposition 1 for $r = 2$ we get

$$d(X(\omega), a)^2 = \inf_{t \in \mathbf{R}^N} \sum_{i=1}^{n} ((X_i(\omega) - t) - a_i)^2 =$$

$$= \sum_{i=1}^{n} \left(\left(X_i(\omega) - \frac{1}{n} \sum_{j=1}^{n} X_j(\omega) \right) - \left(a_i - \frac{1}{n} \sum_{j=1}^{n} a_j \right) \right)^2 .$$

Without loss of generality, we can suppose $\sum_{j=1}^{n} a_j = 0$. Under this condition we have to minimize

$$\mathrm{E}(d(X, a)^2) = \sum_{i=1}^{n} \mathrm{E} \left(\left(X_i - \frac{1}{n} \sum_{j=1}^{n} X_j - a_i \right)^2 \right) .$$

Obviously only $(a_1^0, \ldots, a_n^0)$ with $a_i^0 = \mathrm{E}(X_i) - 1/n \sum_{j=1}^{n} \mathrm{E}(X_j)$, $i = 1, \ldots, n$, fulfils this requirement, and as $(a_1^0, \ldots, a_n^0)$ and $(\mathrm{E}(X_1), \ldots, \mathrm{E}(X_n))$ belong to the same equivalence class modulo d, the proposition is proved.

4. ORDER STATISTICS AS RANDOM FIGURES

Let $X_1, \ldots, X_n$ be n random variables in $\mathbf{R}$ with existing second moments. The order statistics $X_{(r)}$, $r = 1, \ldots, n$, of $X_1, \ldots, X_n$ are defined by the condition that $X_{(1)}(\omega), X_{(2)}(\omega), \ldots, X_{(n)}(\omega)$ is an ascending ordering of $X_1(\omega), X_2(\omega), \ldots, X_n(\omega)$, $\omega \in \Omega$. $X_{(r)}$ is a random variable with existing second moment and is called the rth order statistic of $X_1, \ldots, X_n$, $r = 1, \ldots, n$, [1].

Using the fact that for real numbers $\mathsf{x}_1 \leqq \mathsf{x}_2 \leqq \ldots \leqq \mathsf{x}_n$ and $\mathsf{y}_1 \leqq \mathsf{y}_2 \leqq \ldots \leqq \mathsf{y}_n$ the equality

$$\sum_{i=1}^{n} (\mathsf{x}_i - \mathsf{y}_i)^2 = \min_{\pi \in \mathbf{S}_n} \sum_{i=1}^{n} (\mathsf{x}_{\pi(i)} - \mathsf{y}_i)^2 \tag{13}$$

holds, an easy computation yields the following characterization:

PROPOSITION 4. *Let $X_1, \ldots, X_n$ be random variables in* $\mathbf{R}$ *with existing second moments and let $\mathscr{X} = (\mathbf{R}^1)^n$ be endowed with the quasi-metric $d_{\mathscr{B},\mathbf{S}_n}$, $\mathscr{B}$ containing only the identity. Then* $\mathrm{E}((X_1, \ldots, X_n))$ *equals the equivalence class of* $(\mathrm{E}(X_{(1)}), \ldots$ $\ldots, \mathrm{E}(X_{(n)}))$ *modulo $d_{\mathscr{B},\mathbf{S}_n}$.**

* I am grateful to Professor G. S. Watson for this hint.

Applying furthermore Proposition 3 to $X_{(1)}, \ldots, X_{(n)}$ we get the

COROLLARY. *The assertion of Proposition* 4 *is also valid, if* $\mathscr{B}$ *is the group of all translations in* **R**.

5. RANDOM ELEMENTS IN $(\mathbf{R}^2)^n$

We identify $\mathbf{R}^2$ and the set $\mathbf{C}$ of complex numbers by $(\xi_1, \xi_2) \leftrightarrow \xi_1 + i\xi_2$. In $\mathscr{X} = (\mathbf{R}^2)^n = \mathbf{C}^n$ we shall consider the following two quasi-metrics

$$(14) \qquad d_1(x, y) = d_1((x_1, \ldots, x_n), (y_1, \ldots, y_n)) = \inf_{t \in \mathbf{C}} \Big(\sum_{k=1}^{n} |x_k - t - y_k|^2 \Big)^{1/2} =$$

$$= \left(\sum_{k=1}^{n} \left| \left(x_k - \frac{1}{n} \sum_{j=1}^{n} x_j \right) - \left(y_k - \frac{1}{n} \sum_{j=1}^{n} y_j \right) \right|^2 \right)^{1/2}$$

and

$$(15) \qquad d_2(x, y) = \inf_{\varphi \in [0, 2\pi[} d_1(x e^{i\varphi}, y) \,,$$

i.e. $d_1 = d_{\mathscr{T}_2, \Xi}$ and $d_2 = d_{\mathscr{M}_2, \Xi}$, where in both cases Ξ contains only the identity and $\mathscr{M}_2$ is the group of all Euclidean motions in $\mathbf{R}^2$.

To $x = (x_1, \ldots, x_n) \in \mathbf{C}^n$ we associate the figure $x^* = (x_1^*, \ldots, x_n^*)$ with $x_k^* = x_k - 1/n \sum_{j=1}^{n} x_j$, $k = 1, \ldots, n$, and for $x, y \in \mathbf{C}^n$ we define the figure

$$(16) \qquad h(x, y) = h((x_1, \ldots, x_n), (y_1, \ldots, y_n)) = (x_1^* \overline{y_1^*}, \ldots, x_n^* \overline{y_n^*}) \,,$$

where $\bar{x}$ denotes the conjugate to $x \in \mathbf{C}$.

PROPOSITION 5. *Given* $x, y \in \mathbf{C}^n$, *let* $Re^{i\psi}$, $R \geq 0, \psi \in [0, 2\pi[$, *be the centre of gravity of the figure* $h(x, y)$. *Then*

$$(17) \qquad 4nR = \sup_{\varphi \in [0, 2\pi[} d_1(x e^{i\varphi}, y)^2 - \inf_{\varphi \in [0, 2\pi[} d_1(x e^{i\varphi}, y)^2 =$$

$$= \sup_{\varphi \in [0, 2\pi[} d_1(x e^{i\varphi}, y)^2 - d_2(x, y)^2 \,,$$

$$(18) \qquad d_2(x, y)^2 = d_1(x, 0)^2 + d_1(y, 0)^2 - 2nR \,,$$

and

$$(19) \qquad d_2(x, y) = d_1(x e^{-i\psi}, y) = \sum_{k=1}^{n} |x_k^* e^{-i\psi} - y_k^*|^2 \,,$$

i.e. $x^* e^{-i\psi}$ *and* y^* *are in optimal position to each other with respect to* d_2.

Proof. As $Re^{i\psi} = 1/n \sum_{k=1}^{n} x_k^* \overline{y_k^*}$, we get

$$d_1(xe^{i\varphi}, y)^2 = d_1(x^* e^{i\varphi}, y^*)^2 = \sum_{k=1}^{n} |x_k^* e^{i\varphi} - y_k^*|^2 =$$

$$= \sum_{k=1}^{n} (x_k^* e^{i\varphi} - y_k^*) \overline{(x_k^* e^{i\varphi} - y_k^*)} = \sum_{k=1}^{n} (x_k^* \overline{x_k^*} + y_k^* \overline{y_k^*} - x_k^* \overline{y_k^*} e^{i\varphi} - \overline{x_k^*} y_k^* e^{-i\varphi}) =$$

$$= \sum_{k=1}^{n} (x_k^* \overline{x_k^*} + y_k^* \overline{y_k^*}) - 2 \operatorname{Re} \left(\sum_{k=1}^{n} x_k^* \overline{y_k^*} e^{i\varphi} \right) = \sum_{k=1}^{n} \left(|x_k^*|^2 + |y_k^*|^2 \right) - 2n \operatorname{Re} \left(R e^{i(\psi + \varphi)} \right).$$

This yields

$$d_2(x, y)^2 = \inf_{\varphi \in [0, 2\pi[} d_1(xe^{i\varphi}, y)^2 = \tag{20}$$

$$= \sum_{k=1}^{n} \left(|x_k^*|^2 + |y_k^*|^2 \right) - 2n \sup_{\varphi \in [0, 2\pi[} \operatorname{Re} \left(R e^{i(\psi + \varphi)} \right), \tag{21}$$

and as the supremum is attained for $\varphi = -\psi$, we get from (20) equation (19) and from (21) equation (18). Analogically,

$$\sup_{\varphi \in [0, 2\pi[} d_1(xe^{i\varphi}, y)^2 = d_1(x, 0)^2 + d_1(y, 0)^2 + 2nR. \tag{22}$$

Subtracting (18) from (22) yields (17).

Corollary 1. *The figures $x^*, y^* \in \mathbf{C}^n$ are in optimal position to each other with respect to d_2 if and only if $x^* \overline{y^*} = \sum_{k=1}^{n} x_k^* \overline{y_k^*} \in [0, \infty[$.*

Corollary 2. *If $R > 0$, then $\mathbf{M}(x, y)$, considered in $(\mathbf{C}^n, d_2)$, contains exactly one equivalence class modulo d_2.*

The first of the following two examples will demonstrate that for $x, y \in \mathbf{C}^3$ $\mathbf{M}(x, y)$ can contain infinitely many non-equivalent figures. The second example will show that a relation like $E(E(X \mid \mathscr{G})) = E(X)$ (see e.g. [6], p. 138) cannot be true in general.

Example 1. To $x = (1, i, -1 - i)$ and $y = (1 - 2i, -2 + i, 1 + i)$ formula (16) associates the figure $h(x, y) = (1 + 2i, 1 - 2i, -2)$ with centre of gravity $0 \in \mathbf{C}$. Therefore, by Proposition 5, $xe^{i\psi}$ and y are in optimal position to each other with respect to d_2 for all $\psi \in [0, 2\pi[$. By Proposition 1 we get therefore

$$\tfrac{1}{2}(xe^{i\psi} + y) \in \mathbf{M}(x, y) \quad \text{for all} \quad \psi \in [0, 2\pi[\,.$$

Furthermore these figures are not equivalent with respect to d_2, because an easy computation yields

$$d_2\left(\tfrac{1}{2}(xe^{i\psi_1} + y), \tfrac{1}{2}(xe^{i\psi_2} + y)\right)^2 = 8 - 2\left|3 + e^{i(\psi_1 - \psi_2)}\right|,$$

and this vanishes if and only if $\psi_1 \equiv \psi_2 \mod (2\pi)$.

EXAMPLE 2. We consider $x_1 = (-1, 0, 1)$, $x_2 = (-3, 2, 1)$, and $x_3 = (-3, 3 - 4i, 4i)$ as elements of $\mathbf{C}^n$, $\mathbf{C}^n$ being endowed with d_2. By using Proposition 5 one gets $x' = (-2, 1, 1) \in \mathbf{M}(x_1, x_2)$ and $\tilde{x}' = \frac{1}{10}(-18 - 4i, 15 - 20i, 3 + 24i) \in \mathbf{M}(x_1, x_3)$. These elements are uniquely determined modulo d_2. Further computations yield

$$x'' = \tfrac{1}{3}(-7, 5 - 4i, 2 + 4i) \in \mathbf{M}(x', x', x_3)\,,$$

$$\tilde{x}'' = \frac{1}{3\sqrt{7585}}(-310 - 3\sqrt{7585} - 84i,\ 277 + 2\sqrt{7585} - 336i,\ 33 + \sqrt{7585} + $$

$$+ 420i) \in \mathbf{M}(\tilde{x}', x_2, \tilde{x}')\,,$$

and again both elements are uniquely determined modulo d_2. It is readily seen that x'' and $\tilde{x}''$ are not equivalent modulo d_2 (consider for example the squares of the absolute values of the first components, which must be equal for equivalent figures with the origin as centre of gravity). Therefore $\mathbf{M}(x', x', x_3) \cap \mathbf{M}(\tilde{x}', x_2, \tilde{x}') = \emptyset$, while for a relation like $\mathrm{E}(\mathrm{E}(X \mid \mathscr{G})) = \mathrm{E}(X)$ the equality $\mathbf{M}(x', x', x_3) = \mathbf{M}(x_1, x_2, x_3) = \mathbf{M}(\tilde{x}', x_2, \tilde{x}')$ should hold.

This example also shows that the relation *to be in optimal position to each other* is not transitive, as x_1 and x' as well as x' and x_3 are in optimal position to each other, but x_1 and x_3 are not in optimal position to each other.

REFERENCES

[1] H. A. DAVID: Order statistics. J. Wiley, New York 1970.

[2] S. DOSS: Sur la moyenne d'un élément aléatoire dans un espace distancié. Bull. Sci. Math. *73* (1949), 2, 48–72.

[3] R. FORTET: Lois des grand nombres pour des éléments aléatoires généraux. In: Proceedings of the International Congress of Mathematicians, Amsterdam 1954. Erven P. Nordhoff N. V., Groningen; North Holland Publishing Co, Amsterdam 1956, vol. III, 360–364.

[4] R. FORTET, E. MOURIER: Les fonctions aléatoires comme éléments aléatoires dans un espace de Banach. J. Math. Pures Appl. *38* (1959), 9, 347–364.

[5] M. FRÉCHET: Les éléments aléatoires de nature quelconque dans un espace distancié. Ann. Inst. H. Poincaré *10* (1948), 215–310.

[6] K. KRICKEBERG: Probability Theory. Addison-Wesley, Reading/Mass. 1965.

[7] W. J. PADGETT, R. L. TAYLOR: Laws of large numbers for normed linear spaces and certain Fréchet spaces. Springer, Berlin–Heidelberg–New York 1973.

[8] H. SCHUBERT: Topologie. Teubner, Stuttgart 1964.

UNIVERSITÄT BIELEFELD

FAKULTÄT FÜR MATHEMATIK

MIX
Papier aus verantwortungsvollen Quellen
Paper from responsible sources
FSC® C105338

FSC
www.fsc.org

If you have any concerns about our products,
you can contact us on
ProductSafety@springernature.com

In case Publisher is established outside the EU,
the EU authorized representative is:
Springer Nature Customer Service Center GmbH
Europaplatz 3, 69115 Heidelberg, Germany

Printed by Libri Plureos GmbH
in Hamburg, Germany